The Delivery of Water to Protoplanets, Planets and Satellites

Alessandro Morbidelli • Michel Blanc •
Yann Alibert • Lindy Elkins-Tanton • Paul Estrada •
Keiko Hamano • Helmut Lammer • Sean Raymond •
Maria Schönbächler
Editors

The Delivery of Water to Protoplanets, Planets and Satellites

Previously published in *Space Science Reviews* in the Topical Collection "The Delivery of Water to Protoplanets, Planets and Satellites"

Springer

Editors
Alessandro Morbidelli
Observatoire de la Cote d'Azur
Nice Cedex 4, France

Michel Blanc
International Space Science Institute
Bern, Switzerland

Yann Alibert
Physikalisches Institute
Universitat Bern
Bern, Switzerland

Lindy Elkins-Tanton
School of Earth and Space Exploration
Arizona State University
Tempe, AZ, USA

Paul Estrada
SETI and NASA/Ames Res. Center
Mountain View, CA, USA

Keiko Hamano
Earth-Life Science Institute
Tokyo Institute of Technology
Ookayama, Tokyo, Japan

Helmut Lammer
Space Research Institute
Austrian Academy of Sciences
Graz, Steiermark, Austria

Sean Raymond
Lab. D'Astrophysique de Bordeaux
Floirac, France

Maria Schönbächler
Inst. für Geochemie und Petrologie
ETH Zürich
Zürich, Switzerland

ISSN 1385-7525 Space Sciences Series of ISSI
ISBN 978-94-024-1627-5

Library of Congress Control Number: 2018966393

Cover Image: An extrasolar oceanic moon near a star-forming region. The concept behind this scene is that we know there are "super-Jupiters" in extrasolar planetary systems, and if we scaled up our Jupiter system by a factor two to three, Ganymede would be an Earth-sized world. By that reasoning, a large number—perhaps the majority—of worlds with liquid water oceans could be satellites of giant planets. Heat sources for such a moon could be a host star at an ideal distance, substantial thermal infrared radiation from the planet, or tidal heating, as in the Jupiter-Io system. Acrylic painting copyright William K. Hartmann (Planetary Science Institute, Tucson).

Printed on acid-free paper

This Springer imprint is published by the registered company Springer Nature B.V.
The registered company address is: Van Godewijckstraat 30, 3311 GX Dordrecht, The Netherlands

Contents

Introduction to the Delivery of Water to Proto-Planets, Planets and Satellites

Alessandro Morbidelli[1] · Shun-Ichiro Karato[2] · Masahiro Ikoma[3,4] · Yann Alibert[5] · Michel Blanc[6,7] · Lindy Elkins-Tanton[8] · Paul Estrada[9] · Keiko Hamano[10] · Helmut Lammer[11] · Sean Raymond[12] · Maria Schönbächler[13]

This book is the ultimate result of a workshop on "The Delivery of Water to Proto-Planets, Planets and Satellites" that was organized and hosted by the International Space Science Institute (ISSI) in Bern (CH) from January 11 to 15, 2016. It is organized as a series of review chapters on water in the Solar System, in protoplanetary disks and extrasolar planets, with the ultimate goal of discussing the origin of water on Earth and the likelihood of finding habitable conditions on other worlds. These chapters have been carefully designed to combine in a complementary and consistent way the different oral contributions presented and discussed during the workshop.

The book starts with a chapter by F. Westall and A. Brack on the importance of water for life. It may be possible to theoretically conceive of life based on compound-solvent

✉ A. Morbidelli
morby@obs-nice.fr

1 Observatoire de la Cote d'Azur, Nice, France

2 Department of Geology and Geophysics, Yale University, New Haven, CT, USA

3 Department of Earth and Planetary Science, University of Tokyo, Tokyo, Japan

4 Research Center for the Early Universe (RESCEU), University of Tokyo, Tokyo, Japan

5 Physics Institute, University of Bern, Bern, Switzerland

6 International Space Science Institute ISSI, Bern, Switzerland

7 Institut de Recherche en Astrophysique et Planétologie, IRAP, Toulouse, France

8 School of Earth and Space Exploration, Arizona State University, Tempe, AZ, USA

9 SETI and NASA/Ames Res. Center, Mountain View, CA, USA

10 Earth-Life Science Institute (ELSI), Tokyo Institute of Technology, Tokyo, Japan

11 Space Research Institute, Austrian Academy of Sciences, Graz, Austria

12 Lab. D'Astrophysique de Bordeaux, Floirac, France

13 Institut für Geochemie und Petrologie, ETH Zurich, Zurich, Switzerland

pairs other than carbon-based molecules and liquid water, but the chapter makes clear that organic material and liquid water have physical and chemical properties that make them optimal among all known molecules. This is the reason why water is at the core of exobiology and the focus of considerable research, as described in this book. The chapter argues that hydrothermal environments most likely played a key role in the appearance of life, because hot rock-water interfaces can provide the chemical disequilibria to fuel reactions, and rocks and minerals can provide the reactive surfaces needed to help the formation and stability of prebiotic molecules. Moreover, once life emerged, rocks and minerals provided the required nutriments for life to prosper. This intimate relationship between liquid water, organic material and hot rocks for the sustainability of life comes back later in the book when discussing habitability in the sub-surface oceans of some giant planets' satellites (see chapter by Grasset et al.) or water-rich extrasolar planets (see chapter by Noak et al.).

The book then comprises four chapters that make the inventory of water throughout the Solar System. The chapter by Alexander et al. describes the water budgets (and, more generally, volatile budgets) in small bodies. The chapter by Peslier et al. focuses on the water budget on Earth and its main reservoirs (core, mantle, crust and hydrosphere). The chapter by Greenwood et al. extends the analysis to the other terrestrial planets and the Moon, while the chapter by Grasset et al. describes the giant planets, their satellites, trans-Neptunian objects and the dwarf planet Ceres in the asteroid belt. These chapters establish a clear distinction between a water-poor inner Solar System and a water-rich outer Solar System, with the exception of the asteroid belt, where water-rich and water-poor asteroids coexist in the same region (but might have been separated at origin). The water budget on Earth is uncertain because water concentrations in the lower mantle and the core are poorly known, but it seems likely that our planet is intermediate in terms of bulk water content between water-poor and water-rich asteroids. Venus and Mars have, or started with, water budgets comparable to the Earth's, whereas this is unknown for Mercury. The Moon, long thought to be bone dry, seems now to have a considerable water budget, perhaps 1/10 of that of the Earth. From the isotopic point of view (D/H and $^{14}N/^{15}N$ ratios), the Earth and Mars are very similar to water-rich asteroids, but distinct from comets: they are probably made of a mixture of dryer bodies and about 2% carbonaceous chondrites (Marty 2012 and Greenwood et al. chapter). A signature of cometary bombardment, however, has been found by comparing the Xenon isotope composition in the escape-corrected terrestrial atmosphere with those in asteroids and comets (Marty et al. 2017; see chapter by Alexander et al.), concluding that 20% of atmospheric Xenon should be of cometary origin. The corresponding amount of cometary material would provide a negligible contribution to the terrestrial water budget.

The book continues with three chapters discussing the formation of planetary systems and the fate of water in the various phases of this process. The chapter by Hartmann et al. focuses on protoplanetary disks and in particular it discusses the evolution of the so-called snowline that separates the part of the disk where water is in vapor form from that where water ice is stable. If the detection of the water snowline is still beyond observational capabilities (and therefore remains an issue investigated on the basis of theoretical models), interferometric observations are now providing constraints on the positions of CO snow lines, testing disk models at large scales. The chapter by Paardekooper and Johansen is a complete compendium of all processes leading to giant planet formation: dust coagulation in the form of pebbles, aerodynamic radial drift of these particles, formation of self-gravitating clumps of pebbles leading to the birth of large planetesimals, the subsequent growth of planetesimals due to mutual collisions and continued pebble accretion until forming protoplanetary cores, gas accretion onto these cores and—last but not least—planet migration in the different planetary mass regimes. The chapter of O'Brien et al. instead focuses on the terrestrial planets of the Solar System, describing the leading models that attempt to reproduce their orbital and physical properties. As the giant planets formed before the terrestrial planets, the Paardekooper and Johansen chapter sets the stage for the O'Brien et al. one. A specific emphasis is put on the dispersal of water-rich planetesimals during giant planet growth and migration, eventually "contaminating" the growing terrestrial planets with this molecule so precious for life, as well as other volatiles.

After this description of planet formation at large scale, the book presents four chapters devoted to specific processes related to water that affected the geochemical evolution of planets and their precursors, without which the present properties of the planets cannot be understood in detail. The chapter by Ikoma et al. describes the absorption of water in silicate magma. During the magma ocean stage a planet can store a large amount of water, either produced by oxygen-hydrogen reaction between the silicate and a primitive H-rich atmosphere or delivered by water-rich planetesimals. The silicate-metal segregation in the magma ocean, leading to core formation, partitions the water between the mantle and the core, possibly making H one of its major light elements. As the magma ocean cools off, water exolves and eventually outgasses into the growing atmosphere. The chapter by Monteux et al. describes the flip side of this process—how water influences the differentiation process by increasing the oxygen fugacity of the planet. It is likely that the differentiation observed in/for icy worlds, such as Ganymede, probably occurred in two stages: rock-ice differentiation, then metal silicate differentiation within the rocky core. The chapter discusses also water losses during radioactive heating in planetesimals.

The chapter by Tian et al. describes atmosphere losses due to stellar irradiation, i.e. the process of hydrodynamic escape (the loss of H-molecules dragging heavier species into space). This process is very important for water. In fact, if water is dissociated, the escape of hydrogen is fatal for the final water budget of the atmosphere. This is most likely what happened on Venus (see chapter by Greenwood et al.). But this process could also be important for planets in the so-called habitable zone of M-type stars, due to the strong XUV radiation that these stars emit relative to a G-type, Sun-like star. The chapter by Schlichting and Mukhopadhyay focuses on a different process removing the atmosphere, i.e. the role of impacts. The chapter describes how giant impacts are much less effective in removing an atmosphere than a large number of impacts from small planetesimals for the same total impacting mass. Using geochemical arguments, it also demonstrates that, for the Earth, atmospheric impact losses have been dominant over hydrodynamic escape; moreover, part of the Earth's water survived the Moon-forming giant impact.

The book closes with a chapter by Noak et al. addressing extrasolar planets. It describes the current census of the extrasolar planet population and the present and future techniques for determining the composition of the planets and particularly their water abundance. But it also shows that, contrary to what could be naively expected, water-rich planets are likely *not* favorable for the development of life. The reason is that a large amount of surface water would inevitably produce a layer of high-pressure ice between the liquid layer and the rocky mantle, preventing the interaction between liquid water, rock and minerals that the Westall and Brack chapter argues to be essential for the development and sustainability of life. The same happens on large icy moons of the Solar System, such as Ganymede or Titan (but not on the smaller Europa; see the chapter by Grasset et al.). If this is true, then the habitability of the Earth is due to our planet acquiring some water but not too much in the chaotic game of planet formation. Understanding the conditions that led to this result is then essential to predict the likelihood that other truly habitable worlds exist around other stars.

From all the chapters presented in this book it seems that the origin of terrestrial water, long-time considered to be a major open question, is now almost a closed problem. In fact, geochemical and cosmochemical measurements, astronomical observations and theoretical models converge to a quite coherent common view, although some gray areas still subsist, as described below.

In all evidence, proto-planetary disks are hot in their inner part and cold in their outer part. So there is no surprise that, at first order, the inner Solar System objects are water-poor, while the outer Solar System bodies are water-rich. The problems start when we compare things quantitatively. Water-rich and water-poor asteroids overlap in the asteroid belt, suggesting that the snowline separating water-vapor from water-ice was located between 2 and 3 AU when these asteroids accreted (2–4 My after "time-0"—Villeneuve et al. 2009—defined as the time of accretion of calcium aluminum inclusions: 4.567 Gy ago; Bouvier and Wadhwa 2010). As explained in the chapters by Alexander et al. and O'Brien et al., it is likely that water-rich asteroids formed beyond Jupiter and were subsequently implanted into the asteroid belt. We don't know very well where Jupiter formed (due to planet migration its original location is not necessarily the current one), but if water-rich asteroids formed beyond its orbit, the snowline was probably *beyond* the asteroid belt, possibly somewhere between 3 and 5 AU.

Can disks be so hot to have a snowline there? As explained in Hartman et al. chapter, a passive disk heated only by the central star would not be hot enough (Chiang and Youdin 2010). It was thought that protoplanetary disks are viscous and therefore that the Keplerian sheer in the disk generates frictional heating; thus, a massive, high-opacity disk was expected to have its snowline even beyond 5 AU (Bitsch et al. 2015). But modern models (Turner et al. 2014) suggest that disks are much less turbulent than previously expected, stellar accretion being driven by angular momentum removal in disk-winds at the surface of the disk rather than viscous dissipation. If this is the case, the disk would be almost as cold as a passive disk. So, which source of heat could place the snowline initially beyond 3 AU is unknown.

A second problem is that, as the disk loses mass, it is likely to cool off. So, the snowline, even if originally beyond 3 AU, should migrate towards the Sun, reaching a distance smaller than 1 AU before the disk becomes optically thin (Oka et al. 2011). As ice dust and pebbles drift radially due to aerodynamic drag (see chapter by Paardekooper and Johansen), all planetesimals down to 1 AU or so, including *all* asteroids and many of the planetary embryos precursors of the Earth, should have accreted substantial amounts of water. Signs of water alteration exist in all meteorites, even the driest, but the amount of water inferred is much less than that expected for planetesimals formed beyond the snowline (Alexander et al. chapter). Radioactive heating can erase water alteration features (Monteux et al. chapter), but the recorded peak temperatures are not high enough to explain the limited water alteration of water-poor meteorites, if the original planetesimals had been water rich. So, the cosmochemical analyses suggest that the disk throughout the asteroid belt was indeed cold when asteroids formed (some water could be accreted), but it remained nevertheless water-depleted relative to solar abundance (Lodders 2003). The problem can be solved if Jupiter formed quickly, before the disk cooled. When the mass of Jupiter exceeds 20–30 Earth masses, the flux of pebbles is interrupted beyond Jupiter's orbit. Thus, the bodies inwards of the orbit of Jupiter can accrete very little water, even if the temperature drops below ice-stability values (Morbidelli et al. 2016).

Recent cosmochemical analyses (Kruijer et al. 2017) argue that Jupiter exceeded 20–30 Earth masses within the first My of Solar System history. This conclusion is derived from the observation that meteorites cluster in two distinct isotopic reservoirs (the carbonaceous and the non-carbonaceous groups) and that this separation was already present when the parent bodies of iron meteorites formed, i.e. no later than 1 My from "time-0". Because water-rich meteorites are carbonaceous chondrites, this supports the aforementioned idea that water-rich planetesimals could form only beyond Jupiter. A problem though is that not all carbonaceous chondrites are water rich (see Alexander et al.), and this is not understood so far.

If one accepts that water-rich planetesimals existed originally only beyond Jupiter, the possible sequences of subsequent events are described in the chapter by O'Brien et al. The giant planets experienced two major phases of evolution: when they were still embedded in the gaseous protoplanetary disk they grew in mass and also migrated (although the amount of migration they suffered is still not quantified), acquiring a compact resonant orbital configuration; after the removal of gas, they experienced a final dynamical instability that broke the original resonant chain and brought the planets to their current orbital configuration. The terrestrial planets were still at the stage of planetary embryos (probably with a mass comparable to the mass of Mars) when the gas disappeared and completed their formation in a much longer timescale (tens of My). O'brien et al. identify the planetesimals in the giant planet region with water-rich asteroids and the planetesimals beyond the orbit that Neptune occupied at the end of the gas-phase (roughly 12–15 AU) with comets. Notice however that Alexander et al. argue that water-rich asteroids could only be the planetesimals between the orbits of Jupiter and Saturn, observing that Titan has the same N-isotope composition of comets (i.e. distinct from asteroids) and supposing a radial gradient of N-isotopic composition throughout the disk. Although well taken, this consideration should be retained with the caveat that Titan formed in the circumplanetary disk of Saturn, which might have had different thermal conditions (hence possibly composition) from those of the protoplanetary disk at the same location.

As described in the chapter by O'brien et al., while growing in mass and/or migrating, the giant planets scatter away the planetesimals from their neighborhood. Some planetesimals are ejected or sent in the outer disk, others are sent towards the Sun. Of these, a fraction is captured into the asteroid belt (thus becoming the water-rich asteroids we see today), while others acquire more eccentric orbits that intersect those of the protoplanetary embryos in the terrestrial planet region. The simulations, calibrated on today's population of water-rich asteroids, predict that the Earth accreted about 1–2% of its mass from these bodies. Assuming that these bodies carried 10% water by mass and neglecting losses (thanks to the capability of the magma ocean to absorb water (see Ikoma et al.) and to the relative inefficiency of collisions in devolatilizing the planet (see Schlichting and Mukhopadhyay)), this would correspond to 1,000–2,000 ppm of chondritic water in the Earth. As discussed in Peslier et al., the terrestrial water budget is estimated to be between 600 and 36,000 ppm. So, numbers are in the same ball-park, with the exception of the high-end estimates of Earth's water content, that require huge amounts of water partitioned in the core. The D/H ratio of the Earth's water matches the chondritic mean value, confirming that water-rich asteroids are the most likely source.

An important caveat, though, is made in the Alexander et al. chapter. The D/H ratio measured in chondritic meteorites is probably the result of the interaction between an original water with low D/H ratio (intermediate between the solar and terrestrial values) and a D-rich organic material (Alexander et al. 2012). This is not a problem as long as the Earth accreted only chondritic-like material. But if the Earth also accreted asteroids with a larger water/organic ratio than recorded in chondrites (for instance bodies with large ice reservoirs in their interior), the final D/H ratio of the accreted water should be sub-terrestrial. We know that today there are still asteroids rich in ice in the belt. Ceres, the major body in the asteroid belt, is one of them (see Grasset et al. chapter); the so-called main-belt comets are others. It is unknown what the D/H ratios of these ices are. If measurements in Ceres are not foreseeable in a realistic future, for main-belt comets the prospects are more favorable, because these objects degas when they pass near perihelion. Both NASA and ESA have received proposals for space missions to these objects, but so far none has been selected despite the undisputable scientific interest in isotopically characterizing these bodies, as ESA/Rosetta did for comet 67P.

The dynamical models also show that the Earth is not a singular planet. All terrestrial planets have been affected by the bombardment of water-rich asteroids during their growth. This is in agreement with the indications, reviewed in the Greenwood et al. chapter, that Venus and Mars had originally water budgets comparable to the Earth's. Mercury, closer to the Sun, could have received comparably less water than the other planets, although there is no sharp difference in the model predictions.

Still according to models, comets would have entered into the game of water delivery during the giant planet instability phase, sometime after the disappearance of the gas from the protoplanetary disk. It's expected that the Earth accreted only 7×10^{-6} of its mass from comets (Morbidelli et al. 2018). This is in remarkable agreement with the estimates from the abundances of Argon (Marty et al. 2016) and Xenon (Marty et al. 2017) in the Earth's atmosphere, and the concentration of these gases measured in comet 67P by ESA/Rosetta (see Alexander et al.). This tiny amount of cometary material would have brought to the Earth less than 1% of the Earth's water, i.e. a negligible amount.

Models and observations agree not only for what concerns the total water amount and its origin, but also what concerns the water delivery chronology. In the models (see again O'Brien et al.), the terrestrial planets form from a system of planetary embryos. The latter formed from local material, so they are expected to have been totally dry. The first merging events among embryos therefore should have started to build dry terrestrial planets, but then water-rich planetesimals scattered from beyond Jupiter's orbit became available in the terrestrial planet region. The accretion of some of these objects progressively built up the water's budgets as the terrestrial planets were growing towards their final mass.

On the geochemical side (see chapters by Peslier et al. and Schlichting and Mukhopadhyay), there is evidence that Earth's early accretion history was volatile poor and that water was not incorporated in the Earth in significant quantities until the planet had grown to $\sim$60–90% of its current size. Models and geochemical considerations also agree that water was accreted throughout Earth's formation and not solely during the late veneer (see chapter by Schlichting and Mukhopadhyay). The late veneer is defined as the accretion of material that occurred after the end of core formation; because the Moon-forming giant impact necessarily led to the growth of Earth's core, traditionally the late veneer is identified with the tail of accretion occurred after the Moon forming event. The mass of material accreted during the late veneer is constrained to be $<1\%$ of the Earth's mass by the concentration of the highly siderophile elements (HSE) in the terrestrial mantle (Walker 2009). This mass would be insufficient to deliver the terrestrial water budget. Moreover, the isotope composition of some of the HSEs (e.g. osmium, ruthenium) exclude that the late veneer was dominated by water-rich (i.e. carbonaceous chondritic) material (Drake and Righter 2002; Dauphas et al. 2004). Finally, terrestrial volatile elements of comparable condensation temperature show different levels of partial depletion in the terrestrial mantle, correlated with their affinity to partition in metal, i.e. being trapped into the core. This implies that volatile accretion occurred while the Earth's core was still building up (Wood et al. 2010).

The presence of water on Earth prior to the Moon-forming event can explain the presence of water in the Moon (see Greenwood et al.). In fact, all models of Moon-forming events predict that at least a fraction of the lunar material came from the proto-Earth; so the Moon would have inherited part of the Earth's water. In the recent "synestia" model (Lock et al. 2018), the giant impact produced an extended, hot and fast rotating silicate vapor atmosphere around the Earth, called synestia. The synestia was in rapid equilibration with the terrestrial magma ocean. Upon cooling, the synestia could not contract entirely towards the Earth's surface; because its angular momentum was high, part of it had to form a disk in orbit around the Earth. The Moon formed from such a disk. If the Earth was water rich at the moment of the giant impact, the synestia would have contained water vapor. Due to the high solubility of water in liquid silicate (see Ikoma et al.), it is likely that part of the water remained trapped in the condensing silicates and therefore in the final Moon (Karato 2013; Pahlevan et al. 2016; see Greenwood et al.).

Not just water, but also most volatile elements have isotope compositions that are chondritic. This supports the view that carbonaceous chondritic asteroids dominated the whole terrestrial volatile budget. However, the existence of Ne and possible H with a solar isotopic composition in the deep mantle (see Schlichting and Mukhopadhyay) points that the process of in-gassing of volatiles from a primitive atmosphere (i.e. gas trapped from the protoplanetary disk) described in the Ikoma et al. chapter played some role (see also Lammer et al. 2018). The capability to build a primitive atmosphere and its pressure depend sensitively on the planet's mass. The existence of solar gases in the deep mantle suggests that the Earth reached a few Mars-masses through mutual merging events among embryos before the disappearance of the disk, again in agreement with some dynamical models, but the uncertainty on this mass estimate unfortunately remains large. There is no trace of this primitive atmosphere now on Earth. As discussed in the chapter by Schlichting and Mukhopadhyay, the present-day geochemistry of volatiles shows no evidence of strong hydrodynamic escape. This suggests that impacts played a major role during Earth's formation in removing the original solar gases and delivering the chondritic ones. The chapter also provides geochemical evidence from the $^{3}He/^{22}Ne$ ratio for multiple giant impacts on Earth, again in agreement with accretion models that indicate that most of the terrestrial mass came from merging events with other planetary embryos.

This book is dedicated to the memory of Prof. Yutaka Abe. Throughout his career, Prof. Abe had a long list of outstanding contributions to the subject of this book. Sadly, the ISSI workshop that this book builds upon was the last meeting that Prof. Abe attended. He is well visible on his wheel chair in the first row of the workshop picture. Prof. Abe ran into medical complications during his return flight to Japan, and after a two-year struggle, he passed away on the first day of 2018.

Already in his student days (mid-1980s), Prof. Abe was highly recognized based on a series of papers on the generation of the atmosphere-ocean in the late stage of terrestrial planet formation. During these studies, he recognized the importance of atmosphere-hydrosphere and solid planet interactions to define the "habitable conditions" on a terrestrial planet. To explore such interactions, he used detailed knowledge of materials properties (melting relations, solubility of volatiles in the melts etc.) together with an elegant treatment of mass and energy transfer in the models of planetary formation and evolution. A particularly important notion coming from his works is the recognition of the role of the magma ocean (a molten silicate layer on a growing planet) to control the surface conditions of a terrestrial planet.

In addition to the generation of the atmosphere and ocean, he made great contributions to understanding the stability of aqua-planets. In particular, he found the presence of a radiative-emission limit of wet atmospheres, which is generally known today as a condition that defines the inner edge of the "habitable zone". Through those studies, he also noticed that a difference in ocean mass and water transport brings about a qualitative change to the stability of the atmosphere of aqua-planets. He developed a new concept of the land planet such that water transport is dominated not by surface transport but by atmospheric circulation. In recent years, he clarified the stability of the climate of such land planets by

performing 3D GCM simulations. Even after the onset of ALS, without reducing his research activity, he continued such studies until his premature death at the age of 58.

Another important contribution by Professor Abe is that he has trained so many brilliant students (and post-docs) throughout his career. This is truly remarkable in light of the fact that he needed to fight against ALS during his last ~15 years. We are sure that not only his scientific papers, but also these lucky students/postdocs will maintain his legacy in the planetary science community for years to come.

References

C.M.O. Alexander, R. Bowden, M.L. Fogel, K.T. Howard, C.D.K. Herd, L.R. Nittler, The provenances of asteroids, and their contributions to the volatile inventories of the terrestrial planets. Science **337**, 721 (2012)

B. Bitsch, A. Johansen, M. Lambrechts, A. Morbidelli, The structure of protoplanetary discs around evolving young stars. Astron. Astrophys. **575**, A28 (2015)

A. Bouvier, M. Wadhwa, The age of the Solar System redefined by the oldest Pb–Pb age of a meteoritic inclusion. Nat. Geosci. **3**, 637–641 (2010)

E. Chiang, A.N. Youdin, Forming planetesimals in solar and extrasolar nebulae. Annu. Rev. Earth Planet. Sci. **38**, 493–522 (2010)

N. Dauphas, A.M. Davis, B. Marty, L. Reisberg, The cosmic molybdenum-ruthenium isotope correlation. Earth Planet. Sci. Lett. **226**, 465–475 (2004)

M.J. Drake, K. Righter, Determining the composition of the Earth. Nature **416**, 39–44 (2002)

S-i. Karato, Geophysical constraints on the water content of the lunar mantle and its implications for the origin of the Moon. Earth Planet. Sci. Lett. **384** (2013)

T.S. Kruijer, C. Burkhardt, G. Budde, T. Kleine, Age of Jupiter inferred from the distinct genetics and formation times of meteorites. Proc. Natl. Acad. Sci. **114**, 6712–6716 (2017)

H. Lammer, et al., Origin and evolution of the atmospheres of early Venus, Earth and Mars. Astron. Astrophys. Rev. **26**, 2 (2018)

S.J. Lock, S.T. Stewart, M.I. Petaev, Z. Leinhardt, M.T. Mace, S.B. Jacobsen, M. Cuk, The origin of the moon within a terrestrial synestia. J. Geophys. Res., Planets **123**, 910–951 (2018)

K. Lodders, Solar system abundances and condensation temperatures of the elements. Astrophys. J. **591**, 1220–1247 (2003)

B. Marty, The origins and concentrations of water, carbon, nitrogen and noble gases on Earth. Earth Planet. Sci. Lett. **313**, 56–66 (2012)

B. Marty, G. Avice, Y. Sano, K. Altwegg, H. Balsiger, M. Hassig, A. Morbidelli, O. Mousis, M. Rubin, Origins of volatile elements (H, C, N, noble gases) on Earth and Mars in light of recent results from the ROSETTA cometary mission. Earth Planet. Sci. Lett. **441**, 91–102 (2016)

B. Marty, et al., Xenon isotopes in 67P/Churyumov–Gerasimenko show that comets contributed to Earth's atmosphere. Science **356**, 1069–1072 (2017)

A. Morbidelli, B. Bitsch, A. Crida, M. Gounelle, T. Guillot, S. Jacobson, A. Johansen, M. Lambrechts, E. Lega, Fossilized condensation lines in the Solar System protoplanetary disk. Icarus **267**, 368–376 (2016)

A. Morbidelli, D. Nesvorny, V. Laurenz, S. Marchi, D.C. Rubie, L. Elkins-Tanton, M. Wieczorek, S. Jacobson, The timeline of the lunar bombardment: revisited. Icarus **305**, 262–276 (2018)

A. Oka, T. Nakamoto, S. Ida, Evolution of snow line in optically thick protoplanetary disks: effects of water ice opacity and dust grain size. Astrophys. J. **738**, 141 (2011)

K. Pahlevan, S-i. Karato, B. Fegley, Speciation and dissolution of hydrogen in the proto-lunar disk. Earth Planet. Sci. Lett. **445**, 104–113 (2016)

N.J. Turner, S. Fromang, C. Gammie, H. Klahr, G. Lesur, M. Wardle, X.N. Bai, Transport and accretion in planet-forming disks. Protostars and Planets VI, 411–432 (2014)

J. Villeneuve, M. Chaussidon, G. Libourel, Homogeneous distribution of ^{26}Al in the Solar System from the Mg isotopic composition of chondrules. Science **325**, 985 (2009)

R.J. Walker, Highly siderophile elements in the Earth, Moon and Mars: update and implications for planetary accretion and differentiation. Chem. Erde Geochemistry **69**, 101–125 (2009)

B.J. Wood, A.N. Halliday, M. Rehkamper, Volatile accretion history of the Earth. Nature **467**, E6–E7 (2010)

Space Sci Rev (2018) 214:50
https://doi.org/10.1007/s11214-018-0476-7

The Importance of Water for Life

Frances Westall[1] · **André Brack**[1]

Received: 10 August 2016 / Accepted: 23 January 2018 / Published online: 20 February 2018

Abstract Liquid water is essential for life as we know it, i.e. carbon-based life. Although other compound-solvent pairs that could exist in very specific physical environments could be envisaged, the elements essential to carbon and water-based life are among the most common in the universe. Carbon molecules and liquid water have physical and chemical properties that make them optimised compound-solvent pairs. Liquid water is essential for important prebiotic reactions. But equally important for the emergence of life is the contact of carbon molecules in liquid water with hot rocks and minerals. We here review the environmental conditions of the early Earth, as soon as it had liquid water at its surface and was habitable. Basing our approach to life as a "cosmic phenomenon" (de Duve 1995), i.e. a chemical continuum, we briefly address the various hypotheses for the origin of life, noting their relevance with respect to early environmental conditions. It appears that hydrothermal environments were important in this respect. We continue with the record of early life noting that, by 3.5 Ga, when the sedimentary environment started being well-preserved, anaerobic life forms had colonised all habitable microenvironments from the sea floor to exposed beach environments and, possibly, in the photic planktonic zone of the sea. Life on Earth had also evolved to the relatively sophisticated stage of anoxygenic photosynthesis. We conclude with an evaluation of the potential for habitability and colonisation of other planets and satellites in the Solar System, noting that the most common life forms in the Solar System and probably in the Universe would be similar to terrestrial chemotrophs whose carbon source is either reduced carbon or CO_2 dissolved in water and whose energy would be sourced from oxidized carbon, H_2, or other transition elements.

Keywords Water · Carbon · Rocks · Minerals · Origin of life · Early life · Extraterrestrial life

The Delivery of Water to Protoplanets, Planets and Satellites
Edited by Michel Blanc, Alessandro Morbidelli, Yann Alibert, Lindy Elkins-Tanton, Paul Estrada, Keiko Hamano, Helmut Lammer, Sean Raymond and Maria Schönbächler

✉ F. Westall
frances.westall@cnrs-orleans.fr

[1] CNRS-Centre de Biophysique Moléculaire, CS80054, Rue Charles Sadron, 45071 Orléans cedex 2, France

1 Introduction

The idea that life could be universal can be traced back to the Greek philosophers. In 300 B.C., Epicurus wrote to Herodotus: "There is an infinite number of worlds... There is no reason why these worlds could not contain germs of plants and animals and all the rest of what can be seen on Earth." This age-old dream was further formulated by the Roman philosopher Lucretius: "Confess you must that other worlds exist in other regions of the sky, and different tribes of men, kinds of wild beasts". For agreeing with Lucretius, the defrocked Dominican friar Giordano Bruno was burned at the stake in 1600 on Piazza Campo dei Fiori in Roma, by order of the Inquisition. It is only relatively recently that we have the visionary perception of evolution from some prebiotic compounds floating in a warm pond proposed by Charles Darwin in a 1863 letter to Joseph Hooker (in parallel with the Contributions to the Theory of Natural Selection, 1870 of Alfred Wallace) (see discussion in Peretò et al. 2009). But even today, many are uncomfortable with the idea that we might be descended a simple speck of carbonaceous interstellar dust, let alone an ape or even a bacterium. The hypothesis proposed here was stated very elegantly by Christian de Duve in his seminal 1995 work "Vital Dust", where he proposed the life to be a cosmic phenomenon in which biology is simply a continuation of chemistry. The latter concept will be the guideline throughout this contribution.

The first part of this contribution addresses briefly the definition of life, concluding that, at a minimum, cellular life, irrespective of its ingredients, is characterized by a membrane, energy producing molecules and information transferring molecules. We will see that, although alternative scenarios for life in completely different physico-chemical conditions could be theoretically envisaged, it is life based on carbon molecules and liquid water that could be the most common in the Universe for the simple reason that these components are made up of atoms that are among the most common in the Universe. Understanding of the geological record of the Earth informs us that, apart from liquid water and carbon molecules, there is another essential ingredient to the emergence of life—rocks! And not just any kind of rocks but rocks and the minerals contained in them that were common on the early Earth and very common on rocky planets in general. In fact, in our understanding, life could not have emerged without rocks—in particular, hot rocks in contact with water (or hot water in contact with rocks). The rock-water interface provided the chemical disequilibria to fuel reactions, and rocks and minerals provided the reactive surfaces that most likely helped conformation and stability of prebiotic molecules. Rock formations or their derivatives were thus likely locations for the emergence of the first cells and, once life had emerged, they provided additional nutrients.

Despite the dearth of rocks dating from the earliest period of the Earth, the Hadean (4.56–3.85 billion years (Ga) ago), evidence suggests that there was water on the surface of the Earth by 4.3 Ga (Nemchin et al. 2006). The ancient crustal environment of the Archaean Earth can be interpreted from a well preserved geologic record covering the period 3.8–3.0 Ga. While there is almost no data for the Hadean, we can attempt to make inferences about its environmental situation as soon as it had become habitable (i.e. when there was liquid water). The early Earth was hot, especially at the water/rock interface where, presumably, the important prebiotic reactions were taking place. It hosted a variety of environments that have been suggested as the cradle of life, ranging from hydrothermal vents, the beach environment (including tidal flats) to continental springs or rivers. We will see that hydrothermal environments are possibly the most common and likely locations in which primitive cells could emerge.

When did life emerge? Based on modelling of the impact rate on the Moon and early Earth, It had been hypothesised that life, if it had appeared in the Hadean, would have been wiped out by planet-sterilising impacts at the end of the epoch (Maher and Stephenson 1988; Sleep et al. 1989; Ryder 2003). However a recent re-evaluation of the ages of the lunar rocks, upon which the model was based, suggests that there supposed impact spike in fact did not exist (Boehnke and Harrison 2016). Nevertheless, lack of good preservation of Hadean rocks precludes direct investigation but, considering the ubiquity of the essential ingredients (organic molecules, water, energy and nutrients) probably by ~ 4.3 Ga, it seems reasonable to infer that the chemical reactions leading to a living cell could have started taking place by this time. How long did it take for life to emerge? We do not know but the fact that the process had to be driven by a kinetic momentum such that there could be no turning back, no undoing of the haphazard combination of molecules to form a protocell, means that it had to have been relatively fast. However, given the uncertainty of our understanding of conditions on the early Earth, all that we can state is that, not knowing how an RNA-based life form appeared from abiotic precursors, it is not possible to estimate the time necessary for life to emerge (Orgel 1998). This also depends on exactly which steps in the emergence of life are considered in the processes leading to the appearance of an auto-catalytic entity. Do we include the time required for the creation of the prebiotic building bricks of life or simply the assemblage of the pre-formed molecules? If so, the process could have started very rapidly or even instantaneously, like supersaturated solutions that suddenly crystallise upon the addition of a crystal acting as a germ. Perhaps more time was necessary for the addition of the basic elementary units, but how much?

The oldest, well-preserved traces of life dating back to about 3.5 Ga document life forms inhabiting a variety of microenvironments in a water world—hydrothermal environments, sediment/rock surfaces, intertidal beach surfaces, subaerial hydrothermal environments (geysers), and possibly even the planktonic realm in the water column (Knoll 2003; Westall 2011, 2016; Djokic et al. 2017). Life was already highly diverse, encompassing primitive chemotrophic organisms, as well as more sophisticated photosynthesisers, albeit all living in an anaerobic environment. Indeed, the Hadean/Archaean environment can be considered as an extreme environment and the early microbes as extremophiles, at least from our anthropomorphic point of view. In fact, the inverse is true; the environment of the early microbes is probably far more common in the Universe than the present day, geologically evolved Earth in which we are living. We are the extremophiles!

Since the watery environmental conditions pertaining to the emergence of life on the early Earth were similar to those of other planets in the early Solar System—at the microbial scale—it is possible that life appeared on Venus, Mars, and even the icy satellites having a rocky core. Life may still exist at depth in the subsurface of Mars, as well as in the ice covered oceans of Europa and Enceladus orbiting Jupiter and Saturn, respectively. Considering carbon and water-based life as the "norm" in the Universe and the likely profusion of rocky, habitable planets in the galaxy and Solar System, and considering the emergence of life as a "chemical continuum", primitive life forms, similar to chemotrophs, could be relatively common in the Universe.

2 Definition of Life

It is commonly said that there as many definitions of life as there are people trying to define it. Apart from the basic characteristics, it could be stated that life is more than the sum of its parts. Here are just a few representative definitions:

- The state or quality that identifies living beings, characterised chiefly by growth, reproduction, and response to stimulation (Collins dictionary);

- A self-sustained chemical system capable of undergoing Darwinian evolution (working definition adopted in October 1992 by the NASA Exobiology Programme (Joyce 1995));
- A system that is capable of making more of itself, by itself, and of evolving (Brack and Troublé 2010);
- A chemical entity that consists of bounded microenvironments in chemical disequilibrium with their environment, capable of maintaining a low entropy state by energy and environment transformation, and capable of information encoding and transfer (Schulze-Makuch and Irwin 2004).

One of the reasons why it is difficult to define life is that the definitions generally refer to its processes, e.g. metabolism, growth, reproduction, or adaptation to the environment, that have many characteristics and similarities with processes occurring in non-biological systems. Let us consider the definition by Schulze-Makuch and Irwin (2004) and their discussions which provide some clues as to the essential characteristics of life.

(1) *A bounded environment* is needed to provide order in a chaotic external environment. In living cells, this boundary is an organic membrane but many inorganic features form boundaries, such as porous minerals. Indeed, the porous walls of hydrothermal vent edifices (Russell and Hall 1997) or even pumice (Brasier et al. 2011, 2013) are considered to have been the first cellular membranes before being replaced by organic membranes.
(2) *Chemical disequilibria* are the key to energy transfer for fueling reactions producing prebiotic molecules and conducting the metabolism in living entities. In living cells the energy transfer is driven by chemical or photochemical pathways and the energy is stored in molecules, such as ATP. However, inorganic substances, such as certain minerals, can also store energy, releasing it as luminescence.
(3) *Maintaining a state of low entropy:* One of the specificities of living cells is their ability to maintain a state of low entropy in a disordered environment. This requires transformation of energy and the environment to create and maintain the complexity which characterises the system, including an internal microenvironment with physico-chemical characteristics that differ from the external environment, e.g. Eh, pH, solute composition, etc. An example of abiotic energy transfer is demonstrated by soil weathering profiles, in which an oxidised surface grades into an underlying reduced layer.
(4) *Information encoding and transmission* ensures permanency and succession of the living entity. However, mutations or mistakes in the encoding appear to be a fundamental feature for evolution and survival in changing conditions. We are unfortunately all too familiar with "viral" computer programmes as an example of abiotic information coding and transmission. Minerals are another example of abiotic reproduction.

The fact that a living entity needs some kind of a membrane to separate it from the outer environment and protect the internal processes, a means of transforming external energy to fuel internal functions, and some means of replication, has a number of consequences. In the first place, it needs to be of relatively small size, i.e. having a high surface to volume ratio to ensure sufficient diffusion of external nutrients into the cell, but still large enough to host the machinery driving metabolism. Secondly, such an entity would need to coexist with a liquid medium because it is the latter that can contain and transport the solutes (charged ions) necessary for chemical reactions on a relatively rapid timescale (N.B., these reactions must be unidirectional and fast enough to produce a kinetic "snowball"). Finally, the environmental conditions need to be compatible the molecules making up the living entity in terms of physico-chemical factors, such as temperature, ionic strength, pH, and trace element composition.

3 Life Based on Carbon and Water

Apart from our anthropomorphically skewed perspective, there are specific reasons to believe that life forms based on carbon and water may be the most frequent form of life in the Universe. The elements H, C, and O, together with other elements considered to be life-essential, N, S, and P, are among the most common in the Universe (Fig. 1a). A wide variety of carbon molecules, about 118, has been detected in interstellar space, interstellar dust particles (IDPs), and in primitive Solar System materials, such as asteroids and carbonaceous meteorites, comets and micrometeorites. These molecules range from those with only one carbon atom to huge molecules with more than 60 (fullerines) (Millar 2005) (Fig. 1b). Some molecules in interstellar space are structurally branched, such as i-propyl cyanide (Belloche et al. 2014). This is an important characteristic of the amino acids essential to the formation of prebiotic molecules and life.

Carbon atoms have many advantages. Being tetravalent, they can form complex, stable molecules with millions of possible combinations (Fig. 1c). Life actually only uses a relatively small number of the possible combinations. One of the reasons is that only a small number of these combinations has stable, non-hydrolysable bonds (Westheimer 1987). Moreover, a large amount of energy would be necessary to create and incorporate an infinite variety of molecules. Indeed, there is a far greater variety of carbon molecules in carbonaceous meteorites than there is in a living cell (Summons et al. 2011). Indeed, this is a useful characteristic to distinguish biotic from abiotic carbon. Not only is the number of combinations of carbon huge, but also the variety of structures that can be formed. Another peculiarity of carbon molecules is that for each structure formed, it can have a mirror image, i.e. each molecule can have two structural forms that are mirror images on each other. This is known as chirality, or left and right handedness, and is another significant characteristic of life (Fig. 1d). For each class of component, amino acids, sugars, etc., only one form of the molecule is used since it takes less energy to incorporate a molecule having the same structural orientation than one that does not. Amino acids used by cellular life are characterised by the "L" form whereas abiotic syntheses of amino acids produce an equal mixture of "L" and "D" forms. Finally, another advantage of carbon is that it can be oxidised to CO_2, thus producing energy to fuel living cells.

Carbon molecules are formed not only in the interstellar medium, on asteroids and comets, but also on the Earth. It had originally been hypothesised that abiotic production of organics by photochemistry could have occurred by Strecker synthesis in an very early, reducing (NH_3, CH_4, and H_2O), pre Moon-forming impact (~ 4.5 Ga) atmosphere (Miller 1953), but more recent studies indicate that the early, outgassed atmosphere was not so reducing, consisting instead of H_2O, CO_2, and N_2, with only small amounts of CO and H_2 (Holland 1984).

There was also another source of simple organic molecules from reactions occurring within the hot early crust. Seawater seeping into the crust, heated up to become hydrothermal fluids could react with reduced carbon (of magmatic, extraterrestrial and even possibly atmospheric origin) through Fischer-Tropsch synthesis (Shock et al. 1998; see recent review in Konn et al. 2015) to produce small organic molecules, such as CH_4, alkanes, formic acid, and acetate (Sherwood Lollar et al. 2008; Proskurowski et al. 2008). Moreover, it has recently been proposed that crustal methane and possibly the higher n-alkanes may come from hydrothermal fluid interaction with of volatile-rich fluid (mantle derived) inclusions in crustal rocks (McDermott et al. 2015).

Of the three sources of reduced carbon, extraterrestrial delivery appears to have played the greatest role. About 2.5×1019 kg of complex organic matter was probably accreted

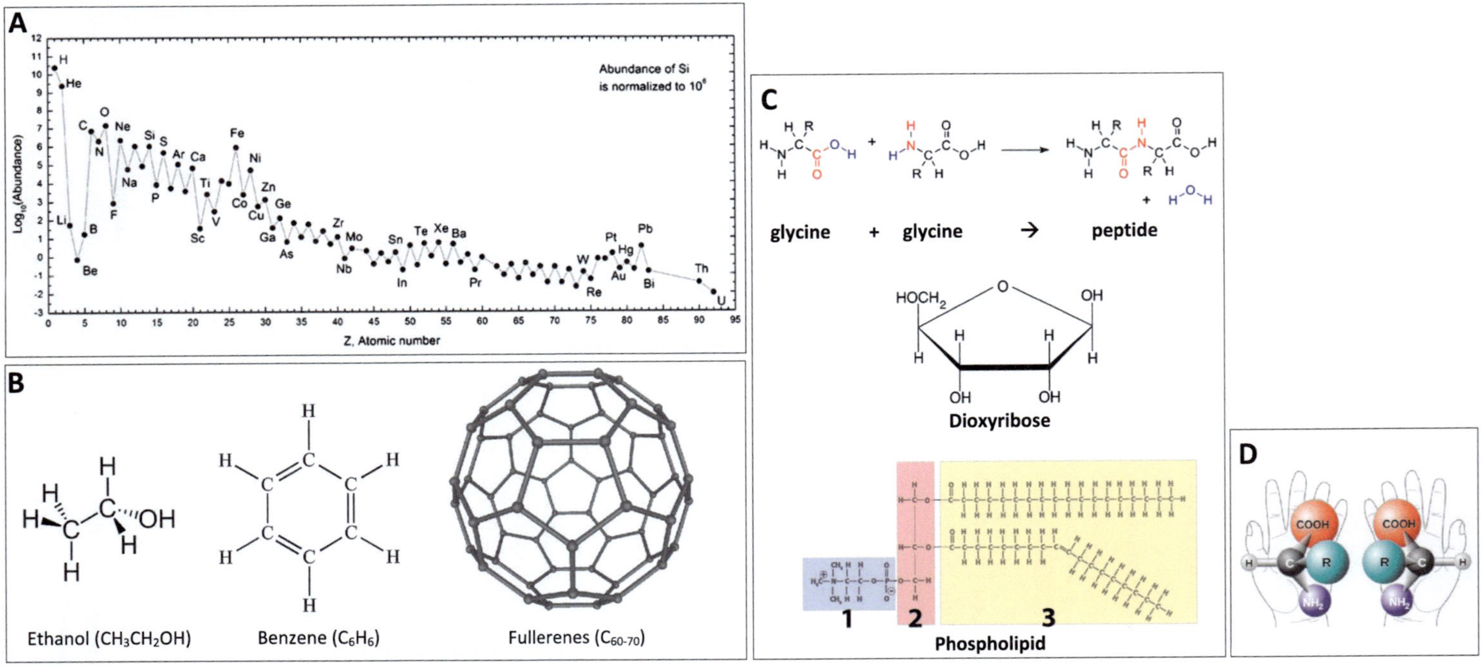

Fig. 1 (**A**) Distribution of elements in the Universe. (**B**) Variety of carbon molecules in interstellar space, ranging from small, two molecules, such as ethanol, up to huge molecules of 60–70 carbon atoms ("fullerines or bucky balls"). (**C**) Variety of structurally complex biomolecules that can be formed in water. (**D**) Chirality, a mirror image structure typical of biomolecules

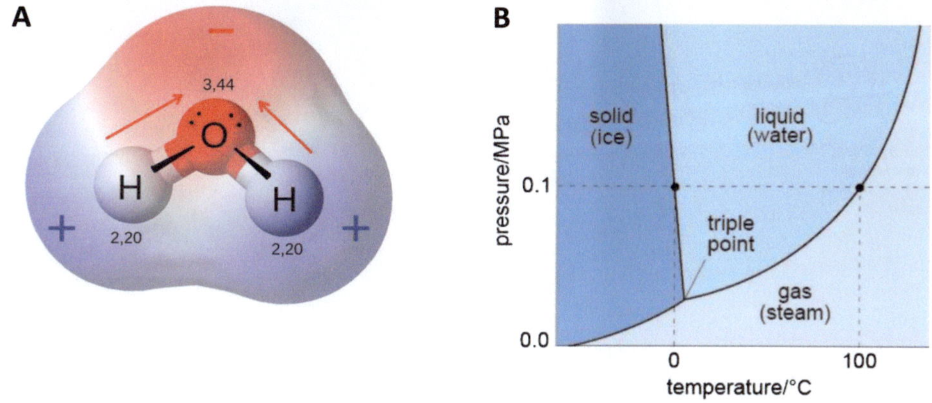

Fig. 2 (**A**) The water molecules (H_2O). (**B**) Phase diagramme for water showing its broad stability range

by the Earth during the first $\sim$ 300 Ma of the post-lunar period, mostly in carbonaceous chondrites and micrometeorites that can contain up to 4% organic matter (Maurette and Brack 2006).

We noted above the necessity of a solvent to accompany the carbon molecules. Liquid water, H_2O has a number of properties that make it an ideal solvent (Brack 2002; Schulze-Makuch and Irwin 2004). First of all, why is a liquid better than a gas or a solid? A liquid provides an environment that allows for the stability of some chemical bonds to maintain macromolecular structure. It should allow dissolution of some chemical bonds to ensure chemical interchange and energy transformations. However, other organic macromolecules need to be insoluble in the liquid medium, thus allowing the formation of boundary surfaces, such as vesicles or other kinds of interfaces, necessary for creating a protected micro-environment having different physico-chemical conditions compared to the external environment. A liquid phase should also promote the stability of macromolecules, e.g. the stability of chiral molecules. It should have a density that can maintain critical concentrations of reactants, e.g. the dissolved salts necessary for cellular life, and transport them to the cell as well as transport effluent away from it. Finally, the liquid phase should be stable in as wide a range of temperatures and pressures as possible in order to function as a buffer against environmental fluctuations, as well as to favour a broad spectrum of biochemical reactions. Note that the physical properties of the solvent need to match those of its host environment.

Liquid water, H_2O, is the perfect solute for carbon-based life (Fig. 2). The hydrogen bonding confers high boiling and low freezing temperatures. Water has additional useful physical properties, such as high heat retention, surface tension, and dielectric constant. The high density of water means that it can contain the dissolved salts necessary for life. Its polarity, the ability to make covalent bonds, confers a polymer-like structure on water. Finally, water favours the polymerisation and structural organisation of organic molecules (Brack 2007) (Fig. 3).

Schulze-Makuch and Irwin (2004) make an excellent discussion about the possibility of life based on other chemical compounds and solutes but none of these combinations presents the same range of properties and advantages as carbon and water and, at least on Earth, most are not stable or are outcompeted by carbon. The most popular, alternative, covalent-bonding element is silicon but other elements could be envisaged, such as boron, nitrogen, phosphorus, sulphur and germanium. Some combinations based on Si include silicon itself;

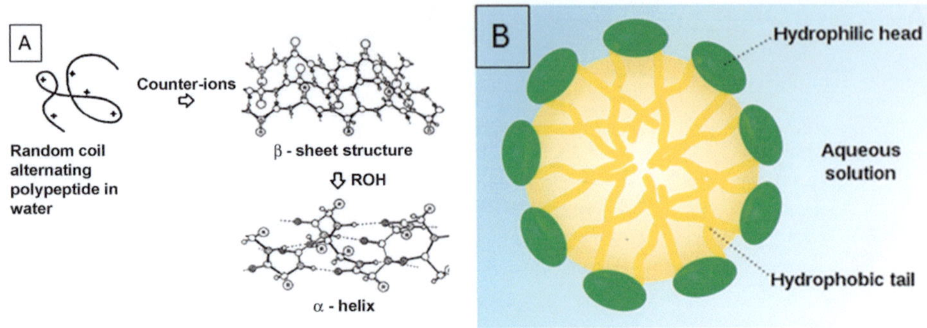

Fig. 3 (**A**) Importance of water containing dissolved ions (salts) for structuring prebiotic molecules. Polypeptides form random structures in plain water but can take on specific structures, such as β sheets or α helices in salty water (Brack 2002). (**B**) Formation of cellular structures due to the spontaneous orientation of carbon molecules having a hydrophilic and a hydrophobic termination

Fig. 4 Examples of some silicon-based molecules that could, under very restricted physical and chemical conditions (after Schulze-Makuch and Irwin 2004)

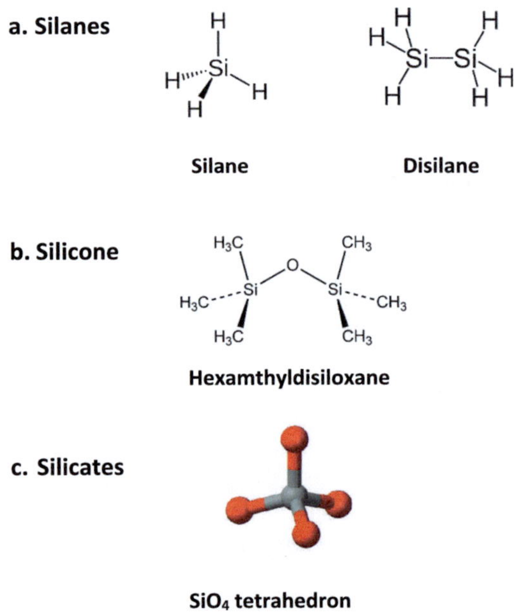

silanes (e.g. Si-H or Si-Si); silicones, i.e. organo-silicon polymers (R_2SiO); and silicates, i.e. salts containing SiO_2 (Fig. 4).

Concluding this section on the carbon and water, the essential ingredients of life, we can also add that other elements are common components in both abiotic and biogenic carbon: N, P, and S. Furthermore, certain transition metals play a critical role in the structure of cellular macromolecules or in their metabolic processes. Finally, there can be no life without energy, either chemical, heat or light energy. And we will see in the next section that rocks and minerals are critical sources of energy and elements—as well as being critical in other respects for the origin of life.

3.1 Early Terrestrial Environmental Conditions

Any discussion of the importance of water for life impinges on theories about how life emerged. These in turn are highly dependent upon the geological and environmental conditions reigning when life emerged. Unfortunately, understanding of the conditions on the early Earth is hampered by the fact that almost all of the Hadean crust (4.4–3.85 Ga) has disappeared, recycled by tectonic activity and most likely large impacts (Kamber 2015). Nevertheless, we can interpret different scenarios from detailed geochemical analysis of both primitive extraterrestrial materials and those ancient terrestrial rocks that still hold clues to the Hadean conditions, from comparison with the situation on our sister planets Venus and Mars, from *in situ* analysis of comets and asteroids, and from modelling. Two recent reviews are particularly pertinent to this brief overview. Shaw (2008, 2016) closely investigated data and models pertaining to the volatile component of the Earth, i.e. its atmosphere, oceans and carbon inventory, while Kamber (2015) was specifically concerned with the evolution of the early crust.

Of crucial importance for the emergence of life is the origin of the volatiles and timing of water condensation on the Earth. Although the volatiles, including water and reduced carbon molecules, appear to have originated on a variety of extraterrestrial bodies, the vast majority on Earth were brought in by carbonaceous chondrites (Pizzarello and Shock 2010), micrometeorites (Maurette 2006), and cometary material (Altwegg et al. 2016). Of particular importance is the "Late Veneer" during which an estimated 1–2% of the Earth's mass (e.g. Halliday 2000) was accreted after the large, Moon-forming impact at about 4.51 Ga (Hartmann and Davis 1975; Nisbet and Sleep 2001). This included the $\sim 0.01\%$ of matter accreted during the hypothesized Late Heavy Bombardment during which there was an apparent peak in extraterrestrial bombardment (4.1–3.85 Ga). The impact that resulted in the formation of the Moon created a planet with a magma ocean to depths of several kilometers which rapidly cooled down such that water could have condensed on the Earth's surface on a geologically-instantaneous time scale (Nisbet and Sleep 2001; Köberl 2006). This implies that, at least by 4.4 Ga, there was liquid water and possibly a global ocean on the Earth (Shaw 2008). There is some indication for the presence of water on the Hadean Earth preserved in the form of crystals of zircon. These refractory and long-lived minerals form through remelting and fractionation of hydrated crust. Some ancient, reworked zircon crystals had been dated at 4.4 Ga (Wilde et al. 2001) and it had been hypothesized that their oxygen isotope compositions reflected the presence of relatively low temperature (< 150 °C) hydrothermal fluids. However, later re-analysis of the minerals established a younger date of formation (Valley et al. 2014). The oldest known zircons are now dated at about 4.36 Ga old (Nemchin et al. 2006; Valley et al. 2014).

Of importance to discussion of the temperature of the early Earth is the 30% lower luminosity of the Sun, the "faint young Sun paradox" of Sagan and Mullen (1972). The temperature of the early oceans is of importance since temperatures > 120 °C break down the prebiotic building bricks of life (Kashefi and Lovley 2003). The Hadean ocean temperatures would have been controlled by two major, global factors, atmospheric composition and heat flux from below, with local perturbation caused by volcanic and hydrothermal activity as well as sporadic impacts, some of which could have been large enough to cause local evaporation of the ocean and remelting of the crust (Nisbet and Sleep 2001; Zahnle and Sleep 2006; Grieve et al. 2006). In fact the composition of the Hadean atmosphere is highly debated. As noted above, the lower luminosity of the young Sun invokes the question as to how the atmosphere could remain at a temperature compatible with liquid water? The various scenarios to explain this have consequences also for the inventory of carbon on the early

Earth. It has been generally accepted that the Hadean atmosphere after the Moon-forming impact consisted mostly of CO_2 at pressures sufficiently dense to keep temperatures above freezing point (e.g. Kasting 1993). If this were the case, where are the thick deposits of carbonate that should have resulted from acid rain and submarine alteration of the rocks at the surface of the Earth? Carbonate rocks from the Early Archaean period are few, consisting of carbonate-altered komatiitic/basaltic crustal rocks and some thin carbonate deposits (siderite, dolomite, all of which were subsequently silicified). Thick carbonate deposits typical of the later Archaean, Proterozoic and Phanerozoic did not form on the early Earth. We will come back to this question below but in the meantime let us address the possibility of additional "greenhouse" gases. It is probable that water vapour and methane (CH_4) were common greenhouse gases, with the addition of other minor species, such as sulphur dioxide (SO_2), Sulphur oxide (SO), ammonia (NH_3), and nitrogen (N_2). A significant amount of methane (and some ammonia) would have been produced from hydrothermal fluid contact with reduced carbon (Seewald et al. 2006), as well as more minor amounts by serpentinisation of ultramafic crust (Kelley et al. 2005), or reactions involving H_2-rich fluid inclusions in these rocks (McDermott et al. 2015). Thus, abiotically-produced methane may have been the key to keeping the Earth's atmosphere warm. Indeed, Shaw (2008) estimates that this effect could have endured for a billion years, i.e. well into the Archaean.

An alternative model posits that the Hadean Earth was frozen and that any liquid water present would have been due to volcanic/hydrothermal activity or impacts (Zahnle and Sleep 2006). Could this have been a possibility (an important question because certain experiments to condense the elusive sugars necessary for a primitive cell require ice Monnard and Ziock 2008)? Indeed, alternating freezing-hot scenarios have been proposed by a number of authors in the origins of life field, e.g. Russell and Arndt (2005).

One way of answering this question is to interpret backwards in time from the period when representative, sedimentary rocks are preserved well enough to provide environmental information, i.e. after 3.5 Ga. Although the terrains containing these rocks are rare, the information they hold is precious. At 3.5 Ga, the Earth was warm and probably hot. It was volcanically and hydrothermally active. All the sedimentary successions preserved document shallow water deposits that clearly indicate that the surface of the Earth was not frozen. Had this been the case, there would have been no evidence for tidal features, such as ripples, or shallow storm dunes (Lowe 1999; Tice and Lowe 2004; Westall et al. 2015a). This situation is evidenced by two Early Archaean terrains, the 3.5–3.3 Ga Pilbara (Australia) and Barberton (South Africa) greenstone belts. In this time period, these terrains were distant from each other although by 3.2 Ga they may have amalgamated into one craton. Although they are only two points of data in time and space, the story they tell is the same.

The Hadean and Early-Mid Archaean Earth was volcanically and hydrothermally active (Fig. 5). Early Archaean rock compositions (including komatiites, Mg-rich basalt-type rocks that only occurred on the hotter early Earth) testify to higher mantle temperatures (Arndt 1994), or simply higher convection plume temperatures according to Kamber (2015). Well-preserved sedimentary successions in the Pilbara (Australia) and Barberton (South Africa) greenstone belts document the prevailing influence of hydrothermal fluid circulation at the critical rock/sediment/water interface (Westall 2005, 2012; Westall et al. 2015a; Van Kranendonk et al. 2015). Scaled back in time to an Earth that was still (and is still) cooling down, after global melting due to the Moon-forming impact, it is clear that heat flow at the rock/water interface would have been even higher. It is highly unlikely that the global ocean, at least in the vicinity of hot rock, could have been frozen and we see below how important the water/mineral/rock connection is for the appearance of organized prebiotic molecules and, probably the first cells.

Fig. 5 Artist's impression of the surface of the early Earth. Note the numerous impact craters, hydrothermal vents and dusty atmosphere. ©Don Dixon

Let us come back to the carbon inventory of the early Earth. If the early atmosphere was indeed composed principally of CO_2 and, if the carbon was not sequestered significantly as carbonate deposits of carbonatised mafic/ultramafic crust, what happened to it? Shaw proposes an elegant solution. It is possible that the early atmosphere was not entirely slightly oxidizing, as could be expected from a CO_2 composition. Instead, Hadean (and Archaean) impactors could have resulted in creating a temporally, hydrogen-rich, reducing atmosphere because of the high input of reducing species, i.e. carbon and iron, into the atmosphere (Shaw 2008) (Fig. 5). In such an atmosphere, photochemical and electric discharge could produce a haze of reduced organics, including CH_4 (Tian et al. 2005), which could be rained out into the ocean as dissolvable and refractory organics. Intriguingly, Shaw (2008) envisages an ocean rich in dissolved organics while the insoluble organic matter formed a layer of "goo" floating on the ocean surface and/or sedimented to the bottom of the ocean by coating the surfaces of silicate grains. The missing carbon inventory could be simply reduced carbon sedimented onto the ocean floor and eventually recycled through tectonic processes to reappear in the atmosphere as volcanic CO_2. In this respect it is interesting to note that volcanic detrital particles in Early Archaean sediments are coated with a thin layer of organics (of either abiotic or biogenic origin, Westall et al. 2015a, 2015b). This scenario has important implications for the availability of organic carbon for prebiotic reactions.

A final word about the pH of the Hadean/Early Archaean oceans. With a relatively dense CO_2 atmosphere, the oceans would be *a priori* somewhat acidic in composition (Morse and MacKenzie 1998), although its pH would have been buffered by alteration of the mafic/ultramafic rocks and sediments on the ocean floor (Kempe and Degons 1985)—at least in the vicinity of rocks and sediments. One of the most important prebiotic reactions, for example the formose reaction, requires an alkaline environment, which would have been abundant at the water/rock interface, despite a probable overlying acidic ocean.

Various hypotheses propose different types of environments for the emergence of life. Hydrothermal vents associated with tectonic spreading ridges are favoured environ-

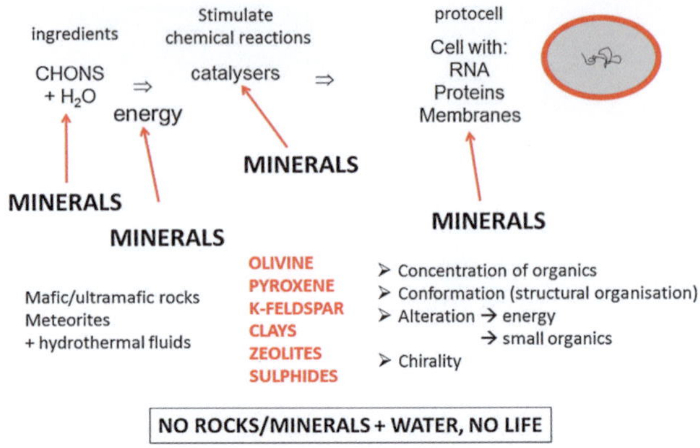

Fig. 6 Schematic view of the emergence of life and the importance of (hot) rock/mineral-water interactions

ments (Baross and Hoffman 1985; Russell and Hall 1997; Martin et al. 2008). This presupposes the existence of conventional plate tectonics (the Willson cycle described by Kamber 2015) during the Hadean about which there is great debate (e.g. Arndt 2013; Kamber 2015). As noted above, no Hadean crust has survived. Even during the Archaean, there is no evidence for Willson style plate tectonics until about 3.2 Ga, although it is likely that tectonic evolution towards this style of crustal recycling could have occurred earlier but was not preserved, its appearance being gradual during the Archaean (Kamber 2015). Hadean tectonic recycling is likely to have been of the kind described as sagduction, sagging of weak crust into the mantle, plus plume tectonics. This means that, most likely, hydrothermal vents aligned along a spreading ridge did not exist. However, this does not imply the demise of the hydrothermal vent scenario for the emergence of life. As noted above, on the hot, volcanically and hydrothermally active crust, there would have been an abundance of hydrothermal vents, especially in the vicinity of volcanic activity.

4 The Origin of Life

For the reasons evoked above, life based on carbon molecules had to have emerged in an aqueous environment. This is not the place to go into detail about the mechanisms for life's origin (still not yet understood; see Dass et al. 2016, for a comprehensive review); we note here the salient points. Simply explained, the essential elements (C, H, N, O, P, and S) form complex prebiotic molecules in water using various forms of energy (chemical, heat, radiation/light) and catalysed by reactive substrates, such as metal sulphides (Fe, Cu, Zn, Pb, Mn). These macromolecules combine in one or more environments to form a primitive protocell consisting of membrane-forming molecules, molecules for metabolism, and molecules for information storage and transcription. We will see below that reactions involving carbon molecules and water at the surfaces of hot rocks and minerals are crucial for the emergence of life (Fig. 6). These protocells were far more primitive than the most primitive microorganisms known today, anaerobic chemotrophs.

Various theoretical models and elegant experiments have been developed to try to better understand the how each class of macromolecule came into being. The chemotron model proposed a minimal cell containing an autocatalytic chemical (metabolism), replication, and a membrane-producing cycle (Ganti 1975). The autopoetic model proposed a system capable of reproducing and maintaining itself (Varela et al. 1974), while computer models follow the evolution of entities by a self-amplifying feedback process, transitioning from autocatalysis to self organisation (e.g. Kauffman 1991; Jessen 1978). Patel et al. (2015) suggested a common origin of RNA, protein, and lipid precursors in a cyanosulphidic protometabolism, although there is no experimental evidence for this.

Many experiments have tried to build parts of a protocell with varying degrees of success. Vesicular membranes of amphiphilic molecules can form readily (e.g. Bachmann et al. 1992) in which it has been possible to incorporate RNA as the information storage and transfer molecule (more stable than DNA and its probable precursor) (Hanczyc et al. 2007). Other molecules have been suggested as precursors to RNA, such as PNA, TNA or GNA (Lindahl 1993) since ribose, the sugar component of the molecule, is inherently unstable.

Various scenarios for the formation of prebiotic molecules have been proposed. One important aspect is the necessity of concentrating or condensing the prebiotic molecules so that they could interact with each other. From this point of view, the apparent importance of mineral surfaces for concentrating, stabilising and/or catalysing prebiotic reactions is becoming increasingly important. Phyllosilicate minerals, such as clays and zeolites have been proposed as playing an essential role in the origin of life. The former are layered, aluminosilicate minerals with charged surfaces that can chelate organics and catalyse reactions (Reid and Orgel 1967). Cairns-Smith (1985) even hypothesised an origin of life based on clay minerals in which complex organic molecules arose gradually on pre-existing, non-organic replication surfaces of silicate crystals in solution. Experiments using the clay montmorillonite showed that this phyllosilicate could catalyse the polymerization of RNA and the formation of membranes from lipids (Wenhua and Ferris 2006).

Any mineral with a porous structure and a reactive surface is of interest for the origins of life because of the possibility of concentrating the building blocks of life in a limited area and for the provision of surfaces that can catalyse reactions and stabilize molecules. Zeolites are porous aluminosilicate minerals of interest because of the potential of the pores for concentrating organics and the catalytic effects of their surfaces. Both clays and zeolites form by aqueous alteration of volcanic rocks and sediments. Furthermore, slightly acidic aqueous corrosion of the surfaces of minerals common in these rocks and sediments, such as pyroxene and feldspar, causes the formation of micropores in which reactions can take place (Westall et al. 2018, in press).

Another theory hypothesised autotrophic life as starting on the surface of iron sulphide minerals whereby the mineral acts as a catalyst for prebiotic reactions. Wächtershäuser (1988, 2007) and Huber and Wächtershäuser (2006) considered that a regular mesh of electric charges at the surface of pyrite may have driven primeval polymerisation of information-containing molecules, such as peptides and RNA. Other sulphides, Zn, Pb, Cu and Mn, have also been considered. These metal sulphides are precipitated by the interaction of H_2S-rich hydrothermal fluids with transition metal containing seawater, the metals leached from the underlying mafic and ultramafic rocks and sediments. Two of the sulphides, spahalerite (ZnS) and alabandite (MnS) can store energy produced by UV light and may have provided the necessary energy to fuel prebiotic reaction on an early Earth bathed in UV radiation Mulkidjanian (2009). If these minerals were important, the hot hydrothermal environments (black smoker types) in which such metals are precipitated would need to form in shallow water, a situation that does not occur on the modern Earth.

The discovery of porous, transition metal-rich hydrothermal vent edifices further fueled ideas about metal sulphide pores acting as membranes-bound compartments in which prebiotic reactions could take place (e.g. Russell and Hall 1997; Martin et al. 2008) (Fig. 7). Genetic studies also suggest that a hydrothermal scenario formed the framework for the emergence of life (Weiss et al. 2016). In all the above-mentioned scenarios, it is the contact of these minerals with seawater that makes them so reactive.

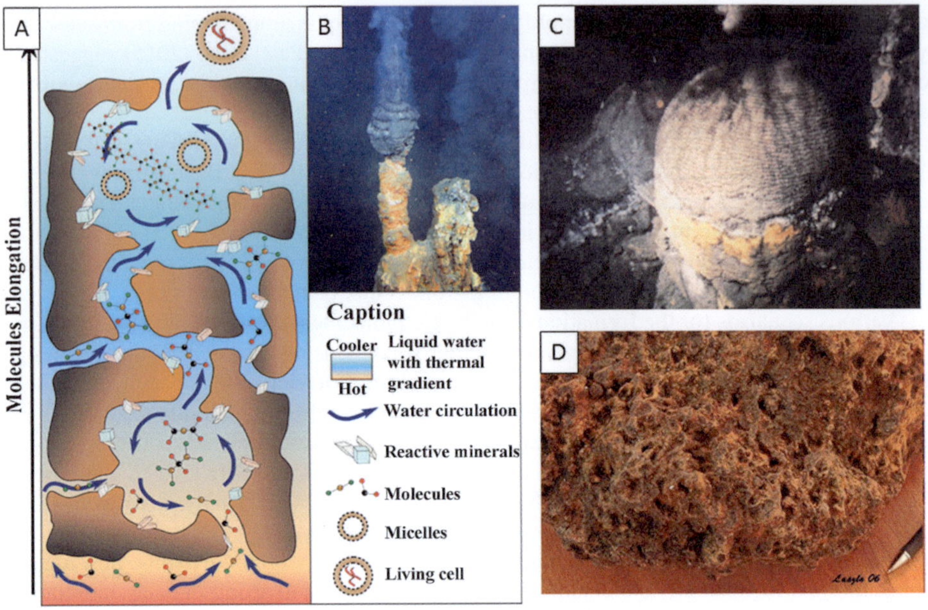

Fig. 7 (**A**) Schematic illustration of the emergence of protocells from hydrothermal vent environments (Westall et al. 2013, after Russell and Hall 1997). (**B**) Black smoker vent. (**C**) Beehive vent structure (image curtesy of Ifremer). (**D**) Porous structure of a hydrothermal beehive

It can be safely stated that, as an ingredient in the processes leading to the emergence of life, rocks and the minerals they contain play as an essential a role as carbon and water. Without rocks there would be no life. With the exception of the small amount of organic molecules formed in the primitive atmosphere, all organic molecules on the early Earth were either imported in dusty or rocky extraterrestrial materials or formed in the subsurface by processes catalysed by aqueous reactions on the surfaces of subsurface rocks. The important catalytic and concentrating role of minerals has been evoked above. Moreover, minerals contribute to the stabilisation of organic molecules by decreasing the degrees of freedom which they have when floating around in a liquid medium (Hazen and Sverjensky 2010). The conformation or structural organisation of organic molecules can be improved because the structure of minerals is inherently regular and repetitive (Sowerby et al. 2002). For the same reason, mineral surfaces can also augment the small chiral excess exhibited in abiotic carbon molecules in carbonaceous meteorites (Pizzarello et al. 2006). In addition, irregularities in mineral surfaces caused by their aqueous alteration produced surface defects aiding concentration and/or attachment of the molecules to the surfaces (Hazen and Sverjensky 2010).

The above discussion leads to the critical question, where did life originate, in what kind of environmental scenario, and when?

Given the fact that the ingredients of life were present, our hypothesis is that life emerged in the Hadean period as soon as environmental conditions were conducive to the assemblage of the macromolecular prebiotic building bricks, i.e. as soon as temperatures were low enough (< 100 °C, Miller and Lazcano 1995). It is probable that water condensed on the surface of the Earth quite soon after the giant impact that formed the Moon, thus we can assume that there were oceans by 4.4 Ga, although we do not know what the global ocean temperature was. Since heating by solar illumination was less efficient than today, perhaps the surfaces of the oceans cooled to clement temperatures ($< 80\,^{\circ}C$) relatively rapidly. However, it is the critical temperature at the water/rock/sediment interface that is of importance for the origin of life.

Life probably emerged in more than one location and possibly over and over again because of impact events (Maher and Stephenson 1988; Sleep et al. 1989; Ryder 2003), volcanic eruptions, eruptions of hot hydrothermal fluids, high UV flux to the surface of the early Earth (Lammer et al. 2008), etc. Eventually, one or more cell types would have dominated and been spread by circulation in the global ocean.

Various hypotheses propose different types of environments for the emergence of life, reviewed in Westall et al. (2018, in press). We have already evoked hydrothermal vent environments with their porous, membrane-like and reactive, sulphide mineral surfaces. Such environments could have been common the Hadean Earth, even without active spreading ridges, although temperatures in black-smoker type environments are far too hot to preserve carbon molecules. However, temperature gradients between the vents and lower temperature seawater would make the outer surfaces of the hydrothermal edifices interesting candidates.

For the same reasons, pumice has been put forward as an alternative location for the emergence of the first cells (Brasier et al. 2011, 2013). Disequilibria in the form of pH, redox and salinity gradients between porous compartments in the pumice could have driven prebiotic reactions while the pore surfaces could act as a template for primitive lipid membranes. Pumice floats on water and forms rafts floating on the ocean surface, which could have foundered at the edge of exposed land masses where there was hydrothermal activity, thus allowing for mixing of prebiotic molecules formed in the hydrothermal environments with those in the pumice compartments. This is certainly a plausible scenario although pumice, a derivative of evolved silicic magmas, would have been less common on the mafic/ultramafic dominated Hadean/Early Archaean Earth than it is today (Cleaves et al. 2012).

A number of origins of life hypotheses call for scenarios occurring in rivers or streams on dry land. The group of John Sutherland (Patel et al. 2015) considered the possibility of a "cyanosulphide" network of reactions based on hydrogen cyanide and hydrogen sulfide in streams of water irradiated by UV light. This network of reactions produced the chemical components of proteins and lipids, as well as those of RNA. Benner and colleagues (Benner et al. 2012; Benner 2013; Gollihar et al. 2014) proposed that the elements boron and molybdenum had important properties pertinent to the stabilization of the sugar ribose. Based on analogy with the modern Earth, borate minerals were proposed to have been relatively common on early continents where they would have been precipitated from altered tourmaline crystals, common in continental granites. Likewise, the usefulness of molybdenum in sugar stabilization led Benner (2013) to propose initially that life arose on early terrestrial continents where Mo was supposed to be present. However, the fact that the early environment was anoxic and that Mo is preferentially made available in significant quantities only in oxygenated conditions pushed Benner (2013) to suggest that, in this case, life must have arisen on the oxidized surface of Mars and been transported to Earth in martian meteorites, thus supporting the Arrhenius' theory of panspermia (1908).

The main difficulty with hypotheses requiring dry land is the paucity and lack of permanence of the early land masses (Kamber 2015). The early Earth was basically an ocean planet, even during the Early Archaean (Shaw 2008; Kamber 2015). Boron, although generally associated with high temperature fractionated rocks, such as granites (early granitoids were less evolved and less common on the Hadean/Early Archaean Earth than today), is also found in lower temperature hydrothermal vent environments and is a component of the mineral tourmaline which, can also occur in hydrothermal as well as hypersaline environments (Henry and Dutrow 2012). Thus, if boron is indeed a key element in the formation

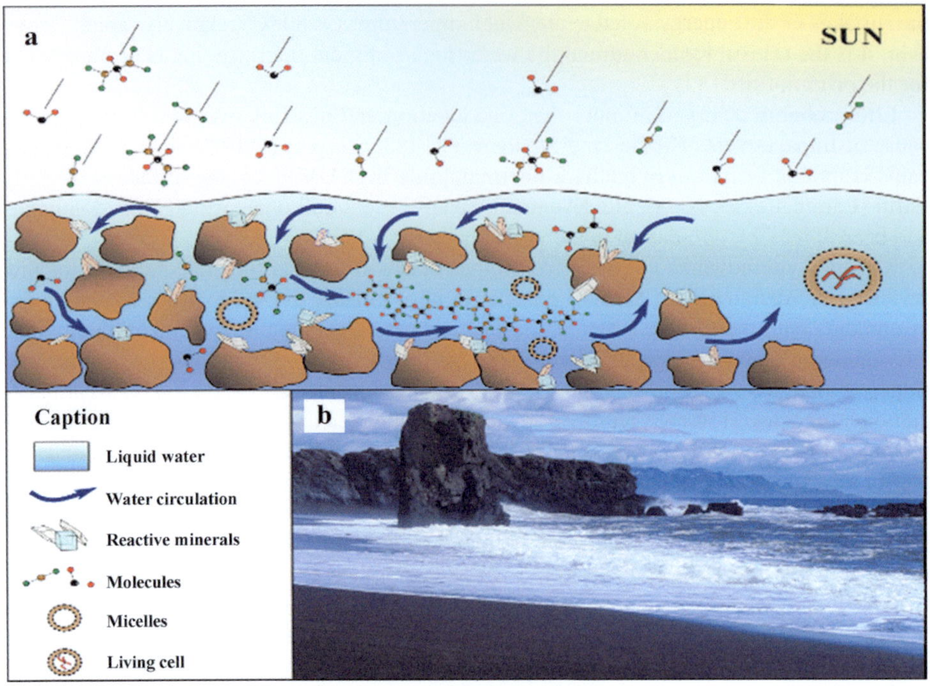

Fig. 8 Schematic view of the emergence of protocells in a tidal environment (Westall et al. 2013)

of a component of an information storage and transmission molecule, it would have been available. Molybdenum is a different story, rare on the early Earth and also rare on an early habitable, anaerobic Mars!

In an attempt to identify a plausible environmental scenario, Damer and Deamer (2015) suggest lipid cell membrane formation in freshwater volcanic pools believing that cell membranes cannot be formed in salty seawater. In their hypothesis, these membranes would be porous and could have contained other essential molecules that could be transferred to other protocells, the latter developing into true cells that would have gradually adapted to saltier environments in the ocean. With acid rain being a characteristic of the early Earth's atmosphere, one wonders whether such a scenario would have much success?

Of importance to most origin of life theories is the necessity of wetting/drying cycles that aid incorporation of essential biomolecules, such as RNA, within lipid vesicles (Damer and Deamer 2015; Higgs 2016). If such a process is indeed essential, it could only have occurred at the edge of the oceans in contact with the land masses. Such cycles have to be driven by intermittent evaporation and wetting. Tidal fluctuations can be invoked although, as noted above, the exposed landmasses would not have been long-lived features (Fig. 8). If tides and drying/wetting cycles were really critical, life could not have appeared on the other, potentially habitable planets in the Solar System. Only Mars and Venus had water lapping onto dry land—but no moon to provoke tides. It is possible that cyclic rainfall and evaporation could have produced the required effect. But the icy moons of Jupiter and Saturn would not have seen life appearing, at least in this manner.

Following the tidal scenario, Parnell (2004) and Adam (2007) have suggested that tidal concentration of radioactive particles, such as uranium, on beaches could have had an influence on the formation of prebiotic molecules. Such radioactive beach sand might have provided sufficient energy to generate organic molecules, such as amino acids and sugars from acetonitrile in water. The major problem with this hypothesis is that radioactive elements are mobilized only in oxidizing conditions. Thus, while present in rocks and minerals on the early Earth, they could not have been concentrated in such a way until the Earth had been oxidized, after 2.2 Ga!

Many prebiotic experiments have made use of ice in at least one or more steps during the fabrication of macromolecules. Alternating freeze-thaw conditions were used to form pyrimidines and triazines (Menor-Salván et al. 2009), while Orgel (2004) showed that freezing temperatures are advantageous for the synthesis of purines, due to the concentrating effect for key precursors such as hydrogen cyanide. Likewise, Robertson and Miller (1995) and Levy et al. (2000) noted the formation of several kinds of amino acids and nucleobases from ammonia and cyanide under freezing conditions.

Recently Westall et al. (2018, in press) have suggested another plausible scenario, the emergence of life in hydrothermally-influenced volcanic sediments. In this scenario, based on direct observations and measurements of porous volcanic sediments from the early Earth, ubiquitous hydrothermal fluids penetrate porous, carbon molecule-containing sediments whose pore spaces contain a matrix of hydrothermal silica gel. Characterised by physical and chemical gradients, with reduced water activity in the silica gels, this environment represented a widespread system of miniature chemical reactors in which the production and complexification of prebiotic molecules could have led to the origin of life.

5 Water and the Earliest Preserved Traces of Life

We have already stated above that almost no Hadean crustal rocks have survived, the whole crust having been recycled. A small, highly metamorphosed enclave in Canada, the Acasta Gneisses originally dated to 4.2 Ga (Bowring et al. 1990), has been shown to have a complicated history in which material from old crust was mixed with a younger component 3.6–3.7 Ga. In terms of the history of life, we are particularly concerned with sediments deposited in water because it is in this environment that life emerged, lived, and was preserved. Indeed, the oldest, well preserved sediments of the Pilbara (Australia and Barberton greenstone terranes (3.5–3.3 Ga, and younger) record thin layers of volcanoclastic sediments deposited between generally thick layers of mafic to ultramafic lavas in a platform type of environment formed by enormous outpourings of lava to create thickened crust above very hot mantle plumes (Van Kranendonk et al. 2015). The high heat flow in these regions contributed not only to extensive volcanic activity but also to extremely rapid lithification by silicification of the sediment layers, as well as the upper layers of the volcanic lavas underlying the sediments. The result of this rapid lithification is superb physical preservation of the sediments and the traces of life they contain.

After more than 30 years of study of the ancient traces of life, the surmounting of numerous pitfalls in the sense of mis-identifications, and the advent of new, fine-scale technology, we now have a fairly good overview of the diversity of early life by $\sim$ 3.5 Ga. It appears that life was as diverse as it could be in an anaerobic environment. Chemotrophic microbes colonized the surfaces of volcanic detrital particles in all aqueous environments forming thin, monolayer biofilms in relatively oligotrophic environments away from nutrient-rich hydrothermal sources, and thick biofilms in the vicinity of these hydrothermal fluids (Westall et al. 2006, 2011a, and 2015a). The chemotrophs would have comprised organisms obtaining their carbon from either reduced carbon, or in the seawater or hydrothermal fluids, and their energy from either redox reactions at the surfaces of reactive minerals or of H_2 in the hydrothermal fluids (lithotrophs), or their energy from oxidation of pre-existing carbon (organotrophs). While the individual chemotrophs are very small (< 1 μm in size), their biofilms can be quite extensive, producing colonies of tens to many hundreds of μm^2 and, where well-developed, completely coating the detrital particles with films up to 20–30 μm thickness (Westall et al. 2006, 2011a, 2015a). Phototrophic microorganisms (obtaining their carbon from CO_2 in the seawater and energy from sunlight) formed films and mats on flat sediment surfaces in the photic zone of shallow the shallow seas, sometimes even competing with the chemotrophs for space! These mats could be flat-lying features or domical in shape (termed tabular or domical stromatolites, cf. Krumbein 1983). The 3.45 Ga-old Strelley Pool Chert in the Pilbara contains a variety of domical shaped stromatolites indicating different microenvironments on the carbonate platform on which they formed (Hofmann et al. 1999; Allwood et al. 2006). Such domical structures have not yet been identified in the Barberton region where the Early to Mid Archaean phototrophic mats are tabular (Byerly et al. 1986; Westall et al. 2006, 2011b, 2015b). The organisms forming the phototrophic biofilms and mats are rarely preserved, it is mostly the polymer-rich mat that is fossilised. However, where visible, the filaments forming the phototrophic films are also very small, < 1 μm in diameter and up to tens of microns in length (Westall et al. 2006). Other enigmatic structures have been described form the Pilbara and Barberton sediments. Sugitani et al. (2015) have been studying relatively large flanged carbonaceous structures up to several 100 μms in size that may have been early planktonic forms in 3.4–3.2 Ga sediments from the Pilbara, while Javaux et al. (2010) described large carbonaceous disks (again up to hundreds of μm in size) that could be the remains of planktonic organisms in 3.2 Ga sediments from Barberton. Again, in 3.2 Ga sands from Barberton, Homann et al. (2016) found the casts of microbes in silica precipitated in a beach vadose zone, indicating that life had already started to colonise aqueous environments above sea-level. Indeed, Westall et al. (2006 and 2011b) documented microbial mats exposed and desiccated on a beach surface in 3.3 Ga sediments from Barberton.

What this brief review tells us is that life had colonized all habitable environments and microenvironments by 3.5–3.3 Ga and that it had already evolved the sophisticated phototrophic metabolism. The degree of metabolic diversity and the distribution exhibited by these oldest preserved traces of life testifies (1) that life already had a long history and most likely appeared as soon as conditions on the Hadean Earth were habitable; (2) that it had probably also colonized crustal fractures (at least where hydrothermal temperatures were supportable); (3) that at least at 3.5 Ga life must have been thermophilic in the hydrothermal fluid-bathed shallow waters. Estimates of seawater temperature from silicon and oxygen isotopes vary widely, from 70–80 °C (Knauth and Lowe 2003), through temperatures around 50–55 °C (van den Boorn et al. 2007), and even to a relatively cool 35 °C (Stefurak et al. 2015). At the seawater/rock/sediment interface where hydrothermal activity dominated, temperatures must have been high but they may have been lower in the overlying water column if it was thick enough and not mixed by tides or storms with the hot, bottom layers.

6 Water and the Emergence of Life Elsewhere?

Water and carbon are key ingredients of life but equally important are the juxtaposition of these ingredients with hot rocks and a source of energy. Although environmental conditions at the surfaces of the other planets and satellites in the Solar System today are not conducive to life, the Earth being the only "habitable planet *sensu stricto*", where water is stable at a planet's surface, this was not the case during the early stages of evolution of the Solar System and is probably not the case with respect to conditions under the surfaces of certain planets and satellites. The subsurface of Mars and below the icy crusts of planets, such as Europa, Ganymede, and Enceladus could be termed "habitable *sensu lato*".

All of these planets were, at one stage, hot, rocky bodies with liquid water directly in contact and with the same extraterrestrial, crustal and possibly atmospheric sources of reduced carbon molecules. If life is the result of the evolution of chemical processes based on the simplest of ingredients but within the constraints of the water/hot rock scenario, as we hypothesise, then life could have appeared on a number of bodies in the Solar System and, indeed, elsewhere. Many Solar System bodies, even those now coated with ice or simply a dead rocky surface, have a metal core coated by a silicate mantle. Study of stony meteorites shows us that many come from bodies, such as Ceres, that had seen liquid water in contact with hot rock. It now appears that, like some of the icy satellites of Jupiter and Saturn, Ceres may have had a subsurface ocean and may have been habitable (Castillo-Rogez and Conrad 2010). This was even the case on young Titan (Raulin 2008). While the interaction of significant water with hot rock at the surface of a planet like Titan may have been long enough to have produced protocells or even cells, this situation is unlikely to have occurred on asteroids, where aqueous contact of water and hot rock was rather within the silicate mantle. Although crustal hydrothermal circulation in the asteroid could have led to the production of H_2 and small carbon molecules, as on Earth, such molecules are not protocells. In the case of Titan, whose hydrosphere contains of ethane and methane, note that life forms using hydrocarbons or ammonia and water as a solvent were proposed by Shapiro and Schulze-Makuch (2009).

If life on Earth is the result of specific, uncommon environmental conditions, such as tidal cycles, or on dry land (which we believe not very likely given their relative rarity), this poses a limit to the appearance of life, similar to that on Earth, elsewhere. Tidal forcing could be evoked for the Jovian and Saturnian planets but not in the sense of producing tidal cycles at the edges of land masses. On the other hand, exposed land masses may have occurred in an ocean, sea or large impact crater setting on Venus and were certainly present in abundance on Mars, which could be considered as a "land-locked" planet with relatively isolated bodies of water during its early history.

For life to be the "cosmic phenomenon" described by Christian de Duve (1995), is should appear simply, without very specific and restricted conditions. As we have seen from the discussion above, the rarity of specific scenarios even on the early Earth lead us to believe that the most common environmental situation would be the most likely, i.e. those involving carbon/water/hot rock/mineral interactions.

The kind of life forms that could have evolved on other bodies during the early part of the Solar System's history would be simple chemotrophic forms, as on Earth; i.e. life forms obtaining their carbon and energy from simple ingredients already in abundance—carbon coming from CO_2 in water and reduced carbon molecules, and energy from oxidation of reduced carbon or H_2 or redox reaction on mineral surfaces in contact with water. Life on Earth further evolved the capability of exploiting another, more powerful source of energy, i.e. sunlight, used in the photosynthetic metabolic pathway. Photosynthesis produces generally more energy than chemosynthesis, thus allowing a buildup of biomass (Des Marais 2001) that, eventually with the mutations that produced oxygenic photosynthesis, completely covered all shallow water locations (i.e. in the photic zone) and exposed landmass surfaces.

However, while photosynthesis appears to have evolved a number of times in different branches of organisms, present understanding of the importance of reactions between organic molecules and mineral surfaces seems to indicate that protocells were chemotrophic, as does the fact that phototrophs do not reach the root of the phylogenetic tree, therefore phototrophic characteristics could not have constituted those attributed to the Last Universal Common Ancestor (LUCA).

Despite probable large impacts that could have vaporized rock and water in specific locations on the Earth during its early history, early terrestrial life was a global phenomenon, redistributed in continuation around the globe by ocean currents and possibly terrestrial impact ejecta containing viable cells. This means that, from the time liquid water condensed on its surface, after the Moon-forming impact, the Earth has been permanently habitable at its surface and certainly in subsurface fractures. No other planet in the Solar System has had a permanently habitable surface. Although Mars was a land-locked planet with isolated surficial habitats, the lack of long lived, relatively stable habitable conditions and connectivity between these habitats resulted in what has been described as "punctuated" habitability (Westall et al. 2013, 2015b; Cockell et al. 2012). It is highly unlikely that a photosynthetic metabolism could have developed on Mars. Perhaps something like photosynthesis developed on Venus. Schulze-Makuch and Irwin (2004) even suggest that phototrophs could have adapted a sulphur-based UV strategy within water droplets in the atmosphere of Venus, although this hypothesis was refuted by Cockell and Westall (2004) on the grounds that habitability requires that the physiological requirements of microorganisms on Earth match the empirically determined combined physical and chemical conditions in the extraterrestrial body (i.e. Venus) and the cloud deck of Venus does not satisfy these conditions.

Thus, simple, chemotrophic life forms could still exist in the subsurface of Mars and in the oceans of some icy satellites if the following preconditions are met: (1) water in contact with hot rock (heated by tidal forcing from Jupiter and Saturn, and due to crustal heat flow on Mars), (2) a means of transport of the reduced carbon molecules still raining down on the surfaces of the icy crusts of the Jovian and Saturnian planets to the underlying liquid water or, in the case of Mars, its subsurface, (3) preservation of the molecules from the intense radiation damage to which they would be exposed at the surfaces of the satellites or through crustal radioactivity on Mars, and (4) transformation of the molecules in such a way as to be available for assimilation by chemotrophs. Similar situations can be evoked for rocky/water planets in other sun systems throughout the Universe. However, noting the relative singularity of conditions giving rise to photosynthesis on Earth—and evolution to technologically-sophisticated life forms, which appears to have depended on geological and biological co-evolution—the most common life forms in the Universe would probably have some kind of chemosynthetic metabolism—which would be invisible from orbital observation (unless perhaps the planet was colonised by an abundance of CH_4-producing methanogens) and thus difficult to identify. However, if life is a cosmic phenomenon, we can surmise that any body with hot rock in contact with water and a source of carbon molecules (ubiquitous) on periods long enough for life to appear, is likely inhabited.

7 Conclusion

Considering life forms based on carbon molecules and water, recent work suggests that rocks and minerals in contact with water and heat were essential ingredients for the emergence of life—at least on Earth under physico-chemical conditions that were and are common on many terrestrial planets in the Universe. Using this premise, and the fact that the ingredients for such life forms were and are common in the Universe, we espouse the concept of Christian de Duve, that carbon-water based life is a cosmic phenomenon. Nevertheless, the constraints imposed the necessity of carbon/water/rock interactions where heat is involved, place limits on the locations where life could appear. Under these conditions, life could have emerged on a number of planets and satellites in the early Solar System, including the Earth, Mars, Venus and possibly the dwarf planet Ceres, as well as on icy satellites, such as Enceladus, Titan, Europa and possibly others. We posit that extraterrestrial life forms in the Solar System were (and possibly still are) likely to be chemotrophic rather than phototrophic on the basis that phototrophy evolved later than chemotrophy on a permanently habitable planet. Chemotrophic biosignatures on exoplanets are likely to be too subtle for detection at distance. The signatures of oxygenic phototrophy, on the other hand, would produce the tell-tale mixture of incompatible gases in the planet's atmosphere, such as oxygen and methane.

References

Z. Adam, Astrobiology **7**(6), 852 (2007)
A.C. Allwood, M.R. Walter, B.S. Kamber, I.W. Burch, Nature **441**, 714 (2006)
K. Altwegg, H. Balsiger, A. Bar-Nun, J-J. Berthelier, A. Bieler, P. Bochsler, C. Briois, U. Calmonte, M.R. Combi, H. Cottin, J. De Keyser, F. Dhooghe, B. Fiethe, S.A. Fuselier, S. Gasc, T. Gombosi, K.C. Hansen, M. Haessig, A. Jäckel, E. Kopp, A. Korth, L. Le Roy, U. Mall, B. Marty, O. Mousis, T. Owen, H. Rème, M. Rubin, T. Sémon, C-Y. Tzou, J.H. Waite, P. Wurz, Sci. Adv. **2** (2016)
N.T. Arndt, in *Archean Crustal Evolution*, ed. by K.C. Condie (Elsevier, Amsterdam, 1994), p. 11
N.T. Arndt, Geochem. Perspect. **2**, 405 (2013)
S. Arrhenius, *Worlds in the Making: The Evolution of the Universe* (Harper & Row, New York, 1908)
P.A. Bachmann, P.L. Luisi, J. Lang, Nature **357** (1992)
J.A. Baross, S.E. Hoffman, Orig. Life Evol. Biosph. **15** (1985)
A. Belloche, R.T. Garrod, H.S.P. Müller, K.M. Menten, Science **345** (2014)
S.A. Benner, H-J. Kim, M.A. Carrigan, Acc. Chem. Res. **45** (2012)
S.A. Benner, Mineral. Mag. **77** (2013)
P. Boehnke, M.T. Harrison, Proc. Natl. Acad. Sci. **113**, 10802–10806 (2016)
S.A. Bowring, T.B. Housh, C.E. Isachsen, in *Origin of the Earth*, ed. by H.E. Newsom, J.H. Jones (Oxford University Press, New York, 1990), p. 319
A. Brack, in *Astrobiology*, ed. by G. Horneck, C. Baulstark-Kahn (Springer, Berlin, 2002), p. 79
A. Brack, Chem. Biodivers. **4** (2007)
A. Brack, M. Troublé, Orig. Life Evol. Biosph. **40**, 31 (2010)
M.D. Brasier, R. Matthewman, S. McMahon, D. Wacey, Astrobiology **11**, 725 (2011)
M.D. Brasier, R. Matthewman, S. McMahon, M.R. Kilburn, D. Wacey, Precambrian Res. **224**, 1 (2013)
G.R. Byerly, D.R. Lowe, M.M. Walsh, Nature **319**, 489 (1986)
A.G. Cairns-Smith, *Seven Clues to the Origin of Life—A Scientific Detective Story* (Cambridge University Press, Cambridge, 1985), pp. 1–144
J.C. Castillo-Rogez, P.G. Conrad, AbSciCon, Abstr. 5302 (2010)
H.J. Cleaves II, A. Michalkova Scott, F.C. Hill, J. Leszczynski, N. Sahai, R. Hazen, Chem. Soc. Rev. **41**, 5502 (2012)
C.S. Cockell, F. Westall, Int. J. Astrobiol. **3**, 157 (2004)
C.S. Cockell, M. Balme, J.C. Bridges, A. Davilad, S.P. Schwenzer, Icarus **217**, 184 (2012)
B. Damer, D. Deamer, Life **5**(1), 872 (2015)
A.V. Dass, K. Hickman-Lewis, A. Brack, T.P. Kee, F. Westall, Chem. Sel. **1**, 4906 (2016)
C. de Duve, *Vital Dust: Life as a Cosmic Imperative* (Basic Books, New York, 1995), p. 362
D.J. Des Marais, Science **291**, 436 (2001)
T. Djokic, M.J. van Kranendonk, K.A. Campbell, M.R. Walter, C.R. Ward, Nat. Commun. (2017). https://doi.org/10.1038/ncomms15263
T. Ganti, Biosystems **7**, 15 (1975)
J. Gollihar, M. Levy, A.D. Ellington, Science **343**, 259 (2014)
R.A.F. Grieve, M.J. Cintala, A.M. Theriault, in *Processes on the Early Earth*, ed. by U.W. Reimold, R.L. Gibson (2006), pp. 23–31. GSA Sp. Paper
A.N. Halliday, Earth Planet. Sci. Lett. **176**, 17 (2000)
M.M. Hanczyc, S.S. Mansy, J.W. Szostak, Orig. Life Evol. Biosph. **37**, 67 (2007)
W.K. Hartmann, D.R. Davis, Icarus **24**, 504 (1975)
R.M. Hazen, D.A. Sverjensky, Cold Spring Harb. Perspect. Biol. **2** (2010)

D.J. Henry, B.L. Dutrow, Lithos **154**, 16 (2012)
P. Higgs, Life **624** (2016)
H.J. Hofmann, K. Grey, A.H. Hickman, R.I. Thorpe, Geol. Soc. Am. Bull. **111**, 1256 (1999)
H.D. Holland, *The Chemical Evolution of the Atmosphere and Oceans* (Princeton University Press, Princeton, 1984), p. 582
M. Homann, C. Heubeck, T.R.R. Bontognali, A-S. Bouvier, L.P. Baumgartner, A. Airo, Geology **44**, 51 (2016)
C. Huber, G. Wächtershäuser, Science **314**, 630 (2006)
E.J. Javaux, C.P. Marshall, A. Bekker, Nature **463**, 934 (2010)
W. Jessen, Naturwissenschaften **65**, 449 (1978)
G.F. Joyce, *The RNA World: Life Before DNA and Protein* (Cambridge University Press, Cambridge, 1995), pp. 139–151
K. Kashefi, D.R. Lovley, Science **301**, 934 (2003)
B.S. Kamber, Precambrian Res. **258**, 48 (2015)
J.F. Kasting, Science **259**, 920 (1993)
S.A. Kauffman, Sci. Am. **64** (1991)
D.S. Kelley, J.A. Karson, G.L. Früh-Green, D.R. Yoerger, T.M. Shank, D.A. Butterfield, J.M. Hayes, M.O. Schrenk, E.J. Olson, G. Proskurowski, M. Jakuba, A. Bradley, B. Larson, K. Ludwig, D. Glickson, K. Buckman, A.S. Bradley, W.J. Brazelton, K. Roe, M.J. Elend, A. Delacour, S.M. Bernasconi, M.D. Lilley, J.A. Baross, R.E. Summons, S.P. Sylva, Science **307**, 1428 (2005)
S. Kempe, E.T. Degons, Chem. Geol. **53**, 95 (1985)
L.P. Knauth, D.R. Lowe, Geol. Soc. Am. Bull. **115**, 566 (2003)
W.E. Krumbein, Dev. Precambrian Geol. **7**, 385 (1983)
C. Köberl, Elements **2**, 211 (2006)
C. Konn, J.L. Charlou, N.G. Holm, O. Mousis, Astrobiology **15**, 381 (2015)
A. Knoll, *Life on a Young Planet: The First Three Billion Years of Evolution on Earth* (Princeton University Press, Princeton, 2003), pp. 1–277
H. Lammer, J.F. Kasting, E. Chassefière, R.E. Johnson, Y.N. Kulikov, F. Tian, Space Sci. Rev. **139**, 399 (2008)
M. Levy, S.L. Miller, K. Brinton, J.L. Bada, Icarus **145**(2), 609 (2000)
T. Lindahl, Nature **362**(6422), 709 (1993)
D.R. Lowe, Spec. Pap., Geol. Soc. Am. **329**, 83 (1999)
K.A. Maher, D.J. Stephenson, Nature **331**, 612 (1988)
W.F. Martin, J. Baross, D. Kelley, M.J. Russell, Nat. Rev. Microbiol. **6**, 805 (2008)
M. Maurette, *Micrometeorites and the Mysteries of Our Origins* (Springer, Berlin, 2006), p. 182
M. Maurette, A. Brack, Meteorit. Planet. Sci. **41**, 5247 (2006)
J.M. McDermott, J.S. Seewald, C.R. German, S.P. Sylva, Proc. Natl. Acad. Sci. USA **112**, 7668 (2015)
C. Menor-Salván, R-B. Marta, G. Marcelo, O-E. Susana, V-V. Sabino, Chemistry **15**(17), 4411 (2009)
T.J. Millar, in *Astrobiology: Future Perspectives*, ed. by P. Ehrenfreund, W.M. Irvine, T. Owen, L. Becker, J. Blank, J.R. Brucato, L. Colangeli, S. Derenne, A. Dutrey, D. Despois, A. Lazcano F. Robert (Springer, Netherlands, 2005), p. 4411
S.L. Miller, Science **117**, 528 (1953)
S.L. Miller, A. Lazcano, J. Mol. Evol. **41**, 689 (1995)
P-A. Monnard, H. Ziock, Chem. Biodivers. **8**, 1521 (2008)
J.W. Morse, F.T. MacKenzie, Aquat. Geochem. **4**(3–4), 301 (1998)
A.Y. Mulkidjanian, Biol. Direct **4**, 26 (2009)
A.A. Nemchin, R.T. Pidgeon, M.J. Whitehouse, Earth Planet. Sci. Lett. **244**, 218 (2006)
E.G. Nisbet, N.H. Sleep, Nature **409**, 1083 (2001)
L.E. Orgel, Orig. Life Evol. Biosph. **28**, 91 (1998)
L.E. Orgel, Orig. Life Evol. Biosph. **34**(4), 361 (2004)
J. Parnell, Orig. Life Evol. Biosph. **34**(6), 533 (2004)
B.H. Patel, C. Percivalle, D.J. Ritson, C.D. Duffy, J.D. Sutherland, Nat. Chem. **7**(4), 301 (2015)
J. Peretò, J.L. Bada, A. Lazcano, Orig. Life Evol. Biosph. **39**, 395 (2009)
S. Pizzarello, G.W. Cooper, G.J. Flynn, in *Meteorites and the Early Solar System II*, ed. by D.S. Lauretta, H.Y. McSween (University of Arizona Press/Lunar and Planetary Institute, Tucson/Houston, 2006), p. 625
S. Pizzarello, E. Shock, Cold Spring Harb. Perspect. Biol. (2010). https://doi.org/10.1101/cshperspect.a002105
G. Proskurowski, M.D. Lilley, J.S. Seewald, G.L. Früh-Green, E.J. Olson, J.E. Lupton, S.P. Sylva, D.S. Kelley, Science **319**, 604 (2008)
F. Raulin, Space Sci. Rev. **135**, 37 (2008)
C. Reid, L.E. Orgel, Nature **216**, 455 (1967)

M.P. Robertson, S.L. Miller, Nature **375**(6534), 772 (1995)
M.J. Russell, A.J. Hall, J. Geol. Soc. Lond. **154**, 377 (1997)
M.J. Russell, N.T. Arndt, Biogeosciences **2**, 97 (2005)
G. Ryder, Astrobiology **3**, 3 (2003)
C. Sagan, G. Mullen, Science **177**, 52 (1972)
D. Schulze-Makuch, L.N. Irwin, *Life in the Universe: Expectations and Constraints* (Springer, Berlin, 2004), p. 172
J.S. Seewald, M.Y. Zolotov, T. McCollom, Geochim. Cosmochim. Acta **70**, 446 (2006)
R. Shapiro, D. Schulze-Makuch, Astrobiology **9**, 335 (2009)
G.H. Shaw, Chem. Erde **68**, 235 (2008)
G.H. Shaw, *Earth's Early Atmosphere and Oceans, and the Origin of Life* (Springer, Switzerland, 2016), p. 113
B. Sherwood Lollar, L-C. Georges, K. Voglesonger, F.S. Gregory, Geochim. Cosmochim. Acta **72**, 4778 (2008)
E.L. Shock, T.M. McCollom, M.D. Schulte, in *The Keys to Molecular Evolution and the Origin of Life*, ed. by J. Wiegel, M.W. Adams (Taylor & Francis, London, 1998), p. 59
N.H. Sleep, K.J. Zahnle, J.F. Kasting, H.J. Morowitz, Nature **342**, 139 (1989)
S.J. Sowerby, N.G. Petersen, N.G. Holm, Orig. Life Evol. Biosph. **32**, 35 (2002)
E.J.T. Stefurak, W.W. Fischer, D.R. Lowe, Geochim. Cosmochim. Acta **150**, 26 (2015)
K. Sugitani, K. Mimura, M. Takeuchi, K. Lepot, S. Ito, E.J. Javaux, Geobiology **13**, 507 (2015)
R.E. Summons, J.P. Amend, D. Bish, R. Buick, G.D. Cody, D.J. Des Marais, G. Dromart, J.L. Eigenbrode, A.H. Knoll, D.Y. Sumner, Astrobiology **11**, 157 (2011)
F. Tian, O.B. Toon, A.A. Pavlov, H. De Steck, Science **308**, 1014 (2005)
M. Tice, D.R. Lowe, Nature **431**, 549 (2004)
J.W. Valley, A.J. Cavosie, T. Ushikubo, D.A. Reinhard, D.F. Lawrence, D.J. Larson, P.H. Clifton, T.F. Kelly, S.A. Wilde, D.E. Moser, M.J. Spicuzza, Nat. Geosci. **7**, 219 (2014)
S.H.J.M. van den Boorn, M.J. van Bergen, W. Nijman, P.Z. Vroon, Geology **35**, 939 (2007)
M. Van Kranendonk, R.H. Smithies, W.L. Griffin, D.L. Huston, A.H. Hickman, D.C. Champion, C.R. Anhaeusser, F. Pirajno, in *Continent Formation Through Time*, ed. by N.M.W. Roberts, M. Van Kranendonk, S. Parman, S. Shirey, P.D. Clift (Geological Society, London, 2015), p. 83. Special Publications
F.J. Varela, H. Maturana, R. Uribe, Biosystems **5**, 187 (1974)
G. Wächtershäuser, Microbiol. Rev. **52**, 452–484 (1988)
G. Wächtershäuser, Chem. Biodivers. **4**, 584 (2007)
M.C. Weiss, F.L. Sousa, N. Mrnjavac, S. Neukirchen, M. Roettger, S. Nelson-Sathi, W.F. Martin, Nat. Microbiol. **1**, 16116 (2016)
H. Wenhua, J.P. Ferris, J. Am. Chem. Soc. **128**(27), 8914 (2006)
F. Westall, in *The Early Earth and the Origin of Life*, ed. by H. Martin, M. Gardaud, G. Reisse, B. Barbier (Springer, Berlin, 2005), p. 195
F. Westall, in *Origins and Evolution of Life, an Astrobiology Perspective*, ed. by M. Gargaud, P. Lopez-Garcia, H. Martin (Cambridge University Press, Cambridge, 2011), p. 391
F. Westall, *Astrobiology*, ed. by C. Impey, J. Lunine, J. Funes (Cambridge University Press, Cambridge, 2012), p. 89
F. Westall, Boll. Soc. Paleontol. Ital. **55**, 85 (2016)
F. Westall, C.E.J. de Ronde, G. Southam, N. Grassineau, M. Colas, C. Cockell, H. Lammer, Philos. Trans. R. Soc. B **361**, 1857 (2006)
F. Westall, B. Cavalazzi, L. Lemelle, Y. Marrocchi, J.N. Rouzaud, A. Simionovici, M. Salome´, S. Mostefaoui, C. Andreazza, F. Foucher, J. Toporski, A. Jauss, V. Thiel, G. Southam, L. MacLean, S. Wirick, A. Hofmann, A. Meibom, F. Robert, C. Défarge, Earth Planet. Sci. Lett. **310**, 468–479 (2011a)
F. Westall, F. Foucher, B. Cavalazzi, S.T. de Vries, W. Nijman, V. Pearson, J. Watson, A. Verchovsky, I. Wright, J-N. Rouzaud, D. Marchesini, S. Anne, Planet. Space Sci. **59**, 1093 (2011b)
F. Westall, D. Loizeau, F. Foucher, N. Bost, M. Bertrand, J. Vago, G. Kminek, Astrobiology **13**, 887 (2013)
F. Westall, K.A. Campbell, J. Bréhéret, F. Foucher, P. Gautret, A. Hubert, S. Sorieul, N. Grassineau, D.M. Guido, Geology **43**, 615 (2015a)
F. Westall, F. Foucher, N. Bost, M. Bertrand, D. Loizeau, J.L. Vago, G. Kminek, F. Gaboyer, K.A. Campbell, J.G. Bréhéret, P. Gautret, C.S. Cockell, Astrobiology **15**, 988 (2015b)
F. Westall, K. Hickman-Lewis, N. Hinman, P. Gautret, K.A. Campbell, J.G. Bréhéret, F. Foucher, A. Hubert, S. Sorieul, A.V. Dass, T.P. Kee, T. Georgelin, A. Brack, Astrobiology (3) (2018, in press)
F.H. Westheimer, Science **235**, 1173 (1987)
S.A. Wilde, J.W. Valley, W.H. Peck, C.M. Graham, Nature **409**, 175 (2001)
K. Zahnle, N.H. Sleep, in *Comets and the Origin and the Evolution of Life*, ed. by P.J. Thomas, R.D. Hicks, C.F. Chyba, C.P. McKay (Springer, Berlin, 2006), pp. 175–208

Space Sci Rev (2018) 214:36
https://doi.org/10.1007/s11214-018-0474-9

Water Reservoirs in Small Planetary Bodies: Meteorites, Asteroids, and Comets

Conel M.O'D. Alexander[1] · Kevin D. McKeegan[2] · Kathrin Altwegg[3]

Received: 10 January 2018 / Accepted: 11 January 2018 / Published online: 23 January 2018

Abstract Asteroids and comets are the remnants of the swarm of planetesimals from which the planets ultimately formed, and they retain records of processes that operated prior to and during planet formation. They are also likely the sources of most of the water and other volatiles accreted by Earth. In this review, we discuss the nature and probable origins of asteroids and comets based on data from remote observations, in situ measurements by spacecraft, and laboratory analyses of meteorites derived from asteroids. The asteroidal parent bodies of meteorites formed ≤ 4 Ma after Solar System formation while there was still a gas disk present. It seems increasingly likely that the parent bodies of meteorites spectroscopically linked with the E-, S-, M- and V-type asteroids formed sunward of Jupiter's orbit, while those associated with C- and, possibly, D-type asteroids formed further out, beyond Jupiter but probably not beyond Saturn's orbit. Comets formed further from the Sun than any of the meteorite parent bodies, and retain much higher abundances of interstellar material. CI and CM group meteorites are probably related to the most common C-type asteroids, and based on isotopic evidence they, rather than comets, are the most likely sources of the H and N accreted by the terrestrial planets. However, comets may have been major sources

The Delivery of Water to Protoplanets, Planets and Satellites
Edited by Michel Blanc, Alessandro Morbidelli, Yann Alibert, Lindy Elkins-Tanton, Paul Estrada, Keiko Hamano, Helmut Lammer, Sean Raymond and Maria Schönbächler

✉ C.M.O'D. Alexander
calexander@carnegiescience.edu

K.D. McKeegan
mckeegan@epss.ucla.edu

K. Altwegg
kathrin.altwegg@space.unibe.ch

1 Department of Terrestrial Magnetism, Carnegie Institution for Science, 5241 Broad Branch Road NW, Washington, DC 20015, USA

2 Department of Earth, Planetary, and Space Sciences, University of California–Los Angeles, Los Angeles, CA 90095-1567, USA

3 Physikalisches Institut, University of Bern, Sidlerstrasse 5, 3012 Bern, Switzerland

of the noble gases accreted by Earth and Venus. Possible constraints that these observations can place on models of giant planet formation and migration are explored.

Keywords Asteroids · Comets · Meteorites · Chondrites · Volatiles · Water · Organics · Rosetta · 67P/Churyumov-Gerasimenko · Hydrogen isotopes · Nitrogen isotopes · Giant planet migration

1 Introduction

Despite water's status as the second most abundant molecule in the Solar System, Earth and the rocky planets are highly desiccated worlds. Even if the deep Earth sequesters an order-of-magnitude more water than is in its ocean (e.g., Marty 2012), which is by no means certain, water would still constitute at most 0.2% of the planet's mass. Nevertheless, this is significantly more than what might be expected based on primary accretion of material from the inner solar nebula (region sunward of Jupiter's orbit). Early-accreted planetesimals from the inner disk would have formed relatively dry because of nebular temperatures that were above the sublimation temperature of water ice. However, even as the temperatures in the region around 1 AU dropped below the ice condensation point later in the evolution of the solar nebula, the local gas is thought to have still remained dry due to the action of proto-Jupiter preventing the inward drift of earlier-condensed icy particles (Morbidelli et al. 2016). Thus, what water exists in the Earth and rocky planets was probably delivered after the dissipation of the solar nebula gas via later addition of volatile-rich planetesimals that had accreted beyond a "fossilized" snowline (radial distance from the Sun at which H_2O starts to condense at the disk midplane) at around 3 AU (Morbidelli et al. 2016).

The principal reservoirs of such volatile-rich outer Solar System materials are comets and, possibly, the parent asteroids of the carbonaceous chondrite meteorites. It is not known precisely when and at what heliocentric radii each of these classes of primitive Solar System bodies accreted, but it is clear that comets could only have accumulated their volatile inventories in very cold environments (e.g., Bar-Nun and Kleinfeld 1989; Rubin et al. 2015a). Dynamical models of giant planet migration, such as the Grand Tack (Walsh et al. 2011), allow for the formation of asteroids beyond the orbit of Jupiter, well beyond the snowline, and then insertion of some fraction of these water-rich bodies into the present Asteroid Belt during the subsequent outgoing motion of Jupiter. Regardless of the details of their formation, volatile-rich asteroids are known to exist through observations of so-called main-belt comets (Jewitt 2012; Jewitt et al. 2015) and as the parent bodies of carbonaceous chondrites. Many of these meteorites contain 'water' at the level of several percent, and in some cases up to 15% by mass, locked mostly as OH in clay minerals. Thus, considering only mass balance, it is readily apparent that the addition of a few percent of any of these materials to an originally completely dry proto-Earth would be sufficient to account for the abundance of water in this planet. The existence of ample sources is therefore not in question, yet our understanding of the proportion of those sources contributing to the volatile inventory of Earth is far from complete (e.g., Marty et al. 2016, 2017).

This chapter discusses our current understanding of the nature of the asteroidal and cometary reservoirs that may have contributed volatile elements to the Earth and other terrestrial planets. Constraints are derived from the laboratory analysis of meteorites, which are (possibly non-representative) samples of asteroids, and of cometary dust. This cometary dust was either captured from the coma of a comet and returned to the Earth by the Stardust mission (Brownlee et al. 2006) or accreted by the Earth in the form of interplanetary dust particles (IDPs) that were derived from unknown sources. Due to the high-velocity capture of both Stardust particles and IDPs, no laboratory measurements of cometary ices are available, but in situ analyses of molecules in the coma of comet 67P/Churyumov-Gerasimenko have been performed by the Rosetta mission (Le Roy et al. 2015; Rubin et al. 2015a, 2015b). Here we assess what has been learned about the nature of the volatile inventories of these sources as they relate to those on the Earth and we discuss implications for the accretion of volatiles in the solar nebula, the architecture and dynamics of the Solar System, and processes affecting the distribution of volatiles that may be observed in planetary systems around other stars.

2 Meteorites

2.1 Classification

The classification of meteorites, recently reviewed in detail by Krot et al. (2014), uses both their physical and chemical characteristics. These characteristics reflect both primary features that were inherited at the time of the accretion of their parent asteroids and secondary features produced by processes that occurred in their parent bodies after accretion. The primary features were set by both ambient conditions (e.g., ice/rock ratios influenced by temperature) at the formation location of a given parent body and by mixing of materials that formed over a wide range of times and radial distances from the Sun. Their secondary features were determined by the extent to which they were heated in their parent bodies and/or interacted with aqueous fluids. The Solar System inherited a number of short-lived radionuclides when it formed, most notably ^{26}Al ($t_{1/2} \approx 720{,}000$ yrs), whose decay drove the internal heating of the meteorite parent bodies. The extent of parent body processing that a given meteorite experienced primarily reflects the time at which its asteroidal parent body accreted and the depths to which it was buried.

The most basic division of meteorites is into unmelted (chondrites) and melted ones (non-chondrites). Based on estimates of the initial abundances of the short-lived ^{26}Al and, possibly, ^{60}Fe ($t_{1/2} \approx 2.53 \times 10^6$ yrs) and thermal modeling, meteorite parent bodies would not have experienced widespread melting (and differentiation) if they accreted later than ~ 2 Ma after the beginning of Solar System formation (see Sects. 2.3 and 2.4). Hence, the chondrite parent bodies should be younger than those of the non-chondrites. Nevertheless, the chondrites provide the more detailed view of the processes that were operating in the early Solar System and will be described first.

2.1.1 *Chondrites*

The chondrites are comprised of three dominant components—refractory inclusions and chondrules formed at high temperatures and are embedded in a fine-grained (< 5–10 μm), volatile-rich matrix. Refractory inclusions are a diverse group of objects that are broadly divided into Ca-Al-rich inclusions (CAIs) and amoeboid olivine aggregates (AOAs). They are the oldest dated Solar System solids and formed by condensation from a hot gas and/or by melting and evaporation of pre-existing solids (MacPherson 2014). Chondrules appear to have typically formed 1–4 Ma after refractory inclusions (see below for more details) as molten silicate-metal droplets in brief (hours to days) heating events. Being high temperature products that formed in low pressure environments, both refractory inclusions and

Table 1 The major chondrite classes, their groups, and their ranges of petrologic types

Class	Groups	Types
Enstatite	EH, EL	3–6
Ordinary	H, L, LL	3–6
Rumaruti	R	3–6
Kakangari	K	3
Carbonaceous	CK	3–6
	CO, CV	3
	CI, CM, CR	1–2
	CH	Mixed chondritic & impact products
	CB	Impact products

chondrules are depleted in volatile elements relative to the bulk solar composition. The majority of the most volatile elements in chondrites were probably accreted in the matrix, along with organic matter and presolar circumstellar grains. The matrix probably also originally contained the ices accreted by most chondrite groups. Because it was so fine grained, the matrix was very susceptible to modification by heating and fluid-rock reaction, thus partially or totally obscuring many of the primary components. Nevertheless, the matrix seems to have originally been dominated by a mixture of crystalline and amorphous silicates.

Almost all chondrites belong to one of five classes that have been further subdivided into a number of groups (Table 1), although a few chondrites remain ungrouped.

Each chondrite group is defined by a relatively narrow range of physical features (e.g., sizes and abundances of chondritic components), major and trace element compositions, and bulk O (and Cr) isotopic compositions. After accretion, the internal heating in the chondrite parent bodies, driven by the decay of ^{26}Al and, possibly, impacts, led to varying degrees of aqueous alteration and/or thermal metamorphism that have modified some or all of the primary features in chondrites. The extent of this modification is reflected in a chondrite's petrologic type—a chondrite that was minimally affected by either aqueous alteration or thermal metamorphism is assigned a petrologic type of 3.0, while petrologic types 3.01–6 reflect increasing degrees of thermal metamorphism and types 2.9–1 reflect increasing extents of aqueous alteration. A meteorite's petrologic type is normally given after its group (e.g., CM1, L3.0 and R6).

While a few chondrites have been assigned to petrologic type 3.0, they are not completely free of the effects of parent body modification. This is because lithification associated with aqueous alteration and/or thermal metamorphism is necessary to produce rocks from originally unconsolidated 'cosmic sediments' that are strong enough to survive impact-driven excavation from their parent bodies and later passage through the Earth's atmosphere. Unless impact heating was more important than is currently generally accepted, we might not expect meteorites to sample asteroids that accreted more than $\sim$ 4–4.5 Ma after Solar System formation because they would not have contained enough ^{26}Al to melt the water-ice that is necessary to begin the lithification process. If they exist, these unlithified bodies could be one of the sources of interplanetary dust particles (IDPs), comets being the other.

2.1.2 Non-chondrites

The non-chondrites are divided into (i) primitive achondrites that experienced relatively low degrees of partial melting and melt extraction, and (ii) chemically differentiated meteorites that come from bodies that underwent extensive melting and segregation of metal from silicates. The differentiated meteorites are subdivided on the basis of their Fe-metal contents into achondrites (metal-poor), stony-irons, and irons.

Primitive achondrites—these meteorites (winonaites, acapulcoites/lodranites and brachinites) occasionally contain relict chondrules (i.e., chondrules that have survived the partial melting of their host rocks). However, they have not been conclusively linked to any known chondrite groups. Nevertheless, they support the widely held belief that most planetesimals originally contained abundant chondrules. Only one primitive achondrite, the ungrouped Tafassasset, has geochemical affinities to the carbonaceous chondrites (Gardner-Vandy et al. 2012).

Achondrites—These are a diverse class of meteorites, although many are broadly basaltic in composition. The aubrites have geochemical affinities to the highly reduced enstatite (E) chondrites. The howardites, eucrites and diogenites (HEDs) are a suite of basaltic and more slowly cooled rocks that, with perhaps a few exceptions, are thought to come from the asteroid 4 Vesta (e.g., McSween et al. 2011). The angrites are also basaltic in nature, but are from a different parent body than the aubrites and HEDs. Ureilites are complex rocks that appear to have been derived from the mantle of their parent body. They are often quite C-rich and exhibit a range of O isotopic compositions that resemble those of carbonaceous chondrites. However, the isotopic compositions of other elements show that they are not related to carbonaceous chondrites. In fact, to date only one achondrite, the ungrouped basaltic achondrite NWA 011, seems to have formed by melting of a carbonaceous chondrite-like body (Warren 2011).

Stony irons—Mesosiderites are impact breccias composed of diverse silicate clasts mixed with roughly equal amounts of Fe,Ni-metal and sulfide. Pallasites are composed of Fe,Ni-metal/sulfide that is intergrown with similar proportions of relatively coarse grained olivine and pyroxene crystals. There are three recognized sub-types that probably derive from at least three different parent bodies. Traditionally, it has been assumed that pallasites are core-mantle boundary samples, but they experienced a range of cooling rates that would be more consistent with mixing of core and mantle material after an impact (Yang et al. 2010).

Irons—Iron meteorites are divided into 14 chemical groups, although $\sim 15\%$ of irons remain ungrouped (for review, see Goldstein et al. 2009). The chemical groups fall into one of two categories, magmatic and non-magmatic. The intragroup chemical variations in the magmatic irons are consistent with crystal-liquid metal fractionation during solidification of planetesimal cores. The non-magmatic irons often contain silicate clasts and the chemical variations of their metal are more difficult to understand in terms of crystal-liquid fractionation in a slowly cooling core (e.g., Worsham et al. 2016). These irons are most likely the products of impact melting. Many of the ungrouped irons are also probably impact melts.

2.2 Compositions—Elemental and Isotopic

While large-scale radial thermal gradients in the disk probably produced some of the gross differences in parent body compositions (e.g., rock/ice ratios), the chondrites show that more transient processes (e.g., chondrule formation) were also important, and that radial transport brought together materials with different thermal histories. The thermal processing of dust in the disk has left its imprint on the major and trace element compositions of almost every inner Solar System object that we have samples of. The most obvious feature in these compositional variations is the strong correlation for many elements between abundance and volatility (Figs. 1 and 2). By convention, the estimates of the volatilities of the elements are based on thermodynamic equilibrium calculations of their 50% condensation temperatures from a gas of solar composition at a total pressure of 10^{-4} bars (e.g., Lodders 2003).

Fig. 1 The CI normalized elemental abundances of bulk CV chondrites as a function of 50% condensation temperatures (Lodders 2003) and divided into their chemical affinities

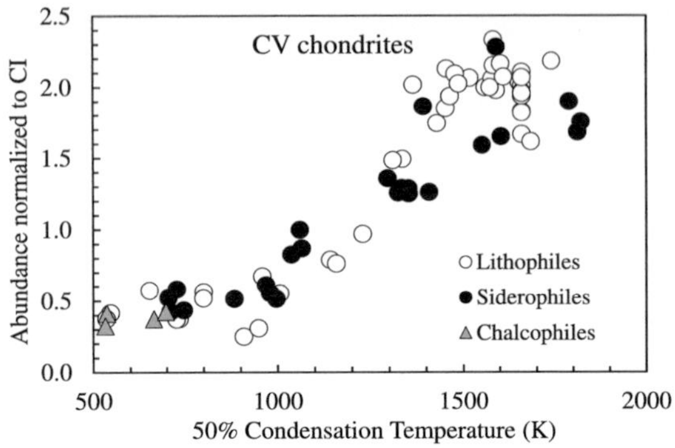

It should always be borne in mind that the absolute and relative condensation temperatures are only as good as the assumptions and thermodynamic data in the models, and will change under different conditions (e.g., total pressure or composition). The CI chondrites occupy a special place in meteoritics as they have compositions that are essentially identical, within error, to the condensable component of the solar photospheric composition (e.g., Lodders 2003; Asplund et al. 2009). Since all Solar System materials presumably ultimately evolved from the solar composition, meteorite elemental compositions are often normalized to (divided by) the CI abundances. In addition to a volatility control on abundance, there is also evidence in meteorites for the fractionation of elements in the nebula according to their chemical affinities (lithophile—rock-loving, siderophile—Fe-metal-loving, and chalcophile—sulfide-loving). Physical processes that separated solid/melt from gas (e.g., condensation, evaporation) and silicates from metal (e.g., gravitational settling, aerodynamic sorting) seem the most likely causes for these variations.

Figure 1 plots the CI normalized elemental composition of the CV chondrites as a function of 50% condensation temperature. It is apparent that for elements with condensation temperatures below ~ 1400 K there is a fairly smooth decrease in abundance, irrespective of their chemical affinities, that then levels off for elements with condensation temperatures below ~ 800 K. There is some scatter that at least in part must reflect uncertainties in the abundances and the condensation temperatures, as well as the influences of processes other than equilibrium condensation (e.g., transport and mixing). The fairly smooth decrease in abundance with condensation temperature is something of a puzzle, as is the leveling off below condensation temperatures of ~ 800 K. The thermodynamic models predict that the degree of condensation of an element should be a steep function of temperature. To produce the continuous fractionations in chondrites (e.g., Fig. 1) seems to require a combination of condensation and gas-solid fractionation, although the environment in which this took place is still a mystery (Ciesla 2008).

The CV abundance pattern is not unique. The CK, CO and CR abundance patterns are quite similar to those of the CVs. The CM and ordinary (OC) chondrites exhibit, respectively, less and more volatile element fractionation than the CV, but the overall shapes are similar (Fig. 2). The abundances of elements with condensation temperatures below ~ 800 K roughly correlate with the relative abundance of matrix. This can be seen more clearly in Figure 3a where Zn abundances are plotted vs. the approximate matrix abundances of the different chondrite groups. Not all of the most volatile elements behave in such a systematic way. Cadmium abundances, for instance, are highly variable even within chondrite groups, probably because this element has been redistributed by parent body processes and shock heating (Wombacher et al. 2008).

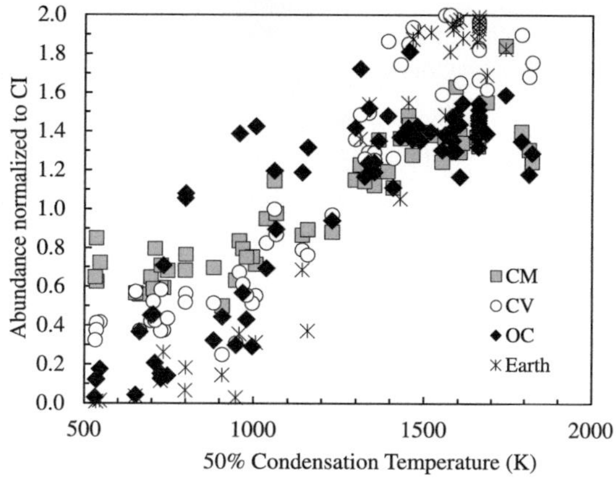

Fig. 2 The CI normalized elemental abundances of bulk CM, CV and OC chondrites, as well as the estimated abundances of lithophile elements in bulk Earth, as a function of their 50% condensation temperatures (Lodders 2003)

The simplest interpretation of the fact that the Zn data in Fig. 3a fall very close to the line that passes through CI chondrite (essentially 100% matrix) is that the matrices of all chondrites are dominated by material that is similar in composition to CI chondrites and that the volatile element fractionations are carried by the chondrules and refractory inclusions (Alexander et al. 2001; Alexander 2005; Zanda et al. 2009, 2012). The enstatite chondrites do not fall on the Zn trend, probably because they formed under highly reducing conditions and as a result the chemical affinities and condensation temperatures for many elements would have been very different to those of the other chondrite groups. It should be pointed out that the dominance of CI-like material in matrix is not universally accepted (e.g., Bland et al. 2005; Hezel and Palme 2010; Palme et al. 2015). Nevertheless, the abundances of circumstellar nanodiamonds (carriers of supernovae derived Xe-HL) (Huss and Lewis 1995) and SiC (Davidson et al. 2014) and bulk C (Fig. 3b) also seem to be in roughly CI-like proportions in chondrite matrices. The bulk of the C (and N) in the most primitive carbonaceous and OC chondrites is in a complex solvent/acid-insoluble, macromolecular organic matter that is generally referred to as IOM (e.g., Alexander et al. 2017a). The organic matter and the presolar circumstellar grains are not Solar System condensates and would not have survived chondrule or refractory inclusion formation, so clearly they are tracers of the presence of very primitive materials in chondrites. Enstatite (E) chondrites do contain circumstellar grains (Huss and Lewis 1995) and C that may once have been organic (Alexander et al. 1998, 2007). However, all the E chondrites studied to date have been heavily metamorphosed, making the direct comparison with the more primitive carbonaceous and OC chondrites difficult.

The Earth shows a similar pattern of volatile element fractionations to the chondrites, at least for lithophile elements that would not have been affected by core formation (Fig. 2), although Earth is more enriched in refractory elements and more depleted in volatile elements than any chondrite. The addition of 2–4 wt.% of CI- or CM-like material can explain the abundances of many of the most volatile elements in the Earth (Alexander et al. 2012; Marty 2012). There are at least two possible explanations for how CI/CM-like material could be added to the Earth. (1) The conventional explanation is that most of the Earth's planetesimal building blocks were volatile-free but included a small fraction of CI/CM-like planetesimals that were scattered into the terrestrial planet region from the outer asteroid belt (Raymond et al. 2009) or beyond (Walsh et al. 2011; Raymond and Izidoro 2017). (2) Alternatively, like the chondrites, most of the building blocks included some CI-like matrix. Which of these explanations is more likely has important implications for the mechanism of volatile

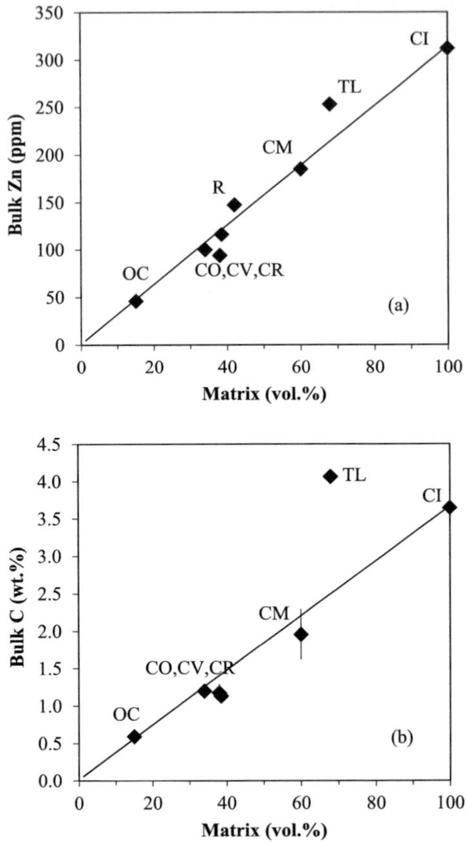

Fig. 3 Matrix abundances vs. bulk Zn and C contents. In general, the abundances of Zn (Wasson and Kallemeyn 1988; Kallemeyn et al. 1991; Brown et al. 2000; Bischoff et al. 2011) and C (Alexander et al. 2012, 2014a) are what would be expected if the matrices in these chondrites were dominated by CI-like material. The exception is the bulk C content of Tagish Lake (TL), but this seems to be because it is unusually rich in carbonate. The C contents for the OC, CO and CV chondrites are for the most primitive members of their groups (Semarkona, DOM 08006, and Kaba, respectively)

delivery to the terrestrial planets and the dynamical evolution of the inner Solar System (see Sect. 5).

There is increasing stable isotopic evidence that the carbonaceous chondrites are distinct from all other inner Solar System materials for which we have samples. For many years, bulk O isotopes were the only isotope system used for the classification of meteorites (e.g., Clayton 1993). It is now apparent that variable excesses in ^{54}Cr and ^{50}Ti abundances can also be used to distinguish between the different meteorite groups (Trinquier et al. 2007, 2009; Qin et al. 2010; Warren 2011). The sizes of certain isotopic anomalies in other elements also hold promise, such as ^{62}Ni, ^{92}Mo and ^{100}Ru (Regelous et al. 2008; Burkhardt et al. 2011). Except for O, these isotopic anomalies are thought to ultimately be nucleosynthetic in origin. However, it remains unclear whether the anomalies reflect spatial/temporal heterogeneities in the material accreted by the Solar System or an 'unmixing' of an originally homogeneous mixture of presolar materials by unidentified physical and/or chemical processes operating in the disk. Heating of dust composed of materials with different thermal stabilities coupled with gas-solid fractionation is one possible mechanism for 'unmixing' in the disk (e.g., Trinquier et al. 2009; Burkhardt et al. 2012), although there appears to be no simple correlation with the volatility trends seen in the chondrites. The isotopic differences between carbonaceous chondrites and all other materials is certainly consistent with the suggestion of the Grand Tack model that the carbonaceous chondrites formed in the outer Solar System, beyond the orbit of Jupiter (Walsh et al. 2011). Indeed, it is possible that the isotopic differences reflect the isolation of the inner and outer Solar Systems by the growth of Jupiter to a size of about 20 Earth masses by ~ 1 Ma after CAIs (Kruijer et al. 2017a). However, this still leaves unexplained the cause(s) of the differences in the inner and outer Solar System compositions.

2.3 Timing of Differentiation, Metamorphism and Aqueous Alteration

While the timing of various parent body processes are important in their own right, they are also a means for indirectly estimating the times of accretion of the various meteorite parent bodies (next section). Constraints on the timing of accretion of meteorite parent bodies would allow inferences regarding variations in the composition of dust in the regions of the disk where the parent asteroids of meteorites formed, including the snowline. In addition, since parent body processes have profoundly modified the H (and C and N) budgets of most planetesimals, understanding the timing of these processes with respect to planet formation timescales is essential.

In recent years there have been tremendous advances in our understanding of the timing of chemical differentiation and core formation in planetesimals. This has been largely based on the short-lived ^{182}Hf-^{182}W system ($t_{1/2} = 8.9 \times 10^6$ yrs). For instance, core formation in the IIAB, IIIAB, IVA, IVB, and IID iron meteorite parent bodies occurred at 0.7 ± 0.3, 1.2 ± 0.3, 1.4 ± 0.54, 2.9 ± 0.5, and 3.1 ± 0.8 Ma after CAIs, respectively (Kruijer et al. 2014). Kleine et al. (2012) were also able to determine that core formation in the angrite parent body occurred ≤ 2 Ma after CAIs. Estimates for the timing of differentiation of the HED parent body, 4 Vesta, range from 2.2 ± 1.1 Ma (Trinquier et al. 2008) based on the short-live ^{53}Mn-^{53}Cr system ($t_{1/2} = 3.7 \times 10^6$ yrs), to ~ 1 Ma using the ^{182}Hf-^{182}W system (Touboul et al. 2015), and to $\leq 0.6^{+0.5}_{-0.4}$ Ma (Schiller et al. 2011) using the short-lived ^{26}Al-^{26}Mg system. Taken together, these results demonstrate that planetesimal accretion and planetary-scale melting and chemical differentiation occurred rapidly, within a few million years (or less) of the formation of the first Solar System solids.

Large, fully differentiated asteroids or their collisional remnants are not likely to be volatile-rich and are generally not considered as important sources of water to the inner Solar System. Instead, it is the smaller, later formed planetesimals, the parent bodies of the chondritic meteorites, that are the best candidates for being the sources of water and other volatiles. The abundances of water in such bodies, and in the meteorites derived therefrom, would be largely dictated by the peak temperatures experienced during early geologic processing in a given parent asteroid. On the one hand, if temperatures never exceeded the melting temperature of ice, there would have been no lithification and samples of these parent bodies are unlikely to be delivered to Earth as meteorites. On the other hand, if peak temperatures exceeded the low-pressure stability limits of common phyllosilicates, most of the water is likely to have been lost to space. There is no single age for such metamorphism since the cycle of heating and cooling will have taken place over a protracted length of time. However, since ^{26}Al with a half-life of ~ 0.72 Ma was the main internal heat source, peak metamorphic temperatures ought to have been achieved within 5–10 Ma of Solar System formation (e.g., Miyamoto et al. 1981; Kleine et al. 2008).

Some idea of when aqueous alteration occurred in the affected chondrites comes from the dating of two products of fluid-rock interaction, fayalitic olivine (Fe_2SiO_4) and carbonates (e.g., $CaCO_3$), that incorporated short-lived ^{53}Mn. Fayalitic olivine in the OC chondrites formed at $2.4^{+1.8}_{-1.3}$ Ma after CAIs, while in CV and CO chondrites it formed at $4.2^{+0.8}_{-0.7}$ Ma and $5.1^{+0.5}_{-0.4}$ Ma after CAIs, respectively (Doyle et al. 2015). The carbonates in CI, CM and CR chondrites and the ungrouped C2 Tagish Lake also formed within this 4–5 Ma interval

(Fujiya et al. 2012, 2013; Jilly-Rehak et al. 2017; Steele et al. 2017). Of course, these ages are only the formation ages of these particular minerals. We have few constraints on the total length of time liquid water was present in these chondrites.

2.4 Timing of Accretion

At present, the accretion ages of chondrites cannot be measured directly, but there are indirect methods for estimating the time of accretion. This subject has recently been reviewed by DeMeo et al. (2015) and is only briefly summarized and updated here and in Table 2.

One way to estimate the accretion ages of the chondrite parent bodies is to assume that they are the same as the ages of their youngest chondrules. Chondrules ages can be dated using both short-lived chronometers, such as the ^{26}Al-^{26}Mg system (e.g., Ushikubo et al. 2013), and the long-lived Pb-Pb system (e.g., Connelly et al. 2012). However, parent body processes can disturb these chronometers producing younger apparent ages that do not date chondrule formation. So, it is essential to study only the most primitive chondrites and to select only those chondrules that can be shown to have undergone no secondary modification. As noted earlier, chronology based on the ^{26}Al-^{26}Mg system depends on the assumption of

Table 2 A summary of average chondrule ages and estimates of the accretion ages of chondrites and non-chondrites. The most complete set of accretion age estimates are from Sugiura and Fujiya (2014) (S + F). The ages are in millions of years after CV CAI formation

Group/meteorite	Chondrules	S + F	Other
Non-chondrites:			
Angrites		0.5 ± 0.4	≤ 1.5[a]
HEDs		0.8 ± 0.3	$\leq 2.2 \pm 1.1$[b], $\leq 0.6^{+0.5}_{-0.4}$[c], < 1[d]
Magmatic irons		0.9 ± 0.3	1.0–1.5[e], 0.1–0.3[f]
Stony irons		0.9 ± 0.3	
Ureilites		1.0 ± 0.3	~ 1.6[g]
Acap.+Lodran.		1.3 ± 0.3	
Aubrites		1.5 ± 0.1	
NWA 011		1.5 ± 0.1	
Tafassasset		1.9 ± 0.2	
Chondrites:			
E		1.8 ± 0.1	~ 2[h]
OC	$2.0^{+0.5}_{-0.3}$[i], 1.7 ± 0.7[j]	2.1 ± 0.1	~ 2[k,h], ~ 1.8[l]
R		2.1 ± 0.1	
CK		2.6 ± 0.2	
CO	$2.0^{+0.3}_{-0.2}$[m]	2.7 ± 0.2	~ 2.5[l]
CV	2–2.5[n]	3.0 ± 0.2	~ 2.5[l]
CI, CM, TL		3.5 ± 0.5	3–4[o]
CR	$3.7^{+0.3}_{-0.2}$[p], $3.7^{+0.9}_{-0.9}$[p]	3.5 ± 0.5	$> 4.0^{+0.5}_{-0.3}$[p]

[a] Kleine et al. (2012). [b] Trinquier et al. (2008). [c] Schiller et al. (2011). [d] Neumann et al. (2014); Touboul et al. (2015). [e] Kruijer et al. (2013). [f] Kruijer et al. (2014). [g] Budde et al. (2015). [h] Polnau and Lugmair (2001); Shukolyukov and Lugmair (2004). [i] Mean age of chondrules from Kita et al. (2000); Villeneuve et al. (2009). [j] Kleine et al. (2008). [k] Kleine et al. (2008); Harrison and Grimm (2010); Henke et al. (2013). [l] Doyle et al. (2015). [m] Selected chondrules from Kurahashi et al. (2008). [n] Budde et al. (2016); Nagashima et al. (2017). [o] Fujiya et al. (2013). [p] Schrader et al. (2016)

an initially uniform $^{26}Al/^{27}Al$ ratio throughout the solar nebula, which is still the subject of debate. Given that the number of chondrules with fairly precisely measured ages for any chondrite group also tends to be small and the possibility that particularly young chondrules are disturbed, DeMeo et al. (2015) used averages of measured chondrules to constrain accretion ages. The most careful ^{26}Al-^{26}Mg study of chondrule ages to date has been for the ungrouped carbonaceous chondrite Acfer 094 in which 9 of 10 chondrule ages are within error of a mean of $2.3^{+0.5}_{-0.3}$ Ma after CAIs (Ushikubo et al. 2013). As can be seen from Table 2, the mean ages for the OC, CO and CV chondrite chondrules are similar to Acfer 094. On the other hand, Schrader et al. (2016) have estimated an average age for 95% of CR chondrite chondrules of $3.7^{+0.3}_{-0.2}$ Ma after CAIs. This compares favorably with a corrected bulk CR chondrule Pb-Pb age of $3.7^{+0.9}_{-0.9}$ Ma after CAIs (Schrader et al. 2016). However, there are some CR chondrules without detectible ^{26}Mg excesses that may be significantly younger than these mean chondrule ages, prompting Schrader et al. (2016) to suggest a limit on the CR accretion age of $> 4.0^{+0.5}_{-0.3}$ Ma after CAIs.

The timing of accretion of chondrites can also be constrained from their thermal histories since accretion time will determine the abundance of the dominant heat source, the short-lived ^{26}Al, at the time of accretion. For instance, several modeling studies of the H ordinary chondrite parent body suggest that it formed ~ 2 Ma after CAIs (Table 2), which is consistent with estimates of the mean OC chondrule ages (Table 2). Modeling of the formation conditions and ages of secondary minerals during aqueous alteration suggests accretion ages after CAIs for the OC parent bodies of ~ 1.8 Ma, for the COs and CVs of ~ 2.5 Ma, and 3–4 Ma for the CI, CM, CR and Tagish Lake chondrites (Table 2). Based on the properties of their insoluble organic matter (IOM) that are indicative of peak temperatures during alteration, the CIs and CMs may have accreted slightly earlier than Tagish Lake and the CRs (Alexander et al. 2014b). To estimate accretion ages, Sugiura and Fujiya (2014) took the maximum peak temperatures experienced by any member of a chondrite group to be the peak central temperature achieved in their parent body. Despite the many assumptions in their approach, the estimates by Sugiura and Fujiya (2014) are broadly consistent with the other constraints (Table 2). The fact that no known chondrites have accretion ages much younger than ~ 4 Ma after CAIs is consistent with our earlier statement that there would not have been enough short-lived radioactivity in bodies that accreted later than this for them to become lithified. It is true that the CB chondrites (and possibly the related CHs) formed 4.6 ± 0.5 Ma after CAIs (Krot et al. 2005; Yamashita et al. 2010), but they seem to be the product of an impact. A gaseous disk protects material near the midplane from the solar wind. The lack of solar wind implanted noble gases in the CRs suggests that the gas disk dissipated later than ~ 4 Ma after Solar System formation.

Further complicating the estimates of chondrite accretion ages has been the suggestion that the CV chondrites, at least, formed in the presence of a magnetic field that was generated by a core dynamo (Weiss and Elkins-Tanton 2013). If correct, this means that the CVs are the unheated crusts of differentiated bodies and that they must have accreted significantly earlier than estimated by Sugiura and Fujiya (2014). This may be more consistent with their ~ 2 Ma chondrule ages (Table 2). However, no achondrites or iron meteorites have yet been found that can be linked to the CV group. The search for evidence in other chondrite groups for formation in the presence of magnetic fields (both disk- and asteroid-generated) is ongoing.

Thermal modeling of differentiated bodies is even more problematic than for chondrites because, for instance, the initial bulk composition is not known and how the melts segregate within the bodies can have a profound effect on internal heating (e.g., Moskovitz and Gaidos 2011; Neumann et al. 2014). Nevertheless, models of varying degrees of sophistication have been used to estimate their accretion times, and range from ~ 0.1–0.3 Ma after CAIs for some iron meteorites, to < 1 Ma after CAIs for 4 Vesta, and ≤ 1.5 Ma and ~ 1.6 Ma after CAIs for the angrite and ureilite parent bodies, respectively.

2.5 Water in Chondrites

The initial abundances and isotopic compositions of water in chondrites can help constrain the ambient conditions when they accreted as well as the origin of the water.

Parent body processes will have modified the initial water abundances of chondrites. Thermal metamorphism, for instance, will have largely dehydrated the interiors of the earlier-formed chondrites (E, OC, R, CK, CV, CO). In the aqueously altered chondrites, much of the water will have been consumed in the formation of phyllosilicates (clay minerals, etc.) and is now stored as OH in the mineral structures. However, the H in water will also have been lost through oxidation reactions mostly involving Fe, such as $3Fe + 4H_2O = Fe_3O_4 + 4H_2$. For chondrites that accreted with high metal/water ratios, a large fraction of the water may have been lost in this way (Sutton et al. 2017).

It is very unlikely that the chondrites in our collections retain much, if any, indigenous water. If any chondrite parent bodies did retain some unreacted water it would have refrozen as they cooled, but it would have subsequently sublimed away when the meteorites were small objects in space. However, many of the more primitive chondrites do contain OH in phyllosilicates, phosphates and hydroxides that are the products of aqueous alteration. Some phyllosilicates can have water in their structures, but this water is easily lost in vacuum or exchanged once in the terrestrial environment. The chondrites also contain H associated with organic matter that when delivered to a growing planet could be oxidized to produce water. Hence, in terms of their potential as sources of 'water' to growing terrestrial planets, it is the bulk H contents of chondrites that are important and are listed in Table 3. Also given in Table 3 are the bulk C and N contents, and the bulk H and N isotopic compositions as they can also be used to constrain the contributions that the different chondrite groups may have made to the volatile inventories of the terrestrial planets.

The H contents in chondrites can be highly variable even between meteorites belonging to the same group. This is due to the variable degrees of alteration and/or thermal metamorphism that they experienced. For the CI, CM and CR chondrites and Tagish Lake the averages of the samples that have been analyzed are given in Table 3, with the uncertainties reflecting the intra group variations. It is not clear why the extent of alteration and H content is particularly variable amongst the CM and CR chondrites. It is possible that the abundance of ice at the time of accretion was heterogeneous at the scales sampled by the meteorites, but it seems more likely that there was some redistribution of water within the parent bodies once the ice melted.

For the CV, CO, and OC chondrites that experienced variable degrees of thermal metamorphism and dehydration, the values in Table 3 are for the most primitive known members of the groups. There are no known very primitive (i.e., type 3.0) E, K, CK, or R chondrites. The H, C, and N contents of the more metamorphosed chondrites fall to near zero by petrologic types 3.6–3.7 and above. Thermal models suggest that the lowest petrologic types will have been confined to relatively thin layers near the surfaces of the metamorphosed chondrite parent bodies (e.g., Miyamoto et al. 1981; Harrison and Grimm 2010; Henke et al. 2012). Hence the H, C, and N abundances in Table 3 should be considered as upper limits for the final bulk compositions of the chondrite parent bodies, although it is possible that at least early on these bodies had ice-rich, near-surface layers as a result of re-freezing of water expelled from their interiors.

Table 3 The bulk H, C and N contents and H and N isotopic compositions of various chondrite groups. For the CI, CM and CR chondrites, the values are the averages for multiple samples and the uncertainties are the standard deviations. The large standard deviations for the CM and CR chondrite H abundances and isotopes reflect the variations in degree of alteration, and in the case of the CM isotopes the anomalous Bells meteorite. For Tagish Lake, the values are averages for several lithologies, except for H abundances and isotopes that are for the most primitive (least heated?) lithology 5b. The OC, CV and CO chondrites are represented by the most primitive and volatile-rich (i.e., least metamorphosed) members of their groups (Semarkona, Kaba, and DOM 08006, respectively), hence no uncertainties are given. The data are from Alexander et al. (2012) and Alexander et al. (2018)

	H (wt.%)	C (wt.%)	N (wt.%)	δD (‰)	δ^{15}N (‰)
OC	0.12	0.59	0.01	1616	39
CV	0.23	1.13	0.13	14	−24
CO	0.46	1.20	0.02	0	8
CM	1.15 ± 0.18	1.95 ± 0.34	0.10 ± 0.03	-53 ± 150	44 ± 66
CI	1.55 ± 0.02	3.65 ± 0.24	0.19 ± 0.03	78 ± 7	42 ± 5
Tagish Lake	0.94	4.07 ± 0.10	0.20 ± 0.04	508	66 ± 9
CR	0.49 ± 0.26	1.17 ± 0.12	0.08 ± 0.01	651 ± 135	175 ± 8

The standard ratios for calculating delta values are: $D/H = 1.5576 \times 10^{-4}$ and $^{15}N/^{14}N = 3.67 \times 10^{-3}$

As we have seen, the peak of metamorphism and aqueous alteration would have been over within the first 5–10 Ma of the Solar System. The Earth did not reach its final mass until ~ 60 Ma after CAIs (Touboul et al. 2007; Barboni et al. 2017). However, there is geochemical evidence that Mars had reached $\sim 50\%$ of its present mass in $\leq 1.8^{+0.9}_{-1.0}$ Ma after CAIs (Dauphas and Pourmand 2011; Tang and Dauphas 2014). Hence, it is possible that the planetary embryos from which the Earth and rocky planets formed accreted some or all of their volatiles very early when many chondrite-like planetesimals were more volatile-rich than they are now.

The H contents of the most primitive members of the CV, CO and OC chondrites provide lower limits on the initial contents of their parent bodies at the time of accretion (i.e., pre-metamorphism/alteration) because of the potential loss of H (as H_2O, H_2, etc.) to space. An alternative estimation of their initial H contents can be made if the earlier argument is correct that the matrices of chondrites were largely CI-like. Figure 4 plots the H contents vs. matrix contents of the various chondrites listed in Table 3.

The H contents of CVs, Tagish Lake and OCs are significantly below what would be predicted if their matrices initially had CI-like H concentrations. This probably reflects a combination of H loss associated with oxidation of Fe by water (Alexander et al. 2010, 2012; Sutton et al. 2017) and thermal dehydration. The CO and CR chondrites fall very close to the CI mixing line, but this may partially be coincidence as the COs are represented by the Antarctic meteorite DOM 08006 that has seen some terrestrial weathering and the CR value is quite uncertain (Table 3). Alteration in the CI chondrites has essentially gone to completion. Since it is very unlikely that the CI chondrites coincidently accreted just enough water to achieve this, they probably accreted significantly more water than they currently contain in their clay minerals. This may explain why the generally less altered CM chondrites fall well above the CI mixing line—unlike for the CIs, in the CMs excess water after alteration of the matrix could then react with chondrules. If correct, the average CM value would suggest that the initial H contents of CI chondrites (and chondrite matrices) could have been at least 50% higher than observed today. Note, however, that even this elevated water content

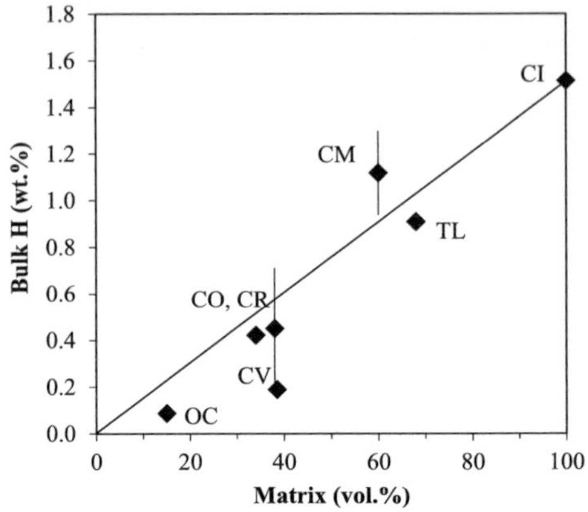

Fig. 4 The bulk H (Table 3) and matrix contents for the more primitive chondrite groups. The CVs, COs and OCs are represented by the most primitive members of their groups (Kaba, DOM 08006, and Semarkona, respectively). The *line* is what would be predicted if the matrices of all chondrites had H contents like those of CI chondrites

is significantly less than the roughly equal masses of water and silicates predicted for a system of solar composition (e.g., Lodders 2003). The fate of water that was not incorporated into clay minerals or consumed by oxidizing Fe is uncertain. It could have simply refrozen in pore spaces—ice-bearing asteroids are present in the asteroid belt (see below). Alternatively, it may have been lost to space while the interiors of the chondrite parent bodies were still warm.

The isotopic composition of water can also be revealing about its origins. It has long been suspected that water was responsible for producing the intragroup and, to some extent, intergroup mass independent O isotopic variations seen in chondrites (e.g., Clayton and Mayeda 1999; Young et al. 1999). Typically, O isotope compositions are given in the geochemical delta notation, for instance where $\delta^{17}O(‰) = 1000 \times (R_{sample}/R_{standard} - 1)$ and $R={}^{17}O/{}^{16}O$, with the scale normalized to Standard Mean Ocean Water or SMOW. Most physical and chemical processes fractionate O isotopes in a mass dependent fashion, such that in a plot of $\delta^{17}O$ vs. $\delta^{18}O$ samples will lie on a line with a slope of ~ 0.52. Unlike almost all terrestrial materials, most extraterrestrial materials exhibit some mass independent fractionation (i.e., they do not fall on the terrestrial mass fractionation line or TFL). This mass independent fractionation is usually expressed as $\Delta^{17}O(‰) = \delta^{17}O - 0.52 \times \delta^{18}O$, i.e., the distance from the TFL for a given value of $\delta^{18}O$. By definition, anything lying on the TFL has a $\Delta^{17}O = 0‰$. Most available Solar System materials (Earth, Moon, Mars, bulk meteorites, chondrules, matrix, and interplanetary dust) have $\Delta^{17}O$ values that fall fairly close to the TFL ($\pm 5‰$), but remarkably the Sun has a $\Delta^{17}O - 28.4 \pm 1.8‰$ (McKeegan et al. 2011). The only known objects with O isotopic compositions that are close to solar are refractory inclusions and a few rare chondrules (Kobayashi et al. 2003).

It has been suggested that an influx of water produced by UV dissociation of CO either in the protosolar molecular cloud (Yurimoto and Kuramoto 2004) or in the disk (Lyons and Young 2005) was responsible for the near-terrestrial O isotopic compositions of almost all inner Solar System solids. The models predict that the water should have large mass independent O isotope anomalies, although the exact values depend on the model parameters. The most extreme ^{16}O depletions observed in any Solar System material is $\Delta^{17}O \approx 86‰$ in a rare mineral association (magnetite-sulfide) found only in the ungrouped meteorite Acfer 094 (Sakamoto et al. 2007). It is often assumed to reflect the isotopic composition of the water responsible for modifying the composition of dust in the inner Solar System.

The O isotopic compositions of the fluids responsible for the alteration in the chondrites can be estimated from the O isotopic compositions of secondary minerals such as magnetite

Table 4 The estimated initial $\Delta^{17}O$ and δD values of water in the various chondrite groups. The $\Delta^{17}O$ values are the highest reported for secondary minerals in any member of the group, but are still probably lower limits as it is likely that there was some isotopic exchange with anhydrous minerals prior to formation of the secondary minerals. The δD values are estimated after subtraction of D-rich organic matter from bulk chondrite compositions, except for the R chondrites that is a direct measurement of hydrated silicates

	$\Delta^{17}O^a$ (‰)	δD^b (‰)	D/H^b
R	3.5	3660 ± 75	$7.26 \pm 0.13 \times 10^{-4}$
OC	4.8 ± 1.7	1002 ± 207	$3.1 \pm 0.3 \times 10^{-4}$
CV	-0.4 ± 0.9		
CO	-2		
CM	1.3	-427 ± 45	$8.7 \pm 0.4 \times 10^{-5}$
CI	1.7	-481 ± 107	$8.1 \pm 1.7 \times 10^{-5}$
Tagish Lake	2	-550 ± 130	$7.0 \pm 2.0 \times 10^{-5}$
CR		96^{+110}_{-65}	$1.7 \pm 0.2 \times 10^{-4}$

(a) R—Greenwood et al. (2000); OC—Choi et al. (1998); Doyle et al. (2015); CV—Choi et al. (1997, 2000); Doyle et al. (2015); CO—Choi et al. (2008); CM—Rowe et al. (1994); Airieau et al. (2005); CI—Rowe et al. (1994); Tagish Lake—Leshin et al. (2001). (b) R—McCanta et al. (2008); others—Alexander et al. (2012)

(Fe_3O_4), fayalite (Fe_2SiO_4) and calcite ($CaCO_3$). It is likely that prior to forming these secondary minerals the fluid interacted with primary, anhydrous silicates that tend to be more ^{16}O-rich. Nevertheless, the most ^{16}O-poor secondary mineral compositions from a chondrite group (Table 4) can be used to place lower limits on the compositions of the altering fluids. As can be seen from Table 4, the compositions of the fluids that produced the secondary minerals in the chondrites do not seem to have been particularly anomalous. Nor is there any obvious evolution in isotopic composition with the inferred accretion ages of the chondrite parent bodies, despite the fact that material and the snowline would have been migrating sunwards during this time (e.g., Oka et al. 2011). The O (and H) isotopic composition of water is likely to have evolved with time and radial location along with the thermal structure of the disk (e.g., Yang et al. 2013).

Hydrogen isotopes can also be informative about the origins of the water in meteorites. The potential extremes for the water H isotopic composition, relative to SMOW ($D/H = 1.5576 \times 10^{-4}$), range from $\delta D = -865‰$ ($D/H = 2.1 \times 10^{-5}$) for water that equilibrated with solar H_2 at high temperatures to $\delta D \approx 5400‰$ ($D/H = 1 \times 10^{-3}$) for interstellar water inherited from the protosolar molecular cloud (e.g., Liu et al. 2011; Coutens et al. 2012, 2013). Unlike molecular clouds, disks are probably unable to make D-rich water even in their cooler regions (Cleeves et al. 2014, 2016). The bulk Earth δD value is estimated to be about $-43‰$ (Lécuyer et al. 1998).

Estimating the initial water isotopic compositions in chondrites is made difficult by the intimate association of phyllosilicates and D-rich organic matter in meteorite matrices, as well as the possibility of isotopic exchange between water and organics during aqueous alteration. Alexander et al. (2012) estimated the initial water compositions (Table 4) by assuming that all chondrites accreted the same D-rich organic matter and subtracting this from the bulk H compositions. There are three major features of the estimated H isotopic compositions. The waters in CIs, CMs and Tagish Lake seem to have had H isotopic compositions that fall between the solar and terrestrial values, the CR chondrite water were slightly heavier than terrestrial, whereas the waters in OC and R chondrites were very D-rich.

If taken at face value, the compositions of the OC and R chondrites imply that they accreted proportionally more interstellar water than any carbonaceous chondrites (Deloule et al. 1998; Piani et al. 2015), although the matrices of the OC chondrites at least are no richer in circumstellar grains or organics than the carbonaceous chondrites. An alternative explanation is that when H_2 was lost as a result of Fe oxidation, the H_2 was very D-poor and left behind increasingly D-rich water (Alexander et al. 2010, 2012; Sutton et al. 2017). On Earth, H_2 generated by Fe oxidation during low temperature serpentinization can have very isotopically light (δD as low as -700 to $-800‰$) compositions (e.g., Coveney et al. 1987; Fritz et al. 1992; Proskurowski et al. 2006). As discussed above, the OC and R chondrites probably had amongst the lowest water/rock ratios of the chondrites, and so the isotopic effect of H_2 loss would have been largest in them. Given this scenario, the initial δD value of water in OC chondrites might have been similar to those estimated for the CIs and CMs (Sutton et al. 2017). This fractionation process should have also affected the carbonaceous chondrites. It should have had a relatively modest effect on the CIs, CMs and Tagish Lake because of their relatively high water/rock ratios, but may explain the higher δD value of the CR chondrites than other carbonaceous chondrites (Sutton et al. 2017). At present, there is little evidence for isotopic fractionation of H associated with Fe oxidation in the CV and CO chondrites, which is a puzzle.

The fact that both the ^{16}O depletions and D-enrichments in chondritic water are much less than predicted for interstellar water, and in the case of O less than predicted for water produced by UV dissociation of CO near the disk surface, suggests that most of the water in chondrites had been reprocessed in the disk prior to accretion by the chondrites (Alexander et al. 2017b).

2.6 Water in Achondrites

While most of the work on water in meteorites has focused on the chondrites, HEDs and angrites are known to contain trace amounts of water in minor phases such as phosphate (Sarafian et al. 2014, 2017a, 2017b; Barrett et al. 2016). They may also contain water, albeit at even lower abundances, in nominally anhydrous minerals like olivine and pyroxene. The other achondrite groups have yet to be studied in detail. Nevertheless, the implication of these observations is that even differentiated bodies should be considered as potential sources of some of the water in the terrestrial planets, although at present there are no accurate estimates for the bulk water contents of even the HED and angrite parent bodies.

The H isotopic compositions of the HEDs and angrites are broadly consistent with those of bulk carbonaceous chondrites (Sarafian et al. 2014, 2017a, 2017b; Barrett et al. 2016). It has been argued that this water was acquired by late accretion of carbonaceous chondrite material while Vesta and the angrite parent body were still at least partially molten (Sarafian et al. 2014, 2017a, 2017b). A few HEDs and angrites have crystallization ages that are < 4 Ma after CAIs, presenting a bit of a conundrum since this is roughly contemporaneous with the formation times of the CI, CM and Tagish Lake parent bodies and before formation of the CRs or the Grand Tack. In which case, what was the source of the water-rich, carbonaceous-chondrite-like bodies that collided with Vesta and the angrite parent bodies? Raymond and Izidoro (2017) have presented one potential solution by arguing that outer Solar System objects would have been continuously scattered into the inner Solar System as the giant planets grew. On the other hand, it has not been definitively demonstrated that differentiated bodies would lose all the water that they may have accreted when they formed. Matrix material with several characteristics that resemble CI chondrites was present in the inner Solar System when the OC and R chondrites formed ~ 2 Ma after CAIs (Sect. 2.2), despite the evidence that Jupiter may have more-or-less sealed off the inner Solar System 1 Ma earlier (Kruijer et al. 2017a). If chondrite matrix material and the snowline were present in the inner Solar System earlier than 2 Ma after CAIs, bodies that differentiated could have accreted some water. How ever they acquired their water, it is possible that differentiated bodies were contributors to the volatile budgets of the terrestrial planets, but to assess this possibility quantitatively requires knowledge of the volatile inventories of differentiated bodies, which remains unknown.

3 Asteroids

3.1 Classification

The Asteroid Belt is often divided into annular regions based on the locations of major resonances. Accordingly, the asteroids can be grouped dynamically based on the semi-major axes of their orbits into: (1) the Hungaria 1.8–2 AU; (2) inner belt bodies 2.1–2.5 AU; (3) the middle belt bodies 2.5–2.8 AU; (4) outer belt bodies 2.8–3.3 AU; (5) Cybeles 3.3–3.6 AU; (6) Hildas 3.9–4 AU; and (7) Jupiter Trojans (5–5.4 AU).

Historically, three spectroscopic classification schemes have been proposed for asteroids (for an overview see DeMeo et al. 2015). All three classification schemes are based on features in the reflectance spectra of asteroids at visible and, when possible, near infrared wavelengths, as well as their albedos, if measured. The asteroid spectra have been divided into three main groups or complexes (S-complex, C-complex and X-complex) that have been subdivided into a number of types. It is important to note that membership in a given complex does not necessarily imply a genetic relationship between bodies. There are also a number of asteroid types that do not belong to any of these complexes.

When asteroids are separated into these broad groups, it is apparent that for the larger (> 50 km diameter) asteroids there is a systematic variation in their relative abundances with heliocentric distance (Gradie and Tedesco 1982; DeMeo and Carry 2013, 2014) with the inner belt being dominated by S-complex asteroids and Vesta, the middle and outer belts by C-complex asteroids, the Cybeles and Hildas by P-types, and the Jupiter Trojans by D-types. Whereas P-types are members of the X-complex, the D-types do not belong to any of the complexes, although both P- and D-types are thought to be amongst the most primitive (least heated and most organic-rich) asteroids.

While particular groups may dominate in one region, it is important to bear in mind that asteroids of that group are not confined to that region. For example, although S-complex asteroids dominate in the inner belt, if one excludes Vesta, more than half the mass of S-complex bodies is found in the middle belt and a further $\sim 15\%$ in the outer belt. Similarly, E-type asteroids (members of the X-complex) are often associated with the Hungaria (interior to 2 AU), but most of the mass of E-types is to be found in the inner and middle belts. Finally, the D-types dominate the Jupiter Trojans but can also be found in the inner belt (DeMeo et al. 2014).

3.2 The Asteroid-Meteorite Connection and the Distribution of Water-Bearing Bodies

Establishing links between meteorite groups and asteroid types can help illuminate the age distribution of asteroids, the abundances of potential sources of water for the terrestrial planets, and constrain how the Asteroid Belt was populated. The subject of asteroid-meteorite links was reviewed in detail by Burbine et al. (2002), and here we highlight more recent developments. The E-asteroids have been linked to the aubrites (or enstatite achondrites) (Zellner et al. 1977; Gaffey et al. 1992; Binzel et al. 2004). The OCs have long been thought to be associated with S-complex and ungrouped Q asteroids, although R chondrites, primitive and basaltic achondrites, and pallasites have also been linked to S-complex asteroids. Specifically, the LL chondrites have been associated with the Flora family (~ 2.2 AU) (Vernazza et al. 2008; Dunn et al. 2013), the L chondrites to the Gefion family (~ 2.8 AU) (Nesvorný et al. 2009), and the H chondrites to Hebe (~ 2.4 AU) (Gaffey and Gilbert 1998), although there are many potential sources of the H chondrites between the Kirkwood gaps at the 3:1 (2.5 AU) and 5:2 (2.82 AU) resonances with Jupiter (Vernazza et al. 2014).

The vast majority of the HED meteorites are thought to derive from Vesta (~ 2.4 AU) and the V-type asteroid members of the Vesta family (McCord et al. 1970; McSween et al. 2011). However, a small number of eucrites are isotopically distinct from the other HEDs and may come from five other differentiated bodies (Scott et al. 2009). This is consistent with the fact that there are a small number of V-type asteroids, unrelated to Vesta, that are distributed throughout the asteroid belt (Lazzaro et al. 2000; Sunshine et al. 2004; Moskovitz et al. 2008; Duffard and Roig 2009), the most distant having semi-major axes of ~ 3.1 AU.

The signature feature of the CI and CM chondrites is that they have undergone extensive aqueous alteration and contain abundant phyllosilicates. This is also true for some CR chondrites. There is spectroscopic evidence that at least 60–70% of C-complex asteroids experienced some aqueous alteration and that this fraction is fairly uniform across the asteroid belt (Jones et al. 1990; Rivkin 2012; Fornasier et al. 2014; Rivkin et al. 2015). The majority of these belong to the Ch- and Chg-types that are spectroscopically similar (having 0.7 μm and 'sharp' 3 μm absorption features) to the CM chondrites (Rivkin et al. 2015; Takir et al. 2015). Those asteroids with a 'sharp' 3 μm absorption feature but lacking the 0.7 μm absorption features could be similar to either CM or CI chondrites. Given the number of asteroids with similar spectra, it is possible that the CMs and CIs come from multiple parent bodies. The sources of the other carbonaceous chondrites are even less certain.

Beyond ~ 3.3 AU, the 3 μm absorption feature of most measured asteroids is rounded rather than sharp (Takir and Emery 2012), and this is thought to be due to the presence of ice rather than phyllosilicates (Campins et al. 2010; Rivkin and Emery 2010). This is observed for several members of the Cybele family (~ 3.4 AU) that are mixed C- and P- types, and the Hilda family (~ 4 AU) that are P-type. One C-type member of the Hildas (but not part of the collisional family) shows the 'sharp' 3 μm absorption feature characteristic of phyllosilicates (Takir and Emery 2012). The same is true for one member of the Cybele family, suggesting that, prior to disruption, the parent body of the Cybele family experienced at least some aqueous alteration but its water was not entirely consumed by this process. Although water ice is most commonly found in bodies beyond 3.3 AU, it is not exclusive to this outer zone. In fact, the first direct detection of ice was made on the asteroid Themis (~ 3.1 AU) (Campins et al. 2010; Rivkin and Emery 2010) and four of the eight active asteroids (or main-belt comets) that appear to be driven by ice sublimation are members of the Themis family, two others belong to the Lixiaohua family (~ 3.2 AU), and of the remaining two, one belongs to a small, young cluster (Jewitt et al. 2015). Thus, it seems likely that disruption of larger asteroids produces fragments with surface ice. Small impacts on the surfaces of the fragments reveals buried ice that then sublimes to space, producing the observed activity, until an insulating layer of dust develops. In this respect, Ceres (~ 2.8 AU) is unique as phyllosilicates are widespread on its surface, it vents water from multiple, localized areas and it has surface deposits that are rich in carbonates that do not appear to be related to recent impacts (Küppers et al. 2014; De Sanctis et al. 2016). It seems possible that internal temperatures in parts of Ceres are sufficient to maintain liquid brines that occasionally escape to the surface. To date, we have no meteorites that can be linked to Ceres.

The D-type asteroids dominate the Trojan population, but < 50 km diameter bodies with D-type spectra are present even in the inner belt. There is no evidence in their spectra for either ice or phyllosilicates. Nevertheless, it has been suggested that one aqueously altered chondrite, the ungrouped C2 Tagish Lake, may be related to D-type asteroids (Hiroi et al. 2001) though this has been challenged (Vernazza et al. 2013).

3.3 The Stratigraphy of the Asteroid Belt

If the association of asteroid spectral classes with specific meteorite groups is correct then we can infer, based on meteorite chronologies, that Vesta, the V-type, E-type and S-complex asteroids all formed ≤ 2 Ma after Solar System formation. They are also relatively volatile-poor, either because they did not accrete many volatiles or because their strong internal heating drove out almost all of the volatiles that they had originally contained. The fact that most C-complex asteroids exhibit affinities to the CM and CI chondrites suggests that they formed ~ 3.5–4 Ma after CAIs. We do not have any recognized samples of P-type asteroids in the meteorite collection, but the absence of evidence for phyllosilicates at their surfaces despite the evidence for ice on some of them seems to imply that they formed later than ~ 4 Ma after CAIs, by which time there was insufficient internal radioactivity to melt the ice. However, D-type asteroids do not show evidence for phyllosilicates at their surface either. Yet if the Tagish Lake chondrite is from a D-type asteroid, they did experience aqueous alteration and they formed at about the same time as the CM and CI chondrites.

To date, it is unclear how abundant CK-, CV- and CO-like parent bodies are in the Asteroid Belt and what their orbital distributions are. The best spectral matches are to the relatively rare K-type asteroids, but the matches are by no means perfect (e.g., Cloutis et al. 2012a, 2012b, 2012c). The timing of the accretion of the CK, CV and CO chondrites seems to have been between the accretion of the OCs and the CI-CM-CR and Tagish Lake chondrites. This would suggest that there may have been a delay in planetesimal formation between the accretion of the V- and S-complex asteroids on the one hand, and the accretion of most C-complex and, possibly, D-type asteroids on the other.

Thus, to first order, it appears that asteroids in the inner belt tend to have formed earlier, experienced higher internal temperatures and accreted (or retained) less volatiles than asteroids in the outer belt, the Hildas and the Trojans. One explanation for this distribution invokes a wave of planetesimal formation that slowly progressed outwards through an Asteroid Belt that was chemically zoned with the snow line located somewhere in the middle belt (e.g., Grimm and McSween 1993; McSween et al. 2002). However, it seems unlikely that conditions would have remained essentially static in the Asteroid Belt for the 1.5–2 Ma that apparently separates the formation of the OC and E chondrites from the CI-CM-CR carbonaceous chondrites. Nor is it clear why there would have been such an orderly and sedate progression through the asteroid belt when bodies as large as Mars were beginning to form within ~ 2 Ma in the inner Solar System and the cores of the giant planets must have been largely formed by ~ 4 Ma if they were to have sufficient time available to acquire their gaseous envelopes before the gas disk dissipated.

The OCs show clear petrologic evidence for aqueous alteration prior to metamorphism. If the LL chondrites are linked to the Flora family, this suggests that by ~ 2 Ma the snowline was sunward of 2.2 AU. This is consistent with model predictions for the evolution of the snowline inwards in a disk as the main heat source, transport of material through the disk and onto the star, decreases with time (e.g., Oka et al. 2011; Bitsch et al. 2015). In principle, dry planetesimals could have accreted ice inside the snowline from ice-bearing bodies that were scattered or drifted sunward (e.g., Morbidelli et al. 2015), but as discussed earlier the chondrites accreted their water and other volatiles in fine-grained matrix, not in large bodies. This then raises the question of why the OCs did not accrete water in roughly its solar abundance (water/rock ≈ 1: Lodders 2003). Morbidelli et al. (2016) argue that in accretion disks gas will flow sunward faster than will condensation fronts. As a result, dry gas that has been depleted of water by ice condensation beyond the snowline will gradually flush any wet gas sunward of the snowline into the Sun. In this scenario, the chondrite matrices could then be dust that followed the gas as it migrated from the outer Solar System into the inner Solar System. The influx of dust into an Asteroid Belt that had been depleted by an earlier phase of planetesimal formation (S-complex and V-types) could potentially explain a second phase of asteroid formation (the C-complex and D-types), the general increase in matrix abundances in chondrites with their formation age after CAIs, and the distinctly different isotopic compositions of carbonaceous chondrites from other inner Solar System materials.

However, even this scenario may too simplistic. There is growing cosmochemical evidence (Sect. 2.2) that the carbonaceous chondrites formed beyond the orbit of Jupiter (Kruijer et al. 2017b). At least two episodes of giant planet migration have been invoked to explain various features of the Solar System that would have scattered bodies from the outer Solar System into the Asteroid Belt (Levison et al. 2009; Walsh et al. 2011). However, giant planet migration may not even be required for scattering of outer Solar System objects into the Asteroid Belt (Raymond and Izidoro 2017). Additionally, it has been proposed that the process of terrestrial planet formation would have scattered bodies from < 2 AU into the Asteroid Belt (Bottke et al. 2006). Thus, it is possible that a large fraction of the bodies in the Asteroid Belt may not have formed there. If correct, the stratigraphy of the Asteroid Belt may be a complex record of the dynamical evolution of large bodies in the Solar System rather than a remnant of planetesimal formation in the belt.

The proposed dynamical event that would have had the most dramatic effect on the Asteroid Belt is the Grand Tack of Jupiter (Walsh et al. 2011), the prime motivation of which was to explain the small mass of Mars. In the Grand Tack, an already formed Jupiter migrates inward to roughly 1.5 AU clearing out most of the previously formed planetesimals from the Asteroid Belt. At this point, a still growing Saturn catches up with Jupiter, they become trapped in a resonance and this causes them to migrate out again in tandem until the disk dissipates. In the process of moving out again, Jupiter and Saturn force Uranus and Neptune outwards as well. One consequence of this action is that the migrating giant planets scatter planetesimals from ~ 5 AU to ~ 13 AU into the inner Solar System. These scattered bodies would have been potent sources of volatiles for the growing terrestrial planets. However, some become trapped in the Asteroid Belt, particularly the outer belt, while the inner belt is dominated by the remnants of the inner Solar System planetesimals that were scattered by Jupiter's inward migration. Based on the estimates of the accretion ages of the CR and CB chondrites, if it happened the Grand Tack must have taken place ≥ 4–5 Ma after Solar System formation. Such a timescale would also be consistent with the absence of evidence for alteration in the P-type asteroids. The sealing off of the inner Solar System by the growth of Jupiter and having the carbonaceous chondrites form in the outer Solar System may also explain why the formation of carbonaceous chondrite bodies continued for up to 2 Ma after any known inner Solar System chondrites—i.e., material migrating inwards through the disk could not pass beyond Jupiter, causing it to accumulate outside of Jupiter's orbit. It also implies that the enigmatic process of chondrule formation was much more widespread in the nebula than is often assumed (i.e., it was not confined to the inner Solar System).

The next dynamical instability, the Nice model, was invoked to explain the architecture of the Kuiper Belt and Scattered Disk, as well as providing a mechanism for the hypothesized Late Heavy Bombardment that may have occurred ~ 600 Ma after Solar System formation. In the Nice model, the orbits of the giant planets become unstable again and, most importantly for the purposes of this discussion, Neptune moves outwards from ~ 15 AU to its current orbit at 30 AU (Levison et al. 2011). In so doing, it scatters bodies from this region into the inner Solar System, some of which are implanted into the Asteroid Belt as well as becoming trapped in the Hildas and Jupiter Trojans. Levison et al. (2009) suggested that most of the D- and P-type asteroids were acquired in this way. The Oort Cloud and Jupiter family comets would also have formed in this region (Brasser and Morbidelli 2013).

Thus, the Nice model predicts a genetic link between comets and D-/P-type asteroids. This potential link will be explored further in Section 5, but petrologically the aqueously altered Tagish Lake meteorite that has been linked spectroscopically to D-type asteroids (Hiroi et al. 2001) appears to be quite distinct from comet Wild 2 (e.g., Zolensky et al. 2002, 2006; Hanner and Zolensky 2010; Blinova et al. 2014). However, the spectroscopic link between Tagish Lake and D-type asteroids has been questioned (Vernazza et al. 2013). The Nice model could not have supplied a significant amount of volatiles to the Earth because: (1) there is clear evidence that Earth had an atmosphere and hydrosphere within 30–300 Ma of Solar System formation (Mojzsis et al. 2001; Wilde et al. 2001; Avice and Marty 2014), and (2) the material accreted by the Earth after the Moon-forming impact, which occurred ~ 50–70 Ma after Solar System formation (Touboul et al. 2007; Barboni et al. 2017), was too little, volatile-depleted and inner Solar System in origin (Bottke et al. 2010, 2012; Walker et al. 2015; Fischer-Gödde and Kleine 2017).

Simulations by Bottke et al. (2006) suggest that significant numbers of bodies would have been scattered into the inner Asteroid Belt from the terrestrial planet region by planetary embryos. Since it seems likely that planetesimal accretion happened faster in the terrestrial planet region than further out, they proposed that all iron meteorites derived from planetesimal cores come from bodies that were implanted into the asteroid belt in this way. They also suggested that Vesta and some fraction of S-complex asteroids could be interlopers from the terrestrial planet region.

4 Comets

Comets are undoubtedly the most primitive and volatile-rich bodies in our Solar System. Thanks to the Rosetta mission, a lot more is known about comets now than a few years ago, and once all the data are properly analysed more results will certainly come forward. At the time of writing, we can already state a few things about comets, mainly concerning their classification and origin, their accretion, and their molecular and isotopic compositions. In the following, we concentrate mostly on new results from the Rosetta mission, which does not mean that previous work on comets should be forgotten. In many ways, Rosetta confirms previous measurements, but in others has prompted the emergence of new paradigms.

4.1 Classification

It has long been known that there are two dynamically distinct families of comets: Oort cloud comets (OCC) and Jupiter family comets (JFC). Their orbits differ by their Tisserand parameters (dynamical quantities that are roughly conserved during an encounter with a planet) with respect to Jupiter (Levison 1996). OCCs generally are long or medium period comets (periods > 50 yr) with more or less isotropic inclinations with respect to the ecliptic plane, whereas JFCs have their aphelion close to Jupiter and are almost in the ecliptic plane. Both classes of comets then are sometimes subdivided into sub-families, like Encke-type comets that do not cross Jupiter's orbit, or Halley-type (HT) comets that are OCCs but have orbital periods of less than 200 yr. Because of the distinct orbits of OCCs and JFCs, it was postulated that their formation zones were also distinct: OCCs close to the giant planets (Dones et al. 2004; Brasser 2008), and JFCs outside of Neptune's orbit (Duncan and Levison 1997). JFCs are thought to have their reservoir in the scattered disk—a dynamical subclass of objects in the Kuiper Belt with highly eccentric obits and perihelia > 30 AU. Once they are deflected into the inner Solar System, comets are captured by Jupiter.

Recently, this paradigm of two distinct formation regions for the two comet families has been challenged. Several authors have tried to classify comets according to their chemical compositions (e.g., A'Hearn et al. 1995; Mumma and Charnley 2011). However, most attempts have failed as OCCs and JFCs both have variable chemistries and their compositional characteristics cannot be tied to a specific comet family. Solar System formation models predicted that the D/H ratio in water should increase with radial distance from the Sun as temperatures decrease and mixing with processed water from the inner Solar System diminishes (e.g., Horner et al. 2007; Jacquet and Robert 2013; Yang et al. 2013). The D/H ratio in water in comets should, therefore, be an indication for their region of formation; hence OCCs should have a lower D/H ratio than JFCs. Table 5 gives an overview of D/H in comets. When a terrestrial D/H ratio was found in the JFC Hartley 2 (Hartogh et al. 2011) that was a factor of two lower than in all OCCs measured up to then, the problem was solved by considering mixing processes in the evolving accretion disk that result in a D/H ratio that does not monotonically increase with distance from the Sun (Yang et al. 2013). However, once Rosetta found a very high D/H in water from the coma of the JFC comet 67P/Churyumov-Gerasimenko (Altwegg et al. 2015), this idea was no longer supported. Meanwhile, more D/H ratios have become available (Biver et al. 2016) showing wider spreads in the ratios for OCCs and JFCs that clearly overlap one another. This suggests that the formation zones for the two comet families overlapped over large radial distances and that they experienced different dynamical histories following accretion (Brasser and Morbidelli 2013). The D/H ratio in water, therefore, may still reflect the region of comet formation, but not the dynamical family a comet belongs to. This picture is supported by the N isotopes of HCN and CN in comets that are similarly ^{15}N-rich in both OCCs and JFCs (Table 6).

4.2 From Dirty Snowballs to Icy Dust

Another paradigm that is changing thanks to the Rosetta mission is the notion of comets as "dirty snowballs". From the Giotto mission to comet Halley, it was already apparent that comets were probably not the dirty snowballs postulated by Oort (Oort 1986), but rather icy dust balls. The dust/gas ratios in comets has continued to be debated (e.g., Singh et al. 1992; Fulle et al. 2000). Measurements of the dust/gas ratio in comet 67P of at least 4:1 (Rotundi et al. 2015) clearly support the icy dust ball picture, although recent observations of a cliff collapse that exposed subsurface ice suggests that 67P may be more ice-rich than this (Pajola et al. 2017). The porosity of 67P is 75% and its density is correspondingly very low ($\sim 530\ \mathrm{kg/m^3}$), and it seems to be homogeneous down to decimeter scales (Kofman et al. 2015; Pätzold et al. 2016). The Giada instrument observed two kinds of grains: compact grains with densities of up to 4000 $\mathrm{kg/m^3}$ and fluffy grains with extremely low effective densities of $< 1\ \mathrm{kg/m^3}$ (Fulle et al. 2016) that take less than 10^3 Pa to disrupt (Hornung et al. 2016). The authors proposed that the compact grains are reprocessed material from the solar nebula, whereas the fluffy grains are primitive material, probably connected to interstellar dust. These ice-free dust grains were also observed with the COSIMA (Langevin et al. 2016) instrument. The existence of these fluffy aggregates means that the cometary material has not experienced any violent collisions; collisional speeds must have been below 1 m/s. It is also postulated that the two lobes of 67P formed separately and that the shape of the comet is due to a very gentle collision of two cometesimals (Jutzi and Asphaug 2015; Massironi et al. 2015; Jutzi and Benz 2017; Jutzi et al. 2017). This bi-lobal shape has previously been found for several comets (e.g., Hartley 2; A'Hearn et al. 2011).

In terms of composition, 67P contains quite a lot of N_2 (Rubin et al. 2015a), CO (Le Roy et al. 2015) and Ar (Balsiger et al. 2015), all highly volatile species that require a low formation temperature of ~ 25 K for the comet and almost no heating of its interior. No signs of hydrated minerals have been found so far in the infrared spectra of the VIRTIS instrument (Quirico et al. 2016), which implies little or no internal radiogenic heating as would have occurred in a bigger parent body that accreted early. The presence of highly volatile species in such a porous and weak body suggests a relatively late accretion of 67P and other volatile-rich comets when the temperatures in the outer disk would have been very cold, and the abundances of radioactive heat sources would have been low.

4.3 Presolar Cloud–Comet Connection

More information about the links between presolar cloud chemistry and cometary material comes from the detections of abundant O_2 (Bieler et al. 2015a), many S-bearing species (Calmonte et al. 2016), as well as of D_2O and HDS (Altwegg et al. 2017). The O_2 in comets 67P and Halley (Rubin et al. 2015b) has an abundance relative to water of ~ 3.8 mol.%. In 67P, the O_2 is very well correlated to water. This cannot be the result of solar nebula chemistry, but has to have come from the presolar cloud, either by radiolysis or by gas phase chemistry (Taquet et al. 2016). According to Mousis et al. (2016b), O_2 can be trapped efficiently in amorphous water ice. In order to keep the correlation between water and O_2, interstellar water ice cannot have sublimed at any stage prior to accretion by 67P. A similar conclusion may be drawn from the presence of S_2. Sulphur in protostellar cores and dense clouds is highly depleted in the gas compared to its cosmic abundance. In diffuse clouds, however, S has a cosmic abundance in the gas (Woods et al. 2015, and references therein). It was postulated that the reason for the depletions in the dense cloud gas is that photolysis/radiolysis converts H_2S, which is predominantly formed on dust grains, into refractory S_n that stays on the grains and cannot be observed by remote sensing. The ROSINA instrument on Rosetta has detected S_2, S_3 and S_4 (Calmonte et al. 2016) in the coma of 67P. Sublimation of S_2 occurred even at large heliocentric distances (> 3 AU) and its relative abundance to the major S-bearing species H_2S remained constant for many months inbound. Once the coma became dustier, ROSINA detected not only S_2, but S_3 and S_4 as well, and the relative abundance of S_2 to H_2S increased by more than a factor of 10. This behaviour shows that S_2 is present both in a very volatile form as well as part of the more refractory S reservoir associated dust. The detection of S_3 and S_4 clearly shows that the pathway described by Woods et al. (2015) also fits cometary material and that, therefore, S in comets and in dense clouds seem to be present in very similar forms. Once released into the gas of a cometary coma, S_2 has a very short photodissociation lifetime of ~ 250 s at 1 AU (de Almeida and Singh 1986). Its lifetime would still be short even at 30 AU as the wavelength needed for the destruction of S_2 is only 280 nm and the flux at this wavelength would probably have been sufficient even near the midplane. The presence of volatile S_2 in 67P shows that most probably the dust grain mantles formed in the dense cloud and survived the accretion of the

Table 5 The D/H ratios of water in Oort Cloud (OCC) and Jupiter Family (JF) comets. The estimated D/H ratios for the solar nebula, bulk Earth and water in Enceladus are also provided for comparison

	D/H	Ref.
OCCs		
1P/Halley	$3.16 \pm 0.34 \times 10^{-4}$	1
1P/Halley	2.12×10^{-4}	2
C/1996 B2 Hyakutake	$2.90 \pm 1.0 \times 10^{-4}$	3
C/1995 O1 Hale-Bopp	$3.30 \pm 0.8 \times 10^{-4}$	4
C/2002 T7 LINEAR	$2.50 \pm 0.4 \times 10^{-4}$	5
8P/Tuttle	$4.09 \pm 1.45 \times 10^{-4}$	6
C/2009 P1 Garradd	$2.06 \pm 0.22 \times 10^{-4}$	7
C/2001 Q4 Neat	$4.60 \pm 1.4 \times 10^{-4}$	8
153P/Ikeya-Zhang	$< 2.50 \times 10^{-4}$	9
C/2014 Q2 Lovejoy	$1.40 \pm 0.4 \times 10^{-4}$	10
C/2012 F6 Lemmon	$6.50 \pm 1.6 \times 10^{-4}$	10
JFCs		
103P/Hartley 2	$1.61 \pm 0.24 \times 10^{-4}$	11
67P/Churyumov-Gerasimenko	$5.30 \pm 0.7 \times 10^{-4}$	12
45P/Honda-Mrkos-Pajdušáková	$< 2.00 \times 10^{-4}$	13
Other		
Solar	$2.1 \pm 0.5 \times 10^{-5}$	14
Earth	$1.49 \pm 0.03 \times 10^{-4}$	15
Angrites	$1.53 \pm 0.04 \times 10^{-4}$	16
Vesta	$1.40 \pm 0.14 \times 10^{-4}$	17
Enceladus	$2.9 + 1.5/ - 0.7 \times 10^{-4}$	18
Titan (CH_4)	$1.59 \pm 0.33 \times 10^{-4}$	19

1 Eberhardt et al. (1995). 2 Brown et al. (2012). 3 Bockelée-Morvan et al. (1998). 4 Meier et al. (1998b). 5 Hutsemékers et al. (2008). 6 Villanueva et al. (2009). 7 Bockelée-Morvan et al. (2012). 8 Altwegg et al. (2015). 9 Biver et al. (2006). 10 Biver et al. (2016). 11 Hartogh et al. (2011). 12 Altwegg et al. (2015). 13 Lis et al. (2013). 14 Geiss and Gloeckler (1998). 15 Lécuyer et al. (1998). 16 Sarafian et al. (2017a). 17 Sarafian et al. (2014); Barrett et al. (2016). 18 Waite Jr. et al. (2009). 19 Nixon et al. (2012)

solar nebula without being heated. It is interesting to note that elemental S is found in CM chondrites and that it has mass independent isotopic fractionations that are most consistent with photodissociation of H_2S (Labidi et al. 2017).

Another link between the presolar cloud, star forming regions, and the comet compositions can be derived from the presence of D_2O and HDS. D_2O has been measured in low-mass star forming regions—the D_2O/HDO ratio is roughly $1.2 \pm 0.5\%$ and the HDO/H_2O ratio is on the order of $0.17 \pm 0.08\%$ (Coutens et al. 2014). Similar values have been found in comet 67P (Altwegg et al. 2017). These high values for D_2O can be explained by dust grain chemistry in presolar clouds. Furuya et al. (2016) have shown that the ratio of [D_2O/HDO] to [HDO/H_2O] is a good diagnostic for the amount of reprocessing that the water experienced in the solar nebula. For values of this ratio that are $\gg 1$, the water is unprocessed, i.e., inherited from the presolar molecular cloud, whereas for reprocessed water this ratio is < 1. In the case of 67P, the ratio is 17 (Altwegg et al. 2017), which suggests that the comet mostly contains primordial material inherited from the presolar molecular cloud. The same conclusion holds for the high HDS/H_2S ratio measured in 67P of 0.12% (Altwegg et al. 2017) that is indicative of dust grain chemistry (Hatchell et al. 1999).

Overall, the Rosetta data suggests that comet 67P has incorporated a significant amount of material that was formed in the presolar cloud and was little processed in the protoplanetary nebula. This would be consistent with the high D/H ratio of its water (Altwegg et al. 2015). The large range of D/H ratios amongst comets (Table 5) may then be the result

Table 6 The N isotopic compositions of HCN and/or CN in Oort Cloud (OCC) and Jupiter Family (JFC) comets. Comet 8P/Tuttle is included in both as it has mixed orbital properties. The estimated bulk N isotopic compositions for the solar nebula, the terrestrial planets and Titan's atmosphere are also included for comparison

	$^{14}N/^{15}N$	Ref.
OCCs		
Hale-Bopp	140 ± 35	1
C/2002 T7 (LINEAR)	160 ± 25	2
8P/Tuttle	150 ± 20	3
153P/Ikeya-Zhang	140 ± 50	4
122P/de Vico	145 ± 20	4
C/1999 S4 (LINEAR)	150 ± 50	5
C/1999 T1 (McNaught-Hartley)	160 ± 50	2
C/2000 WM1 (LINEAR)	150 ± 30	6
C/2002 X5 (Kudo-Fujikawa)	130 ± 20	2
C/2002 V1 (NEAT)	160 ± 35	2
C/2002 Y1 (Juels-Holvorcem)	150 ± 35	2
C/2001 Q4 (NEAT)	135 ± 20	1
C/2003 K4 (LINEAR)	145 ± 25	1
C/2006 M4 (SWAN)	145 ± 50	2
C/2007 N3 (Lulin)	150 ± 50	2
C/2012 S1 (ISON)	139 ± 38	7
C/2012 F6 (Lemmon)	152 ± 72	8
C/2014 Q2 (Lovejoy)	145 ± 12	8
JFCs		
88P/Howell	140 ± 20	5
9P/Temple 1	145 ± 20	9
73P/Schwassmann–Wachmann 3	215 ± 32	10
17P/Holmes	139 ± 25	11
8P/Tuttle	150 ± 30	3
Other		
Venus	273 ± 56	12
Earth	272	
Mars	280 ± 5	13
Angrites	272 ± 1	14
Vesta	277	15
Titan	168 ± 2	16
Solar	441 ± 6	17

1 Manfroid et al. (2005). 2 Manfroid et al. (2009). 3 Jehin et al. (2009). 4 Jehin et al. (2004). 5 Hutsemékers et al. (2005). 6 Arpigny et al. (2003). 7 NH_3, Shinnaka et al. (2014). 8 Biver et al. (2016). 9 Jehin et al. (2006). 10 Jehin et al. (2008). 11 Bockelée-Morvan et al. (2008). 12 Hoffman et al. (1979). 13 Mathew and Marti (2001). 14 Abernethy et al. (2013). 15 Miura and Sugiura (1993). 16 Niemann et al. (2010). 17 Marty et al. (2011)

of mixing of presolar material with water formed at high temperatures in the inner Solar System.

Further evidence for a link between cometary ices and the ISM, or at least evidence for low temperature ion-molecule and gas-grain chemistry, can be seen in the ^{15}N-rich compositions of HCN and CN (Table 6). These ^{15}N enrichments in HCN are similar to what is seen in dense, prestellar cores in molecular clouds (Ikeda et al. 2002; Milam and Charnley 2012; Hily-Blant et al. 2013). The D/H ratio ($2.3 \pm 0.4 \times 10^{-3}$) in the HCN of Hale-Bopp (Meier et al. 1998a) is almost an order of magnitude higher than its water (Table 5), but still an order of magnitude lower than observed for HCN in low mass star forming regions (Roberts et al. 2002).

4.4 Bulk Compositions of Comets

On arrival of Rosetta at comet 67P, several instruments detected a large heterogeneity in the coma composition (Bockelée-Morvan et al. 2015; Hässig et al. 2015; Luspay-Kuti et al. 2015). At that time, the northern hemisphere was in summer. This hemisphere experiences a long but cold summer during aphelion passage, whereas the southern hemisphere has a short intense summer of only 10 months during perihelion passage. This difference in insolation for the two hemispheres leads to different outgassing behaviours (Hässig et al. 2015; Le Roy et al. 2015) and results, over time, in different morphological changes (Feaga et al. 2007). Such heterogeneities have been observed before, e.g., for comets Hartley 2 and Temple 1 (A'Hearn et al. 2011). The heterogeneities make it difficult to derive a bulk composition for 67P. However, as the comet sheds more than 2 m from the southern hemisphere during its perihelion passage (Keller et al. 2015) and the seasonal heat wave penetrates more slowly than the erosion rate on the southern part of the comet, it is probably safe to say that the composition during the few months around perihelion best represents the bulk abundance at least of the southern hemisphere. For other comets, whose rotational axes are generally not known, any variations in composition may well be due to their seasonal evolution. The relative abundances of the major volatile species change dramatically during the orbit of 67P, even close to perihelion (Fougere et al. 2016). The same has been shown for many other comets. In addition, observing comets by remote sensing can introduce additional biases. One result from Rosetta is that for UV (Feldman et al. 2015) and optical spectroscopy (Bodewits et al. 2016) solar wind electron impact excitation of molecules plays a major role in the intensity of what is observed remotely, and the characteristics of the solar wind can be quite variable. Due to its irregular shape, 67P also shows a very pronounced diurnal variation, which can be mostly explained by variable illumination conditions (Bieler et al. 2015b). It is, therefore, not straightforward to compare abundances of different comets as their rotational state, shape, heliocentric distance and interaction with a dynamic solar wind all strongly influence what is observed.

For many comets water is the most abundant species ($> 50\%$ of the volatiles), followed by CO and CO_2 (Mumma and Charnley 2011) and, at least in the case of 67P and Halley, O_2. The inventory of organics is very rich as can be seen from measurements at 67P (Goesmann et al. 2015; Le Roy et al. 2015; Wright et al. 2015; Altwegg et al. 2016; Calmonte et al. 2016). The data analysis is still ongoing and the assignment of species to the low-resolution mass spectra of the Ptolemy and COSAC instruments are not unique. Nevertheless, it is clear that 67P contains many hydrocarbons, oxygenated species, a complex S chemistry, as well as N-bearing species. The C-chain molecules detected by ROSINA (Le Roy et al. 2015) match the semi-volatile component of surface material detected by the Virtis instrument (Capaccioni et al. 2015; Quirico et al. 2016). The volatile inventories of 67P and other comets are similar to what is measured in presolar clouds (Mumma and Charnley 2011; Le Roy et al. 2015) and is close to what is expected based on laboratory experiments with interstellar ice analogues (for an overview see Linnartz et al. 2015 and references therein).

While the coma is clearly heterogeneous in composition, no firm statement can be made about the compositional heterogeneity of the nucleus at any given time. Some authors claim that the ice of 67P contains clathrates (Luspay-Kuti et al. 2016; Mousis et al. 2016a), but this seems to be very unlikely in view of the presence of O_2 and S_2. In order to form clathrates, all volatile species must be expelled from the amorphous ice first (Schmitt et al. 1989), even if the ice transformed without sublimation from amorphous to clathrates as postulated by Mousis et al. (2016b). This would have destroyed the very good correlation between O_2 and H_2O in the coma (Bieler et al. 2015a), and it would also have destroyed S_2 by photodissociation (Calmonte et al. 2016). We have to wait for further analysis to see if any nucleus heterogeneity exists that cannot be explained by the evolution of the outer layers of the comet. This could then shed light on the formation conditions of 67P, especially regarding whether the two lobes have different compositions because they formed separately.

4.5 Summary

With the new results from Rosetta, some paradigms about comets are changing. The most important one is that the regions where OCCs and JFCs formed probably overlapped and extended over large radial distances from the Sun. This may explain the range of D/H ratios found in cometary water, as well as some of the other compositional variations. However, care has to be taken as cometary comas can be very heterogeneous and vary with time. Evidence for a connection between the presolar cloud composition and the observed cometary chemical inventories is seen in the highly deuterated species and the complex organic chemistry, especially of S-bearing molecules. There are even indications that some presolar ice layers on dust grains have survived comet accretion. Clear signs of grain-surface chemistry, as well as of radiolysis/photolysis, are evident in the volatile inventory of 67P. There are no significant traces of mixing of the volatile material with processed material from the inner Solar System, at least for 67P. This might be somewhat different for comets with lower D/H ratios in their water, as they may have formed closer to the Sun. So far, there have been no firm detections of CAIs or chondrules by the COSIMA instrument, although Fulle et al. (2015) claims that the compact grains they observe must have undergone modification in the solar nebula.

The very porous nature and the physical homogeneity of the nucleus down to the decimeter level, as well as the extremely fluffy, low density, low strength properties of emitted dust aggregates from 67P make violent collisions between dust grains or cometesimals in the early history of comets unlikely. These features are more compatible with comet formation by the streaming instabilities (Blum et al. 2014; Davidsson et al. 2016) or by forming in dust traps (van der Marel et al. 2015). However, it cannot be ruled out that small comets are the products of collisional disruption of larger bodies (Farinella and Davis 1996; Morbidelli and Rickman 2015). This is certainly one possible explanation for the generally elongated shapes of known JFCs (Jewitt et al. 2003). Whatever the mechanisms were for forming comets and ultimately producing their current shapes, what has become increasingly evident is that comets like 67P preserve very primitive materials that have undergone little, if any, reprocessing in the protosolar nebula.

The average D/H of water measured in comets so far is well above terrestrial. This clearly makes the notion that comets delivered all of the terrestrial water unlikely. However, it was shown by Marty et al. (2016) that, while not required, the amount of ^{36}Ar in the terrestrial atmosphere is compatible with its delivery by comets like 67P and that this could be done without changing the terrestrial D/H ratio and other volatile element abundances by much. In addition, there is isotopic evidence that $\sim 20\%$ of atmospheric Xe could have come from comets like 67P (Marty et al. 2017). It has been suggested that comets are also responsible for the large enrichments in ^{22}Ne and ^{36}Ar in Venus's atmosphere (Owen and Bar-Nun 2000), although other explanations are possible (Bogard 1988). In this context, the Earth could also have acquired complex organics from comets that, if accreted after the Moon-forming impact, helped to spark life.

5 Tests of the Dynamical Models and Constraints on the Sources of Terrestrial Planet Volatiles

5.1 Testing the Dynamical Models

The predictions of giant planet formation models, whether or not they involve orbital migration (e.g., the Grand Tack and Nice models), are that C-complex and P-/D-type asteroids formed in the outer Solar System, and a significant fraction will have formed in the same regions as comets. Since these asteroids are thought to be the sources of the carbonaceous chondrites, the chondrites can potentially be used to test the models, but to do this requires indicators of formation distances from the Sun. At present, the H isotopes of water are the most promising indicators of formation distance (Alexander et al. 2012). This is because there is likely to have been a radial gradient in the H isotopic composition of water in the disk. This gradient would have been the result of radial mixing between (1) water in the outer Solar System that retains a large D-rich interstellar component inherited from the presolar molecular cloud, and (2) D-poor (solar D/H) water in the warm inner Solar System produced by isotopic re-equilibration between H_2 and H_2O. While some models predict a simple monotonic increase in D/H in water with radial distance (e.g., Horner et al. 2007), this is almost certainly a gross oversimplification (e.g., Yang et al. 2013). Nevertheless, the basic prediction that planetesimals that formed further from the Sun should tend to be more D-rich is at least consistent with the H isotopic compositions of water ice in carbonaceous chondrites, most comets and Saturn's moon Enceladus (Fig. 5, as well as Tables 4 and 5).

As previously discussed (Sect. 2.5), the estimates of the H isotopic compositions of OH/water in CI and CM chondrites, as well as Tagish Lake, are all intermediate between the terrestrial and solar values (Table 4), and quite distinct from those of any comets (Table 5). The estimate for the CR chondrite water is slightly above terrestrial and overlaps with the compositions of the least D enriched comets. However, perhaps most surprising are the OC and R chondrite compositions that overlap with those of comets. Taken at face value, this is the reverse of what was expected from the dynamical models as it would seem to imply that the OC and R chondrites formed further from the Sun than the carbonaceous chondrites and even many comets. At the very least, and counter intuitively, it is possible that the OC and R chondrites accreted more interstellar water than the carbonaceous chondrites (Deloule et al. 1998; Piani et al. 2015). However, there may be a more mundane explanation for the OC and R chondrite results related to the oxidation of Fe (Alexander et al. 2010, 2012; Bonal et al. 2013). This process will have affected all the chondrites to varying degrees and it may partly explain why the CR chondrites have higher δD values than the other carbonaceous chondrites (Sutton et al. 2017).

Figure 5 plots the estimated water isotopic compositions of the carbonaceous chondrites, comets, and Saturn's moon Enceladus as a function of the radial distance from the Sun, either in their current location (Saturn's moons and chondrites) or approximately where they are thought to have formed (comets). The H isotopic composition of CH_4 in Titan's atmosphere (Nixon et al. 2012) is also plotted, although it is unclear how closely it reflects the isotopic composition of the water that Titan initially accreted. The formation locations of the two classes of comets are unknown, but are likely to have been in the region of the current orbit of Neptune (e.g., Brasser and Morbidelli 2013). Hence, they have both been given nominal formation locations of ~ 25 AU. If Enceladus and Titan formed from Saturn's subdisk, this would presumably have been towards the end of Saturn's growth that in the Grand Tack model would have been when Saturn was between roughly 3 AU and 7 AU (arrowed) rather than at its current orbital radius of ~ 10 AU. Again, it is clear that,

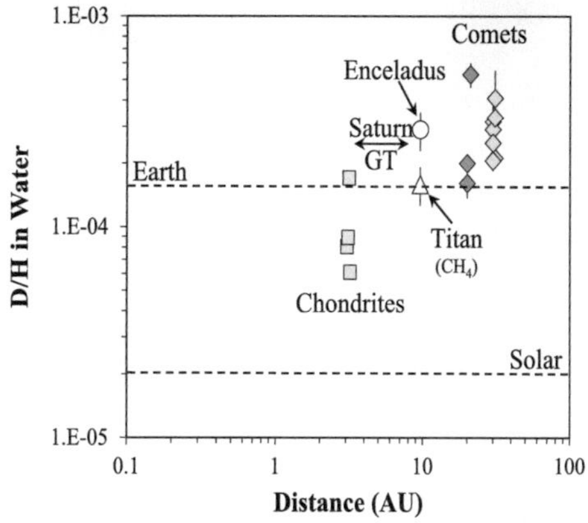

Fig. 5 The H isotopic compositions of OH/water in various objects as a function of radial distance from the Sun (modified after Alexander 2017). The H isotopic composition of methane in Titan's atmosphere is also shown, although it is not known how closely it reflects that of the water that Titan accreted. The locations of the asteroidal parent bodies of the carbonaceous chondrites are not known and so they have been given nominal radial distances of 3 AU. Saturn's moons Titan and Enceladus have been given their current orbital distances, although if there was a Grand Tack they may have formed between ~ 3 AU and ~ 7 AU (*double arrowed*). The formation locations of the comets are unknown, but are thought to have been between ~ 20 AU and ~ 30 AU (e.g., Brasser and Morbidelli 2013). The Oort (*light grey*) and Jupiter family (*dark grey*) comets are plotted at 30 AU and 20 AU, respectively, but their overlapping D/H ratios suggest that they formed in similar regions of the disk. With the possible exception of the CRs, the carbonaceous chondrite with the most D-rich water, the carbonaceous chondrites have compositions that are distinct from any measured outer Solar System body. Data sources are given in Tables 4 and 5

with the possible exception of the CRs, the carbonaceous chondrites had significantly lighter initial water H isotopic compositions than those of comets and Enceladus. Consequently, it is tempting to infer that water beyond somewhere between 3–7 AU was enriched in D relative to the bulk Earth and somewhere between 3–7 AU there was a steep decrease in the isotopic composition of the water in the disk.

There are at least two caveats to this line of reasoning. First, if Saturn's moons formed in its subdisk, their compositions will reflect the composition of the subdisk when they formed and not necessarily that of the surrounding nebula. During the main phase of Saturn's growth, the subdisk will have been very hot, erasing any intermolecular isotopic fractionations inherited from the nebula material it accreted. The subdisk will have cooled as accretion slowed and the protoplanetary disk dissipated. Whether at this stage conditions (e.g., temperature) and timescales would have produced larger isotopic fractionations than were present in the disk at Saturn's location is not known but seems doubtful. If correct, Saturn's moons provide lower limits for the isotopic compositions of planetesimals that formed in its neighborhood. However, it is also possible that Enceladus formed from the debris of a tidally disrupted planetesimal of unknown origin that was captured by Saturn (Crida and Charnoz 2012), in which case it cannot be used to constrain the H isotopic gradient in water in the outer Solar System.

Nitrogen isotopes might provide some further constraints on the formation distances of planetesimals (Alexander 2017). The N_2-NH_3 system in the nebula might have behaved in an analogous way to the H_2-H_2O system, although N_2 is less chemically reactive than H_2 and there has been no detailed isotopic modeling of the N system in the nebula. Nevertheless, the N isotopic compositions of the bulk chondrites are less ^{15}N enriched than HCN or NH_3 in comets, as well as Titan's atmospheric N_2 (Niemann et al. 2010) that may have

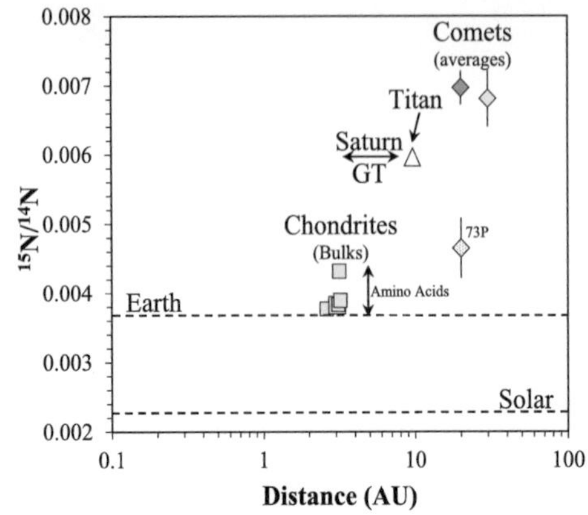

Fig. 6 The same as for Fig. 5 but for N isotopes (modified after Alexander 2017). The N isotopic compositions of individual comets are generally very consistent but the uncertainties can be large (Table 6), so they have been averaged, except for the JFC outlier comet 73P. Titan's atmospheric N isotopic composition is thought to reflect that of the NH_3 it accreted (Mandt et al. 2014). The carbonaceous chondrite compositions are for bulk meteorites (Table 3). However, amino acids and other N-bearing soluble organic compounds (*vertical double arrow*) probably formed from HCN/NH_3 and exhibit a similar range of isotopic compositions to the bulk meteorites (see text and Table 6 for sources)

originally been accreted as NH_3 (Mandt et al. 2014). Hence, when the N isotopes are plotted against radial distance (Fig. 6) there is a similar pattern to that for H isotopes (Fig. 5), with outer Solar System bodies being more enriched than the chondrites in ^{15}N. However, to truly compare like with like we should be using HCN/NH_3 compositions in chondrites not their bulk compositions. There is little HCN, NH_3 or ammonium salts in chondrites, and what there is has not been extensively studied. However, the CI, CM, and CR chondrites do contain amino acids that are thought to have formed in their parent bodies via Strecker-cyano synthesis that would have involved HCN and NH_3. The N isotopic compositions of these amino acids, as well as a few measurements of NH_3 and other N-bearing soluble organic compounds, have a similar range to the bulk chondrites (Epstein et al. 1987; Pizzarello et al. 1994; Engel and Macko 1997; Pizzarello and Holmes 2009; Elsila et al. 2012).

The amino acids and other N-bearing compounds that probably formed from HCN/NH_3 show that, like comets, at least some fraction of the ice accreted by chondrites contained highly volatile compounds. The ubiquitous presence of carbonates in aqueously altered chondrites suggests that the ices, as in comets, also contained CO/CO_2 (Alexander et al. 2015). However, on balance, the combination of H and N isotopes in carbonaceous chondrites suggest that they did not accrete in the formation locations of most comets and probably not beyond the formation locations of Saturn's moons, which in the Grand Tack model would have been between ~ 3 AU and ~ 7 AU. It should always be borne in mind that our meteorite collections probably do not sample all asteroid types. Nevertheless, if there was a Grand Tack it appears that it did not scatter many bodies into the asteroid belt from as far out as envisaged in the model. On the other hand, the meteoritic data are equally consistent with the chondrites having formed more-or-less where their parent bodies are presently located in the asteroid belt. Nor is there evidence in the meteorite collection, particularly from the Tagish Lake meteorite, for scattering of P-/D-type asteroids from the comet forming regions beyond 15 AU as required by the Nice model. Raymond and Izidoro (2017) have argued that scattering of outer Solar System planetesimals from as far out as 20 AU into the inner Solar System, including implantation into the Asteroid Belt, is an inevitable consequence of

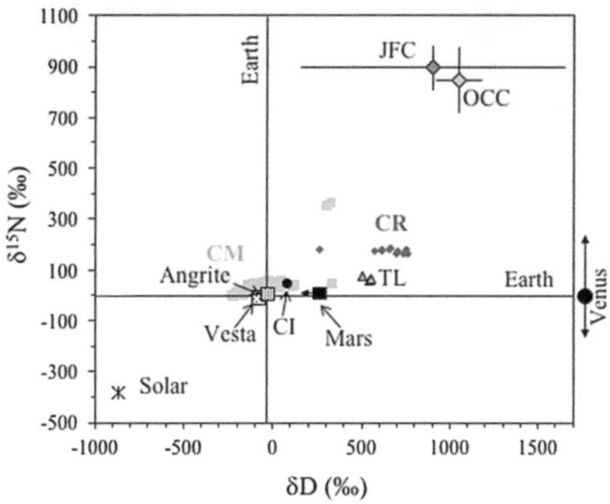

Fig. 7 Comparison of the estimated bulk H and N isotopic compositions of the major inner Solar System bodies with those of the average Jupiter family (JFC) and Oort Cloud (OCC) comets and individual members of the most volatile-rich chondritic meteorites. The H isotopic composition given for Mars is probably an upper limit. The H isotopes of the comets, based on their water compositions, are probably lower limits for their bulk compositions, whereas their N isotopes are probably upper limits (see text for details). Venus has lost almost all of its original water and its initial H isotopic composition is unknown. The sources are given in Tables 5 and 6

giant planet formation, irrespective of whether there was a Grand Tack. Again, the isotopic evidence does not seem to be consistent with a large fraction of objects that were implanted in the asteroid belt coming from beyond the orbit of Saturn, unless we are being misled by the moons of Saturn about the nature of the H and N isotopic gradients in the nebula.

5.2 Sources of Volatiles for the Terrestrial Planets

When the terrestrial planets formed, they will have accreted bulk planetesimals and comets, not just their water. So, in trying to constrain the sources of the volatiles in the Earth it is important to compare bulk compositions. For the comets, this is problematic as what can be measured in their comas remotely may not be representative of their bulk compositions. For instance, refractory organic matter may make significant contributions to the bulk H and N contents of comets, and N_2 may also be an important component of the N budgets. The abundances and elemental and isotopic compositions of the refractory organics and N_2 in comets are poorly known. For the organics, the similarities between Halley CHON particles, and IOM in primitive meteorites and IDPs suggest that the cometary organics are probably significantly more D-rich than their water, but will not be as ^{15}N-rich as their HCN and NH_3. Mass balance would suggest that as the major N-bearing component in the nebula, the N_2 in comets should be close to the bulk solar value, which is highly depleted in ^{15}N relative to the Earth (Marty et al. 2011). Hence, the H isotopic compositions measured in cometary water should be regarded as minimum estimates for their bulk D/H, while the N isotopic compositions of HCN/NH_3 should be regarded as upper limits for their bulk $^{15}N/^{14}N$. Even with the chondrites one has to be somewhat cautious because there is no guaranty that they faithfully represent the bulk compositions of their parent bodies. For instance, as discussed earlier, their parent bodies may have retained some ice and other volatiles that the meteorites will have lost when they were small bodies prior to being captured by the Earth.

There are also considerable uncertainties and disagreements in the estimates of the volatile budget of the Earth (e.g., Marty 2012; Halliday 2013). Nevertheless, Marty (2012) has argued that the volatile budget of the Earth (H, C, noble gases, Cl, and I) can be explained by addition of $\sim$ 2–4 wt.% CI/CM-like material. Hydrogen and N isotopes (Fig. 7) also suggest that CI/CM-like material, rather than comets or other chondrites, are the dominant source of Earth's volatiles (Alexander et al. 2012). The bulk CI and CM H and N isotopic compositions are not identical to the Earth's, which could be explained either by the CI/CM-like planetesimals retaining some isotopically light ices and other volatiles, or by accretion of $\sim$10% of solar H and N. The interior H and N isotopic compositions of Mars (Mathew and Marti 2001; Usui et al. 2012; Hallis et al. 2017) are also consistent with CI/CM-like sources.

That CI/CM-like bodies were the major sources of volatiles (except perhaps some of the noble gases) for the terrestrial planets is at least consistent with the fact that CI/CM-like bodies are the most abundant type of asteroid. The terrestrial planets probably acquired their volatiles either by: (a) scattering of volatile-rich objects from the Asteroid Belt into their formation regions, as in the 'classical' model (Chambers 2001; Raymond et al. 2009), or (b) the accretion of outer Solar System bodies scattered into the inner Solar System (Walsh et al. 2011; Raymond and Izidoro 2017). In either case, the sources of the terrestrial planets' volatiles must be the most abundant of the volatile-rich asteroid types.

6 Summary and Conclusions

The evidence from meteorites is that planetesimal formation in the inner Solar System began very early ($\sim$ 0.1–0.3 Ma after CAIs) and continued for at least 2 Ma (formation of OC and R chondrites). Planetesimals that formed earlier than $\sim$ 2 Ma after CAIs would have contained sufficient short-lived ^{26}Al to have experienced extensive melting and in most cases differentiation into a silicate crust/mantle (achondrites) and iron core (magmatic iron meteorites). Spectroscopic comparisons suggest that the OC and R chondrites come from S-complex asteroids that dominate the inner Asteroid Belt. The evidence for water in the OC and R chondrites indicate that the snowline had migrated to within $\sim$ 2 AU of the Sun by $\sim$ 2 Ma after CAIs, assuming that the S-complex asteroids formed roughly where they are found today.

The carbonaceous chondrites have been spectroscopically linked to the C-complex and P-/D-type asteroids. The various carbonaceous chondrites parent bodies formed between roughly 2 Ma and 4 Ma after CAIs. At present, it has yet to be firmly established whether they formed in the inner or outer Solar System, although there is new isotopic evidence that suggests that they formed beyond the orbit of a growing Jupiter. Planetesimal formation may have continued beyond $\sim$ 4 Ma after CAIs. However, the lack of a sufficiently strong radiogenic heat source to drive lithification (at a minimum this requires melting of ice) means that these bodies would not be consolidated enough to produce meteorites. Most C-complex asteroids show spectroscopic evidence for the presence of phyllosilicates, indicating that they, like the carbonaceous chondrites, formed within $\sim$ 4 Ma of CAIs. There are asteroids without clear spectroscopic evidence for phyllosilicates, most notably the P- and D-types, that could have formed after $\sim$ 4 Ma, but spectroscopic linking of the heavily altered ungrouped Tagish Lake chondrite to the D-types calls this into question. The absence of implanted solar wind in unbrecciated chondrites indicates that the solar nebula dissipated later than $\sim$ 4 Ma after CAIs.

Primitive anhydrous IDPs may come from bodies that formed later than $\sim$ 4 Ma after CAIs, but at present we have no way to date when they formed. It has often been suggested that primitive IDPs come from comets. Certainly, the abundances of highly volatile species in comets suggest that they formed late and/or as small bodies. There also seems to be many common features between chondrite matrices, IDPs and comet dust. These include an abundance of fine-grained crystalline and amorphous silicates, refractory organic matter, presolar grains and water. The presence of carbonates and amino acids in chondrites suggests that the ices they accreted contained CO/CO_2 and NH_3/HCN, just like comets. However, the lower abundance of organic matter in chondrite matrices suggests that there is a continuum between it and comet dust rather than there being a direct genetic relationship. The fact that matrix abundance tends to increase with accretion age after Solar System formation could reflect the transport of more primitive dust into the inner Solar System whose dust content was being progressively depleted by planetesimal formation and accretion onto the Sun. Alternatively, it might reflect progressive planetesimal formation at increasing distances from the Sun in a compositionally zoned disk.

The elemental fractionations in chondrites can be explained by mixing of volatile-depleted chondrules and refractory inclusions with volatile-rich matrix that had a CI-like bulk composition. To first order, this also accounts for the bulk C, N and H abundances of chondrites at the time of accretion, although subsequent parent body processes have modified these abundances in most chondrites. Interestingly, water abundances in chondrite matrices and comets appear to be less than would be expected assuming a bulk solar composition. The cause of this depletion is unclear but would have implications for planetesimal and planet formation if it is representative of the dust composition beyond the snowline since it would imply a lower than predicted dust surface density in the disk.

The H isotopes of water and N isotopes of HCN and NH_3 support the evidence for a continuum between chondrites and comets rather than a direct genetic relationship. The D/H ratio in water in the nebula is thought to have increased with distance from the Sun due to mixing between interstellar ice and water that had been re-equilibrated with H_2 in the inner Solar System. Comets exhibit a range of water D/H ratios but, with a few exceptions, they are all enriched in D relative to the bulk Earth and solar compositions. On the other hand, the CI and CM chondrites, as well as Tagish Lake, have estimated water D/H ratios that are significantly depleted relative to Earth. Nevertheless, they are still enhanced relative to the bulk solar composition, indicating that these chondrites accreted some interstellar water but most of it had been re-equilibrated in the inner Solar System. The H isotopic composition of water in the CR chondrites does overlap with the most D depleted comets, but this elevated D/H ratio may have been the product of parent body processes. Parent body processes may also explain the very D-rich water compositions of the OC and R chondrites.

As with the D/H in water, the N isotopes of the HCN and NH_3 in comets are enhanced in ^{15}N relative to the Earth and the bulk solar composition, and this enhancement is thought to be an interstellar signature. Amino acids in chondrites are thought to be the products of reactions involving HCN and NH_3, and are also enriched in ^{15}N although not as much as in comets. Thus, the isotopic compositions of H in water and N in HCN/NH_3 indicate that chondrites did not form in the same regions as comets, but probably formed sunward of comets where the abundance of unprocessed interstellar material was lower.

In terms of the major sources of volatiles to the terrestrial planets, the addition of 2–4 wt.% CI and CM chondrites give the best fits to the abundances and isotopic compositions of H, C and N. This would also account for much of the Earth's noble gases, but comets may have been a significant source of Ar and Xe for the Earth, and Ar and Ne for Venus. A minor contribution to the Earth from solar H and N may be required to explain their isotopic compositions, although how this material was accreted is unclear. Differentiated planetesimals are often not considered as significant sources of Earth's volatiles. However, two groups of achondrites have been shown to contain water that was either retained during melting and differentiation, or was accreted late as their parent bodies cooled and began to solidify.

Acknowledgements This manuscript was greatly improved by the thoughtful comments from two anonymous reviewers and the editor A. Morbidelli. C.A. was partially funded by the National Aeronautics and Space Administration Cosmochemistry program grant NNX14AJ54G, and K.M. by National Aeronautics and Space Administration Emerging Worlds program grant NNX17AE77G.

References

F.A.J. Abernethy, A.B. Verchovsky, N.A. Starkey, M. Anand, I.A. Franchi, M.M. Grady, Stable isotope analysis of carbon and nitrogen in angrites. Meteorit. Planet. Sci. **48**, 1590–1606 (2013)

M.F. A'Hearn, R.C. Millis, D.O. Schleicher, D.J. Osip, P.V. Birch, The ensemble properties of comets: results from narrowband photometry of 85 comets, 1976–1992. Icarus **118**, 223–270 (1995)

M.F. A'Hearn, M.J.S. Belton, W.A. Delamere, L.M. Feaga, D. Hampton, J. Kissel, K.P. Klaasen, L.A. McFadden, K.J. Meech, H.J. Melosh, P.H. Schultz, J.M. Sunshine, P.C. Thomas, J. Veverka, D.D. Wellnitz, D.K. Yeomans, S. Besse, D. Bodewits, T.J. Bowling, B.T. Carcich, S.M. Collins, T.L. Farnham, O. Groussin, B. Hermalyn, M.S. Kelley, M.S. Kelley, J.-Y. Li, D.J. Lindler, C.M. Lisse, S.A. McLaughlin, F. Merlin, S. Protopapa, J.E. Richardson, J.L. Williams, EPOXI at comet Hartley 2. Science **332**, 1396–1400 (2011)

S.A. Airieau, J. Farquhar, M.H. Thiemens, L.A. Leshin, H. Bao, E. Young, Planetesimal sulfate and aqueous alteration in CM and CI carbonaceous chondrites. Geochim. Cosmochim. Acta **69**, 4167–4172 (2005)

C.M.O'D. Alexander, Re-examining the role of chondrules in producing the volatile element fractionations in chondrites. Meteorit. Planet. Sci. **40**, 943–965 (2005)

C.M.O'D. Alexander, The origin of inner solar system water. Philos. Trans. R. Soc., Math. Phys. Eng. Sci. **375**, 20150384 (2017)

C.M.O'D. Alexander, S.S. Russell, J.W. Arden, R.D. Ash, M.M. Grady, C.T. Pillinger, The origin of chondritic macromolecular organic matter: A carbon and nitrogen isotope study. Meteorit. Planet. Sci. **33**, 603–622 (1998)

C.M.O'D. Alexander, A.P. Boss, R.W. Carlson, The early evolution of the inner solar system: a meteoritic perspective. Science **293**, 64–68 (2001)

C.M.O'D. Alexander, M. Fogel, H. Yabuta, G.D. Cody, The origin and evolution of chondrites recorded in the elemental and isotopic compositions of their macromolecular organic matter. Geochim. Cosmochim. Acta **71**, 4380–4403 (2007)

C.M.O'D. Alexander, S.N. Newsome, M.L. Fogel, L.R. Nittler, H. Busemann, G.D. Cody, Deuterium enrichments in chondritic macromolecular material—implications for the origin and evolution of organics, water and asteroids. Geochim. Cosmochim. Acta **74**, 4417–4437 (2010)

C.M.O'D. Alexander, R. Bowden, M.L. Fogel, K.T. Howard, C.D.K. Herd, L.R. Nittler, The provenances of asteroids, and their contributions to the volatile inventories of the terrestrial planets. Science **337**, 721–723 (2012)

C.M.O'D. Alexander, R. Bowden, K. Howard, A multi-technique search for the most primitive CO chondrites. Lunar Planet. Sci. **45**, #2667 (2014a)

C.M.O'D. Alexander, G.D. Cody, Y. Kebukawa, R. Bowden, M.L. Fogel, A.L.D. Kilcoyne, L.R. Nittler, C.D.K. Herd, Elemental, isotopic and structural changes in Tagish Lake insoluble organic matter produced by parent body processes. Meteorit. Planet. Sci. **49**, 503–525 (2014b)

C.M.O'D. Alexander, R. Bowden, M.L. Fogel, K.T. Howard, Carbonate abundances and isotopic compositions in chondrites. Meteorit. Planet. Sci. **50**, 810–833 (2015)

C.M.O'D. Alexander, G.D. Cody, B.T. De Gregorio, L.R. Nittler, R.M. Stroud, The nature, origin and modification of insoluble organic matter in chondrites, the major source of Earth's C and N. Chem. Erde **77**, 227–256 (2017a)

C.M.O'D. Alexander, L.R. Nittler, J. Davidson, F.J. Ciesla, Measuring the level of interstellar inheritance in the solar protoplanetary disk. Meteorit. Planet. Sci. **52**, 1797–1821 (2017b)

C.M.O'D. Alexander, R.C. Greenwood, R. Bowden, J.M. Gibson, K.T. Howard, I. Franchi, A mutli-technique search for the most primitive CO chondrites. Geochim. Cosmochim. Acta **221**, 406–420 (2018)

K. Altwegg, H. Balsiger, A. Bar-Nun, J.J. Berthelier, A. Bieler, P. Bochsler, C. Briois, U. Calmonte, M. Combi, J. De Keyser, P. Eberhardt, B. Fiethe, S. Fuselier, S. Gasc, T.I. Gombosi, K.C. Hansen, M. Hässig, A. Jäckel, E. Kopp, A. Korth, L. LeRoy, U. Mall, B. Marty, O. Mousis, E. Neefs, T. Owen, H. Rème, M. Rubin, T. Sémon, C.-Y. Tzou, H. Waite, P. Wurz, 67P/Churyumov-Gerasimenko, a Jupiter family comet with a high D/H ratio. Science **347**, 1261952 (2015)

K. Altwegg, H. Balsiger, A. Bar-Nun, J.-J. Berthelier, A. Bieler, P. Bochsler, C. Briois, U. Calmonte, M.R. Combi, H. Cottin, J. De Keyser, F. Dhooghe, B. Fiethe, S.A. Fuselier, S. Gasc, T.I. Gombosi, K.C. Hansen, M. Haessig, A. Jäckel, E. Kopp, A. Korth, L. Le Roy, U. Mall, B. Marty, O. Mousis, T. Owen, H. Rème, M. Rubin, T. Sémon, C.-Y. Tzou, J.H. Waite, P. Wurz, Prebiotic chemicals—amino acid and phosphorus—in the coma of comet 67P/Churyumov-Gerasimenko. Sci. Adv. **2**, e1600285 (2016)

K. Altwegg, H. Balsiger, J.J. Berthelier, A. Bieler, U. Calmonte, J. De Keyser, B. Fiethe, S.A. Fuselier, S. Gasc, T.I. Gombosi, T. Owen, L. Le Roy, M. Rubin, T. Sémon, C.-Y. Tzou, D_2O and HDS in the coma of 67P/Churyumov-Gerasimenko. Philos. Trans. R. Soc. A **375**, 20160253 (2017)

C. Arpigny, E. Jehin, J. Manfroid, D. Hutsemékers, R. Schulz, J.A. Stüwe, J.-M. Zucconi, I. Ilyin, Anomalous nitrogen isotope ratio in comets. Science **301**, 1522–1524 (2003)

M. Asplund, N. Grevesse, A.J. Sauval, P. Scott, The chemical composition of the Sun. Annu. Rev. Astron. Astrophys. **47**, 481–522 (2009)

G. Avice, B. Marty, The iodine–plutonium–xenon age of the Moon–Earth system revisited. Philos. Trans. R. Soc. A **372**, 20130260 (2014)

H. Balsiger, K. Altwegg, A. Bar-Nun, J.-J. Berthelier, A. Bieler, P. Bochsler, C. Briois, U. Calmonte, M. Combi, J. De Keyser, P. Eberhardt, B. Fiethe, S.A. Fuselier, S. Gasc, T.I. Gombosi, K.C. Hansen, M. Hässig, A. Jäckel, E. Kopp, A. Korth, L. Le Roy, U. Mall, B. Marty, O. Mousis, T. Owen, H. Rème, M. Rubin, T. Sémon, C.-Y. Tzou, J.H. Waite, P. Wurz, Detection of argon in the coma of comet 67P/Churyumov-Gerasimenko. Sci. Adv. **1**, e1500377 (2015)

A. Bar-Nun, I. Kleinfeld, On the temperature and gas composition in the region of comet formation. Icarus **80**, 243–253 (1989)

M. Barboni, P. Boehnke, B. Keller, I.E. Kohl, B. Schoene, E.D. Young, K.D. McKeegan, Early formation of the Moon 4.51 billion years ago. Sci. Adv. **3**, e1602365 (2017)

T.J. Barrett, J.J. Barnes, R. Tartèse, M. Anand, I.A. Franchi, R.C. Greenwood, B.L.A. Charlier, M.M. Grady, The abundance and isotopic composition of water in eucrites. Meteorit. Planet. Sci. **51**, 1110–1124 (2016)

A. Bieler, K. Altwegg, H. Balsiger, A. Bar-Nun, J.J. Berthelier, P. Bochsler, C. Briois, U. Calmonte, M. Combi, J. De Keyser, E.F. van Dishoeck, B. Fiethe, S.A. Fuselier, S. Gasc, T.I. Gombosi, K.C. Hansen, M. Hassig, A. Jackel, E. Kopp, A. Korth, L. Le Roy, U. Mall, R. Maggiolo, B. Marty, O. Mousis, T. Owen, H. Reme, M. Rubin, T. Semon, C.Y. Tzou, J.H. Waite, C. Walsh, P. Wurz, Abundant molecular oxygen in the coma of comet 67P/Churyumov-Gerasimenko. Nature **526**, 678–681 (2015a)

A. Bieler, K. Altwegg, H. Balsiger, J.-J. Berthelier, U. Calmonte, M. Combi, J. De Keyser, B. Fiethe, N. Fougere, S. Fuselier, S. Gasc, T. Gombosi, K. Hansen, M. Hässig, Z. Huang, A. Jäckel, X. Jia, L. Le Roy, U.A. Mall, H. Rème, M. Rubin, V. Tenishev, G. Tóth, C.-Y. Tzou, P. Wurz, Comparison of 3D kinetic and hydrodynamic models to ROSINA-COPS measurements of the neutral coma of 67P/Churyumov-Gerasimenko. Astron. Astrophys. **583**, A7 (2015b)

R.P. Binzel, A.S. Rivkin, J.S. Stuart, A.W. Harris, S.J. Bus, T.H. Burbine, Observed spectral properties of near-Earth objects: results for population distribution, source regions, and space weathering processes. Icarus **170**, 259–294 (2004)

A. Bischoff, N. Vogel, J. Roszjar, The Rumuruti chondrite group. Chem. Erde **71**, 101–133 (2011)

B. Bitsch, A. Johansen, M. Lambrechts, A. Morbidelli, The structure of protoplanetary discs around evolving young stars. Astron. Astrophys. **575**, A28 (2015)

N. Biver, D. Bockelée-Morvan, J. Crovisier, D.C. Lis, R. Moreno, P. Colom, F. Henry, F. Herpin, G. Paubert, M. Womack, Radio wavelength molecular observations of comets C/1999 T1 (McNaught-Hartley), C/2001 A2 (LINEAR), C/2000 WM1 (LINEAR) and 153P/Ikeya-Zhang. Astron. Astrophys. **449**, 1255–1270 (2006)

N. Biver, R. Moreno, D. Bockelée-Morvan, A. Sandqvist, P. Colom, J. Crovisier, D.C. Lis, J. Boissier, V. Debout, G. Paubert, S. Milam, A. Hjalmarson, S. Lundin, T. Karlsson, M. Battelino, U. Frisk, D. Murtagh, O. Team, Isotopic ratios of H, C, N, O, and S in comets C/2012 F6 (Lemmon) and C/2014 Q2 (Lovejoy). Astron. Astrophys. **589**, A78 (2016)

P.A. Bland, O. Alard, G.K. Benedix, A.T. Kearsley, O.N. Menzies, L.E. Watt, N.W. Rogers, Volatile fractionations in the early solar system and chondrule/matrix complimentarity. Proc. Natl. Acad. Sci. **102**, 13755–13760 (2005)

A.I. Blinova, T. Zega, C.D.K. Herd, R. Stroud, Testing variations within the Tagish Lake meteorite, I: mineralogy and petrology of pristine samples. Meteorit. Planet. Sci. **49**, 473–502 (2014)

J. Blum, B. Gundlach, S. Mühle, J.M. Trigo-Rodriguez, Comets formed in solar-nebula instabilities! An experimental and modeling attempt to relate the activity of comets to their formation process. Icarus **235**, 156–169 (2014)

D. Bockelée-Morvan, D. Gautier, D.C. Lis, K. Young, J. Keene, T. Phillips, T. Owen, J. Crovisier, P.F. Goldsmith, E.A. Bergin, D. Despois, A. Wootten, Deuterated water in comet C/1996 B2 (Hyakutake) and its implications for the origin of comets. Icarus **133**, 147–162 (1998)

D. Bockelée-Morvan, N. Biver, E. Jehin, A.L. Cochran, H. Wiesemeyer, J. Manfroid, D. Hutsemékers, C. Arpigny, J. Boissier, W. Cochran, P. Colom, J. Crovisier, N. Milutinovic, R. Moreno, J.X. Prochaska, I. Ramirez, R. Schulz, J.-M. Zucconi, Large excess of heavy nitrogen in both hydrogen cyanide and cyanogen from comet 17P/Holmes. Astrophys. J. Lett. **679**, L49 (2008)

D. Bockelée-Morvan, N. Biver, B. Swinyard, M. de Val-Borro, J. Crovisier, P. Hartogh, D.C. Lis, R. Moreno, S. Szutowicz, E. Lellouch, M. Emprechtinger, G.A. Blake, R. Courtin, C. Jarchow, M. Kidger, M. Küppers, M. Rengel, G.R. Davis, T. Fulton, D. Naylor, S. Sidher, H. Walker, Herschel measurements of the D/H and $^{16}O/^{18}O$ ratios in water in the Oort-cloud comet C/2009 P1 (Garradd). Astron. Astrophys. **544**, L15 (2012)

D. Bockelée-Morvan, V. Debout, S. Erard, C. Leyrat, F. Capaccioni, G. Filacchione, N. Fougere, P. Drossart, G. Arnold, M. Combi, B. Schmitt, J. Crovisier, M.-C. de Sanctis, T. Encrenaz, E. Kührt, E. Palomba, F.W. Taylor, F. Tosi, G. Piccioni, U. Fink, G. Tozzi, A. Barucci, N. Biver, M.-T. Capria, M. Combes, W. Ip, M. Blecka, F. Henry, S. Jacquinod, J.-M. Reess, A. Semery, D. Tiphene, First observations of H_2O and CO_2 vapor in comet 67P/Churyumov-Gerasimenko made by VIRTIS onboard Rosetta. Astron. Astrophys. **583**, A6 (2015)

D. Bodewits, L.M. Lara, M.F. A'Hearn, F. La Forgia, A. Gicquel, G. Kovacs, J. Knollenberg, M. Lazzarin, Z.-Y. Lin, X. Shi, C. Snodgrass, C. Tubiana, H. Sierks, C. Barbieri, P.L. Lamy, R. Rodrigo, D. Koschny, H. Rickman, H.U. Keller, M.A. Barucci, J.-L. Bertaux, I. Bertini, S. Boudreault, G. Cremonese, V. Da Deppo, B. Davidsson, S. Debei, M. De Cecco, S. Fornasier, M. Fulle, O. Groussin, P.J. Gutiérrez, C. Güttler, S.F. Hviid, W.-H. Ip, L. Jorda, J.-R. Kramm, E. Kührt, M. Küppers, J.J. López-Moreno, F. Marzari, G. Naletto, N. Oklay, N. Thomas, I. Toth, J.-B. Vincent, Changes in the physical environment of the inner coma of 67P/Churyumov-Gerasimenko with decreasing heliocentric distance. Astron. J. **152**, 130 (2016)

D.D. Bogard, On the origin of Venus' atmosphere—possible contributions from simple component mixtures and fractionated solar wind. Icarus **74**, 3–20 (1988)

L. Bonal, C.M.O'D. Alexander, G.R. Huss, K. Nagashima, E. Quirico, P. Beck, Hydrogen isotopic composition of the water in CR chondrites. Geochim. Cosmochim. Acta **106**, 111–133 (2013)

W.F. Bottke, D. Nesvorný, R.E. Grimm, A. Morbidelli, D.P. O'Brien, Iron meteorites as remnants of planetesimals formed in the terrestrial planet region. Nature **439**, 821–824 (2006)

W.F. Bottke, R.J. Walker, J.M.D. Day, D. Nesvorny, L. Elkins-Tanton, Stochastic late accretion to Earth, the Moon, and Mars. Science **330**, 1527–1530 (2010)

W.F. Bottke, D. Vokrouhlický, D. Minton, D. Nesvorný, A. Morbidelli, R. Brasser, B. Simonson, H.F. Levison, An Archaean heavy bombardment from a destabilized extension of the asteroid belt. Nature **485**, 78–81 (2012)

R. Brasser, A two-stage formation process for the Oort comet cloud and its implications. Astron. Astrophys. **492**, 251–255 (2008)

R. Brasser, A. Morbidelli, Oort cloud and scattered disc formation during a late dynamical instability in the solar system. Icarus **225**, 40–49 (2013)

P.G. Brown, A.R. Hildebrand, M.E. Zolensky, M. Grady, R.N. Clayton, T.K. Mayeda, E. Tagliaferri, R. Spalding, N.D. MacRae, E.L. Hoffman, D.W. Mittlefehldt, J.F. Wacker, J.A. Bird, M.D. Cambell, R. Carpenter, H. Gingerich, M. Glatiotis, E. Greiner, M.J. Mazur, P.J. McCausland, H. Plotkin, T.R. Mazur, The fall, recovery, orbit and composition of the Tagish Lake meteorite: a new type of carbonaceous chondrite. Science **290**, 320–325 (2000)

R.H. Brown, D.S. Lauretta, B. Schmidt, J. Moores, Experimental and theoretical simulations of ice sublimation with implications for the chemical, isotopic, and physical evolution of icy objects. Planet. Space Sci. **60**, 166–180 (2012)

D.E. Brownlee, P. Tsou, J. Aléon, C.M.O'D. Alexander, T. Araki, S. Bajt, G.A. Baratta, R. Bastien, P.A. Bland, P. Bleuet, J. Borg, J.P. Bradley, A.J. Brearley, F. Brenker, S. Brennan, J.C. Bridges, N.D. Browning, J.R. Brucato, E. Bullock, M.J. Burchell, H. Busemann, A. Butterworth, M. Chaussidon, A. Cheuvront, M. Chi, M.J. Cintala, B.C. Clark, S.J. Clemett, G. Cody, L. Colangeli, G. Cooper, P. Cordier, C. Daghlian, Z.R. Dai, L. d'Hendecourt, Z. Djouadi, G. Dominguez, T. Duxbury, J.P. Dworkin, D.S. Ebel, T.E. Economou, S. Fakra, S.A.J. Fairey, S. Fallon, G. Ferrini, T. Ferroir, H. Fleckenstein, C. Floss, G.J. Flynn, I.A. Franchi, M. Fries, Z. Gainsforth, J.P. Gallien, M.J. Genge, M.K. Gilles, P. Gillet, J.D. Gilmour, D.P. Glavin, M. Gounelle, M.M. Grady, G.A. Graham, P.G. Grant, S.F. Green, F. Grossemy, L. Grossman, J.N. Grossman, Y. Guan, K. Hagiya, R. Harvey, P.R. Heck, G.F. Herzog, P. Hoppe, F. Hörz, J. Huth, I.D. Hutcheon, K. Ignatyev, H. Ishii, M. Ito, D. Jacob, C. Jacobsen, S.B. Jacobsen, S. Jones, D.J. Joswiak, A. Jurewicz, A.T. Kearsley, L.P. Keller, H. Khodja, A.L. Kilcoyne, J. Kissel, A.N. Krot, F. Langenhorst, A. Lanzirotti, L. Le, L.A. Leshin, J. Leitner, L. Lemelle, H. Leroux, M.-C. Liu, K. Luening, I.C. Lyon et al., Comet 81P/Wild 2 under a microscope. Science **314**, 1711–1716 (2006)

G. Budde, T.S. Kruijer, M. Fischer-Gödde, A.J. Irving, T. Kleine, Planetesimal differentiation revealed by the Hf–W systematics of ureilites. Earth Planet. Sci. Lett. **430**, 316–325 (2015)

G. Budde, T. Kleine, T.S. Kruijer, C. Burkhardt, K. Metzler, Tungsten isotopic constraints on the age and origin of chondrules. Proc. Natl. Acad. Sci. **113**, 2886–2891 (2016)

T.H. Burbine, T.J. McCoy, A. Meibom, B. Gladman, K. Kiel, in *Asteroids III*, ed. by W.F. Bottke Jr., A. Cellino, P. Paolicchi, R.P. Binzel (Univ. of Arizona Press, Tuscon, 2002), pp. 653–667

C. Burkhardt, T. Kleine, F. Oberli, A. Pack, B. Bourdon, R. Wieler, Molybdenum isotope anomalies in meteorites: constraints on solar nebula evolution and origin of the Earth. Earth Planet. Sci. Lett. **312**, 390–400 (2011)

C. Burkhardt, T. Kleine, N. Dauphas, R. Wieler, Origin of isotopic heterogeneity in the solar nebula by thermal processing and mixing of nebular dust. Earth Planet. Sci. Lett. **357–358**, 298–307 (2012)

U. Calmonte, K. Altwegg, H. Balsiger, J.J. Berthelier, A. Bieler, G. Cessateur, F. Dhooghe, E.F. van Dishoeck, B. Fiethe, S.A. Fuselier, S. Gasc, T.I. Gombosi, M. Hässig, L. Le Roy, M. Rubin, T. Sémon, C.-Y. Tzou, S.F. Wampfler, Sulphur-bearing species in the coma of comet 67P/Churyumov-Gerasimenko. Mon. Not. R. Astron. Soc. **462**, S253–S273 (2016)

H. Campins, K. Hargrove, N. Pinilla-Alonso, E.S. Howell, M.S. Kelley, J. Licandro, T. Mothé-Diniz, Y. Fernández, J. Ziffer, Water ice and organics on the surface of the asteroid 24 Themis. Nature **464**, 1320–1321 (2010)

F. Capaccioni, A. Coradini, G. Filacchione, S. Erard, G. Arnold, P. Drossart, M.C. De Sanctis, D. Bockelee-Morvan, M.T. Capria, F. Tosi, C. Leyrat, B. Schmitt, E. Quirico, P. Cerroni, V. Mennella, A. Raponi, M. Ciarniello, T. McCord, L. Moroz, E. Palomba, E. Ammannito, M.A. Barucci, G. Bellucci, J. Benkhoff, J.P. Bibring, A. Blanco, M. Blecka, R. Carlson, U. Carsenty, L. Colangeli, M. Combes, M. Combi, J. Crovisier, T. Encrenaz, C. Federico, U. Fink, S. Fonti, W.H. Ip, P. Irwin, R. Jaumann, E. Kuehrt, Y. Langevin, G. Magni, S. Mottola, V. Orofino, P. Palumbo, G. Piccioni, U. Schade, F. Taylor, D. Tiphene, G.P. Tozzi, P. Beck, N. Biver, L. Bonal, J.-P. Combe, D. Despan, E. Flamini, S. Fornasier, A. Frigeri, D. Grassi, M. Gudipati, A. Longobardo, K. Markus, F. Merlin, R. Orosei, G. Rinaldi, K. Stephan, M. Cartacci, A. Cicchetti, S. Giuppi, Y. Hello, F. Henry, S. Jacquinod, R. Noschese, G. Peter, R. Politi, J.M. Reess, A. Semery, The organic-rich surface of comet 67P/Churyumov-Gerasimenko as seen by VIRTIS/Rosetta. Science **347**, aaa0628 (2015)

J.E. Chambers, Making more terrestrial planets. Icarus **152**, 205–224 (2001)

B.-G. Choi, K.D. McKeegan, L.A. Leshin, J.T. Wasson, Origin of magnetite in oxidized CV chondrites: in situ measurement of oxygen isotope compositions of Allende magnetite and olivine. Earth Planet. Sci. Lett. **146**, 337–349 (1997)

B.-G. Choi, K.D. McKeegan, A.N. Krot, J.T. Wasson, Extreme oxygen-isotope compositions in magnetite from unequilibrated ordinary chondrites. Nature **392**, 577–579 (1998)

B.-G. Choi, A.N. Krot, J.T. Wasson, Oxygen-isotopes in magnetite and fayalite in CV chondrites Kaba and Mokoia. Meteorit. Planet. Sci. **35**, 1239–1248 (2000)

B.G. Choi, S. Itoh, H. Yurimoto, A.E. Rubin, J.T. Wasson, J.N. Grossman, Oxygen-isotopic composition of magnetite in the DOM 03238 CO3.1 chondrite. Meteorit. Planet. Sci. Suppl. **43**, A32 (2008)

F.J. Ciesla, Radial transport in the solar nebula: implications for moderately volatile element depletions in chondritic meteorites. Meteorit. Planet. Sci. **43**, 639–655 (2008)

R.N. Clayton, Oxygen isotopes in meteorites. Annu. Rev. Earth Planet. Sci. **21**, 115–149 (1993)

R.N. Clayton, T.K. Mayeda, Oxygen isotope studies of carbonaceous chondrites. Geochim. Cosmochim. Acta **63**, 2089–2104 (1999)

L.I. Cleeves, E.A. Bergin, C.M.O'D. Alexander, F. Du, D. Graninger, K.I. Öberg, T.J. Harries, The ancient heritage of water ice in the solar system. Science **345**, 1590–1593 (2014)

L.I. Cleeves, E.A. Bergin, C.M.O'D. Alexander, F. Du, D. Graninger, K.I. Öberg, T.J. Harries, Exploring the origins of deuterium enrichments in solar nebular organics. Astrophys. J. **819**, 13 (2016)

E.A. Cloutis, P. Hudon, T. Hiroi, M.J. Gaffey, Spectral reflectance properties of carbonaceous chondrites, 7: CK chondrites. Icarus **221**, 911–924 (2012a)

E.A. Cloutis, P. Hudon, T. Hiroi, M.J. Gaffey, P. Mann, Spectral reflectance properties of carbonaceous chondrites, 5: CO chondrites. Icarus **220**, 466–486 (2012b)

E.A. Cloutis, P. Hudon, T. Hiroi, M.J. Gaffey, P. Mann, J.F. Bell Iii, Spectral reflectance properties of carbonaceous chondrites, 6: CV chondrites. Icarus **221**, 328–358 (2012c)

J.N. Connelly, M. Bizzarro, A.N. Krot, Å. Nordlund, D. Wielandt, M.A. Ivanova, The absolute chronology and thermal processing of solids in the solar protoplanetary disk. Science **338**, 651–655 (2012)

A. Coutens, C. Vastel, E. Caux, C. Ceccarelli, S. Bottinelli, L. Wiesenfeld, A. Faure, Y. Scribano, C. Kahane, A study of deuterated water in the low-mass protostar IRAS 16293-2422. Astron. Astrophys. **539**, A132 (2012)

A. Coutens, C. Vastel, S. Cabrit, C. Codella, L.E. Kristensen, C. Ceccarelli, E.F. van Dishoeck, A.C.A. Boogert, S. Bottinelli, A. Castets, E. Caux, C. Comito, K. Demyk, F. Herpin, B. Lefloch, C. McCoey,

J.C. Mottram, B. Parise, V. Taquet, F.F.S. van der Tak, R. Visser, U.A. Yıldız, Deuterated water in the solar-type protostars NGC 1333 IRAS 4A and IRAS 4B⋆⋆⋆. Astron. Astrophys. **560**, A39 (2013)
A. Coutens, J.K. Jørgensen, M.V. Persson, E.F. van Dishoeck, C. Vastel, V. Taquet, High D_2O/HDO ratio in the inner regions of the low-mass protostar NGC 1333 IRAS2A. Astrophys. J. Lett. **792**, L5 (2014)
R.M. Coveney Jr., E.D. Goebel, E.J. Zeller, G.A.M. Dreschhoff, E.E. Angino, Serpentinization and the origin of hydrogen gas in Kansas. AAPG Bull. **71**, 39–48 (1987)
A. Crida, S. Charnoz, Formation of regular satellites from ancient massive rings in the solar system. Science **338**, 1196–1199 (2012)
N. Dauphas, A. Pourmand, Hf-W-Th evidence for rapid growth of Mars and its status as a planetary embryo. Nature **473**, 489–492 (2011)
J. Davidson, H. Busemann, L.R. Nittler, C.M.O'D. Alexander, F.-R. Orthous-Daunay, I.A. Franchi, P. Hoppe, Abundances of presolar silicon carbide grains in primitive meteorites determined by NanoSIMS. Geochim. Cosmochim. Acta **139**, 248–266 (2014)
B.J.R. Davidsson, H. Sierks, C. Güttler, F. Marzari, M. Pajola, H. Rickman, M.F. A'Hearn, A.-T. Auger, M.R. El-Maarry, S. Fornasier, P.J. Gutiérrez, H.U. Keller, M. Massironi, C. Snodgrass, J.-B. Vincent, C. Barbieri, P.L. Lamy, R. Rodrigo, D. Koschny, M.A. Barucci, J.-L. Bertaux, I. Bertini, G. Cremonese, V. Da Deppo, S. Debei, M. De Cecco, C. Feller, M. Fulle, O. Groussin, S.F. Hviid, S. Höfner, W.-H. Ip, L. Jorda, J. Knollenberg, G. Kovacs, J.-R. Kramm, E. Kührt, M. Küppers, F. La Forgia, L.M. Lara, M. Lazzarin, J.J. Lopez Moreno, R. Moissl-Fraund, S. Mottola, G. Naletto, N. Oklay, N. Thomas, C. Tubiana, The primordial nucleus of comet 67P/Churyumov-Gerasimenko. Astron. Astrophys. **592**, A63 (2016)
A.A. de Almeida, P.D. Singh, Photodissociation lifetime of $^{32}S_2$ molecule in comets. Earth Moon Planets **36**, 117–125 (1986)
M.C. De Sanctis, A. Raponi, E. Ammannito, M. Ciarniello, M.J. Toplis, H.Y. McSween, J.C. Castillo-Rogez, B.L. Ehlmann, F.G. Carrozzo, S. Marchi, F. Tosi, F. Zambon, F. Capaccioni, M.T. Capria, S. Fonte, M. Formisano, A. Frigeri, M. Giardino, A. Longobardo, G. Magni, E. Palomba, L.A. McFadden, C.M. Pieters, R. Jaumann, P. Schenk, R. Mugnuolo, C.A. Raymond, C.T. Russell, Bright carbonate deposits as evidence of aqueous alteration on (1) Ceres. Nature **536**, 54–57 (2016)
E. Deloule, F. Robert, J.C. Doukhan, Interstellar hydroxyl in meteoritic chondrules: implications for the origin of water in the inner solar system. Geochim. Cosmochim. Acta **62**, 3367–3378 (1998)
F.E. DeMeo, B. Carry, The taxonomic distribution of asteroids from multi-filter all-sky photometric surveys. Icarus **226**, 723–741 (2013)
F.E. DeMeo, B. Carry, Solar system evolution from compositional mapping of the asteroids. Nature **505**, 629–634 (2014)
F.E. DeMeo, R.P. Binzel, B. Carry, D. Polishook, N.A. Moskovitz, Unexpected D-type interlopers in the inner main belt. Icarus **229**, 392–399 (2014)
F.E. DeMeo, C.M.O'D. Alexander, K.J. Walsh, R.P. Binzel, C.R. Chapman, in *Asteroids IV*, ed. by P. Michel, F.E. DeMeo, W.F. Bottke (Univ. of Arizona Press, Tucson, 2015), pp. 13–41
L. Dones, P.R. Weissman, H.F. Levison, M.J. Duncan, H.U. Keller, H.A. Weaver, in *Comets II*, ed. by M.C. Festou (2004), pp. 153–174
P.M. Doyle, K. Jogo, K. Nagashima, A.N. Krot, S. Wakita, F.J. Ciesla, I.D. Hutcheon, Early aqueous activity on the ordinary and carbonaceous chondrite parent bodies recorded by fayalite. Nat. Commun. **6**, 7444 (2015)
R. Duffard, F. Roig, Two new V-type asteroids in the outer Main Belt? Planet. Space Sci. **57**, 229–234 (2009)
M.J. Duncan, H.F. Levison, A scattered comet disk and the origin of Jupiter family comets. Science **276**, 1670–1672 (1997)
T.L. Dunn, T.H. Burbine, W.F. Bottke Jr., J.P. Clark, Mineralogies and source regions of near-Earth asteroids. Icarus **222**, 273–282 (2013)
P. Eberhardt, M. Reber, D. Krankowsky, R.R. Hedges, The D/H and $^{18}O/^{16}O$ ratios in water from comet Halley. Astron. Astrophys. **302**, 301–316 (1995)
J.E. Elsila, S.B. Charnley, A.S. Burton, D.P. Glavin, J.P. Dworkin, Compound-specific carbon, nitrogen, and hydrogen isotopic ratios for amino acids in CM and CR chondrites and their use in evaluating potential formation pathways. Meteorit. Planet. Sci. **47**, 1517–1536 (2012)
M.H. Engel, S.A. Macko, Isotopic evidence for extraterrestrial non-racemic amino acids in the Murchison meteorite. Nature **389**, 265–268 (1997)
S. Epstein, R.V. Krishnamurthy, J.R. Cronin, S. Pizzarello, G.U. Yuen, Unusual stable isotope ratios in amino acid and carboxylic acid extracts from the Murchison meteorite. Nature **326**, 477–479 (1987)
P. Farinella, D.R. Davis, Short-period comets: primordial bodies or collisional fragments? Science **273**, 938–941 (1996)
L.M. Feaga, M.F. A'Hearn, J.M. Sunshine, O. Groussin, T.L. Farnham, Asymmetries in the distribution of H_2O and CO_2 in the inner coma of Comet 9P/Tempel 1 as observed by Deep Impact. Icarus **190**, 345–356 (2007)

P.D. Feldman, M.F. A'Hearn, J.-L. Bertaux, L.M. Feaga, J.W. Parker, E. Schindhelm, A.J. Steffl, S.A. Stern, H.A. Weaver, H. Sierks, J.-B. Vincent, Measurements of the near-nucleus coma of comet 67P/Churyumov-Gerasimenko with the Alice far-ultraviolet spectrograph on Rosetta. Astron. Astrophys. **583**, A8 (2015)

M. Fischer-Gödde, T. Kleine, Ruthenium isotopic evidence for an inner solar system origin of the late veneer. Nature **541**, 525–527 (2017)

S. Fornasier, C. Lantz, M.A. Barucci, M. Lazzarin, Aqueous alteration on main belt primitive asteroids: results from visible spectroscopy. Icarus **233**, 163–178 (2014)

N. Fougere, K. Altwegg, J.-J. Berthelier, A. Bieler, D. Bockelée-Morvan, U. Calmonte, F. Capaccioni, M.R. Combi, J. De Keyser, V. Debout, S. Erard, B. Fiethe, G. Filacchione, U. Fink, S.A. Fuselier, T.I. Gombosi, K.C. Hansen, M. Hässig, Z. Huang, L. Le Roy, C. Leyrat, A. Migliorini, G. Piccioni, G. Rinaldi, M. Rubin, Y. Shou, V. Tenishev, G. Toth, C.-Y. Tzou, Three-dimensional direct simulation Monte-Carlo modeling of the coma of comet 67P/Churyumov-Gerasimenko observed by the VIRTIS and ROSINA instruments on board Rosetta. Astron. Astrophys. **588**, A134 (2016)

P. Fritz, I.D. Clark, J.-C. Fontes, M.J. Whiticar, E. Faber, Deuterium and ^{13}C evidence for low temperature production of hydrogen and methane in a highly alkaline groundwater environment in Oman. Proc. Int. Symp. Water-Rock Interact. **7**, 793–796 (1992)

W. Fujiya, N. Sugiura, H. Hotta, K. Ichimura, Y. Sano, Evidence for the late formation of hydrous asteroids from young meteoritic carbonates. Nat. Commun. **3**, 627 (2012)

W. Fujiya, N. Sugiura, Y. Sano, H. Hiyagon, Mn–Cr ages of dolomites in CI chondrites and the Tagish Lake ungrouped carbonaceous chondrite. Earth Planet. Sci. Lett. **362**, 130–142 (2013)

M. Fulle, A.C. Levasseur-Regourd, N. McBride, E. Hadamcik, In situ dust measurements from within the coma of 1P/Halley: first-order approximation with a dust dynamical model. Astron. J. **119**, 1968–1977 (2000)

M. Fulle, V. Della Corte, A. Rotundi, P. Weissman, A. Juhasz, K. Szego, R. Sordini, M. Ferrari, S. Ivanovski, F. Lucarelli, M. Accolla, S. Merouane, V. Zakharov, E. Mazzotta Epifani, J.J. López-Moreno, J. Rodríguez, L. Colangeli, P. Palumbo, E. Grün, M. Hilchenbach, E. Bussoletti, F. Esposito, S.F. Green, P.L. Lamy, J.A.M. McDonnell, V. Mennella, A. Molina, R. Morales, F. Moreno, J.L. Ortiz, E. Palomba, R. Rodrigo, J.C. Zarnecki, M. Cosi, F. Giovane, B. Gustafson, M.L. Herranz, J.M. Jerónimo, M.R. Leese, A.C. López-Jiménez, N. Altobelli, Density and charge of pristine fluffy particles from comet 67P/Churyumov-Gerasimenko. Astrophys. J. Lett. **802**, L12 (2015)

M. Fulle, F. Marzari, V. Della Corte, S. Fornasier, H. Sierks, A. Rotundi, C. Barbieri, P.L. Lamy, R. Rodrigo, D. Koschny, H. Rickman, H.U. Keller, J.J. López-Moreno, M. Accolla, J. Agarwal, M.F. A'Hearn, N. Altobelli, M.A. Barucci, J.-L. Bertaux, I. Bertini, D. Bodewits, E. Bussoletti, L. Colangeli, M. Cosi, G. Cremonese, J.-F. Crifo, V. Da Deppo, B. Davidsson, S. Debei, M. De Cecco, F. Esposito, M. Ferrari, F. Giovane, B. Gustafson, S.F. Green, O. Groussin, E. Grün, P. Gutierrez, C. Güttler, M.L. Herranz, S.F. Hviid, W. Ip, S.L. Ivanovski, J.M. Jerónimo, L. Jorda, J. Knollenberg, R. Kramm, E. Kührt, M. Küppers, L. Lara, M. Lazzarin, M.R. Leese, A.C. López-Jiménez, F. Lucarelli, E. Mazzotta Epifani, J.A.M. McDonnell, V. Mennella, A. Molina, R. Morales, F. Moreno, S. Mottola, G. Naletto, N. Oklay, J.L. Ortiz, E. Palomba, P. Palumbo, J.-M. Perrin, F.J.M. Rietmeijer, J. Rodríguez, R. Sordini, N. Thomas, C. Tubiana, J.-B. Vincent, P. Weissman, K.-P. Wenzel, V. Zakharov, J.C. Zarnecki, Evolution of the dust size distribution of comet 67P/Churyumov-Gerasimenko from 2.2 au to perihelion. Astrophys. J. **821**, 19 (2016)

K. Furuya, E.F. van Dishoeck, Y. Aikawa, Reconstructing the history of water ice formation from HDO/H_2O and D_2O/HDO ratios in protostellar cores. Astron. Astrophys. **586**, A127 (2016)

M.J. Gaffey, S.L. Gilbert, Asteroid 6 Hebe: the probable parent body of the H-type ordinary chondrites and the IIE iron meteorites. Meteorit. Planet. Sci. **33**, 1281–1295 (1998)

M.J. Gaffey, K.L. Reed, M.S. Kelley, Relationship of E-type Apollo asteroid 3103 (1982 BB) to the enstatite achondrite meteorites and the Hungaria asteroids. Icarus **100**, 95–109 (1992)

K.G. Gardner-Vandy, D.S. Lauretta, R.C. Greenwood, T.J. McCoy, M. Killgore, I.A. Franchi, The Tafassasset primitive achondrite: insights into initial stages of planetary differentiation. Geochim. Cosmochim. Acta **85**, 142–159 (2012)

J. Geiss, G. Gloeckler, Abundances of deuterium and helium-3 in the protosolar cloud. Space Sci. Rev. **84**, 239–250 (1998)

F. Goesmann, H. Rosenbauer, J.H. Bredehöft, M. Cabane, P. Ehrenfreund, T. Gautier, C. Giri, H. Krüger, L. Le Roy, A.J. MacDermott, S. McKenna-Lawlor, U.J. Meierhenrich, G.M.M. Caro, F. Raulin, R. Roll, A. Steele, H. Steininger, R. Sternberg, C. Szopa, W. Thiemann, S. Ulamec, Organic compounds on comet 67P/Churyumov-Gerasimenko revealed by COSAC mass spectrometry. Science **349**, aab0689 (2015)

J.I. Goldstein, E.R.D. Scott, N.L. Chabot, Iron meteorites: crystallization, thermal history, parent bodies, and origin. Chem. Erde **69**, 293–325 (2009)

J. Gradie, E. Tedesco, Compositional structure of the asteroid belt. Science **216**, 1405–1407 (1982)

J.P. Greenwood, A.E. Rubin, J.T. Wasson, Oxygen isotopes in R-chondrite magnetite and olivine: links between R chondrites and ordinary chondrites. Geochim. Cosmochim. Acta **64**, 3897–3911 (2000)

R.E. Grimm, H.Y. McSween Jr., Heliocentric zoning of asteroid belt by ^{26}Al heating. Science **259**, 653–655 (1993)

A.N. Halliday, The origins of volatiles in the terrestrial planets. Geochim. Cosmochim. Acta **105**, 146–171 (2013)

L.J. Hallis, G.R. Huss, K. Nagashima, G.J. Taylor, D. Stöffler, C.L. Smith, M.R. Lee, Effects of shock and Martian alteration on Tissint hydrogen isotope ratios and water content. Geochim. Cosmochim. Acta **200**, 280–294 (2017)

M.S. Hanner, M.E. Zolensky, in *Astromineralogy*, ed. by T. Henning (Springer, Berlin, 2010), pp. 203–232

K.P. Harrison, R.E. Grimm, Thermal constraints on the early history of the H-chondrite parent body reconsidered. Geochim. Cosmochim. Acta **74**, 5410–5423 (2010)

P. Hartogh, D.C. Lis, D. Bockelée-Morvan, M. de Val-Borro, N. Biver, M. Küppers, M. Emprechtinger, E.A. Bergin, J. Crovisier, M. Rengel, R. Moreno, S. Szutowicz, G.A. Blake, Ocean-like water in the Jupiter-family comet 103P/Hartley 2. Nature **478**, 218–220 (2011)

M. Hässig, K. Altwegg, H. Balsiger, A. Bar-Nun, J.J. Berthelier, A. Bieler, P. Bochsler, C. Briois, U. Calmonte, M. Combi, J. De Keyser, P. Eberhardt, B. Fiethe, S.A. Fuselier, M. Galand, S. Gasc, T.I. Gombosi, K.C. Hansen, A. Jäckel, H.U. Keller, E. Kopp, A. Korth, E. Kührt, L. Le Roy, U. Mall, B. Marty, O. Mousis, E. Neefs, T. Owen, H. Rème, M. Rubin, T. Sémon, C. Tornow, C.-Y. Tzou, J.H. Waite, P. Wurz, Time variability and heterogeneity in the coma of 67P/Churyumov-Gerasimenko. Science **347**, aaa0276 (2015)

J. Hatchell, H. Roberts, T.J. Millar, Limits on HDS/H_2S abundance ratios in hot molecular cores. Astron. Astrophys. **346**, 227–232 (1999)

S. Henke, H.-P. Gail, M. Trieloff, W.H. Schwarz, T. Kleine, Thermal history modelling of the H chondrite parent body. Astron. Astrophys. **545**, A135 (2012)

S. Henke, H.P. Gail, M. Trieloff, W.H. Schwarz, Thermal evolution model for the H chondrite asteroid-instantaneous formation versus protracted accretion. Icarus **226**, 212–228 (2013)

D.C. Hezel, H. Palme, The chemical relationship between chondrules and matrix and the chondrule matrix complementarity. Earth Planet. Sci. Lett. **294**, 85–93 (2010)

P. Hily-Blant, L. Bonal, A. Faure, E. Quirico, The ^{15}N-enrichment in dark clouds and solar system objects. Icarus **223**, 582–590 (2013)

T. Hiroi, M.E. Zolensky, C.M. Pieters, The Tagish Lake meteorite: a possible sample from a D-type asteroid. Science **293**, 2234–2236 (2001)

J.H. Hoffman, R.R. Hodges, M.B. McElroy, T.M. Donahue, M. Koplin, Composition and structure of the Venus atmosphere: results from Pioneer Venus. Science **205**, 49–52 (1979)

J. Horner, O. Mousis, F. Hersant, Constraints on the formation regions of comets from their D: H ratios. Earth Moon Planets **100**, 43–56 (2007)

K. Hornung, S. Merouane, M. Hilchenbach, Y. Langevin, E.M. Mellado, V. Della Corte, J. Kissel, C. Engrand, R. Schulz, J. Ryno, J. Silen, C.t. the, A first assessment of the strength of cometary particles collected in-situ by the COSIMA instrument onboard ROSETTA. Planet. Space Sci. **133**, 63–75 (2016)

G.R. Huss, R.S. Lewis, Presolar diamond, SiC, and graphite in primitive chondrites: abundances as a function of meteorite class and petrologic type. Geochim. Cosmochim. Acta **59**, 115–160 (1995)

D. Hutsemékers, J. Manfroid, E. Jehin, C. Arpigny, A. Cochran, R. Schulz, J.A. Stüwe, J.-M. Zucconi, Isotopic abundances of carbon and nitrogen in Jupiter-family and Oort Cloud comets. Astron. Astrophys. **440**, L21–L24 (2005)

D. Hutsemékers, J. Manfroid, E. Jehin, J.-M. Zucconi, C. Arpigny, The ^{16}OH/^{18}OH and OD/OH isotope ratios in comet C/2002 T7 (LINEAR). Astron. Astrophys. **490**, L31–L34 (2008)

M. Ikeda, T. Hirota, S. Yamamoto, The H^{13}CN/HC^{15}N abundance ratio in dense cores: possible source-to-source variation of isotope abundances? Astrophys. J. **575**, 250–256 (2002)

E. Jacquet, F. Robert, Water transport in protoplanetary disks and the hydrogen isotopic composition of chondrites. Icarus **223**, 722–732 (2013)

E. Jehin, J. Manfroid, A.L. Cochran, C. Arpigny, J.-M. Zucconi, D. Hutsemékers, W.D. Cochran, M. Endl, R. Schulz, The anomalous ^{14}N/^{15}N Ratio in comets 122P/1995 S1 (de Vico) and 153P/2002 C1 (Ikeya-Zhang). Astrophys. J. Lett. **613**, L161–L164 (2004)

E. Jehin, J. Manfroid, D. Hutsemékers, A.L. Cochran, C. Arpigny, W.M. Jackson, H. Rauer, R. Schulz, J.-M. Zucconi, Deep Impact: high-resolution optical spectroscopy with the ESO VLT and the Keck I telescope. Astrophys. J. Lett. **641**, L145–L148 (2006)

E. Jehin, J. Manfroid, H. Kawakita, D. Hutsemékers, M. Weiler, C. Arpigny, A. Cochran, O. Hainaut, H. Rauer, R. Schulz, J.-M. Zucconi, Optical spectroscopy of the B and C fragments of comet 73P/Schwassmann-Wachmann 3 at the ESO VLT. LPI Contrib. **1405**, 8319 (2008)

E. Jehin, D. Bockelée-Morvan, N. Dello Russo, J. Manfroid, D. Hutsemékers, H. Kawakita, H. Kobayashi, R. Schulz, A. Smette, J. Stüwe, M. Weiler, C. Arpigny, N. Biver, A. Cochran, J. Crovisier, P. Magain, H. Rauer, H. Sana, R.J. Vervack, H. Weaver, J.-M. Zucconi, A multi-wavelength simultaneous study of the composition of the Halley family comet 8P/Tuttle. Earth Moon Planets **105**, 343–349 (2009)

D. Jewitt, The active asteroids. Astron. J. **143**, 66 (2012)

D. Jewitt, S. Sheppard, Y. Fernández, 143P/Kowal-Mrkos and the shapes of cometary nuclei. Astron. J. **125**, 3366–3377 (2003)

D. Jewitt, H. Hsieh, J. Agarwal, in *Asteroids IV*, ed. by P. Michel, F.E. DeMeo, W.F. Bottke (Univ. of Arizona Press, Tucson, 2015), pp. 221–241

C.E. Jilly-Rehak, G.R. Huss, K. Nagashima, ^{53}Mn-^{53}Cr radiometric dating of secondary carbonates in CR chondrites: timescales for parent body aqueous alteration. Geochim. Cosmochim. Acta **201**, 224–244 (2017)

T.D. Jones, L.A. Lebofsky, J.S. Lewis, M.S. Marley, The composition and origin of the C, P, and D asteroids—water as a tracer of thermal evolution in the outer belt. Icarus **88**, 172–192 (1990)

M. Jutzi, E. Asphaug, The shape and structure of cometary nuclei as a result of low-velocity accretion. Science **348**, 1355–1358 (2015)

M. Jutzi, W. Benz, Formation of bi-lobed shapes by sub-catastrophic collisions. A late origin of comet 67P's structure. Astron. Astrophys. **597**, A62 (2017)

M. Jutzi, W. Benz, A. Toliou, A. Morbidelli, R. Brasser, How primordial is the structure of comet 67P? Combined collisional and dynamical models suggest a late formation. Astron. Astrophys. **597**, A61 (2017)

G.W. Kallemeyn, A.E. Rubin, J.T. Wasson, The compositional classification of chondrites, V: the Karoonda (CK) group of carbonaceous chondrites. Geochim. Cosmochim. Acta **55**, 881–892 (1991)

H.U. Keller, S. Mottola, B. Davidsson, S.E. Schröder, Y. Skorov, E. Kührt, O. Groussin, M. Pajola, S.F. Hviid, F. Preusker, F. Scholten, M.F. A'Hearn, H. Sierks, C. Barbieri, P. Lamy, R. Rodrigo, D. Koschny, H. Rickman, M.A. Barucci, J.-L. Bertaux, I. Bertini, G. Cremonese, V. Da Deppo, S. Debei, M. De Cecco, S. Fornasier, M. Fulle, P.J. Gutiérrez, W.-H. Ip, L. Jorda, J. Knollenberg, J.R. Kramm, M. Küppers, L.M. Lara, M. Lazzarin, J.J. Lopez Moreno, F. Marzari, H. Michalik, G. Naletto, L. Sabau, N. Thomas, J.-B. Vincent, K.-P. Wenzel, J. Agarwal, C. Güttler, N. Oklay, C. Tubiana, Insolation, erosion, and morphology of comet 67P/Churyumov-Gerasimenko. Astron. Astrophys. **583**, A34 (2015)

N.T. Kita, H. Nagahara, S. Togashi, Y. Morishita, A short duration of chondrule formation in the solar nebula: evidence from ^{26}Al in Semarkona ferromagnesian chondrules. Geochim. Cosmochim. Acta **64**, 3913–3922 (2000)

T. Kleine, M. Touboul, J.A. van Orman, B. Bourdon, C. Maden, K. Mezger, A.N. Halliday, Hf-W thermochronometry: closure temperature and constraints on the accretion and cooling history of the H chondrite parent body. Earth Planet. Sci. Lett. **270**, 106–118 (2008)

T. Kleine, U. Hans, A.J. Irving, B. Bourdon, Chronology of the angrite parent body and implications for core formation in protoplanets. Geochim. Cosmochim. Acta **84**, 186–203 (2012)

S. Kobayashi, H. Imai, H. Yurimoto, New extreme ^{16}O-rich reservoir in the early solar system. Geochem. J. **37**, 663–669 (2003)

W. Kofman, A. Herique, Y. Barbin, J.-P. Barriot, V. Ciarletti, S. Clifford, P. Edenhofer, C. Elachi, C. Eyraud, J.-P. Goutail, E. Heggy, L. Jorda, J. Lasue, A.-C. Levasseur-Regourd, E. Nielsen, P. Pasquero, F. Preusker, P. Puget, D. Plettemeier, Y. Rogez, H. Sierks, C. Statz, H. Svedhem, I. Williams, S. Zine, J. Van Zyl, Properties of the 67P/Churyumov-Gerasimenko interior revealed by CONSERT radar. Science **349**, aab0639 (2015)

A.N. Krot, Y. Amelin, P. Cassen, A. Meibom, Young chondrules in CB chondrites from a giant impact in the early solar system. Nature **436**, 989–992 (2005)

A.N. Krot, K. Keil, E.R.D. Scott, C.A. Goodrich, M.K. Weisberg, in *Meteorites and Cosmochemical Processes*, ed. by A.M. Davis (Elsevier–Pergamon, Oxford, 2014), pp. 1–63

T.S. Kruijer, M. Fischer-Gödde, T. Kleine, P. Sprung, I. Leya, R. Wieler, Neutron capture on Pt isotopes in iron meteorites and the Hf–W chronology of core formation in planetesimals. Earth Planet. Sci. Lett. **361**, 162–172 (2013)

T.S. Kruijer, M. Touboul, M. Fischer-Gödde, K.R. Bermingham, R.J. Walker, T. Kleine, Protracted core formation and rapid accretion of protoplanets. Science **344**, 1150–1154 (2014)

T.S. Kruijer, C. Burkhardt, G. Budde, T. Kleine, Age of Jupiter inferred from the distinct genetics and formation times of meteorites. Proc. Natl. Acad. Sci. **114**, 6712–6716 (2017a)

T.S. Kruijer, T. Kleine, C. Burkhardt, G. Budde, Dating the formation of Jupiter through W and Mo isotope analyses of meteorites. Lunar Planet. Sci. **48**, #1386 (2017b)

M. Küppers, L. O'Rourke, D. Bockelée-Morvan, V. Zakharov, S. Lee, P. von Allmen, B. Carry, D. Teyssier, A. Marston, T. Müller, J. Crovisier, M.A. Barucci, R. Moreno, Localized sources of water vapour on the dwarf planet (1) Ceres. Nature **505**, 525–527 (2014)

E. Kurahashi, N.T. Kita, H. Nagahara, Y. Morishita, ^{26}Al–^{26}Mg systematics of chondrules in a primitive CO chondrite. Geochim. Cosmochim. Acta **72**, 3865–3882 (2008)

J. Labidi, J. Farquhar, C.M.O'D. Alexander, D.L. Eldridge, H. Oduro, Mass independent sulfur isotope signatures in CMs: implications for sulfur chemistry in the early solar system. Geochim. Cosmochim. Acta **196**, 326–350 (2017)

Y. Langevin, M. Hilchenbach, N. Ligier, S. Merouane, K. Hornung, C. Engrand, R. Schulz, J. Kissel, J. Rynö, P. Eng, Typology of dust particles collected by the COSIMA mass spectrometer in the inner coma of 67P/Churyumov Gerasimenko. Icarus **271**, 76–97 (2016)

D. Lazzaro, T. Michtchenko, J.M. Carvano, R.P. Binzel, S.J. Bus, T.H. Burbine, T. Mothé-Diniz, M. Florczak, C.A. Angeli, A.W. Harris, Discovery of a basaltic asteroid in the outer Main Belt. Science **288**, 2033–2035 (2000)

L. Le Roy, K. Altwegg, H. Balsiger, J.-J. Berthelier, A. Bieler, C. Briois, U. Calmonte, M.R. Combi, J. De Keyser, F. Dhooghe, B. Fiethe, S.A. Fuselier, S. Gasc, T.I. Gombosi, M. Hässig, A. Jäckel, M. Rubin, C.-Y. Tzou, Inventory of the volatiles on comet 67P/Churyumov-Gerasimenko from Rosetta/ROSINA. Astron. Astrophys. **583**, A1 (2015)

C. Lécuyer, P. Gillet, F. Robert, The hydrogen isotope composition of seawater and the global water cycle. Chem. Geol. **145**, 249–261 (1998)

L.A. Leshin, J. Farquhar, Y. Guan, S. Pizzarello, T.L. Jackson, M.H. Thiemens, Oxygen isotopic anatomy of Tagish Lake: relationship to primary and secondary minerals in CI and CM chondrites. Lunar Planet. Sci. **32**, #1843 (2001)

H.F. Levison, *Comet Taxonomy* (Astronomical Society of the Pacific, San Francisco, 1996), pp. 173–191

H.F. Levison, W.F. Bottke, M. Gounelle, A. Morbidelli, D. Nesvorný, K. Tsiganis, Contamination of the asteroid belt by primordial trans-Neptunian objects. Nature **460**, 364–366 (2009)

H.F. Levison, A. Morbidelli, K. Tsiganis, D. Nesvorný, R. Gomes, Late orbital instabilities in the outer planets induced by interaction with a self-gravitating planetesimal disk. Astron. J. **142**, 152 (2011)

H. Linnartz, S. Ioppolo, G. Fedoseev, Atom addition reactions in interstellar ice analogues. Int. Rev. Phys. Chem. **34**, 205–237 (2015)

D.C. Lis, N. Biver, D. Bockelée-Morvan, P. Hartogh, E.A. Bergin, G.A. Blake, J. Crovisier, M. de Val-Borro, E. Jehin, M. Küppers, J. Manfroid, R. Moreno, M. Rengel, S. Szutowicz, A Herschel study of D/H in water in the Jupiter-family comet 45P/Honda-Mrkos-Pajdušáková and prospects for D/H measurements with CCAT. Astrophys. J. Lett. **774**, L3 (2013)

F.-C. Liu, B. Parise, L. Kristensen, R. Visser, E.F. van Dishoeck, R. Güsten, Water deuterium fractionation in the low-mass protostar NGC1333-IRAS2A. Astron. Astrophys. **527**, A19 (2011)

K. Lodders, Solar System abundances and condensation temperatures of the elements. Astrophys. J. **591**, 1220–1247 (2003)

A. Luspay-Kuti, M. Hässig, S.A. Fuselier, K.E. Mandt, K. Altwegg, H. Balsiger, S. Gasc, A. Jäckel, L. Le Roy, M. Rubin, C.-Y. Tzou, P. Wurz, O. Mousis, F. Dhooghe, J.J. Berthelier, B. Fiethe, T.I. Gombosi, U. Mall, Composition-dependent outgassing of comet 67P/Churyumov-Gerasimenko from ROSINA/DFMS. Implications for nucleus heterogeneity? Astron. Astrophys. **583**, A4 (2015)

A. Luspay-Kuti, O. Mousis, M. Hässig, S.A. Fuselier, J.I. Lunine, B. Marty, K.E. Mandt, P. Wurz, M. Rubin, The presence of clathrates in comet 67P/Churyumov-Gerasimenko. Sci. Adv. **2**, e1501781 (2016)

J.R. Lyons, E.D. Young, CO self-shielding as the origin of oxygen isotope anomalies in the early solar nebula. Nature **435**, 317–320 (2005)

G.J. MacPherson, in *Meteorites and Cosmochemical Processes*, ed. by A.M. Davis (Elsevier–Pergamon, Oxford, 2014), pp. 139–179

K.E. Mandt, O. Mousis, J. Lunine, D. Gautier, Protosolar ammonia as the unique source of Titan's nitrogen. Astrophys. J. Lett. **788**, L24 (2014)

J. Manfroid, E. Jehin, D. Hutsemékers, A. Cochran, J.-M. Zucconi, C. Arpigny, R. Schulz, J.A. Stüwe, Isotopic abundance of nitrogen and carbon in distant comets. Astron. Astrophys. **432**, L5–L8 (2005)

J. Manfroid, E. Jehin, D. Hutsemékers, A. Cochran, J.-M. Zucconi, C. Arpigny, R. Schulz, J.A. Stüwe, I. Ilyin, The CN isotopic ratios in comets. Astron. Astrophys. **503**, 613–624 (2009)

B. Marty, The origins and concentrations of water, carbon, nitrogen and noble gases on Earth. Earth Planet. Sci. Lett. **313–314**, 56–66 (2012)

B. Marty, M. Chaussidon, R.C. Wiens, A.J.G. Jurewicz, D.S. Burnett, A ^{15}N-poor isotopic composition for the solar system as shown by Genesis solar wind samples. Science **332**, 1533–1536 (2011)

B. Marty, G. Avice, Y. Sano, K. Altwegg, H. Balsiger, M. Hässig, A. Morbidelli, O. Mousis, M. Rubin, Origins of volatile elements (H, C, N, noble gases) on Earth and Mars in light of recent results from the ROSETTA cometary mission. Earth Planet. Sci. Lett. **441**, 91–102 (2016)

B. Marty, K. Altwegg, H. Balsiger, A. Bar-Nun, D.V. Bekaert, J.-J. Berthelier, A. Bieler, C. Briois, U. Calmonte, M. Combi, J. De Keyser, B. Fiethe, S.A. Fuselier, S. Gasc, T.I. Gombosi, K.C. Hansen, M. Hässig, A. Jäckel, E. Kopp, A. Korth, L. Le Roy, U. Mall, O. Mousis, T. Owen, H. Rème, M. Rubin,

T. Sémon, C.-Y. Tzou, J.H. Waite, P. Wurz, Xenon isotopes in 67P/Churyumov-Gerasimenko show that comets contributed to Earth's atmosphere. Science **356**, 1069–1072 (2017)

M. Massironi, E. Simioni, F. Marzari, G. Cremonese, L. Giacomini, M. Pajola, L. Jorda, G. Naletto, S. Lowry, M.R. El-Maarry, F. Preusker, F. Scholten, H. Sierks, C. Barbieri, P. Lamy, R. Rodrigo, D. Koschny, H. Rickman, H.U. Keller, M.F. A'Hearn, J. Agarwal, A.-T. Auger, M.A. Barucci, J.-L. Bertaux, I. Bertini, S. Besse, D. Bodewits, C. Capanna, V. da Deppo, B. Davidsson, S. Debei, M. de Cecco, F. Ferri, S. Fornasier, M. Fulle, R. Gaskell, O. Groussin, P.J. Gutiérrez, C. Güttler, S.F. Hviid, W.-H. Ip, J. Knollenberg, G. Kovacs, R. Kramm, E. Kührt, M. Küppers, F. La Forgia, L.M. Lara, M. Lazzarin, Z.-Y. Lin, J.J. Lopez Moreno, S. Magrin, H. Michalik, S. Mottola, N. Oklay, A. Pommerol, N. Thomas, C. Tubiana, J.-B. Vincent, Two independent and primitive envelopes of the bilobate nucleus of comet 67P. Nature **526**, 402–405 (2015)

K.J. Mathew, K. Marti, Early evolution of Martian volatiles: nitrogen and noble gas components in ALH84001 and Chassigny. J. Geophys. Res. **106**, 1401–1422 (2001)

M.C. McCanta, A.H. Treiman, M.D. Dyar, C.M.O'D. Alexander, D. Rumble III., E.J. Essene, The LaPaz Icefield 04840 meteorite: mineralogy, metamorphism, and origin of an amphibole- and biotite-bearing R chondrite. Geochim. Cosmochim. Acta **72**, 5757–5780 (2008)

T.B. McCord, J.B. Adams, T.V. Johnson, Asteroid Vesta: spectral reflectivity and compositional implications. Science **168**, 1445–1447 (1970)

K.D. McKeegan, A.P.A. Kallio, V.S. Heber, G. Jarzebinski, P.H. Mao, C.D. Coath, T. Kunihiro, R.C. Wiens, J.E. Nordholt, R.W. Moses, D.B. Reisenfeld, A.J.G. Jurewicz, D.S. Burnett, The oxygen isotopic composition of the Sun inferred from captured solar wind. Science **332**, 1528–1532 (2011)

H.Y. McSween Jr., A. Ghosh, R.E. Grimm, L. Wilson, E.D. Young, in *Asteroids III*, ed. by W.F. Bottke Jr., A. Cellino, P. Paolicchi, R.P. Binzel (Univ. of Arizona Press, Tucson, 2002), pp. 559–571

H.Y. McSween, D.W. Mittlefehldt, A.W. Beck, R.G. Mayne, T.J. McCoy, HED meteorites and their relationship to the geology of Vesta and the Dawn mission. Space Sci. Rev. **163**, 141–174 (2011)

R. Meier, T.C. Owen, D.C. Jewitt, H.E. Matthews, M. Senay, N. Biver, D. Bockelée-Morvan, J. Crovisier, D. Gautier, Deuterium in comet C/1995 O1 (Hale-Bopp): detection of DCN. Science **279**, 1707–1710 (1998a)

R. Meier, T.C. Owen, H.E. Matthews, D.C. Jewitt, D. Bockelée-Morvan, N. Biver, J. Crovisier, D. Gautier, A determination of the HDO/H_2O ratio in comet C/1995 O1 (Hale-Bopp). Science **279**, 842–844 (1998b)

S.N. Milam, S.B. Charnley, Observations of nitrogen fractionation in prestellar cores: nitriles tracing interstellar chemistry. Lunar Planet. Sci. **43**, #2618 (2012)

Y. Miura, N. Sugiura, Nitrogen isotopic compositions in three Antarctic and two non-Antarctic eucrites. Antarct. Meteor. Res. **6**, 338 (1993)

M. Miyamoto, N. Fujii, H. Takeda, Ordinary chondrite parent body: an internal heating model, Proc. Lunar Planet. Sci. Conf. **12B**, 1145–1152 (1981)

S.J. Mojzsis, T.M. Harrison, R.T. Pidgeon, Oxygen-isotope evidence from ancient zircons for liquid water at the Earth's surface 4,300 Myr ago. Nature **409**, 178–181 (2001)

A. Morbidelli, H. Rickman, Comets as collisional fragments of a primordial planetesimal disk. Astron. Astrophys. **583**, A43 (2015)

A. Morbidelli, K.J. Walsh, D.P. O'Brien, D.A. Minton, W.F. Bottke, in *Asteroids IV*, ed. by P. Michel, F.E. DeMeo, W.F. Bottke (Univ. of Arizona Press, Tucson, 2015), pp. 493–507

A. Morbidelli, B. Bitsch, A. Crida, M. Gounelle, T. Guillot, S. Jacobson, A. Johansen, M. Lambrechts, E. Lega, Fossilized condensation lines in the solar system protoplanetary disk. Icarus **267**, 368–376 (2016)

N. Moskovitz, E. Gaidos, Differentiation of planetesimals and the thermal consequences of melt migration. Meteorit. Planet. Sci. **46**, 903–918 (2011)

N.A. Moskovitz, R. Jedicke, E. Gaidos, M. Willman, D. Nesvorný, R. Fevig, Ž. Ivezić, The distribution of basaltic asteroids in the Main Belt. Icarus **198**, 77–90 (2008)

O. Mousis, J.I. Lunine, A. Luspay-Kuti, T. Guillot, B. Marty, M. Ali-Dib, P. Wurz, K. Altwegg, A. Bieler, M. Hässig, M. Rubin, P. Vernazza, J.H. Waite, A protosolar nebula origin for the ices agglomerated by comet 67P/Churyumov-Gerasimenko. Astrophys. J. Lett. **819**, L33 (2016a)

O. Mousis, T. Ronnet, B. Brugger, O. Ozgurel, F. Pauzat, Y. Ellinger, R. Maggiolo, P. Wurz, P. Vernazza, J.I. Lunine, A. Luspay-Kuti, K.E. Mandt, K. Altwegg, A. Bieler, A. Markovits, M. Rubin, Origin of molecular oxygen in comet 67P/Churyumov–Gerasimenko. Astrophys. J. Lett. **823**, L41 (2016b)

M.J. Mumma, S.B. Charnley, The chemical composition of comets—emerging taxonomies and natal heritage. Annu. Rev. Astron. Astrophys. **49**, 471–524 (2011)

K. Nagashima, A.N. Krot, M. Komatsu, ^{26}Al-^{26}Mg systematics in chondrules from Kaba and Yamato 980145 CV3 carbonaceous chondrites. Geochim. Cosmochim. Acta **201**, 303–319 (2017)

D. Nesvorný, D. Vokrouhlický, A. Morbidelli, W.F. Bottke, Asteroidal source of L chondrite meteorites. Icarus **200**, 698–701 (2009)

W. Neumann, D. Breuer, T. Spohn, Differentiation of Vesta: implications for a shallow magma ocean. Earth Planet. Sci. Lett. **395**, 267–280 (2014)
H.B. Niemann, S.K. Atreya, J.E. Demick, D. Gautier, J.A. Haberman, D.N. Harpold, W.T. Kasprzak, J.I. Lunine, T.C. Owen, F. Raulin, Composition of Titan's lower atmosphere and simple surface volatiles as measured by the Cassini-Huygens probe gas chromatograph mass spectrometer experiment. J. Geophys. Res., Planets **115**, 12006 (2010)
C.A. Nixon, B. Temelso, S. Vinatier, N.A. Teanby, B. Bézard, R.K. Achterberg, K.E. Mandt, C.D. Sherrill, P.G.J. Irwin, D.E. Jennings, P.N. Romani, A. Coustenis, F.M. Flasar, Isotopic ratios in Titan's methane: measurements and modeling. Astrophys. J. **749**, 159 (2012)
A. Oka, T. Nakamoto, S. Ida, Evolution of snow line in optically thick protoplanetary disks: effects of water ice opacity and dust grain size. Astrophys. J. **738**, 141 (2011)
J.H. Oort, The origin and dissolution of comets. Observatory **106**, 186–193 (1986)
T.C. Owen, A. Bar-Nun, in *Origin of the Earth and Moon*, ed. by R.M. Canup, K. Righter (2000), pp. 459–471
M. Pajola, S. Höfner, J.B. Vincent, N. Oklay, F. Scholten, F. Preusker, S. Mottola, G. Naletto, S. Fornasier, S. Lowry, C. Feller, P.H. Hasselmann, C. Güttler, C. Tubiana, H. Sierks, C. Barbieri, P. Lamy, R. Rodrigo, D. Koschny, H. Rickman, H.U. Keller, J. Agarwal, M.F. A'Hearn, M.A. Barucci, J.-L. Bertaux, I. Bertini, S. Besse, S. Boudreault, G. Cremonese, V. da Deppo, B. Davidsson, S. Debei, M. de Cecco, J. Deller, J.D.P. Deshapriya, M.R. El-Maarry, S. Ferrari, F. Ferri, M. Fulle, O. Groussin, P. Gutierrez, M. Hofmann, S.F. Hviid, W.-H. Ip, L. Jorda, J. Knollenberg, G. Kovacs, J.R. Kramm, E. Kührt, M. Küppers, L.M. Lara, Z.-Y. Lin, M. Lazzarin, A. Lucchetti, J.J. Lopez Moreno, F. Marzari, M. Massironi, H. Michalik, L. Penasa, A. Pommerol, E. Simioni, N. Thomas, I. Toth, E. Baratti, The pristine interior of comet 67P revealed by the combined Aswan outburst and cliff collapse. Nat. Astron. **1**, 0092 (2017)
H. Palme, D.C. Hezel, D.S. Ebel, The origin of chondrules: constraints from matrix composition and matrix-chondrule complementarity. Earth Planet. Sci. Lett. **411**, 11–19 (2015)
M. Pätzold, T. Andert, M. Hahn, S.W. Asmar, J.-P. Barriot, M.K. Bird, B. Häusler, K. Peter, S. Tellmann, E. Grün, P.R. Weissman, H. Sierks, L. Jorda, R. Gaskell, F. Preusker, F. Scholten, A homogeneous nucleus for comet 67P/Churyumov-Gerasimenko from its gravity field. Nature **530**, 63–65 (2016)
L. Piani, F. Robert, L. Remusat, Micron-scale D/H heterogeneity in chondrite matrices: a signature of the pristine solar system water? Earth Planet. Sci. Lett. **415**, 154–164 (2015)
S. Pizzarello, W. Holmes, Nitrogen-containing compounds in two CR2 meteorites: ^{15}N composition, molecular distribution and precursor molecules. Geochim. Cosmochim. Acta **73**, 2150–2162 (2009)
S. Pizzarello, X. Feng, S. Epstein, J.R. Cronin, Isotopic analyses of nitrogenous compounds from the Murchison meteorite: ammonia, amines, amino acids, and polar hydrocarbons. Geochim. Cosmochim. Acta **58**, 5579–5588 (1994)
E. Polnau, G.W. Lugmair, Mn–Cr isotope systematics in the two ordinary chondrites Richardton (H5) and Ste. Marguerite (H4). Lunar Planet. Sci. Conf. Abstr. **XXXII**, 1527 (2001)
G. Proskurowski, M.D. Lilley, D.S. Kelley, E.J. Olson, Low temperature volatile production at the Lost City hydrothermal field, evidence from a hydrogen stable isotope geothermometer. Chem. Geol. **229**, 331–343 (2006)
L. Qin, C.M.O'D. Alexander, R.W. Carlson, M.F. Horan, T. Yokoyama, Contributors to chromium isotope variation of meteorites. Geochim. Cosmochim. Acta **74**, 1122–1145 (2010)
E. Quirico, L.V. Moroz, B. Schmitt, G. Arnold, M. Faure, P. Beck, L. Bonal, M. Ciarniello, F. Capaccioni, G. Filacchione, S. Erard, C. Leyrat, D. Bockelée-Morvan, A. Zinzi, E. Palomba, P. Drossart, F. Tosi, M.T. Capria, M.C. De Sanctis, A. Raponi, S. Fonti, F. Mancarella, V. Orofino, A. Barucci, M.I. Blecka, R. Carlson, D. Despan, A. Faure, S. Fornasier, M.S. Gudipati, A. Longobardo, K. Markus, V. Mennella, F. Merlin, G. Piccioni, B. Rousseau, F. Taylor, Refractory and semi-volatile organics at the surface of comet 67P/Churyumov-Gerasimenko: insights from the VIRTIS/Rosetta imaging spectrometer. Icarus **272**, 32–47 (2016)
S.N. Raymond, A. Izidoro, Origin of water in the inner solar system: planetesimals scattered inward during Jupiter and Saturn's rapid gas accretion. Icarus **297**, 134–148 (2017)
S.N. Raymond, D.P. O'Brien, A. Morbidelli, N.A. Kaib, Building the terrestrial planets: constrained accretion in the inner Solar System. Icarus **203**, 644–662 (2009)
M. Regelous, T. Elliott, C.D. Coath, Nickel isotope heterogeneity in the early solar system. Earth Planet. Sci. Lett. **272**, 330–338 (2008)
A.S. Rivkin, The fraction of hydrated C-complex asteroids in the asteroid belt from SDSS data. Icarus **221**, 744–752 (2012)
A.S. Rivkin, J.P. Emery, Detection of ice and organics on an asteroidal surface. Nature **464**, 1322–1323 (2010)
A.S. Rivkin, C.A. Thomas, E.S. Howell, J.P. Emery, The Ch-class asteroids: connecting a visible taxonomic class to a 3 μm band shape. Astron. J. **150**, 198 (2015)

H. Roberts, G.A. Fuller, T.J. Millar, J. Hatchell, J.V. Buckle, A survey of [HDCO]/[H_2CO] and [DCN]/[HCN] ratios towards low-mass protostellar cores. Astron. Astrophys. **381**, 1026–1038 (2002)

A. Rotundi, H. Sierks, V. Della Corte, M. Fulle, P.J. Gutierrez, L. Lara, C. Barbieri, P.L. Lamy, R. Rodrigo, D. Koschny, H. Rickman, H.U. Keller, J.J. López-Moreno, M. Accolla, J. Agarwal, M.F. A'Hearn, N. Altobelli, F. Angrilli, M.A. Barucci, J.-L. Bertaux, I. Bertini, D. Bodewits, E. Bussoletti, L. Colangeli, M. Cosi, G. Cremonese, J.-F. Crifo, V. Da Deppo, B. Davidsson, S. Debei, M. De Cecco, F. Esposito, M. Ferrari, S. Fornasier, F. Giovane, B. Gustafson, S.F. Green, O. Groussin, E. Grün, C. Güttler, M.L. Herranz, S.F. Hviid, W. Ip, S. Ivanovski, J.M. Jerónimo, L. Jorda, J. Knollenberg, R. Kramm, E. Kührt, M. Küppers, M. Lazzarin, M.R. Leese, A.C. López-Jiménez, F. Lucarelli, S.C. Lowry, F. Marzari, E.M. Epifani, J.A.M. McDonnell, V. Mennella, H. Michalik, A. Molina, R. Morales, F. Moreno, S. Mottola, G. Naletto, N. Oklay, J.L. Ortiz, E. Palomba, P. Palumbo, J.-M. Perrin, J. Rodríguez, L. Sabau, C. Snodgrass, R. Sordini, N. Thomas, C. Tubiana, J.-B. Vincent, P. Weissman, K.-P. Wenzel, V. Zakharov, J.C. Zarnecki, Dust measurements in the coma of comet 67P/Churyumov-Gerasimenko inbound to the Sun. Science **347**, aaa3905 (2015)

M.W. Rowe, R.N. Clayton, T.K. Mayeda, Oxygen isotopes in separated components of CI and CM meteorites. Geochim. Cosmochim. Acta **58**, 5341–5348 (1994)

M. Rubin, K. Altwegg, H. Balsiger, A. Bar-Nun, J.-J. Berthelier, A. Bieler, P. Bochsler, C. Briois, U. Calmonte, M. Combi, J. De Keyser, F. Dhooghe, P. Eberhardt, B. Fiethe, S.A. Fuselier, S. Gasc, T.I. Gombosi, K.C. Hansen, M. Hässig, A. Jäckel, E. Kopp, A. Korth, L. Le Roy, U. Mall, B. Marty, O. Mousis, T. Owen, H. Rème, T. Sémon, C.-Y. Tzou, J.H. Waite, P. Wurz, Molecular nitrogen in comet 67P/Churyumov-Gerasimenko indicates a low formation temperature. Science **348**, 232–235 (2015a)

M. Rubin, K. Altwegg, E.F. van Dishoeck, G. Schwehm, Molecular oxygen in Oort Cloud comet 1P/Halley. Astrophys. J. Lett. **815**, L11 (2015b)

N. Sakamoto, Y. Seto, S. Itoh, K. Kuramoto, K. Fujino, K. Nagashima, A.N. Krot, H. Yurimoto, Remnants of the early solar system water enriched in heavy oxygen isotopes. Science **317**, 231–233 (2007)

A.R. Sarafian, S.G. Nielsen, H.R. Marschall, F.M. McCubbin, B.D. Monteleone, Early accretion of water in the inner solar system from a carbonaceous chondrite-like source. Science **346**, 623–626 (2014)

A.R. Sarafian, E.H. Hauri, F.M. McCubbin, T.J. Lapen, E.L. Berger, S.G. Nielsen, H.R. Marschall, G.A. Gaetani, K. Righter, E. Sarafian, Early accretion of water and volatile elements to the inner solar system: evidence from angrites. Philos. Trans. R. Soc., Math. Phys. Eng. Sci. **375**, 20160209 (2017a)

A.R. Sarafian, S.G. Nielsen, H.R. Marschall, G.A. Gaetani, E.H. Hauri, K. Righter, E. Sarafian, Angrite meteorites record the onset and flux of water to the inner solar system. Geochim. Cosmochim. Acta **212**, 156–166 (2017b)

M. Schiller, J. Baker, J. Creech, C. Paton, M.-A. Millet, A. Irving, M. Bizzarro, Rapid timescales for magma ocean crystallization on the Howardite-Eucrite-Diogenite parent body. Astrophys. J. Lett. **740**, L22 (2011)

B. Schmitt, S. Espinasse, R.J.A. Grim, J.M. Greenberg, J. Klinger, T.D. Guyenne, Laboratory studies of cometary ice analogues, in *Physics and Mechanics of Cometary Materials*, ed. by J.J. Hunt (European Space Agency, Münster, 1989), pp. 65–69

D.L. Schrader, K. Nagashima, A.N. Krot, R.C. Ogliore, Q.-Z. Yin, Y. Amelin, C.H. Stirling, A. Kaltenbach, Distribution of ^{26}Al in the CR chondrite chondrule-forming region of the protoplanetary disk. Geochim. Cosmochim. Acta **201**, 275–302 (2016)

E.R.D. Scott, R.C. Greenwood, I.A. Franchi, I.S. Sanders, Oxygen isotopic constraints on the origin and parent bodies of eucrites, diogenites, and howardites. Geochim. Cosmochim. Acta **73**, 5835–5853 (2009)

Y. Shinnaka, H. Kawakita, H. Kobayashi, M. Nagashima, D.C. Boice, $^{14}NH_2/^{15}NH_2$ ratio in comet C/2012 S1 (ISON) observed during its outburst in 2013 November. Astrophys. J. Lett. **782**, L16 (2014)

A. Shukolyukov, G.W. Lugmair, Manganese-chromium isotope systematics of enstatite meteorites. Geochim. Cosmochim. Acta **68**, 2875–2888 (2004)

P.D. Singh, A.A. de Almeida, W.F. Huebner, Dust release rates and dust-to-gas mass ratios of eight comets. Astron. J. **104**, 848–858 (1992)

R.C.J. Steele, V.S. Heber, K.D. McKeegan, Matrix effects on the relative sensitivity factors for manganese and chromium during ion microprobe analysis of carbonate: implications for early solar system chronology. Geochim. Cosmochim. Acta **201**, 245–259 (2017)

N. Sugiura, W. Fujiya, Correlated accretion ages and e^{54}Cr of meteorite parent bodies and the evolution of the solar nebula. Meteorit. Planet. Sci. **49**, 772–787 (2014)

J.M. Sunshine, S.J. Bus, T.J. McCoy, T.H. Burbine, C.M. Corrigan, R.P. Binzel, High-calcium pyroxene as an indicator of igneous differentiation in asteroids and meteorites. Meteorit. Planet. Sci. **39**, 1343–1357 (2004)

S. Sutton, C.M.O'D. Alexander, A. Bryant, A. Lanzirotti, M. Newville, E.A. Cloutis, The bulk valence state of Fe and the origin of water in chondrites. Geochim. Cosmochim. Acta **211**, 115–132 (2017)

D. Takir, J.P. Emery, Outer Main Belt asteroids: identification and distribution of four 3-mm spectral groups. Icarus **219**, 641–654 (2012)
D. Takir, J.P. Emery, H.Y. McSween Jr., Toward an understanding of phyllosilicate mineralogy in the outer main asteroid belt. Icarus **257**, 185–193 (2015)
H. Tang, N. Dauphas, ^{60}Fe–^{60}Ni chronology of core formation in Mars. Earth Planet. Sci. Lett. **390**, 264–274 (2014)
V. Taquet, K. Furuya, C. Walsh, E.F. van Dishoeck, A primordial origin for molecular oxygen in comets: a chemical kinetics study of the formation and survival of O_2 ice from clouds to discs. Mon. Not. R. Astron. Soc. **462**, S99–S115 (2016)
M. Touboul, T. Kleine, B. Bourdon, H. Palme, R. Wieler, Late formation and prolonged differentiation of the Moon inferred from W isotopes in lunar metals. Nature **450**, 1206–1209 (2007)
M. Touboul, P. Sprung, S.M. Aciego, B. Bourdon, T. Kleine, Hf–W chronology of the eucrite parent body. Geochim. Cosmochim. Acta **156**, 106–121 (2015)
A. Trinquier, J.-L. Birck, C.J. Allègre, Widespread ^{54}Cr heterogeneity in the inner Solar System. Astrophys. J. **655**, 1179–1185 (2007)
A. Trinquier, J.L. Birck, C.J. Allègre, C. Göpel, D. Ulfbeck, ^{53}Mn–^{53}Cr systematics of the early Solar System revisited. Geochim. Cosmochim. Acta **72**, 5146–5163 (2008)
A. Trinquier, T. Elliott, D. Ulfbeck, C. Coath, A.N. Krot, M. Bizzarro, Origin of nucleosynthetic isotope heterogeneity in the solar protoplanetary disk. Science **324**, 374–376 (2009)
T. Ushikubo, D. Nakashima, M. Kimura, T.J. Tenner, N.T. Kita, Contemporaneous formation of chondrules in distinct oxygen isotope reservoirs. Geochim. Cosmochim. Acta **109**, 280–295 (2013)
T. Usui, C.M.O'D. Alexander, J. Wang, J.I. Simon, J.H. Jones, Origin of water and mantle-crust interactions on Mars i nferred from hydrogen isotopes and volatile element abundances of olivine-hosted melt inclusions of primitive shergottites. Earth Planet. Sci. Lett. **357–358**, 119–129 (2012)
N. van der Marel, E.F. van Dishoeck, S. Bruderer, L. Pérez, A. Isella, Gas density drops inside dust cavities of transitional disks around young stars observed with ALMA. Astron. Astrophys. **579**, A106 (2015)
P. Vernazza, R.P. Binzel, C.A. Thomas, F.E. DeMeo, S.J. Bus, A.S. Rivkin, A.T. Tokunaga, Compositional differences between meteorites and near-Earth asteroids. Nature **454**, 858–860 (2008)
P. Vernazza, D. Fulvio, R. Brunetto, J.P. Emery, C.A. Dukes, F. Cipriani, O. Witasse, M.J. Schaible, B. Zanda, G. Strazzulla, R.A. Baragiola, Paucity of Tagish Lake-like parent bodies in the asteroid belt and among Jupiter Trojans. Icarus **225**, 517–525 (2013)
P. Vernazza, B. Zanda, R.P. Binzel, T. Hiroi, F.E. DeMeo, M. Birlan, R. Hewins, L. Ricci, P. Barge, M. Lockhart, Multiple and fast: the accretion of ordinary chondrite parent bodies. Astrophys. J. **791**, 120 (2014)
G.L. Villanueva, M.J. Mumma, B.P. Bonev, M.A. Di Santi, E.L. Gibb, H. Böhnhardt, M. Lippi, A sensitive search for deuterated water in comet 8P/Tuttle. Astrophys. J. Lett. **690**, L5–L9 (2009)
J. Villeneuve, M. Chaussidon, G. Libourel, Homogeneous distribution of ^{26}Al in the solar system from the Mg isotopic composition of chondrules. Science **325**, 985–988 (2009)
J.H. Waite Jr., W.S. Lewis, B.A. Magee, J.I. Lunine, W.B. McKinnon, C.R. Glein, O. Mousis, D.T. Young, T. Brockwell, J. Westlake, M.J. Nguyen, B.D. Teolis, H.B. Niemann, R.L. McNutt Jr., M. Perry, W.H. Ip, Liquid water on Enceladus from observations of ammonia and ^{40}Ar in the plume. Nature **460**, 487–490 (2009)
R.J. Walker, K. Bermingham, J. Liu, I.S. Puchtel, M. Touboul, E.A. Worsham, In search of late-stage planetary building blocks. Chem. Geol. **411**, 125–142 (2015)
K.J. Walsh, A. Morbidelli, S.N. Raymond, D.P. O'Brien, A.M. Mandell, A low mass for Mars from Jupiter's early gas-driven migration. Nature **475**, 206–209 (2011)
P.H. Warren, Stable-isotopic anomalies and the accretionary assemblage of the Earth and Mars: a subordinate role for carbonaceous chondrites. Earth Planet. Sci. Lett. **311**, 93–100 (2011)
J.T. Wasson, G.W. Kallemeyn, Composition of chondrites. Philos. Trans. R. Soc. Lond. Ser. A **325**, 535–544 (1988)
B.P. Weiss, L.T. Elkins-Tanton, Differentiated planetesimals and the parent bodies of chondrites. Annu. Rev. Earth Planet. Sci. **41**, 529–560 (2013)
S.A. Wilde, J.W. Valley, W.H. Peck, C.M. Graham, Evidence from detrital zircons for the existence of continental crust and oceans on Earth 4.4 Gyr ago. Nature **409**, 175–178 (2001)
F. Wombacher, M. Rehkämper, K. Mezger, A. Bischoff, C. Münker, Cadmium stable isotope cosmochemistry. Geochim. Cosmochim. Acta **72**, 646–667 (2008)
P.M. Woods, A. Occhiogrosso, S. Viti, Z. Kaňuchová, M.E. Palumbo, S.D. Price, A new study of an old sink of sulphur in hot molecular cores: the sulphur residue. Mon. Not. R. Astron. Soc. **450**, 1256–1267 (2015)
E.A. Worsham, K.R. Bermingham, R.J. Walker, Siderophile element systematics of IAB complex iron meteorites: new insights into the formation of an enigmatic group. Geochim. Cosmochim. Acta **188**, 261–283 (2016)

I.P. Wright, S. Sheridan, S.J. Barber, G.H. Morgan, D.J. Andrews, A.D. Morse, CHO-bearing organic compounds at the surface of 67P/Churyumov-Gerasimenko revealed by Ptolemy. Science **349**, aab0673 (2015)

K. Yamashita, S. Maruyama, A. Yamakawa, E. Nakamura, ^{53}Mn-^{53}Cr chronometry of CB chondrite: evidence for uniform distribution of ^{53}Mn in the early solar system. Astrophys. J. **723**, 20–24 (2010)

J. Yang, J.I. Goldstein, E.R.D. Scott, Main-group pallasites: thermal history, relationship to IIIAB irons, and origin. Geochim. Cosmochim. Acta **74**, 4471–4492 (2010)

L. Yang, F.J. Ciesla, C.M.O'D. Alexander, The D/H ratio of water in the solar nebula during its formation and evolution. Icarus **226**, 256–267 (2013)

E.D. Young, R.D. Ash, P. England, D. Rumble III, Fluid flow in chondritic parent bodies: deciphering the compositions of planetesimals. Science **286**, 1331–1335 (1999)

H. Yurimoto, K. Kuramoto, Molecular cloud origin for the oxygen isotope heterogeneity in the Solar System. Science **305**, 1763–1766 (2004)

B. Zanda, P.A. Bland, C. Le Guillou, R.H. Hewins, Volatile element distribution in matrix and chondrules of carbonaceous and ordinary chondrites. Lunar Planet. Sci. **40**, #1810 (2009)

B. Zanda, M. Humayun, R.H. Hewins, Chemical composition of matrix and chondrules in carbonaceous chondrites: implications for disk transport. Lunar Planet. Sci. **43**, #2413 (2012)

B. Zellner, M. Leake, D. Morrison, J.G. Williams, The E asteroids and the origin of the enstatite achondrites. Geochim. Cosmochim. Acta **41**, 1759–1767 (1977)

M.E. Zolensky, K. Nakamura, M. Gounelle, T. Mikouchi, T. Kasama, O. Tachikawa, E. Tonui, Mineralogy of Tagish Lake: an ungrouped type 2 carbonaceous chondrite. Meteorit. Planet. Sci. **37**, 737–761 (2002)

M.E. Zolensky, T.J. Zega, H. Yano, S. Wirick, A.J. Westphal, M.K. Weisberg, I. Weber, J.L. Warren, M.A. Velbel, A. Tsuchiyama, P. Tsou, A. Toppani, N. Tomioka, K. Tomeoka, N. Teslich, M. Taheri, J. Susini, R. Stroud, T. Stephan, F.J. Stadermann, C.J. Snead, S.B. Simon, A. Simionovici, T.H. See, F. Robert, F.J.M. Rietmeijer, W. Rao, M.C. Perronnet, D.A. Papanastassiou, K. Okudaira, K. Ohsumi, I. Ohnishi, K. Nakamura-Messenger, T. Nakamura, S. Mostefaoui, T. Mikouchi, A. Meibom, G. Matrajt, M.A. Marcus, H. Leroux, L. Lemelle, L. Le, A. Lanzirotti, F. Langenhorst, A.N. Krot, L.P. Keller, A.T. Kearsley, D. Joswiak, D. Jacob, H. Ishii, R. Harvey, K. Hagiya, L. Grossman, J.N. Grossman, G.A. Graham, M. Gounelle, P. Gillet, M.J. Genge, G. Flynn, T. Ferroir, S. Fallon, D.S. Ebel, Z.R. Dai, P. Cordier, B. Clark, M. Chi, A.L. Butterworth, D.E. Brownlee, J.C. Bridges, S. Brennan, A. Brearley, J.P. Bradley, P. Bleuet, P.A. Bland, R. Bastien, Mineralogy and petrology of comet 81P/Wild 2 nucleus samples. Science **314**, 1735–1739 (2006)

Space Sci Rev (2017) 212:743–810
DOI 10.1007/s11214-017-0387-z

Water in the Earth's Interior: Distribution and Origin

Anne H. Peslier[1] · Maria Schönbächler[2] · Henner Busemann[2] · Shun-Ichiro Karato[3]

Received: 21 November 2016 / Accepted: 3 June 2017 / Published online: 9 August 2017

Abstract The concentration and distribution of water in the Earth has influenced its evolution throughout its history. Even at the trace levels contained in the planet's deep interior (mantle and core), water affects Earth's thermal, deformational, melting, electrical and seismic properties, that control differentiation, plate tectonics and volcanism. These in turn influenced the development of Earth's atmosphere, oceans, and life. In addition to the ubiquitous presence of water in the hydrosphere, most of Earth's "water" actually occurs as trace amounts of hydrogen incorporated in the rock-forming silicate minerals that constitute the planet's crust and mantle, and may also be stored in the metallic core. The heterogeneous distribution of water in the Earth is the result of early planetary differentiation into crust, mantle and core, followed by remixing of lithosphere into the mantle after plate-tectonics

This article has been corrected. Figure 3 was initially published with erroneous data because in Fig. 3B and 3D the *x* axis should be Cpx H_2O, instead of Grt H_2O.

The Delivery of Water to Protoplanets, Planets and Satellites
Edited by Michel Blanc, Alessandro Morbidelli, Yann Alibert, Lindy Elkins-Tanton, Paul Estrada, Keiko Hamano, Helmut Lammer, Sean Raymond and Maria Schönbächler

Electronic supplementary material The online version of this article (doi:10.1007/s11214-017-0387-z) contains supplementary material, which is available to authorized users.

✉ A.H. Peslier
anne.h.peslier@nasa.gov

M. Schönbächler
mariasc@ethz.ch

H. Busemann
henner.busemann@erdw.ethz.ch

S.-I. Karato
shun-ichiro.karato@yale.edu

[1] Jacobs, NASA-Johnson Space Center, Mail Code XI3, Houston, TX 77058, USA

[2] Institute of Geochemistry and Petrology, Department of Earth Sciences, ETH Zurich, Clausiusstrasse 25, 8092 Zurich, Switzerland

[3] Department of Geology and Geophysics, Yale University, New Haven, CT 06520, USA

started. The Earth's total water content is estimated at 18^{+81}_{-15} times the equivalent mass of the oceans (or a concentration of 3900^{+32700}_{-3300} ppm weight H_2O). Uncertainties in this estimate arise primarily from the less-well-known concentrations for the lower mantle and core, since samples for water analyses are only available from the crust, the upper mantle and very rarely from the mantle transition zone (410–670 km depth). For the lower mantle (670–2900 km) and core (2900–4500 km), the estimates rely on laboratory experiments and indirect geophysical techniques (electrical conductivity and seismology).

The Earth's accretion likely started relatively dry because it mainly acquired material from the inner part of the proto-planetary disk, where temperatures were too high for the formation and accretion of water ice. Combined evidence from several radionuclide systems (Pd-Ag, Mn-Cr, Rb-Sr, U-Pb) suggests that water was not incorporated in the Earth in significant quantities until the planet had grown to $\sim$ 60–90% of its current size, while core formation was still on-going. Dynamic models of planet formation provide additional evidence for water delivery to the Earth during the same period by water-rich planetesimals originating from the asteroid belt and possibly beyond. This early delivered water may have been partly lost during giant impacts, including the Moon forming event: magma oceans can form in their aftermath, degas and be followed by atmospheric loss. More water may have been delivered and/or lost after core formation during late accretion of extraterrestrial material ("late-veneer"). Stable isotopes of hydrogen, carbon, nitrogen and some noble gases in Earth's materials show similar compositions to those in carbonaceous chondrites, implying a common origin for their water, and only allowing for minor water inputs from comets.

Keywords Water · Hydrogen · Earth · Crust · Mantle · Core · Delivery · Origin · Solar system

1 Introduction

The importance of water for the development of life on Earth renders this compound one of the most intriguing for investigations aimed at understanding our origins. Water in the Earth is rarely present as H_2O molecules but is most abundant in the form of hydrogen (H) dissolved in various minerals of the planet's interior. Because of the ample abundance of H in the universe and its small size, it is likely present in nearly all phases, from the Earth's atmosphere to its core. Although water typically exists in trace quantities in the deep Earth, it influences tremendously how the Earth evolved from its formation, to its so far unique plate tectonic regime and with liquid water oceans on its surface.

This chapter summarizes the current knowledge of the water content of the Earth, the water distribution between our planet's different layers, and the water fluxes between these reservoirs. To this end, a vast range of data is utilized, encompassing actual measurements of water concentrations in crustal and mantle rock samples, experiments simulating the Earth's interior conditions for chemistry and rheology, geophysical data such as seismic wave propagation, electrical and thermal conductivities, ab initio calculations of molecular dynamics, and geodynamical and planetary accretion models.

This chapter also examines when water was incorporated into the Earth throughout its history, i.e. during and/or after the planet's accretion from the protoplanetary disk, using the latest findings from the isotopes of moderately volatile elements. Finally, we address from where in the solar system the Earth's water possibly originated, based on constraints from H isotopes and the behavior of other volatile elements and isotopes.

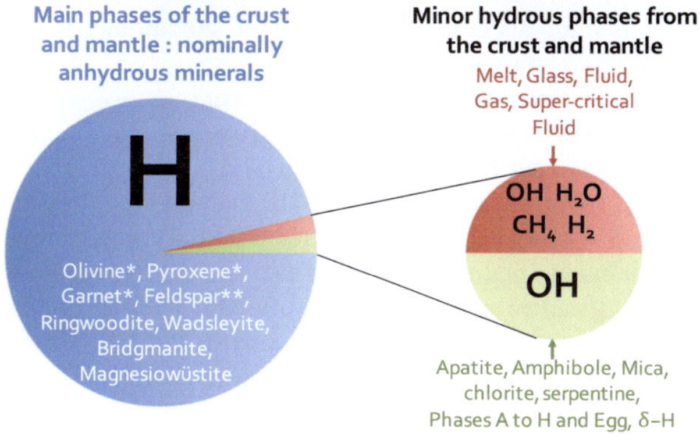

Fig. 1 Sketch illustrating the "water" species present in the various phases of the Earth's mantle and crust. The size of the pie slices represents the approximate volume percentage of the phases. The main reservoir of water in the mantle is nominally anhydrous minerals where hydrogen (H) enters their lattice in defects, and bonds to structural oxygen (*blue field*) (Bell and Rossman 1992b). *Olivine, pyroxene and garnet can incorporate water as H_2 under reduced conditions (Yang et al. 2016). **K-feldspar can sometimes include water as H_2O and NH_4 (Johnson and Rossman 2004)

2 Background

2.1 Definition of "Water"

Although water is primarily in the form of H_2O molecules in the hydrosphere (atmosphere and oceans), deeper in the Earth, what is generally referred to as "water" is actually H incorporated as various species in minerals, melts and fluids (Fig. 1). In the latter, i.e. *silicate/carbonate melts or volatile-rich fluids*, H can occur as water H_2O, hydroxyl OH, hydrogen gas H_2, methane CH_4, hydrogen sulfide H_2S, and possibly more complex H-bearing molecules at high pressures (e.g., Stolper 1982; Dixon et al. 1988; Carroll and Webster 1994; Kohn 2000; Luth 2003; Mookherjee et al. 2008; Song et al. 2009; Zhang and Duan 2009; Hirschmann et al. 2012; Bali et al. 2013; Armstrong et al. 2015). Melts and aqueous fluids are generally thought to become miscible at high pressure (> 6 GPa) in the upper mantle and the resulting super-critical fluids may contain water as both H_2O and OH (e.g., Kennedy et al. 1962; Bureau and Keppler 1999; Mibe et al. 2007). The relative proportions of these H-bearing species in fluids depend on melt composition and structure, the concentrations of other volatiles (C, F, Cl, S), pressure, temperature, water and oxygen fugacities (Holloway and Blank 1994). For example, methane and H_2 are present in fluids in reduced environments while OH and H_2O are prominent in oxidized conditions (e.g., Matveev et al. 1997; Frost and McCammon 2008; Mysen et al. 2009). When a melt is quenched at the surface in a volcanic eruption or in an experiment at high temperature (> 700 °C), the relative proportions of water species may change and the glass incorporates H principally as H_2O with minor OH (Shen and Keppler 1995; Zhang 1999; Kohn 2000).

Hydrous minerals incorporate "water" as hydroxyl OH. These hydrous phases such as mica, amphibole, apatite, serpentine and chlorite are only stable at relatively low mantle temperatures (e.g., Luth 2003). Consequently, these minerals play an important role in carrying water into the deep mantle in subducting slabs, which are cold relative to the surrounding mantle. Once in the deeper mantle, these hydrous minerals transform into high-pressure minerals such as phases B, D, H, Egg and AlOOH (e.g., Frost 2006;

Smyth 2006). Hydrous minerals contain H as a major stoichiometric element, but occupy only a small fraction of the total mantle.

The main phases of the mantle and crust are *nominally anhydrous minerals* (NAM) in which water is only present as a trace element (< 0.1 weight %). The main mineral phases are olivine ($(Mg, Fe)_2SiO_4$), pyroxene ($(Mg, Fe, Ca)SiO_3$), feldspar ($(K, Na, Ca_{0.5})AlSi_3O_8$) and garnet ($(Mg, Fe, Ca)_3Al_2Si_3O_{12}$) at depth < 410 km and high pressure forms of olivine called ringwoodite and wadsleyite as well as perovskite (of compositions $(Mg, Fe, Al)(Si, Al)O_3$ and $CaSiO_3$) at deeper mantle levels. In these phases, "water" is in reality H located in defects of the crystal lattice and bonded to structural oxygen (e.g., Beran 1976; Bell 1992; Rossman 1996; Johnson and Rossman 2004; Smyth 2006; Wright 2006). Interestingly, a recent study has evidenced that water could also be present as H_2 molecules in olivine, pyroxene and garnet under the reduced conditions prevailing in parts of the upper mantle (Yang et al. 2016). A significant amount of Earth's water is located in nominally anhydrous minerals, probably amounting to more than one mass of the Earth's oceans stored in this manner in the Earth's mantle (see Sect. 3.2). Several additional oceans' mass equivalents could also be stored in the metallic core (see Sect. 3.3). As a result, this chapter uses a very wide definition for the word "water", in most cases referring to H incorporated in silicate minerals. Water concentrations are reported here in part per million weight H_2O (ppm wt H_2O) in silicate phases and in ppm wt H in metals.

2.2 Chemical Behavior of Water in the Earth's Interior

To understand the water distribution in the Earth's interior, four key behaviors of water are relevant: (1) *incompatible partitioning behavior between melt and solid*, (2) *fast diffusion*, (3) *degassing from magmas and* (4) *sub-solidus re-equilibration*. These four topics are discussed in more detail below as well as how these features can affect the final water content measured in geological samples.

First, during melting of rocks or crystallization of melts, *water behaves as an incompatible element*, i.e. preferentially partitions into a silicate melt relative to a silicate solid (Moore 1970; Dixon et al. 1988; Michael 1988; Koga et al. 2003; Aubaud et al. 2004, 2008; Hauri et al. 2006; Grant et al. 2007b; Tenner et al. 2009; O'Leary et al. 2010; Hamada et al. 2013; Novella et al. 2014; Rosenthal et al. 2015). In contrast to previous assumptions, growing evidence shows that H is less incompatible than K or the light rare earth elements (LREE, e.g., Ce) (discussion in Sect. 3.2.2). Regarding conditions relevant to the Earth's core, H is incompatible in a solid-melt iron system, and may even still partition into Fe melt in the presence of silicate melt at these high pressures (Okuchi 1997; Okuchi and Takahashi 1998).

Second, *H is one of the fastest diffusing elements* in silicate melts and nominally anhydrous minerals at typical pressures and temperatures in the Earth's interior (e.g., Kohlstedt and Mackwell 1998; Demouchy and Mackwell 2006; Ingrin and Blanchard 2006; Farver 2010; Zhang and Ni 2010; Johnson and Rossman 2013; Padrón-Navarta et al. 2014; Sun et al. 2015; Ferriss et al. 2016). Careful consideration is therefore necessary to ascertain the significance of measured concentrations in natural minerals of mantle rocks and lavas because water may be lost or gained by diffusion during transit from the mantle to the surface. Mantle samples can be brought up to the surface as chunks, called xenoliths, ripped from the mantle as magma ascends through it. In these, olivine can sometimes experience H loss by diffusion even if a typical host-magma (alkali basalt or kimberlite) takes less than a day to travel from the mantle to the surface (Demouchy et al. 2006; Peslier and Luhr 2006; Peslier et al. 2008, 2015; Denis et al. 2013; Thoraval and Demouchy 2014). Recently, H loss has also been evidenced in orthopyroxene from mantle xenoliths (Tian et al. 2017). In contrast, transfer of water from a host magma to the minerals of a xenolith has not yet been observed and experimental data indicates no water addition from a kimberlite melt to olivine in less than a day (Baptiste et al. 2015). On the other hand, H loss by diffusion is likely pervasive in minerals from erupted basaltic magmas, whose sources are located in the mantle (Jamtveit et al. 2001; Wade et al. 2008; Hamada et al. 2011; Nazzareni et al. 2011; Xia et al. 2013b; Liu et al. 2015; Weis et al. 2015; Edmonds et al. 2016). Finally, on the Earth's surface, alteration of minerals or glasses can occur through interaction with meteoritic and oceanic water or hydrothermal fluids, and this needs to be carefully assessed. Modification of water concentration and/or speciation by such alteration has been evidenced in silicate glass, feldspar and apatite (e.g., Friedman et al. 1966; Johnson and Rossman 2004; Webster and Piccoli 2015; Seligman et al. 2016).

The third behavior of water that can affect the water content of natural samples is *magma degassing*. As the magmas reach upper crustal levels and pressure decreases, volatiles become less soluble in silicate melts and exsolve as a gas phase (e.g., Sparks et al. 1994). Degassing is a strong concern for most analyses of silicate glasses, because their water concentration may not reflect that of their parent melt (e.g., Dixon et al. 1991; Dixon 1997; Clay et al. 2015). Some silicate glasses, which were quenched fast enough or erupted under significant pressure, for example at the bottom of the ocean (> 500 m of water column) or under a significant volume of ice, experience only minimal degassing (Moore and Schilling 1973; Jambon and Zimmermann 1987; Dixon and Stolper 1995; Saal et al. 2002). The water content of silicate melts can also be inferred from that of melt inclusions, drops of melt trapped early in the crystallization sequence of a magma in a growing mineral such as olivine, spinel, garnet, or diamond. Encapsulation by these host minerals may prevent the inclusion from losing its water during magma ascent (Danyushevsky 2002; Wallace 2005; Kent 2008). However, disturbances of the water content in melt inclusions in olivine are documented and need careful evaluation prior to data interpretation (Chen et al. 2013; Lloyd et al. 2013, 2014; Le Voyer et al. 2014; Hartley et al. 2015).

Finally, the fourth important process affecting water in mantle rocks is *sub-solidus re-equilibration*. In the case of the slow (several Ma) tectonic rise of mantle units (i.e., orogenic massifs or abyssal peridotites), re-distribution of water (H) among co-existing minerals likely occurs because the hydrogen partition coefficient between minerals is a function of the thermodynamic conditions and mineral composition (Katayama et al. 2006; Warren and Hauri 2014). Even peridotites found as xenoliths experienced sub-solidus re-equilibration. Subsequent re-distribution of their H among minerals occurs after peridotites, the residues of melting events at relatively high-temperatures, equilibrate at the lower temperatures of the ambient upper mantle after melting ceases (Schaffer et al. 2016).

In this chapter, all data are filtered to exclude any measurements of a phase affected by water loss or gain, and we only take into account initial (mantle or pre-degassing magmatic) water concentrations. All water concentrations and ancillary data for lower crustal and upper mantle samples used in this chapter are available in the online electronic supplementary Table 1.

2.3 Methods of Water Determination in the Earth's Interior

A brief summary of the techniques used to estimate the amount of water in the Earth's interior is given here. A comprehensive review of analytical techniques available to measure water in minerals and glasses of natural samples or experiments is given by Rossman (2006). Briefly, the two main techniques are Fourier transform infrared spectrometry (FTIR) and secondary ion mass spectrometry (SIMS). Both now achieve similar detection limits (< 0.5 ppm wt H_2O) and have their own advantages and disadvantages. FTIR is a non-destructive technique that allows detection of the speciation of water and H-defect location (e.g., Libowitzky and Beran 2004), but requires cumbersomely made doubly-polished samples for transmission quantitative analysis. SIMS, on the other hand, can be performed on samples polished on a single side but is destructive. Its advantage is the simultaneous measurement of water concentration and H isotopes or other volatiles (Cl, F, C, S). Nano-SIMS can also achieve a smaller spatial resolution ($< 1 \times 1$ μm) compared to FTIR ($> 20 \times 20$ μm) (Hauri 2002; Hauri et al. 2011). SIMS, however, requires suitable standards (Hauri et al. 2002; Aubaud et al. 2007; Mosenfelder et al. 2011), while FTIR necessitates composition-specific absorption coefficients (Bell et al. 1995, 2003a; Johnson and Rossman 2003; King et al. 2004; Koch-Müller and Rhede 2010; Withers et al. 2012; Thomas et al. 2014; Mosenfelder et al. 2015). Geophysical methods such as seismology and electrical conductivity can be employed to infer the amount of water in the deep Earth (Jacobsen 2006; Karato 2006b, 2011). Among various geophysical observations, electrical conductivity is particularly useful because it is highly sensitive to the water content but relatively insensitive to other parameters such as major element chemistry, temperature and oxygen fugacity (Karato 2006a, 2011; Gardés et al. 2014).

3 Concentration and Distribution of Water in the Earth's Interior

3.1 Water in Crustal Minerals

A detailed account of the distribution of water in the crust is beyond the scope of this chapter. The crust contains many water-bearing minerals such as mica, amphibole and clay minerals in its igneous and metamorphic units and water-rich sediments (Taylor and McLennan 1985; Rudnick and Fountain 1995; Bodnar et al. 2013). Water is also present in pores and fractures. In the upper crustal lithologies, originating from basaltic to evolved igneous or metamorphic processes, trace amounts of water are incorporated in pyroxene, feldspar, quartz and olivine (Aines and Rossman 1984; Rossman and Smyth 1990; Nakashima et al. 1995; Kronenberg et al. 1996; Herd et al. 2000; Ito and Nakashima 2002; Johnson and Rossman 2003, 2004; Filip et al. 2006; Andrut et al. 2007; Müller and Koch-Müller 2009; Stalder and Konzett 2012; Seaman et al. 2013). The lower continental crust, primarily consists of granulite, and also contains water in nominally anhydrous minerals (Xia et al. 2006; Koch-Müller et al. 2007; Yang et al. 2008a; Seaman et al. 2013; Schmädicke et al. 2015). Finally, water can be found in accessory phases from crustal units of igneous or metamorphic origin like rutile, zircon, quartz and kyanite (Vlassopoulos et al. 1993; Bell et al. 2004a; Beran and Libowitzky 2006; Lucassen et al. 2013; De Hoog et al. 2014). The average water content of the continental crust (amphibolite + granulite) is estimated to be ~ 1.3 wt% H_2O (0.1–5 wt% H_2O), while that of the oceanic crust (basalt + gabbro) is ~ 1.5 wt% H_2O (Bodnar et al. 2013).

3.2 Water in the Mantle

3.2.1 Water in the Continental Mantle Lithosphere

The Earth's lithosphere is on average thickest under continents, particularly beneath the oldest crustal units, that can date back to the Archean (Jordan 1978; Boyd et al. 1985;

Artemieva and Mooney 2001; Sleep 2005; Priestley and McKenzie 2006; Eaton et al. 2008; Rychert and Shearer 2009; Fisher et al. 2010; Yuan and Romanowicz 2010). These old parts of the continental lithosphere are called cratons and can be 200–300 km thick. In that respect, these cratonic roots resemble the submerged part of icebergs in the ocean (analog of the asthenosphere). They are isolated from asthenospheric convection and are thus colder than the surrounding asthenosphere with heat transmitted through conduction (e.g., Jaupart and Mareschal 1999; Michaut et al. 2009). The amount and distribution of water in the mantle lithosphere below the continental crust is especially well documented because of the abundance of mantle xenoliths and other mantle samples (megacrysts, diamonds) found in these settings.

Water in Olivine and Lithospheric Strength The amount of water stored in the continental lithosphere has implications on how strong and rigid tectonic plates are. The old ages of cratonic roots are a particularly puzzling case (e.g., Walker et al. 1989; Pearson et al. 1995a; Griffin et al. 2004). They are surrounded by hotter convecting asthenosphere and therefore would be expected to be eroded and destroyed from below, i.e. "delaminated", over > 3 billion year (Ga). Water plays a fundamental role in how the mantle lithosphere resists delamination over time by influencing its deformation properties (e.g., Mackwell et al. 1985; Pollack 1986; Karato 1993, 2010; Hirth et al. 2000; Sleep 2003; Lee et al. 2010; Wang 2010).

The key to evaluating the strength of the mantle lithosphere relative to that of the asthenosphere is the mineral olivine, the main mineral of the upper mantle ($> 50\%$ in volume) in its dominant lithology, peridotite (Boyd 1989; Griffin et al. 2009). The presence of trace amounts of water (tens of ppm wt H_2O) in its structure weakens olivine (e.g., Mackwell et al. 1985; Demouchy et al. 2012; Girard et al. 2013; Faul et al. 2016; Tielke et al. 2017). The distribution of olivine with different water contents in the lithospheric mantle (lateral and with depth) is thus of particular importance for assessing lithospheric strength. In contrast, water does not weaken clinopyroxene, and the weakening effect on orthopyroxene depends on mineral orientation (Gavrilenko et al. 2010; Ohuchi et al. 2010). The fact that olivine is the most interconnected phase of the upper mantle causes its hydrolytic weakening properties to control the weakening of peridotite, even though olivine typically contains about 10 times less water than pyroxenes (e.g., Peslier 2010; Demouchy and Bolfan-Casanova 2016). Figure 2A shows water concentrations in olivine as a function of depth for peridotite xenoliths from the Kaapvaal (Southern Africa), Siberian (Russia) and Tanzanian (Eastern Africa) cratons (Kurosawa et al. 1997; Grant et al. 2007a; Peslier et al. 2008, 2010; Baptiste et al. 2012; Doucet et al. 2014; Hui et al. 2015). The depth, i.e. the pressure at which equilibration occurred, and the equilibration temperature are obtained by comparing the major element composition of peridotite minerals (for example Al exchange between orthopyroxene and garnet) to those of laboratory experiments calibrated at known pressures (e.g., Brey and Köhler 1990). Each craton has a different water concentration distribution with depth (Fig. 2A). The Tanzanian craton displays the lowest concentrations (1–39 ppm wt H_2O), that of the Kaapvaal craton is characterized by water-poor olivines (< 10 ppm wt H_2O) at > 190 km and the Siberian craton exhibits the largest range of water contents (7–223 ppm wt H_2O).

These water contents, along with equilibration temperatures and pressures, can be inserted into a flow law for the deformation of olivine aggregates (Li et al. 2008). Viscosity variation with depth in each craton (i.e., along the cratonic geotherm) can then be calculated for a constant water content (curves in Fig. 2B) or for individual peridotite samples (symbols in Fig. 2B). The majority of the calculated viscosities for individual peridotites are

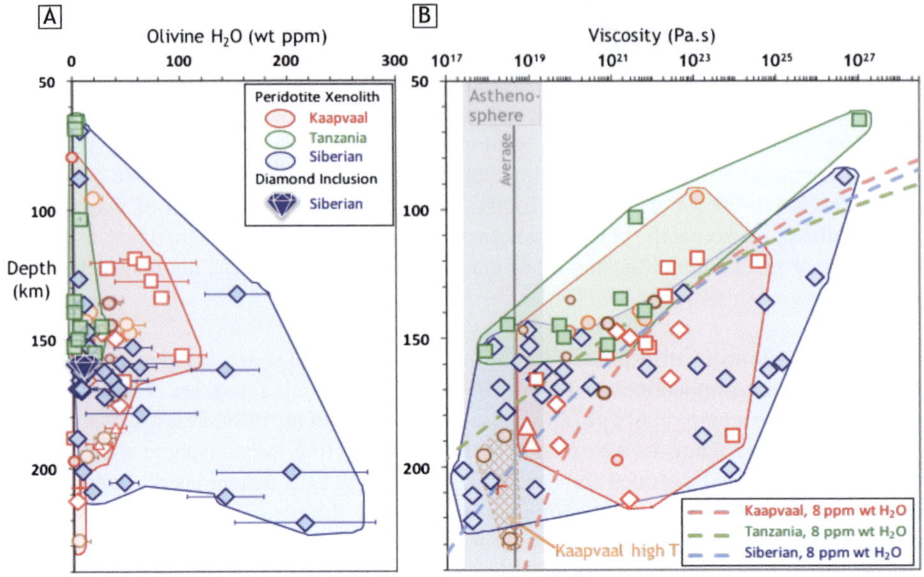

Fig. 2 **A**: Water contents of olivine versus depth of equilibration for peridotites from the mantle lithosphere of the Kaapvaal, Siberian and Tanzanian cratons. Each symbol represents one peridotite xenolith (Kurosawa et al. 1997; Grant et al. 2007a; Peslier et al. 2008, 2010; Baptiste et al. 2012; Doucet et al. 2014; Hui et al. 2015; Kolesnichenko et al. 2016), except the diamond which represents the average of the water contents of 29 olivine inclusions found in diamonds from the Siberian craton (Novella et al. 2015; Jean et al. 2016; Taylor et al. 2016). Although the pressure of equilibration of the olivine inclusions in diamonds is not precisely known, they are shown here at 5 GPa ($\sim$ 160 km depth), a formation pressure determined for a Siberian craton diamond from the same location, the Udachnaya kimberlite field (Nestola et al. 2011). The *red symbols* correspond to different kimberlite locations where the xenoliths were found in the Kaapvaal (*circle*: various Lesotho locations, *square*: Kimberley, *triangle*: Finsh, *cross*: Premier). The Siberian and Tanzanian craton xenoliths come from only one location in each craton (Udachnaya and Labait, respectively). Water contents obtained from FTIR data are calculated with the absorption coefficient of Withers et al. (2012). **B**: Viscosity versus depth of equilibration calculated for each xenolith using the water content measured in olivine and its equilibration depth entered in a flow law for olivine aggregates. Viscosity calculated for 8 ppm wt H_2O in olivine along the pressure-temperature variation of the geotherm of each craton is also shown (*dashed lines*). The Tanzanian craton has a hotter geotherm than the two others. The range and average viscosity of the asthenosphere is shown in grey. At depths $<$ 200 km, peridotites from the Kaapvaal and Siberian craton record viscosities higher than that of the asthenosphere, consistent with a strong mantle lithosphere. At depths $>$ 200 km, the peridotites from the Kaapvaal that have anomalous high temperatures (Kaapvaal high T field, their high temperatures being linked to a Mesozoic heating event (Thaba Putsoa) or plume event (Premier); Pearson et al. 1995a; Bell et al. 2003b) and the peridotites from the Siberian craton have similar viscosities to that of the asthenosphere

consistent with the mantle lithosphere at depths $<$ 180 km having a higher viscosity than the asthenosphere (Fig. 2B). This indicates that the water contents and equilibration temperatures are consistent with a rigid lithosphere, as expected (Pollack 1986; Karato 1993; Hirth et al. 2000; Sleep 2003).

The viscosities at the bottom of the lithosphere ($>$ 200 km depth), however, are the ones that warrant closer inspection to understand why cratonic roots do not get delaminated by the surrounding asthenosphere. A high viscosity contrast between the lithosphere and the asthenosphere ($>$ 3 to $>$ 50 times depending on estimated conditions) is necessary for the survival of a cratonic root for several billion years (Shapiro et al. 1999; Sleep 2003; O'Neill et al. 2008; Karato 2010; Wang et al. 2014). Viscosity estimates for the asthenosphere can be inferred from post-glacial rebound and earthquake data (Pollitz et al. 1998; Sjöberg et al. 2000; Larsen et al. 2005; Fleming et al. 2007; Masuti et al. 2016). The viscosity calculated for the deepest xenoliths from the Kaapvaal craton is high enough compared to the asthenosphere average viscosity to suggest that the low water content of olivines at the base of that craton could play a role to maintain its strength and help prevent its delamination (Fig. 2B, Peslier et al. 2010).

On the other hand, the viscosities calculated for the deepest Siberian craton xenoliths or for the anomalous high-temperature Kaapvaal peridotites are similar to those of the asthenosphere (Fig. 2B, Doucet et al. 2014). If these xenoliths were representative of the cratonic root, that of the Siberian craton should have been delaminated. However, we know that the Siberian craton still has a deep cratonic root as evidenced by the presence of xenoliths equilibrated down to > 200 km (7 GPa) and by seismic data (Priestley and Debayle 2003). In contrast, olivine inclusions in diamonds from the Siberian craton, all display low water contents (average 8 ± 8 ppm wt H_2O; Novella et al. 2015; Jean et al. 2016; Taylor et al. 2016). Consequently, the xenoliths with high water contents in olivine cannot be representative for the overall Siberian craton. They instead represent a biased sampling of local water-rich zones in the mantle (see next paragraph).

The kimberlites, the host magma of xenoliths in cratonic settings, may use some of these water-rich zones, perhaps shear zones, in the mantle lithosphere as conduits to pass through (Doucet et al. 2014). Interestingly, there may be a link between metasomatic conduits in the mantle and the formation of kimberlite melts. Modification of the cratonic mantle by melts and fluids (metasomatism) is often characterized by the formation of orthopyroxene (Kelemen et al. 1998). It has also been proposed that melts of kimberlitic composition may be generated by carbonatite melts dissolving orthopyroxene (Russell et al. 2012). In this scenario, melts ultimately forming kimberlites would preferentially follow metasomatic zones with their abundance of orthopyroxenes and their relative high water concentrations. These melts also pick up mantle samples on their way to the surface (xenoliths, diamonds), but a large proportion of the xenoliths are thereby unrepresentative of the whole cratonic mantle, with compositions dominated by water-rich metasomatized peridotite. The water-poor olivine from diamond inclusions are likely more representative because encapsulation in diamond protected the olivines from being re-enriched in water by localized circulation of water-rich melts or fluids (Novella et al. 2015; Jean et al. 2016; Taylor et al. 2016). In particular, the viscosity-pressure curve calculated for 8 ppm wt H_2O and the Siberian craton geotherm intersects with the viscosity range of the asthenosphere at ~ 200 km depth, i.e. close to the lithosphere-asthenosphere boundary. This suggests that the water content of olivine in the Siberian craton, as represented by that of olivine inclusions in diamond, have low enough water contents to provide the high viscosities that are necessary for preserving a thick (200 km) cratonic root from delamination.

Additional evidence for the "dry" nature of cratonic mantle lithosphere comes from the interpretation of geophysical data: electrical conductivity and seismic anisotropy. The presence of hydrogen in minerals enhances their *electrical conductivity* (Karato 1990; Karato and Wang 2013; Pommier 2014). Electrical conductivity profiles across the Kaapvaal craton are best explained with olivine having < 20 ppm wt H_2O at a depth of 100 km (Jones et al. 2013b). For the Slave craton in Canada, another thick Archean craton, electrical conductivity data is best modeled with olivine water contents decreasing from 150 to mostly < 20 ppm wt H_2O from 100 to 200 km depth (Jones et al. 2013a). In the center of the Tanzanian craton, < 100 ppm wt H_2O has been inferred for the mantle lithosphere between 90 km and the lithosphere-asthenosphere boundary. At the eastern edge of the craton, decreasing water contents from 10 ppm wt H_2O at 60 km depth to < 1 ppm wt H_2O at 150 km depth match best electrical conductivity data (Selway et al. 2014), i.e. the water contents are mostly lower than those measured in mantle xenoliths from this region (20 ± 12 ppm wt H_2O, Fig. 2A; Hui et al. 2015). Overall, electrical conductivity measurements reveal regional and depth variations of water concentrations, likely linked to local metasomatic enrichments (see next section on Water and Metasomatism). Interestingly, electrical conductivity also predicts low water contents at the base of the lithosphere of the Slave and Tanzanian cratons. These observations provide additional evidence that the deepest xenoliths with water-rich olivines are not representative of the overall cratonic lithosphere.

Observations of the dominant *micro-structure of cratonic mantle xenoliths* can also provide constraints on the amount of water present when the fabric was formed (Karato 2006a; Karato et al. 2008). Laboratory experiments on olivine-aggregate deformation constrain which type of fabric (lattice-preferred orientation, "LPO") forms as a function of water content, temperature and pressure (Jung and Karato 2001a, 2001b; Karato et al. 2008; Ohuchi et al. 2011; Ohuchi and Irifune 2013). However, results from these experiments are contradictory. For example, experiments at high pressures (7 GPa, i.e., $\sim$ 223 km depth) obtain the so-called "Type-A" fabric, i.e. a coarse ($>$ 1 mm grains) granular fabric, with high water contents ($>$ 127 ppm H_2O) in olivine. But at low pressures ($<$ 2 GPa or $<$ 64 km depth), it is experiments with low olivine water contents ($<$ 10 ppm H_2O) that generate the Type A fabric (Karato 2006a; Ohuchi and Irifune 2013). Moreover, the link between water content and fabric is not always obvious in natural samples. Observations in a tectonically-uplifted mantle section (Josephine peridotite) and in a xenolith suite from Udachnaya (Siberian craton) are nevertheless consistent with the more deformed fabrics associated with the highest water concentrations (Skemer et al. 2013; Kolesnichenko et al. 2016). However, no correlation was observed in Kaapvaal mantle xenoliths between the dominant fabric of peridotite (Type-A; Ben-Ismail et al. 2001) and their water content, suggesting that water could have been added locally after the main fabric was set, after melting and early deformation and annealing events (Baptiste et al. 2012). Water could also be lost from fabrics characteristic of formation in wet conditions (so called "E-type" fabric) if the final equilibration temperature of such a peridotite is low ($<$ 800 °C). This is because the solubility of water in olivine decreases with decreasing temperature (Chin et al. 2016). Deformation fabrics (LPO) are directly connected to the observed *seismic anisotropy*. Seismological observations of the cratonic lithosphere are most consistent with the dominance of the Type-A fabric (Mainprice and Silver 1993; Vinnik et al. 1996; Gung et al. 2003; Karato 2006b; Baptiste and Tommasi 2014). If further studies of natural mantle samples confirm that this fabric is generally associated with dry olivine it would imply that the cratonic mantle lithosphere has primarily olivines with $<$ 10 ppm wt H_2O at depths $<$ 190 km.

Water and Metasomatism Cratonic peridotites formed by partial melting in the Archean and Proterozoic, and are the solid residues of these melting processes (e.g., Walker et al. 1989; Pearson et al. 1995a, 1995b; Carlson et al. 1999; Doucet et al. 2015). Since then, however, their compositions have been modified by melts or fluids percolating through the cratonic lithosphere, i.e. they have been metasomatized. We now further investigate further the hypothesis of the previous section that the range of their water content is primarily the result of various degrees of metasomatism. To this end, we compare the variations of water content with those of other trace elements and with modal composition in olivine, pyroxene and garnet. Evidence for metasomatism in peridotites vary with the type of metasomatic agent involved (for example fluids, silicate melts of alkalitic, kimberlitic and ultramafic compositions, carbonatitic melts). It includes (i) addition of "secondary" clinopyroxene or orthopyroxene, (ii) increase of the Ti content of garnet, (iii) increase of incompatible trace element concentrations like rare-earth elements (REE) and Na in clinopyroxene and (iv) oxidation (increase of $Fe^{3+}/\Sigma Fe$ in clinopyroxene, spinal and garnet) (e.g., Harte 1983; Menzies and Hawkesworth 1987; Kesson and Ringwood 1989; Boyd et al. 1997; Kelemen et al. 1998; Griffin et al. 1999; Lee and Rudnick 1999; Grégoire et al. 2003; Burgess and Harte 2004; Bell et al. 2005; Creighton et al. 2009; Ionov et al. 2010; Doucet et al. 2013).

Water contents of the peridotite minerals for individual cratonic xenolith suites exhibit positive trends with these metasomatism indicators as illustrated in Fig. 3. For example, the water content of garnet correlates with its Ti contents in two Kaapvaal xenoliths suites and also roughly in that of Udachnaya from the Siberian craton (Fig. 3A). The water content of clinopyroxene correlates with REE, Na and $Fe^{3+}/\Sigma Fe$ in clinopyroxene from Liqhobong xenoliths in the Kaapvaal craton (Fig. 3B). Importantly, the highest water contents in olivine of the Siberian xenoliths are associated with the highest clinopyroxene modal proportions (Fig. 3C) and the highest REE concentrations in clinopyroxene (not shown), highlighting the role of water-bearing metasomatic melts in the most water-rich Udachnaya xenoliths. Overall, these trends signify that the range of water contents in cratonic mantle xenoliths is linked to water addition by melts or fluids (Bell and Rossman 1992a; Peslier et al. 2012; Doucet et al. 2014). As discussed in the previous section, circulation of metasomatic agents through the lithosphere is probably facilitated in zones of preexisting weaknesses such as faults, ancient sutures or magma conduits. In addition, these zones are also likely the preferred path of kimberlite melt ascent, the host-magma that brings up xenoliths in cratonic settings. It is thus no surprise that the kimberlites would provide us with a biased sample of the cratonic lithosphere, with an overabundance of metasomatized and water-rich peridotite xenoliths.

On the other hand, olivine inclusions in diamonds from the Siberian craton are relatively unaffected by metasomatism. This is indicated by the relatively high forsterite contents (Mg/(Mg+Fe)) of these olivines combined with their low Ni, Co and Ti concentrations compared to those of most olivines from Udachnaya xenoliths (Fig. 3E and F) (Kelemen et al. 1998; Jean et al. 2016). The diamonds may thus have acted as a protective shell around their inclusions, preventing late interactions with water-bearing metasomatic agents. The inclusions, therefore, preserve a more representative cratonic water content of peridotite minerals (Novella et al. 2015; Taylor et al. 2016).

Similar conclusions can be derived from garnet and pyroxene inclusions in diamonds. Water contents of two peridotitic clinopyroxene from diamond inclusions are at the low end of the range (< 100 ppm H_2O) of their counterparts found in xenolith peridotites (Fig. 4B). In contrast, peridotitic garnet inclusions from diamonds display water contents that span the whole range of water contents of garnets from peridotite xenoliths (Fig. 4D). Garnet, however, is the phase with the least amount of water in peridotite (< 30 ppm H_2O, Fig. 4). Eclogitic garnet inclusions in diamond, on the other hand, have lower water contents than garnet in eclogite xenoliths (Fig. 4C). These observations are based on limited data, but are overall consistent with the conclusion that diamonds protect their pyroxene and garnet inclusions from interaction with water-bearing metasomatizing agents, as we discussed above for olivine inclusions in diamond.

Two small cratons stand out compared to the larger Kaapvaal and Siberian cratons in terms of pyroxene water contents: the Tanzanian craton peridotites have low water contents, and those from the Colorado Plateau (USA) have high water contents compared to those of the Kaapvaal craton (Fig. 4B and 4C; Li et al. 2008; Hui et al. 2015). The high water contents of the Colorado peridotites were explained by metasomatism induced by subduction-related

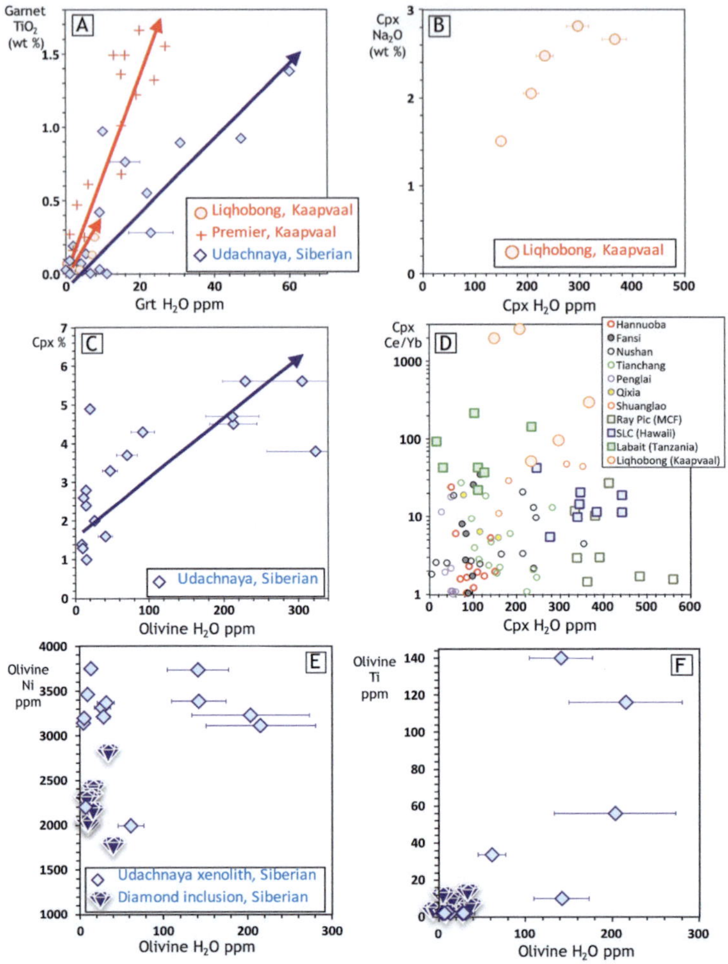

Fig. 3 Examples of water content variation with indices of metasomatism in mantle minerals. **A**: Garnet Ti concentrations increase with increasing water contents in three cratonic xenoliths suites (Premier, Kaapvaal craton, Bell and Rossman 1992a, Liqhobong, Kaapvaal craton, Peslier et al. 2012 and Udachnaya, Siberian craton, Doucet et al. 2014). *Arrows* indicate the trends. **B**: The Na_2O and water contents of clinopyroxene are positively correlated in Liqhobong clinopyroxenes (Kaapvaal craton; Peslier et al. 2012). **C**: Water contents of Udachnaya olivine increase with the modal proportion of clinopyroxene in peridotites from the Siberian craton (Doucet et al. 2014). These trends illustrate that the highest water contents measured in peridotite minerals from cratonic xenoliths can be explained by the addition of water by metasomatic fluid or melts. **D**: Water content versus Ce/Yb ratio of clinopyroxene from various xenolith suites in plume-lithosphere tectonic settings (*square symbols*), cratonic setting (*large round symbols*, Liqhobong, Kaapvaal), and Phanerozoic continental lithosphere beneath China (*small round symbols*) (Yang et al. 2008b; Xia et al. 2010, 2013a; Peslier et al. 2012; Denis et al. 2015; Hui et al. 2015; Hao et al. 2016). Cerium is more incompatible than Yb resulting in Yb primarily reflecting melting processes, while Ce is sensitive to metasomatism by REE-bearing melts. The lack of correlation between water content and Ce/Yb ratio indicates that water and LREE like Ce are decoupled during metasomatism of nominally anhydrous minerals. Error bars on the clinopyroxene water contents are omitted in **D** for clarity. Cpx = clinopyroxene, MCF = Massif Central in France, SLC = Salt Lake Crater. **E** and **F**: Water versus Ni and Ti concentrations in Siberian craton mantle olivines. The low water, Ni and Ti concentrations of the olivines from diamond inclusions relative to those found in xenoliths may indicate that diamonds protected the inclusions from being metasomatized and enriched in water. In contrast, most olivines found in peridotites have high Ni, Ti and water contents derived from metasomatism (Jean et al. 2016)

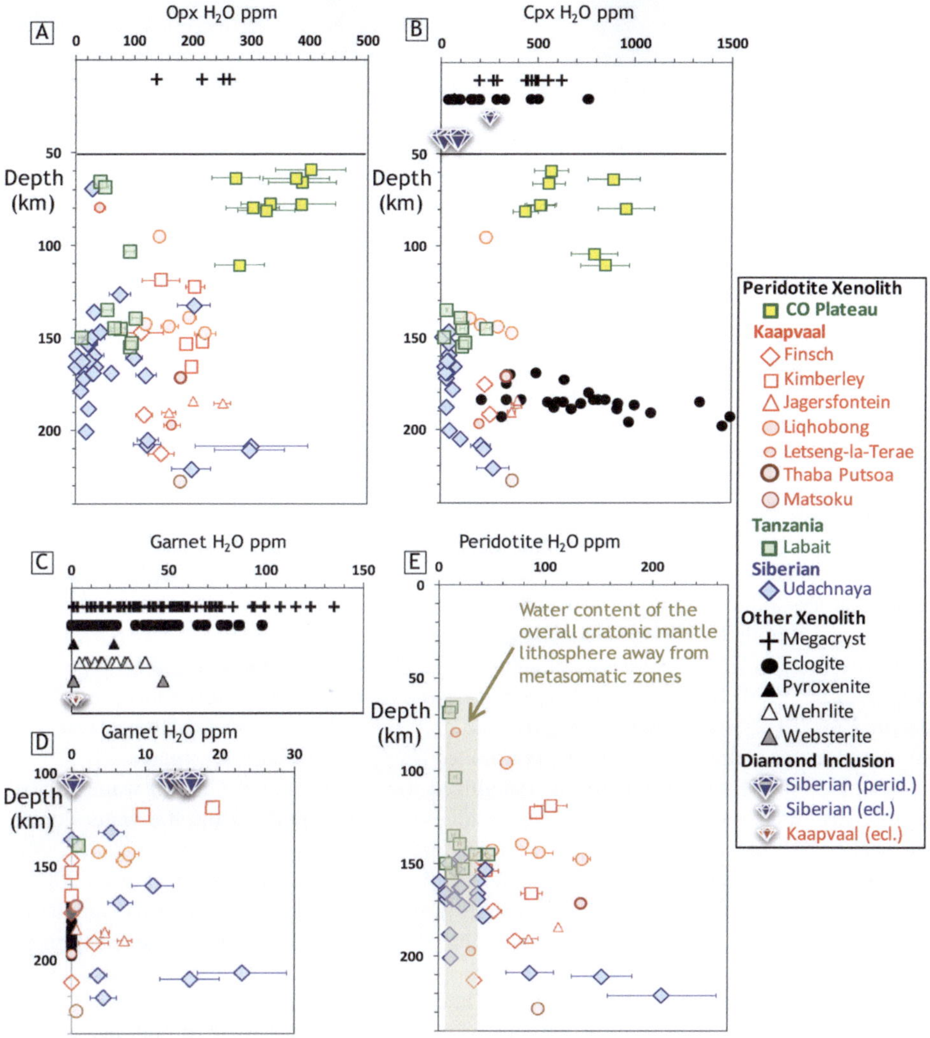

Fig. 4 Water contents of orthopyroxene (Opx), clinopyroxene (Cpx), garnet and those calculated for bulk-rock peridotites versus depth in the cratonic mantle lithosphere for mantle xenoliths from the Kaapvaal, Siberian, Tanzanian, and Colorado (CO) plateau cratons (Bell and Rossman 1992b; Grant et al. 2007a; Li et al. 2008; Sundvall and Stalder 2011; Peslier et al. 2012; Doucet et al. 2014; Hui et al. 2015; Kolesnichenko et al. 2016). Water contents in minerals from non-peridotitic lithologies from the Kaapvaal and Siberian cratons are shown at the top of the diagrams when their depth of equilibration is not known (Smyth et al. 1991; Bell and Rossman 1992a; Matsyuk et al. 1998; Bell et al. 2004b; Koch-Müller et al. 2004, 2007; Matsyuk and Langer 2004; Sundvall and Stalder 2011; Huang et al. 2014). The water contents of clinopyroxene and garnet inclusions in diamonds from the Siberian and Kaapvaal cratons are also shown in **B**, **C** and **D** (Bell and Rossman 1992a; Novella et al. 2015; Jean et al. 2016; Taylor et al. 2016)

water-rich melts, which were initiated by the Farallon plate subducting beneath the southwestern US (Usui et al. 2003; Dixon et al. 2004; Li et al. 2008; Behr and Smith 2016). The relative "dryness" of the peridotite minerals from the Tanzanian craton, however, is puzzling because this craton is placed on top of one of the largest deep mantle upwelling (plume) of the Earth and has been extensively metasomatized by it (Lee and Rudnick 1999; Aulbach et al. 2008). Although no water measurements in melts erupted from that plume exist, plume melts are generally considered water-rich (see section on oceanic mantle below). The interpretation for the low water content of the Tanzanian xenoliths is that either the plume melts that metasomatized that craton were water-poor, or water and other incompatible elements were decoupled during the metasomatizing process (Hui et al. 2015).

Decoupling of incompatible elements and H during metasomatism has also been observed in peridotites from off cratonic settings and is illustrated by the lack of correlations between water content and the Ce/Yb ratio of clinopyroxene in individual xenolith suites (Fig. 3D) (Yang et al. 2008b; Xia et al. 2010, 2013a; Peslier et al. 2012; Denis et al. 2015; Hui et al. 2015; Hao et al. 2016). Finally, melt inclusions in olivine that are particularly water-rich (2–4 wt% H_2O for undegassed ones) from Columbia River basalts (USA, a plume setting impinging on continental lithosphere) were interpreted as their source being a mantle enriched in water by prior subduction events and re-melted during plume upwelling (Cabato et al. 2015).

Other cratonic mantle lithologies besides peridotites have been analyzed for water (Smyth et al. 1991; Bell and Rossman 1992a; Matsyuk et al. 1998; Bell et al. 2004b; Koch-Müller et al. 2004, 2007; Matsyuk and Langer 2004; Huang et al. 2014). Megacrysts represent crystallized minerals from melts in the mantle lithosphere, pyroxenites represent crystallized cumulate melts or metasomatic veins at depth, while eclogites may represent either crystallized melts or remnants of subducted slabs (e.g., MacGregor and Manton 1986; Smyth et al. 1989; Bodinier and Godard 2003; Foley et al. 2003; Bell and Moore 2004; Jacob 2004). The water content of pyroxene and garnet from these lithologies extend to more water-rich values than those of peridotites (Fig. 4). This can be explained by the incompatible behavior of water; these crystallized melts become enriched in water relative to the peridotites which are the melting residues. For the same reason, the water content of megacrysts correlates with major and trace elements concentrations which reflect crystallization processes of melts in the mantle (Bell and Rossman 1992a; Bell et al. 2004b). More exotic and rare lithologies with a high proportion of hydrous minerals such as the PIC and MARID xenolith suites (for mica (phlogopite)–ilmenite–clinopyroxene and mica–amphibole–rutile–ilmenite–clinopyroxene (diopside), respectively) may represent locally crystallized water-rich metasomatic melts (Dawson and Smith 1977; Grégoire et al. 2002; Konzett et al. 2014). Interestingly, the type of volcanism characteristic of cratonic settings, kimberlites, is characterized by volatile-rich (mainly CO_2 and H_2O) compositions as evidenced from a high proportion of hydrous minerals and the explosive character of their eruption (Mitchell 1986; Russell et al. 2012; Sparks 2013). There is a controversy, however, on when the kimberlite magmas acquired their water. One possibility is that kimberlite parent melts are water-rich, and they could be the surface expression of the PIC and MARID mantle melts (Dawson and Smith 1977; Sweeney et al. 1993; Konzett et al. 1998; Grégoire et al. 2002). However, there is growing evidence that most of the water contained in kimberlites was acquired during hydrothermal processes at the surface while their source was carbonate-rich rather than water-rich (Russell et al. 2012; Giuliani et al. 2014; Kamenetsky et al. 2014).

Summary of Water Content Distribution in the Continental Lithosphere The bulk-rock water contents of cratonic peridotite xenoliths, calculated from the water content measured in individual minerals combined with the modal mineralogy, range from 7 to 209 ppm wt H_2O (average 65 ± 47 ppm wt H_2O, Fig. 4E). However, the cratonic lithosphere is mostly dry, as indicated by mineral inclusions in diamonds. There are localized water-rich zones linked to melt/fluid infiltration which are over-represented in the xenolith population. Eight ppm wt H_2O is a reasonable estimation for the water content of cratonic olivine, and this amounts to $\sim 24 \pm 21$ ppm wt H_2O in the overall cratonic lithosphere when pyroxene is taken into account (Table 1). The low water content of olivine in the overall cratonic lithosphere contributes to the strength and longevity of cratonic root for billions of years.

Water contents vary regionally in the off-craton continental mantle lithosphere. For example, peridotite xenolith suites with low or no water content in their minerals (Letseng-la-Terae in Lesotho in the Kaapvaal craton or Nuomin in northeast China) are found < 100 km away from peridotite xenoliths containing several hundreds ppm H_2O (Peslier et al. 2012; Hao et al. 2016). Similarly, water-rich pyroxenes from the Colorado plateau (USA) or Jiande (southwest China) lithospheric mantle are localized occurrences (Li et al. 2008; Hao et al. 2014).

The water content of the lithospheric mantle is therefore highly heterogeneous laterally on a regional scale but also vertically with depth. These water content variations originated from the local melting and metasomatic history of the mantle, that was sampled as xenoliths or diamond inclusions by each alkali basalt or kimberlite host-magma. Overall they depend on the pressure, the temperature and oxygen fugacity conditions at any given depth, and thereby water fugacity.

Water Solubility in Minerals Laboratory experiments equilibrating a water-bearing fluid with a silicate mineral provide estimates of solubility for a variety of temperature, pressure, oxygen and water fugacity conditions and fluid composition. Such experiments are particularly useful to constrain the maximum amount of water (i.e. "storage capacity", Hirschmann et al. 2005) that could be present at depths in the Earth from which samples are not available (see Sect. 3.2.3 on the transition zone and Sect. 3.2.4 on the lower mantle).

The first important observation is that mantle minerals from xenoliths have lower water contents compared to their maximum storage capacity estimated from experiments (Bali et al. 2008; Ardia et al. 2012; Férot and Bolfan-Casanova 2012; Tenner et al. 2012; Yang et al. 2014; Yang 2015). This implies that the lithospheric mantle is under-saturated in water (Fig. 5).

The second observation is that the solubility of water in olivine increases with increasing pressure, and these predictions from experiments are in agreement with the maximum amounts of water reported for olivine from mantle xenoliths at a given depth (Demouchy and Bolfan-Casanova 2016). Water solubility strongly depends on the volatile species present in the fluid: pure H_2O fluid results in a higher solubility of water in olivine compared to that of a fluid with less water, but containing other volatile species like CO_2, H_2 or CH_4 in addition to H_2O (Fig. 5A). The amount of water that will enter mantle phases consequently depends strongly on the water fugacity of the melt or fluid it equilibrates with (Gaetani et al. 2014; Baptiste et al. 2015; Yang 2016). In cratonic lithospheres, where conditions are increasingly reducing with depth (Woodland and Koch 2003; McCammon and Kopylova 2004), the proportion of CH_4 in the metasomatic agent could play a role in the amount of water present. For example, the Kaapvaal is slightly more reduced at 160–223 km depths than the Siberian craton, which could explain the lower maximum water contents recorded in the Kaapvaal craton olivines compared to those of the Siberian ones (Fig. 2A; Doucet et al. 2014).

Water solubility in pyroxene, in contrast, is expected to decrease with increasing pressure at depths < 300 km. This is because the more Al, the more water can enter the crystal structure of pyroxene, and the amount of Al in pyroxene decreases with depth due to the enhanced stability of garnet, an Al-rich mineral (Brey et al. 1990; Stalder and Skogby 2002;

Table 1 Estimates of water concentration, mass of water, relative mass to that of the Earth's oceans and H isotopes of the main reservoirs of the Earth. See text for references

			Volume[a] 10^9 km^3	Mass[a] 10^{24} g	Concentration[b] H_2O ppm wt	Concentration[b] H ppm wt	Mass water[b] 10^{23} g H_2O	X Mass oceans[b]	δD_{SMOW} ‰
Hydrosphere	Rivers, Ice, Atmosphere						0.4	0.03	
	Oceans						13.7	1	0
	Total Hydrosphere			1.41			14.1		
Crust	Continental Crust		8.16	21.78	20000	2238	2.8–3.6	0.2–0.3	
	Oceanic Crust		2.14	6.34	15000	1679	1.2–1.4	0.1	−34 to 46
Mantle	Upper Mantle	Continental Mantle Lithosphere Craton[c]			24–100	3–11			
		Continental Mantle Lithosphere off-Craton			50–100	6–11			−92 to −113
		Oceanic Mantle Lithosphere			50–100	6–11			−60 to −130
		Asthenosphere[d]			55–400 (200)	6–45 (22)	1.2	0.1	−20 to −100
		Total Upper Mantle	16.10	584.00	200	22	1.2	0.1	
	Transition Zone[e]		10.70	400.00	3140–5100 (3350)	351–571 (375)	12.6–20 (13.5)	0.9–1.4 (0.96)	
	Lower Mantle[f]		60.60	3030.00	100–2000? (1340)	11–224? (168)	3.0–61.0? (40.6)	0.2–4.5? (2.9)	
	Subduction Zone	Mantle Wedge (Arc Basalt Source)			600–8000	67–895			−55 to −12
		Slab in Upper Mantle[g]			300–3000	37–336			−33 to −126
	Primitive Mantle				330–1200	37–134			< −218
	Total mantle		**87.40**	**4014.00**	**382–2564 (1374)**	**43–287 (167)**	**15.4–103.0 (55.3)**	**1.1–7.4 (4)**	

Table 1 (*Continued*)

		Volume[a] 10^9 km^3	Mass[a] 10^{24} g	Concentration[b] H_2O ppm wt	Concentration[b] H ppm wt	Mass water[b] 10^{23} g H_2O	X Mass oceans[b]	δD_{SMOW} ‰
Core[f]		17.70	1928.70		20–7300? (1000)	2.4–960? (17.4)	0.2–90.0? (12)	
Total	Without Hydrosphere			404–36442 (3898)	45–4078 (436)	24.1–1370 (234)	1.7–97.7 (16.7)	−43
	Including Hydrosphere			638–36668 (**3921**)	71–4103 (**439**)	38.6–1380 (**248**)	2.8–98.8 (**17.6**)	

[a]Reservoir volume and mass estimates from McDonough (2005) and Bodnar et al. (2013)

[b]When possible, a likely reasonable average is given in parentheses. See text for details. X mass oceans = times the mass of the oceans

[c]The overall water content of the cratonic mantle lithosphere is probably low ($\sim$ 24 ppm H_2O) while localized metasomatized zones have more water

[d]The source of depleted MORB has $\sim$ 20–250 ppm H_2O but the deepest part of the upper mantle may have higher water content due to the increasing solubility of water in olivine and less depletion with depth

[e]The amount of water in the transition zone is heterogeneous. The ranges are calculated with 2–30% water-rich zones (at 10,000 ppm H_2O) and the rest being drier (3000–6000 ppm H_2O). The 3350 ppm H_2O is calculated with 5% at 10,000 ppm H_2O and 95% at 3000 ppm H_2O

[f]The amounts of water in the lower mantle and core are controversial, thereby the question mark next to the range provided

[g]Amount of water measured in pyroxenite and eclogite samples

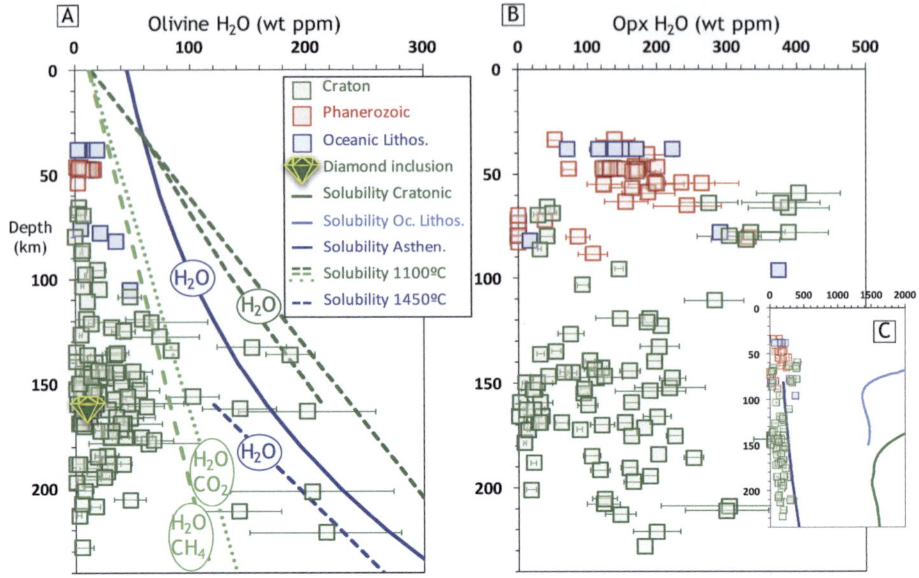

Fig. 5 Water contents of olivine (**A**) and orthopyroxene (**B**) in mantle peridotites compared to solubility curves obtained experimentally. H_2O, H_2O-CH_4 or H_2O-CO_2 indicate the volatiles present in the fluid the minerals equilibrated with in the experiments. The *dotted* and *dashed solubility curves* for olivine are calculated from experiments at a set temperature: H_2O, H_2O-CH_4 or H_2O-CO_2 at 1100 °C (*in green*, Yang et al. 2014; Yang 2015), H_2O at 1450 °C (*in blue*, Ardia et al. 2012; Tenner et al. 2012). The H_2O *blue solid curve* is calculated with a geotherm for the convective mantle (Asthen., Férot and Bolfan-Casanova 2012). All concentrations for olivine, natural or experimental, obtained from FTIR data are calculated with the absorption coefficient of Withers et al. (2012). *Inset* figure C shows the contrasting estimates for the solubility of water in orthopyroxene (Opx): for Al-saturated and Fe-free Opx (Mierdel et al. 2007), the water storage capacity of the lithosphere is high (> 1000 ppm wt H_2O, *light blue* and *green curve*), while the solubility curve of water in Opx for the convective mantle (Férot and Bolfan-Casanova 2012) provides lower water contents (< 250 ppm wt H_2O, *dark blue curve*). The oceanic lithosphere (Oc. Lithos.) geotherm used in the opx solubility calculation (*light blue*) is for a 40 Ma old lithosphere moving at a rate of 10 mm per year (i.e. Pacific plate conditions). Peridotite data from the literature (Kurosawa et al. 1997; Peslier et al. 2002, 2008, 2010, 2012; Grant et al. 2007a; Baptiste et al. 2012; Doucet et al. 2014; Demouchy et al. 2015; Hui et al. 2015; Peslier and Bizimis 2015; Kolesnichenko et al. 2016; Schaffer et al. 2016). Peridotites are grouped by tectonic setting: cratonic, continental Phanerozoic and oceanic. Oceanic peridotites include peridotites from Hawaii and the Ontong-Java plateau (Demouchy et al. 2015; Peslier and Bizimis 2015). The peridotite xenoliths from Hawaii do not contain any garnet and a pressure of equilibration could thus not be calculated. They are shown here, however, at 38 km depth, a reasonable assumption for these spinel peridotite lithospheric samples (Bizimis et al. 2004). The main conclusion from these figures is that natural olivine and pyroxene from the mantle lithosphere are under-saturated in water

Stalder 2004; Hirschmann et al. 2009; Stalder et al. 2015). The amount of Al in pyroxene at depths > 300 km is relatively constant and low, limiting the maximum water intake of pyroxene. This effect may cause the orthopyroxenes at depths < 100 km to record the highest water contents in natural peridotites (> 300 ppm wt H_2O; Fig. 5B). Experiment-based estimations for the water storage capability of orthopyroxene, however, vary by almost an order of magnitude (Fig. 5C), with Al- and water-saturated experiments predicting > 1000 ppm wt H_2O and experiments in undersaturated conditions predicting < 300 ppm wt H_2O. The latter is close to the maximum water content recorded in orthopyroxenes from natural peridotites (Mierdel et al. 2007; Férot and Bolfan-Casanova 2012).

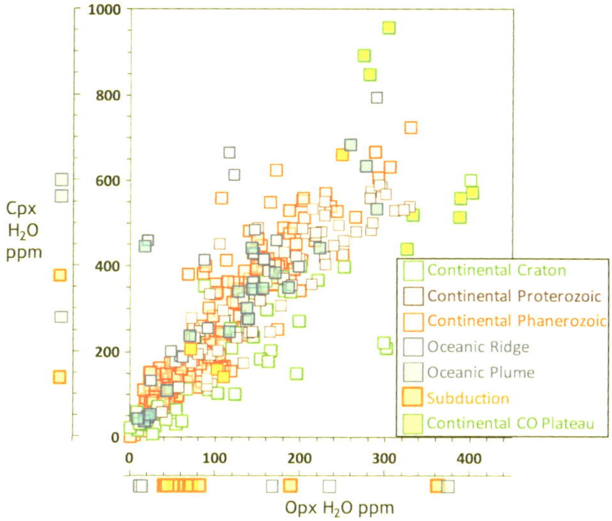

Fig. 6 Water content in clinopyroxene (Cpx) and orthopyroxene (Opx) of peridotites from mantle xenoliths, abyssal peridotites, and tectonically exhumed massifs grouped by tectonic setting (see online supplementary Table 1 for data and references). Uncertainties on water contents are omitted for clarity and are typically 10 to 35% depending on the dataset. When only one mineral was analyzed for water, these are shown to the *left* or *bottom* of the figure

3.2.2 Water in the Asthenosphere and Oceanic Mantle Lithosphere

Mantle xenoliths in oceanic settings are rare, but abyssal peridotites, mantle zones tectonically lifted to the ocean floor at mid-oceanic ridges, also offer samples of the shallow oceanic mantle. Water concentration data from these *direct samples of the oceanic mantle lithosphere*, however, are limited (Gose et al. 2009, 2011; Schmädicke et al. 2011; Warren and Hauri 2014; Bizimis and Peslier 2015; Demouchy et al. 2015; Peslier and Bizimis 2015; Peslier et al. 2015). The water content of these oceanic lithospheric peridotites (blue symbols in Fig. 6) is similar to that of continental peridotites (red and green symbols in Fig. 6). Oceanic pyroxenites contain three to four times more water than peridotites (Fig. 7, Bizimis and Peslier 2015). Two reasons may be proposed for this difference. First pyroxenites primarily consist of pyroxene in which water is more compatible than in olivine, the main mineral of peridotites. Second, they represent crystallized melts at depth, and melts should be water-rich due to the incompatible behavior of water during mantle melting.

On the other hand, abundant water data exists on *indirect samples of the mantle below the lithosphere*, i.e. basalts from mid-oceanic ridge (MORB) and oceanic island (OIB) settings from which the water content of the mantle can be inferred. The water content of their sources in the mantle can be calculated from the analyses of their un-degassed glasses and melt inclusions using reasonable assumptions for the degree and style of partial melting and partition coefficients (Dixon and Clague 2001; Dixon et al. 2002; Saal et al. 2002; Wallace 2002; Workman et al. 2006; Cabral et al. 2014; Kendrick et al. 2014, 2015; Jackson et al. 2015). Basalts from oceanic islands are generally more water-rich than MORB, which implies that their sources are more water-rich than those of MORB (Fig. 7). This is explained by MORB sampling the shallow asthenosphere, depleted in incompatible elements such as H (DM for depleted mantle in Fig. 7; e.g., Dixon et al. 1988; Michael 1988). The source of normal MORB typically has 50–230 ppm wt H_2O (Fig. 7, Dixon et al. 2002; Schaffer et al. 2016). These estimates are consistent with electrical conductivity results that were measured for the oceanic mantle away from any plume influence. These analyses reveal a dry lithosphere (< 50 ppm wt H_2O) underlain by a progressively wetter asthenosphere with depth (400 ppm wt H_2O at 300 km depth; Hirth and Kohlstedt 1996; Evans et al. 2005; Baba et al. 2006; Dai and Karato 2009; Naif et al. 2013; Gardés et al. 2014; Pommier 2014;

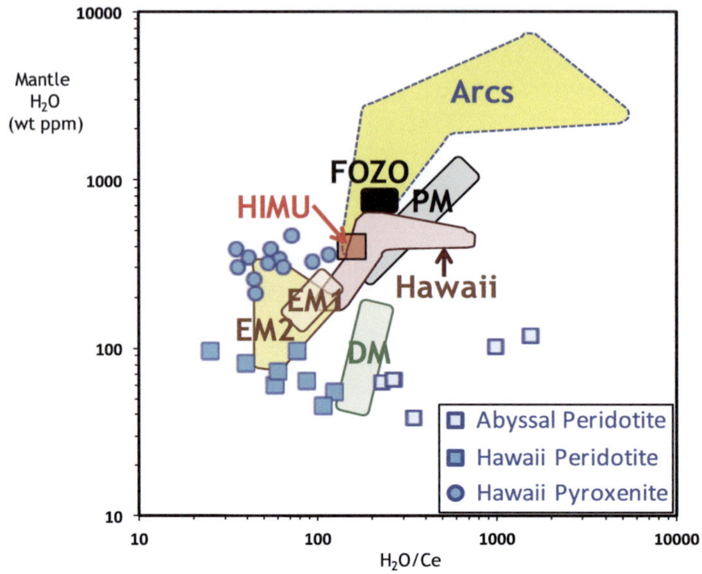

Fig. 7 Water and water/Ce ratio in the oceanic mantle and in subduction zones. Water in subduction zones is obtained from basalts from oceanic arc settings (Arcs; Plank et al. 2009, 2013). Water in the oceanic mantle can be obtained using actual mantle samples (peridotites and pyroxenites) or oceanic basalts which are melts from the mantle (DM, EM1 and 2, HIMU, Hawaii). For peridotites, the water contents are calculated from water analyses of their minerals and each symbol represents an individual sample (Warren and Hauri 2014; Bizimis and Peslier 2015; Peslier and Bizimis 2015). For basalts, the water content of their sources is shown and is calculated from numerous measurements in glasses and melt inclusions. Oceanic basalt end-members are: DM = depleted mantle, the source of mid-oceanic ridges (MOR in Fig. 9) (Michael 1988), and oceanic island (OI in Fig. 9) basalts: EM = enriched mantle (Wallace 2002; Workman and Hart 2005; Workman et al. 2006; Dixon et al. 2008; Kendrick et al. 2014, 2015); and HIMU = "high-mu," with mu being defined as $(^{238}U/^{204}Pb)_{t=0}$ (Cabral et al. 2014; Jackson et al. 2014, 2015). The H_2O/Ce ratio is that of the bulk peridotites, pyroxenites and basalts. Theoretical mantle reservoirs are also shown: PM = primitive mantle (Sun and McDonough 1989; O'Neill and Palme 1998; Dixon and Clague 2001) and FOZO = focal zone (Dixon et al. 2002). The large variations of the water/Ce ratio for a given mantle water content in peridotites and oceanic island basalts indicates that water and Ce are decoupled during mantle processes. See text in Sect. 3.2.2 for details

Sarafian et al. 2015; Khan 2016; Masuti et al. 2016). Such a dry oceanic lithosphere is rigid enough to be passively transported above the hotter convecting asthenosphere from its formation at ridges towards subduction zones or passive continental margins.

Oceanic island basalts sample a broader range of sources, which is indicated by their large range of Sr, Nd, Hf, Pb and noble gases isotopic compositions. Some of these sources are deeper than those of MORB, and OIB are the surface expression of deep mantle up-welling, generally called "plume". The sources of OIB are not well constrained but likely include deep fertile reservoirs, recycled subducted slabs and the depleted upper mantle. Several compositional isotopic end-members have been defined (e.g., Sun 1982; Zindler and Hart 1986; Hart et al. 1992; Brandon et al. 1998; Hofmann 2003; Stracke et al. 2005; Jackson et al. 2007; Delavault et al. 2016). The enriched mantle 1 (EM1) could involve recycled continental mantle or Archean sediments, while EM2 may incorporate recycled continental crust into its source. Another end-member is high-μ (HIMU, with μ being defined as the present-day $^{238}U/^{204}Pb$ ratio of a basalt) and may represent recycled subducted oceanic crust. Primitive mantle (PM) is a theoretical reservoir representing the mantle prior the melt extraction that formed the crust; some remnants of this original mantle may still exist in the deep upper mantle or lower mantle. Finally, focal zone (FOZO) is defined by a convergence of isotopic trends to a relatively undegassed reservoir. Inclusion of deep mantle material in OIB plumes seem inevitable because they originate from the deep mantle, possibly from the core-mantle boundary at the edges of so called large low shear wave velocity provinces (LLSVP, Burke et al. 2008; Garnero et al. 2016; Mundl et al. 2017). The high water content of OIB thus is linked to the incompatible element enriched nature of their sources, either because they sample (i) undegassed reservoirs (lower mantle, core-mantle boundary), (ii) water-rich reservoirs (for example the transition zone, see Sect. 3.2.3) or (iii) water-rich lithologies (for example pyroxene-rich lithologies like pyroxenites and eclogites) that result from recycling via subduction zones (see Sect. 4; Dixon et al. 2002; Bizimis and Peslier 2015). The water-rich nature of plume sources is also confirmed by analyses of another type of rare primitive (Mg-rich) magma called komatiites, whose melt inclusions are as rich in water as those of OIB (Gurenko et al. 2016), and by electromagnetic imaging of plumes (Tada et al. 2016). Finally, the inventory of noble gas isotopic systems of K-Ar, Xe and Ne are best explained if a primordial volatile-rich reservoir is still present in the mantle and is being sampled by plumes (Marty 2012; Mukhopadhyay 2012; Caracausi et al. 2016; Mundl et al. 2017). The sources of OIB contain from 80 to 800 ppm wt H_2O (Dixon et al. 2002; Cabral et al. 2014).

Finally, we would like to caution the reader regarding the use of a popular alternative method for estimating the amount of water in oceanic basalt sources, which is to use concentration ratios of water over an element of presumed similar incompatibility during partitioning between silicate melt and solid (e.g., Ce, K, La). This idea derives from the observation that these ratios (e.g., water/Ce in Fig. 7) span a narrow range for MORB in each ocean and thus likely represent those of the basalt source (Michael 1995). However, several observations have questioned this approach, primarily the difficulty in explaining the lack of correlation between Ce enrichments and water enrichment in OIB and oceanic peridotites (Fig. 7) (Workman et al. 2006; Kendrick et al. 2014; Warren and Hauri 2014; Hartley et al. 2015; Peslier and Bizimis 2015). Moreover, mantle lithosphere that was pervasively metasomatized by plume-related melts, and thereby enriched in incompatible elements such as LREE, does not exhibit enriched water contents, as evidence from Hawaii, the Tanzanian craton, and many other locations demonstrates (Fig. 3D) (Hui et al. 2015; Peslier and Bizimis 2015). This is best interpreted by decoupling of water from incompatible elements like Ce during mantle processes, with several mechanisms proposed. One is that the fast diffusion of H relative to other slower incompatible elements (e.g., REE) either dilutes water contents in the upper mantle and in the oceanic mantle lithosphere or generates preferential loss of water in subduction zones (Workman et al. 2006; Kendrick et al. 2014; Jackson et al. 2015; Peslier and Bizimis 2015). Subducted material in the mantle could thus be drier than previously thought and its incorporation in a plume source would not generate water-rich OIB. Other mechanisms invoke more fundamental differences of behavior between water and other incompatible elements. One hypothesis is that the large sizes of REE atoms could hinder their incorporation into mantle minerals when pressure increases because of the shrinking radii of the lattice sites. This results in lower REE concentration in the mantle minerals from greater depths. The small H atom, on the other hand, is less prone to this effect. This could affect the water/Ce ratio of the various sources of basalts (Karato 2016). Moreover, melts from the deepest sources will have lower water/Ce ratios then those from shallower levels because Ce will preferentially partitions into the melt at the higher pressures (Adam et al. 2016). Another mechanism that could explain water and Ce decoupling relates on the dependency of H partitioning on water speciation in the melt resulting in H being less incompatible than Ce (Adam et al. 2016;

Schaffer et al. 2016). It is therefore suggested that water estimates that are based on constant ratios of water over other incompatible elements, such as LREE or K, are not reliable.

3.2.3 Water in the Transition Zone

At 410 km depth, pyroxene is no longer stable and transforms into garnet while olivine changes into the denser wadsleyite structure, which in turn is converted into ringwoodite at 520 km depth. The transition zone thus comprises 56% wadsleyite and 44% majorite, a high pressure garnet of formula $(Mg, Fe, Ca)_3(Mg, Si, Al)_2Si_3O_{12}$ from 410 to 520 km, and 56% ringwoodite and 44% majorite from 520 to 660 km depth (e.g., Frost 2008). Experiments show that the water storage capacity of wadsleyite, ringwoodite and majorite is high (0.5–2, 0.3–2, and 0.07–0.2 wt% H_2O respectively) (Smyth 1994; Inoue et al. 1995; Kohlstedt et al. 1996; Bolfan-Casanova et al. 2000; Katayama et al. 2003; Demouchy et al. 2005; Panero et al. 2013; Thomas et al. 2014; Ohtani 2015; Pigott et al. 2015; Thomas et al. 2015). We have discussed earlier that the storage capacity does not necessarily reflect the actual water content with the example of the upper mantle being undersaturated in water. However, additional evidence from natural samples suggests that some regions of the transition zone are indeed particularly rich in water. Two diamonds were found from these depths, each with a mineral inclusion containing water (Wirth et al. 2007; Pearson et al. 2014): a ringwoodite inclusion containing 1.5 wt% H_2O, and a hydrous mineral inclusion called Phase EGG ($AlSiO_3(OH)$), named after its discoverer Eggleton (Eggleton et al. 1978). Although these samples provide very important direct evidence for water in the transition zone, it is not clear whether these two samples represent the typical water content of this mantle layer.

Geophysical studies used to remotely map the mantle are indeed more consistent with regionally variable water contents in the transition zone (Karato et al. 2001; Huang et al. 2005; Karato 2011; Koyama et al. 2013; Khan 2016). Electrical conductivity and global seismic data are consistent with a heterogeneous transition zone that is on average undersaturated in water (i.e. $\sim$3000–6000 ppm wt H_2O), but still about 10 times more water-rich than the asthenosphere (Karato et al. 2001; Huang et al. 2005; Smyth and Jacobsen 2006; Houser 2016; Khan 2016). Some areas of the transition zone, for example beneath China or southwestern USA, are strongly hydrated with $>$ 10,000 ppm wt H_2O (Karato et al. 2001; Courtier and Revenaugh 2006; Khan et al. 2011; Khan and Shankland 2012; Houser 2016; Khan 2016). Water-rich inclusions in diamonds might have come from such regions. The total amount of water that the transition zone could hold can be expressed in masses of present Earth's oceans. Given that the mass of the Earth's oceans is $\sim 13.7 \times 10^{23}$ g, estimating the mass of the transition zone (4×10^{26} g) from its average density and its volume, and assuming the water-rich zones constitute 5–30% (Houser 2016) of the transition zone, the transition zone contains from 1 to 2 Earth's oceans in mass (Fig. 8, Table 1, Bodnar et al. 2013; Nestola and Smyth 2015).

3.2.4 Water in the Lower Mantle

The main phase (80% volume) of the lower mantle is a dense Mg-perovskite called brigmanite ($(Mg, Fe, Ca)SiAlO_3$) and the minor phases are magnesiowüstite ($(Mg, Fe)O$) and calcium silicate perovskite ($CaSiO_3$) (e.g., Frost 2008). The results of experiments attempting to determine their water storage capacity are contradictory. However, except possibly Al-rich bridgmanite and calcium perovskite, the water solubility in major lower mantle minerals is generally low ($<$ 1000 ppm wt H_2O; Bolfan-Casanova 2005; Ohtani 2015;

Panero et al. 2015; Walter et al. 2015; Townsend et al. 2016). Minor hydrous phases are stable under lower mantle conditions, and could be brought down by subducting material all the way to the core-mantle boundary (Sano et al. 2008; Ohira et al. 2014; Walter et al. 2015; Hu et al. 2016, 2017; Nishihara and Matsukage 2016).

There are some diamond inclusions that show evidence for a lower mantle origin (e.g., Stachel et al. 2000; Kaminsky et al. 2001; Tappert et al. 2008; Walter et al. 2011), but the water content in these samples has not been measured. The hydrous mineral assemblage (brucite and $Mg(OH)_2$) found in one of these diamond inclusions, however, is best explained by the presence of hydrous melt in the lower mantle (Palot et al. 2016). In addition, there are no geophysical constraints on the water content of the lower mantle because the existing laboratory data cannot be used to infer the water content from geophysical observations at these high pressure and temperature conditions.

However, there is some indirect evidence for the transport of water into the lower mantle. This is the observation of low velocity regions near the top of the lower mantle and at the core-mantle boundary (Revenaugh and Meyer 1997; Garnero and McNamara 2008; Schmandt et al. 2014; Garnero et al. 2016). One possible interpretation is that these low velocity regions correspond to regions of partial melting. Partial melting in the lower mantle is impossible without water (Walter et al. 2015). Therefore, if these low velocity regions are caused by partial melting, some water is likely present. Alternatively, water could be stored in these lower most parts of the mantle in Al-postperovskite (Townsend et al. 2016).

In summary, the water content of the lower mantle is not well constrained and could vary from 0.2 to 4.3 times the mass of the ocean (Bodnar et al. 2013; Nestola and Smyth 2015). We estimated that if the lower mantle consists of 80% bridgmanite with $\sim$ 800 ppm H_2O, 11% magnesiowüstite with $\sim$ 2000 ppm H_2O, 8% $CaSiO_2$ with 3500 ppm H_2O and 1% of hydrous minerals with $\sim$ 20000 ppm H_2O (numbers from the review of Ohtani 2015), then it contains on average 1340 ppm H_2O, i.e. 2.9 times the mass of the oceans (Fig. 8A, Table 1).

3.3 Hydrogen in the Core

The core mainly consists of an Fe-Ni alloy and is liquid from 2900 to 5150 km depth and solid to the center of the Earth (e.g., Birch 1952; Dziewonski and Anderson 1981; Fiquet et al. 2008). The presence of a few % of light elements in the inner core has been invoked to explain core densities inferred from seismology (e.g., Birch 1952; Poirier 1994). One candidate is H because it is ubiquitous in planetary interiors and highly compatible in Fe liquid metal (Okuchi 1997; Abe et al. 2000). The solubility of H in Fe far exceeds that in the transition zone minerals at high pressures, and the core is potentially the largest reservoir of hydrogen in the Earth (e.g., Fukai 1984; Kuramoto and Matsui 1996; Ohtani et al. 2005; Shibazaki et al. 2009; Terasaki et al. 2012; Karato 2015; Iizuka-Oku et al. 2017). A recent temperature estimate of the outer core is lower than previously suggested and would necessitate a large amount of a light element to maintain its liquid state (Nomura et al. 2014). In this case and assuming the only light element is H, then as much as 80 times the mass of oceans could be stored in the core (Genda 2016). Mass balance constraints and planetary accretion models, however, predict relatively low concentrations of H in the core, from 20 to $<$ 200 ppm H (McDonough 2005; Rubie et al. 2015). Moreover, first principles calculations of the effect of H on the properties of iron suggest that seismological observations cannot be explained by the presence of H in the solid core (Caracas 2015; Umemoto and Hirose 2015). In summary, and emphasizing the enormous uncertainties on these numbers, the core could contain 20 to 7300 ppm wt H, i.e. the equivalent of 0.2 to 90 times the present mass of the Earth's oceans (Table 1). The higher estimates ($>$ 2300 ppm wt% H in the core; e.g., Shibazaki et al. 2012;

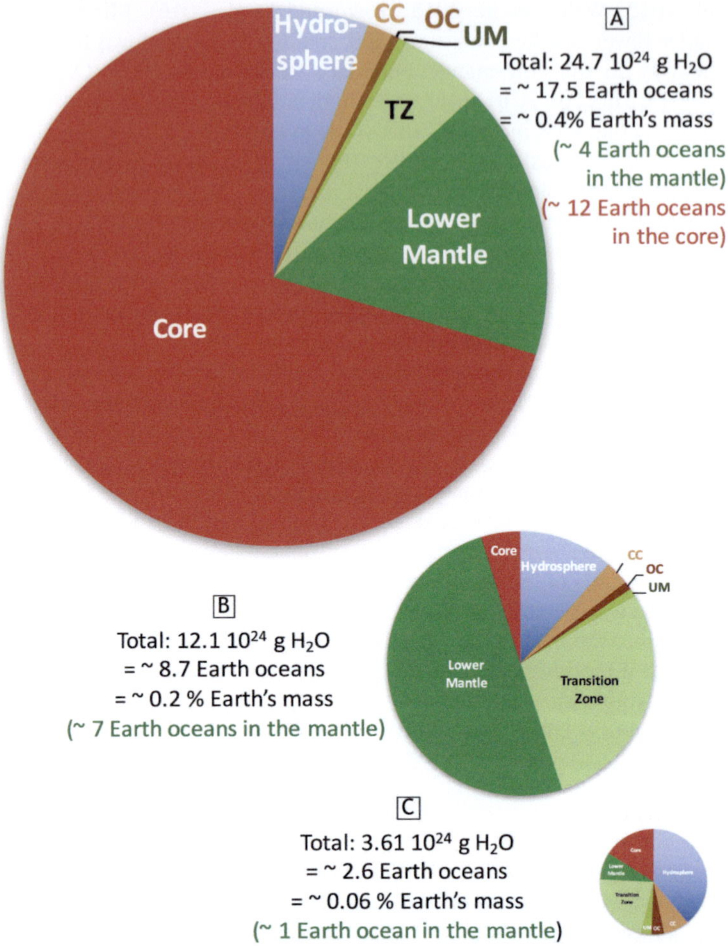

Fig. 8 Sketch of the relative amount of water (in mass %) in Earth's layers from our estimate (**A**) and two recent compilations, **B**: Nestola and Smyth (2015) and **C**: Bodnar et al. (2013). The size of the pie charts corresponds to the total mass of water calculated. The key difference between the three estimations is the amount of water in the lower mantle and core which are not well constrained. The estimate of water in the core in **B** and **C** is from the planetary accretion models of Rubie et al. (2015), while in **A**, it is based on the compatibility of H into iron derived from silicate-iron partitioning experiments. The mass of water in the core was calculated as H_2O for comparison with the silicate oxide layers of the Earth. However, "water" in the core is H in metal and this calculation does not necessary imply that oxygen is present in the metallic core. CC = continental crust, OC = oceanic crust, UM = upper mantle, TZ = transition zone

Genda 2016; Iizuka-Oku et al. 2017) are upper limits because they assume the unlikely scenario that H is the only volatile in the core. Other light elements (for example O, S, C or Si) are likely present in the core (e.g., Poirier 1994; Wood et al. 2013; Badro et al. 2014; Chabot et al. 2015). A more conservative estimate is to be within one order of magnitude of the planetary accretion models ($10\times \sim 100$ ppm H) and about half the amount necessary to explain seismic wave velocity of the inner core and core density with H as the only volatile element (2300/2), i.e. $\sim$ 1000 ppm H in the core, equivalent to 12 times the mass of the oceans (Fig. 8A, Table 1).

3.4 Amount of Water in the Earth

The largest uncertainties on the total amount of water of the Earth are the water contents of the lower mantle and core (Table 1). Based on experiments that show that H is strongly compatible in metal relative to silicate at high pressures (e.g., Ohtani et al. 2005), our model assumes that the core contains the equivalent of 12 masses of the Earth's present oceans. The mantle in our model contains the equivalent of 4 present masses of the Earth's oceans and this is based on a water rich but heterogeneous transition zone and a lower mantle with 1340 ppm H_2O. Including the hydrosphere, this results in a total amount of water for the Earth of 0.4 wt% H_2O (Fig. 8A). A conservative model suggests $<$ 1 present mass of the oceans in the lower mantle (Bodnar et al. 2013) and, combined with an estimate of the water content in the core of 0.4 the oceans mass from planetary accretion modeling (Rubie et al. 2015), results in the total amount of water contained in the Earth, including the hydrosphere, at 0.06 wt% H_2O (Fig. 8C). An intermediary estimate calculates about 7 times the present ocean masses in the mantle (Nestola and Smyth 2015), and combined with the water content in the core of Rubie et al. (2015), results in 0.2 wt% water for the Earth, including the hydrosphere, or 8.7 times the present mass of the oceans (Fig. 8B). Our estimate for the concentration of water in the Earth ($\sim$ 3900 ppm H_2O, Table 1) is within uncertainties to that inferred from noble gas systematics (2700 ± 1350 ppm H_2O) (Marty 2012).

4 The Water Cycle in the Earth

4.1 Water in Subduction Zones

Subduction zones are the main location of water exchange between the deep Earth and the atmosphere, with the highest fluxes estimated between any terrestrial layers. The fate of water in subduction zones is not only unique to Earth, but also plays a crucial role in the origin of its volcanism, mantle geodynamics and plate tectonics. A detailed account is beyond the scope of this chapter and several reviews are available (Hacker 2008; Grove et al. 2012; Faccenda 2014; Ohtani 2015). A short summary of essential observations is given here.

As subducted slabs are entrained into the mantle, they carry with them water-logged lithologies such as sediments, with structural (i.e., incorporated in mineral lattices) and pore water, and basalt, gabbro and mantle lithologies altered by sea water (e.g., Peacock 1990; Plank and Langmuir 1998; Rüpke et al. 2004; Hacker 2008). They also carry down unaltered crustal and mantle lithologies which, as discussed in the previous sections, contain hundreds to tens of thousands ppm wt H_2O. The input of water is also facilitated by the presence of hydrated faults generated when the plate is stretched through bending downwards towards the subduction trench (Garth and Rietbrock 2014). About 60–90% of this water returns to the oceans via faults and diffusion along the bottom of accumulated material in subduction trenches, called the accretionary prism, because this material is being compacted and clays start to dehydrate (e.g., Jarrard 2003). As the remaining material is transported to higher pressures, metamorphic processes dehydrate the slab minerals and water is transferred to the mantle wedge. This is evidenced by dehydration experiments on hydrous phases like serpentine and amphibole at relevant pressure and temperature, and low seismic velocities and high electrical conductivities observed in the mantle wedge (e.g., Schmidt and Poli 1998; Jarrard 2003; Luth 2003; Kawamoto 2006; Soyer and Unsworth 2006; Kawakatsu and Watada 2007; Rondenay et al. 2008; Ichiki et al. 2009; Frezzotti and Ferrando 2015). These fluids or melts are typically oxidized, which likely increases their water fugacity (Wood et al. 1990; Brandon and Draper 1996; Debret et al. 2015). Finally, water solubility in subduction zone fluids could also be enhanced by their relatively high Cl content derived from the contribution of sea water to subducted material (Bénard et al. 2017; Keppler 2017).

Given that the presence of water facilitates melting by triggering it at lower temperatures (e.g., Green 1973; Hirose and Kawamoto 1995; Médard and Grove 2008), this water-rich zone (dehydrating slab and mantle-wedge above) is the source of extensive volcanism (McGary et al. 2014). Subduction zone volcanoes therefore produce the most water-rich lavas on Earth (Danyushevsky et al. 1993; Sisson and Layne 1993; Stolper and Newman 1994; Sobolev and Chaussidon 1996; Ruscitto et al. 2012; Plank et al. 2013). Based on the measured water content in melt inclusions, the amount of water in the source of arc volcanoes is higher than that of any other basalt sources (Fig. 7). Note that for subduction zone lavas in Fig. 7 (Arcs), only data from oceanic arc volcanoes have been utilized. Oceanic arc volcanoes have the advantage that their interaction with crustal units is minimal because they only traverse relatively thin oceanic lithosphere (Wallace 2005). Therefore their water content may reflect more accurately the water content of their mantle sources than lavas from subduction zones along continents. Another evidence for the water-rich nature of the mantle wedge may be found in fluid inclusions, which are trapped droplets of fluids embedded in nominally anhydrous minerals. Typical fluid inclusions from upper mantle peridotites not related to subduction contain primarily CO_2 and little to no water (Roedder 1984). This has been interpreted as the result of (i) metasomatism by an ascending volatile-rich melt with CO_2 that exsolves more extensively than water, or (ii) diffusive loss of water from the inclusion during xenolith ascent, or (iii) as evidence that fluids in the shallow upper mantle are not hydrous (Pasteris 1987; Andersen and Neumann 2001; Frezzotti et al. 2012). Fluid inclusions in subduction zones, however, can contain significant amounts of water, which are best explained by particularly water-rich subduction mantle wedges (Roedder 1965; Schiano et al. 1995; McInnes et al. 2001).

It should be noted that some arc basalts are water poor (< 0.5 wt% H_2O; Sisson and Bronto 1998; Wade et al. 2008). In addition, the few subduction zone peridotites (xenoliths from Simcoe (WA, USA), Kamtchatka (Russia) and Papua-New Guinea; exhumed peridotites in China) analyzed so far for water do not have pyroxenes more water-rich than those from continental or oceanic peridotites (yellow symbols in Fig. 6; Peslier et al. 2002; Soustelle et al. 2010, 2013; Cao et al. 2015; Tollan et al. 2015; Wang et al. 2016). The only evidence for possible enrichment of mantle peridotite by water-rich subduction metasomatism is provided by xenoliths from the Colorado plateau in southwestern USA (Fig. 4A and B; Dixon et al. 2004; Li et al. 2008). Occasionally dry zones in the mantle wedge above subduction zones could result from (i) prior melting events that depleted the mantle wedge, (ii) upwelling of comparatively water-poor asthenospheric mantle, perhaps related to back-arc initiation or slab breaking occurrences, (iii) the low water fugacity of rare reduced slab fluids or melts, potentially generated via the melting of subducted organic matter, or by segregation of immiscible C-H-O fluids with low water fugacity (Karato 1986; Sisson and Bronto 1998; Wang et al. 2007; Song et al. 2009; Li 2017). Recent results from thermal modeling also suggest that the majority of mantle fore-arcs are not water-rich because most slabs are too cold to dehydrate rapidly enough over an arc life span (Abers et al. 2017).

Finally, despite the extensive melting that occurs in subduction zones, the mantle residue from this process is probably still relatively rich in water (i.e. it contains the same amount of water as the MORB source or upper asthenosphere; Hirth and Kohlstedt 2004). Therefore, water in the peridotitic mantle wedge likely contributes to the rehydration of the asthenosphere via downward passive entrainment above a subducted slab ("corner flow", Fig. 9B), in addition to the water-rich slab lithologies described in the next paragraph. This mantle flow is facilitated by the low viscosity (similar to that of the asthenosphere) of this relatively water-rich mantle (Billen and Gurnis 2001; Hirth and Kohlstedt 2004; Behr and Smith 2016).

4.2 Water Transport to the Deep Earth

Not all subducted water is returned to the surface in subduction-related volcanoes. Some water is entrained deeper, locked in slab minerals that do not dehydrate completely despite the increasing pressures, in part due to the colder temperature of the sinking slab compared to the surrounding mantle (Dixon et al. 2002; Rüpke et al. 2006; Ohtani 2015; Chang et al. 2017). Eclogites, metamorphic rocks of subduction origin, contain significant amounts of water locked in their minerals amounting to 460–3000 ppm wt H_2O in the bulk-rock (Katayama and Nakashima 2003; Katayama et al. 2006). At higher pressures, phases like hydrous Mg or Al silicates, or Al or Si oxides, have been synthesized in laboratory experiments. Such minerals could carry water to the transition zone (hydrous ringwoodite, akimotoite, stishovite, Phases A, B, C, D, E, X, EGG, δ-H, $(FeOOH)TiO_2$), the lower mantle, and all the way down to the core-mantle boundary (δ-AlOOH, phase H) (e.g., Panero et al. 2003; Bolfan-Casanova 2005; Komabayashi 2006; Lawrence and Wysession 2006; Sano et al. 2008; Ohira et al. 2014; Ohtani 2015; Nishihara and Matsukage 2016).

In particular, the main reason why the transition zone (Sect. 3.2.3) comprises water-rich zones might be that cold subducted slabs seem to stagnate as these depths (Fig. 9). Imaging by seismic tomography (van der Hilst et al. 1991; Fukao et al. 2009) shows these slabs, which entails that subducted water could accumulate locally in the transition zone via this process (e.g., Braunmiller et al. 2006; Suetsugu et al. 2006; Richards and Bercovici 2009; van Mierlo et al. 2013). Although water-rich minerals (hydrous wadsleyite and ringwoodite) have been invoked to explain seismic velocities within the transition zone and the seismic contrast seen at 410 km (Smyth and Frost 2002; van der Meijde et al. 2003; Smyth et al. 2004; Hirschmann et al. 2009), widespread presence of water-rich solid material is not supported by geophysical observations (Houser 2016). Another hypothesis is that water-rich melts are generated at the top and at the bottom of the transition zone by material flowing through the transition zone (Fig. 9A). In the framework of this hypothesis, partial melting could occur at the top of the transition zone, linked to the transformation of olivine into wadsleyite at cold slab temperatures or the stagnation of upwelling transition zone dense melts (Bercovici and Karato 2003; Sakamaki et al. 2006; Frost and Dolejš 2007). A melt layer at 410 km depth is consistent with both seismic and electrical conductivity data (Revenaugh and Sipkin 1994; Song et al. 2004; Yoshino et al. 2006). Downward flow of transition zone material to the lower mantle involves the transformation of hydrous ringwoodite to bridgmanite, which could generate water-rich melts at the top of the lower mantle and could explain the sudden decrease in seismic velocities at 660 km (Ghosh and Schmidt 2014; Schmandt et al. 2014).

4.3 Fluxes of Water Between Earth's Geological Reservoirs

The water cycle in the Earth can be constrained by combining the estimated amount of water in each reservoir (hydrosphere, continental crust, oceanic crust, upper mantle, transition zone, lower mantle and core, Fig. 9A, Table 1) with the estimated net flux at the main locations of water transfer in and out of each reservoir for a given period of time (e.g.,

A

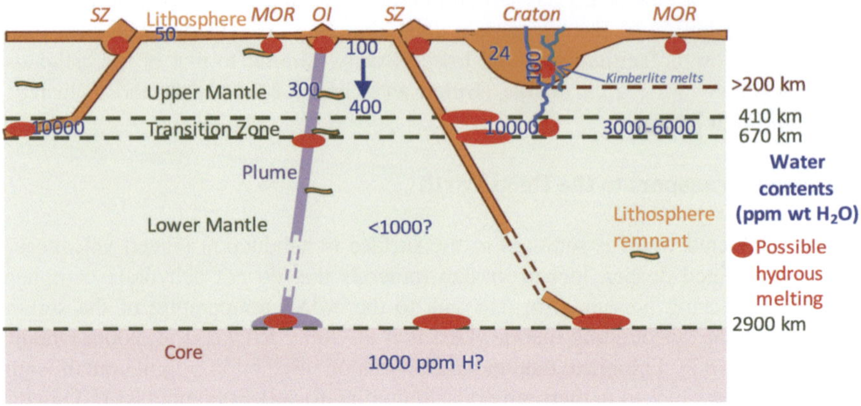

B

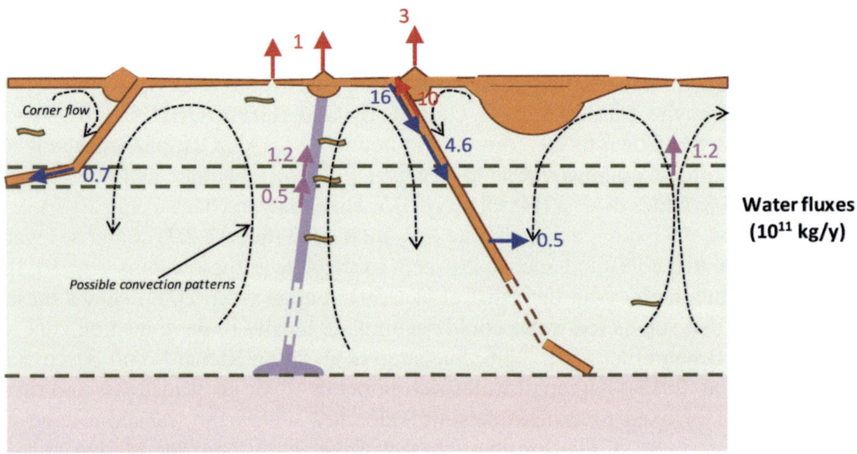

C

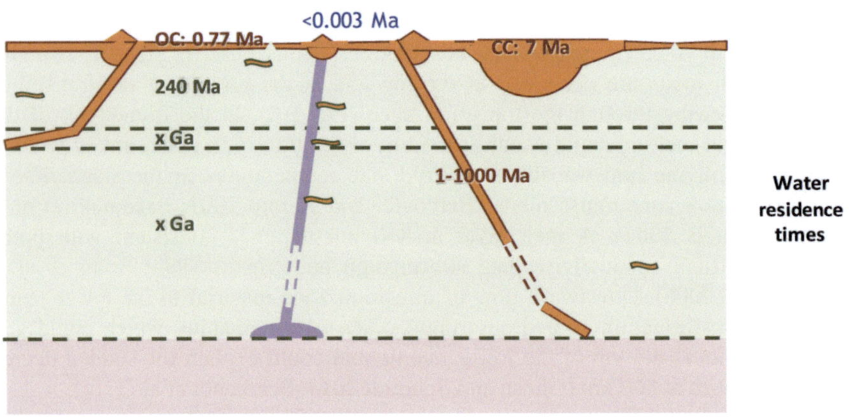

Fig. 9 Sketch of the key layers, tectonic settings and processes of a dynamic Earth interior in relation to water with **A**: average water concentrations (ppm weight H_2O) and zones where partial melting occurs (*red*), **B**: present water fluxes between reservoirs in 10^{11} kg/y (*solid colored arrows*) and schematic mantle convection patterns (*dotted black arrows*), and **C**: residence time of water for various reservoirs. MOR = mid oceanic ridge, OI = oceanic island, SZ = subduction zone, × Ga = several billion years, OC = oceanic crust, CC = continental crust. See text for references

Bodnar et al. 2013; Magni et al. 2014). Figure 9B shows averages of water fluxes in the Earth in 10^{11} kg/y from estimations of the last 25 years (Peacock 1990; Bebout 1996; Schmidt and Poli 1998; Wallmann 2001; Hilton et al. 2002; Bercovici and Karato 2003; Rüpke et al. 2004; Wallace 2005; Rüpke et al. 2006; Hacker 2008; van Keken et al. 2011; Faccenda et al. 2012; Parai and Mukhopadhyay 2012; Bodnar et al. 2013; Faccenda 2014; Magni et al. 2014). Details can be found in online supplementary Table 2.

The primary regions of water exchange between Earth's interior and hydrosphere are subduction zones (Fig. 9B). It is estimated that ~25% of the water entering subduction zones reaches the transition zone and ~3% reaches the lower mantle (Bodnar et al. 2013). Compared to subduction volcanism, mid-oceanic ridge and oceanic island volcanisms contribute a minor percentage of the total water loss from the Earth's interior to the hydrosphere. Yet, oceanic basaltic volcanism constitutes the main exhaust of water from the deep mantle to the surface via mantle convection and plume upwelling processes. The net balance of input versus output of water results in addition of water to the mantle over time. Assuming that plate tectonics has operated over the last 3 Ga (Farquhar et al. 2002; Shirey and Richardson 2011; Debaille et al. 2013), a first order calculation results in about half the mass of the Earth's present oceans being added to the mantle over that time period. It should be emphasized, however, that the uncertainty on the estimated fluxes presented here is high (~50%).

4.4 Residence Time and Origin of Distribution of Water in the Earth

The highly heterogeneous distribution of water in the Earth is a remarkable conclusion. Particularly important is the inferred layered structure of water content in the mantle. Although diffusion of water (i.e., hydrogen) in silicates is fast compared to that of other elements in silicates, diffusion is ineffective to modify the water content at a large scale: for typical asthenospheric conditions, diffusion of H over a distance of ~10 km takes ~1 Ga (Karato 2007; Peslier and Bizimis 2015). However, because of mantle convection, water contents should be homogeneous among regions of more than ~100 km size, not considering partial melting and resultant melt-solid segregation.

Heterogeneity in the water concentrations within the terrestrial reservoirs was first established during the early planetary differentiation processes, i.e. the formation of the metal core and the first silicate reservoirs. The processes include magma ocean degassing, as well as silicate differentiation in the magma ocean and water extraction from the mantle via crustal building over time (e.g., McKenzie 1985; Rudnick 1995; Hirth and Kohlstedt 1996; Hirschmann 2006; Labrosse et al. 2007; Elkins-Tanton 2008). Evidence for a water-rich reservoir in the Archean mantle comes from the water analyses of melt inclusions in 2.7 Ga old komatiites, which are primitive magmas that erupted primarily in the Archean (Sobolev et al. 2016). Heterogeneity is sustained by (i) recycling of water-bearing material via subduction processes, (ii) partial melting via solid material vertical transport, with resulting melt ascent or sinking depending on density, and (iii) inefficient re-homogenization by convection processes over Earth's history. Further evidence for this inefficient mixing stems from H isotopes. Melt inclusions of primitive basalts have low D/H isotopic ratios that were inherited from their source, possibly a primordial reservoir deep in the mantle that may still carry solar nebula H isotopic ratios (see Sect. 5.2.1; Hallis et al. 2015). Basalt glasses with OIB-like geochemical signatures also carry H isotopic signatures consistent with preservation of subduction-derived H, signifying that water in slabs was not diluted over time in the ambient mantle but stayed isolated during the recycling process over Ma to 1 Ga (Shaw et al. 2012).

Residence times for water shown in Fig. 9C are taken from Bodnar et al. (2013) and calculated using the amount of water in each reservoir and the estimated fluxes between

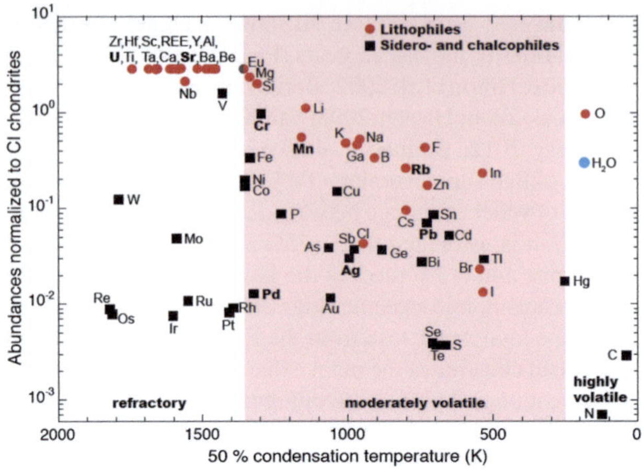

Fig. 10 The elemental abundances in the Earth's mantle normalized relative to CI chondrites as a function of the 50% condensation temperature (T_C) (from Lodders 2003; Palme and O'Neill 2014b). At this temperature, 50% of an element is in solid form, during condensation from a gas of solar system composition at low pressure conditions and under the assumption of complete chemical equilibrium. Lithophile elements define a distinct trend that correlates with condensation temperatures, while siderophile and chalcophile elements are depleted relative to this trend due to sequestration into the Fe-Ni core

them. Shortest residence times are in the hydrosphere (< 3000 years) and lithosphere (0.77–7 Ma), and water could stay locked in the transition zone and lower mantle for billions of years. If we assume that plate tectonics started at least 3 Ga ago (Farquhar et al. 2002; Shirey and Richardson 2011; Debaille et al. 2013), a H atom may have travelled from the oceans to the deeper mantle and back 3 times since then. You could be drinking water from the mantle right now!

5 Acquisition of the Earth's Water

5.1 Timing of Water Delivery to Earth

The total water budget of the Earth today (Table 1) reflects the integrated history of a long evolution starting with the formation of the Earth and the water present in the building blocks, followed by processes that modified its water content. An example for the latter is magma ocean degassing (Chap. 12) that resulted in an atmosphere, which in turn experienced losses through impacts, thermal processing and interaction with the radiation from the active young Sun (Chaps. 13 and 14). Moreover, water may also have been acquired by the Earth through a later delivery by water-rich asteroids and comets after the Earth had already reached > 99% of its current size.

To assess the potential presence of water in the building blocks of the Earth, we first evaluate the general volatile depletion trend of the Earth. In this context, the cosmochemical classification of elements becomes important (e.g., Lodders 2003; Palme and O'Neill 2014a). This classification is based on the condensation sequence of the chemical elements from a gas with solar composition at pressures of 10^{-4} bar (Fig. 10). The first elements to condensate when the solar gas cools are refractory elements (half mass condensation temperatures (T_c) = 1850–1355 K, Fig. 10). They are followed by the so-called main component elements (e.g., Si, Fe and Mg; $T_c = 1355$–1250 K, not highlighted in Fig. 10), then the moderately volatile elements ($T_c = 1250$–250 K), and, at the end of the sequence, the highly volatile elements ($T_c < 250$ K, e.g., H), which condense after the nebular gas had cooled to below 250 K. Of particular interest for our purpose are the moderately volatile elements, which provide powerful constraints on the presence of volatiles. Subsequently, we establish the link to the highly volatile compounds and in the next section (Sect. 5.2), we summarize attempts to use isotopic constraints of the highly volatile elements to assess the origin of Earth's water.

5.1.1 The Formation of the Earth

The birth of our solar system started with the gravitational collapse of an interstellar molecular cloud that contained gas and dust from a variety of stellar sources. The cloud collapsed and most of the mass accumulated in the center to form the young Sun, surrounded by a protoplanetary disk. Within this disk, dusty materials collided and accreted into small planetary bodies (planetesimals, planetary embryos), most of which further accumulated to build the terrestrial planets including the Earth (Chaps. 2 and 10).

The combination of heat released from radioactive nuclides and impacts on the growing proto-Earth likely triggered the formation of a deep magma ocean or potentially even several generations of magma oceans (Chap. 12). In these, metal segregation and sinking occurred resulting in core formation (e.g., Wood et al. 2008; Rubie et al. 2011). Dynamic modeling indicates that the last giant impact at the end of terrestrial accretion created the Moon (e.g., Benz et al. 1986; Canup 2012). This last potentially Mars-sized impactor, and the subsequent magma ocean, most likely concluded the terrestrial core segregation, and it also could have brought volatile elements to the Earth (Halliday 2004; Schönbächler et al. 2010). After this event, only a small amount of material (< 1%, Chou 1978; Walker 2009) has been accreted to the Earth until today. This late material was added to the mantle only and did not experience metal segregation to the core. This latest episode of extraterrestrial material addition is referred to as "late veneer" (e.g., Wang and Becker 2013), "late accretion" (e.g., Walker 2009), or terminal bombardment (Willbold et al. 2011). In this work, we use the term "late veneer". Its existence was originally proposed based on the elevated highly siderophile element (HSE) abundances in chondritic proportions in the Earth's mantle (Chou 1978), which could not be explained by core formation in a magma ocean that would have almost exclusively sequestered these elements into the core.

5.1.2 Volatile-Poor Early Accretion

The Sun, which comprises > 99% of the mass in the solar system, and CI chondrites display nearly identical elemental compositions, with a few exceptions (e.g., Li, Be, the highly volatile elements H, O, N, C and the noble gases). This implies that CI chondrites reflect the average solar system composition for most elements and are therefore usually taken as a reference (Chap. 3). The Earth and other terrestrial planets are depleted in moderately and highly volatile elements relative to CI chondrites and the Sun (Fig. 10). Some of the depleted elements (e.g., Ag, Mn, Rb and Pb) are part of a radioactive decay system such as the short-lived decay systems ^{53}Mn-^{53}Cr (half-life of 3.5 ± 0.4 Ma) and ^{107}Pd-^{107}Ag (half-life of 6.5 ± 0.3 Ma), the long-lived chronometer ^{87}Rb-^{87}Sr (half-life of 49 Ga), and the U-Pb double decay system, where ^{238}U decays to ^{206}Pb with a half-life of 4.47 Ga and ^{235}U to ^{207}Pb with a half-life of 710 Ma. Manganese, Ag, Rb and Pb possess lower condensation temperatures than Cr, Pd, Sr and U (Lodders 2003, Fig. 10), and for this reason they are fractionated from each other by volatile depletion events. This fractionation can be dated by the chronometers. It is noteworthy that Pd-Ag, Mn-Cr and U-Pb also fractionate during core formation (Ballhaus et al. 2013; Kiseeva and Wood 2013; Rubie et al. 2015), because the involved elements (except U) partition to different degrees into the core. This needs to be taken into account for the interpretation.

The short-lived decay systems provide very powerful constraints because their signatures are established within the first 10–30 Ma of our solar system and are not modified by later radiogenic ingrowth. It can be shown that the Cr isotope composition of the bulk silicate Earth (= crust and mantle) is relatively unradiogenic, i.e. contains relatively small amounts of radiogenic ^{53}Cr from the decay of ^{53}Mn, compared to the volatile-rich CI chondrites (e.g., Trinquier et al. 2008; Qin et al. 2010). This implies that the volatile ^{53}Mn was lost early, presumably within the first 2 Ma after the start of the solar system dated by oldest known solar system materials—calcium-aluminum-rich inclusion (CAI) (Trinquier et al. 2008; Qin et al. 2010; Palme and O'Neill 2014a). A longer presence of the volatile ^{53}Mn would have generated a more radiogenic ^{53}Cr-rich composition. The Mn-Cr data was incorporated into different accretion models, i.e. simple two-stage models, but also models that take core formation into account and use exponential growth curves for the Earth (Trinquier et al. 2008; Schönbächler et al. 2010; Carlson et al. 2015). These models agree that a large amount of the Earth's building blocks were significantly depleted in the moderately volatile element Mn. Evidence from the long-lived Rb-Sr system also argues for early volatile depletion (Halliday and Porcelli 2001; Schönbächler et al. 2010; Nebel et al. 2011; Hans et al. 2013; Carlson et al. 2015), despite some uncertainties regarding the initial Sr isotope composition of the solar system (e.g., Hans et al. 2013).

Similarly, evidence from the U-Pb system is also in agreement with an early depletion (Wood and Halliday 2010; Nebel et al. 2011; Albarède et al. 2013; Ballhaus et al. 2013). However, the interpretation of the U-Pb system has long been a matter of debate. There are some uncertainties regarding the estimate of the Pb isotope composition of bulk silicate Earth, partly due to the high concentration of Pb in the crust (Hofmann 2001). Other works illustrated the potential sensitivity of volatile elements such as Pb to the late veneer and proposed that all Pb was delivered by the late veneer (Albarède et al. 2013; Ballhaus et al. 2013). Most studies agree, however, that the Earth started with a high U/Pb ratio (volatile depleted) and that this ratio was further enhanced around 100–150 Ma after the start of the solar system formation, either by Pb loss through volatilization (e.g., during the giant impact, Connelly and Bizzarro 2016) or partitioning into the core (Wood and Halliday 2010).

In summary, radioactive decay systems, which involve moderately volatile elements, provide strong evidence that the Earth accreted on average from volatile depleted materials. Since moderately volatile elements display higher condensation temperatures than the highly volatile elements and compounds (Fig. 10), it is likely that the highly volatiles (e.g., H) were even more depleted in the early building blocks of the Earth. Selective adsorption on the surface of olivine in the protoplanetary disk could be a potential mechanism to incorporate water into the terrestrial building blocks without significant amounts of moderately volatile elements (e.g., Vattuone et al. 2013). However, the amount of water (H_2O) that can be tolerated in the early building blocks of the Earth is limited. This is because models of metal-silicate partitioning in a deep magma ocean, which are based on high-pressure experiments, generally require reduced material in the early stages of terrestrial accretion to successfully match the metal-silicate partitioning behavior of the elements in the Earth (e.g., Wood et al. 2008; Rubie et al. 2011; Rubie et al. 2015; Chap. 11), although a more oxidized scenario has also been suggested (Siebert et al. 2013). Hydrogen instead of H_2O could have been present under early reducing conditions (Hirschmann et al. 2012;

Sharp et al. 2013). While the proto-Earth is small, and because H is easily lost from cooling magmas (Sharp et al. 2013), a significant proportion of H was presumably lost to space. Given that there is ample evidence for the addition of volatile-rich/bearing material later (see below), this early signal is likely swamped by the later addition.

Another limiting factor for water retention for the early proto-Earth is the small size of the proto-Earth and the impacting bodies, which does not allow gravitational retention of an atmosphere. To a first order, the proto-Earth must be massive enough such that its Bondi radius is larger than its physical radius to retain an atmosphere. The Bondi radius is the radius at which the escape velocity from the planets gravitational potential is equal to the thermal velocity of the gas molecules. If the average thermal velocity of the gas molecules is larger than the escape velocity of the planet, it cannot hold on to any primordial atmosphere. For the conditions relevant for the Earth (Hayashi 1981), the proto-Earth at 1 AU needs to be at least 0.002 of its current mass to retain some H- and He-bearing envelope from the protoplanetary disk (Inamdar and Schlichting 2015). At the same time, the accreted material is heated by radioactive isotopes and accretion, which will lead to degassing of the interior and volatile loss. The meteorite record shows that small differentiated bodies (e.g., Vesta or the Angrite parent body) are generally volatile-poor (e.g., Hans et al. 2013), but they are not completely devoid of water (Sarafian et al. 2013, 2017; Scully et al. 2015).

In addition, the concept of the snowline further supports water-poor building blocks for the Earth. The snowline marks the boundary between condensation and vaporization of water in the protoplanetary disk. In the framework of the snowline model, rocky planetesimals form inside and icy planetesimals outside the snowline, where ice particles can exist and accrete. The temperature in the protoplanetary disk at 1 AU distance from the Sun is estimated to 270 K by Hayashi (1981). In a protoplanetary disk of solar composition, water ice forms below $\sim$180 K (Lodders 2003). Therefore, the calculated warm temperature at 1 AU would not allow condensation of water ice. Water remained in the gas phase and could not be accreted as ice. This resulted in the formation of water-poor planetesimals in the 1 AU region, which is the main "feeding zone" for the early accreted material of Earth (e.g., O'Brien et al. 2006, 2014; Raymond et al. 2006).

5.1.3 How Late Did the Volatiles Arrive?

While there is powerful isotopic evidence that the Earth's accretion started volatile poor as discussed in Sect. 5.1.2 above, the timing of volatile addition to the Earth is more uncertain. Small amounts of H may have been accreted early (before the Earth reached $\sim$60–80% of its current size), but larger amounts of moderately volatile elements can be excluded. Hence, a later addition for these elements is required and they are likely coupled with significant amounts of highly volatile elements as, for example, provided by carbonaceous chondrites (Alexander 2017).

The combined modeling of several radionuclide systems (Pd-Ag, Mn-Cr, Rb-Sr and Hf-W) shows that the Earth most likely accreted heterogeneously; first, materials depleted in moderately volatile elements accreted, while the later accretion history was dominated by volatile-rich materials (Schönbächler et al. 2010). Their preferred model encompasses that the Earth first grew to 86% of its present mass from volatile-depleted material, with similar or slightly lower volatile content than the Earth today. Considering the uncertainties, the model allows for volatile-depleted accretion of the first 60–70% of the Earth's mass, before the volatile—rich material arrives (Fig. 11). This result is in good agreement with several other recent studies that employed very different approaches and techniques (O'Brien et al. 2006; Wood et al. 2008; Rubie et al. 2011, 2015; Marty 2012;

Hirschmann 2016; Fischer-Gödde and Kleine 2017). While Schönbächler et al. (2010) based their conclusions on the combined modeling of radioactive decay systems, other studies (Wood et al. 2008; Rubie et al. 2011, 2015) used high-pressure experiments to model metal-silicate partitioning in a deep magma ocean. Others (Marty 2012; Hirschmann 2016) modeled C/H, C/N and C/S ratios or their element concentrations (water, C, N and noble gases) in the bulk silicate Earth and concluded that some of the volatile budget (in particular some H; Hirschmann 2016) were acquired prior to the late veneer. The nucleosynthetic Ru isotope composition of the Earth compared to those of chondrites supports this conclusion and even indicates that the late veneer could have been water poor and similar to enstatite chondrites in composition (Fischer-Gödde and Kleine 2017). O'Brien et al. (2006) employed numerical simulations to determine the final stages of accretion of the terrestrial planets and concluded that volatile-rich material was added late during the Earth's accretion history, but that a part of it was incorporated when core formation was still proceeding. In this so-called "classical" model (O'Brien et al. 2006; Raymond et al. 2006; Chap. 10), the volatile-rich material is delivered by bodies from the primordial asteroid belt beyond $\sim$2.5 AU. In the more recently proposed "Grand Tack" model, which takes gas giant planet migration into account, some volatile-rich material is also delivered during ongoing core formation—but in this model the water-rich bodies originate from between and beyond the present orbits of the gas giant planets (O'Brien et al. 2014; Chap. 10).

The model of substantial volatile delivery starting during core formation stands in contrast to other studies (Albarède 2009; Albarède et al. 2013; Ballhaus et al. 2013), which argue that the Earth's volatiles were exclusively delivered by the late veneer. Although it is possible that the late veneer delivered volatile-rich material (e.g., similar to CM chondrites; Wang and Becker 2013), delivering *all* the volatiles during the late veneer would lead to significantly higher HSE abundances than observed in the bulk silicate Earth. Moreover, this scenario would generate a strong radiogenic signal in the Ag isotope composition of the Earth's mantle due to the decay of the refractory isotope ^{107}Pd, which is very abundant relative to Ag in volatile depleted material. Such a radiogenic signature is not observed (Schönbächler et al. 2010; Schönbächler and Nimmo 2011). Mixing of previously accreted materials with volatile-rich material during the late veneer would modify and dilute such a strong radiogenic signal. In addition, admixture of more than 10% by mass of volatile-rich material during the late veneer event would be required (Schönbächler and Nimmo 2011). This is not realistic because of the relatively low HSE content in the Earth's mantle (e.g., Wood et al. 2010). Special scenarios invoked to explain the missing high HSE abundances (e.g., late sulphide or metal segregation; Albarède et al. 2013; Ballhaus et al. 2013) fail to relax the Ag isotope argument.

5.1.4 Summary

In summary, geo- and cosmochemical evidence indicates that the last third of the Earth's growth involved volatile-rich (water containing) material (Fig. 11). The amount of water added before the late veneer is, however, uncertain. The Earth's total water budget is estimated at 2–18 Earth ocean masses (Fig. 8), but large uncertainties on the core and lower-mantle water contents extend the range from 0.2 to 90 (Table 1). Because the inferred amount of water in Earth is small compared to that of carbonaceous chondrites or planetary bodies from the outer part of the solar system, it is conceivable that some fraction of the Earth's water might have been acquired very early, before water-rich materials were accreted later from regions beyond the snowline. Incorporation of

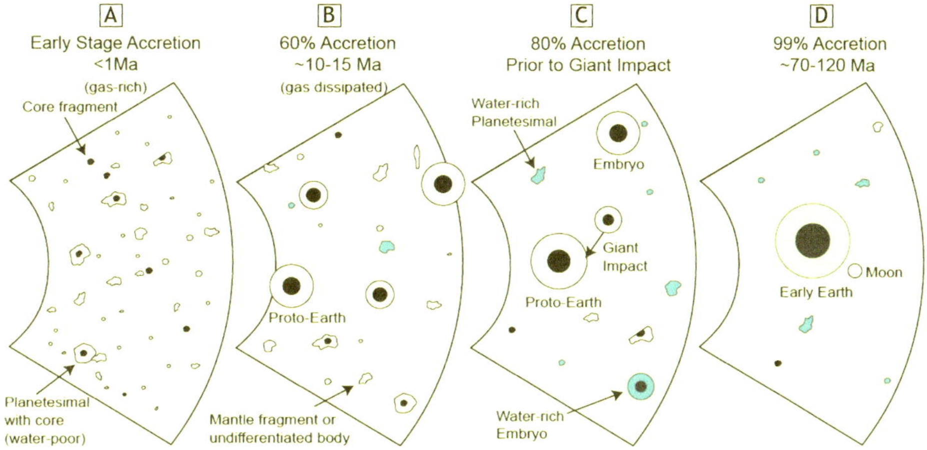

Fig. 11 Schematic evolution of water delivery to the Earth. The view looks down onto the accreting disk of the solar system and shows snapshots of the section of the disk where the Earth formed (around 1 AU) at different times. It illustrates the planetary materials that likely were near the Earth's orbit, i.e. the material available at each time period to be accreted to the Earth (not everything will be accreted). The amount of water-rich bodies in the "feeding zone" of the Earth increases from *left* to *right*. The Earth started relatively dry, i.e. water-poor, and received the first significant volatile delivery after it reached 60–80% of its current size. Later accretion is dominated by volatile-rich bodies (*blue*). The accretion percentage is given relative to today's mass of the Earth. *White bodies*: volatile-poor mantle fragments or undifferentiated material originally from the 1 AU region. *Blue bodies*: water-rich mantle fragments or undifferentiated material originally from (i) the primordial asteroid belt (e.g., carbonaceous chondrites for classical model; Raymond et al. 2006) or (ii) from between and beyond the giant planets (Grand Tack model; O'Brien et al. 2014). In *black*: core material. Figure is based on a sketch in Nimmo and Kleine (2015)

1% CI chondrite material (with > 10% H_2O; e.g., King et al. 2015) can already deliver ~4.5 times the mass of the present Earth's oceans. However, up to 30% of volatile-rich material may have been added before the late veneer (Schönbächler et al. 2010; Rubie et al. 2011), which implies that the water mass delivered to the Earth—during ongoing core formation—was likely orders of magnitude higher than today's water budget. This requires significant water loss from the terrestrial atmosphere during impact-driven accretion, the magma ocean state, and/or during the late veneer (Chaps. 12–14), to match the current Earth's water budget. In addition, water was also likely delivered through the late veneer, which is discussed in the next section.

5.2 Origins of the Earth's Water

As discussed above (Sects. 3 and 5), there are two first-order conclusions from the studies on the water distribution in the Earth.

(i) The total amount of water in the Earth is small (0.06 wt% to ~3.9 wt% H_2O or ~3 to ~18 ocean masses; Fig. 8), compared to that of the most primitive meteorites, carbonaceous chondrites, which can contain up to 10–20 wt% H_2O (e.g., Alexander et al. 2012; King et al. 2015; Stephant et al. 2017). Hence, moderately water-rich materials, such as CI chondrites are sufficient to supply the Earth's total water budget, including the ocean and interior water. Other materials such as comets are even more water-rich and thus less such material would be required to explain the Earth's water content.

The amount of water in other—more water-poor—materials that were potentially accreted by the Earth is not well-constrained. Although 0.1 to 1 wt% H_2O were estimated earlier (Jarosewich 1990; Robert 2002; Alexander et al. 2012; Barnes et al. 2016), ordinary chondrites are still likely water-poor. This is based on the fact that their apatites appear generally devoid of water, except one LL chondrite that contains the driest apatite ever measured in solar system materials (Jones et al. 2014, 2016). In agreement with this observation, ordinary chondrites exhibit an overall strong depletion trend in moderately volatile elements (Schönbächler et al. 2008; Schaefer and Fegley 2010). Enstatite chondrites may have < 0.5 wt% H_2O (Robert et al. 1987; Barnes et al. 2016). Moreover, the differentiated parent bodies of angrites and eucrites contain some water (Sarafian et al. 2013, 2017; Barrett et al. 2016), but still 2–3 orders of magnitude less than CI chondrites.

(ii) Much of the initial mass of the Earth, ranging from at least 60% to up to almost 90% of present Earth mass (based on the moderately volatile elements, Sect. 5.1.), accreted from volatile-depleted precursor material (e.g., Albarède 2009; Schönbächler et al. 2010; Rubie et al. 2011). It has been suggested that only the last 1% of accreting mass, the "late veneer" (Sect. 5.1), provided the bulk of the terrestrial water. The initial material that formed the Earth (and the other terrestrial planets) may have lost some of its volatiles by vaporization on impact. However, the main reason for the volatile depletion of the precursor material is that most volatile elements were likely not present in the first place in the earlier building blocks (Albarède 2009; Schönbächler et al. 2010; Rubie et al. 2015, and references therein).

The origin of the water and the nature of its source(s), whether accreted early or late in Earth's evolution, is not entirely determined. The highly volatile elements H, C, N and the noble gases can potentially provide new constraints on the origin of Earth's volatiles. Particularly helpful are their isotopes, as the Earth and many of the reservoirs that are potential sources, show distinct isotopic signatures. In the following, we aim to briefly summarize these efforts with the main focus on *isotopic* characterization. More comprehensive models and considerations beyond the major realm of this chapter can be found in recent, dedicated reviews (e.g., Robert 2001; Drake 2005; Marty and Yokochi 2006; Alexander et al. 2012; Marty 2012; Halliday 2013; Dauphas and Morbidelli 2014; Genda 2016; Marty et al. 2016; Hallis 2017; Chaps. 5, 6, 11–14). First, we will summarize the D/H and $^{15}N/^{14}N$ ratios detected in various inner solar system bodies that then can be compared with terrestrial values. This is followed by the discussion of noble gas isotopes that provide further constraints on the origin of Earth's water.

5.2.1 Constraints from Hydrogen Isotopes

The D/H ratio might be the most diagnostic isotope ratio to assess the origin of water, due to its large observed range in solar system reservoirs (Fig. 12; e.g., Alexander et al. 2012; Bockelée-Morvan et al. 2015; Alexander 2017; Hallis 2017). We express here the D/H ratio by the delta notation, which describes the variation of the measured ratio relative to a standard in permil, i.e. the D/H ratio of terrestrial oceanic water ("SMOW"). Most reservoirs including the Earth ($\delta D \sim 0$ to $< -220‰$ in its interior, Table 1), are enriched—to different degrees—relative to the initial value of the solar nebula ($\delta D = -872 \pm 22‰$; Geiss and Gloeckler 2003). Nebular H, with the lowest D/H ratio in the solar system, is sampled essentially unmodified by the giant planets Jupiter and Saturn (e.g., Alexander et al. 2012; Bockelée-Morvan et al. 2015; Chaps. 3, 6 and 8).

Comets Cometary isotope compositions were detected by remote sensing (e.g., Bockelée-Morvan et al. 2015), in-situ study by spacecraft (Balsiger et al. 1995; Eberhardt et al. 1995; Altwegg et al. 2015) and in the laboratory—by examining returned samples (comet Wild 2 dust, McKeegan et al. 2006; Sandford et al. 2006) or bona-fide cometary interplanetary dust particles (IDPs, e.g., dust from comet Grigg-Skjellerup; Busemann et al. 2009;

Bradley et al. 2014 and references therein). Comets display the highest D/H ratios, in H_2O, and C- and N-bearing organic molecules (e.g., Altwegg et al. 2015; Bockelée-Morvan et al. 2015; Chaps. 3 and 6). These high D/H ratios are usually assumed to be inherited from D enrichments produced in the cold interstellar medium and the slightly warmer solar nebula, in low-temperature ion-derived reactions (e.g., Aikawa and Herbst 1999; Cleeves et al. 2014; Bockelée-Morvan et al. 2015). Incomplete mixing of the nebula, heliocentric temperature gradients, and migrating planets affecting the snowline, cause the H isotope variations between inner and outer solar system, and the distinct compositions of all its planetary bodies (Fig. 12; Brownlee 2014; Bockelée-Morvan et al. 2015; Chaps. 2 and 8).

Earth The Earth ocean has per definition $\delta D = 0‰$. The silicate Earth reservoirs (Table 1, Fig. 12) display lower δD values, e.g., for the convecting mantle δD of $-60 \pm 5‰$ (Clog et al. 2013), δD of -60 to $-130‰$ for the oceanic and continental mantle lithosphere (Deloule et al. 1991; Bell and Ihinger 2000), $\delta D < -126‰$ for subducting material (Shaw et al. 2012) and a bulk Earth δD of $-43 \pm 19‰$ (Lécuyer et al. 1998).

Moon The Moon is, in contrast to early assumptions, not entirely dry and may contain significant amounts of water (the bulk silicate Moon may have 133–292 ppm H_2O; Saal et al. 2008; Hauri et al. 2015). This includes water in volcanic glass, mare basalt melt inclusions and minerals such as apatite, olivine and plagioclase. From these, up to a few 100 ppm H_2O is calculated for the mantle source rocks, with a $\delta D \sim -100‰$ (Saal et al. 2013; Füri et al. 2014; Hui et al. 2017). Similarly, most lunar highland rocks show Earth-like δD values (δD between -281 and $-27‰$, Barnes et al. 2014). A much higher ratio compared to the Earth ($\delta D = 310 \pm 110‰$) was observed in highland plagioclase, which has been interpreted as being the result of degassing of the lunar magma ocean (Hui et al. 2017).

However, the bulk interior of the Moon could also contain only little water (< 10 ppm H_2O), with the D-rich water being supplied to the magma ocean late and more superficially by impact (Elkins-Tanton and Grove 2011). In any case, the range of the various Earth-like isotopic H compositions observed in the Moon is suggestive of a provenance of the volatiles similar to those in the Earth (Saal et al. 2013).

Mars In order to determine the mantle H isotopic composition for Mars, the D/H was examined in various phases from Martian meteorites (e.g., apatite, amphibole, melt inclusions and impact melts, Chap. 5). Tucker et al. (2015) measured a range between -300 to $+100‰$ in various Martian meteorites. An upper limit for the D/H in the Martian mantle has been given as $\delta D < 275‰$ (Usui et al. 2012; Hallis et al. 2012). Much higher values ($\delta D \sim$ 900–5500‰) have been detected elsewhere in Martian meteorites (e.g., Bogard et al. 2001; Hu et al. 2014; Usui et al. 2015). These ratios represent samples affected by atmospheric or surficial water components and fractionating losses.

Achondrites In addition to contributions from primitive chondrites, differentiated (achondritic) planetesimals could also have brought volatiles to Earth (Chaps. 11 and 12), because even this material, which is severely depleted in volatile elements relative to CI chondrites, may contain significant indigenous water (Chaps. 11 and 12). The D/H isotope examinations of achondrites are still rare. Recent studies of the D/H composition of the water in eucritic phosphates show similarities to that in Earth and carbonaceous chondrites ($\delta D \sim -162 \pm 127‰$, Sarafian et al. 2014; $\delta D \sim -34 \pm 67‰$, Barrett et al. 2016; $\delta D \sim -210$–$50‰$, Guan et al. 2016). Eucrites, supposed to originate from Vesta, formed early (e.g., 4561 ± 13 Ma; Hopkins et al. 2015), which implies that water similar to that in

Earth, the Moon and Vesta could have been incorporated into all inner solar system objects, from a similar source such as, e.g., carbonaceous chondrites (see below). On the other hand, phosphates in angrites, another particularly old achondrite class (4564.42±0.12 Ma, Amelin 2008), appear to contain water with significant enrichments in D (δD > 500‰, > 400 ppm H_2O, Sarafian et al. 2017). The same authors also report 20–60 ppm of water in olivine, which may suggest, albeit deduced with some reservation, that the angrites' parent melt was water-rich and contained > 1 wt% water (Sarafian et al. 2017). The high D/H ratio could be the result of a D-rich, potentially cometary source or the result of strong fractionation during degassing (Sarafian et al. 2017). Similar to the work discussed in the preceding paragraph, Stephant et al. (2016) found low D/H ratios in phosphate (δD = −173 ± 72‰), but also in pyroxene (δD = −241 ± 64‰) of eucrites. Hydrogen in pyroxenes may better represent the parent melt composition than phosphates. The pyroxenes show lower H_2O content and D/H ratios than phosphates. The ratios are similar to an upper limit recently reported for the terrestrial deep mantle based on melt inclusions in olivine phenocrysts (δD < −218‰, Hallis et al. 2015). The similarity in low D/H ratios of the terrestrial deep mantle and of pyroxenes in eucrites was interpreted as potential evidence for the presence of a primordial, nebula-inherited (low D, see Sect. 5.2.1) component present in both Vesta and Earth.

A Carbonaceous Chondrite Origin of Earth's Water The general similarity of the Earth's D/H ratios measured in ocean water and silicate Earth reservoirs with ratios reported for the Moon, Mars, Vesta and, perhaps, the angrite parent body (Fig. 12), strongly implies a common source for the delivery of the water. This was suggested in numerous publications (e.g., Marty and Yokochi 2006; Saal et al. 2008). The most likely source are the carbonaceous chondrites, and more precisely CI chondrites (Alexander et al. 2012; Chap. 3). Bulk carbonaceous chondrites show a range of D/H ratios (δD between −230 and +766‰; Alexander et al. 2012). However, much of the D enrichments in primitive chondrites is carried by organic matter (bulk OM δD reaches up to 3500‰, whereas small organic aggregates show ratios up to 19000‰; e.g., Busemann et al. 2006; Alexander et al. 2007). The D/H ratios of chondritic water can be deduced by comparing measurements on bulk meteorite and organic matter. The calculated and extrapolated D/H ratios of the water bound in the hydrous phases of carbonaceous chondrites show a characteristic variability between the various carbonaceous chondrite groups (δD ~ −587 to +98‰; Alexander et al. 2012). This can be used, together with other isotope ratios, particularly N (see below), to further constrain the most probable chondritic provider of Earth's water. Isotopic similarities with the Earth for many elements may argue for a close link of enstatite chondrites to Earth and, hence, the origin of the terrestrial water through enstatite chondrites (Javoy et al. 2010). A recent publication supports a significant contribution to the Earth from enstatite chondrites (Dauphas 2017). However, the Si isotopic system (Fitoussi and Bourdon 2012), as well as other geochemical evidence (Wang and Becker 2013), do not support this.

In principle, the rough isotopic match of the terrestrial δD and δD in the other bodies of the inner solar system, with carbonaceous chondrite δD could be fortuitous. Mixing of low δD primordial nebula water with other sources such as cometary δD contributions could also result in chondritic δD values. The δD values of cometary water measured in 8 out of 12 comets are so far consistent with a high value of δD ≥ 1050‰ (Bockelée-Morvan et al. 2015). Vice versa, the δD values in comet Hartley 2 (δD = 34 ± 154‰, Hartogh et al. 2011) and three further comets (only upper limits given, Bockelée-Morvan et al. 2015) are consistent with the D/H of Earth's water. However, the cometary organic matter would still be more D-rich than the water ice and would deliver too much D to Earth (Alexander et al. 2012).

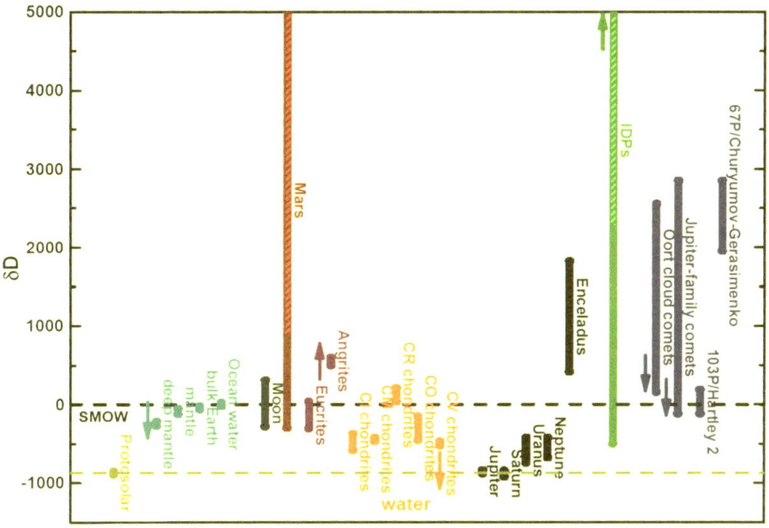

Fig. 12 Hydrogen isotope ratios, i.e. D/H ratios, in Earth and solar system reservoirs, given in delta notation (δD). The delta notation for H gives the variation of the measured ratio relative to a standard in permil. Standards are terrestrial mean ocean water for H ("SMOW"). Most data represent the D/H ratio as measured in water or hydroxyl groups in minerals. The protosolar value is observed essentially unaltered in the atmospheres of Jupiter and Saturn. Elevated ratios can be the result of D-enrichments during ion–molecule reactions in cold environments and can be found in pristine water and organic phases in comets, the outer icy planets and IDPs. The water of only a few comets including Hartley 2 may match the D/H of the Earth's water (further comets show upper limits consistent with SMOW). The high D/H in IDPs is mostly carried in organic matter in so-called "hotspots", although IDPs with "cometary" D/H, residing most likely in hydrous phases, were also observed (Mukhopadhyay and Nittler 2003). Organic matter is also responsible for most of the D enrichments in bulk carbonaceous chondrites (not shown). Here only D/H ratios of non-organic phases of carbonaceous chondrites (CI, CM, CR, CO, CV) are shown (labelled "water"). The high D/H ratios of many Martian samples and reservoirs are the result of incorporation of Martian surface material with atmospheric signatures because fractionation of H isotopes occurs during H escape from the planet's atmosphere with preferential loss of H over D (Watson et al. 1994). *Arrows* illustrate/symbolize that a precise upper and lower limits, respectively, for a given range is not known. References are given in the main text, apart from those for the giant planets and Enceladus (Feuchtgruber et al. 1999; Lellouch et al. 2001; Waite et al. 2009)

Most importantly, the N isotopic composition of comets probably excludes a significant cometary input (see below). In this context, it is also important to keep in mind that the volatile-depleted material initially accreted to the Earth (pre-late-veneer), or some of the volatile-rich (up to 30%, see above) Earth-forming material, will contain some primordial water. Another mean to compensate the heavy H isotopic composition of cometary water to achieve terrestrial ocean D/H was suggested to be IDP, which are exposed to the solar wind in space and would carry isotopically light water (Pavlov et al. 1999). Water-bearing rims in anhydrous IDPs due to space-weathering by solar wind have indeed been observed (Bradley 2015) and dust might have been an important source of volatiles during the early evolution (100 Ma) of the Earth (Marty and Yokochi 2006).

Finally, an alternative view is that indeed a substantial amount of Earth's water was accreted early from the nebula (Halliday 2013; Hallis 2017), e.g., by water adsorption (Drake 2005; Vattuone et al. 2013) and incorporation into olivine grains. This water would have been strongly isotopically fractionated during condensation in the nebula (Asaduzzaman et al. 2015; Ganguly et al. 2016). The low D/H ratio of nebular water could have been increased by mass fractionation during atmospheric hydrogen loss (Genda and Ikoma 2008).

Indeed, the large D/H variations reported for various minerals in the eucrite and angrite achondrites, Martian meteorites and the Earth, indicate that comparing D/H ratios alone might not be conclusive, without a full understanding of melting and impact processing on the planetary/parent body that could cause D/H fractionation during partitioning, diffusion and degassing.

5.2.2 *Constraints from Nitrogen and Carbon Isotopes*

The discussion on H isotopes above illustrates the need for additional constraints to determine the possible origin(s) for Earth's water. The volatile element N is commonly the first choice to be added. Similar to H, the N isotopic compositions of the reservoirs in the solar system show a significant range. It is given here in the delta notation, i.e. the difference to the N isotopic ratio of the Earth atmosphere ($\delta^{15}N = 0‰$). The Sun, assumed to represent nebular N, has a strongly negative $\delta^{15}N$ value of $-389 \pm 5‰$, which is, as for H isotopes, supported by measurements of N isotope data of Jupiter ($\delta^{15}N = -396 \pm 141‰$, Fouchet et al. 2004). In contrast, comets and carbonaceous chondrites are strongly enriched: carbonaceous chondrites bulk $\delta^{15}N$ range between -20 to up to $+350‰$ (Alexander et al. 2012). Almost all N-bearing molecules detected in comets show $\delta^{15}N$ in the range 650 to 2000‰, with only two detections roughly consistent with terrestrial N (Bockelée-Morvan et al. 2015). The N isotopic compositions of terrestrial silicate reservoirs scatter around $\delta^{15}N \sim 0‰$. For example, the $\delta^{15}N$ for bulk-silicate Earth + atmosphere, MORB-source, and continental crust are 2.7‰, $-5‰$ and 7.3‰, respectively (Johnson and Goldblatt 2015). Others reported a larger range between $\delta^{15}N \sim -40‰$ and $-5‰$ for the terrestrial mantle (Cartigny and Marty 2013). An even larger range between -46 and $+27‰$ was reported for indigenous lunar N, which is challenging to determine (Füri et al. 2015), and suggests a link to N in Earth's pristine mantle and in carbonaceous chondrites. The $\delta^{15}N$ in eucrites, i.e. Vesta, is similar to the Earth, Moon and carbonaceous chondrites (Sarafian et al. 2014). In agreement with the high δD in most comets, their high $\delta^{15}N$, however, excludes them as the dominant N source for the Earth. On the other hand, similarly to the interpretation of H isotopic values, a mixture of ^{15}N-enriched cometary material and nebula N, either primordially accreted by the Earth, or trapped in cometary ices (Marty et al. 2016), could in principle provide the terrestrial N composition.

The $^{13}C/^{12}C$ ratios in all examined solar system reservoirs, the Earth, carbonaceous chondrites and comets (with large uncertainties) are within $< 10\%$ similar (e.g., Marty et al. 2013). Their values are slightly higher than the protosolar value, consistent with any of the currently proposed origins of Earth's water, and not sufficiently well constrained to distinguish subtle differences in the source reservoirs (Marty et al. 2013).

5.2.3 *Constraints from the Isotopes of Noble Gases*

While the study of H and N isotopes are most instructive, isotope compositions of the chemically inert noble gases He to Xe will add additional constraints to the question of the origin of Earth's water, and we will summarize here the major points. The moon-forming giant impact, as many preceding impacts on Earth, may have erased most signatures of a pre-existing primary atmosphere (Chap. 13). These impacts, combined with global magma oceans over an extended period, led to partial degassing of the Earth's mantle. Outgassing of mostly radiogenic noble gases, together with incoming volatile-rich chondritic or cometary material, replenished the terrestrial atmosphere (e.g., Pepin and Porcelli 2002; Porcelli and Ballentine 2002).

Protosolar Cloud and Planetary Atmospheres The noble gas composition of the protosolar cloud is best represented by the Sun. The exception is solar ^{3}He, which significantly increased immediately after the Sun's accretion due to solar D-burning (e.g., Geiss and Gloeckler 2003). The protosolar pre-D-burning ^{3}He/^{4}He ratio might be best represented by Jovian He or, perhaps, primitive meteorites (Mahaffy et al. 1998; Busemann et al. 2000, 2001). All other noble gas isotopic ratios of the solar nebula are best sampled by the solar wind (e.g., Heber et al. 2009; Vogel et al. 2011; Meshik et al. 2012). So-called "planetary" noble gases (the atmospheres of Earth and Mars, but also certain components in chondrites) are isotopically and elementally heavy compared to solar gases. The exception is Xe in the present terrestrial atmosphere, which is more depleted compared to Kr and the solar composition (e.g., Dauphas and Morbidelli 2014). These planetary patterns, i.e. fractionation relative to the protoplanetary cloud can be the result of processes that favour the depletion of the lighter species such as (i) incomplete or selective trapping by adsorption and dissolution, or (ii) diffusive, evaporative or gravitationally-controlled losses from reservoirs of an originally solar composition.

Earth's Mantle Helium and Ne isotopes in the mantle, measured in oceanic basalts in numerous locations, show a mixing range between atmosphere and solar end-members. The presence of solar He and Ne in the Earth's mantle (e.g., Pepin and Porcelli 2002; Yokochi and Marty 2004; Moreira and Charnoz 2016, and references therein) suggests either that (i) an early magma ocean has dissolved parts of a thick proto-atmosphere of solar nebula gas (e.g., Hayashi et al. 1979), (ii) dust precursors delivered elementally fractionated adsorbed nebula gas (e.g., Pepin and Porcelli 2002), or (iii) small dust grains incorporated solar wind by irradiation, perhaps by an early active Sun after dissipation of the nebula (Podosek et al. 2000; Trieloff et al. 2000; Moreira and Charnoz 2016). In all cases, the noble gases were likely accompanied by abundant H_2 that then formed H_2O in the Earth by oxidization (e.g., Marty and Yokochi 2006). In contrast to these hints for solar gases, chondritic (non-solar) Xe and Kr are observed in mantle gases (Holland et al. 2009; Caracausi et al. 2016). This supports the general idea that most water (see above) is of chondritic, i.e. asteroidal origin.

Most (~85%; e.g., Allègre et al. 1995) of the Earth's ^{129}I-derived ^{129}Xe was possibly lost within the first 90–110 Ma, either by transferal into a "hidden Xe" reservoir in the deep mantle (e.g., Podosek and Ozima 2000), or during continued magma ocean degassing and subsequent escape from the atmosphere into space (e.g., Marty and Yokochi 2006). In the latter case, water could have been similarly lost. However, the relatively low D/H ratio in the Earth does not support such losses, which implies that the water remained dissolved in the mantle (Marty and Yokochi 2006). Sophisticated considerations of the Ne and Xe isotopes in the Earth provide evidence for two distinct noble gas reservoirs in the mantle, isolated since ~100 Ma after the formation of the solar system (Mukhopadhyay 2012). In particular, these reservoirs show (i) distinct abundances of ^{129}Xe (produced by decay of once extant ^{129}I), (ii) the distinct presence of primordial ^{22}Ne relative to nucleogenic ^{21}Ne, and (iii) different primordial ^{20}Ne/^{22}Ne endmember compositions. These mantle reservoirs, which could also have distinct water contents, are individually sampled by OIB and MORB (Sect. 3.2.2; Caracausi et al. 2016).

The present, secondary atmosphere likely contains mostly radiogenic noble gas isotopes replenished from the interior and from the "late veneer". The elemental noble gas pattern shows the characteristic depletion in Xe relative to lighter elements (if normalized to solar composition), which was taken as evidence that the noble gases were delivered from volatile-rich comets (see above; Dauphas 2003). Comets would then have delivered abundant cometary water. However, the predominantly high D/H ratios reported for comets limit this contribution (Sect. 5.2.1). In the following, we discuss briefly further potential source reservoirs.

Chondrites Bulk noble gas isotopes in primitive chondrites are carried in (i) presolar grains, dominating He and Ne (e.g., the "HL" and "Ne-E" components) with significant "exotic" isotopic anomalies related to stellar nucleosynthesis (e.g., Wieler et al. 2006; Ott 2014, and references therein), (ii) most likely organic matter (the component called "phase Q" dominating Ar-Xe; e.g., Busemann et al. 2000; Ott 2014, and references therein) and (iii)—not negligible but often neglected—less abundant bulk silicates (Wieler et al. 2006, and references therein). Apart from the rare but important observations of chondritic Kr and Xe signatures in the terrestrial mantle (Holland et al. 2009; Caracausi et al. 2016), other typical isotopic signatures of primitive meteorites (see above) are not observed in the atmosphere or interior. However, if substantiated, the observation of chondritic Kr and Xe in the Earth's mantle supports an origin of the Earth's water from asteroidal sources.

Achondrites and Moon *Achondrites:* Noble gas concentrations in achondrites are, as expected in view of the experienced elevated temperatures, severely depleted, by orders of magnitude compared to primitive carbonaceous chondrites (e.g., Busemann and Eugster 2002). Only Ar-Xe in the most primitive achondrites (e.g., lodranites) survived the increased temperatures during metamorphism and incipient differentiation (Busemann and Eugster 2002), likely in the most thermally-resistant phase Q component.

Moon: In contrast to hydrogen isotopes (Sect. 5.2.1), primordial "indigenous" noble gases in lunar samples that survived the harsh process of lunar formation have not been detected yet (Füri et al. 2015).

Comets and IDPs Noble gases in comets are notoriously difficult to observe and only fragmentary information could be collected so far. For the first time, cometary Ar could clearly be detected ($^{36}Ar/^{38}Ar = 5.4 \pm 1.4$) in the coma of comet 67P/Churyumov-Gerasimenko (Balsiger et al. 2015). If the volatiles did not fractionate during emission from the comet, the Ar/H_2O ratio implies that comets are less likely than carbonaceous chondrites to have contributed Ar and water to Earth, because the ratio is, as the D/H ratio (Altwegg et al. 2015), much higher than the terrestrial values. Helium and Ne in comet Wild 2 dust seem very abundant (Marty et al. 2008). The Ne is isotopically consistent with "Phase Q" (Ne found in minor amounts in meteorites), while He is rich in ^3He compared to phase Q, suggesting some exchange from inner and outer solar system material. Most likely He and Ne represent a component incorporated by strong irradiation (Marty et al. 2008).

First attempts to detect trapped Kr and Xe in IDPs show large concentrations, perhaps similar to those found in primitive meteorites (Kehm et al. 2009; Busemann et al. 2010; Spring et al. 2014). However, the Kr and Xe isotopic compositions in bona-fide cometary dust could not be analyzed sufficiently precisely yet, although they are essential to clarify the origin and character of the cometary noble gases. The light noble gases are dominated by solar wind and cosmogenic components, and the primordially trapped He and Ne isotopic compositions are hence still unknown (e.g., Nier and Schlutter 1992). Altogether, evidence from cometary noble gases is currently inconclusive regarding the origin of Earth's water. Major advances might arise from future results on the noble gas isotopic compositions in comet Churyumov-Gerasimenko. Until then, trapping experiments on noble gases in ices at various temperatures (e.g., Owen et al. 1992; Bar-Nun and Owen 1998) are the best tool for a better understanding of the crucial cometary noble gases.

5.2.4 Summary

In summary, isotopic considerations alone cannot reveal the origin of Earth's water. Generally, large contributions from chondritic asteroidal sources fit better than cometary material. However, at least for the noble gases, two distinct sources for mantle and atmosphere are probable. More sophisticated modeling involving relative abundances of many volatile elements leads to a variety of detailed conclusions: Alexander et al. (2012) suggest that Earth's water is essentially delivered with carbonaceous chondrites, in this case mainly (90%) CI chondrites. This is in agreement with considerations from platinum group elements and elemental abundances of S, Se and Te, which argue for a CM-like late veneer (Dauphas 2003; Marty and Yokochi 2006; Wang and Becker 2013). Other work pleads for an enstatite chondrite-like late veneer (Fischer-Gödde and Kleine 2017). Marty et al. (2016) suggest a significant cometary contribution (possibly as IDPs) to the noble gases at least in the terrestrial atmosphere during the late veneer or late heavy bombardment phase. However, Marty (2012) and Marty et al. (2016) limit cometary contributions to the other volatiles in the Earth to a few percent. Halliday (2013) envisions 10–30% of cometary material mixing with the CI chondritic late veneer. CI-CM chondritic contributions of ~2–4% of the Earth's mass could have delivered volatiles to the dry proto-Earth (Marty 2012; Alexander 2017). While most volatiles in the Earth may be present in chondritic relative proportions, N is depleted, possibly residing in the core (Marty 2012). Halliday (2013) also includes H, C and Xe for possible elements residing in the core. All, Marty (2012), Halliday (2013) and Dauphas and Morbidelli (2014) favour volatile-rich outer asteroid belt sources to deliver the Earth's water.

In order to finally clarify the origin of Earth's water, it will be essential to determine (i) the H and N isotopic compositions of more comets, and (ii) the noble gas elemental and isotopic compositions of cometary solids and ices, as well as of the terrestrial core and mantle.

6 Conclusion

The present consensus on the origin of Earth's water seems to converge on the following scenario (Fig. 11), although many details are still debated. The Earth initially accreted water-poor from the inner part of the solar nebula 4.56 Ga ago. After the Earth had reached ~60–90% of its current size, water-rich bodies from further away from the Sun were delivered to the Earth, while core segregation was well underway. This delivery likely already brought more water than needed to explain the Earth's water content. Evidence suggests that the late veneer, the last $< 1\%$ of accretion of material to the Earth after core formation, predominantly delivered carbonaceous chondrites from the asteroid belt to the Earth. The late veneer may have brought 20 to 100% of the H and C to the Earth's mantle depending on which estimates are used for the water and C contents of the bulk silicate Earth (Wang and Becker 2013). Accretion through giant impacts and radioactive decay of isotopes released sufficient heat such that deep magma oceans were generated. While these lasted for several million years (Elkins-Tanton 2008), significant degassing of the oceans occurred and resulted in water transport from the Earth's interior to the atmosphere, which in turn could be lost by interaction with the irradiation of the early active Sun. Differentiation of our planet into core, mantle and crust likely initially generated a layered distribution of water. The early Earth probably operated in a stagnant lid regime before the onset of plate tectonics and this could have resulted in the upper part of the mantle being depleted in water by crust formation events, leaving deeper parts of the mantle comparatively water-rich. Degassing and cooling allowed the formation of water oceans early in the Earth's history, > 4 Ga ago (Wilde et al. 2001). The onset of plate tectonics ~3 Ga ago started to rehydrate the mantle by subducting water-rich lithologies of the crust back into the mantle. This resulted in a heterogeneously hydrated mantle. Despite fast H diffusion in minerals at mantle temperatures, large (> km in size) water-rich zones could not be erased on billion-year time scales. The heterogeneous mantle also implies that convection was not efficient enough to homogenize water in the mantle.

This story line has many uncertainties. The deeper the Earth, the less certain are the concentration and distribution estimates for water. In particular, estimates for H in the lower mantle and core are very uncertain, while also fluxes of water between reservoirs, including at the crucial subduction zones, have large errors. When considering these uncertainties, the total water amount in the Earth could range from the equivalent of 3 to 18 times the present mass of the oceans. This uncertainty also affects the determination of the origin and amount of water-rich material from which the Earth accreted or the material that was delivered later over the Earth history. Moreover, solar nebula condensation models have large uncertainties on condensation mechanisms and the position of the snowline, the temperature-driven boundary of water condensation. These uncertainties strongly affect the initial conditions that are used for dynamic models (e.g., Grand Tack), which describe the delivery of water-rich material to the inner solar system and the Earth. Carbonaceous chondrites are the most promising candidates to date for supplying the Earth's water, but we have not yet sampled and analyzed all potential water-rich materials of the solar system, in particular data are missing for comets and bodies from the outer solar system.

These open questions require more comprehensive future studies of terrestrial and extra-terrestrial samples, with continuously improved analytical techniques, including robot or human solar-system exploration. For example, two robotic missions, OSIRIS-REx (NASA, USA) and Hayabusa 2 (JAXA, Japan) are presently travelling towards asteroids and are planned to return samples of carbonaceous chondrite composition in 2023 and 2020, respectively. These missions are vital to complement the meteorite collection and Apollo samples used to study the formation of the Earth and our solar system in general. Observation of protoplanetary disks and planetary systems around other stars will further constrain our planetary formation scenarios. For Earth, progress can also be expected from geophysical methods such as seismology and electrical conductivity, globally by increasing geographical coverage of measurements, and in the laboratory by better constraining physical properties of deep Earth phases. Increased computational power will also improve models of Earth's formation and geodynamics, as well as ab initio calculations of water incorporation in high pressure phases. Water plays such a key role in planetary differentiation, volcanism, plate-tectonics, mantle convection, and the origin of life that we expect the flowering of studies of the last 10-years on water in the Earth to be sustained by a high rate of discoveries.

Acknowledgements The authors are very grateful to Rosie Jones and an anonymous reviewer for careful detailed comments that greatly improved this manuscript. MS also thanks Hilke Schlichting for inspiring discussions that helped to improve the manuscript. Thanks to editor Michel Blanc and ISSI in Bern (Switzerland) for organizing in February 2016 the workshop on Water delivery to the Solar System from which this book originates. This work was supported by NSF grant #OCE1624310 to AHP and, in part (HB and MS), has been carried out within the frame of the National Centre for Competence in Research 'PlanetS' supported by the Swiss National Science Foundation (SNSF).

References

N. Abe, E. Ohtani, T. Okuchi, K. Righter, M.J. Drake, Water in the early Earth, in *Origin of the Earth and Moon*, ed. by R.M. Canup, K. Righter (University of Arizona Press, Tucson, 2000), pp. 413–433

G.A. Abers, P.E. van Keken, B.R. Hacker, The cold and relatively dry nature of mantle forearcs in subduction zones. Nat. Geosci. **10**, 333–337 (2017)

J. Adam, M. Turner, E.H. Hauri, S. Turner, Crystal/melt partitioning of water and other volatiles during the near-solidus melting of mantle peridotite: comparisons with non-volatile incompatible elements and implications for the generation of intraplate magmatism. Am. Mineral. **101**, 876–888 (2016)

Y. Aikawa, E. Herbst, Deuterium fractionation in protoplanetary disks. Astrophys. J. **526**, 314–326 (1999)

R.D. Aines, G.R. Rossman, The hydrous component in garnets: pyralspites. Am. Mineral. **69**, 1116–1126 (1984)

F. Albarède, Volatile accretion history of the terrestrial planets and dynamic implications. Nature **461**, 1227–1233 (2009)

F. Albarède, C. Ballhaus, J. Blichert-Toft, C.-T. Lee, B. Marty, F. Moynier, Q.-Z. Yin, Asteroidal impacts and the origin of terrestrial and lunar volatiles. Icarus **222**, 44–52 (2013)

C.M.O.D. Alexander, R. Bowden, M.L. Fogel, K.T. Howard, C.D.K. Herd, L.R. Nittler, The provenances of asteroids, and their contributions to the volatile inventories of the terrestrial planets. Science **337**, 721–723 (2012)

C.M.O.D. Alexander, M. Fogel, H. Yabuta, G.D. Cody, The origin and evolution of chondrites recorded in the elemental and isotopic compositions of their macromolecular organic matter. Geochim. Cosmochim. Acta **71**, 4380–4403 (2007)

C.M.O.D. Alexander, The origin of inner Solar System water. Philos. Trans. R. Soc. Lond. A **375**, 20150384 (2017)

C.J. Allègre, G. Manhès, C. Göpel, The age of the Earth. Geochim. Cosmochim. Acta **59**, 1445–1456 (1995)

K. Altwegg, H. Balsiger, A. Bar-Nun, J.J. Berthelier, A. Bieler, P. Bochsler, C. Briois, U. Calmonte, M. Combi, J. De Keyser, P. Eberhardt, B. Fiethe, S. Fuselier, S. Gasc, T.I. Gombosi, K.C. Hansen, M. Hässig, A. Jäckel, E. Kopp, A. Korth, L. LeRoy, U. Mall, B. Marty, O. Mousis, E. Neefs, T. Owen, H. Rème, M. Rubin, T. Sémon, C.-Y. Tzou, H. Waite, P. Wurz, 67P/Churyumov-Gerasimenko, a Jupiter family comet with a high D/H ratio. Science **347**, 1261952 (2015)

Y. Amelin, U–Pb ages of angrites. Geochim. Cosmochim. Acta **72**, 221–232 (2008)

T. Andersen, E.-R. Neumann, Fluid inclusions in mantle xenoliths. Lithos **55**, 301–320 (2001)

M. Andrut, M. Wildner, J. Ingrin, A. Beran, Mechanisms of OH defect incorporation in naturally occurring, hydrothermally formed diopside and jadeite. Phys. Chem. Miner. **34**, 543–549 (2007)

P. Ardia, M.M. Hirschmann, A.C. Withers, T.J. Tenner, H_2O storage capacity of olivine at 5–8 GPa and consequences for dehydration partial melting of the upper mantle. Earth Planet. Sci. Lett. **345–348**, 104–116 (2012)

L.S. Armstrong, M.M. Hirschmann, B.D. Stanley, E.G. Falksen, S.D. Jacobsen, Speciation and solubility of reduced C-O-H-N volatiles in mafic melt: implications for volcanism, atmospheric evolution, and deep volatile cycles in terrestrial planets. Geochim. Cosmochim. Acta **171**, 283–302 (2015)

I.M. Artemieva, W.D. Mooney, Thermal thickness and evolution of Precambrian lithosphere: a global study. J. Geophys. Res. **106**, 16387–16414 (2001)

A. Asaduzzaman, K. Muralidharan, J. Ganguly, Incorporation of water into olivine during nebular condensation: insights from density functional theory and thermodynamics, and implications for phyllosilicate formation and terrestrial water inventory. Meteorit. Planet. Sci. **50**, 578–589 (2015)

C. Aubaud, E. Hauri, M.M. Hirschmann, Hydrogen partition coefficients between nominally anhydrous minerals and basaltic melts. Geophys. Res. Lett. **31**, L20611 (2004). doi:10.1029/2004GL021341

C. Aubaud, M.M. Hirschmann, A.C. Withers, R.L. Hervig, Hydrogen partitioning between melt, clinopyroxene, and garnet at 3 GPa in a hydrous MORB with 6 wt.% H_2O. Contrib. Mineral. Petrol. **156**, 607–625 (2008)

C. Aubaud, A.C. Withers, M.M. Hirschmann, Y. Guan, L.A. Leshin, S.J. Mackwell, D.R. Bell, Intercalibration of FTIR and SIMS for hydrogen measurements in glasses and nominally anhydrous minerals. Am. Mineral. **92**, 811–828 (2007)

S. Aulbach, R.L. Rudnick, W.F. McDonough, Li-Sr-Nd isotope signatures of the plume and cratonic lithospheric mantle beneath the margin of the rifted Tanzanian craton (Labait). Contrib. Mineral. Petrol. **155**, 79–92 (2008)

K. Baba, A. Chave, R.L. Evans, G. Hirth, R.L. Mackie, Mantle dynamics beneath the East Pacific Rise at 17 S: insights from the Mantle Electromagnetic and Tomography (MELT) experiment. J. Geophys. Res. **111**, B02101 (2006)

J. Badro, A.S. Côté, J.P. Brodholt, A seismologically consistent compositional model of Earth's core. Proc. Natl. Acad. Sci. **111**, 7542–7545 (2014)

E. Bali, A. Audédat, H. Keppler, Water and hydrogen are immiscible in Earth's mantle. Nature **495**, 220–222 (2013)

E. Bali, N. Bolfan-Casanova, K.T. Koga, Pressure and temperature dependence of H solubility in forsterite: an implication to water activity in the Earth interior. Earth Planet. Sci. Lett. **268**, 354–363 (2008)

C. Ballhaus, V. Laurenz, C. Münker, R.O.C. Fonseca, F. Albarède, A. Rohrbach, M. Lagos, M.W. Schmidt, K.-P. Jochum, B. Stoll, U. Weis, H.M. Helmy, The U/Pb ratio of the Earth's mantle—a signature of late volatile addition. Earth Planet. Sci. Lett. **362**, 237–245 (2013)

H. Balsiger, K. Altwegg, J. Geiss, D/H and $^{18}O/^{16}O$ ratio in the hydronium ion and in neutral water from in situ ion measurements in comet Halley. J. Geophys. Res. **100**, 5827–5834 (1995)

H. Balsiger, K. Altwegg, A. Bar-Nun, J.-J. Berthelier, A. Bieler, P. Bochsler, C. Briois, U. Calmonte, M. Combi, J. De Keyser, P. Eberhardt, B. Fiethe, S.A. Fuselier, S. Gasc, T.I. Gombosi, K.C. Hansen, M. Hässig, A. Jäckel, E. Kopp, A. Korth, L. Le Roy, U. Mall, B. Marty, O. Mousis, T. Owen, H. Rème, M. Rubin, T. Sémon, C.-Y. Tzou, J.H. Waite, P. Wurz, Detection of argon in the coma of comet 67P/Churyumov–Gerasimenko. Sci. Adv. **1** (2015)

V. Baptiste, S. Demouchy, S. Keshav, F. Parat, N. Bolfan-Casanova, P. Condamine, P. Cordier, Decrease of hydrogen incorporation in forsterite from CO_2-H_2O-rich kimberlitic liquid. Am. Mineral. **100**, 1912–1920 (2015)

V. Baptiste, A. Tommasi, Petrophysical constraints on the seismic properties of the Kaapvaal craton mantle root. Solid Earth **5**, 1–19 (2014)

V. Baptiste, A. Tommasi, S. Demouchy, Deformation and hydration of the lithospheric mantle beneath the Kaapvaal craton, South Africa. Lithos **149**, 31–50 (2012)

A. Bar-Nun, T. Owen, Trapping of gases in water ice and consequences to comets and the atmospheres of the inner planets, in *Solar System Ices*, ed. by B. Schmitt, C. De Bergh, M. Festou (Springer, Dordrecht, 1998), pp. 353–366

J.J. Barnes, D.A. Kring, R. Tartèse, I.A. Franchi, M. Anand, S.S. Russell, An asteroidal origin for water in the Moon. Nat. Commun. **7**, 11684 (2016)

J.J. Barnes, R. Tartèse, M. Anand, F.M. McCubbin, I.A. Franchi, N.A. Starkey, S.S. Russell, The origin of water in the primitive Moon as revealed by the lunar highlands samples. Earth Planet. Sci. Lett. **390**, 244–252 (2014)

T.J. Barrett, J.J. Barnes, R. Tartèse, M. Anand, I.A. Franchi, R.C. Greenwood, B.L.A. Charlier, M.M. Grady, The abundance and isotopic composition of water in eucrites. Meteorit. Planet. Sci. **51**, 1110–1124 (2016)

G.E. Bebout, Volatile transfer and recycling at convergent margins: mass-balance and insights from high-P/T metamorphic rocks, in *Subduction Top to Bottom*, ed. by G.E. Bebout, D.W. Scholl, S.H. Kirby, J.P. Platt (American Geophysical Union, Washington, 1996), pp. 179–193

W. Behr, D. Smith, Deformation in the mantle wedge associated with Laramide flat-slab subduction. Geochem. Geophys. Geosyst. **17**, 1–18 (2016)

D.R. Bell, Water in mantle minerals. Nature **357**, 646–647 (1992)

D.R. Bell, M. Grégoire, T.L. Grove, N. Chatterjee, R.W. Carlson, P.R. Buseck, Silica and volatile-element metasomatism of Archean mantle: a xenolith-scale example from the Kaapvaal craton. Contrib. Mineral. Petrol. **150**, 251–267 (2005)

D.R. Bell, P.D. Ihinger, The isotopic composition of hydrogen in nominally anhydrous mantle minerals. Geochim. Cosmochim. Acta **64**, 2109–2118 (2000)

D.R. Bell, P.D. Ihinger, G.R. Rossman, Quantitative analysis of trace OH in garnet and pyroxenes. Am. Mineral. **80**, 465–474 (1995)

D.R. Bell, R.O. Moore, Deep chemical structure of the southern African mantle from kimberlite megacrysts. S. Afr. J. Geol. **107**, 59–80 (2004)

D.R. Bell, G.R. Rossman, The distribution of hydroxyl in garnets from the subcontinental mantle in southern Africa. Contrib. Mineral. Petrol. **111**, 161–178 (1992a)

D.R. Bell, G.R. Rossman, Water in Earth's mantle: the role of nominally anhydrous minerals. Science **255**, 1391–1397 (1992b)

D.R. Bell, G.R. Rossman, J. Maldener, D. Endisch, F. Rauch, Hydroxide in olivine: a quantitative determination of the absolute amount and calibration of the IR spectrum. J. Geophys. Res. **108**(B2), 2105 (2003a)

D.R. Bell, M.D. Schmitz, P.E. Janney, Mesozoic thermal evolution of the southern African mantle lithosphere. Lithos **71**, 273–287 (2003b)

D.R. Bell, G.R. Rossman, J. Maldener, D. Endisch, F. Rauch, Hydroxide in kyanite; a quantitative determination of the absolute amount and calibration of the IR spectrum. Am. Mineral. **89**, 998–1003 (2004a)

D.R. Bell, G.R. Rossman, R.O. Moore, Abundance and partitioning of OH in a high pressure magmatic system: megacrysts from the Monastery kimberlite, South Africa. J. Petrol. **45**, 1539–1564 (2004b)

W. Ben-Ismail, G. Barruol, D. Mainprice, The Kaapvaal craton seismic anisotropy: petrophysical analyses of upper mantle kimberlite nodules. Geophys. Res. Lett. **28**, 2497–2500 (2001)

A. Bénard, K.T. Koga, N. Shimizu, M.A. Kendrick, D.A. Ionov, O. Nebel, R.J. Arculus, Chlorine and fluorine partition coefficients and abundances in sub-arc mantle xenoliths (Kamchatka, Russia): implications for melt generation and volatile recycling processes in subduction zones. Geochim. Cosmochim. Acta **199**, 324–350 (2017)

W. Benz, W.L. Slattery, A.G.W. Cameron, The origin of the moon and the single-impact hypothesis I. Icarus **66**, 515–535 (1986)

A. Beran, Messung des Ultrarot-Pleochroismus von Mineralen. XIV. Der Pleochroismus der OH-Streckfrequenz in Diopsid. Tschermak's Mineral. Petrogr. Mitt. **23**, 79–85 (1976)

A. Beran, E. Libowitzky, Water in natural mantle minerals II: olivine, garnet and accessory minerals, in *Water in Nominally Anhydrous Minerals*, ed. by H. Keppler, J.R. Smyth (Mineralogical Society of America, Chantilly, 2006), pp. 169–191

D. Bercovici, S.-I. Karato, Whole-mantle convection and the transition-zone water filter. Nature **425**, 39–44 (2003)

M.I. Billen, M. Gurnis, A low viscosity wedge in subduction zones. Earth Planet. Sci. Lett. **193**, 227–236 (2001)

F. Birch, Elasticity and constitution of the Earth's interior. J. Geophys. Res. **57**, 227–286 (1952)

M. Bizimis, A.H. Peslier, Water in Hawaiian garnet pyroxenites: implications for water heterogeneity in the mantle. Chem. Geol. **397**, 61–75 (2015)

M. Bizimis, G. Sen, V.J.M. Salters, Hf-Nd isotope decoupling in the oceanic lithosphere: constraints from spinel peridotites from Oahu, Hawaii. Earth Planet. Sci. Lett. **217**, 43–58 (2004)

D. Bockelée-Morvan, U. Calmonte, S. Charnley, J. Duprat, C. Engrand, A. Gicquel, M. Hässig, E. Jehin, H. Kawakita, B. Marty, S. Milam, A.D. Morse, P. Rousselot, S. Sheridan, E. Wirström, Cometary isotopic measurements. Space Sci. Rev. **197**, 47–83 (2015)

J.-L. Bodinier, M. Godard, Orogenic, ophiolitic, and abyssal peridotites, in *Treatise on Geochemistry*, ed. by R. Carlson (Elsevier, Oxford, 2003), pp. 103–170

R.J. Bodnar, T. Azbej, S.P. Becker, C. Cannatelli, A. Fall, M. Severs, Whole Earth geohydrologic cycle, from the clouds to the core: the distribution of water in the dynamic Earth system. Spec. Pap., Geol. Soc. Am. **500**, 431–461 (2013)

D.D. Bogard, R.N. Clayton, K. Marti, T. Owen, G. Turner, Martian volatiles: isotopic composition, origin, and evolution. Space Sci. Rev. **96**, 425–458 (2001)

N. Bolfan-Casanova, Water in the Earth's mantle. Mineral. Mag. **69**, 229–257 (2005)

N. Bolfan-Casanova, H. Keppler, D.C. Rubie, Water partitioning between nominally anhydrous minerals in the MgO-SiO_2-H_2O system up to 24 GPa: implications for the distribution of water in the Earth's mantle. Earth Planet. Sci. Lett. **182**, 209–221 (2000)

F.R. Boyd, Compositional distinction between oceanic and cratonic lithosphere. Earth Planet. Sci. Lett. **96**, 15–26 (1989)

F.R. Boyd, J.J. Gurney, S.H. Richardson, Evidence for a 150–200 km thick Archean lithosphere from diamond inclusion thermobarometry. Nature **315**, 387–389 (1985)

F.R. Boyd, N.P. Pokhilenko, D.G. Pearson, S.A. Mertzman, N.V. Sobolev, L.W. Finger, Composition of the Siberian cratonic mantle: evidence from Udachnaya peridotite xenoliths. Contrib. Mineral. Petrol. **128**, 228–246 (1997)

J.P. Bradley, *Water and organics in interplanetary dust particles* (American Astronomical Society Assembly, Honolulu, 2015)

J.P. Bradley, H.A. Ishii, J.J. Gillis-Davis, J. Ciston, M.H. Nielsen, H.A. Bechtel, M.C. Martin, Detection of solar wind-produced water in irradiated rims on silicate minerals. Proc. Natl. Acad. Sci. **111**, 1732–1735 (2014)

A.D. Brandon, D.S. Draper, Constraints on the origin of the oxidation state of mantle overlying subduction zones: an example from Simcoe, Washington, USA. Geochim. Cosmochim. Acta **60**, 1739–1749 (1996)

A.D. Brandon, R.J. Walker, J.W. Morgan, M.D. Norman, H.M. Prichard, Coupled ^{186}Os and ^{187}Os evidence for core-mantle interaction. Science **280**, 1570–1573 (1998)

J. Braunmiller, S. van der Lee, L. Doermann, Mantle transition zone thickness in the central South-American subduction zone, in *Earth's Deep Water Cycle*, ed. by S. Jacobsen, S. Lee (AGU Geophysical Monograph, Washington D.C., 2006), pp. 215–224

G.P. Brey, T. Köhler, Geothermobarometry in four-phase lherzolites II. New thermobarometers, and practical assessment of existing thermobarometers. J. Petrol. **31**, 1353–1378 (1990)

G.P. Brey, T. Köhler, K.G. Nickel, Geothermobarometry in four-phase lherzolites I. Experimental results from 10 to 60 kb. J. Petrol. **31**, 1313–1352 (1990)

D.E. Brownlee, Comets, in *Treatise on Geochemistry*, ed. by A.S. Davis 2nd edn. (Elsevier, Amsterdam, 2014), pp. 335–363

H. Bureau, H. Keppler, Complete miscibility between silicate melts and hydrous fluids in the upper mantle: experimental evidence and geochemical implication. Earth Planet. Sci. Lett. **165**, 187–196 (1999)

S.R. Burgess, B. Harte, Tracing lithosphere evolution through the analysis of heterogeneous G9–G10 garnets in peridotite xenoliths, II: REE chemistry. J. Petrol. **43**, 609–634 (2004)

K. Burke, B. Steinberger, T.H. Torsvik, M.A. Smethurst, Plume generation zones at the margins of Large Low Shear Velocity Provinces on the core-mantle boundary. Earth Planet. Sci. Lett. **265**, 49–60 (2008)

H. Busemann, H. Baur, R. Wieler, Primordial noble gases in "phase Q" in carbonaceous and ordinary chondrites studied by closed-system stepped etching. Meteorit. Planet. Sci. **35**, 949–973 (2000)

H. Busemann, H. Baur, R. Wieler, Helium isotopic ratios in carbonaceous chondrites: significant for the early solar nebula and circumstellar diamonds? in *32nd Lunar and Planetary Science Conference*, The Woodlands, TX (2001)

H. Busemann, O. Eugster, The trapped noble gas component in achondrites. Meteorit. Planet. Sci. **37**, 1865–1891 (2002)

H. Busemann, A.N. Nguyen, G.D. Cody, P. Hoppe, A.L.D. Kilcoyne, R.M. Stroud, T.J. Zega, L.R. Nittler, Ultra-primitive interplanetary dust particles from the comet 26P/Grigg-Skjellerup dust stream collection. Earth Planet. Sci. Lett. **288**, 44–57 (2009)

H. Busemann, N. Spring, S. Crowther, J. Claydon, J. Gilmour, L. Nittler, Abundant primordial xenon in interplanetary dust particles from the comet Grigg-Skjellerup collection, in *Lunar Planet. Sci. Conf.*, The Woodlands, TX (2010), p. 1947

H. Busemann, A.F. Young, C.M.O.D. Alexander, P. Hoppe, S. Mukhopadhyay, L.R. Nittler, Interstellar chemistry recorded in organic matter from primitive meteorites. Science **312**, 727–730 (2006)

J.A. Cabato, C.J. Stefano, S.B. Mukasa, Volatile concentrations in olivine-hosted melt inclusions from the Columbia River flood basalts and associated lavas of the Oregon Plateau: implications for magma genesis. Chem. Geol. **392**, 59–73 (2015)

R.A. Cabral, M.G. Jackson, K.T. Koga, E.F. Rose-Koga, E.H. Hauri, M.J. Whitehouse, A.A. Price, J.M.D. Day, N. Shimizu, K.A. Kelley, Volatile cycling of H_2O, CO_2, F, and Cl in the HIMU mantle: a new window provided by melt inclusions from oceanic hot spot lavas at Mangaia, Cook Islands. Geochem. Geophys. Geosyst. **15**, 4445–4467 (2014). doi:10.1002/2014GC005473

R.M. Canup, Forming a Moon with an Earth-like composition via a giant impact. Science **338**, 1052–1055 (2012)

Y. Cao, H. Jung, S. Song, M. Park, S. Jung, J. Lee, Plastic deformation and seismic properties in fore-arc mantles: a petrofabric analysis of the Yushigou harzburgites, North Qilian Suture Zone, NW China. J. Petrol. **56**, 1897–1944 (2015)

R. Caracas, The influence of hydrogen on the seismic properties of solid iron. Geophys. Res. Lett. **42**, 3780–3785 (2015)

A. Caracausi, G. Avice, P.G. Burnard, E. Füri, B. Marty, Chondritic xenon in the Earth's mantle. Nature **533**, 82–85 (2016)

R.W. Carlson, M. Boyet, H. Rizo, R.J. Walker, Early differentiation and its long-term consequences for Earth evolution, in *The Early Earth: Accretion and Differentiation*, ed. by J. Badro, M.J. Walter (Wiley, New Jersey, 2015), pp. 143–172

R.W. Carlson, D.G. Pearson, F.R. Boyd, S.B. Shirey, G. Irvine, A.H. Menzies, J.J. Gurney, Re-Os systematics of lithospheric peridotites: implications for lithosphere formation and preservation, in *Proc. VIIth International Kimberlite Conference, B.J. Dawson Volume*, ed. by J.J. Gurney, J.L. Gurney, M.D. Pascoe, S.R. Richardson (1999), pp. 99–108

M.R. Carroll, J.D. Webster, Solubilities of sulfur, noble gases, nitrogen, chlorine, and fluorine in magmas, in *Volatiles in Magmas*, ed. by M.R. Carroll, J.R. Holloway (Mineralogical Society of America, Washington D.C., 1994), pp. 231–279

P. Cartigny, B. Marty, Nitrogen isotopes and mantle geodynamics: the emergence of life and the atmosphere-crust-mantle connection. Elements **9**, 359–366 (2013)

N.L. Chabot, E.A. Wollack, M. Humayun, E.M. Shank, The effect of oxygen as a light element in metallic liquids on partitioning behavior. Meteorit. Planet. Sci. **50**, 530–546 (2015)

Y.-Y. Chang, W.-P. Hsieh, E. Tan, J. Chen, Hydration-reduced lattice thermal conductivity of olivine in Earth's upper mantle. Proc. Natl. Acad. Sci. **114**, 4078–4081 (2017)

Y. Chen, A. Provost, P. Schiano, N. Cluzel, Magma ascent rate and initial water concentration inferred from diffusive water loss from olivine-hosted melt inclusions. Contrib. Mineral. Petrol. **165**, 525–541 (2013)

E.J. Chin, V. Soustelle, G. Hirth, A.E. Saal, S.C. Kruckenberg, J.M. Eiler, Microstructural and geochemical constraints on the evolution of deep arc lithosphere. Geochem. Geophys. Geosyst. **17**, 2497–2521 (2016)

C.-L. Chou, Fractionation of siderophile elements in the Earth's upper mantle, in *Proc. Lunar Planet. Sci. Conf*, vol. 9 (1978), pp. 219–230

P.L. Clay, H. Busemann, S.C. Sherlock, T.L. Barry, S.P. Kelley, D.W. McGarvie, $^{40}Ar/^{39}Ar$ ages and residual volatile contents in degassed subaerial and subglacial glassy volcanic rocks from Iceland. Chem. Geol. **403**, 99–110 (2015)
L.I. Cleeves, E.A. Bergin, C.M.O.D. Alexander, F. Du, D. Graninger, K.I. Öberg, T.J. Harries, The ancient heritage of water ice in the solar system. Science **345**, 1590–1593 (2014)
M. Clog, C. Aubaud, P. Cartigny, L. Dosso, The hydrogen isotopic composition and water content of southern Pacific MORB: a reassessment of the D/H ratio of the depleted mantle reservoir. Earth Planet. Sci. Lett. **381**, 156–165 (2013)
J.N. Connelly, M. Bizzarro, Lead isotope evidence for a young formation age of the Earth-Moon system. Earth Planet. Sci. Lett. **452**, 36–43 (2016)
A.M. Courtier, J. Revenaugh, A water-rich transition zone beneath the Eastern United States and Gulf of Mexico from multiple ScS reverberations, in *Earth's Deep Water Cycle*, ed. by S. Jacobsen, S. Lee (AGU Geophysical Monograph, Washington D.C., 2006), pp. 181–194
S. Creighton, T. Stachel, S. Matveev, H. Höfer, C.A. McCammon, R.W. Luth, Oxidation of the Kaapvaal lithospheric mantle derived by metasomatism. Contrib. Mineral. Petrol. **157**, 491–504 (2009)
L. Dai, S.-I. Karato, Electrical conductivity of orthopyroxene: implications for the water content of the asthenosphere. Proc. Jpn. Acad. Ser. B, Phys. Biol. Sci. **85**, 466–475 (2009)
L.V. Danyushevsky, Experimental and petrological studies of melt inclusions in phenocrysts from mantle-derived magmas: an overview of techniques, advantages and complications. Chem. Geol. **183**, 5–24 (2002)
L.V. Danyushevsky, T.J. Fallon, A.V. Sobolev, A.J. Crawford, M. Carroll, R.C. Price, The H_2O content of basalt glasses from southwest Pacific back-arc basins. Earth Planet. Sci. Lett. **117**, 347–362 (1993)
N. Dauphas, The dual origin of the terrestrial atmosphere. Icarus **165**, 326–339 (2003)
N. Dauphas, A. Morbidelli, Geochemical and planetary dynamical views on the origin of Earth's atmosphere and oceans, in *Treatise on Geochemistry*, ed. by A.S. Davis 2nd edn. (Elsevier, Amsterdam, 2014), pp. 1–35
N. Dauphas, The isotopic nature of the Earth's accreting material through time. Nature **541**, 521–524 (2017)
J.B. Dawson, J.V. Smith, The MARID (mica-amphibole-rutile-ilmenite-diopside) suite of xenoliths in kimberlite. Geochim. Cosmochim. Acta **41**, 309–323 (1977)
J.C.M. De Hoog, C.J. Lissenberg, R.A. Brooker, R. Hinton, D. Trail, E. Hellebrand, Hydrogen incorporation and charge balance in natural zircon. Geochim. Cosmochim. Acta **141**, 472–486 (2014)
V. Debaille, C. O'Neill, A.D. Brandon, P. Haenecour, Q.-Z. Yin, N. Mattielli, A.H. Treiman, Stagnant-lid tectonics in early Earth revealed by ^{142}Nd variations in late Archean rocks. Earth Planet. Sci. Lett. **373**, 83–92 (2013)
B. Debret, N. Bolfan-Casanova, J.A. Padrón-Navarta, F. Martin-Hernandez, M. Andreani, C.J. Garrido, V. López Sánchez-Vizcaíno, M.T. Gómez-Pugnaire, M. Muñoz, N. Trcera, Redox state of iron during high-pressure serpentinite dehydration. Contrib. Mineral. Petrol. **169**, 1–18 (2015)
H. Delavault, C. Chauvel, E. Thomassot, C. Devey, B. Dazas, *Relicts of Archean Sediment in the Present Earth Mantle* (Goldschmidt, Yokohama, 2016), p. 633
E. Deloule, F. Albarède, S.M.F. Sheppard, Hydrogen isotope heterogeneities in the mantle from ion probe analysis of amphibole from ultramafic rocks. Earth Planet. Sci. Lett. **105**, 543–553 (1991)
S. Demouchy, N. Bolfan-Casanova, Distribution and transport of hydrogen in the lithospheric mantle: a review. Lithos **240–243**, 402–425 (2016)
S. Demouchy, E. Deloule, D.J. Frost, H. Keppler, Pressure and temperature dependence of water solubility in Fe-free wadsleyite. Am. Mineral. **90**, 1084–1091 (2005)
S. Demouchy, A. Ishikawa, A. Tommasi, O. Alard, S. Keshav, Characterization of hydration in the mantle lithosphere: peridotite xenoliths from the Ontong Java Plateau as an example. Lithos **212–215**, 189–201 (2015)
S. Demouchy, S.D. Jacobsen, F. Gaillard, C.R. Stern, Rapid magma ascent recorded by water diffusion profiles in olivine from Earth's mantle. Geology **34**, 429–432 (2006)
S. Demouchy, S. Mackwell, Mechanisms of hydrogen incorporation and diffusion in iron-bearing olivine. Phys. Chem. Miner. **33**, 347 (2006)
S. Demouchy, A. Tommasi, F. Barou, D. Mainprice, P. Cordier, Deformation of olivine in torsion under hydrous conditions. Phys. Earth Planet. Inter. **202–203**, 56–70 (2012)
C.M.M. Denis, O. Alard, S. Demouchy, Water content and hydrogen behaviour during metasomatism in the uppermost mantle beneath Ray Pic volcano (Massif Central France). Lithos **236–237**, 256–274 (2015)
C.M.M. Denis, S. Demouchy, C.S.J. Shaw, Evidence of dehydration in peridotites from Eifel volcanic field and estimates of the rate of magma ascent. J. Volcanol. Geotherm. Res. **258**, 85–99 (2013)
J.E. Dixon, Degassing of alkalic basalts. Am. Mineral. **82**, 368–378 (1997)
J.E. Dixon, D.A. Clague, Volatiles in basaltic glasses from Loihi seamount, Hawaii: evidence for a relatively dry plume component. J. Petrol. **42**, 627–634 (2001)

J.E. Dixon, D.A. Clague, B. Cousens, M.L. Monsalve, J. Uhl, Carbonatite and silicate melt metasomatism of the mantle surrounding the Hawaiian plume: evidence from volatiles, trace elements, and radiogenic isotopes in rejuvenated-stage lavas from Niihau, Hawaii. Geochem. Geophys. Geosyst. **9**, 1–34 (2008)

J.E. Dixon, D.A. Clague, E.M. Stolper, Degassing history of water, sulfur, and carbon in submarine lavas from Kilauea volcano, Hawaii. J. Geol. **99**, 371–394 (1991)

J.E. Dixon, T.H. Dixon, D.R. Bell, R. Malservisi, Lateral variation in upper mantle viscosity: role of water. Earth Planet. Sci. Lett. **222**, 451–467 (2004)

J.E. Dixon, L. Leist, C. Langmuir, J.-G. Schilling, Recycled dehydrated lithosphere observed in plume-influenced mid-ocean-ridge basalt. Nature **420**, 385–389 (2002)

J.E. Dixon, E. Stolper, J.R. Delaney, Infrared spectroscopic measurements of CO_2 and H_2O in Juan de Fuca basaltic glasses. Earth Planet. Sci. Lett. **90**, 87–104 (1988)

J.E. Dixon, E.M. Stolper, An experimental study of water and carbon dioxide solubilities in mid-ocean ridge basaltic liquids. Part II: application to degassing. J. Petrol. **36**, 1633–1646 (1995)

L.S. Doucet, D.A. Ionov, A.V. Golovin, The origin of coarse garnet peridotites in cratonic lithosphere: new data on xenoliths from the Udachnaya kimberlite, central Siberia. Contrib. Mineral. Petrol. **165**, 1225–1242 (2013)

L.S. Doucet, D.A. Ionov, A.V. Golovin, Paleoproterozoic formation age for the Siberian cratonic mantle: Hf and Nd isotope data on refractory peridotite xenoliths from the Udachnaya kimberlite. Chem. Geol. **391**, 42–55 (2015)

L.S. Doucet, A.H. Peslier, D.A. Ionov, A.D. Brandon, A.V. Golovin, A.G. Goncharov, I.V. Ashchepkov, High water contents in the Siberian cratonic mantle linked to metasomatism: an FTIR study of Udachnaya peridotite xenoliths. Geochim. Cosmochim. Acta **137**, 159–187 (2014)

M.J. Drake, Origin of water in the terrestrial planets. Meteorit. Planet. Sci. **40**, 519–527 (2005). doi:10.1111/j.1945-5100.2005.tb00960.x

A.M. Dziewonski, D.L. Anderson, Preliminary reference Earth model. Phys. Earth Planet. Inter. **25**, 297–356 (1981)

D.W. Eaton, F. Darbyshire, R.L. Evans, H. Grütter, A.G. Jones, X. Yuan, The elusive lithosphere-asthenosphere boundary (LAB) beneath cratons. Lithos **109**, 1–22 (2008)

P. Eberhardt, M. Reber, D. Krankowsky, R.R. Hodges, The D/H and $^{18}O/^{16}O$ ratios in water from comet P/Halley. Astron. Astrophys. **302**, 301–316 (1995)

M. Edmonds, S.C. Kohn, E.H. Hauri, M.C.S. Humphreys, M. Cassidy, Extensive, water-rich magma reservoir beneath southern Montserrat. Lithos **252–253**, 216–233 (2016)

R.A. Eggleton, J.N. Boland, A.E. Ringwood, High pressure synthesis of a new aluminium silicate: $Al_5Si_5O_{17}(OH)$. Geochem. J. **12**, 191–194 (1978)

L.T. Elkins-Tanton, Linked magma ocean solidification and atmospheric growth for Earth and Mars. Earth Planet. Sci. Lett. **271**, 181–191 (2008)

L.T. Elkins-Tanton, T.L. Grove, Water (hydrogen) in the lunar mantle: results from petrology and magma ocean modeling. Earth Planet. Sci. Lett. **307**, 173–179 (2011)

R.L. Evans, G. Hirth, K. Baba, D.W. Forsyth, A. Chave, R. Mackie, Geophysical evidence from the MELT area for compositional controls on oceanic plates. Nature **437**, 249–252 (2005)

M. Faccenda, Water in the slab: a trilogy. Tectonophysics **614**, 1–30 (2014)

M. Faccenda, T.V. Gerya, N.S. Mancktelow, L. Moresi, Fluid flow during slab unbending and dehydration: implications for intermediate-depth seismicity, slab weakening and deep water recycling. Geochem. Geophys. Geosyst. **13**, 1–23 (2012)

J. Farquhar, B. Wing, K.D. McKeegan, J.W. Harris, P. Cartigny, M.H. Thiemens, Mass-independent sulfur of inclusions in diamond and sulfur recycling on early Earth. Science **298**, 2369–2372 (2002)

J.R. Farver, Oxygen and hydrogen diffusion in minerals, in *Diffusion in Minerals and Melts*, ed. by Y. Zhang, D.J. Cherniak (Mineralogical Society of America, Geochemical Society, Chantilly, 2010), pp. 447–507

U.H. Faul, C.J. Cline Ii, E.C. David, A.J. Berry, I. Jackson, Titanium-hydroxyl defect-controlled rheology of the Earth's upper mantle. Earth Planet. Sci. Lett. **452**, 227–237 (2016)

A. Férot, N. Bolfan-Casanova, Water storage capacity in olivine and pyroxene to 14 GPa: implications for the water content of the Earth's upper mantle and nature of seismic discontinuities. Earth Planet. Sci. Lett. **349–350**, 218–230 (2012)

E. Ferriss, T. Plank, D. Walker, Site-specific hydrogen diffusion rates during clinopyroxene dehydration. Contrib. Mineral. Petrol. **171**, 1–24 (2016)

H. Feuchtgruber, E. Lellouch, B. Bézard, T. Encrenaz, T. de Graauw, G. Davis, Detection of HD in the atmospheres of Uranus and Neptune: a new determination of the D/H ratio. Astron. Astrophys. **341**, L17–L21 (1999)

J. Filip, M. Novák, A. Beran, R. Zbořil, Crystal chemistry and OH defect concentrations in spodumene from different granitic pegmatites. Phys. Chem. Miner. **32**, 733–746 (2006)

G. Fiquet, F. Guyot, J. Badro, The Earth's lower mantle and core. Elements **4**, 177–182 (2008)

M. Fischer-Gödde, T. Kleine, Ruthenium isotopic evidence for an inner Solar System origin of the late veneer. Nature **541**, 525–527 (2017)
K.M. Fisher, H.A. Ford, D.L. Abt, C.A. Rychert, The lithosphere-asthenosphere boundary. Annu. Rev. Earth Planet. Sci. **38**, 551–575 (2010)
C. Fitoussi, B. Bourdon, Silicon isotope evidence against an enstatite chondrite Earth. Science **335**, 1477–1480 (2012)
K. Fleming, Z. Martinec, D. Wolf, Glacial-isostatic adjustment and the viscosity structure underlying the Vatnajökull ice cap, Iceland. Pure Appl. Geophys. **164**, 751–768 (2007)
S.F. Foley, S. Buhre, D.E. Jacob, Evolution of the Archaean crust by delamination and shallow subduction. Nature **421**, 249–252 (2003)
T. Fouchet, P.G.J. Irwin, P. Parrish, S.B. Calcutt, F.W. Taylor, C.A. Nixon, T. Owen, Search for spatial variation in the jovian $^{15}N/^{14}N$ ratio from Cassini/CIRS observations. Icarus **172**, 50–58 (2004)
M.L. Frezzotti, S. Ferrando, The chemical behavior of fluids released during deep subduction based on fluid inclusions. Am. Mineral. **100**, 352–377 (2015)
M.L. Frezzotti, S. Ferrando, F. Tecce, D. Castelli, Water content and nature of solutes in shallow-mantle fluids from fluid inclusions. Earth Planet. Sci. Lett. **351–352**, 70–83 (2012)
I. Friedman, R.L. Smith, W.D. Long, Hydration of natural glass and formation of perlite. Geol. Soc. Am. Bull. **77**, 323–328 (1966)
D.J. Frost, The stability of hydrous mantle phases, in *Water in Nominally Anhydrous Minerals*, ed. by H. Keppler, J.R. Smyth (Mineralogical Society of America, Chantilly, 2006), pp. 243–271
D.J. Frost, The upper mantle and transition zone. Elements **4**, 171–176 (2008)
D.J. Frost, D. Dolejš, Experimental determination of the effect of H_2O on the 410-km seismic discontinuity. Earth Planet. Sci. Lett. **256**, 182–195 (2007)
D.J. Frost, C.A. McCammon, The redox state of the Earth's mantle. Annu. Rev. Earth Planet. Sci. **36**, 389–420 (2008)
Y. Fukai, The iron-water reaction and the evolution of the Earth. Nature **308**, 174–175 (1984)
Y. Fukao, M. Obayashi, T. Nakakuki, Stagnat slab: a review. Annu. Rev. Earth Planet. Sci. **37**, 19–46 (2009)
E. Füri, E. Deloule, A.A. Gurenko, B. Marty, New evidence for chondritic lunar water from combined D/H and noble gas analyses of single Apollo 17 volcanic glasses. Icarus **229**, 109–120 (2014)
E. Füri, P.H. Barry, L.A. Taylor, B. Marty, Indigenous nitrogen in the Moon: constraints from coupled nitrogen–noble gas analyses of mare basalts. Earth Planet. Sci. Lett. **431**, 195–205 (2015)
G.A. Gaetani, J.A. O'Leary, K.T. Koga, E.H. Hauri, E.F. Rose-Koga, B.D. Monteleone, Hydration of mantle olivine under variable water and oxygen fugacity conditions. Contrib. Mineral. Petrol. **167**, 965–979 (2014)
J. Ganguly, A. Asaduzzaman, K. Muralidharan, Origin of water in Earth with high D/H ratio relative to protosolar nebula, and an explanation of its similarity with the isotopic ratios of carbonaceous chondrites and asteroid Vesta, in *79th Meteoritical Society Meeting*, Berlin, Germany (2016), p. 6055
E. Gardés, F. Gaillard, P. Tarits, Toward a unified hydrous olivine electrical conductivity law. Geochem. Geophys. Geosyst. **15**, 4984–5000 (2014)
E.J. Garnero, A.K. McNamara, Structure and dynamics of Earth's lower mantle. Science **320**, 626–628 (2008)
E.J. Garnero, A.K. McNamara, S.-H. Shim, Continent-sized anomalous zones with low seismic velocity at the base of Earth's mantle. Nat. Geosci. **9**, 481–489 (2016)
T. Garth, A. Rietbrock, Order of magnitude increase in subducted H_2O due to hydrated normal faults within the Wadati-Benioff zone. Geology **42**, 207–210 (2014)
P. Gavrilenko, T. Boffa Ballaran, H. Keppler, The effect of Al and water on the compressibility of diopside. Am. Mineral. **95**, 608–616 (2010)
J. Geiss, G. Gloeckler, Isotopic composition of H, He and Ne in the protosolar cloud, in *Solar System History from Isotopic Signatures of Volatile Elements*, ed. by R. Kallenbach, T. Encrenaz, J. Geiss, K. Mauersberger, T.C. Owen, F. Robert (Springer, Dordrecht, 2003), pp. 3–18
H. Genda, Origin of Earth's oceans: an assessment of the total amount, history and supply of water. Geochem. J. **50**, 27–42 (2016)
H. Genda, M. Ikoma, Origin of the ocean on the Earth: early evolution of water D/H in a hydrogen-rich atmosphere. Icarus **194**, 42–52 (2008)
S. Ghosh, M.W. Schmidt, Melting of Phase D in the lower mantle and implications for recycling and storage of H_2O in the deep mantle. Geochim. Cosmochim. Acta **145**, 72–88 (2014)
J. Girard, J. Chen, P.C. Raterron, C.W. Holyoke, Hydrolytic weakening of olivine at mantle pressure: evidence of [1 0 0](0 1 0) slip system softening from single-crystal deformation experiments. Phys. Earth Planet. Inter. **216**, 12–20 (2013)
A. Giuliani, D. Phillips, V.S. Kamenetsky, M.L. Fiorentini, J. Farquhar, M.A. Kendrick, Stable isotope (C, O, S) compositions of volatile-rich minerals in kimberlites: a review. Chem. Geol. **374–375**, 61–83 (2014)

J. Gose, E. Schmädicke, A. Beran, Water in enstatite from Mid-Atlantic Ridge peridotite: evidence for water content of suboceanic mantle? Geology **37**, 543–546 (2009)

J. Gose, E. Schmädicke, R. Stalder, Water in mantle orthopyroxene—no visible change in defect water during serpentinization. Eur. J. Mineral. **23**, 529–536 (2011)

K. Grant, J. Ingrin, J.P. Lorand, P. Dumas, Water partitioning between mantle minerals from peridotite xenoliths. Contrib. Mineral. Petrol. **154**, 15–34 (2007a)

K.J. Grant, S.C. Kohn, R.A. Brooker, The partitioning of water between olivine, orthopyroxene and melt synthesised in the system albite-forsterite-H_2O. Earth Planet. Sci. Lett. **260**, 227–241 (2007b)

D.H. Green, Experimental melting studies on a model upper mantle composition at high pressure under H_2O-saturated and H_2O undersaturated conditions. Earth Planet. Sci. Lett. **19**, 37–45 (1973)

M. Grégoire, D.R. Bell, A.P. Le Roex, Trace element geochemistry of phlogopite-rich mafic mantle xenoliths: their classification and relationship to phlogopite-bearing peridotites and to kimberlites revisited. Contrib. Mineral. Petrol. **142**, 603–625 (2002)

M. Grégoire, D.R. Bell, A.P. Le Roex, Garnet lherzolites from the Kaapvaal craton (South Africa): trace element evidence for a metasomatic history. J. Petrol. **44**, 629–657 (2003)

W.L. Griffin, S. Graham, S.Y. O'Reilly, N.J. Pearson, Lithosphere evolution beneath the Kaapvaal craton: Re-Os systematics of sulfides in mantle-derived peridotites. Chem. Geol. **208**, 89–118 (2004)

W.L. Griffin, S.Y. O'Reilly, J.C. Afonso, G.C. Begg, The composition and evolution of lithospheric mantle: a re-evaluation and its tectonic implications. J. Petrol. **50**, 1185–1204 (2009)

W.L. Griffin, S.R. Shee, C.G. Ryan, T.T. Win, B.A. Wyatt, Harzburgite to lherzolite and back again: metasomatic processes in ultramafic xenoliths from the Wesselton kimberlite, Kimberley, South Africa. Contrib. Mineral. Petrol. **134**, 232–250 (1999)

T.L. Grove, C.B. Till, M.J. Krawczynski, The role of H_2O in subduction zone magmatism. Annu. Rev. Earth Planet. Sci. **40**, 413–439 (2012)

Y. Guan, Y. Wang, W. Hsu, J.M. Eiler, Analysis of OH and D/H of Apatites from Eucrites, in *79th Meteoritical Society Meeting*, Berlin, Germany (2016)

Y. Gung, M. Panning, B. Romanowicz, Global anisotropy and the thickness of continents. Nature **422**, 707–711 (2003)

A.A. Gurenko, V.S. Kamenetsky, A.C. Kerr, Oxygen isotopes and volatile contents of the Gorgona komatiites, Colombia: a confirmation of the deep mantle origin of H_2O. Earth Planet. Sci. Lett. **454**, 154–165 (2016)

B.R. Hacker, H_2O subduction beyond arcs. Geochem. Geophys. Geosyst. **9**, 1–24 (2008)

A.N. Halliday, Mixing, volatile loss and compositional change during impact-driven accretion of the Earth. Nature **427**, 505–509 (2004)

A.N. Halliday, The origins of volatiles in the terrestrial planets. Geochim. Cosmochim. Acta **105**, 146–171 (2013)

A.N. Halliday, D. Porcelli, In search of lost planets—the paleocosmochemistry of the inner solar system. Earth Planet. Sci. Lett. **192**, 545–559 (2001)

L.J. Hallis, G.R. Huss, K. Nagashima, G.J. Taylor, S.A. Halldórsson, D.R. Hilton, M.J. Motti, K.J. Meech, Evidence for primordial water in Earth's deep mantle. Science **350**, 795–797 (2015)

L.J. Hallis, G.J. Taylor, K. Nagashima, G.R. Huss, Magmatic water in the martian meteorite Nakhla. Earth Planet. Sci. Lett. **359–360**, 84–92 (2012)

L.J. Hallis, D/H ratios of the inner Solar System. Philos. Trans. R. Soc. Lond. A **375**, 20150390 (2017)

M. Hamada, T. Kawamoto, E. Takahashi, T. Fujii, Polybaric degassing of island arc low-K tholeiitic basalt magma recorded by OH concentrations in Ca-rich plagioclase. Earth Planet. Sci. Lett. **308**, 259–266 (2011)

M. Hamada, M. Ushioda, T. Fujii, E. Takahashi, Hydrogen concentration in plagioclase as a hygrometer of arc basaltic melts: approaches from melt inclusion analyses and hydrous melting experiments. Earth Planet. Sci. Lett. **365**, 253–262 (2013)

U. Hans, T. Kleine, B. Bourdon, Rb–Sr chronology of volatile depletion in differentiated protoplanets: BABI, ADOR and ALL revisited. Earth Planet. Sci. Lett. **374**, 204–214 (2013)

Y. Hao, Q.-K. Xia, Q. Li, H. Chen, M. Feng, Partial melting control of water contents in the Cenozoic lithospheric mantle of the Cathaysia block of South China. Chem. Geol. **380**, 7–19 (2014)

Y.-T. Hao, Q.-K. Xia, Z.-B. Jia, Q.-C. Zhao, P. Li, M. Feng, S.-C. Liu, Regional heterogeneity in the water content of the Cenozoic lithospheric mantle of Eastern China. J. Geophys. Res. **121**, 517–537 (2016)

S.H. Hart, E.H. Hauri, L.A. Oschmann, J.A. Whitehead, Mantle plumes and entrainment: isotopic evidence. Science **256**, 517–519 (1992)

B. Harte, Mantle peridotites and processes—the kimberlite sample, in *Continental Basalts and Mantle Xenoliths*, ed. by C.J. Hawkesworth, M.J. Norry (Shiva Publishing Limited, Cambridge, 1983), p. 272

M.E. Hartley, D.A. Neave, J. MacLennan, M. Edmonds, Diffusive over-hydration of olivine-hosted melt inclusions. Earth Planet. Sci. Lett. **425**, 168–178 (2015)

P. Hartogh, D.C. Lis, D. Bockelee-Morvan, M. de Val-Borro, N. Biver, M. Kuppers, M. Emprechtinger, E.A. Bergin, J. Crovisier, M. Rengel, R. Moreno, S. Szutowicz, G.A. Blake, Ocean-like water in the Jupiter-family comet 103P/Hartley 2. Nature **478**, 218–220 (2011)

E. Hauri, SIMS analysis of volatiles in silicate glasses, 2: isotopes and abundances in Hawaiian melt inclusions. Chem. Geol. **183**, 115–141 (2002)

E. Hauri, J. Wang, J.E. Dixon, P.L. King, C. Mandeville, S. Newman, SIMS analysis of volatiles in silicate glasses 1. Calibration, matrix effects and comparison with FTIR. Chem. Geol. **183**, 99–114 (2002)

E.H. Hauri, G.A. Gaetani, T.H. Green, Partitioning of water during melting of the Earth's upper mantle at H_2O-undersaturated conditions. Earth Planet. Sci. Lett. **248**, 715–734 (2006)

E.H. Hauri, A.E. Saal, M.C. Rutherford, J.A. Van Orman, Water in the Moon's interior: truth and consequences. Earth Planet. Sci. Lett. **409**, 252–264 (2015)

E.H. Hauri, T. Weinreich, A.E. Saal, M.C. Rutherford, J.A. Van Orman, High pre-eruptive water contents preserved in lunar melt inclusions. Science **333**, 213–215 (2011)

C. Hayashi, Structure of the solar nebula, growth and decay of magnetic fields and effects of magnetic and turbulent viscosities on the nebula. Prog. Theor. Phys. Suppl. **70**, 35–53 (1981)

C. Hayashi, K. Nakazawa, H. Mizuno, Earth's melting due to the blanketing effect of the primordial dense atmosphere. Earth Planet. Sci. Lett. **43**, 22–28 (1979)

V.S. Heber, R. Wieler, H. Baur, C. Olinger, T.A. Friedmann, D.S. Burnett, Noble gas composition of the solar wind as collected by the Genesis mission. Geochim. Cosmochim. Acta **73**, 7414–7432 (2009)

C.D.K. Herd, R.C. Peterson, G.R. Rossman, Violet-colored diopside from southern Baffin island, Nunavut, Canada. Can. Mineral. **38**, 1193–1199 (2000)

D.R. Hilton, T.P. Fischer, B. Marty, Noble gases and volatile recycling at subduction zones. Rev. Mineral. Geochem. **47**, 319–370 (2002)

K. Hirose, T. Kawamoto, Hydrous partial melting of lherzolite at 1 GPa: the effect of H_2O on the genesis of basaltic magmas. Earth Planet. Sci. Lett. **133**, 463–473 (1995)

M.M. Hirschmann, Water, melting, and the deep Earth H_2O cycle. Annu. Rev. Earth Planet. Sci. **34**, 629–653 (2006)

M.M. Hirschmann, Constraints on the early delivery and fractionation of Earth's major volatiles from C/H, C/N, and C/S ratios. Am. Mineral. **101**, 540–553 (2016)

M.M. Hirschmann, C. Aubaud, A.C. Withers, Storage capacity of H_2O in nominally anhydrous minerals in the upper mantle. Earth Planet. Sci. Lett. **236**, 167–181 (2005)

M.M. Hirschmann, T.J. Tenner, C. Aubaud, A.C. Withers, Dehydration melting of nominally anhydrous mantle: the primacy of partitioning. Phys. Chem. Earth **176**, 54–68 (2009)

M.M. Hirschmann, A.C. Withers, P. Ardia, N.T. Foley, Solubility of molecular hydrogen in silicate melts and consequences for volatile evolution of terrestrial planets. Earth Planet. Sci. Lett. **345–348**, 38–48 (2012)

G. Hirth, R.L. Evans, A.D. Chave, Comparison of continental and oceanic mantle electrical conductivity: is the Archean lithosphere dry? Geochem. Geophys. Geosyst. **1**, 1030 (2000)

G. Hirth, D.L. Kohlstedt, Water in the oceanic upper mantle; implications for rheology, melt extraction and the evolution of the lithosphere. Earth Planet. Sci. Lett. **144**, 93–108 (1996)

G. Hirth, D.L. Kohlstedt, Rheology of the upper mantle and the mantle wedge: a view from the experimentalists, in *Inside the Subduction Factory*, ed. by J. Eiler (American Geophysical Union, Washington, 2004), pp. 83–106

A.W. Hofmann, Lead isotopes and the age of the Earth—a geochemical accident. Geol. Soc. (Lond.) Spec. Publ. **190**, 223–236 (2001)

A.W. Hofmann, Sampling mantle heterogeneity through oceanic basalts: isotopes and trace elements, in *The Mantle and Core*, ed. by H.D. Holland, K.K. Turekian (Elsevier, Oxford, 2003), pp. 1–44

G. Holland, M. Cassidy, C.J. Ballentine, Meteorite Kr in Earth's mantle suggests a late accretionary source for the atmosphere. Science **326**, 1522–1525 (2009)

D. Holloway, J.G. Blank, Application of experimental results to C-O-H species in natural melts, in *Volatiles in Magmas*, ed. by M.R. Carroll, J.R. Holloway (Mineralogical Society of America, Washington D.C., 1994), pp. 187–230

M.D. Hopkins, S.J. Mojzsis, W.F. Bottke, O. Abramov, Micrometer-scale U–Pb age domains in eucrite zircons, impact re-setting, and the thermal history of the HED parent body. Icarus **245**, 367–378 (2015)

C. Houser, Global seismic data reveal little water in the mantle transition zone. Earth Planet. Sci. Lett. **448**, 94–101 (2016)

Q. Hu, D.Y. Kim, J. Liu, Y. Meng, L. Yang, D. Zhang, W.L. Mao, H-k. Mao, Dehydrogenation of goethite in Earth's deep lower mantle. Proc. Natl. Acad. Sci. **114**, 1498–1501 (2017)

Q. Hu, D.Y. Kim, W. Yang, L. Yang, Y. Meng, L. Zhang, H.-K. Mao, FeO_2 and FeOOH under deep lower-mantle conditions and Earth's oxygen–hydrogen cycles. Nature **534**, 241–244 (2016)

S. Hu, Y. Lin, J. Zhang, J. Hao, L. Feng, L. Xu, W. Yang, J. Yang, NanoSIMS analyses of apatite and melt inclusions in the GRV 020090 Martian meteorite: hydrogen isotope evidence for recent past underground hydrothermal activity on Mars. Geochim. Cosmochim. Acta **140**, 321–333 (2014)

J.-X. Huang, P. Li, W.L. Griffin, Q.-K. Xia, Y. Gréau, N.J. Pearson, S.Y. O'Reilly, Water contents of Roberts Victor xenolithic eclogites: primary and metasomatic controls. Contrib. Mineral. Petrol. **168**, 1–13 (2014)

X. Huang, Y. Xu, S.-I. Karato, Water content in the transition zone from electrical conductivity of wadsleyite and ringwoodite. Nature **434**, 746–749 (2005)

H. Hui, Y. Guan, Y. Chen, A.H. Peslier, Y. Zhang, Y. Liu, R.L. Flemming, G.R. Rossman, J.M. Eiler, C.R. Neal, G.R. Osinsky, A heterogeneous lunar interior for hydrogen isotopes as revealed by the lunar highlands samples. Earth Planet. Sci. Lett. **473**, 14–23 (2017). doi:10.1016/j.epsl.2017.05.029

H. Hui, A.H. Peslier, R.L. Rudnick, A. Simonetti, C.R. Neal, Plume-cratonic lithosphere interaction recorded by water and other trace elements in peridotite xenoliths from the Labait volcano, Tanzania. Geochem. Geophys. Geosyst. **16**, 1–24 (2015)

M. Ichiki, K. Baba, H. Toh, K. Fuji-ta, An overview of electrical conductivity structures of the crust and upper mantle beneath the northwestern Pacific, the Japanese Islands, and continental East Asia. Gondwana Res. **16**, 545–562 (2009)

R. Iizuka-Oku, T. Yagi, H. Gotou, T. Okuchi, T. Hattori, A. Sano-Furukawa, Hydrogenation of iron in the early stage of Earth's evolution. Nat. Commun. **8**, 14096 (2017)

N.K. Inamdar, H.E. Schlichting, The formation of super-Earths and mini-Neptunes with giant impacts. Mon. Not. R. Astron. Soc. **448**, 1751–1760 (2015)

J. Ingrin, M. Blanchard, Diffusion of hydrogen in minerals, in *Water in Nominally Anhydrous Minerals*, ed. by H. Keppler, J.R. Smyth (Mineralogical Society of America, Chantilly, 2006), pp. 291–320

T. Inoue, H. Yurimoto, Y. Kudoh, Hydrous modified spinel, $Mg_{1.75}SiH_{0.5}O_4$: a new water reservoir in the mantle transition region. J. Geophys. Res. **22**, 117–120 (1995)

D.A. Ionov, L.S. Doucet, I.V. Ashchepkov, Composition of the lithospheric mantle in the Siberian craton: new constraints from fresh peridotites in the Udachnaya-East kimberlite. J. Petrol. **51**, 2177–2210 (2010)

Y. Ito, S. Nakashima, Water distribution in low-grade siliceous metamorphic rocks by micro-FTIR and its relation to grain size: a case from the Kanto Mountain region, Japan. Chem. Geol. **189**, 1–18 (2002)

M.G. Jackson, S.R. Hart, J.G. Konter, M.D. Kurz, J. Blusztajn, K.A. Farley, Helium and lead isotopes reveal the geochemical geometry of the Samoan plume. Nature **514**, 355–358 (2014)

M.G. Jackson, S.R. Hart, A.A.P. Koppers, H. Staudigel, J. Konter, J. Blusztajn, M.D. Kurz, J.A. Russell, The return of subducted continental crust in Samoan lavas. Nature **448**, 684–687 (2007)

M.G. Jackson, K.T. Koga, A. Price, J.G. Konter, A.A.P. Koppers, V.A. Finlayson, K. Konrad, E.H. Hauri, A. Kylander-Clark, K.A. Kelley, M.A. Kendrick, Deeply dredged submarine HIMU glasses from the Tuvalu Islands, Polynesia: implications for volatile budgets of recycled oceanic crust. Geochem. Geophys. Geosyst. **16**, 3210–3234 (2015)

D.E. Jacob, Nature and origin of eclogite xenoliths from kimberlites. Lithos **77**, 295–316 (2004)

S.D. Jacobsen, Effect of water on the equation of state of nominally anhydrous minerals, in *Water in Nominally Anhydrous Minerals*, ed. by H. Keppler, J.R. Smyth (Mineralogical Society of America, Chantilly, 2006), pp. 321–342

A. Jambon, J.L. Zimmermann, Major volatiles from a North Atlantic MORB glass and calibration to He: a size fraction analysis. Chem. Geol. **62**, 177–189 (1987)

B. Jamtveit, R. Brooker, K. Brooks, L.M. Larsen, T. Pedersen, The water content of olivines from the North Atlantic volcanic province. Earth Planet. Sci. Lett. **186**, 401–415 (2001)

E. Jarosewich, Chemical analyses of meteorites: a compilation of stony and iron meteorite analyses. Meteoritics **25**, 323–337 (1990)

R.D. Jarrard, Subduction fluxes of water, carbon dioxide, chlorine, and potassium. Geochem. Geophys. Geosyst. **4**, 1–50 (2003)

C. Jaupart, J.C. Mareschal, The thermal structure and thickness of continental roots. Lithos **48**, 93–114 (1999)

M. Javoy, E. Kaminski, F. Guyot, D. Andrault, C. Sanloup, M. Moreira, S. Labrosse, A. Jambon, P. Agrinier, A. Davaille, C. Jaupart, The chemical composition of the Earth: enstatite chondrite models. Earth Planet. Sci. Lett. **293**, 259–268 (2010)

M.M. Jean, L.A. Taylor, G.H. Howarth, A.H. Peslier, L. Fedele, R.J. Bodnar, Y. Guan, L.S. Doucet, D.A. Ionov, A.M. Logvinova, A.V. Golovin, N.V. Sobolev, Contrasting water and trace elements in olivines from Siberian diamonds versus mantle xenoliths. Lithos **265**, 31–41 (2016). doi:10.1016/j.lithos.2016.07.023

B. Johnson, C. Goldblatt, The nitrogen budget of Earth. Earth-Sci. Rev. **148**, 150–173 (2015)

E.A. Johnson, G.R. Rossman, The concentration and speciation of hydrogen in feldspars using FTIR and ^{1}H MAS NMR spectroscopy. Am. Mineral. **88**, 901–911 (2003)

E.A. Johnson, G.R. Rossman, A survey of hydrous species and concentrations in igneous feldspars. Am. Mineral. **89**, 586–600 (2004)

E.A. Johnson, G.R. Rossman, The diffusion behavior of hydrogen in plagioclase feldspar at 800–1000 °C: implications for re-equilibration of hydroxyl in volcanic phenocrysts. Am. Mineral. **98**, 1779–1787 (2013)

A.G. Jones, J. Ledo, I.J. Ferguson, J.A. Craven, M.J. Unsworth, M. Chouteau, J.E. Spratt, The electrical resistivity of Canada's lithosphere and correlation with other parameters: contributions from LITHOPROBE and other programmes. Can. J. Earth Sci. **51**, 573–617 (2013a)

A.G. Jones, M.R. Muller, S. Fishwick, R.L. Evans, J. Fullea, Velocity-conductivity relations for cratonic lithosphere and their application: example of southern Africa. Geochem. Geophys. Geosyst. **14**, 806–827 (2013b)

R.H. Jones, F.M. McCubbin, L. Dreeland, Y. Guan, P.V. Burger, C.K. Shearer, Phosphate minerals in LL chondrites: a record of the action of fluids during metamorphism on ordinary chondrite parent bodies. Geochim. Cosmochim. Acta **132**, 120–140 (2014)

R.H. Jones, F.M. McCubbin, Y. Guan, Phosphate minerals in the H group of ordinary chondrites, and fluid activity recorded by apatite heterogeneity in the Zag H3-6 regolith breccia. Am. Mineral. **101**, 2452–2467 (2016)

T.H. Jordan, Composition and development of the continental tectosphere. Nature **274**, 544–548 (1978)

H. Jung, S. Karato, Effects of water on dynamically recrystallized grain-size of olivine. J. Struct. Geol. **23**, 1337–1344 (2001a)

H. Jung, S. Karato, Water-induced fabric transitions in olivine. Science **293**, 1460–1463 (2001b)

V.S. Kamenetsky, A.V. Golovin, R. Maas, A. Giuliani, M.B. Kamenetsky, Y. Weiss, Towards a new model for kimberlite petrogenesis: evidence from unaltered kimberlites and mantle minerals. Earth-Sci. Rev. **139**, 145–167 (2014)

F. Kaminsky, O. Zakharchenko, R. Davies, W. Griffin, G. Khachatryan-Blinova, A. Shiryaev, Superdeep diamonds from the Juina area, Mato Grosso State, Brazil. Contrib. Mineral. Petrol. **140**, 734–753 (2001)

S. Karato, Does partial melting reduce the creep strength of the upper mantle? Nature **319**, 309–310 (1986)

S. Karato, The role of hydrogen in the electrical conductivity of the upper mantle. Nature **347**, 272–273 (1990)

S. Karato, Rheology of the upper mantle: a synthesis. Science **260**, 771–778 (1993)

S. Karato, Influence of hydrogen-related defects on the electrical conductivity and plastic deformation of mantle minerals: a critical review, in *Earth's Deep Water Cycle*, ed. by S.D. Jacobsen, S. van der Lee (American Geophysical Union, Washington, 2006a), pp. 113–130

S. Karato, Remote sensing of hydrogen in Earth's mantle, in *Water in Nominally Anhydrous Minerals*, ed. by H. Keppler, J.R. Smyth (Mineralogical Society of America, Chantilly, 2006b), pp. 343–375

S. Karato, Rheology of the deep upper mantle and its implications for the preservation of the continental roots: a review. Tectonophysics **481**, 82–98 (2010)

S. Karato, Water distribution across the mantle transition zone and its implications for global material circulation. Earth Planet. Sci. Lett. **301**, 413–423 (2011)

S. Karato, M.R. Riedel, D.A. Yuen, Rheological structure and deformation of subducted slabs in the mantle transition zone: implications for mantle circulation and deep earthquakes. Phys. Earth Planet. Inter. **127**, 83–108 (2001)

S.-I. Karato, Microscopic models for the effects of hydrogen on physical and chemical properties of Earth materials, in *Superplumes: Beyond Plate Tectonics*, ed. by D.A. Yuen, S. Maruyama, S.-I. Karato, B.F. Windley (Springer, Dordrecht, 2007), pp. 321–356

S.-I. Karato, Water in the evolution of Earth and other terrestrial planets, in *Treatise on Geophysics*, ed. by G. Schubert (Elsevier, Amsterdam, 2015), pp. 105–144

S.-I. Karato, Physical basis of trace element partitioning: a review. Am. Mineral. **101**, 2577–2593 (2016)

S.-I. Karato, H. Jung, I. Katayama, P. Skemer, Geodynamic significance of seismic anisotropy of the upper mantle: new insights from laboratory studies. Annu. Rev. Earth Planet. Sci. **36**, 59–95 (2008)

S.-I. Karato, D. Wang, Electrical conductivity of minerals and rocks, in *Physics and Chemistry of the Deep Earth*, ed. by S.-I. Karato (Wiley, New York, 2013), pp. 145–182

I. Katayama, K. Hirose, H. Yurimoto, S. Nakashima, Water solubility in majoritic garnet in subducting oceanic crust. Geophys. Res. Lett. **30**, 2155 (2003)

I. Katayama, S. Nakashima, Hydroxyl in clinopyroxene from the deep subducted crust: evidence for H_2O transport into the mantle. Am. Mineral. **88**, 229–234 (2003)

I. Katayama, S. Nakashima, H. Yurimoto, Water content in natural eclogite and implication for water transport into the deep upper mantle. Lithos **86**, 245–259 (2006)

H. Kawakatsu, S. Watada, Seismic evidence for deep-water transportation in the mantle. Science **316**, 1468–1471 (2007)

T. Kawamoto, Hydrous phases and water transport in the subducting slab, in *Water in Nominally Anhydrous Minerals*, ed. by H. Keppler, J.R. Smyth (Mineralogical Society of America, Chantilly, 2006), pp. 273–289

K. Kehm, S. Crowther, J.D. Gilmour, R.K. Mohapatra, C.M. Hohenberg, Upper limit concentrations of trapped xenon in individual interplanetary dust particles from the stratosphere. Meteorit. Planet. Sci. **44**, 249–259 (2009)

P.B. Kelemen, S.R. Hart, S. Bernstein, Silica enrichment in the continental upper mantle via melt/rock reaction. Earth Planet. Sci. Lett. **164**, 387–406 (1998)

M.A. Kendrick, M. Jackson, E.H. Hauri, D. Phillips, The halogen (F, Cl, Br, I) and H_2O systematics of Samoan lavas: assimilated-seawater, EM2 and high-$^3He/^4He$ components. Earth Planet. Sci. Lett. **410**, 197–2069 (2015)

M.A. Kendrick, M.G. Jackson, A.J.R. Kent, E.H. Hauri, P.J. Wallace, J. Woodhead, Contrasting behaviours of CO_2, S, H_2O and halogens (F, Cl, Br and I) in enriched-mantle melts from Pitcairn and Society seamounts. Chem. Geol. **370**, 69–81 (2014)

G.C. Kennedy, G.J. Wasserburg, H.C. Heard, R.C. Newton, The upper-three phase region in the system SiO_2-H_2O. Am. J. Sci. **260**, 501–521 (1962)

A.J.R. Kent, Melt inclusions in basaltic and related volcanic rocks, in *Minerals, Inclusions and Volcanic Processes*, ed. by K.D. Purtika, F.J. Tepley III. (Mineralogical Society of America, Chantilly, 2008), pp. 273–331

H. Keppler, Fluids and trace element transport in subduction zones. Am. Mineral. **102**, 5–20 (2017)

S.E. Kesson, A.E. Ringwood, Slab-mantle interactions: 2. The formation of diamonds. Chem. Geol. **78**, 97–118 (1989)

A. Khan, On Earth's mantle constitution and structure from joint analysis of geophysical and laboratory-based data: an example. Surv. Geophys. **37**, 149–189 (2016)

A. Khan, A. Kushinov, A. Semenov, On the heterogeneous electrical conductivity structure of the Earth's mantle with implications for transition zone water content. J. Geophys. Res. **116**, B01103 (2011)

A. Khan, T.J. Shankland, A geophysical perspective on mantle water content and melting: inverting electromagnetic sounding data using laboratory-based electrical conductivity profiles. Earth Planet. Sci. Lett. **317–318**, 27–43 (2012)

A.J. King, J.R. Solomon, P.F. Schofield, S.S. Russell, Characterising the CI and CI-like carbonaceous chondrites using thermogravimetric analysis and infrared spectroscopy. Earth Planets Space **67**, 198 (2015)

P.L. King, P.F. McMillan, G.M. Moore, Infrared spectroscopy of silicate glasses with application to natural systems, in *Infrared Spectroscopy in Geochemistry, Exploration Geochemistry, and Remote Sensing*, ed. by P.L. King, M.S. Ramsey, G.A. Swayze (Mineralogical Association of Canada, London, 2004), pp. 92–133

E.S. Kiseeva, B.J. Wood, A simple model for chalcophile element partitioning between sulphide and silicate liquids with geochemical applications. Earth Planet. Sci. Lett. **383**, 68–81 (2013)

M. Koch-Müller, I. Abs-Wurmbach, D. Rhede, V. Kahlenberg, S. Matsyuk, Dehydration experiments on natural omphacites: qualitative and quantitative characterization by various spectroscopic methods. Phys. Chem. Miner. **34**, 663–678 (2007)

M. Koch-Müller, S.S. Matsyuk, R. Wirth, Hydroxyl in omphacite and omphacitic clinopyroxenes of upper mantle to lower crustal origin beneath the Siberian platform. Am. Mineral. **89**, 921–931 (2004)

M. Koch-Müller, D. Rhede, IR absorption coefficients for water in nominally anhydrous high-pressure minerals. Am. Mineral. **95**, 770–775 (2010)

K.T. Koga, E. Hauri, M.M. Hirschmann, D. Bell, Hydrogen concentration analyses using SIMS and FTIR: comparison and calibration for nominally anhydrous minerals. Geochem. Geophys. Geosyst. **4**, 1019 (2003). doi:10.1029/2002GC000378

D.L. Kohlstedt, H. Keppler, D.C. Rubie, Solubility of water in the α, β and γ phases of $(Mg, Fe)_2SiO_4$. Contrib. Mineral. Petrol. **123**, 345–357 (1996)

D.L. Kohlstedt, S.J. Mackwell, Diffusion of hydrogen and intrinsic point defects in olivine. Z. Phys. Chem. **207**, 147–162 (1998)

S.C. Kohn, The dissolution mechanism of water in silicate melts; a synthesis of recent data. Mineral. Mag. **64**, 389–408 (2000)

M.V. Kolesnichenko, D.A. Zedgenizov, K.D. Litasov, I.Y. Safonova, A.L. Ragozin, Heterogeneous distribution of water in the mantle beneath the central Siberian Craton: implications from the Udachnaya Kimberlite Pipe. Gondwana Res. **47**, 249–266 (2016)

T. Komabayashi, Phase relations of hydrous peridotite: implications for water circulation in the Earth's mantle, in *Earth's Deep Water Cycle*, ed. by S. Jacobsen, S. Lee (AGU Geophysical Monograph, Washington D.C., 2006), pp. 29–43

J. Konzett, R.A. Armstrong, R.J. Sweeney, W. Compston, The timing of MARID metasomatism in the Kaapvaal mantle: an ion probe study of zircons from MARID xenoliths. Earth Planet. Sci. Lett. **160**, 133–145 (1998)

J. Konzett, K. Krenn, D. Rubatto, C. Hauzenberger, R. Stalder, The formation of saline mantle fluids by open-system crystallization of hydrous silicate-rich vein assemblages—evidence from fluid inclusions and their host phases in MARID xenoliths from the central Kaapvaal Craton, South Africa. Geochim. Cosmochim. Acta **147**, 1–25 (2014)

T. Koyama, H. Shimizu, H. Utada, M. Ichiki, E. Ohtani, R. Hae, Water content in the mantle transition zone beneath the North Pacific derived from the electrical conductivity anomaly, in *Earth's Deep Water Cycle* (American Geophysical Union, Washington, 2013), pp. 171–179

A.K. Kronenberg, R.A. Yund, G.R. Rossman, Stationary and mobile hydrogen defects in potassium feldspar. Geochim. Cosmochim. Acta **60**, 4075–4094 (1996)

K. Kuramoto, T. Matsui, Partitioning of H and C between the mantle and core during the core formation in the Earth: its implications for the atmospheric evolution and redox state of early mantle. J. Geophys. Res. **101**, 14909–14932 (1996)

M. Kurosawa, H. Yurimoto, S. Sueno, Patterns in the hydrogen and trace element compositions of mantle olivines. Phys. Chem. Miner. **24**, 385–395 (1997)

S. Labrosse, J.W. Hernlund, N. Coltice, A crystallizing dense magma ocean at the base of the Earth's mantle. Nature **450**, 866–869 (2007)

C.F. Larsen, R.J. Motyka, J.T. Freymueller, K.A. Echelmeyer, E.R. Ivins, Rapid viscoelastic uplift in southeast Alaska caused by post-Little Ice Age glacial retreat. Earth Planet. Sci. Lett. **237**, 548–560 (2005)

J.F. Lawrence, M.E. Wysession, Seismic evidence for subduction-transported water in the lower mantle, in *Earth's Deep Water Cycle*, ed. by S. Jacobsen, S. Lee (AGU Geophysical Monograph, Washington D.C., 2006), pp. 251–262

M. Le Voyer, P.D. Asimow, J.L. Mosenfelder, Y. Guan, P.J. Wallace, P. Schiano, E.M. Stolper, J.M. Eiler, Zonation of H_2O and F concentrations around melt inclusions in olivines. J. Petrol. **55**, 685–707 (2014)

C. Lécuyer, P. Gillet, F. Robert, The hydrogen isotope composition of seawater and the global water cycle. Chem. Geol. **145**, 249–261 (1998)

C.-T. Lee, P. Luffi, E.J. Chin, Building and destroying continental mantle. Annu. Rev. Earth Planet. Sci. **39**, 59–90 (2010)

C.-T. Lee, R.L. Rudnick, Compositionally stratified cratonic lithosphere: petrology and geochemistry of peridotite xenoliths from the Labait Volcano, Tanzania, in *Proc. VIIth International Kimberlite Conference, B.J. Dawson volume*, ed. by J.J. Gurney, J.L. Gurney, M.D. Pascoe, S.R. Richardson (1999), pp. 503–521

E. Lellouch, B. Bézard, T. Fouchet, H. Feuchtgruber, T. Encrenaz, T. de Graauw, The deuterium abundance in Jupiter and Saturn from ISO-SWS observations. Astron. Astrophys. **370**, 610–622 (2001)

Y. Li, Immiscible C-H-O fluids formed at subduction zone conditions. Geochem. Perspect. Lett. **3**, 12–21 (2017)

Z.-X.A. Li, C.-T. Lee, A.H. Peslier, A. Lenardic, S.J. Mackwell, Water contents in mantle xenoliths from the Colorado Plateau and vicinity: implications for the mantle rheology and hydration-induced thinning of continental lithosphere. J. Geophys. Res. **113**, 1–22 (2008)

E. Libowitzky, A. Beran, IR spectroscopic characterization of hydrous species in minerals, in *Spectroscopic Methods in Mineralogy. European Mineralogical Union*, ed. by A. Beran, E. Libowitzky (2004), pp. 227–279

J. Liu, Q.-K. Xia, E. Deloule, J. Ingrin, H. Chen, M. Feng, Water content and oxygen isotopic composition of alkali basalts from the Taihang Mountains, China: recycled oceanic components in the mantle source. J. Petrol. **56**, 681–702 (2015)

A.S. Lloyd, T. Plank, P. Ruprecht, E.H. Hauri, W. Rose, Volatile loss from melt inclusions in pyroclasts of differing sizes. Contrib. Mineral. Petrol. **165**, 129–153 (2013)

A.S. Lloyd, P. Ruprecht, E.H. Hauri, W. Rose, H.M. Gonnerman, T. Plank, NanoSIMS results from olivine-hosted melt embayments: magma ascent rate during explosive basaltic eruptions. J. Volcanol. Geotherm. Res. **283**, 1–18 (2014)

K. Lodders, Solar system abundances and condensation temperatures of the elements. Astrophys. J. **591**, 1220–1247 (2003)

F. Lucassen, M. Koch-Müller, M. Taran, G. Franz, Coupled H and Nb, Cr, and V trace element behavior in synthetic rutile at 600 °C, 400 MPa and possible geological application. Am. Mineral. **98**, 7–18 (2013)

R.W. Luth, Mantle volatiles—distribution and consequences, in *Treatise on Geochemistry*, ed. by R.W. Carlson (Elsevier, Amsterdam, 2003), pp. 319–361

I.D. MacGregor, W.I. Manton, Roberts Victor eclogites: ancient oceanic crust. J. Geophys. Res. **91**, 14063–14079 (1986)

S.J. Mackwell, D.L. Kohlstedt, M.S. Paterson, The role of water in the deformation of olivine single crystals. J. Geophys. Res. **90**, 11319–11333 (1985)

V. Magni, P. Bouilhol, J. van Hunen, Deep water recycling through time. Geochem. Geophys. Geosyst. **15**, 4203–4216 (2014)

P.R. Mahaffy, T.M. Donahue, S.K. Atreya, T.C. Owen, H.B. Niemann, Galileo probe measurements of D/H and $^3He/^4He$ in Jupiter's atmosphere. Space Sci. Rev. **84**, 251–263 (1998)
D. Mainprice, P.G. Silver, Interpretation of SKS-waves using samples from the subcontinental lithosphere. Phys. Earth Planet. Inter. **78**, 257–280 (1993)
B. Marty, The origins and concentrations of water, carbon, nitrogen and noble gases on Earth. Earth Planet. Sci. Lett. **313–314**, 56–66 (2012)
B. Marty, C.M.O.D. Alexander, S.N. Raymond, Primordial Origins of Earth's Carbon. Rev. Mineral. Geochem. **75**, 149–181 (2013)
B. Marty, G. Avice, Y. Sano, K. Altwegg, H. Balsiger, M. Hässig, A. Morbidelli, O. Mousis, M. Rubie, Origins of volatile elements (H, C, N, noble gases) on Earth and Mars in light of recent results from the ROSETTA cometary mission. Earth Planet. Sci. Lett. **441**, 91–102 (2016)
B. Marty, R.L. Palma, R.O. Pepin, L. Zimmermann, D.J. Schlutter, P.G. Burnard, A.J. Westphal, C.J. Snead, S. Bajt, R.H. Becker, J.E. Simones, Helium and neon abundances and compositions in cometary matter. Science **319**, 75–78 (2008)
B. Marty, R. Yokochi, Water in the early Earth. Rev. Mineral. Geochem. **62**, 421–450 (2006)
S. Masuti, S.D. Barbot, S.-I. Karato, L. Feng, P. Banerjee, Upper-mantle water stratification inferred from observations of the 2012 Indian Ocean earthquake. Nature **538**, 373–377 (2016)
S.S. Matsyuk, K. Langer, Hydroxyl in olivines from mantle xenoliths in kimberlites of the Siberian platform. Contrib. Mineral. Petrol. **147**, 413–437 (2004)
S.S. Matsyuk, K. Langer, A. Hösch, Hydroxyl defects in garnets from mantle xenoliths in kimberlites of the Siberian platform. Contrib. Mineral. Petrol. **132**, 163–179 (1998)
S. Matveev, C. Ballhaus, K. Fricke, J. Truckenbrodt, D. Ziegenbein, Volatiles in the Earth's mantle: synthesis of CHO fluids at 1273 K and 2.4 GPa. Geochim. Cosmochim. Acta **61**, 3081–3088 (1997)
C.A. McCammon, M.G. Kopylova, A redox profile of the Slave mantle and oxygen fugacity control in the cratonic mantle. Contrib. Mineral. Petrol. **148**, 55–68 (2004)
W.F. McDonough, Compositional model for the Earth's core, in *Treatise on Geochemistry; The Mantle and Core*, ed. by R.W. Carlson (Elsevier, Amsterdam, 2005), pp. 547–568
R.S. McGary, R.L. Evans, P.E. Wannamaker, J. Elsenbeck, S. Rondenay, Pathway from subducting slab to surface for melt and fluids beneath Mount Rainier. Nature **511**, 338–340 (2014)
B.I.A. McInnes, M. Gregoire, R.A. Binns, P.M. Herzig, M.D. Hannington, Hydrous metasomatism of oceanic sub-arc mantle, Lihir, Papua New Guinea: petrology and geochemistry of fluid-metasomatised mantle wedge xenoliths. Earth Planet. Sci. Lett. **188**, 169–183 (2001)
K.D. McKeegan, J. Aléon, J.P. Bradley, D.E. Brownlee, H. Busemann, A. Butterworth, M. Chaussidon, S. Fallon, C. Floss, J. Gilmour, M. Gounelle, G. Graham, Y.N. Guan, P.R. Heck, P. Hoppe, I.D. Hutcheon, J. Huth, H.A. Ishii, M. Ito, S.B. Jacobsen, A. Kearsley, L.A. Leshin, M.-C. Liu, I.C. Lyon, K. Marhas, B. Marty, G. Matrajt, A. Meibom, S.M. Messenger, S. Mostefaoui, S. Mukhopadhyay, K. Nakamura-Messenger, L.R. Nittler, R.L. Palma, R.O. Pepin, D.A. Papanastassiou, F. Robert, D.J. Schlutter, C.J. Snead, F.J. Stadermann, R.M. Stroud, P. Tsou, A. Westphal, E.D. Young, K. Ziegler, L. Zimmermann, E. Zinner, Isotopic compositions of cometary matter returned by Stardust. Science **314**, 1724–1728 (2006)
D. McKenzie, The extraction of magma from the crust and mantle. Earth Planet. Sci. Lett. **74**, 81–91 (1985)
E. Médard, T.L. Grove, The effect of H_2O on the olivine liquidus of basaltic melts: experiments and thermodynamic models. Contrib. Mineral. Petrol. **155**, 417–432 (2008)
M.A. Menzies, C.J. Hawkesworth (eds.), *Mantle Metasomatism* (Academic Press Inc., London, 1987). 472 pp.
A. Meshik, C.M. Hohenberg, D.S. Burnett, O. Pravdivtseva, Measuring the isotopic composition of solar wind noble gases, in *Exploring the Solar Wind*, ed. by M. Lazar (INTECH Open Access Publisher, Winchester, 2012), pp. 93–120
K. Mibe, M. Kanzaki, T. Kawamoto, K.N. Matsukage, Y. Fei, S. Ono, Second critical endpoint in the peridotite-H_2O system. J. Geophys. Res. **112**, 1–8 (2007)
P. Michael, Regionally distinctive sources of depleted MORB: evidence from trace elements and H_2O. Earth Planet. Sci. Lett. **131**, 301–320 (1995)
P.J. Michael, The concentration, behavior and storage of H_2O in the suboceanic upper mantle: implications for mantle metasomatism. Geochim. Cosmochim. Acta **52**, 555–566 (1988)
C. Michaut, C. Jaupart, J.-C. Mareshal, Thermal evolution of cratonic roots. Lithos **109**, 47–60 (2009)
K. Mierdel, H. Keppler, J.R. Smyth, F. Langenhorst, Water solubility in aluminous orthopyroxene and the origin of Earth's asthenosphere. Science **315**, 364–368 (2007)
R.H. Mitchell, *Kimberlites. Mineralogy, Geochemistry, and Petrology* (Springer, New York, 1986)
M. Mookherjee, L. Stixrude, B. Karki, Hydrous silicate melt at high pressure. Nature **452**, 983–986 (2008)
J.G. Moore, Water content of basalt erupted on the ocean floor. Contrib. Mineral. Petrol. **28**, 272–279 (1970)
J.G. Moore, J.-G. Schilling, Vesicles, water, and sulfur in Reykjanes Ridge basalts. Contrib. Mineral. Petrol. **41**, 105–118 (1973)

M. Moreira, S. Charnoz, The origin of the neon isotopes in chondrites and on Earth. Earth Planet. Sci. Lett. **433**, 249–256 (2016)

J.L. Mosenfelder, M. Le Voyer, G.R. Rossman, Y. Guan, D.R. Bell, P.D. Asimow, J.M. Eiler, Analysis of hydrogen in olivine by SIMS: evaluation of standards and protocol. Am. Mineral. **96**, 1725–1741 (2011)

J.L. Mosenfelder, G.R. Rossman, E.A. Johnson, Hydrous species in feldspars: a reassessment based on FTIR and SIMS. Am. Mineral. **100**, 1209–1221 (2015)

S. Mukhopadhyay, Early differentiation and volatile accretion recorded in deep-mantle neon and xenon. Nature **486**, 101–104 (2012)

S. Mukhopadhyay, L. Nittler, D-rich water in interplanetary dust particles, in *Lunar Planet. Sci. Conf.*, League City, TX (2003), p. 1941

A. Müller, M. Koch-Müller, Hydrogen speciation and trace element contents of igneous, hydrothermal and metamorphic quartz from Norway. Mineral. Mag. **73**, 569–583 (2009)

A. Mundl, M. Touboul, M.G. Jackson, J.M.D. Day, M.D. Kurz, V. Lekic, R.T. Helz, R.J. Walker, Tungsten-182 heterogeneity in modern ocean island basalts. Science **356**, 66–69 (2017)

B.O. Mysen, M.L. Fogel, P.L. Morril, G.D. Cody, Solution behavior of reduced C-O-H volatiles in silicate melts at high pressure and temperature. Geochim. Cosmochim. Acta **73**, 1696–1710 (2009)

S. Naif, K. Key, S. Constable, R.L. Evans, Melt-rich channel observed at the lithosphere-asthenosphere boundary. Nature **495**, 356–359 (2013)

S. Nakashima, H. Matayoshi, T. Yuko, K. Michibayashi, T. Masuda, N. Kuroki, H. Yamagishi, Y. Ito, A. Nakamura, Infrared microspectroscopy analysis of water distribution in deformed and metamorphosed rocks. Tectonophysics **245**, 263–276 (1995)

S. Nazzareni, H. Skogby, P.F. Zanazzi, Hydrogen content in clinopyroxene phenocrysts from Salina mafic lavas (Aeolian arc, Italy). Contrib. Mineral. Petrol. **162**, 275–288 (2011)

O. Nebel, K. Mezger, W. van Westrenen, Rubidium isotopes in primitive chondrites: constraints on Earth's volatile element depletion and lead isotope evolution. Earth Planet. Sci. Lett. **305**, 309–316 (2011)

F. Nestola, P. Nimis, L. Ziberna, M. Longo, A. Marzoli, J.W. Harris, M.H. Manghnani, Y. Fedortchouk, First crystal-structure determination of olivine in diamond: composition and implications for provenance in the Earth's mantle. Earth Planet. Sci. Lett. **305**, 249–255 (2011)

F. Nestola, J.R. Smyth, Diamonds and water in the deep Earth: a new scenario. Int. Geol. Rev. **58**, 1–14 (2015). doi:10.1080/00206814.2015.1056758

A.O. Nier, D.J. Schlutter, Extraction of helium from individual interplanetary dust particles by step-heating. Meteoritics **27**, 166–173 (1992)

F. Nimmo, T. Kleine, Early differentiation and core formation: processes and timescales, in *The Early Earth: Accretion and Differentiation*, ed. by J. Badro, A.M. Walker (Wiley, New York, 2015), pp. 83–102

Y. Nishihara, K.N. Matsukage, Iron-titanium oxyhydroxides as water carriers in the Earth's deep mantle. Am. Mineral. **101**, 919–927 (2016)

R. Nomura, K. Hirose, K. Uesugi, Y. Ohishi, A. Tsuchiyama, A. Miyake, Y. Ueno, Low core-mantle boundary temperature inferred from the solidus of pyrolite. Science **343**, 522–525 (2014)

D. Novella, N. Bolfan-Casanova, F. Nestola, J.W. Harris, H_2O in olivine and garnet inclusions still trapped in diamonds from the Siberian craton: implications for the water content of cratonic lithosphere peridotites. Lithos **230**, 180–183 (2015)

D. Novella, D.J. Frost, E.H. Hauri, H. Bureau, C. Raepsaet, M. Roberge, The distribution of H_2O between silicate melt and nominally anhydrous peridotite and the onset of hydrous melting in the deep upper mantle. Earth Planet. Sci. Lett. **400**, 1–13 (2014)

D.P. O'Brien, A. Morbidelli, H.F. Levison, Terrestrial planet formation with strong dynamical friction. Icarus **184**, 39–58 (2006)

D.P. O'Brien, K.J. Walsh, A. Morbidelli, S.N. Raymond, A.V. Mandell, Water delivery and giant impacts in the 'Grand Tack' scenario. Icarus **239**, 74–84 (2014)

J.A. O'Leary, G.A. Gaetani, E.H. Hauri, The effect of tetrahedral Al^{3+} on the partitioning of water between clinopyroxene and silicate melt. Earth Planet. Sci. Lett. **297**, 111–120 (2010)

C.J. O'Neill, A. Lenardic, W.L. Griffin, S.Y. O'Reilly, Dynamics of cratons in an evolving mantle. Lithos **102**, 12–24 (2008)

H.S.C. O'Neill, H. Palme, Composition of the silicate Earth: implications for accretion and core formation, in *The Earth's Mantle: Composition, Structure, and Evolution*, ed. by I. Jackson (Cambridge University Press, Cambridge, 1998), pp. 3–126

I. Ohira, E. Ohtani, T. Sakai, M. Miyahara, N. Hirao, Y. Ohishi, M. Nishijima, Stability of a hydrous δ-phase, $AlOOH–MgSiO_2(OH)_2$, and a mechanism for water transport into the base of lower mantle. Earth Planet. Sci. Lett. **401**, 12–17 (2014)

E. Ohtani, Hydrous minerals and the storage of water in the deep mantle. Chem. Geol. **418**, 6–15 (2015)

E. Ohtani, N. Hirao, T. Kondo, M. Ito, T. Kikegawa, Iron-water reaction at high pressure and temperature, and hydrogen transport into the core. Phys. Chem. Miner. **32**, 77–82 (2005)

T. Ohuchi, T. Irifune, Development of A-type olivine fabric in water-rich deep upper mantle. Earth Planet. Sci. Lett. **362**, 20–30 (2013)

T. Ohuchi, S.-I. Karato, K. Fujino, Strength of single-crystal orthopyroxene under lithospheric conditions. Contrib. Mineral. Petrol. **161**, 961–975 (2010)

T. Ohuchi, T. Kawazoe, Y. Nishihara, N. Nishiyama, T. Irifune, High pressure and temperature fabric transitions in olivine and variations in upper mantle seismic anisotropy. Earth Planet. Sci. Lett. **304**, 55–63 (2011)

T. Okuchi, Hydrogen partitioning into molten iron at high pressure: implications for Earth's core. Science **278**, 1781–1784 (1997)

T. Okuchi, E. Takahashi, Hydrogen in molten iron at high pressure: the first measurement, in *Properties of Earth and Planetary Materials at High Pressure and Temperature* (American Geophysical Union, Washington, 1998), pp. 249–260

U. Ott, Planetary and pre-solar noble gases in meteorites. Chem. Erde **74**, 519–544 (2014)

T. Owen, A. Bar-Nun, I. Kleinfeld, Possible cometary origin of heavy noble gases in the atmospheres of Venus, Earth and Mars. Nature **358**, 43–46 (1992)

J.A. Padrón-Navarta, J. Hermann, H.S.C. O'Neill, Site-specific hydrogen diffusion rates in forsterite. Earth Planet. Sci. Lett. **392**, 100–112 (2014)

H. Palme, H.S.C. O'Neill, Cosmochemical estimates of mantle composition, in *The Mantle and Core*, ed. by A.M. Davis (Elsevier, Amsterdam, 2014a), pp. 1–39

H. Palme, H.S.C. O'Neill, Solar system abundances of the elements, in *The Mantle and Core*, ed. by A.M. Davis (Elsevier, Amsterdam, 2014b), pp. 15–36

M. Palot, S.D. Jacobsen, J.P. Townsend, F. Nestola, K. Marquardt, J.W. Harris, T. Stachel, C.A. McCammon, D.G. Pearson, Evidence for H_2O-bearing fluids in the lower mantle from diamond inclusion. Lithos **265**, 237–243 (2016). doi:10.1016/j.lithos.2016.06.023

W.R. Panero, L.R. Benedetti, R. Jeanloz, Transport of water into the lower mantle: role of stishovite. J. Geophys. Res. **108**, 2039 (2003)

W.R. Panero, J.S. Pigott, D.M. Reaman, J.E. Kabbes, Z. Liu Dry, (Mg, Fe)SiO_3 perovskite in the Earth's lower mantle. J. Geophys. Res. **120**, 894–908 (2015)

W.R. Panero, J.R. Smyth, J.S. Pigott, Z. Liu, D.J. Frost, Hydrous ringwoodite to 5 K and 35 GPa: multiple hydrogen bonding sites resolved with FTIR spectroscopy. Am. Mineral. **98**, 637–642 (2013)

R. Parai, S. Mukhopadhyay, How large is the subducted water flux? New constraints on mantle regassing rates. Earth Planet. Sci. Lett. **317–318**, 396–406 (2012)

J.D. Pasteris, Fluid inclusions in mantle xenoliths, in *Mantle Xenoliths*, ed. by P.H. Nixon (Wiley, New York, 1987), pp. 691–707

A.A. Pavlov, A.K. Pavlov, J.F. Kasting, Irradiated interplanetary dust particles as a possible solution for the deuterium/hydrogen paradox of Earth's oceans. J. Geophys. Res. **104**, 30725–30728 (1999)

S.M. Peacock, Fluid processes in subduction zones. Science **248**, 329–337 (1990)

D.G. Pearson, F.E. Brenker, F. Nestola, J. McNeill, L. Nasdala, M.T. Hutchison, S. Matveev, K. Mather, G. Silversmit, S. Schmitz, B. Vekemans, L. Vincze, Hydrous mantle transition zone indicated by ringwoodite included within diamond. Nature **507**, 221–224 (2014)

D.G. Pearson, R.W. Carlson, S.B. Shirey, F.R. Boyd, P.H. Nixon, Stabilization of Archean lithospheric mantle: a Re-Os isotope study of peridotite xenoliths from the Kaapvaal craton. Earth Planet. Sci. Lett. **134**, 341–357 (1995a)

D.G. Pearson, S.B. Shirey, R.W. Carlson, F.R. Boyd, N.P. Pokhilenko, N. Shimizu, Re-Os, Sm-Nd, and Rb-Sr isotope evidence for thick Archaean lithospheric mantle beneath the Siberian craton modified by multistage metasomatism. Geochim. Cosmochim. Acta **59**, 959–977 (1995b)

R.O. Pepin, D. Porcelli, Origin of noble gases in the terrestrial planets. Annu. Rev. Earth Planet. Sci. **47**, 191–246 (2002)

A.H. Peslier, A review of water contents of nominally anhydrous natural minerals in the mantles of Earth, Mars and the Moon. J. Volcanol. Geotherm. Res. **197**, 239–258 (2010)

A.H. Peslier, M. Bizimis, Water in Hawaiian peridotite minerals: a case for a dry metasomatized oceanic mantle lithosphere. Geochem. Geophys. Geosyst. **16**, 1–22 (2015)

A.H. Peslier, M. Bizimis, M. Matney, Water disequilibrium in olivines from Hawaiian peridotites: recent metasomatism, H diffusion and magma ascent rates. Geochim. Cosmochim. Acta **154**, 98–117 (2015)

A.H. Peslier, J.F. Luhr, Hydrogen loss from olivines in mantle xenoliths from Simcoe (USA) and Mexico: mafic alkalic magma ascent rates and water budget of the sub-continental lithosphere. Earth Planet. Sci. Lett. **242**, 302–319 (2006)

A.H. Peslier, J.F. Luhr, J. Post, Low water contents in pyroxenes from spinel-peridotites of the oxidized, sub-arc mantle wedge. Earth Planet. Sci. Lett. **201**, 69–86 (2002)

A.H. Peslier, A.B. Woodland, D.R. Bell, M. Lazarov, Olivine water contents in the continental lithosphere and the longevity of cratons. Nature **467**, 78–81 (2010)

A.H. Peslier, A.B. Woodland, D.R. Bell, M. Lazarov, T.J. Lapen, Metasomatic control of water contents in the Kaapvaal cratonic mantle. Geochim. Cosmochim. Acta **97**, 213–246 (2012)

A.H. Peslier, A.B. Woodland, J.A. Wolff, Fast kimberlite ascent rates estimated from hydrogen diffusion profiles in xenolithic olivines from Southern Africa. Geochim. Cosmochim. Acta **72**, 2711–2722 (2008)

J.S. Pigott, K. Wright, J.D. Gale, W.R. Panero, Calculation of the energetics of water incorporation in majorite garnet. Am. Mineral. **100**, 1065–1075 (2015)

T. Plank, L.B. Cooper, C.E. Manning, Emerging geothermometers for estimating slab surface temperatures. Nat. Geosci. **614**, 611–615 (2009)

T. Plank, K.A. Kelley, M.M. Zimmer, E.H. Hauri, P.J. Wallace, Why do mafic magmas contain ~4 wt% water on average? Earth Planet. Sci. Lett. **364**, 168–179 (2013)

T. Plank, C.H. Langmuir, The chemical composition of subducted sediment and its consequences for the crust and mantle. Chem. Geol. **145**, 325–394 (1998)

F.A. Podosek, M. Ozima, The xenon age of the Earth, in *Origin of the Earth and Moon*, ed. by R.M. Canup, K. Righter (University of Arizona Press, Tucson, 2000), pp. 63–72

F.A. Podosek, D.S. Woolum, P. Cassen, R.H.J. Nichols, Solar gases in the Earth by solar wind irradiation, in *Goldschmidt Conf.*, Oxford, UK (2000), p. 804

J.-P. Poirier, Light elements in the Earth's outer core: a critical review. Phys. Earth Planet. Inter. **85**, 319–337 (1994)

H.N. Pollack, Cratonization and thermal evolution of the mantle. Earth Planet. Sci. Lett. **80**, 175–182 (1986)

F.F. Pollitz, R. Bürgmann, B. Romanowicz, Viscosity of oceanic asthenosphere inferred from remote triggering of earthquakes. Science **280**, 1245–1249 (1998)

A. Pommier, Interpretation of magnetotelluric results using laboratory measurements. Surv. Geophys. **35**, 41–84 (2014)

D. Porcelli, C.J. Ballentine, Models for distribution of terrestrial noble gases and evolution of the atmosphere. Rev. Mineral. Geochem. **47**, 411–480 (2002)

K. Priestley, E. Debayle, Seismic evidence for a moderately thick lithosphere beneath the Siberian platform. Geophys. Res. Lett. **30**, 1118 (2003)

K. Priestley, D. McKenzie, The thermal structure of the lithosphere from shear wave velocities. Earth Planet. Sci. Lett. **244**, 285–301 (2006)

L. Qin, C.M.O.D. Alexander, R.W. Carlson, M.F. Horan, T. Yokoyama, Contributors to chromium isotope variation of meteorites. Geochim. Cosmochim. Acta **74**, 1122–1145 (2010)

S.N. Raymond, T. Quinn, J.I. Lunine, High-resolution simulations of the final assembly of Earth-like planets I. Terrestrial accretion and dynamics. Icarus **183**, 265–282 (2006)

J. Revenaugh, R. Meyer, Seismic evidence of partial melt within a possibly ubiquitous low-velocity layer at the base of the mantle. Science **277**, 670–673 (1997)

J. Revenaugh, S.A. Sipkin, Seismic evidence for silicate melt atop the 410-km mantle discontinuity. Nature **369**, 474–476 (1994)

G.G. Richards, D. Bercovici, Water-induced convection in the Earth's mantle transition zone. J. Geophys. Res. **114**, B01205 (2009)

F. Robert, The origin of water on Earth. Science **293**, 1056–1058 (2001)

F. Robert, The D/H ratio in chondrites, in *Solar System History from Isotopic Signatures of Volatile Elements: Volume Resulting from an ISSI Workshop*, 14–18 January, 2003, Bern, Switzerland, ed. by R. Kallenbach, T. Encrenaz, J. Geiss, K. Mauersberger, T.C. Owen, F. Robert (Springer, Dordrecht, 2002), pp. 87–101.

F. Robert, M. Javoy, J. Halbout, B. Dimon, L. Merlivat, Hydrogen isotope abundances in the solar system. Part II: meteorites with terrestrial-like DH ratio. Geochim. Cosmochim. Acta **51**, 1807–1822 (1987)

E. Roedder, Liquid CO_2 inclusions in olivine-bearing nodules and phenocrysts from basalts. Am. Mineral. **10**, 1746–1773 (1965)

E. Roedder, *Fluid Inclusions* (Mineralogical Society of America, Chantilly, 1984). 644 pp.

S. Rondenay, G.A. Abers, P.E. van Keken, Seismic imaging of subduction zone metamorphism. Geology **36**, 275–278 (2008)

A. Rosenthal, E.H. Hauri, M.M. Hirschmann, Experimental determination of C, F, and H partitioning between mantle minerals and carbonated basalt, CO_2/Ba and CO_2/Nb systematics of partial melting, and the CO_2 contents of basaltic source regions. Earth Planet. Sci. Lett. **412**, 77–87 (2015)

G.R. Rossman, Studies of OH in nominally anhydrous minerals. Phys. Chem. Miner. **23**, 299–304 (1996)

G.R. Rossman, Analytical methods for measuring water in nominally anhydrous minerals, in *Water in Nominally Anhydrous Minerals*, ed. by H. Keppler, J.R. Smyth (Mineralogical Society of America, Chantilly, 2006), pp. 1–28

G.R. Rossman, J.R. Smyth, Hydroxyl contents of accessory minerals in mantle eclogites and related rocks. Am. Mineral. **75**, 775–780 (1990)

D.C. Rubie, D.J. Frost, U. Mann, Y. Asahara, F. Nimmo, K. Tsuno, P. Kegler, A.H. Holzheid, H. Palme, Heterogeneous accretion, composition and core–mantle differentiation of the Earth. Earth Planet. Sci. Lett. **301**, 31–42 (2011)

D.C. Rubie, S.A. Jacobson, A. Morbidelli, D.P. O'Brien, E.D. Young, J. de Vries, F. Nimmo, H. Palme, D.J. Frost, Accretion and differentiation of the terrestrial planets with implications for the compositions of early-formed Solar System bodies and accretion of water. Icarus **248**, 89–108 (2015)

R.L. Rudnick, Making continental crust. Nature **378**, 571–578 (1995)

R.L. Rudnick, D.M. Fountain, Nature and composition of the continental crust: a lower crustal perspective. Rev. Geophys. **33**, 267–309 (1995)

L. Rüpke, J.P. Morgan, J.E. Dixon, Implications of subduction rehydration for Earth's deep water cycle, in *Earth's Deep Water Cycle*, ed. by S.D. Jacobsen, S. van der Lee (American Geophysical Union, Washington, 2006), pp. 263–276

L.H. Rüpke, J.P. Morgan, M. Hort, J.A.D. Connolly, Serpentine and the subduction zone water cycle. Earth Planet. Sci. Lett. **223**, 17–34 (2004)

D.M. Ruscitto, P.J. Wallace, L.B. Cooper, T. Plank, Global variations in H_2O/Ce: 2. Relationships to arc magma geochemistry and volatile fluxes. Geochem. Geophys. Geosyst. **13**, 1–27 (2012)

J.K. Russell, L.A. Porritt, Y. Lavallée, D.B. Dingwell, Kimberlite ascent by assimilation-fuelled buoyancy. Nature **481**, 352–356 (2012)

C.A. Rychert, P.M. Shearer, A global view of the lithosphere-asthenosphere boundary. Science **324**, 495–498 (2009)

A.E. Saal, E.H. Hauri, C.H. Langmuir, M.R. Perfit, Vapour undersaturation in primitive mid-ocean-ridge basalt and the volatile content of Earth's upper mantle. Nature **419**, 451–455 (2002)

A.E. Saal, E.H. Hauri, M. Lo Cascio, J.A. Van Orman, M.C. Rutherford, R.F. Cooper, Volatile content of lunar volcanic glasses and the presence of water in the Moon's interior. Nature **454**, 192–196 (2008)

A.E. Saal, E.H. Hauri, J.A. Van Orman, M.C. Rutherford, Hydrogen isotopes in lunar volcanic glasses and melt inclusions reveal a carbonaceous chondrite heritage. Science **340**, 1317–1320 (2013)

T. Sakamaki, A. Suzuki, E. Ohtani, Stability of hydrous melt at the base of the Earth's upper mantle. Nature **439**, 192–194 (2006)

S.A. Sandford, J. Aléon, C.M.O.D. Alexander, T. Araki, S. Bajt, G.A. Baratta, J. Borg, J.P. Bradley, D.E. Brownlee, J.R. Brucato, M.J. Burchell, H. Busemann, A. Butterworth, S.J. Clemett, G.D. Cody, L. Colangeli, G. Cooper, L. D'Hendecourt, Z. Djouadi, J.P. Dworkin, G. Ferrini, H. Fleckenstein, G.J. Flynn, I.A. Franchi, M. Fries, M.K. Gilles, D.P. Glavin, M. Gounelle, F. Grossemy, C. Jacobsen, L.P. Keller, A.L.D. Kilcoyne, J. Leitner, G. Matrajt, A. Meibom, V. Mennella, S. Mostefaoui, L.R. Nittler, M.E. Palumbo, D.A. Papanastassiou, F. Robert, A. Rotundi, C.J. Snead, M.K. Spencer, F.J. Stadermann, A. Steele, T. Stephan, P. Tsou, T. Tyliszczak, A.J. Westphal, S. Wirick, B. Wopenka, H. Yabuta, R.N. Zare, M.E. Zolensky, Organics captured from comet 81P/Wild 2 by the Stardust Spacecraft. Science **314**, 1720–1724 (2006)

A. Sano, E. Ohtani, T. Kondo, N. Hirao, T. Sakai, N. Sata, Y. Ohishi, T. Kikegawa, Aluminous hydrous mineral δ-AlOOH as a carrier of hydrogen into the core-mantle boundary. Geophys. Res. Lett. **35**, L03303 (2008)

A.R. Sarafian, S.G. Nielsen, H.R. Marschall, F.M. McCubbin, B.D. Monteleone, Early accretion of water in the inner solar system from a carbonaceous chondrite-like source. Science **346**, 623–626 (2014)

A.R. Sarafian, M.F. Roden, A.E. Patiño-Douce, The volatile content of Vesta: clues from apatite in eucrites. Meteorit. Planet. Sci. **48**, 2135–2154 (2013)

A.R. Sarafian, E.H. Hauri, F.M. McCubbin, T.J. Lapen, E.L. Berger, S.G. Nielsen, H.R. Marschall, G.A. Gaetani, K. Righter, E. Sarafian, Early accretion of water and volatile elements to the inner Solar System: evidence from angrites. Philos. Trans. R. Soc. Lond. A **375**, 1–27 (2017)

E. Sarafian, R.L. Evans, J.A. Collins, J. Elsenbeck, G.A. Gaetani, J.B. Gaherty, G. Hirth, D. Lizarralde, The electrical structure of the central Pacific upper mantle constrained by the NoMelt experiment. Geochem. Geophys. Geosyst. **16**, 1115–1132 (2015)

L. Schaffer, A.H. Peslier, A.D. Brandon, M. Bizimis, M. Matney, *Why Are Mantle Melting Residues Still Hydrous?* (Goldschmidt, Yokohama, 2016)

L. Schaefer, B. Fegley Jr., Volatile element chemistry during metamorphism of ordinary chondritic material and some of its implications for the composition of asteroids. Icarus **205**, 483–496 (2010)

P. Schiano, R. Clocchiatti, N. Shimizu, R.C. Maury, K.P. Jochum, A.W. Hofmann, Hydrous, silica-rich melts in the sub-arc mantle and their relationship with erupted arc lavas. Nature **377**, 595–600 (1995)

E. Schmädicke, J. Gose, J. Reinhardt, T.M. Will, R. Stalder, Garnet in cratonic and non-cratonic mantle and lower crustal xenoliths from southern Africa: composition, water incorporation and geodynamic constraints. Precambrian Res. **270**, 285–299 (2015)

E. Schmädicke, J. Göse, T.M. Will, Heterogeneous mantle underneath the North Atlantic: evidence from water in orthopyroxene, mineral composition and equilibrium conditions of spinel peridotite from different locations at the Mid-Atlantic ridge. Lithos **125**, 308–320 (2011)

B. Schmandt, S.D. Jacobsen, T.W. Becker, Z. Liu, K. Ducker, Dehydration melting at the top of the lower mantle. Science **344**, 1265–1268 (2014)
M.W. Schmidt, S. Poli, Experimentally based water budgets for dehydrating slabs and consequences for arc magma generation. Earth Planet. Sci. Lett. **163**, 361–379 (1998)
M. Schönbächler, R.W. Carlson, M.F. Horan, T.D. Mock, E.H. Hauri, Silver isotope variations in chondrites: volatile depletion and the initial 107Pd abundance of the solar system. Geochim. Cosmochim. Acta **72**, 5330–5341 (2008)
M. Schönbächler, R.W. Carlson, M.F. Horan, T.D. Mock, E.H. Hauri, Heterogeneous accretion and the moderately volatile element budget of Earth. Science **328**, 884–887 (2010)
M. Schönbächler, F. Nimmo, Heterogeneous accretion of the Earth and the timing of volatile element depletion, in *AGU Fall Meeting*, San Francisco (2011)
J.E.C. Scully, C.T. Russell, A. Yin, R. Jaumann, E. Carey, J. Castillo-Rogez, H.Y. McSween, C.A. Raymond, V. Reddy, L. Le Corre, Geomorphological evidence for transient water flow on Vesta. Earth Planet. Sci. Lett. **411**, 151–163 (2015)
S.J. Seaman, M.L. Williams, M.J. Jercinovic, G.C. Koteas, L.B. Brown, Water in nominally anhydrous minerals: implications for partial melting and strain localization in the lower crust. Geology **41**, 1051–1054 (2013)
A.N. Seligman, I.N. Bindeman, J.M. Watkins, A.M. Ross, Water in volcanic glass: from volcanic degassing to secondary hydration. Geochim. Cosmochim. Acta **191**, 216–238 (2016)
K. Selway, J. Yi, S.-I. Karato, Water content of the Tanzanian lithosphere from magnetotelluric data: implications for cratonic growth and stability. Earth Planet. Sci. Lett. **388**, 175–186 (2014)
S.S. Shapiro, B.H. Hager, T.H. Jordan, Stability and dynamics of the continental tectosphere. Lithos **48**, 135–152 (1999)
Z.D. Sharp, F.M. McCubbin, C.K. Shearer, A hydrogen-based oxidation mechanism relevant to planetary formation. Earth Planet. Sci. Lett. **380**, 88–97 (2013)
A.M. Shaw, E.H. Hauri, M.D. Behn, D.R. Hilton, C.G. McPherson, J.M. Sinton, Long-term preservation of slab signatures in the mantle inferred from hydrogen isotopes. Nat. Geosci. **5**, 224–228 (2012)
A. Shen, H. Keppler, Infrared spectroscopy of hydrous silicate melts to 1000 °C and 10 kbars: direct observation of H_2O speciation in a diamond-anvil cell. Am. Mineral. **80**, 1335–1338 (1995)
Y. Shibazaki, E. Ohtani, H. Fukui, T. Sakai, S. Kamada, D. Ishikawa, S. Tsutsui, A.Q.R. Baron, N. Nishitani, N. Hirao, K. Takemura, Sound velocity measurements in dhcp-FeH up to 70 GPa with inelastic X-ray scattering: implications for the composition of the Earth's core. Earth Planet. Sci. Lett. **313–314**, 79–85 (2012)
Y. Shibazaki, E. Ohtani, H. Terasaki, A. Suzuki, K-i. Funakoshi, Hydrogen partitioning between iron and ringwoodite: implications for water transport into the Martian core. Earth Planet. Sci. Lett. **287**, 463–470 (2009)
S. Shirey, S.H. Richardson, Start of the Wilson cycle at 3 Ga shown by diamonds from subcontinental Mantle. Science **333**, 434–436 (2011)
J. Siebert, J. Badro, D. Antonangeli, F.J. Ryerson, Terrestrial accretion under oxidizing conditions. Science **339**, 1194–1197 (2013)
T.W. Sisson, S. Bronto, Evidence for pressure-release melting beneath magmatic arcs from basalt at Galunggung, Indonesia. Nature **391**, 883–886 (1998)
T.W. Sisson, G.D. Layne, H_2O in basalt and basaltic andesite glass inclusions from four subduction-related volcanoes. Earth Planet. Sci. Lett. **117**, 619–635 (1993)
L.E. Sjöberg, M. Pan, E. Asenjo, S. Erlingsson, Glacial rebound near Vatnajökull, Iceland, studied by GPS campaigns in 1992 and 1996. J. Geodyn. **29**, 63–70 (2000)
P. Skemer, J.M. Warren, L. Hansen, G. Hirth, P.B. Kelemen, The influence of water and LPO on the initiation and evolution of mantle shear zones. Earth Planet. Sci. Lett. **375**, 222–233 (2013)
N.H. Sleep, Survival of Archean cratonal lithosphere. J. Geophys. Res. **108**, 2302 (2003). doi:10.1029/2001JB000169
N.H. Sleep, Evolution of the continental lithosphere. Annu. Rev. Earth Planet. Sci. **33**, 369–393 (2005)
J.R. Smyth, A crystallographic model for hydrous wadsleyite (β-Mg_2SiO_4): an ocean in the Earth's interior? Am. Mineral. **79**, 1021–1024 (1994)
J.R. Smyth, Hydrogen in high pressure silicate and oxide mineral structures, in *Water in Nominally Anhydrous Minerals*, ed. by H. Keppler, J.R. Smyth (Mineralogical Society of America, Chantilly, 2006), pp. 85–115
J.R. Smyth, D.R. Bell, G.R. Rossman, Incorporation of hydroxyl in upper-mantle clinopyroxenes. Nature **351**, 732–735 (1991)
J.R. Smyth, F.A. Caporuscio, T.C. McCormick, Mantle eclogite: evidence of igneous fractionation in the mantle. Earth Planet. Sci. Lett. **93**, 133–141 (1989)

J.R. Smyth, D.J. Frost, The effect of water on the 410-km discontinuity: an experimental study. Geophys. Res. Lett. **29**, 123-121–123-124 (2002)

J.R. Smyth, C.M. Holl, D.J. Frost, S.D. Jacobsen, High pressure crystal chemistry of hydrous ringwoodite and water in the Earth's interior. Phys. Earth Planet. Inter. **143–144**, 271–278 (2004)

J.R. Smyth, S.D. Jacobsen, Nominally anhydrous minerals and Earth's deep water cycle, in *Earth's Deep Water Cycle*, ed. by S. Jacobsen, S. Lee (AGU Geophysical Monograph, Washington D.C., 2006), pp. 1–11

A.V. Sobolev, E.V. Asafov, A.A. Gurenko, N.T. Arndt, V.G. Batanova, M. Portnyagin, D. Garbe-Schönberg, S.P. Krashenninnikov, Komatiites reveal a hydrous Archaean deep-mantle reservoir. Nature **531**, 628–632 (2016)

A.V. Sobolev, M. Chaussidon, H_2O concentrations in primary melts from supra-subduction zones and mid-ocean ridges: implications for H_2O storage and recycling in the mantle. Earth Planet. Sci. Lett. **137**, 45–55 (1996)

S. Song, L. Su, Y. Niu, Y. Lai, L. Zhang, CH_4 inclusions in orogenic harzburgite: evidence for reduced slab fluids and implications for redox melting in mantle wedge. Geochim. Cosmochim. Acta **73**, 1737–1754 (2009)

T.-R.A. Song, D.V. Helmberger, S.P. Grand, Low-velocity zone atop the 410-km seismic discontinuity in the northwestern United States. Nature **427**, 530–533 (2004)

V. Soustelle, A. Tommasi, S. Demouchy, L. Franz, Melt-rock interactions, deformation, hydration and seismic properties in the sub-arc lithospheric mantle inferred from xenoliths from seamounts near Lihir, Papua New Guinea. Tectonophysics **608**, 330–345 (2013)

V. Soustelle, A. Tommasi, S. Demouchy, D.A. Ionov, Deformation and fluid-rock interaction in the supra-subduction mantle: microstructures and water contents in peridotite xenoliths from the Avacha volcano, Kamchatka. J. Petrol. **51**, 363–394 (2010)

W. Soyer, M. Unsworth, Deep electrical structure of the northern Cascadia (British Columbia, Canada) subduction zone: implications for the distribution of fluids. Geology **34**, 53–56 (2006)

R.S.J. Sparks, Kimberlite volcanism. Annu. Rev. Earth Planet. Sci. **41**, 497–528 (2013)

R.S.J. Sparks, J. Barclay, C. Jaupart, H.M. Mader, J.C. Phillips, Physical aspects of magmatic degassing I. Experimental and theoretical constraints on vesiculation, in *Volatiles in Magmas*, ed. by M.R. Carroll, J.R. Holloway (Mineralogical Society of America, Washington D.C., 1994), pp. 413–446

N.H. Spring, H. Busemann, S.A. Crowther, J.D. Gilmour, C. Engrand, L.R. Nittler, Xenon in Antarctic micrometeorites (AMMs) and interplanetary dust particles (IDPs), in *45th Lunar and Planetary Science Conference*, The Woodland, TX (2014), p. 2923

T. Stachel, W.J. Harris, P.G. Brey, W. Joswig, Kankan diamonds (Guinea) II: lower mantle inclusion parageneses. Contrib. Mineral. Petrol. **140**, 16–27 (2000)

R. Stalder, Influence of Fe, Cr and Al on hydrogen incorporation in orthopyroxene. Eur. J. Mineral. **16**, 703–711 (2004)

R. Stalder, A. Karimova, J. Konzett, OH-defects in multiple-doped orthoenstatite at 4–8 GPa: filling the gap between pure and natural systems. Contrib. Mineral. Petrol. **169**, 38 (2015)

R. Stalder, J. Konzett, OH defects in quartz in the system quartz–albite–water and granite–water between 5 and 25 kbar. Phys. Chem. Miner. **39**, 817–827 (2012)

R. Stalder, H. Skogby, Hydrogen incorporation in enstatite. Eur. J. Mineral. **14**, 1139–1144 (2002)

A. Stephant, R.L. Hervig, M. Bose, M. Wadhwa, D/H ratios and water contents in eucrite minerals: implications for the source and abundance of water on Vesta, in *79th Meteoritical Society Meeting* (2016), p. 6212

A. Stephant, L. Remusat, F. Robert, Water in type I chondrules of Paris CM chondrite. Geochim. Cosmochim. Acta **199**, 75–90 (2017)

E. Stolper, The speciation of water in silicate melts. Geochim. Cosmochim. Acta **46**, 2609–2620 (1982)

E. Stolper, S. Newman, The role of water in the petrogenesis of Mariana trough magmas. Earth Planet. Sci. Lett. **293**(325), 293–325 (1994)

A. Stracke, A.W. Hofmann, S.R. Hart, FOZO, HIMU, and the rest of the mantle zoo. Geochem. Geophys. Geosyst. **6**, 1–20 (2005)

D. Suetsugu, T. Inoue, A. Yamada, D. Zhao, M. Obayashi, Towards mapping the three-dimentional distribution of water in the transition zone from P-velocity tomography and 660-km discontinuity depths, in *Earth's Deep Water Cycle*, ed. by S. Jacobsen, S. Lee (AGU Geophysical Monograph, Washington D.C., 2006), pp. 237–250

S.-S. Sun, Chemical composition and origin of the Earth's primitive mantle. Geochim. Cosmochim. Acta **46**, 179–192 (1982)

S.S. Sun, W.F. McDonough, Chemical and isotopic systematics of oceanic basalts: implications for mantle composition and processes, in *Magmatism in Ocean Basins*, ed. by A.D. Saunders, M.J. Norry (Geol. Soc. London Spec. Publ., London, 1989), pp. 313–345

W. Sun, T. Yoshino, N. Sakamoto, H. Yurimoto, Hydrogen self-diffusivity in single crystal ringwoodite: implications for water content and distribution in the mantle transition zone. Geophys. Res. Lett. **42**, 6582–6589 (2015)

R. Sundvall, R. Stalder, Water in upper mantle pyroxene megacrysts and xenocrysts: a survey study. Am. Mineral. **96**, 1215–1227 (2011)

R.J. Sweeney, A.B. Thompson, P. Ulmer, Phase relations of a natural MARID composition and implications for MARID genesis, lithospheric melting and mantle metasomatism. Contrib. Mineral. Petrol. **115**, 225–241 (1993)

N. Tada, P. Tarits, K. Baba, H. Utada, T. Kasaya, D. Suetsugu, Electromagnetic evidence for volatile-rich upwelling beneath the society hotspot, French Polynesia. Geophys. Res. Lett. **43**, 12021–12026 (2016)

R. Tappert, J. Foden, T. Stachel, K. Muehlenbachs, M. Tappert, K. Wills, Deep mantle diamonds from South Australia: a record of Pacific subduction at the Gondwanan margin. Geology **37**, 43–46 (2008)

L.A. Taylor, A.M. Logvinova, G.H. Howarth, Y. Liu, A.H. Peslier, G.R. Rossman, Y. Chen, N.V. Sobolev, Low water in diamond mineral inclusions: proto-genetic origin in a dry cratonic lithosphere. Earth Planet. Sci. Lett. **433**, 125–132 (2016)

S.R. Taylor, S.M. McLennan, *The Continental Crust: Its Composition and Evolution* (Blackwell, Oxford, 1985)

T.J. Tenner, M.M. Hirschmann, A.C. Withers, P. Ardia, H_2O storage capacity of olivine and low-Ca pyroxene from 10 to 13 GPa: consequences for dehydration melting above the transition zone. Contrib. Mineral. Petrol. **163**, 297–316 (2012)

T.J. Tenner, M.M. Hirschmann, A.C. Withers, R.L. Hervig, Hydrogen partitioning between nominally anhydrous upper mantle minerals and melt between 3 and 5 GPa and applications to hydrous peridotite partial melting. Chem. Geol. **262**, 42–56 (2009)

H. Terasaki, E. Ohtani, T. Sakai, S. Kamada, H. Asanuma, Y. Shibazaki, N. Hirao, N. Sata, Y. Ohishi, T. Sakamaki, A. Suzuki, K.-I. Funakoshi, Stability of Fe–Ni hydride after the reaction between Fe–Ni alloy and hydrous phase (δ-AlOOH) up to 1.2 Mbar: possibility of H contribution to the core density deficit. Phys. Earth Planet. Inter. **194–195**, 18–24 (2012)

S.-M. Thomas, S.D. Jacobsen, C.R. Bina, P. Reichart, M. Moser, E.H. Hauri, M. Koch-Müller, J.R. Smyth, G. Dollinger, Quantification of water in hydrous ringwoodite. Front. Earth Sci. **2**, 1–19 (2014)

S.-M. Thomas, K. Wilson, M. Koch-Müller, E.H. Hauri, C.A. McCammon, S.D. Jacobsen, J. Lazarz, D. Rhede, M. Ren, N. Blair, S. Lenz, Quantification of water in majoritic garnet. Am. Mineral. **100**, 1084–1092 (2015)

C. Thoraval, S. Demouchy, Numerical models of ionic diffusion in one and three dimensions: application to dehydration of mantle olivine. Phys. Chem. Miner. **41**, 709–723 (2014)

Z.Z. Tian, J. Liu, Q.-K. Xia, J. Ingrin, Y.T. Hao, D. Christophe, Water concentration profiles in natural mantle orthopyroxenes: a geochronometer for long annealing of xenoliths within magma. Geology **45**, 87–90 (2017)

J.A. Tielke, M.E. Zimmerman, D.L. Kohlstedt, Hydrolytic weakening in olivine single crystals. J. Geophys. Res. **122**, 3465–3479 (2017)

P.M.E. Tollan, H.S.C. O'Neill, J. Hermann, A. Benedictus, R.J. Arculus, Frozen melt-rock reaction in a peridotite xenolith from the sub-arc mantle recorded by diffusion of trace elements and water in olivine. Earth Planet. Sci. Lett. **422**, 169–181 (2015)

J.P. Townsend, J. Tsuchiya, C.R. Bina, S.D. Jacobsen, Water partitioning between bridgmanite and postperovskite in the lowermost mantle. Earth Planet. Sci. Lett. **454**, 20–27 (2016)

M. Trieloff, J. Kunz, D.A. Clague, D. Harrison, C.J. Allègre, The nature of pristine noble gases in mantle plumes. Science **288**, 1036–1038 (2000)

A. Trinquier, J.L. Birck, C.J. Allègre, C. Göpel, D. Ulfbeck, ^{53}Mn–^{53}Cr systematics of the early Solar System revisited. Geochim. Cosmochim. Acta **72**, 5146–5163 (2008)

K. Tucker, R.L. Hervig, M. Wadhwa, Hydrogen isotope systematics of nominally anhydrous phases in martian meteorites, in *46th Lunar and Planetary Science Conference*, The Woodlands, TX (2015)

K. Umemoto, K. Hirose, Liquid iron-hydrogen alloys at outer core conditions by first-principles calculations. Geophys. Res. Lett. **42**, 7513–7520 (2015)

T. Usui, C.M.O.D. Alexander, J. Wang, J.I. Simon, J.H. Jones, Origin of water and mantle-crust interactions on Mars inferred from hydrogen isotopes and volatile element abundances of olivine-hosted melt inclusions of primitive shergottites. Earth Planet. Sci. Lett. **357–358**, 119–129 (2012)

T. Usui, C.M.O.D. Alexander, J. Wang, J.I. Simon, J.H. Jones, Meteoritic evidence for a previously unrecognized hydrogen reservoir on Mars. Earth Planet. Sci. Lett. **410**, 140–151 (2015)

T. Usui, E. Nakamura, K. Kobayashi, S. Maruyama, H. Helmstaedt, Fate of the subducted Farallon plate inferred from eclogite xenoliths in the Colorado Plateau. Geology **31**, 589–592 (2003)

R. van der Hilst, E.R. Engdahl, W. Spakman, G. Nolet, Tomographic imaging of subducted lithosphere below northwest Pacific island arcs. Nature **353**, 37–43 (1991)

M. van der Meijde, F. Marone, D. Giardini, S. van der Lee, Seismic evidence for water deep in Earth's upper mantle. Science **300**, 1556–1558 (2003)

P.E. van Keken, B.R. Hacker, E.M. Syracuse, G.A. Abers, Subduction factory: 4. Depth-dependent flux of H_2O from subducting slabs worldwide. J. Geophys. Res. **116**, B01401 (2011)

W.L. van Mierlo, F. Langenhorst, D.J. Frost, D.C. Rubie, Stagnation of subducting slabs in the transition zone due to slow diffusion in majoritic garnet. Nat. Geosci. **6**, 400–403 (2013)

L. Vattuone, M. Smerieri, L. Savio, A.M. Asaduzzaman, K. Muralidharan, M.J. Drake, M. Rocca, Accretion disc origin of the Earth's water. Philos. Trans. R. Soc. Lond. A **371**, 20110585 (2013)

L.P. Vinnik, R.W.E. Green, L.O. Nicolaysen, Seismic constraints on dynamics of the mantle of the Kaapvaal craton. Phys. Earth Planet. Inter. **95**, 139–151 (1996)

D. Vlassopoulos, G.R. Rossman, S.E. Haggerty, Coupled substitution of H and minor elements in rutile and the implications of high OH contents in Nb- and Cr-rich rutile from the upper mantle. Am. Mineral. **78**, 1181–1191 (1993)

N. Vogel, V.S. Heber, H. Baur, D.S. Burnett, R. Wieler, Argon, krypton, and xenon in the bulk solar wind as collected by the Genesis mission. Geochim. Cosmochim. Acta **75**, 3057–3071 (2011)

J.A. Wade, T. Plank, E.H. Hauri, K.A. Kelley, K. Roggensack, M. Zimmer, Prediction of magmatic water contents measurements of H_2O in clinopyroxene phenocrysts. Geology **36**, 799–802 (2008)

J.H. Waite Jr., W.S. Lewis, B.A. Magee, J.I. Lunine, W.B. McKinnon, C.R. Glein, O. Mousis, D.T. Young, T. Brockwell, J. Westlake, M.J. Nguyen, B.D. Teolis, H.B. Niemann, R.L. McNutt Jr., M. Perry, W.H. Ip, Liquid water on Enceladus from observations of ammonia and ^{40}Ar in the plume. Nature **460**, 487–490 (2009)

R.J. Walker, Highly siderophile elements in the Earth, Moon and Mars: update and implications for planetary accretion and differentiation. Chem. Erde **69**, 101–125 (2009)

R.J. Walker, R.W. Carlson, S.B. Shirey, F.R. Boyd, Os, Sr, Nd, and Pb isotope systematics of southern African peridotite xenoliths: implications for the chemical evolution of subcontinental mantle. Geochim. Cosmochim. Acta **53**, 1583–1595 (1989)

P.J. Wallace, Volatiles in submarine basaltic glasses from the Northern Kerguelen Plateau (Site 1140): implications for source region compositions, magmatic processes, and plateau subsidence. J. Petrol. **43**, 1311–1326 (2002)

P.J. Wallace, Volatiles in subduction zone magmas: concentrations and fluxes based on melt inclusion and volcanic gas data. J. Volcanol. Geotherm. Res. **140**, 217–240 (2005)

K. Wallmann, The geological water cycle and the evolution of marine $\delta^{18}O$ values. Geochim. Cosmochim. Acta **65**, 2469–2485 (2001)

M.J. Walter, S.C. Kohn, D. Araujo, G.P. Bulanova, C.B. Smith, E. Gaillou, J. Wang, A. Steele, S.B. Shirey, Deep mantle cycling of oceanic crust: evidence from diamonds and their mineral inclusions. Science **334**, 54–57 (2011)

M.J. Walter, A.R. Thompson, W. Wang, O.T. Lord, J.L. Ross, S.C. McMahon, M.A. Baron, E. Melekhova, A.K. Kleppe, S.C. Kohn, The stability of hydrous silicates in Earth's lower mantle: experimental constraints from the systems $MgO–SiO_2–H_2O$ and $MgO–Al_2O_3–SiO_2–H_2O$. Chem. Geol. **418**, 16–29 (2015)

H. Wang, J. van Hunen, D.G. Pearson, M.B. Allen, Craton stability and longevity: the roles of composition-dependent rheology and buoyancy. Earth Planet. Sci. Lett. **391**, 224–233 (2014)

J. Wang, K.H. Hattori, R. Kilian, C.R. Stern, Metasomatism of sub-arc mantle peridotites below southernmost South America: reduction of fO_2 by slab-melt. Contrib. Mineral. Petrol. **153**, 607–624 (2007)

Q. Wang, A review of water contents and ductile deformation mechanisms of olivine: implications for the lithosphere-asthenosphere boundary of continents. Lithos **120**, 30–41 (2010)

Y. Wang, H. Ren, Z. Jin, Water and fabric in an ophiolitic peridotite from a supra-subduction zone. Contrib. Mineral. Petrol. **171**, 1–17 (2016)

Z. Wang, H. Becker, Ratios of S, Se and Te in the silicate Earth require a volatile-rich late veneer. Nature **499**, 328–331 (2013)

J.M. Warren, E.H. Hauri, Pyroxenes as tracers of mantle water variations. J. Geophys. Res. **119**, 1851–1881 (2014)

L.L. Watson, I.D. Hutcheon, S. Epstein, E.M. Stolper, Water on Mars: clues from deuterium/hydrogen and water contents of hydrous phases in SNC meteorites. Science **265**, 86–90 (1994)

J.D. Webster, P.M. Piccoli, Magmatic apatite: a powerful, yet deceptive, mineral. Elements **11**, 177–182 (2015)

F.A. Weis, H. Skogby, V.R. Troll, F.M. Deegan, B. Dahren, Magmatic water contents determined through clinopyroxene: examples from the Western Canary Islands, Spain. Geochem. Geophys. Geosyst. **16**, 2127–2146 (2015)

R. Wieler, H. Busemann, I.A. Franchi, Trapping and modification processes of noble gases and nitrogen in meteorites and their parent bodies, in *Meteorites and the Early Solar System II*, ed. by D.S. Lauretta, H.Y. McSween Jr. (University of Arizona Press, Tucson, 2006), pp. 499–521

S.A. Wilde, J.W. Valley, W.H. Peck, C.M. Graham, Evidence from detrital zircons for the existence of continental crust and oceans on the Earth 4.4 Gyr ago. Nature **409**, 175–178 (2001)

M. Willbold, T. Elliott, S. Moorbath, The tungsten isotopic composition of the Earth's mantle before the terminal bombardment. Nature **477**, 195–198 (2011)

R. Wirth, C. Vollmer, F. Brenker, S.S. Matsyuk, F. Kaminsky, Inclusions of nanocrystalline hydrous aluminium silicate "Phase Egg" in superdeep diamonds from Juina (Mato Grosso State, Brazil). Earth Planet. Sci. Lett. **259**, 384–399 (2007)

A.C. Withers, H. Bureau, C. Raepsaet, M.M. Hirschmann, Calibration of infrared spectroscopy by elastic recoil detection analysis of H in synthetic olivine. Chem. Geol. **334**, 92–98 (2012)

B.J. Wood, L.T. Bryndzia, K.E. Johnson, Mantle oxidation state and its relationship to tectonic environment and fluid speciation. Science **248**, 337–345 (1990)

B.J. Wood, A.N. Halliday, The lead isotopic age of the Earth can be explained by core formation alone. Nature **465**, 767–770 (2010)

B.J. Wood, A.N. Halliday, M. Rehkamper, Volatile accretion history of the Earth. Nature **467**, E6–E7 (2010)

B.J. Wood, J.-Y. Li, A. Shahar, Carbon in the core: its influence on the properties of core and mantle. Rev. Mineral. Geochem. **75**, 231–250 (2013)

B.J. Wood, J.A. Wade, M.R. Kilburn, Core formation and the oxidation state of the Earth: additional constraints from Nb, V and Cr partitioning. Geochim. Cosmochim. Acta **72**, 1415–1426 (2008)

A.B. Woodland, M. Koch, Variation in oxygen fugacity with depth in the upper mantle beneath the Kaapvaal craton, Southern Africa. Earth Planet. Sci. Lett. **214**, 295–310 (2003)

R.K. Workman, S.H. Hart, Major and trace element composition of the depleted MORB mantle (DMM). Earth Planet. Sci. Lett. **231**, 53–72 (2005)

R.K. Workman, E. Hauri, S.R. Hart, J. Wang, J. Blusztajn, Volatile and trace elements in basaltic glasses from Samoa: implication for water distribution in the mantle. Earth Planet. Sci. Lett. **241**, 932–951 (2006)

K. Wright, Atomistic models of OH defects in nominally anhydrous minerals, in *Water in Nominally Anhydrous Minerals*, ed. by H. Keppler, J.R. Smyth (Mineralogical Society of America, Chantilly, 2006), pp. 67–83

Q.-K. Xia, Y. Hao, P. Li, E. Deloule, M. Coltorti, L. Dallai, X. Yang, M. Feng, Low water content of the Cenozoic lithospheric mantle beneath the eastern part of the North China Craton. J. Geophys. Res. **115**, B07207 (2010)

Q.-K. Xia, Y.-T. Hao, S.C. Liu, X.-Y. Gu, M. Feng, Water contents of the Cenozoic lithospheric mantle beneath the western part of the North China Craton: peridotite xenolith constraints. Gondwana Res. **23**, 108–118 (2013a)

Q.-K. Xia, J. Liu, S.-C. Liu, I. Kovács, M. Feng, L. Dang, High water content in Mesozoic primitive basalts of the North China Craton and implications on the destruction of cratonic mantle lithosphere. Earth Planet. Sci. Lett. **361**, 85–97 (2013b)

Q.-K. Xia, X.-Z. Yang, E. Deloule, Y.-M. Sheng, Y.-T. Hao, Water in the lower crustal granulite xenoliths from Nushan, eastern China. J. Geophys. Res. **111**, B11202 (2006)

X. Yang, OH solubility in olivine in the peridotite-COH system under reducing conditions and implications for water storage and hydrous melting in the reducing upper mantle. Earth Planet. Sci. Lett. **432**, 199–209 (2015)

X. Yang, Effect of oxygen fugacity on OH dissolution in olivine under peridotite-saturated conditions: an experimental study at 1.5–7 GPa and 1100–1300 °C. Geochim. Cosmochim. Acta **173**, 319–336 (2016)

X. Yang, H. Keppler, Y. Li, Molecular hydrogen in mantle minerals. Geochem. Perspect. Lett. **2**, 160–168 (2016)

X. Yang, D. Liu, Q.-K. Xia, CO_2-induced small water solubility in olivine and implications for properties of the shallow mantle. Earth Planet. Sci. Lett. **403**, 37–47 (2014)

X.-Z. Yang, E. Deloule, Q.-K. Xia, Q.-C. Fan, M. Feng, Water contrast between Precambrian and Phanerozoic continental lower crust in eastern China. J. Geophys. Res. **113**, B08207 (2008a)

X.-Z. Yang, Q.-K. Xia, E. Deloule, L. Dallai, Q.C. Fan, M. Feng, Water in minerals of the continental lithospheric mantle and overlying lower crust: a comparative study of peridotite and granulite xenoliths from the North China craton. Chem. Geol. **256**, 33–45 (2008b)

R. Yokochi, B. Marty, A determination of the neon isotopic composition of the deep mantle. Earth Planet. Sci. Lett. **225**, 77–88 (2004)

T. Yoshino, T. Matsuzaki, S. Yamashita, T. Katsura, Hydrous olivine unable to account for conductivity anomaly at the top of the asthenosphere. Nature **443**, 973–976 (2006)

H. Yuan, B. Romanowicz, Lithospheric layering in the North American craton. Nature **466**, 1063–1068 (2010)

C. Zhang, Z. Duan, A model for C-O-H fluid in the Earth's mantle. Geochim. Cosmochim. Acta **73**, 2089–2102 (2009)

Y. Zhang, H_2O in rhyolitic glasses and melts: measurement, speciation, solubility, and diffusion. Rev. Geophys. **37**, 493–516 (1999)
Y. Zhang, H. Ni, Diffusion of H, C, and O components in silicate melts, in *Diffusion in Minerals and Melts*, ed. by Y. Zhang, D.J. Cherniak (Mineralogical Society of America, Geochemical Society, Chantilly, 2010), pp. 171–225
A. Zindler, S. Hart, Chemical geodynamics. Annu. Rev. Earth Planet. Sci. **14**, 493–571 (1986)

Space Sci Rev (2018) 214:92
https://doi.org/10.1007/s11214-018-0526-1

Water and Volatile Inventories of Mercury, Venus, the Moon, and Mars

James P. Greenwood[1] · Shun-ichiro Karato[2] · Kathleen E. Vander Kaaden[3] · Kaveh Pahlevan[4] · Tomohiro Usui[5]

Received: 16 March 2018 / Accepted: 10 July 2018 / Published online: 26 July 2018

Abstract We review the geochemical observations of water, D/H and volatile element abundances of the inner Solar System bodies, Mercury, Venus, the Moon, and Mars. We focus primarily on the inventories of water in these bodies, but also consider other volatiles when they can inform us about water. For Mercury, we have no data for internal water, but the reducing nature of the surface of Mercury would suggest that some hydrogen may be retained in its core. We evaluate the current knowledge and understanding of venusian water and volatiles and conclude that the venusian mantle was likely endowed with as much water as Earth of which it retains a small but non-negligible fraction. Estimates of the abundance of the Moon's internal water vary from Earth-like to one to two orders of magnitude more depleted. Cl, K, and Zn isotope anomalies for lunar samples argue that the giant impact left a unique geochemical fingerprint on the Moon, but not the Earth. For Mars, an early magma ocean likely generated a thick crust; this combined with a lack of crustal recycling mechanisms would have led to early isolation of the Martian mantle from later delivery of water and volatiles from surface reservoirs or late accretion. The abundance estimates of Martian mantle water are similar to those of the terrestrial mantle, suggesting some similarities in the water and volatile inventories for the terrestrial planets and the Moon.

Keywords Water · Hydrogen isotopes · Mercury · Venus · Mars · Moon

The Delivery of Water to Protoplanets, Planets and Satellites
Edited by Michel Blanc, Alessandro Morbidelli, Yann Alibert, Lindy Elkins-Tanton, Paul Estrada, Keiko Hamano, Helmut Lammer, Sean Raymond and Maria Schönbächler

✉ J.P. Greenwood
jgreenwood@wesleyan.edu

[1] Department of Earth and Environmental Sciences, Wesleyan University, Middletown, CT 06459, USA

[2] Department of Geology and Geophysics, Yale University, New Haven, CT 06520, USA

[3] Jacobs, NASA Johnson Space Center, Mail Code XI3, Houston, TX 77058, USA

[4] School of Earth & Space Exploration, Arizona State University, Tempe, AZ 85281, USA

[5] Earth-Life Science Institute, Tokyo Institute of Technology, 2-12-1 Ookayama, Meguro, Tokyo 152-8550, Japan

1 Introduction

Where did the Earth's oceans come from? This question is at the core of what motivates the study of the sources of water for inner Solar System bodies, and its answer will ultimately inform us on the sources and inventories of water in extrasolar planets. In this article, we will review the current knowledge regarding the abundances and sources of water for Mercury, Venus, the Moon, and Mars. We will primarily consider H and D/H, but will use other highly volatile elements (C, N, O, S, Cl) and their isotopes when they can inform us on the origins of water in the inner Solar System. Use of other common highly volatile elements is especially important when data on water for a body is poor or non-existent (as also occurs in the case of chondrites, e.g., the enstatite chondrites). This brings us to the problem of datasets for these four bodies. For Mercury, H data does not exist. For Venus, H data are only available for the upper atmosphere. For the Moon, H data exists for both the surface and interior, and for Mars a relative plethora of datasets on H abundance exist for the atmosphere, the surface, and the interior. Yet a glut of data does not guarantee an answer to the source of water delivery, as both the sources and inventory of Earth's water continue to be debated (e.g., Peslier et al. 2017; Genda and Ikoma 2008; Halliday 2013; Marty 2012; Marty et al. 2016; Hallis et al. 2015). In this article we will use 'water' to refer to total H, but H should be understood to be the sum of all species of hydrogen (H_2, H_2O, and OH). We will also discuss measurements of hydrogen using weight percent (wt.%) or parts per million (ppm wt) H_2O. The sources of water to the inner Solar System are discussed in the article by Alexander et al. (2018), and the delivery of water to the terrestrial planets is the subject of the article by O'Brien et al. (2018). The abundance and sources of terrestrial water are discussed in the complementary article by Peslier et al. (2017).

1.1 Gamma-Ray Spectroscopy of Mercury, Venus, the Moon, and Mars and Their Moderately Volatile Elements

Gamma-Ray Spectroscopy (GRS) collected by an orbiting spacecraft is the only measurement technique that has been employed on all four bodies, and its dataset is a useful starting point for this article. The GRS measurements acquired in the last decade at Mercury completed this dataset for the four bodies (Fig. 1 adapted from McCubbin et al. 2012a). In Fig. 1, remote GRS data for the surfaces of Mercury, Venus, the Moon, and Mars are compared to estimates of the bulk silicate Earth (Kitts and Lodders 1998; Lodders and Fegley 1998; Lucey et al. 2006; McDonough and Sun 1995; Peplowski et al. 2011; Taylor 2013). Figure 1 shows an important result: while K, U, Th are all lithophile elements, K is also moderately volatile, while U and Th are refractory elements. An element's volatility is a direct reflection of its condensation temperature from a gas of solar composition (Wood et al. 2010). Under highly reducing conditions, K can become less volatile, but still more volatile than U and Th (Ebel and Sack 2013). All three elements behave as similarly incompatible elements during planetary differentiation, meaning that they are not expected to fractionate from one another in the processes of magmatism and volcanism in the parent bodies. Figure 2a shows an abundance plot of another highly incompatible lithophile element, La, vs. U for Mars, Earth, the Moon, and the HED (howardite-eucrite-diogenite) meteorites that are believed to originate from 4 Vesta (McSween et al. 2013) demonstrating U and La are not fractionated from each other during planetary igneous processes. In Fig. 2b, K is plotted vs. La for these same bodies. There is a systematic

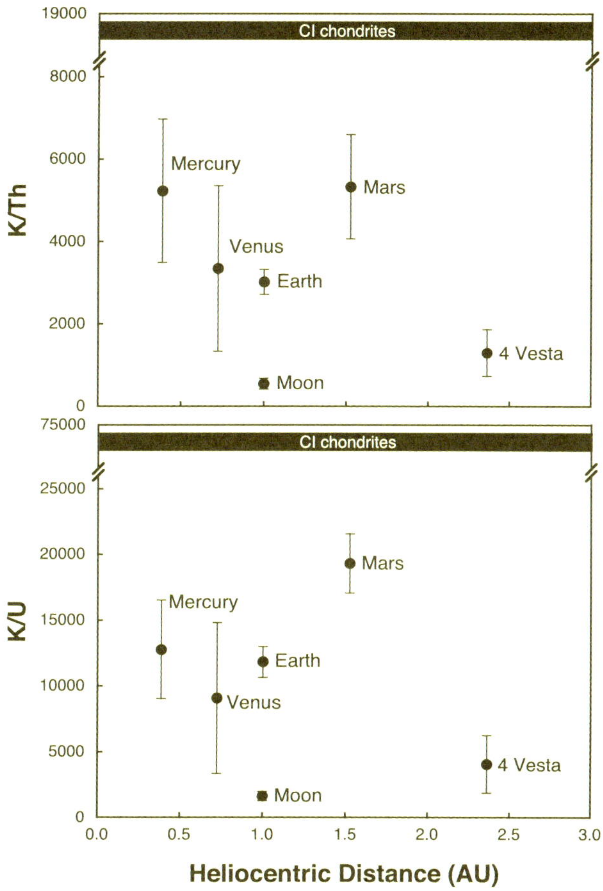

Fig. 1 K/Th and K/U vs. heliocentric distance for the inner Solar System bodies showing the surprising result of Mercury being more enriched in the moderately volatile K than the Moon, Vesta, and possibly the Earth. (Data from Kitts and Lodders 1998; Lodders and Fegley 1998; Lucey et al. 2006; McDonough and Sun 1995; Peplowski et al. 2011; Taylor 2013)

depletion in K in the order Mars, Earth, Vesta, and the Moon from the CI and LL chondrites (Fig. 2b). The depletions in K in the CV and CO chondrites are similar to Earth and Mars. This suggests that the depletion in K could be used as a proxy for a systematic depletion in all the volatile lithophile elements in these bodies relative to the Sun and CI chondrites.

In Fig. 1, Mercury is as enriched in K as Earth, Mars, and Venus. Yet, Earth has oceans, and Mercury clearly does not. While this contrast may be related to the history of redox evolution, volatile processing, and atmospheric loss, it does raise the question: Do moderately volatile lithophile elements give us any information about the inventory of highly volatile elements, such as H, C, and N? When one considers the systematic depletion of K for the Moon and Vesta relative to Earth and Mars shown in Figs. 1 and 2b, it would appear that K was subsequently lost to space or that these bodies accreted different amounts of K. Subsequent loss of K is quite plausible for the Moon and Vesta, which are believed to have had magma oceans that could have led to depletions in K (Wang and Jacobsen 2016; Mandler and Elkins-Tanton 2013). Loss of K during the Earth-Moon giant impact event might also explain why Mars is enriched relative to Earth. Mercury is at least similar in moderately volatile elements to Earth, which argues strongly against models such as the one proposed by Albarède (2009) that predict that the volatile element chemistry of a body is

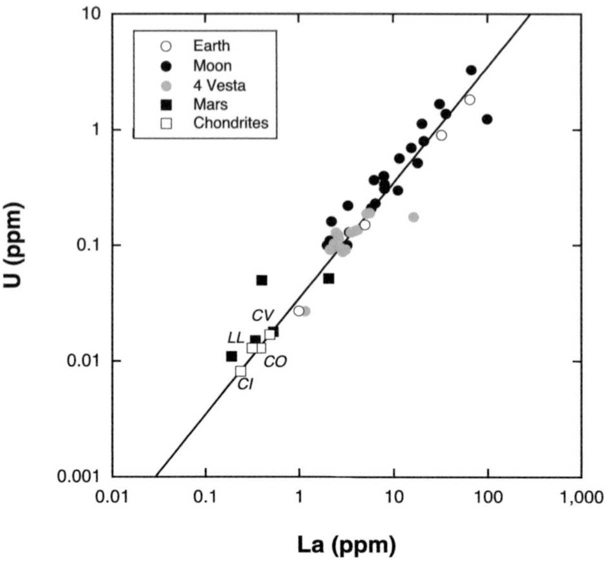

(a)

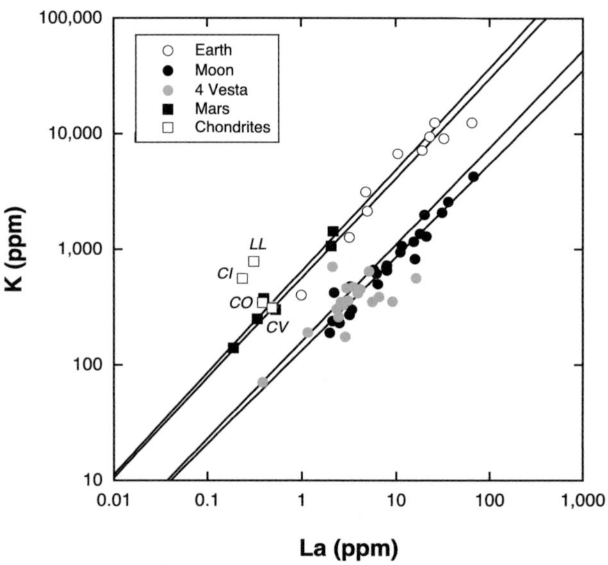

(b)

Fig. 2 (**a**) U vs. La and (**b**) K vs. La for chondrites and the Earth, Moon, Vesta, and Mars. U and La are both refractory incompatible elements and show that planetary differentiation does not alter this primordial ratio. The depletion in K of the Earth is very similar to that of CV chondrites. Data from Halliday et al. (1995); Jochum et al. (1991); BVSP (1981); Newsom et al. (1986); Haskin and Warren (1991); Jerde et al. (1990); Jolliff et al. (2003); Korotev (1987); Korotev and Kremser (1992); Warren and Kallemeyn (1986); Warren (2005); Kitts and Lodders (1998)

related to its heliocentric distance when it formed. Thus, if Mercury is as enriched in K and other moderately volatile elements as the Earth, Venus, and Mars, it raises the question of whether Mercury accreted as much water as the Earth? Did Venus acquire as much water as Earth? And for that matter, what about Mars? Did Mars have *more* water than the Earth? Can moderately volatile elements be used as an indication of abundances of the highly volatile elements H, C, and N?

The similarity between Earth and Mars and CV and CO chondrites in Fig. 2b argues that this depletion in K could be due to accretion of K depleted material, rather than loss of K to space. Chondritic meteorites also show systematic depletions in moderately volatile elements as shown in Fig. 3a–f. Here the moderately and highly volatile elements are shown vs. 50% condensation temperature for three types of carbonaceous chondrites (CM, CO, CV), the LL ordinary chondrites, and EL and EH enstatite chondrites, normalized to CI chondrite and Si. All classes of chondrite show depletions in the moderately and highly volatile elements relative to CI chondrite. The cosmochemical affinities of the elements are also shown, and the three carbonaceous chondrites exhibit a systematic depletion of elements with decreasing condensation temperature, regardless of whether the elements are siderophile, chalcophile, or lithophile, suggesting that these volatile element depletions are nebular in origin, as planetary differentiation would fractionate siderophile and chalcophile from lithophile elements during core formation (Wood et al. 2010).

1.2 Highly Volatile and Moderately Volatile Element Depletions of the Chondrites

The volatile elements H, C, and N are also plotted in Fig. 3, and the abundances of these three highly volatile elements are also strongly depleted in the chondrites. While some of this depletion may be due to redistribution of elements during metamorphism, there appears to be a systematic depletion trend in the volatile elements for each type of chondrite shown in Fig. 3. The CM chondrites have overall higher abundances of volatiles than the other chondrite groups shown, and appear slightly enriched compared to their moderately volatile element depletion trend. The other three chondrite groups have H, C, and N abundances that appear to plot close to an extension of their moderately volatile element depletion trend. The LL chondrites have a strong depletion in the siderophile and chalcophile elements, while being quite high in H, C, and N (Alexander et al. 2012), as well as being high in moderately volatile lithophile elements, such as K. There are no primitive enstatite chondrites (EC), and thus they are also affected by metamorphism. The ECs retain significant C and N demonstrating that they have not suffered complete loss of volatiles as is sometimes thought (H has not been reliably measured in ECs; e.g., Jarosewich 1990). If the volatile element abundances of the chondrites plot on an extension of their moderately volatile element depletion trends in Fig. 3, this would suggest that whatever processes caused the moderately volatile element depletions in chondrites and the inner Solar System bodies may also be responsible for the depleted abundances of the highly volatile elements (H, C, N) in these bodies.

Thus, the chondrites point to the commonality of moderately volatile element depletions between the differentiated inner Solar System bodies Mercury, Venus, the Moon, Mars, and Vesta and the undifferentiated chondrites, and demonstrates an early loss of volatile elements *prior* to planetary differentiation. The chondrites also demonstrate that the highly volatile element abundances may reflect the processes that caused moderately volatile element depletions of these different inner Solar System bodies and materials. The processes that caused the systematic depletion trends in the moderately volatile lithophile elements are

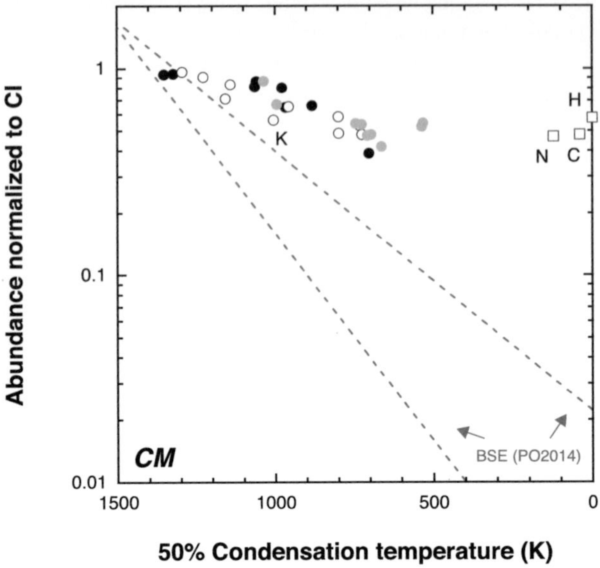

(a)

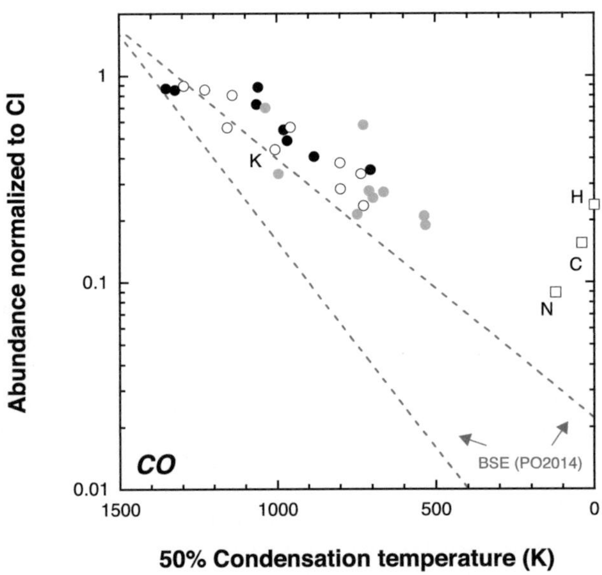

(b)

Fig. 3 Abundances of the moderately volatile elements and volatile elements vs. 50% condensation temperature for chondrites: (**a**) CM, (**b**) CO, (**c**) CV, (**d**) LL, (**e**) EH, (**f**) EL. The abundances are normalized to CI chondrites and silicon. Black circles indicate siderophile elements, open circles are lithophile, grey circles are chalcophile, and open squares are volatile. The chalcophile and siderophile elements are already mobilized due to early parent body processes on the LL parent body. Metamorphism has also affected the EH and EL chondrites. Volatile data from Alexander et al. (2012) for Murchison (CM), Mokoia (CO), Kaba (CV), Semarkona (LL). C and N from Grady et al. (1986) for (EH) and (EL), with data for other elements from Wasson and Kallemeyn (1988) and Lodders and Fegley (1998). Dashed grey lines are the moderately volatile element trend of the Bulk Silicate Earth from Palme and O'Neill (2014)

Fig. 3 (*Continued*)

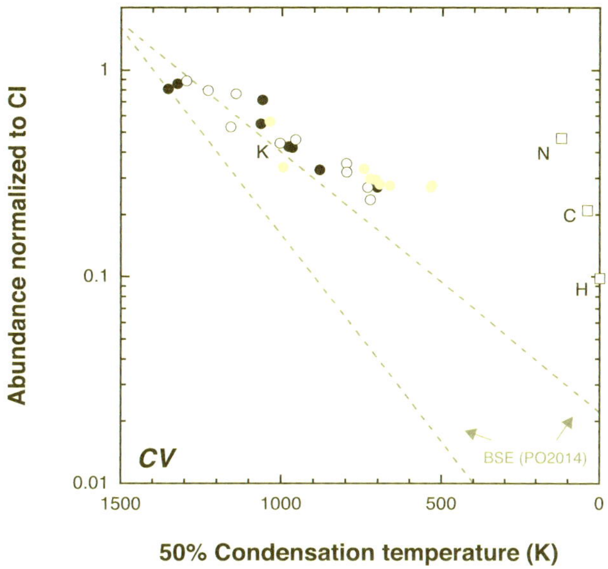

(c)

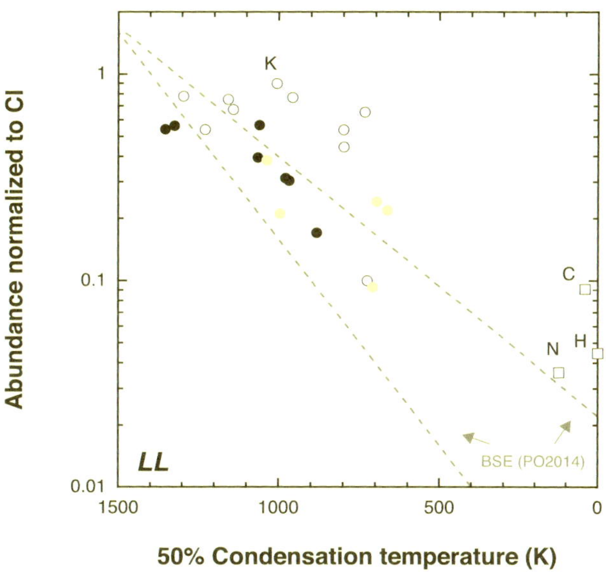

(d)

likely to be high temperature processes, such as condensation or evaporation of rock components (Young 2017). These high temperature processes are likely to significantly deplete the volatile elements H, C, N, as well as the noble gases of the precursor materials for Mercury, Venus, Earth-Moon, and Mars, and have led to the prevailing ideas that highly volatile elements were completely lost during the processes that caused the moderately volatile element depletions, and then were added back to the Earth and the other inner Solar System bodies at some later time (e.g. Albarède 2009). By contrast, the abundances of H, C, and N

Fig. 3 (*Continued*)

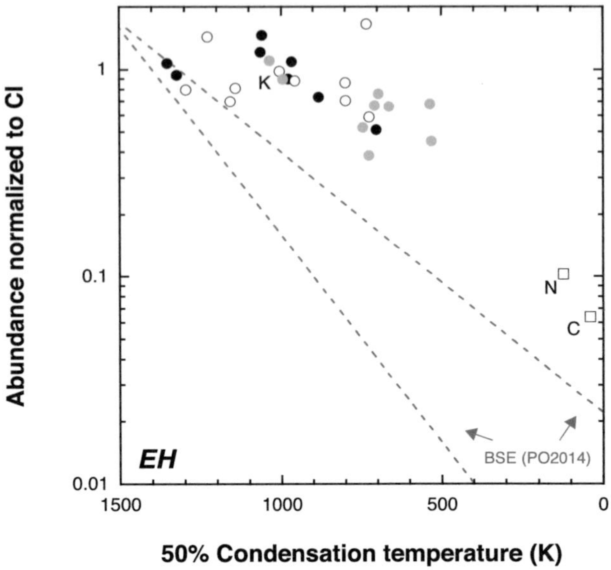

(e)

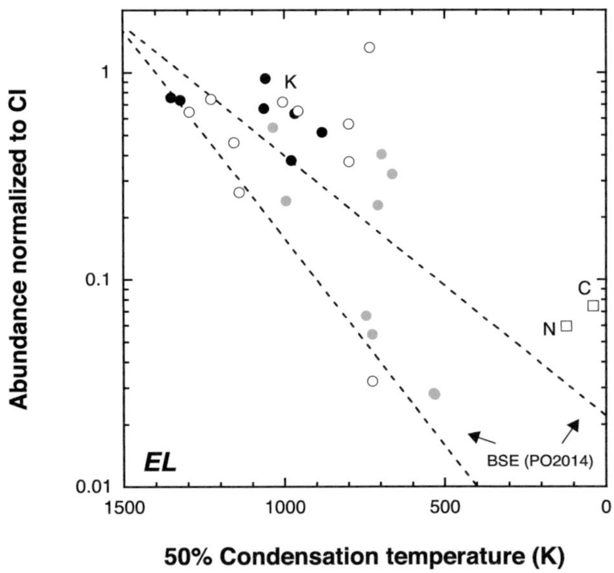

(f)

in primitive CV, CO, and CM chondrites appear to be similar to their expected depletions from an extension of the trend of their depletions in moderately volatile elements to lower temperatures. The enstatite chondrites have quite substantial abundances of C and N (Grady et al. 1986). If Earth's precursor materials were similar to ECs, this would imply that some fraction of the highly volatile elements were inherited from the local nebular environment of inner Solar System bodies, rather than being completely lost during planetary accretion, and then delivered later by chondritic asteroids or comets.

2 Water, D/H, and Volatile Elements of Mercury, Venus, Moon, and Mars

2.1 Mercury

2.1.1 Introduction

Prior to the return of data from the MErcury Surface, Space ENvironment, GEochemistry and Ranging (MESSENGER) spacecraft (Solomon et al. 2001), the planet Mercury was thought to be depleted in volatile elements primarily due to its close proximity to the Sun (Albarède 2009). Initial analyses conducted during flybys from the Mariner 10 mission confirmed the presence of exospheric H, He, and O (Broadfoot et al. 1974, 1976). Furthermore, ground-based discoveries enhanced our knowledge about Na, K, and Ca in the exosphere (Bida et al. 2000; Potter and Morgan 1985, 1986), as well as polar volatiles found in radar-reflective deposits (Harmon and Slade 1992; Slade et al. 1992). Despite this evidence for the presence of volatiles on the surface of the planet and in the exosphere, it was not until the return of data from MESSENGER that it was discovered just how volatile-rich Mercury is (Evans et al. 2015; McCubbin et al. 2012a; Murchie et al. 2015; Nittler et al. 2011; Peplowski et al. 2012, 2014, 2015, 2016; Weider et al. 2012).

Among the numerous instruments onboard MESSENGER, the X-Ray Spectrometer (XRS), Gamma-Ray Spectrometer (GRS), and Neutron Spectrometer (NS) have provided the greatest amount of information regarding the volatile content of Mercury (Evans et al. 2012, 2015; Lawrence et al. 2013; Nittler et al. 2011; Peplowski et al. 2012, 2014, 2015, 2016). These data have shown that the surface of the planet has elevated abundances of C (Murchie et al. 2015; Peplowski et al. 2015, 2016), Na (Peplowski et al. 2014), S (Nittler et al. 2011; Weider et al. 2012), Cl (Evans et al. 2015), and K (Peplowski et al. 2012), likely indicating that Mercury has high levels of indigenous volatiles. Additionally, although a likely exogenic component, elevated H has been seen in the polar regions compared to other portions of the planet's surface (Lawrence et al. 2013). Furthermore, elevated K/Th and near chondritic K/Cl ratios at Mercury's surface lends support to the volatile-rich nature of the planet (Evans et al. 2015; McCubbin et al. 2012a; Peplowski et al. 2011, 2012). Although not considered a volatile element, MESSENGER XRS data has also shown that the surface of the planet is depleted in Fe (Nittler et al. 2011; Weider et al. 2012). These characteristic elemental abundances have facilitated calculation of the oxygen fugacity (fO_2) characterizing silicate Mercury, which is between 2.6 and 7.3 $\log_{10}$ units below the iron-wustite buffer (McCubbin et al. 2012a, 2017; Zolotov et al. 2011, 2013), making Mercury the most reduced terrestrial planet (Herd 2008; Sharp et al. 2013; Wadhwa 2008).

2.1.2 Pyroclastic Deposits

Pyroclastic deposits across the surface of Mercury have been identified by their characteristic diffuse borders, high reflectance compared to nearby units, and a steeper slope in their reflectance spectrum from visible to near-infrared wavelengths (Blewett et al. 2009; Kerber et al. 2011, 2009; Rava and Hapke 1987; Robinson et al. 2008). On Mercury, 51 pyroclastic deposits have been identified from the MESSENGER data, covering $> 111{,}000\ \text{km}^2$ of the mercurian surface, or $\sim 0.15\%$ of the total surface (Goudge et al. 2014; Kerber et al. 2011). Of these 51 deposits, $> 90\%$ of them are located within impact craters and basins (Goudge et al. 2014). The pyroclastic deposits on Mercury are larger than pyroclastic deposits found on the Moon, suggesting that they involved higher energy eruptions with more abundant volatiles from the interior of the planet than what was typically associated with lunar pyroclastic eruptions (Kerber et al. 2011; Thomas et al. 2015). Furthermore, although the majority of pyroclastic deposits on Mercury are located away from the smooth plains, there are pyroclastic deposits associated with the margins of some smooth plains units, analogous to lunar pyroclastic deposits that are adjacent to lunar maria (e.g., Goudge et al. 2014; Head and Wilson 1979). Thomas et al. (2014) presented evidence suggesting that the pyroclastic eruptions initially occurred ~ 3.9 Ga and continued until less than a billion years ago. This long duration of explosive volcanism suggests that large amounts of volatiles are present in the mercurian interior. Kerber et al. (2011) calculated CO abundances of 1600–16000 ppm wt are required if the pyroclastic eruptions on Mercury are similar to Hawaiian-style eruptions on Earth. Given the reduced nature of the planet, as revealed by MESSENGER, Kerber et al. (2011) proposed S_2Cl, Cl, Cl_2, and COS as the possible driving forces behind mercurian pyroclastic eruptions. Given the highly reducing nature of the planet, it is also possible that H_2 or H_2S are present and could have aided in driving these pyroclastic eruptions.

2.1.3 Magmatic Volatiles and the Evolution of Mercury

The role that volatiles played in the thermal and magmatic evolution of Mercury are critical to understanding the present-day planet as investigated by Mariner 10 and MESSENGER. Although all of the chemical analyses currently obtained for Mercury are solely of surface compositions, these data have been used to infer the abundance of volatiles in the mercurian interior (Evans et al. 2015; Malavergne et al. 2004, 2010; Namur et al. 2016; Stockstill-Cahill et al. 2012; Vander Kaaden and McCubbin 2015, 2016; Zolotov et al. 2013, 2011).

In Mercury, C is thought to be present in the form of elemental C as opposed to CO or CO_2. Initially, GRS measurements placed C abundance of Mercury's surface at 1.4 ± 0.9 wt.% (Peplowski et al. 2015). At the 3-sigma standard deviation level, this allowed for C abundances on Mercury to be between 0 wt.% and 4.1 wt.%. Additional analyses using the NS on MESSENGER showed C abundances on Mercury in the low reflectance material to be as high as 1–3 wt.% greater than the surrounding, higher reflectance materials (Peplowski et al. 2016). This elevated C signature found in the low reflectance material on the planet's surface is likely due to the presence of graphite, possibly from an ancient primary floatation crust (Murchie et al. 2015; Vander Kaaden and McCubbin 2015).

Other magmatic volatiles present in the mercurian interior include Na, Cl, and K. These three elements are all higher in the north polar region of Mercury, where the largest volcanic unit on the planet is located. This suggests that these volatiles played a role in the transport and eruption of flood volcanics on Mercury. Given the low fO_2 of Mercury, S is likely present in the form of sulfides instead of sulfates. Analogue and experimental studies have suggested the presence of Fe-, Cr-, Mn-, Ti-, Mg-, Ca-bearing sulfides on Mercury (Malavergne et al. 2014; McCoy et al. 1999; Namur et al. 2016; Vander Kaaden and McCubbin 2016; Vander Kaaden et al. 2017; Zolotov et al. 2013). It has also been suggested that a sulfide-rich conductive lid could have formed during a mercurian magma ocean event (Parman et al. 2016). However, it is unlikely that pure CaS, which is the only sulfide solid or immiscible melt that has been shown to be buoyant in a mercurian magma ocean, would form given the preferential formation of MgS to CaS as well as the likely formation of multi-component sulfides as opposed to single component sulfide phases (Vander Kaaden and McCubbin 2016).

Although the exact accretion history of Mercury is still debated, it is generally believed that Mercury experienced a giant impact that stripped much of its silicate mantle (Benz et al. 1988) and led to a magma ocean (Brown and Elkins-Tanton 2009;

Riner et al. 2009). The highly reducing conditions that were present in this magma ocean may have further contributed to the large mass of the metallic core reforming after the giant impact. However, given the geophysical data from MESSENGER, the density of the mercurian core—like the terrestrial core—requires addition of a light element component in addition to Fe (Hauck et al. 2013; Smith et al. 2012). The most likely light element to make up this density difference, due to the low fO_2 and experimental results, is Si (Chabot et al. 2014a; Hauck et al. 2013). With increasing Si contents in the mercurian core, C and S would both be excluded from it. Although S would preferentially partition into the silicate portion of the planet, C is also not soluble in the mercurian mantle, adding to exotic crust formation processes on the planet (Chabot et al. 2014a; Vander Kaaden and McCubbin 2015; Vander Kaaden et al. 2016). Furthermore, based on experimental results at reducing conditions similar to Mercury, Na, Cl, and K are expected to preferentially partition into the silicate portion of the planet as well (Vander Kaaden and McCubbin 2016). Following the results of Vander Kaaden and McCubbin (2015), this graphite would float to the surface of the magma ocean producing a primary graphite floatation crust, similar to the anorthositic crust exposed on the Moon.

Continued cooling of the planet would have resulted in the crystallization of the magma ocean. Analogous to the lunar magma ocean (Warren 1985), the crystallizing cumulates would sink to the base of the magma ocean, enriching the melt in highly incompatible elements. However, due to the low fO_2 of the planet, and associated low FeO contents of mantle cumulates, solid-state overturn, which occurs in the lunar scenario, is not expected on Mercury (Vander Kaaden and McCubbin 2015). Instead, the stratified mantle would reflect the bottom-up crystallization of the magma ocean, including volatile-rich phases, such as albitic plagioclase and possibly sodalite.

After crystallization of the magma ocean, continued volcanism on Mercury would have resurfaced the planet and buried the evidence of the primary graphite floatation crust (Vander Kaaden and McCubbin 2015). The elevated abundances of S on the surface of Mercury, therefore, are the direct result of transport within reducing magmas that have a higher carrying capacity for S compared to moderately reduced to oxidized magmas (e.g., Tuff et al. 2013). Furthermore, the elevated abundances of Na, Cl, and K on the surface were most likely produced in a similar manner given their lithophilic nature under mercurian oxygen fugacities (Vander Kaaden and McCubbin 2016). Continued resurfacing of the planet through impacts, flood volcanism, and pyroclastic volcanism resulted in partial to total destruction of the primary graphite floatation crust and mixing with surface materials, darkening the reflection of the planet. Additional impacts have exposed evidence of this primary graphite floatation crust in the low reflectance material on the planet (Peplowski et al. 2016; Vander Kaaden and McCubbin 2015). Finally, additional impacts from volatile-rich asteroidal material as well as comets have resulted in polar volatile deposits, which are mainly composed of water ice (Chabot et al. 2014b; Lawrence et al. 2013).

2.2 Venus

2.2.1 Introduction

Although Venus and the Earth have long been considered planetary "twins", Venus does not have plate tectonics. Instead, 92% of its surface has the same crater density, suggesting the same age of $\sim$ 300–700 Ma (Schaber et al. 1992; McKinnon et al. 1997). Models suggest that episodic overturn of the mantle on this timescale leads to catastrophic resurfacing of the planet (Parmentier and Hess 1992). The venusian atmosphere has Earth-like inventories of C and N, with terrestrial isotopic compositions, suggesting a primordial volatile inventory endowment not unlike Earth. In contrast, the modern atmosphere only has ~ 30 ppm wt H_2O, with a $D/H \sim 150$ times greater than the Earth's oceans (Fegley 2014). This high D/H led to the suggestion that Venus originally had Earth-like levels of water, with a similar isotopic composition to the Earth's oceans, and that the majority of this water was subsequently lost to space (Ingersoll 1969; Donahue et al. 1982). Models of this process suggest that such early loss scenarios would leave Venus with an initial water endowment of ~a few percent of a terrestrial ocean mainly held in the primitive venusian mantle (Hamano et al. 2013).

Elkins-Tanton et al. (2007) show that volcanism and volatile element recycling will likely never fully deplete the venusian mantle of volatiles. Thus, if atmospheric N and C could serve as proxies for primordial planetary water content, and if current scenarios for venusian evolution have predictive value, then Venus can have ~ 0.01–0.1 times the terrestrial water inventory in its interior. If the majority of the interior water is released episodically during catastrophic resurfacing events, the high D/H and low water content of the atmosphere could be explained via loss of water with an initially terrestrial isotopic composition to space on a 0.5–1.0 Ga timescale (Grinspoon 1993).

2.2.2 Volatiles in the Venusian Atmosphere

The Venus atmosphere has high levels of CO_2, N_2, SO_2, HCl, and HF (Fegley 2014). The abundance of C in Venus' atmosphere is similar to the crustal C inventory of the Earth (Fegley 2014). The N abundance of Venus's atmosphere is several times higher than that of Earth's atmosphere, and two times higher than the total inventory of N in Earth's atmosphere, biosphere, oceans, and crust (Lodders and Fegley 1998). The high level of SO_2 in the venusian atmosphere is 220 times lower than the Earth's total sulfur inventory in the crust and oceans, suggesting that there is abundant sulfur in the venusian crust and/or mantle. Hydrogen chloride and HF appear to be buffered by reactions with surface minerals (Fegley 2014), most likely phosphate minerals such as apatite [$Ca_5(PO_4)_3$F,Cl,OH] (Treiman et al. 2016a). The presence of these atmospheric gases are believed to be the product of volcanic outgassing.

2.2.3 Water and Volatile Elements in the Venusian Mantle

Elkins-Tanton et al. (2007) showed that volatiles, such as water and CO_2, can partition into anhydrous mantle minerals during partial melting processes on Venus and other planetary bodies, and thus be partially retained in the mantle regardless if the mantle region is unmelted or has been previously involved in a melting event. They predict that some fraction of volatile elements will be retained in the venusian mantle, ranging from 100% in mantle regions that have not been melted at all, to 1–2% in regions that have experienced melting. They also argue that volatile-enriched lithosphere would be involved in crustal delamination processes in the thick venusian crust, and envision a way to reintroduce volatiles back into the mantle much the same way that plate tectonics on Earth re-fertilizes the mantle in volatiles.

High levels of unstable sulfur species in the venusian atmosphere argue for a replenishment mechanism by volatile input via volcanic degassing (Fegley 2014). As Venus is similar in size to the Earth, it is likely to be just as volcanically active as the Earth, providing a reasonable explanation for the venusian atmospheric sulfur contents and species (e.g., Fegley 2014). Thus, if volcanic outgassing is responsible for the input of sulfur, then it is likely that

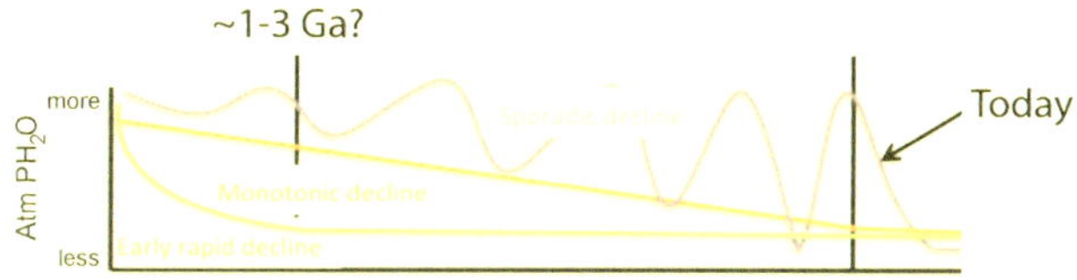

Fig. 4 A pH_2O vs. time schematic showing three possible evolutionary scenarios for water on Venus. The green curve denotes an early ocean-rich Venus which was lost early likely due to hydrodynamic escape (Kasting and Pollack 1983). The orange curve represents a monotonic decline due to continuous loss of hydrogen from the upper atmosphere. The purple curve represents episodic input of volatiles, which takes into account the ability of the mantle to retain non-zero levels of water, and releases mantle volatiles during catastrophic mantle overturn events. This large release would likely release low D/H water (similar to other inner Solar System materials) on a ~ 500 Ma timescale, which then would be fractionated by atmospheric loss processes to the observed D/H of $\sim 150\times$ terrestrial. Modified from Gilmore et al. (2017)

other volatiles, such as C, N, and water are being supplied to the venusian atmosphere as well. The input of volatiles to the venusian atmosphere may be episodic, and at maximum abundance levels at the end of mantle overturn events that resurface the planet. The abundances of some volatiles in the atmosphere would then decline over time due to weathering reactions with crustal rocks (e.g., S, Cl) and loss and escape processes from the upper atmosphere (e.g., H), until the next mantle overturn event (on a timescale of ~ 0.5 to 1 Ga). Atmospheric species that do not react with surface rocks or are lost to space would accumulate (e.g., C, N, and Ne). Alternatively, volcanic outgassing is mostly quiescent, and steadily supplying volatiles to the venusian atmosphere over time (albeit with higher levels during and after mantle overturn events). A schematic of these various processes is shown for water in Fig. 4 (modified from Gilmore et al. 2017).

2.2.4 Timing of Water and Volatile Delivery and Sources of Water to Venus

The similarity in the isotopic ratios and abundances of the major volatiles C and N between the venusian atmosphere and Earth argues for a similar origin of volatiles for these two bodies. The extraordinarily high D/H of water in the venusian atmosphere argues strongly for loss of extensive amounts of water, such as an ocean's worth (Donahue et al. 1982) or more. The only other source of hydrogen known to have D/H similar to Venus are water and organics of the interstellar medium. It has been argued that delivery of cometary water followed by loss of water from the atmosphere through time could explain the elevated D/H of the Venus atmosphere (Grinspoon and Lewis 1988). While comets are undoubtedly a source of water to the inner Solar System bodies, it is not necessary to advocate for this model today due to the expected high levels of volatiles in the venusian mantle (Elkins-Tanton et al. 2007). Loss of atmospheric water either from an early ocean world (Ingersoll 1969; Donahue et al. 1982; Kasting and Pollack 1983), or through episodic inputs of water due to mantle overturn with subsequent fractionation of D/H and loss of hydrogen from the upper atmosphere (Grinspoon 1993), can explain the D/H systematics of Venus.

2.3 The Moon

A major paradigm shift in our understanding of the lunar volatile inventory occurred with a Secondary Ion Mass Spectrometry (SIMS) study of H_2O, CO_2, F, S, and Cl in lunar volcanic glasses by Saal et al. (2008). They measured 4–46 ppm wt H_2O in lunar volcanic glasses. While these are low values compared to terrestrial arc volcanic glasses, the lunar glasses are believed to have undergone significant degassing during lunar fire-fountaining activity. Saal et al. (2008) modeled a diffusive loss of volatiles during eruption to arrive at pre-eruptive volatile estimates of 260–745 ppm wt H_2O for a lunar green glass bead (a very low-Ti glass). These levels of water are similar to water contents in mantle source regions of the Earth, such as mid-ocean ridge basalts (MORBs) and Hawaiian ocean island basalts (OIBs) (Luth 2005). This remarkable result suggests that the Moon was able to retain water after the giant impact, or that water was accreted shortly thereafter (Saal et al. 2008).

Much higher levels of indigenous lunar water have been found in the mineral apatite with three separate SIMS studies finding 1000's of ppm wt OH in this phosphate mineral from rock samples returned by the Apollo astronauts (Boyce et al. 2010; McCubbin et al. 2010a, 2010b; Greenwood et al. 2011). These high hydroxyl contents were far above background levels of at most 10's of ppm, and could not be ascribed to terrestrial contamination. The three groups all used different SIMS instruments, techniques, and standard terrestrial apatite minerals, but obtained similar results for the H contents of lunar apatites (Boyce et al. 2010; McCubbin et al. 2010a, 2010b; Greenwood et al. 2011). The high OH contents of lunar apatites also confirmed the results of Saal et al. (2008) that the Moon and its samples still retain some indigenous water (Boyce et al. 2010; McCubbin et al. 2010a, 2010b; Greenwood et al. 2011).

Since these early pioneering studies, a number of other research groups have embarked on measuring the water contents and D/H of the Moon. A comprehensive review of lunar volatile elements and their isotopes is provided by McCubbin et al. (2015a), and will not be restated here. We will primarily review the evidence for 'water' and D/H in the Moon, but will also discuss other volatiles as appropriate. The current state of our understanding of water and D/H in different lunar reservoirs and sample types will be discussed after we consider issues associated with the measuring of hydrogen contents in nominally anhydrous minerals from extraterrestrial samples.

2.3.1 Considerations for SIMS Measurements of D/H and H in Nominally Anhydrous Extraterrestrial Materials

SIMS SIMS measurements allow for accurate and precise measurements of small sample volumes to be attained for D/H, if terrestrial contamination can be minimized, and is the primary means by which meteorite and returned sample D/H is analyzed today. Detailed reviews of the use of SIMS can be found in Hauri et al. (2002), Greenwood et al. (2011), Barnes et al. (2013), Slodzian et al. (2014), and references therein. In brief, SIMS measurements involve the use of a primary ion beam to sputter/excavate a small sample mass from a material's surface. The 'secondary ions' sputtered from the surface are then accelerated and collimated by electrostatic lenses and filters before being separated according to mass using a large magnet, and then collected/counted on one or more electronic multipliers or Faraday cups, or with a SCAPS (Stacked CMOS-type Activated Pixel Sensor) detector (Hokkaido University and University of Hawaii SIMS machines) (Yurimoto et al. 2003). Removal of terrestrial adsorbed water from the surfaces and near-surface region of the samples is accomplished by pumping of the sample under high vacuum ($> 10^{-8}$ Torr) for several days prior to analysis, and pre-sputtering the area to be analyzed by rastering the primary ion beam over it (a form of ion milling) to remove the topmost exposed surface of the sample ($\lesssim 1$ μm). For D/H measurements, the secondary ions must be extracted from the center of the secondary ion beam, to avoid contaminant water that will condense in the outer regions on the edge of the secondary ion beam. A liquid N_2 cold trap can also be used to trap contaminant water in the vacuum of the sample chamber. Contaminant water will condense

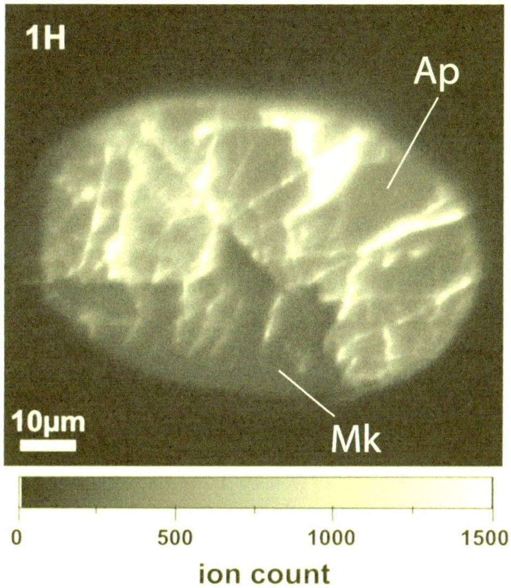

Fig. 5 ^{1}H image of Shergotty meteorite obtained using SCAPS. The upper region is the mineral apatite (Ap); the lower region is maskelynite (Mk), and a sharp grain boundary can be seen between the two phases. Apatite is brighter because it has several thousand ppm wt H_2O, while the maskelynite is nominally anhydrous ($\lesssim$ 100 ppm wt H_2O). The most water-rich phase in this image is terrestrial contamination along cracks, fractures, and micro-cracks. Typically, a $\sim$ 20 μm diameter crack-free area is needed for SIMS analysis. This illustrates the extreme difficulty in obtaining 'true' martian D/H values in shocked extraterrestrial samples (source: J. Greenwood, unpubl. data)

if any instrument components contact the secondary ion beam. This points to difficulty of using NanoSIMS instruments to make D/H measurements of low-water containing materials, as this instrument's immersion lens contacts the secondary ion beam, and will likely lead to an endemically higher background of contaminant water during analyses. For these reasons, in this article we will adopt a 'red light/yellow light/green light' approach to SIMS data collected on extraterrestrial materials via mass spectrometry or SIMS for the interpretation of D/H measurements. Below 50 ppm wt H_2O, this is the "red zone" wherein D/H measurements are likely to be affected most strongly by terrestrial contamination. We will not consider D/H data below 50 ppm wt H_2O in this article. We define the "yellow zone" as 50–150 ppm wt H_2O, wherein caution should be applied when considering the interpretations of this data. We will alert the reader when we consider data in this range of water contents. Above 150 ppm wt H_2O is the "green zone", and data above this water content is generally excellent, though SIMS analyses that overlap onto cracks loaded with terrestrial contamination can still give extraterrestrial δD values (Greenwood et al. 2008). Thus, even above 150 ppm wt H_2O, caution must be exercised in the interpretation of D/H values in heavily shocked meteorites and returned samples, as the density of fractures can make finding an analytical area free of cracks (and thus reasonably free of terrestrial contamination) in the phase of interest quite challenging. An example of the difficulty in measuring water in extraterrestrial materials can be seen in a ^{1}H image of apatite from the Shergotty meteorite in Fig. 5.

Cosmic Ray Exposure (CRE) Corrections for D/H Friedman et al. (1971) first measured elevated δD of 307‰ in basalt 12051, and proposed this to be a mixture of D-rich spallogenic H generated by cosmic rays and terrestrial contamination. Mare basalt 12051 has large grains of apatite (J. Greenwood, unpubl. data), and most apatite in mare basalts have 100's to 1000's of ppm wt OH with elevated δD (McCubbin et al. 2010a, 2010b; Boyce et al. 2010; Greenwood et al. 2011; Tartèse and Anand 2013; Barnes et al. 2013, 2014; Tartèse et al. 2013, 2014a, 2014b), suggesting that rather than spallogenic deuterium, Friedman et al. (1971) measured a mixture of indigenous lunar hydroxyl from apatite and terrestrial contamination. This is relevant, because most of the H isotopic data in the literature (Saal et al. 2013; Barnes et al. 2013, 2014; Tartèse and Anand 2013; Tartèse et al. 2013, 2014a, 2014b; Füri et al. 2014; Robinson and Taylor 2014; Robinson et al. 2016) have been corrected for CRE using the deuterium production rate of Merlivat et al. (1976). Merlivat et al. (1976) derived their deuterium production rate from their stepped heating of basalt 70215, arguing that the majority of the hydrogen released at low temperature was terrestrial contamination, but that spallogenic H dominated the high temperature steps. Mare basalt 70215 is a fine-grained, rapidly cooled sample, and the apatite in it is submicron in size (J. Greenwood, unpubl. data). This suggests that Merlivat et al. (1976) did not measure spallogenic deuterium, but rather high δD associated with OH in apatite (e.g., Greenwood et al. 2011).

Füri and Deloule (2016) have derived a new D production rate for lunar samples, from careful analysis of olivine grains in lunar samples with varying CRE ages. The D-production rate they define is $3\times$ higher than that of Merlivat et al. (1976), but is in good agreement with the theoretical D-production rate of Reedy (1981). The much higher D-production rate has important consequences for a number of published lunar studies of D/H. For example, Hauri et al. (2017) point out that in their own dataset for the pyroclastic glass beads of Saal et al. (2013), the positive correlation of δD with decreasing water content they report (the expected trend of D/H fractionation during degassing of H_2 gas (e.g., Sharp et al. 2013)), would produce a *negative* correlation of δD with decreasing water content if they use the D-production rate of Füri and Deloule (2016). This has profound implications if true: a negative correlation of δD with decreasing water content can only occur by the degassing of molecular water (Taylor 1986). As molecular water is not the dominant hydrogen species at the low fO_2's expected of lunar magmas (Hirschmann et al. 2012), this would imply that the pyroclastic beads were degassing at higher fO_2.

Hauri et al. (2017) suggested that data below ~ 20 ppm wt H_2O were dominated by spallogenic H, and suggested magmatic water dominates in lunar samples above this level. This further emphasizes the importance of the "red light/yellow light/green light zones" of water content for D/H interpretation of extraterrestrial materials that we introduced above.

2.3.2 Water and D/H of Mare Basalts

Our knowledge of water and D/H of mare basalts comes primarily from SIMS analyses of two phases: silicate glasses and apatite. These two phases have benefits and pitfalls for inferring hydrogen and D/H values. Apatite is slow to exchange its hydrogen under most conditions (Higashi et al. 2017), thus making it a potentially strong container for preserving lunar water information. Apatite is a common rock-forming phosphate mineral that can contain F, Cl, and OH as essential structural constituents that occupy a single crystallographic site (Pan and Fleet 2002). The relative proportions of F, Cl, and OH in apatite may be used to extract relative fugacity ratios of F, Cl, and OH in the silicate melts or aqueous fluids from which they formed if thermodynamic models are developed for the mixing of F, Cl, and OH components in apatite and for F, Cl, and OH components in multicomponent melt and fluid systems (e.g., Candela 1986; Hovis and Harlov 2010). In the absence of such models, the partitioning behavior of F, Cl, and OH between apatite and melt or fluid can be used to approximate the relative abundances of F, Cl, and OH in fluid and melt systems (e.g., Brenan 1993; McCubbin et al. 2015b; Boyce et al. 2014; Riker et al. 2018). However, the partitioning cannot be treated with simple Nernst-like partition coefficients given the stoichiometric constraints on F, Cl, and OH abundances in apatite and the lack of constraints on the abundances of F, Cl, and OH in coexisting silicate melts or fluids (Boyce et al. 2014; McCubbin et al. 2015b). This complicates the use of apatite as a quantitative melt hygrometer, especially for mare basalts. Lastly, apatite is a late-crystallizing mineral in lunar magmas, thus only recording information in the last stages of magmatic history.

Glasses record magmatic information directly, without the need for partition coefficients, thus making them ideal for understanding a magma's volatile history, if they do not undergo degassing or water and hydrogen isotope exchange with external reservoirs during cooling. As silicate melts do not cool through the glass transition until they have reached $\sim 2/3$ of their liquidus temperatures, magmatic liquids can be susceptible to extensive exchange during cooling. Melts that are trapped in minerals can potentially be armored against further exchange with the magma (Kent 2008). Yet, even olivine-hosted melt inclusions can undergo substantial H exchange with magmas via diffusive exchange through the crystals, either at magmatic temperatures or during cooling (Gaetani et al. 2012), suggesting that even these records may not be pristine. CO_2 and S can also be affected by this process in olivine-hosted melt inclusions, but F and Cl are not (Bucholz et al. 2013). Furthermore, both apatite and glasses will give poor information on D/H below 50 ppm wt H_2O, due to terrestrial contamination and spallogenic H.

Apatite Measurements of water contents and/or D/H of apatite in mare basalts have been published by Boyce et al. (2010, 2015), McCubbin et al. (2010a, 2010b), Greenwood et al. (2011), Tartèse and Anand (2013), Barnes et al. (2013), Tartèse et al. (2013, 2014a, 2014b), and Treiman et al. (2016b). Early studies used apatite to estimate the water content of the Moon, but Boyce et al. (2014) proposed a model for the partitioning of water in apatite, which predicted that OH-rich apatite would only be produced in volatile-poor magmas, due to the strong preference of F over OH or Cl in apatite. This suggests that F-rich, OH-poor apatite would be a better indicator of higher volatile contents in lunar magmas. Thus, while lunar apatite can contain copious amounts of water (over 1 wt.% OH; Tartèse et al. 2013), until other evidence of water-rich magmas in lunar samples with OH-rich apatite are found, this mineral should not be used to estimate water contents of lunar magmas or the bulk Moon.

The D/H of apatite in lunar samples is not affected by the model of Boyce et al. (2014), but would be affected by degassing of lunar magmas before the onset of apatite crystallization. Greenwood et al. (2011) found a δD range of $-215‰$ to $1010‰$ in mare basalts, and proposed a cometary source for elevated D/H, and mixing with either solar wind or mantle hydrogen with lower D/H. Tartèse and Anand (2013) proposed degassing of H_2 from lunar magmas originally with a $\delta D \sim 100‰$ to explain elevated D/H of lunar apatite. They revised this value down to $-100‰$ for KREEP (K-,Rare Earth Element-,P-rich) samples (Tartèse et al. 2014a). KREEP samples have a component of the last 1–2% percent crystallization of the lunar magma ocean (called urKREEP), and are rich in incompatible elements (Warren and Wasson 1979), and should be volatile-rich. They proposed delivery of chondritic water after the Moon-forming giant impact through crust-breaching impacts, and this idea was further elaborated upon in Barnes et al. (2016).

Pyroclastic Glasses and Their Olivine-Hosted Melt Inclusions Pyroclastic glasses have been studied by Saal et al. (2008, 2013), Hauri et al. (2011), Füri et al. (2014) and Chen et al. (2015). Saal et al. (2008) suggested that the Moon had primitive material below the magma ocean that was inherited from the Earth. The olivine-hosted melt inclusions studied by Hauri et al. (2011) provided the first direct evidence for magmatic volatile contents above 1000 ppm wt H_2O in the Moon. Saal et al. (2013) found the most water-rich melt inclusions have slightly elevated δD ($\delta D = 270 \pm 56‰$). Chen et al. (2015) conducted a more extensive study of the volatile contents of pyroclastic glasses and their olivine-hosted melt inclusions, and generally confirmed the results and interpretations of Saal et al. (2008) and Hauri et al. (2011) for the 74220 orange glasses. Importantly, Chen et al. (2015) found high levels of F, Cl, and S in other types of pyroclastic glasses, showing that 74220 is not an anomalous sample, and that the Moon is likely not completely dry (meaning 1 ppm wt H_2O or less). Chen et al. (2015) estimated from H_2O/Ce that the Moon is depleted in water by only a factor of three relative to Earth, similar to the depletion in K for the Moon (Fig. 2) (Warren and Taylor 2014).

Secondary Processes Affecting Water and D/H of Lunar Samples Singer et al. (2017) conducted measurements of H and D/H of olivine-hosted melt inclusions of the slowly cooled samples (12018, 12035, and 12040) of the Apollo 12 olivine basalt suite (Walker et al. 1976). These are very different samples to the pyroclastic glasses that were probably produced by fire-fountain activity on the lunar surface (e.g., McCubbin et al. 2015a). Walker et al. (1976) proposed that the Apollo 12 olivine basalts sampled ~ 35 m of the basal portion of a magma body, such as a sill, thick lava flow, or lava pond, and that became olivine-rich by gravitationally driven crystal settling. In contrast to the water-rich olivine-hosted melt inclusions of 74220 (average $\delta D = 270 \pm 56‰$), the melt inclusions in Apollo 12 olivine basalts with > 150 ppm wt H_2O were characterized by lower δD values ranging from $-183 \pm 212‰$ to $6 \pm 185‰$. Diffusion calculations show hydrogen and D/H in the melt inclusions should be exchangeable with external hydrogen reservoirs in these slowly cooled samples (Singer et al. 2017). In agreement with this idea, apatite δD is low ($\sim 0‰$) in the most slowly cooled sample (12040), while the δD of apatite in the more rapidly cooled 12018 is very high ($\delta D = 1439 \pm 103‰$). This suggests that D/H underwent exchange with a low D/H reservoir during cooling. These results also lend credence to the model proposed by Treiman et al. (2016b) for secondary processes affecting D/H of apatite in slowly cooled mare basalts. Another implication of this study is that olivine-hosted melt inclusions of 74220 (Saal et al. 2013) may have been undergone exchange with solar wind derived hydrogen in near-surface magma chambers (e.g., Gaetani et al. 2012). This suggests caution should be applied when considering hydrogen and D/H in olivine-hosted melt inclusions.

2.3.3 Water and D/H of Highlands Samples

The lunar highlands are the most ancient portions of the lunar surface, and samples of them that have been analyzed have very low water contents and D/H ratios include ferroan anorthosite crust (FAN) and later intrusive rocks (Mg-suite, alkali-suite). Hui et al. (2013) used FTIR to measure 5–6 ppm wt H_2O in plagioclase from FAN, and predicted ~ 320 ppm wt H_2O for the lunar magma ocean. Hui et al. (2017) measured D/H of this very low water content plagioclase, but these measurements are in the 'red zone' of SIMS analyses and will not be discussed further.

Greenwood et al. (2011) found $\delta D = 290 \pm 73‰$ for apatite with 80 ppm wt H_2O in an alkali anorthosite clast, and suggested this is the value for the hydrogen isotope composition of the lunar mantle. More recent work has found values that overlap with those of mare basalt apatite (Barnes et al. 2014; Robinson et al. 2016). Measurements of QMD clasts from

15403 and 15404 are much lighter, with apatite in 15404 having a $\delta D = -344 \pm 319‰$ (2σ) and 15403 apatite with $\delta D = -721 \pm 48‰$ (2σ). The extremely light value of apatite in 15403 is anomalous for lunar magmatic samples, but in the expected range of solar wind in the lunar regolith (e.g., Liu et al. 2012).

2.3.4 Isotopes of Volatile Elements that Bear on the Question of Water and D/H of the Moon

Cl, Zn and K Isotopes Sharp et al. (2010) showed that the Moon had high $\delta^{37}Cl$, unlike other bodies in the Solar System. They suggested that elevated Cl isotope ratios could only be produced in extremely water-poor conditions, and concluded that the Moon had < 1 ppm wt H_2O. The highest $\delta^{37}Cl$ was measured using SIMS in apatite from KREEP basalt 72275. Boyce et al. (2015) studied δD and $\delta^{37}Cl$ of apatite in mare basalts, and found a strong positive correlation of $\delta^{37}Cl$ and Th or La/Lu (both indicators of KREEP signature). They found no correlation between δD and $\delta^{37}Cl$. They argued that the Cl isotopic composition of urKREEP is $\sim 30‰$ (the mantle reservoir of KREEP; also the last 1–2% of lunar magma ocean crystallization) and was set during lunar magma ocean outgassing after the loss of water. This would imply that urKREEP is very dry (McCubbin et al. 2015a). Barnes et al. (2016) measured $\delta^{37}Cl$ in apatite from KREEP basalt 15386 and found a positive correlation of $\delta^{37}Cl$ with apatite Cl content. They argued that the lunar magma ocean had a $\delta^{37}Cl$ of $\sim 30‰$, and suggested that late crust-breaching impacts could have allowed the urKREEP reservoir to outgas.

The Moon is the only body that currently has resolvable K isotope anomalies (Wang and Jacobsen 2016). Heavy Zn isotope anomalies have also been measured in lunar materials, which have been interpreted as representing loss of light Zn isotopes during first the protolunar disk phase and then during outgassing of the lunar magma ocean (Paniello et al. 2012; Kato et al. 2015). The combination of heavy isotope enrichments for the moderately volatile lithophile element K and the more volatile Zn and Cl argue that the Moon experienced a unique event in the inner Solar System, which is most likely the giant impact event.

N Isotopes Füri et al. (2015) and Mortimer et al. (2015) measured $\delta^{15}N$ in lunar basalts. Mortimer et al. (2015) measured six lunar basalts and found $\delta^{15}N = 0.35‰$, similar to terrestrial atmosphere and earlier measurements of lunar materials (Mortimer et al. 2015 and references therein). Füri et al. (2015) reported a range of $\delta^{15}N$ of -0.2 to $26.7‰$, with the higher values similar to some carbonaceous chondrites. They also reported values of $\delta^{15}N < -30‰$, which would be similar to the isotopic composition of Earth's primordial mantle. On the basis of N isotopes, Füri et al. (2015) argued that indigenous lunar volatiles could have been inherited from a proto-lunar disk sourced from the terrestrial mantle or implanted into the lunar mantle during the subsequent late accretion phase.

2.3.5 Water Abundances of the Moon

Geochemical estimates of the water abundance of the Moon primarily come from three datasets: highland samples (Hui et al. 2013, 2017), pyroclastic glasses (Hauri et al. 2015; Chen et al. 2015), KREEP samples (McCubbin et al. 2015a; Greenwood et al. 2017). There are also geophysical and selenogenetic estimates (Karato 2013; Elkins-Tanton and Grove 2011; Pahlevan et al. 2016). The range of water abundances for the Moon runs the gamut, from Earthlike levels of 100–300 ppm wt H_2O (Hui et al. 2013; Chen et al. 2015; Hauri et al. 2015; Greenwood et al. 2017), to slightly lower abundances (10–100 ppm wt H_2O; Karato 2013; Pahlevan et al. 2016), to order of magnitude lower abundances of 3–13 ppm wt H_2O (McCubbin et al. 2015a) and $\lesssim 10$ ppm wt H_2O (Elkins-Tanton and Grove 2011).

Highland Samples Hui et al. (2017) used their earlier measurements of water in anorthite from FAN 60015 (Hui et al. 2013) to estimate the water content of the lunar magma ocean to be ~ 136 ppm wt H_2O, without outgassing. If outgassing of H_2 from the magma ocean is called upon to raise the δD of the initial hydrogen reservoir from $-250‰$ to $300‰$ (the lower value is an estimate of the δD of the undegassed magma ocean, and the higher value is similar to values in olivine-hosted melt inclusions in pyroclastic glasses (Saal et al. 2013) and highlands apatite (Greenwood et al. 2011)), estimates of the water content for the initial lunar magma ocean ranges from 1320 to 5000 ppm wt H_2O (Hui et al. 2017).

Pyroclastic Glass Bead Estimates The pyroclastic glass beads have been used to estimate the water contents of their parent magmas, as well as the whole Moon. Chen et al. (2015) estimated 110 ppm wt H_2O for the mantle source of the pyroclastic glasses. Hauri et al. (2015) estimated that the bulk silicate Moon (BSM) contains 100–300 ppm wt H_2O primarily using the sulfur concentrations and H_2O/Ce of the pyroclastic glass beads.

urKREEP Estimates McCubbin et al. (2015a) postulated that the late residual liquid of the lunar magma ocean would be Cl-enriched relative to H_2O and F, due to magma ocean volatile abundances being controlled by partitioning with anhydrous minerals. They estimated the lunar mantle to have much less water than the Earth, with only 0.15 ppm wt to 5.3 ppm wt H_2O. They predicted that the majority of the Moon's volatiles would be in urKREEP (which they defined as the last 1% residual liquid from magma ocean crystallization). They predicted 300–1250 ppm wt H_2O for urKREEP and 3–13 ppm wt H_2O for the BSM, an order of magnitude less than estimates using pyroclastic glasses (Chen et al. 2015; Hauri et al. 2015).

Greenwood et al. (2017) also predicted that the bulk of the Moon's volatiles would be in the urKREEP from their finding of Cl-rich late-stage glasses in KREEP basalts 15382 and 15386. Due to an order of magnitude depletion in F in lunar samples relative to Earth, they used an H_2O/Ce of 17–30 to estimate lunar water abundances, instead of the higher $H_2O/Ce = 64$ or 77 of orange glasses (Hauri et al. 2015; Chen et al. 2015). They obtained ~ 1.3 wt.% to 2.4 wt.% H_2O for urKREEP; if urKREEP represents the last 1% of a lunar magma ocean that originated from a 100% molten Moon, then the BSM would contain 130–240 ppm wt H_2O, similar to estimates obtained from the pyroclastic glasses (Hauri et al. 2015). The estimate for urKREEP of Greenwood et al. (2017) is also similar to an estimate for urKREEP of 1.4 wt.% H_2O using highlands FAN plagioclase (Hui et al. 2013).

Models for High Water Content of the Moon As summarized above, there is a wide range of water content estimates for the Moon. But, the most recent estimates have begun to converge on a water and volatile content for the lunar mantle that is depleted relative to Earth's mantle by an order of magnitude (Greenwood et al. 2017), or less (Chen et al. 2015; Hauri et al. 2015; Hui et al. 2013, 2017).

In contrast, a model of volatile content based on condensation temperature (e.g., Albarède et al. 2015) predicts much higher degrees of depletion. Such a model assumes condensation to solids, but condensation could occur to liquids when pressure of the gas is high. Karato (2013) showed that for condensation to liquid in a high-pressure environment in the Moon-forming disk, the solubility in the liquids is not controlled by the bond strength (e.g., Karato 2016) and consequently, the degree of depletion of highly volatile elements such as H is much less when condensation occurs to liquids. Similarly, Pahlevan et al. (2016) modeled the behavior of hydrogen during lunar origin process and concluded that proto-lunar disks with terrestrial abundances of hydrogen would dissolve ~ 10–100 ppm wt H_2O into proto-lunar liquids (see also Canup et al. 2015a, 2015b). Those models are consistent with the

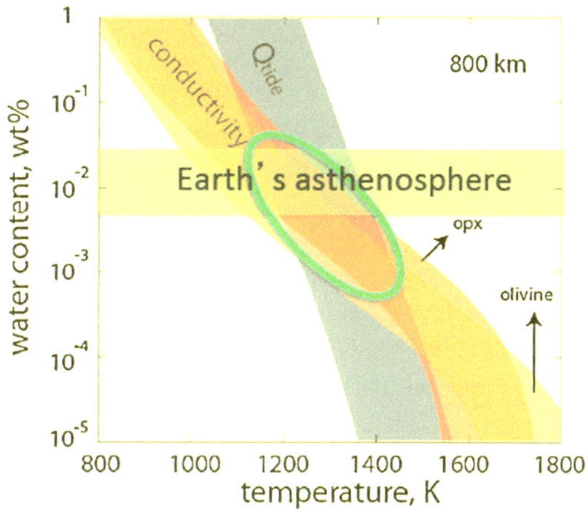

Fig. 6 Geophysical estimates of water content and temperature in the lunar interior at the depth of 800 km. The inferred water content is from 0.01 to 0.001 wt.% that is comparable to, or only a factor of ~ 10 less than, that of Earth's asthenosphere (0.01 wt.%). Modified from Karato (2013)

K isotope observations suggesting a high-pressure environment in the moon-forming disk (Wang and Jacobsen 2016).

2.3.6 Geophysical Constraints on Lunar Water Abundances

Although the extensive study of lunar basaltic inclusions by Chen et al. (2015) strongly suggests that the modest degree of depletion in volatile elements such as water (hydrogen) in the lunar mantle is global, it is also important to develop independent methods to constrain the lunar water abundance. To circumvent the ambiguities inherent in geochemical approaches, Karato (2013) used geophysical observations to infer global water content of the lunar interior. Among various geophysical properties, Karato (2013) chose electrical conductivity and tidal Q; both of these properties are controlled by the motion of crystalline defects that depend strongly on water content (for a review on electrical conductivity see Karato and Wang 2013; for tidal Q, Karato 2013).

Electrical conductivity of the Moon was investigated during the Apollo mission using the measurements of time-dependent electromagnetic fields (e.g., Hood et al. 1982; Sonett et al. 1971). In these studies, the depth-variation of electrical conductivity was inferred assuming a one-dimensional model. Tidal deformation occurs on time scales of months-to-years and it involves not only elastic but also viscous deformation. The tidal Q is a measure of the non-elastic part of the deformation. The tidal Q was determined by radar observations of the shape of the Moon (Williams et al. 2001) and is, therefore, known to rather high precision (to within ~ 10–20%).

Both electrical conductivity and tidal Q depend strongly on the water content and temperature of the body of interest. Consequently, if one uses only one of these observations, then the trade-off between the water content and temperature is strong and the inference of the water content is non-unique. Karato (2013) simultaneously used both properties to more uniquely infer the water content and temperature in the lunar interior. The results show that the lunar interior is cooler (at a depth of 800 km, temperature is $\sim 1350 \pm 150$ K) than Earth's interior (at a depth of 800 km, temperature is $\sim 2200 \pm 100$ K), while the water content is similar to, or slightly less than, that of the asthenosphere of Earth (the Moon: 10–100 ppm, Earth: 100 ppm) (Fig. 6). The results support the notion of a not-so-dry Moon inferred from geochemical studies (Saal et al. 2008; Hauri et al. 2011, 2015; Chen et al. 2015).

2.3.7 *D/H and Sources of the Moon's Water*

Earth Source Saal et al. (2008) suggested that since the Moon has some water in pyroclastic glasses, parts of the deep Moon may have remained unmelted during the giant impact event. Saal et al. (2013) argue that the elevated D/H of olivine-hosted melt inclusions ($\delta D = 270 \pm 56‰$) was due to degassing of H_2 from values that were similar to the terrestrial Earth mantle ($\sim -60‰$, Clog et al. 2013). This terrestrial signature could also be due to equilibrium dissolution of terrestrial hydrogen in the proto-lunar disk with negligible high-temperature isotopic fractionation (Pahlevan et al. 2016). Robinson et al. (2016) argue that extremely low δD they measured in highland rocks represent a solar-like reservoir in the mantle, and call for parts of the Moon to carry solar-like hydrogen, as has been postulated for reservoirs in the Earth's deep mantle (Hallis et al. 2015).

Carbonaceous Chondrite Source The majority of researchers favor a carbonaceous chondrite origin for terrestrial volatiles, due to similarities in δD and $\delta^{15}N$ (e.g., Alexander et al. 2012, 2018; Halliday 2013; Marty 2012; Marty et al. 2016). Saal et al. (2013) called for a carbonaceous chondrite origin for lunar water, though they implied that this may also have been a terrestrial relic of the giant impact event. Tartèse and Anand (2013) argued for delivery of chondritic impactors to the Moon after the giant impact, and Barnes et al. (2016) developed this idea further. Füri et al. (2015) argued that N isotopes for the Moon were best matched by carbonaceous chondrites, but with additional identifiable components.

Cometary Source Due to the elevated D/H of most lunar apatites, Greenwood et al. (2011) suggested cometary delivery of water. This would need to occur most likely in the time after the giant impact event and before the lunar crust had undergone significant thickening. Barnes et al. (2016) considered this process for chondritic impactors, but the constraints on the timing of impact delivery are similar for the cometary impactors. An additional constraint on the cometary contribution comes from the dynamical delivery efficiency of outer Solar System bodies to the inner solar system (O'Brien et al. 2018) which suggests that only a small fraction of terrestrial water could be delivered by cometary sources.

2.3.8 *Timing of Water Delivery to the Moon*

Pre-Giant Impact Saal et al. (2008) first suggested that parts of the Moon may have survived the Moon-forming impact relatively unscathed, and still retain their abundances of terrestrial volatiles. Saal et al. (2013) and Hauri et al. (2015) also favored this idea for the majority of the lunar volatiles to have survived the giant impact event.

Giant Impact Volatile gain and loss accompanying the giant impact would most likely occur during thermal processing in the immediate aftermath of the event. Such post-impact evolution can be separated into two distinct regimes: the proto-lunar disk and magma ocean regimes, the former having received the most attention. Early thinking about the giant impact considered that the proto-lunar disk could have been open with respect to thermal escape of hydrogen (Stevenson 1987). However, since hydrogen constitutes only a small fraction of the proto-lunar disk, it was subsequently recognized that thermal escape required that the light hydrogen atoms separate from heavier, gravitationally bound silicate vapor species (SiO, O_2) in order to undergo significant escape (Genda 2004). The process of condensation (Pahlevan and Stevenson 2008) and molecular diffusion (Nakajima and Stevenson 2018) were found to be inefficient as means of separating the heavy bound gases from the light unbound hydrogen. Thermal escape of the heavier silicate vapor species requires very distended ($>$ 10 Earth radii) disks (Desch and Taylor 2011), much more distended than found in models of lunar accretion (Salmon and Canup 2012). Hence, recent thinking on the evolution of the proto-lunar disk considers the proposition that the disk was nearly closed with respect to thermal escape of gases to space (Canup et al. 2015b; Charnoz and Michaut 2015; Hauri et al. 2017; Karato 2013; Lock et al. 2018; Pahlevan et al. 2016). In these studies, the observed abundances of lunar volatiles are due to liquid-vapor partitioning followed by preferential accretion of the Moon from the liquid fraction. The complementary vapor could, in this setting, be lost to the silicate Earth without violating terrestrial abundance constraints.

A second physical setting that can alter the lunar volatile abundances is the lunar magma ocean, either during a short "naked" magma ocean phase contemporaneous with lunar accretion or after accretion and formation of a flotation crust due to crust-breaching impacts (Barnes et al. 2016). In both cases, the escape of a proto-atmosphere would have been facilitated by the weaker gravity field of the Moon that makes the conditions necessary for hydrodynamic escape easier to achieve. Two lines of evidence for volatile loss from the lunar magma ocean have recently come forward. First, identification of two Cl isotopic reservoirs in the Moon, one associated with a "normal" lunar mantle (δ^{37}Cl $\sim$ 0‰), and the other an isotopically enriched urKREEP end-member (δ^{37}Cl $\sim$ 30‰), suggests ongoing evaporation and fractionation during the formation and closure of these two reservoirs (Boyce et al. 2015), processes that would also have influenced the lunar hydrogen abundances. Secondly, after decades of expended efforts, a K isotopic offset has been observed between the silicate Earth and Moon (Wang and Jacobsen 2016). This is an important observation because K isotopes do not significantly fractionate during petrologic processes or during liquid-vapor equilibrium and, therefore, constrain the kinetics of volatile loss during an episode that presumably would also have affected lunar hydrogen abundances and isotopic compositions.

Post-Giant Impact Greenwood et al. (2011) argued for exogeneous delivery of volatiles to the Moon, favoring delivery of cometary water after the giant impact event, and before solidification of a thick lunar crust, likely 10–200 Ma after Moon formation (Elkins-Tanton and Grove 2011). Tartèse and Anand (2013) suggested a variant on this idea, by calling for delivery of chondritic water via asteroidal delivery. Barnes et al. (2016) undertook detailed modeling of this process and called for a crust-breaching impact to expose the lunar magma ocean and cause it to degas and fractionate Cl isotopes. The high water contents measured in KREEP materials (e.g., Tartèse et al. 2014a) suggest that urKREEP had significant water, as suggested by Greenwood et al. (2017) and Hui et al. (2017), which would likely preclude outgassing under water-poor conditions. Finally, Ustunisik et al. (2015) showed that lunar magmas would degas H before Cl, thus allowing for the possibility of outgassing the magma ocean sequentially, but the dynamics of how that would occur in a crust-breaching impact have yet to be the subject of detailed modeling.

2.4 Mars

2.4.1 Introduction

The cadre of missions to Mars in the last two decades has led to an evolving picture of martian water that argues strongly for the presence of liquid water on the Mars surface in the Noachian and pre-Noachian ($\sim$ 3.7–4.5 Ga) from the presence of fluvial landforms and hydrous minerals (e.g., Carr 2006; Ehlmann and Edwards 2014). What is not known is how much water was involved in creating these aqueous environments. Was it global-scale oceans or intermittent wetting with mostly subsurface water? Recent climate models do not require a warm, thick atmosphere to satisfy the current constraints on water availability (e.g., Wordsworth 2016). During the Hesperian (~ 3.0–3.7 Ga) there was a change to the current dry atmosphere of Mars, and evaporite deposits mark this transition (Bibring et al. 2006). The current dry climate suggests that surface water has not been an important feature of martian landscape evolution for several billion years. Following orbital and landed observations of ground ice near the poles (Boynton et al. 2002; Smith et al. 2009), large reservoirs of subsurface ice have been found in mid-latitude to near-equatorial regions (Mouginot et al. 2012; Dundas et al. 2018), suggesting that there may be large ice reservoirs on Mars that are not part of the current hydrologic cycle, but could be activated during warming due to martian obliquity cycles. Here we discuss the water abundances and D/H of the three recognized reservoirs for water on Mars: the atmosphere, crust, and mantle.

When Mariner 9 radioed back to Earth the first photographs of Mars obtained from orbit, scientists were disappointed to find a desert-like planet, instead of the potential oasis of life imagined by the public, due to the early ideas of martian canals promoted by Lowell (1906). The Viking landers in 1976 had mass spectrometers, but only found trace water in the martian atmosphere (Owen 1992). The Viking mass spectrometers did find elevated levels of ^{15}N (McElroy et al. 1976; Nier et al. 1976), suggesting that there was a loss of an early atmosphere. Owen et al. (1988) measured highly elevated D/H using emission lines of H_2O and HDO. The Viking lander measurements confirmed that the Shergottite-Nakhlite-Chassigny (SNC) meteorite clan were from Mars because trapped gases in impact glasses in the meteorites proved to be almost identical in composition to the current martian atmosphere (Bogard and Johnson 1983). Watson et al. (1994) first measured D/H in the martian meteorites using SIMS, and found values similar to the martian atmosphere. A review of early measurements of D/H in SNC meteorites can be found in Usui (2018).

2.4.2 Water Abundances and D/H of Martian Reservoirs

Atmospheric Water Current atmospheric water is dominantly sourced from the polar caps, which dominate the recognized surface reservoir of water, and has a total volume of 20–30 mGEL (meters of global equivalent layer) (Kurokawa et al. 2014). The precipitable water column in the present-day atmosphere is < 50 μm (Fisher 2007). The D/H of the present atmospheric water is ~ 5000‰, but varies on a seasonal basis between ~ 2000–9000‰ (Fig. 7, Villanueva et al. 2015; Montmessin et al. 2005). The Curiosity rover at Gale Crater using the SAM-TLS also measured $\delta D = 4950 \pm 1080$‰ for the present martian atmosphere, and a much more precise measurement of water from soil components released at 230 °C to 430 °C of $\delta D = 5880 \pm 60$‰ (Webster et al. 2013). The highest value measured using the SAM-TLS for H_2O was for Rocknest fines during stepped heating of $\delta D = 7010 \pm 66$‰ (Leshin et al. 2013).

Crustal Water The martian meteorites are samples of the martian crust that have interacted with martian atmospheric water (e.g., Watson et al. 1994). Watson et al. (1994) suggested that the atmospheric water was introduced into the meteorites by alteration of hydrous minerals by later aqueous fluids, possibly during subsolidus cooling. Greenwood et al. (2008) showed from a correlation of zoning of D/H with igneous crystallization of apatite that the D/H was set during the late-stage crystallization of apatite, and that atmospherically-derived water with a high D/H had mixed with magmatic water of low D/H in the magma, either by assimilation or an igneous hydrothermal system. Hu et al. (2014) showed that an olivine-hosted melt inclusion in GRV 020090 had undergone aqueous alteration by infiltration of high D/H water, suggesting that apatite in martian meteorites needs

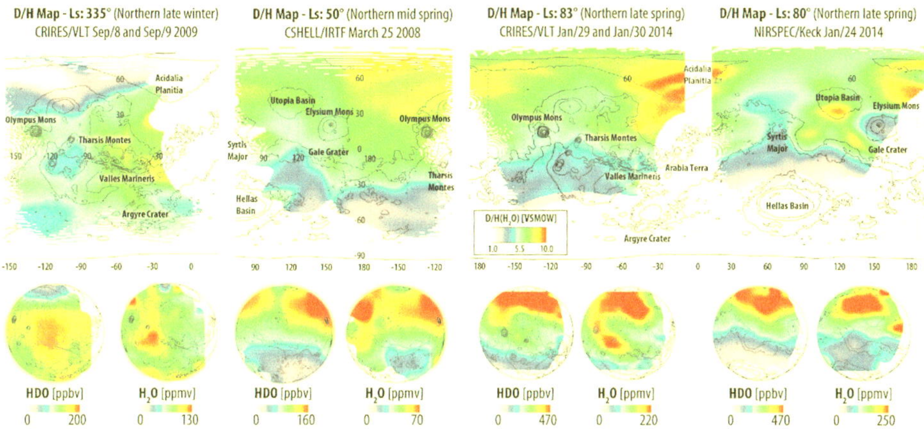

Fig. 7 Seasonal changes in atmospheric water species on Mars. D/H maps on top with abundance maps of HDO and H_2O below. Figure adapted from Villanueva et al. (2015)

to be studied more closely with 2-D isotope imaging (isotopography) to better understand how atmospheric waters alter the presumably lower D/H of the magma or protolith.

Ancient water is likely preserved in the oldest known sample of Mars, the ALH 84001 meteorite which underwent an episode of carbonate deposition 3.9–4.1 Ga (Borg et al. 1999). The δD of apatite associated with carbonate can provide an ancient value for D/H of water, if it is not altered by later fluids. One measurement of apatite in ALH 84001 is $\sim 3000‰$; measurement of δD in associated carbonate reach maximum $\delta D \sim 2000‰$ (Sugiura and Hoshino 2000; Boctor et al. 2003), suggesting that water on the martian surface was as low as 2000–3000‰ at 3.9–4.1 Ga.

The NWA 7034 martian meteorite ("Black Beauty") is a breccia and is the second sample of old martian crust in our collections. The rock has a bulk age of ~ 2 Ga (Agee et al. 2013), but igneous clasts in the meteorite record ancient ages of ~ 4.44 Ga (Nyquist et al. 2016), confirming the early separation of the mantle and crustal reservoirs on Mars. The clasts and matrix of the breccia also host martian water measured during stepped heating and mass spectrometry for δD and oxygen isotopes (Agee et al. 2013). The main martian water-bearing phases in the meteorite are a mélange of hydrous Fe-oxide phases, which suggest the importance of these phases for hosting water in the martian crust (Muttik et al. 2014). Evidence for fluid-rock interactions in the martian crust have also been found from phosphate minerals in NWA 7034 (Liu et al. 2016). The wide range of ages obtained for NWA 7034 require that apatite D/H measurements come from clasts that can be dated, in order to assess the D/H of martian crustal water in different epochs (e.g., Greenwood et al. 2008; Villanueva et al. 2015).

A third distinct reservoir of crustal water on Mars has been suggested from water trapped in impact glasses of martian meteorites that were excavated from the martian crust (Usui et al. 2015). Usui et al. (2015) proposed that there is a reservoir in the martian crust with a $\delta D \sim 1000$–$2000‰$ that is sampled by the impact glasses, and is similar to Hesperian clays measured by the Curiosity rover using the SAM-TLS ($\delta D = 2000 \pm 200‰$) (Mahaffy et al. 2015). This intermediate D/H water reservoir (to distinguish it from the low primordial D/H of the mantle (see below) and the more D-enriched atmospheric water) could be ice or hydrated crustal minerals that have been isolated from exchange with the atmospheric water and represents ancient water, similar to that which deposited the Hesperian-aged clays measured by Curiosity rover. The existence of this putative reservoir in the impact glasses of martian meteorites has been called into question, as Hallis (2017) argue that low water-containing glasses in martian meteorites have likely been affected by shock implantation of the martian atmosphere during the shock event that liberated the samples from the martian crust. It remains to be seen if this reservoir exists in martian meteorites, but the reservoir does exist in Hesperian-aged clays on the martian surface (Mahaffy et al. 2015).

Mantle Water Mars is thought to have had a different early history to that of the Earth, and would have started as a planetary embryo and then underwent early igneous differentiation during a magma ocean phase (Elkins-Tanton et al. 2005). It is generally believed that the crust and mantle did not undergo recycling or melting events that would later significantly mix mantle and crustal components (e.g., Brandon et al. 2000; Debaille et al. 2007; Foley et al. 2005; Kleine et al. 2009), suggesting that mantle water may represent a very early episode of volatile acquisition by Mars (Brasser 2013).

Usui et al. (2012) estimated the primordial mantle water to have $\delta D < 275‰$ from SIMS measurements of water in olivine-hosted melt inclusions of Yamato (Y-) 980459. This meteorite is believed to represent an early melt from a geochemically depleted source in the martian mantle, and its melt inclusions are interpreted to represent undegassed mantle water. Other estimates of the δD of the martian mantle are 900‰ from apatite in QUE 94201 (Leshin 2000), and $-100‰$ from apatite in Nakhla (Hallis et al. 2012). This may indicate significant mantle heterogeneity, or terrestrial contamination during analysis of apatite (e.g., Greenwood et al. 2008; Boctor et al. 2003). As discussed earlier, all H isotope measurements of martian meteorites are challenging, due to the prevalence of fractures which are the result of high shock pressures experienced by these rocks as they left Mars (Fig. 6).

Water abundances for the martian mantle of 73–290 ppm wt H_2O have been estimated from apatite in both geochemically enriched and depleted shergottites (McCubbin et al. 2012b) using experimentally determined partition coefficients for water and apatite in shergottite magmas (McCubbin et al. 2015b). Due to the early formation of Mars, and the lack of crustal recycling, water in the mantle is likely to be primordial, and not delivered by later additions of chondrites or comets. This suggests significant reservoirs for primordial H in the martian interior (McCubbin et al. 2012b).

2.4.3 Evolution of Water Through Time on Mars

Based on the low δD of water in apatite from the ancient ALH 84001 meteorite, it was argued that Mars suffered a two-stage loss of water: significant early loss of water up to 3.9–4.1 Ga, followed by only modest loss since (Greenwood et al. 2008). This significant early loss of water would likely be due to hydrodynamic escape (Hunten et al. 1987). Using the mantle estimate of Usui et al. (2012) for D/H of primordial water, Kurokawa et al. (2014) modeled the loss of water due to atmospheric escape as a two-stage process. Their model suggested loss of > 41–99 mGEL before 4 Ga, and a lower loss rate of > 10–53 mGEL since then. Usui (2018) reports that an updated version of the model of Kurokawa et al. (2014) with atmospheric D/H of 7000‰ (Leshin et al. 2013; Villanueva et al. 2015) suggests $\sim 40\%$ higher loss of water. Regardless, these estimates are much lower than amounts of water loss based on geomorphology of ~ 100–1000 mGEL (Carr and Head 2003; Di Achille and Hynek 2010; Head et al. 1999; Ormö et al. 2004). If Mars had oceans, then a great deal of water is missing and could be buried water ice (Clifford and Parker 2001; Grimm et al. 2017). Usui et al. (2015) predict that this ice would have δD of ~ 1000–2000‰.

2.4.4 Timing of Water Delivery and Sources of Water to Mars

Crust The lack of viable crustal recycling mechanisms on Mars, combined with its early accretion and early magma ocean crystallization, suggests that the ancient crust of Mars (the Noachian-aged southern highlands) holds the outgassed inventories of volatile elements unable to escape from the atmosphere (e.g., S and Cl). The likely early crystallization of the magma ocean would suggest that the Noachian highlands would have witnessed later delivery of volatiles from chondritic or cometary material, as well as volatile-bearing material during the late heavy bombardment. This suggests that early martian crustal water may reflect these sources of water.

Mantle There are four estimates for the H isotopic composition of the martian mantle. Usui et al. (2012) estimate the martian mantle $\delta D < 275‰$, as they argue that water in the olivine-hosted melt inclusion they measured in Y-980459 may have been affected by early primordial degassing that would have led to elevated D/H. The early formation of Mars and lack of crustal recycling argue that martian mantle water may preserve one of the earliest records of accretion of volatiles in the Solar System. A δD of 275‰ is not similar to any classes of carbonaceous chondrites, as this is heavier than CI, CM, and CV chondrites, but lighter than CR and Tagish Lake (Alexander et al. 2012, 2018 in the same special volume), suggesting that a chondritic origin for martian water requires multiple components.

Three other values for the martian mantle come from apatite in the depleted shergottite QUE 94201 ($\delta D = 900 \pm 250‰$; Leshin 2000), the fall Nakhla ($\delta D = -111 \pm 62‰$; Hallis et al. 2012), and the recent fall Tissint ($\delta D = 116 \pm 94‰$) (Mane et al. 2016). If these values are representative of the martian mantle, and are not a result of terrestrial contamination causing significant lowering of D/H (e.g., Greenwood et al. 2008; Boctor et al. 2003), then this would suggest the martian mantle is quite heterogeneous and poorly mixed. If the martian mantle δD ranges from $-100‰$ to 900‰, this would be similar to the range of D/H expected for a mixture of chondritic and cometary material, and also similar to the D/H of asteroidal water (δD: $-400‰$ to 1300‰; Yurimoto et al. 2014).

2.4.5 Geophysical Constraints on Water in the Martian Mantle

Similar to the Moon, the tidal dissipation of Mars is well constrained from the analysis of the motion of its satellites Phobos and Deimos (Lainey et al. 2007). Electrical conductivity of martian interior was also estimated from electromagnetic field observations (Civet and Tarits 2014). Consequently, one can use these results to infer temperature and water content in Mars similar to the case on the Moon (Karato 2013).

Relatively small tidal Q (80 ± 0.7) (i.e., high degree of energy loss) and high conductivity (~ 1 S/m at 1000 km depth) suggest the relatively "wet" interior of Mars. In contrast to this expectation, using the laboratory data on electrical conductivity by Yoshino et al. (2012), Verhoeven and Vasher (2016) argued that the electrical conductivity of the martian mantle can be explained by the high FeO content without invoking water. Note, however that Yoshino et al. (2012) did not measure the water content in their samples, and their results on the influence of FeO on electrical conductivity of olivine at high pressure are in marked contrast to the well-established results on the influence of FeO on electrical conductivity at room pressure. Since the influence of pressure on electrical conductivity is small (Dai and Karato 2014a; Xu et al. 2000), this large difference is likely due to the high water content in their samples (Karato 2017). Indeed, by measuring the electrical conductivity as a function of both water content and FeO content, Dai and Karato (2014b) showed that the influence of FeO on electrical conductivity in olivine is much smaller than that reported by Yoshino et al. (2012). If one uses the results by Dai and Karato (2014b), one would conclude that a substantial amount of water is present in the martian mantle. However, there have been no detailed studies on the water content in Mars using geophysical observations.

3 Summary

The abundances of water and other volatiles in the silicate mantles of the inner Solar System bodies are not well constrained at this time. The detection of Cl-bearing lava flows, as well as evidence for C and S, show that even Mercury has more volatiles than might be expected from volatile-poor precursor materials that are predicted this close to the Sun (e.g., Albarède 2009). Abundances for water in the mercurian mantle are expected to be low, due to the highly reducing nature of Mercury (e.g., Hirschmann et al. 2012). Venus has Earth-like levels of C and N in its atmosphere, suggesting Venus may have accreted Earth-like levels of water, having lost the bulk of it during the magma ocean (Hamano et al. 2013) or water ocean (Ingersoll 1969; Kasting and Pollack 1983) phases, with some small fraction still extant in the mantle (Elkins-Tanton et al. 2007). The low abundance of high D/H water in the venusian atmosphere strongly suggest that significant amounts of water have been lost to space over time.

The interior water budget of the Moon is highly controversial and a wide range of values have been estimated. Volatile/refractory element ratios of olivine-hosted melt inclusions (Chen et al. 2015; Hauri et al. 2015, 2017) and KREEP basalt (Greenwood et al. 2017) are in agreement with estimates from H_2O in plagioclase from ferroan anorthosite (Hui et al. 2013), which all suggest $\sim$ 100–300 ppm wt H_2O for the lunar interior. This range is very similar to that postulated for the Earth's upper mantle (e.g., Peslier et al. 2017). These high, Earth-like values are in contrast to geophysical and selenogenetic studies which suggest 10–100 ppm wt H_2O for the lunar interior (Karato 2013; Pahlevan et al. 2016). Much lower abundances of water for the lunar interior are suggested from Cl isotopes of lunar materials (Sharp et al. 2010) and magma ocean modeling studies which argue for $\lesssim$ 10 ppm wt H_2O.

The amount of water on Mars has been a controversial subject for decades (Carr 2006). Current estimates suggest the martian mantle contains 73–290 ppm wt H_2O (McCubbin et al. 2012b), which is similar to estimates for Earth's upper mantle. This amount of water would argue that the martian interior can be a significant source of water and could have supported an early, active hydrosphere on Mars.

The inner Solar System is believed to have been volatile-depleted at the time of terrestrial planet formation (Albarède 2009; Dauphas and Morbidelli 2014). Mechanisms for the timing and delivery of water and other volatiles to the inner Solar System are reviewed in the article in this volume by O'Brien et al. (2018). The currently favored mechanism involves gravitational scattering of carbonaceous chondrite planetesimals from $\gtrsim$ 3 AU into the inner Solar System and their accretion during the latter parts of the Earth's growth, but before the end of the Earth's core formation (Walsh et al. 2011; O'Brien et al. 2018), since they can be scattered inwards very early, but due to their eccentric orbits, they might not hit the Earth until relatively late in its growth. Spectral similarities between volatile-rich C-type asteroids and carbonaceous chondrites support this theory (e.g., Morbidelli et al. 2012). Carbonaceous chondrites are also favored sources of water for the inner Solar System, due to similarities in volatile isotope and element ratios (e.g., Alexander et al. 2012, 2018; Halliday 2013; Marty 2012; Marty et al. 2016).

Carbonaceous chondrites have also been suggested as the source of water for differentiated meteorites, such as the parent bodies of the eucrites and angrites (Sarafian et al. 2014, 2017). The angrites and eucrites record melting events within ~ 1–2 Ma of the formation of CAIs, and are thus younger than the carbonaceous chondrite parent bodies (e.g., Alexander et al. 2018). The differentiated meteorites require that at least some water and volatiles were delivered inside the orbit of Jupiter within 2 Ma of CAI formation, and before the formation of carbonaceous chondrite parent bodies.

Enstatite chondrites or material like the EC's has been suggested to be an important component of the Earth's building blocks (Wänke and Dreibus 1988; Rubie et al. 2011), and could be an important component for Mars (Dreibus and Wanke 1985; Sanloup et al. 1999) and the other inner Solar System bodies. The water abundance of the enstatite chondrites is not well known due to the confounding influence of terrestrial weathering (Jarosewich 1990) but EC's have abundant C and N (Fig. 3; Grady et al. 1986), suggesting that the material in the terrestrial planet forming regions may not have been as volatile-poor as is currently advocated (e.g., Albarède 2009; Dauphas and Morbidelli 2014).

Acknowledgements We thank the organizing committee for inviting us to write this article. J.P.G. also thanks K. Abe for assistance with figure preparation, M. Gilmore for discussions regarding Venus, and S. Itoh, N. Sakamoto and H. Yurimoto for discussions of water measurements using SIMS and for discussions on volatile delivery mechanisms to the Solar System. This work was supported by NASA-LASER (NNX14AQ76G) and NASA-EW (NNX17AE26G) to J.P.G., and by JSPS (16H04073, 15KK0153, 17H06459) to T.U.

References

B. Agee, N.V. Wilson, F.M. McCubbin, K. Ziegler, V.J. Polyak, Z.D. Sharp, Y. Asmerom, M.H. Numm, R. Shaheen, M.H. Thiemens, A. Steele, M.L. Fogel, R. Bowden, M. Glamoclija, Z. Zhang, S.M. Elardo, Unique meteorite from Early Amazonian Mars: water-rich basaltic breccia Northwest Africa 7034. Nature **339**(6121), 780–785 (2013). https://doi.org/10.1126/science.1228858

F. Albarède, Volatile accretion history of the terrestrial planets and dynamic implications. Nature **461**, 1227–1233 (2009)

F. Albarède, E. Albalat, C.-T.A. Lee, An intrinsic volatility scale relevant to the Earth and Moon and the status of water in the Moon. Meteorit. Planet. Sci. **50**, 568–577 (2015)

C.M.O'D. Alexander, R. Bowden, M.L. Fogel, K.T. Howard, C.D.K. Herd, L.R. Nittler, The provenances of asteroids, and their contributions to the volatile inventories of the terrestrial planets. Science **337**, 721 (2012). https://doi.org/10.1126/science.1223474

C.M.O'D. Alexander, K.D. McKeegan, K. Altwegg, Water reservoirs in small planetary bodies: meteorites, asteroids, and comets. Space Sci. Rev. **214**, 36 (2018) (this book). https://doi.org/10.1007/s11214-018-0474-9.

J.J. Barnes, I.A. Franchi, M. Anand, R. Tartèse, N.A. Starkey, M. Koike, Y. Sano, S.S. Russell, Accurate and precise measurements of the D/H ratio and hydroxyl content in lunar apatites using NanoSIMS. Chem. Geol. **337**, 48–55 (2013)

J.J. Barnes, R. Tartèse, M. Anand, F.M. McCubbin, I.A. Franchi, N.A. Starkey, S.S. Russell, The origin of water in the primitive Moon as revealed by the lunar highlands samples. Earth Planet. Sci. Lett. **390**, 244–252 (2014)

J.J. Barnes, D.A. Kring, R. Tartèse, I.A. Franchi, M. Anand, S.S. Russell, An asteroidal origin for water in the Moon. Nat. Commun. **7**, 11684 (2016)

Basaltic Volcanism Study Project: *Basaltic Volcanism on the Terrestrial Planets*. Pergamon, New York (1981), 1286 pp.

W. Benz, W.L. Slattery, A.G.W. Cameron, Collisional stripping of Mercury's mantle. Icarus **74**, 516–528 (1988)

J.P. Bibring, Y. Langevin, J.F. Mustard, F. Poulet, R. Arvidson, A. Gendrin, B. Gondet, N. Mangold, P. Pinet, F. Forget, M. Berthé, C. Gomez, D. Jouglet, A. Soufflot, M. Vincendon, M. Combes, P. Drossart, T. Encrenaz, T. Fouchet, R. Merchiorri, G. Belluci, F. Altieri, V. Formisano, F. Capaccioni, P. Cerroni, A. Coradini, S. Fonti, O. Korablev, V. Kottsov, N. Ignatiev, V. Moroz, D. Titov, L. Zasova, D. Loiseau,

N. Mangold, P. Pinet, S. Douté, B. Schmitt, C. Sotin, E. Hauber, H. Hoffmann, R. Jaumann, U. Keller, T. Duxbury, G. Neukum, Global mineralogical and aqueous Mars history derived from OMEGA/Mars Express data. Science **312**, 400–404 (2006)

T.A. Bida, R.M. Killen, T.H. Morgan, Discovery of calcium in Mercury's atmosphere. Nature **404**, 159–161 (2000)

D.T. Blewett, M.S. Robinson, B.W. Denevi, J.J. Gillis-Davis, J.W. Head, S.C. Solomon, G.M. Holsclaw, W.E. McClintock, Multispectral images of Mercury from the first MESSENGER flyby: analysis of global and regional color trends. Earth Planet. Sci. Lett. **285**, 272–282 (2009)

N.Z. Boctor, C.M.O'D. Alexander, J. Wang, E. Hauri, The sources of water in Martian meteorites: clues from hydrogen isotopes. Geochim. Cosmochim. Acta **67**, 3971–3989 (2003)

D.D. Bogard, P. Johnson, Martian gases in an Antarctic meteorite? Science **221**, 651–654 (1983)

L.E. Borg, J.N. Connelly, L.E. Nyquist, C.Y. Shih, H. Wiesmann, Y. Reese, The age of the carbonates in martian meteorite ALH84001. Science **286**, 90–94 (1999)

J.W. Boyce, Y. Liu, G.R. Rossman, Y. Guan, J.M. Eiler, E.M. Stolper, L.A. Taylor, Lunar apatite with terrestrial volatile abundances. Nature **466**, 466–470 (2010)

J.W. Boyce, S.M. Tomlinson, F.M. McCubbin, J.P. Greenwood, A.H. Treiman, The Lunar Apatite Paradox. Science **344**, 400–402 (2014)

J.W. Boyce, A.H. Treiman, Y. Guan, C. Ma, J.M. Eiler, J. Gross, J.P. Greenwood, E.M. Stolper, The chlorine isotope fingerprint of the lunar magma ocean. Sci. Adv. **1**, e1500380 (2015)

W.V. Boynton, W.C. Feldman, S.W. Squyres, T.H. Prettyman, J. Brückner, L.G. Evans, R.C. Reedy, R. Starr, J.R. Arnold, D.M. Drake, P.A.J. Englert, A.E. Metzger, I. Mitrofanov, J.I. Trombka, C. d'Uston, H. Wänke, O. Gasnault, D.K. Hamara, D.M. Janes, R.L. Marcialis, S. Maurice, I. Mikheeva, G.J. Taylor, R. Tokar, C. Shinohara, Distribution of hydrogen in the near surface of Mars: evidence for subsurface ice deposits. Science **297**, 81–85 (2002). https://doi.org/10.1126/science.1073722

A.D. Brandon, R.J. Walker, J.W. Morgan, G.G. Goles, Re-Os isotopic evidence for early differentiation of the Martian mantle. Geochim. Cosmochim. Acta **64**, 4083–4095 (2000)

R. Brasser, The formation of Mars: building blocks and accretion time scale. Space Sci. Rev. **174**, 11–25 (2013)

J.M. Brenan, Partitioning of fluorine and chlorine between apatite and aqueous fluids at high-pressure and temperature—implications for the F and Cl content of high P-T fluids. Earth Planet. Sci. Lett. **117**, 251–263 (1993)

A.L. Broadfoot, S. Kumar, M.J.S. Belton, Mercury's atmosphere from Mariner 10: preliminary results. Science **185**, 166–169 (1974)

A.L. Broadfoot, D.E. Shemansky, S. Kumar, Mariner 10: Mercury atmosphere. Geophys. Res. Lett. **3**, 577–580 (1976)

S.M. Brown, L.T. Elkins-Tanton, Compositions of Mercury's earliest crust from magma ocean models. Earth Planet. Sci. Lett. **286**, 446–455 (2009)

C.E. Bucholz, G.A. Gaetani, M.D. Behn, N. Shimizu, Post-entrapment modification of volatiles and oxygen fugacity in olivine-hosted melt inclusions. Earth Planet. Sci. Lett. **374**, 145–155 (2013)

P.A. Candela, Toward a thermodynamic model for the halogens in magmatic systems—an application to melt vapor apatite equilibria. Chem. Geol. **57**, 289–301 (1986)

R. Canup, C. Visscher, J. Salmon, B. Fegley, Protolunar disk evolution and the depletion of volatile elements in the Moon, in *Lunar Planet. Sci. Conf. XLVI*, abstract #2304 (2015a)

R. Canup, C. Visscher, J. Salmon, B. Fegley, Lunar volatile depletion due to incomplete accretion within an impact-generated disk. Nat. Geosci. **8**, 918–921 (2015b)

M.H. Carr, *The Surface of Mars* (Cambridge University Press, Cambridge, 2006), 322 pp.

M.H. Carr, J.W. Head, Oceans on Mars: an assessment of the observational evidence and possible fate. J. Geophys. Res. **108**, 5042 (2003). https://doi.org/10.1029/2002JE001963

N.L. Chabot, E.A. Wollack, R.L. Klima, M.E. Minitti, Experimental constraints on Mercury's core composition. Earth Planet. Sci. Lett. **390**, 199–208 (2014a)

N.L. Chabot, C.M. Ernst, B.W. Denevi, H. Nair, A.N. Deutsch, D.T. Blewett, S.L. Murchie, G.A. Neumann, E. Mazarico, D.A. Paige, J.K. Harmon, J.W. Head, S.C. Solomon, Images of surface volatiles in Mercury's polar craters acquired by the MESSENGER spacecraft. Geology **42**, 1051–1054 (2014b)

S. Charnoz, C. Michaut, Evolution of the protolunar disk: dynamics, cooling timescale and implantation of volatiles onto the Earth. Icarus **260**, 440–463 (2015)

Y. Chen, Y. Zhang, Y. Liu, Y. Guan, J. Eiler, E.M. Stolper, Water, fluorine, and sulfur concentrations in the lunar mantle. Earth Planet. Sci. Lett. **427**, 37–46 (2015)

F. Civet, P. Tarits, Electrical conductivity of the mantle of Mars from MGS magnetic observations. Earth Planets Space **66**, 66–85 (2014)

S.M. Clifford, T.J. Parker, The evolution of the Martian hydrosphere: implications for the fate of a primordial ocean and the current state of the northern plains. Icarus **154**, 40–79 (2001)

M. Clog, C. Aubaud, P. Cartigny, L. Dosso, The hydrogen isotopic composition and water content of southern Pacific MORB: a reassessment of the D/H ratio of the depleted mantle reservoir. Earth Planet. Sci. Lett. **381**, 156–165 (2013)

L. Dai, S. Karato, The effect of pressure on hydrogen-assisted electrical conductivity of olivine. Phys. Earth Planet. Inter. **232**, 51–56 (2014a)

L. Dai, S. Karato, The influence of FeO and H content on the electrical conductivity of olivine: applications to terrestrial planets with different FeO content. Earth Planet. Sci. Lett. **237**, 73–79 (2014b)

N. Dauphas, A. Morbidelli, Geochemical and planetary dynamical views on the origin of Earth's atmosphere and oceans, in *Treatise on Geochemistry*, 2nd edn., ed. by A.S. Davis (Elsevier, Amsterdam, 2014), pp. 1–35

V. Debaille, A.D. Brandon, Q.Z. Yin, B. Jacobsen, Coupled 142Nd–143Nd evidence for a protracted magma ocean in Mars. Nature **450**, 525–528 (2007)

S.J. Desch, G.J. Taylor, A model of the Moon's volatile depletion, in *Lunar Plan. Sci. Conf., XLII*, abstract #2005 (2011)

G. Di Achille, B.M. Hynek, Ancient ocean on Mars supported by global distribution of deltas and valleys. Nat. Geosci. **3**, 459–463 (2010)

T.M. Donahue, J.H. Hoffman, R.R. Hodges, A.J. Watson, Venus was wet—a measurement of the ratio of deuterium to hydrogen. Science **216**, 630–633 (1982)

G. Dreibus, H. Wanke, Mars, a volatile-rich planet. Meteoritics **20**, 367–381 (1985)

C.M. Dundas, A.M. Bramson, L. Ojha, J.J. Wray, M.T. Mellon, S. Byrne, A.S. McEwen, N. Putzig, D. Viola, S. Sutton, E. Clark, J.W. Holt, Exposed subsurface ice sheets in the Martian mid-latitudes. Science **359**, 199–201 (2018)

D.S. Ebel, R.O. Sack, Djerfisherite: nebular source of refractory potassium. Contrib. Mineral. Petrol. **166**, 923–934 (2013)

B.L. Ehlmann, C.S. Edwards, Mineralogy of the Martian surface. Annu. Rev. Earth Planet. Sci. **42**, 291–315 (2014)

L.T. Elkins-Tanton, T.L. Grove, Water (hydrogen) in the lunar mantle: results from petrology and magma ocean modeling. Earth Planet. Sci. Lett. **307**, 173–179 (2011)

L.T. Elkins-Tanton, P.C. Hess, E.M. Parmentier, Possible formation of ancient crust on Mars through magma ocean processes. J. Geophys. Res. **110**, E12S01 (2005). https://doi.org/10.1029/2005JE002480

L.T. Elkins-Tanton, S.E. Smrekar, P.C. Hess, E.M. Parmentier, Volcanism and volatile recycling on a one-plate planet: applications to Venus. J. Geophys. Res. **112**, E04S06 (2007)

L.G. Evans, P.N. Peplowski, E.A. Rhodes, D.J. Lawrence, T.J. McCoy, L.R. Nittler, S.C. Solomon, A.L. Sprague, K.R. Stockstill-Cahill, R.D. Starr, S.Z. Weider, W.V. Boynton, D.K. Hamara, J.O. Goldsten, Major-element abundances on the surface of Mercury: results from the MESSENGER Gamma-Ray Spectrometer. J. Geophys. Res. **117**, E00L07 (2012)

L.G. Evans, P.N. Peplowski, F.M. McCubbin, T.J. McCoy, L.R. Nittler, M.Y. Zolotov, D.S. Ebel, D.J. Lawrence, R.D. Starr, S.Z. Weider, S.C. Solomon, Chlorine on the surface of Mercury: MESSENGER Gamma-Ray measurements and implications for the planet's formation and evolution. Icarus **257**, 417–427 (2015)

B. Fegley Jr., Venus, in *Meteorites, Comets, and Planets*, ed. by A.M. Davis, Treatise on Geochemistry, 2nd edn. (Elsevier-Pergamon, Oxford, 2014), pp. 127–146. https://doi.org/10.1016/B978-0-08-095975-7.00122-4

D.A. Fisher, Mars' water isotope (D/H) history in the strata of the North Polar Cap: inferences about the water cycle. Icarus **187**, 430–441 (2007)

C.N. Foley, M. Madhwa, L.E. Borg, P.E. Janney, R. Hines, T.L. Grove, The early differentiation history of Mars from 182W–142Nd isotope systematics in the SNC meteorites. Geochim. Cosmochim. Acta **69**, 4557–4571 (2005)

I. Friedman, J.R. O'Neil, J.D. Gleason, K. Hardcastle, The carbon and hydrogen content and isotopic composition of some Apollo 12 materials, in *Proc. Lunar Sci. Conf.*, vol. 2 (1971), pp. 1407–1415

E. Füri, E. Deloule, New constraints on the production rate of cosmogenic deuterium at the Moon's surface, in *47th Lunar Planet. Sci. Conf.*, abstract #1351 (2016)

E. Füri, E. Deloule, A.A. Gurenko, B. Marty, New evidence for chondritic lunar water from combined D/H and noble gas analyses of single Apollo 17 volcanic glasses. Icarus **229**, 109–120 (2014)

E. Füri, P.H. Barry, L.A. Taylor, B. Marty, Indigenous nitrogen in the Moon: constraints from coupled nitrogen-noble gas analyses of mare basalts. Earth Planet. Sci. Lett. **431**, 195–205 (2015)

G.A. Gaetani, J.A. O'Leary, N. Shimizu, C.E. Bucholz, M. Newville, Rapid re-equilibration of H_2O and oxygen fugacity in olivine-hosted melt inclusions. Geology **40**, 915–918 (2012)

H. Genda, *Effects of Giant Impacts on the Atmosphere Formation of Terrestrial Planets* (University of Tokyo, Tokyo, 2004)

H. Genda, M. Ikoma, Origin of the ocean on the Earth: early evolution of water D/H in a hydrogen-rich atmosphere. Icarus **194**, 42–52 (2008). https://doi.org/10.1016/j.icarus.2007.09.007

M. Gilmore, A. Treiman, J. Helbert, S. Smrekar, Venus surface composition constrained by observation and experiment. Space Sci. Rev. **212**(3–4), 1511–1540 (2017). https://doi.org/10.1007/s11214-017-0370-8

T.A. Goudge, J.W. Head, L. Kerber, D.T. Blewett, B.W. Denevi, D.L. Domingue, J.J. Gillis-Davis, K. Gwinner, J. Helbert, G.M. Holsclaw, N.R. Izenberg, R.L. Klima, W.E. McClintock, S.L. Murchie, G.A. Neumann, D.E. Smith, R.G. Strom, Z. Xiao, M.T. Zuber, S.C. Solomon, Global inventory and characterization of pyroclastic deposits on Mercury: new insights into pyroclastic activity from MESSENGER orbital data. J. Geophys. Res. **119**, 635–658 (2014)

M.M. Grady, I.P. Wright, L.P. Carr, C.T. Pillinger, Compositional differences in enstatite chondrites based on carbon and nitrogen stable isotope measurements. Geochim. Cosmochim. Acta **50**, 2799–2813 (1986)

J.P. Greenwood, S. Itoh, N. Sakamoto, E. Vicenzi, H. Yurimoto, Hydrogen isotope evidence for loss of water from Mars through time. Geophys. Res. Lett. **35**, L05203 (2008). https://doi.org/10.1029/2007GL032721

J.P. Greenwood, S. Itoh, N. Sakamoto, P.H. Warren, L.A. Taylor, H. Yurimoto, Hydrogen isotope ratios in lunar rocks indicate delivery of cometary water to the Moon. Nat. Geosci. **4**, 79–82 (2011)

J.P. Greenwood, N. Sakamoto, S. Itoh, P.H. Warren, J.A. Singer, K. Yanai, H. Yurimoto, The lunar magma ocean volatile signature recorded in chlorine-rich glasses in KREEP basalts 15382 and 15386. Geochem. J. **51**, 105–114 (2017)

R.E. Grimm, K.P. Harrison, D.E. Stillman, M.R. Kirchoff, On the secular retention of ground water and ice on Mars. J. Geophys. Res., Planets **122**, 94–109 (2017). https://doi.org/10.1002/2016JE005132

D.H. Grinspoon, Implications of the high deuterium-to-hydrogen ratio for the sources of water in Venus' atmosphere. Nature **363**, 428–431 (1993)

D.H. Grinspoon, J.S. Lewis, Cometary water on Venus: implications of stochastic impacts. Icarus **74**, 21–35 (1988)

A.N. Halliday, The origins of volatiles in the terrestrial planets. Geochim. Cosmochim. Acta **105**, 146–171 (2013)

A.N. Halliday, D. Lee, S. Tommasini, G.R. Davies, C.R. Paslick, J.G. Fitton, D.E. James, Incompatible trace elements in OIB and MORB and source enrichment in the sub-oceanic mantle. Earth Planet. Sci. Lett. **133**, 379–395 (1995)

L.J. Hallis, D/H ratios of the inner solar system. Philos. Trans. R. Soc. Lond. A **375**, 20150390 (2017)

L.J. Hallis, G.J. Taylor, K. Nagashima, G.R. Huss, Magmatic water in the martian meteorite Nakhla. Earth Planet. Sci. Lett. **359–360**, 84–92 (2012)

L.J. Hallis, G.R. Huss, K. Nagashima, G.J. Taylor, S.A. Halldórsson, D.R. Hilton, M.J. Mottl, K.J. Meech, Evidence for primordial water in Earth's deep mantle. Science **350**, 795–797 (2015). https://doi.org/10.1126/science.aac4834

K. Hamano, Y. Abe, H. Genda, Emergence of two types of terrestrial planet on solidification of magma ocean. Nature **497**, 607–610 (2013). https://doi.org/10.1038/nature12163

J.K. Harmon, M.A. Slade, Radar mapping of Mercury: full-disk images and polar anomalies. Science **258**, 640–642 (1992)

L.A. Haskin, P.H. Warren, Lunar chemistry, in *Lunar Sourcebook: A User's Guide to the Moon*, ed. by G. Heiken, D. Vaniman, B. French (Cambridge University Press, New York, 1991), pp. 357–474

S.A. Hauck II, J.L. Margot, S.C. Solomon, R.J. Phillips, C.L. Johnson, F.G. Lemoine, E. Mazarico, T.J. McCoy, S. Padovan, S.J. Peale, M.E. Perry, D.E. Smith, M.T. Zuber, The curious case of Mercury's internal structure. J. Geophys. Res., Planets **118**, 1204–1220 (2013)

E. Hauri, J. Wang, J.E. Dixon, P.L. King, C. Mandeville, S. Newman, SIMS analysis of volatiles in silicate glasses 1. Calibration, matrix effects, and comparisons with FTIR. Chem. Geol. **183**, 99–114 (2002)

E.H. Hauri, T. Weinreich, A.E. Saal, M.C. Rutherford, J.A. Van Orman, High pre-eruptive water contents preserved in Lunar melt inclusions. Science **333**, 213–215 (2011)

E.H. Hauri, A.E. Saal, M.C. Rutherford, J.A. Van Orman, Water in the Moon's interior: truth and consequences. Earth Planet. Sci. Lett. **409**, 252–264 (2015)

E.H. Hauri, A.E. Saal, M. Nakajima, M. Anand, M.J. Rutherford, J.A. Van Orman, M. Le Voyer, Origin and evolution of water in the Moon's interior. Annu. Rev. Earth Planet. Sci. **45**, 89–111 (2017)

J.W. Head III, L. Wilson, Alphonsus-type dark-halo craters: morphology, morphometry, and eruption conditions, in *Lunar Planet. Sci. Conf. XI*, vol. 418 (1979)

J.W. Head III, H. Hiesinger, M.A. Ivanov, M.A. Kreslavsky, S. Pratt, B.J. Thomson, Possible ancient oceans on Mars: evidence from Mars Orbiter Laser Altimeter data. Science **286**, 2134–2137 (1999)

C.D.K. Herd, Basalts as probes of planetary interior redox state. Rev. Mineral. Geochem. **68**, 527–553 (2008). *Oxygen in the Solar System*, ed. by G.J. MacPherson

Y. Higashi, S. Itoh, M. Hashiguchi, S. Sakata, T. Hirata, K. Watanabe, I. Sakaguchi, Hydrogen diffusion in the apatite-water system: fluorapatite parallel to the c-axis. Geochem. J. **51**, 115–122 (2017)

M.M. Hirschmann, A.C. Withers, P. Ardia, N.T. Foley, Solubility of molecular hydrogen in silicate melts and consequences for volatile evolution of terrestrial planets. Earth Planet. Sci. Lett. **345–348**, 38–48 (2012)

L.L. Hood, F. Herbert, C.P. Sonett, The deep lunar electrical conductivity profile: structural and thermal inferences. J. Geophys. Res. **87**, 5311–5326 (1982)

G.L. Hovis, D.E. Harlov, Solution calorimetric investigation of fluor-chlorapatite crystalline solutions. Am. Mineral. **95**, 946–952 (2010)

S. Hu, Y. Lin, J. Zhang, J. Hao, L. Feng, L. Xu, W. Yang, J. Yang, NanoSIMS analyses of apatite and melt inclusions in the GRV 020090 Martian meteorite: hydrogen isotope evidence for recent past underground hydrothermal activity on Mars. Geochim. Cosmochim. Acta **140**, 321–333 (2014)

H. Hui, A.H. Peslier, Y. Zhang, C.R. Neal, Water in lunar anorthosites and evidence for a wet early Moon. Nat. Geosci. **6**, 177–180 (2013). https://doi.org/10.1038/ngeo1735

H. Hui, Y. Guan, Y. Chen, A.H. Peslier, Y. Zhang, Y. Liu, R.L. Flemming, G.R. Rossman, J.M. Eiler, C.R. Neal, G.R. Osinsky, A heterogeneous lunar interior for hydrogen isotopes as revealed by the lunar highlands samples. Earth Planet. Sci. Lett. **473**, 14–23 (2017). https://doi.org/10.1016/j.epsl.2017.05.029

D. Hunten, R.O. Pepin, J.C.G. Walker, Mass fractionation in hydrodynamic escape. Icarus **69**, 532–549 (1987)

A.P. Ingersoll, The runaway greenhouse: a history of water on Venus. J. Atmos. Sci. **26**(6), 1191–1198 (1969)

E. Jarosewich, Chemical analyses of meteorite: a compilation of stony and iron meteorite analyses. Meteoritics **25**, 323–337 (1990)

E.A. Jerde, R.V. Morris, P.H. Warren, In quest of lunar regolith breccias of exotic provenance—a uniquely anorthositic sample from the Fra Mauro (Apollo 14) highlands. Earth Planet. Sci. Lett. **98**, 90–108 (1990)

K.P. Jochum, N.T. Arndt, A.W. Hofmann, Nb-Th-La in komatiites and basalts: constraints on komatiite petrogenesis and mantle evolution. Earth Planet. Sci. Lett. **107**, 272–289 (1991)

B.L. Jolliff, R.L. Korotev, R.A. Zeigler, C. Floss, Northwest Africa 773: lunar mare breccia with a shallow-formed olivine-cumulate component, inferred very-low-Ti (VLT) heritage, and a KREEP connection. Geochim. Cosmochim. Acta **67**, 4857–4879 (2003)

S.-I. Karato, Geophysical constraints on the water content of the lunar mantle and its implications for the origin of the Moon. Earth Planet. Sci. Lett. **384**, 144–153 (2013)

S.-I. Karato, Physical basis of trace element partitioning: a review. Am. Mineral. **101**, 2577–2593 (2016)

S.-I. Karato, Some remarks on hydrogen-assisted electrical conductivity in olivine and other minerals, in *Water in the Earth*, ed. by K. Mibe, J. Kasahara (Springer, Tokyo, 2017)

S.-I. Karato, D. Wang, Electrical conductivity of minerals and rocks, in *Physics and Chemistry of the Deep Earth*, ed. by S.-I. Karato (Wiley, New York, 2013), pp. 145–182

J.F. Kasting, J.B. Pollack, Loss of water from Venus. I. Hydrodynamic escape of hydrogen. Icarus **63**, 479–508 (1983)

C. Kato, F. Moynier, M.C. Valdes, J.K. Dhaliwal, J.M.D. Day, Extensive volatiles loss during the formation and differentiation of the Moon. Nat. Commun. **6**, 7617 (2015). https://doi.org/10.1038/ncomms8617

A. Kent Jr., Melt inclusions in basaltic and related volcanic rocks. Rev. Mineral. Geochem. **69**, 273–331 (2008). *Minerals, Inclusions, and Volcanic Processes*, ed. by K.D. Putirka, and F.J. Tepley III. MSA

L. Kerber, J.W. Head, S.C. Solomon, S.L. Murchie, D.T. Blewett, L. Wilson, Explosive volcanic eruptions on Mercury: eruption conditions, magma volatile content, and implications for interior volatile abundances. Earth Planet. Sci. Lett. **285**, 263–271 (2009)

L. Kerber, J.W. Head, D.T. Blewett, S.C. Solomon, L. Wilson, S.L. Murchie, M.S. Robinson, B.W. Denevi, D.L. Domingue, The global distribution of pyroclastic deposits on Mercury: the view from MESSENGER flybys 1–3. Planet. Space Sci. **59**, 1895–1909 (2011)

K. Kitts, K. Lodders, Survey and evaluation of eucrite bulk compositions. Meteorit. Planet. Sci. **33**, A197–A213 (1998)

T. Kleine, M. Touboul, B. Bourdon, F. Nimmo, K. Mezger, H. Palme, S.B. Jacobsen, Q-Z. Yin, A.N. Halliday, Hf–W chronology of the accretion and evolution of asteroids and terrestrial planets. Geochim. Cosmochim. Acta **73**, 5150–5188 (2009)

R.L. Korotev, Mixing levels, the Apennine Front soil component, and compositional trends in the Apollo 15 soils, in *Proc. Lunar Plan. Sci. Conf.*, vol. 17 (1987), pp. E411–E431

R.L. Korotev, D.T. Kremser, Compositional variations in Apollo 17 soils and their relationship to the geology of the Taurus-Littrow site, in *Proc. Lunar Plan. Sci. Conf.*, vol. 22 (1992), pp. 275–301

H. Kurokawa, M. Sato, M. Ushioda, T. Matsuyama, R. Moriwaki, J.M. Dohm, T. Usui, Evolution of water reservoirs on Mars: constraints from hydrogen isotopes in martian meteorites. Earth Planet. Sci. Lett. **394**, 179–185 (2014)

V. Lainey, V. Dehant, M. Pätzold, First numerical ephemerides of the Martian moons. Astron. Astrophys. **465**, 1075–1084 (2007)

D.J. Lawrence, W.C. Feldman, J.O. Goldsten, S. Maurice, P.N. Peplowski, B.J. Anderson, D. Bazell, R.L. McNutt Jr., L.R. Nittler, T.H. Prettyman, D.J. Rodgers, S.C. Solomon, S.Z. Weider, Evidence for water ice near Mercury's north pole from MESSENGER neutron spectrometer measurements. Science **339**, 292–298 (2013)

L.A. Leshin, Insights into martian water reservoirs from analyses of martian meteorite QUE94201. Geophys. Res. Lett. **27**, 2017–2020 (2000)

L.A. Leshin, P.R. Mahaffy, C.R. Webster, M. Cabane, P. Coll, P.G. Conrad, P.D. Archer Jr., S.K. Atreya, A.E. Brunner, A. Buch, J.L. Eigenbrode, G.J. Flesch, H.B. Franz, C. Freissinet, D.P. Glavin, A.C. Mcadam, K.E. Miller, D.W. Ming, R.V. Morris, R. Navarro-González, P.B. Niles, T. Owen, R.O. Pepin, S. Squyres, A. Steele, J.C. Stern, R.E. Summons, D.Y. Sumner, B. Sutter, C. Szopa, S. Teinturier, M.G. Trainer, J.J. Wray, J.P. Grotzinger (MSL Science Team), Volatile, isotope, and organic analysis of Martian fines with the Mars Curiosity rover. Science **341**, 1238937 (2013)

Y. Liu, Y. Guan, Y. Zhang, G.R. Rossman, J.M. Eiler, L.A. Taylor, Direct measurement of hydroxyl in the lunar regolith and the origin of lunar surface water. Nat. Geosci. **5**, 779–782 (2012)

Y. Liu, C. Ma, J.R. Beckett, Y. Chen, Y. Guan, Rare-Earth element minerals in martian meteorites NWA 7034 and 7533: implications for fluid-rock interaction in the martian crust. Earth Planet. Sci. Lett. **451**, 251–262 (2016)

S.J. Lock, S.T. Stewart, M.I. Petaev, Z.M. Leinhardt, M.T. Mace, S.B. Jacobsen, M. Ćuk, The origin of the Moon within a terrestrial synestia. J. Geophys. Res. **123**, 910–951 (2018). https://doi.org/10.1002/2017JE005333

K. Lodders, B. Fegley, *The Planetary Scientist's Companion* (Oxford University Press, New York, 1998)

P. Lowell, *Mars and Its Canals* (Macmillan & Co., London, 1906)

P. Lucey, R.L. Korotev, J.J. Gillis, L.A. Taylor, D. Lawrence, B.A. Campbell, R. Elphic, B. Feldman, L.L. Hood, D. Hunten, M. Mendillo, S. Noble, J.J. Papike, R.C. Reedy, S. Lawson, T. Prettyman, O. Gasnault, S. Maurice, Understanding the lunar surface and Space-Moon interactions. Rev. Mineral. Geochem. **60**, 83–219 (2006). *New Views of the Moon*, ed. by B.L. Joliff, M.A. Weiczorek, C.K. Shearer, C.R. Neal

R.W. Luth, Mantle volatiles-distributions and consequences, in *The Mantle and Core*, ed. by R.W. Carlson. Treatise on Geochemistry (Elsevier-Pergamon, Oxford, 2005), pp. 319–361

P.R. Mahaffy, C.R. Webster, J.C. Stern, A.E. Brunner, S.K. Atreya, P.G. Conrad, S. Domagal-Goldman, J.L. Eigenbrode, G.J. Flesch, L.E. Christensen, H.B. Franz, C. Freissinet, D.P. Glavin, J.P. Grotzinger, J.H. Jones, L.A. Leshin, C. Malespin, A.C. Mcadam, D.W. Ming, R. Navarro-Gonzalez, P.B. Niles, T. Owen, A.A. Pavlov, A. Steele, M.G. Trainer, K.H. Williford, J.J. Wray (MSL Science Team), The imprint of atmospheric evolution in the D/H of Hesperian clay minerals on Mars. Science **347**, 412–414 (2015)

V. Malavergne, J. Siebert, F. Guyot, L. Gautron, R. Combes, T. Hammouda, S. Borensztajn, D. Frost, I. Martinez, Si in the core? New high-pressure and high-temperature experimental data. Geochim. Cosmochim. Acta **68**, 4201–4211 (2004)

V. Malavergne, M.J. Toplis, S. Berthet, J. Jones, Highly reducing conditions during core formation on Mercury: implications for internal structure and the origin of a magnetic field. Icarus **206**, 199–209 (2010)

V. Malavergne, P. Cordier, K. Righter, F. Brunet, B. Zanda, A. Addad, T. Smith, H. Bureau, S. Surblé, C. Raepsaet, E. Charon, R.H. Hewins, How Mercury can be the most reduced terrestrial planet and still store iron in its mantle. Earth Planet. Sci. Lett. **394**, 186–197 (2014)

B.E. Mandler, L.T. Elkins-Tanton, The origin of eucrites, diogenites, and olivine diogenites: magma ocean crystallization and shallow magma chamber processes on Vesta. Meteorit. Planet. Sci. **48**, 2333–2349 (2013)

P. Mane, R. Hervig, M. Wadhwa, L.A.J. Garvie, J.B. Balta, H.Y. McSween Jr., Hydrogen isotopic composition of the Martian mantle inferred from the newest Martian meteorite fall, Tissint. Meteorit. Planet. Sci. **51**, 2073–2091 (2016). https://doi.org/10.1111/maps.12717

B. Marty, The origins and concentrations of water, carbon, nitrogen and noble gases on Earth. Earth Planet. Sci. Lett. **313–314**, 56–66 (2012)

B. Marty, G. Avice, Y. Sano, K. Altwegg, H. Balsiger, M. Hässig, A. Morbidelli, O. Mousis, M. Rubie, Origins of volatile elements (H, C, N, noble gases) on Earth and Mars in light of recent results from the ROSETTA cometary mission. Earth Planet. Sci. Lett. **441**, 91–102 (2016)

T.J. McCoy, T.L. Dickinson, G.E. Lofgren, Partial melting of the Indarch (EH4) meteorite: a textural, chemical, and phase relations view of melting and melt migration. Meteorit. Planet. Sci. **34**, 735–746 (1999)

F.M. McCubbin, A. Steele, E.H. Hauri, H. Nekvasil, S. Yamashita, R.J. Hemley, Nominally hydrous magmatism on the Moon. Proc. Natl. Acad. Sci. **107**(25), 11223–11228 (2010a)

F.M. McCubbin, A. Steele, H. Nekvasil, A. Schnieders, T. Rose, M. Fries, P.K. Carpenter, B.L. Jolliff, Detection of structurally bound hydroxyl in fluorapatite from Apollo Mare basalt 15058,128 using TOF-SIMS. Am. Mineral. **95**, 1141–1150 (2010b)

F.M. McCubbin, M.A. Riner, K.E. Vander Kaaden, L.K. Burkemper, Is Mercury a volatile-rich planet? Geophys. Res. Lett. **39**, L09202 (2012a)

F.M. McCubbin, E.H. Hauri, S.M. Elardo, K.E. Vander Kaaden, J. Wang, C.K. Shearer Jr., Hydrous melting of the martian mantle produced both depleted and enriched shergottites. Geology **40**(8), 683–686 (2012b). https://doi.org/10.1130/G33242.1

F.M. McCubbin, K.E. Vander Kaaden, R. Tartèse, R.L. Klima, Y. Liu, J. Mortimer, J.J. Barnes, C.K. Shearer, A.H. Treiman, D.J. Lawrence, S.M. Elardo, D.M. Hurley, J.W. Boyce, M. Anand, Magmatic volatiles (H, C, N, F, S, Cl) in the lunar mantle, crust, and regolith: abundances, distributions, processes, and reservoirs. Am. Mineral. **100**, 1668–1707 (2015a)

F.M. McCubbin, K.E. Vander Kaaden, R. Tartèse, J.W. Boyce, S. Mikhail, E.S. Whitson, A.S. Bell, M. Anand, I.A. Franchi, J. Wang, E.H. Hauri, Experimental investigation of F, Cl, and OH partitioning between apatite and Fe-rich basaltic melt at 1.0–1.2 GPa and 950–1000 °C. Am. Mineral. **100**, 1790–1802 (2015b)

F.M. McCubbin, K.E. Vander Kaaden, P.N. Peplowski, A.S. Bell, L.R. Nittler, J.W. Boyce, L.G. Evans, L.P. Keller, S.M. Elardo, T.J. McCoy, A low O/Si ratio on the surface of Mercury: evidence for silicon smelting? J. Geophys. Res. **122**, 2053–2076 (2017)

W.F. McDonough, S.-S. Sun, The composition of the Earth. Chem. Geol. **120**, 223–253 (1995)

M.B. McElroy, Y.L. Yung, A.O. Nier, Isotopic composition of nitrogen: implications for the past history of Mars' atmosphere. Science **194**, 70–72 (1976)

W.B. McKinnon, K.J. Zahnle, B.A. Ivanov, H.J. Melosh, Cratering on Venus: models and observations, in *Venus II*, ed. by S.W. Bougher, D.M. Hunten, R.J. Phillips (University of Arizona Press, Tucson, 1997), pp. 969–1014

H.Y. McSween Jr., R.P. Binzel, M.C. De Sanctis, E. Ammannito, T.H. Prettyman, A.W. Beck, V. Reddy, L. Le Corre, M.J. Gaffey, T.B. McCord, C.A. Raymond, C.T. Russell (The Dawn Science Team), Dawn; the Vesta-HED connection; and the geologic context for eucrites, diogenites, and howardites. Meteorit. Planet. Sci. **48**, 2090–2104 (2013)

L. Merlivat, M. Lelu, G. Nief, E. Roth, Spallation deuterium in rock 70215, in *Proc. Lunar Sci. Conf*, vol. 7 (1976), pp. 649–658

F. Montmessin, T. Fouchet, F. Forget, Modeling the annual cycle of HDO in the Martian atmosphere. J. Geophys. Res. **110**, E03006 (2005)

A. Morbidelli, J.I. Lunine, D.P. O'Brien, S.N. Raymond, K.J. Walsh, Building terrestrial planets. Annu. Rev. Earth Planet. Sci. **40**, 251–275 (2012). https://doi.org/10.1146/annurevearth042711-105319

J. Mortimer, A.B. Verchovsky, M. Anand, I. Gilmour, C.T. Pillinger, Simultaneous analysis of abundance and isotopic composition of nitrogen, carbon, and noble gases in lunar basalts: insights into interior and surface processes on the Moon. Icarus (2015). https://doi.org/10.1016/j.icarus.2014.10.006

J. Mouginot, A. Pommerol, P. Beck, W. Kofman, S.M. Clifford, Dielectric map of the Martian northern hemisphere and the nature of plain filling materials. Geophys. Res. Lett. **39**, L02202 (2012)

S.L. Murchie, R.L. Klima, B.W. Denevi, C.M. Ernst, M.R. Keller, D.L. Domingue, D.T. Blewett, N.L. Chabot, C. Hash, E. Malaret, N.R. Izenberg, F. Vilas, L.R. Nittler, J.W. Head, Orbital multispectral mapping of Mercury using the MESSENGER Mercury Dual Imaging System: evidence for the origins of plains units and low-reflectance material. Icarus **254**, 287–305 (2015)

N. Muttik, F.M. McCubbin, L.P. Keller, A.R. Santos, W.A. McCutcheon, P.P. Provencio, Z. Rahman, C.K. Shearer, J.W. Boyce, C.B. Agee, Inventory of H_2O in the ancient martian regolith from Northwest Africa 7034: the important role of iron oxides. Geophys. Res. Lett. **41**, 8235–8244 (2014). https://doi.org/10.1002/2014GL062533

M. Nakajima, D.J. Stevenson, Inefficient volatile loss from the Moon-forming disk: reconciling the giant impact hypothesis and a wet Moon. Earth Planet. Sci. Lett. **487**, 117–126 (2018)

O. Namur, M. Collinet, B. Charlier, T.L. Grove, F. Holtz, C. McCammon, Melting processes and mantle sources of surface lavas on Mercury. Earth Planet. Sci. Lett. **439**, 117–128 (2016)

H.E. Newsom, W.M. White, K.P. Jochum, A.W. Hofmann, Siderophile and chalcophile element abundances in oceanic basalts, Pb isotope evolution and growth of the Earth's core. Earth Planet. Sci. Lett. **80**, 299–313 (1986)

A.O. Nier, M.B. McElroy, Y.L. Yung, Isotopic composition of the Martian atmosphere. Science **194**, 68–70 (1976)

L.R. Nittler, R.D. Starr, S.Z. Weider, T.J. McCoy, W.V. Boynton, D.S. Ebel, C.M. Ernst, L.G. Evans, J.O. Goldsten, D.K. Hamara, D.J. Lawrence, R.L. McNutt, C.E. Schlemm, S.C. Solomon, A.L. Sprague, The major-element composition of Mercury's surface from MESSENGER X-ray spectrometry. Science **333**, 1847–1850 (2011)

L.E. Nyquist, C.-Y. Shih, F.M. McCubbin, A.R. Santos, C.K. Shearer, Z.X. Peng, P.V. Burger, C.B. Agee, Rb-Sr and Sm-Nd isotopic and REE studies of igneous components in the bulk matrix domain of Martian breccia Northwest Africa 7034. Meteorit. Plan. Sci. **51**, 483–498 (2016)

D.P. O'Brien, A. Izidoro, S.A. Jacobson, S.N. Raymond, D.C. Rubie, The delivery of water during terrestrial planet formation. Space Sci. Rev. **214**, 47 (2018) (this book). https://doi.org/10.1007/s11214-018-0475-8

J. Ormö, J.M. Dohm, J.C. Ferris, A. Lepinette, A.G. Fairén, Marine-target craters on Mars? An assessment study. Meteorit. Planet. Sci. **39**, 333–346 (2004)
T. Owen, The composition and early history of the atmosphere of Mars, in *Mars*, ed. by H.H. Kieffer, B.M. Jakosky, C.W. Snyder (University of Arizona Press, Tucson, 1992)
T. Owen, J.P. Maillard, C. de Bergh, B.L. Lutz, Deuterium on Mars: the abundance of HDO and the value of D/H. Science **240**, 1767–1770 (1988)
K. Pahlevan, D.J. Stevenson, Volatile loss Following the Moonforming giant impact. Geochim. Cosmochim. Acta, Suppl. **72**, A716 (2008)
K. Pahlevan, S.-I. Karato, B. Fegley, Speciation and dissolution of hydrogen in the proto-lunar disk. Earth Planet. Sci. Lett. **445**, 104–113 (2016)
H. Palme, H.S.C. O'Neill, Cosmochemical estimates of mantle composition, in *The Mantle and Core*, ed. by R.W. Carlson. Treatise on Geochemistry (Elsevier-Pergamon, Oxford, 2014), pp. 1–39
Y.M. Pan, M.E. Fleet, Compositions of the apatite-group minerals: substitution mechanisms and controlling factors. Rev. Mineral. Geochem. **48**, 13–49 (2002). *Phosphates: Geochemical, Geobiological, and Materials Importance*, ed. by M.J. Kohn, J. Rakovan, J.M. Hughes
R.C. Paniello, J.M.D. Day, F. Moynier, Zinc isotope evidence for the origin of the Moon. Nature **490**, 376–380 (2012)
S.W. Parman, E.M. Parmentier, S. Wang, Crystallization of Mercury's sulfur-rich magma ocean, in *Lunar Planet. Sci. Conf. XLVII*, abstract# 2990 (2016)
E.M. Parmentier, P.C. Hess, Chemical differentiation of a convecting planetary interior: consequences for a one plate planet such as Venus. Geophys. Res. Lett. **19**, 2015–2018 (1992)
P.N. Peplowski, L.G. Evans, S.A. Hauck, T.J. McCoy, W.V. Boynton, J.J. Gillis-Davis, D.S. Ebel, J.O. Goldsten, D.K. Hamara, D.J. Lawrence, R.L. McNutt, L.R. Nittler, S.C. Solomon, E.A. Rhodes, A.L. Sprague, R.D. Starr, K.R. Stockstill-Cahill, Radioactive elements on Mercury's surface from MESSENGER: implications for the planet's formation and evolution. Science **333**, 1850–1852 (2011)
P.N. Peplowski, D.J. Lawrence, E.A. Rhodes, A.L. Sprague, T.J. McCoy, B.W. Denevi, L.G. Evans, J.W. Head, L.R. Nittler, S.C. Solomon, K.R. Stockstill-Cahill, S.Z. Weider, Variations in the abundances of potassium and thorium on the surface of Mercury: results from the MESSENGER Gamma-Ray Spectrometer. J. Geophys. Res. **117**, E00L04 (2012). https://doi.org/10.1029/2012JE004141
P.N. Peplowski, L.G. Evans, K.R. Stockstill-Cahill, D.J. Lawrence, J.O. Goldsten, T.J. McCoy, L.R. Nittler, S.C. Solomon, A.L. Sprague, R.D. Starr, S.Z. Weider, Enhanced sodium abundance in Mercury's north polar region revealed by the MESSENGER Gamma-Ray Spectrometer. Icarus **228**, 86–95 (2014)
P.N. Peplowski, D.J. Lawrence, L.G. Evans, R.L. Klima, D.T. Blewett, J.O. Goldsten, S.L. Murchie, T.J. McCoy, L.R. Nittler, S.C. Solomon, R.D. Starr, S.Z. Weider, Constraints on the abundance of carbon in near-surface materials on Mercury: results from the MESSENGER Gamma-Ray Spectrometer. Planet. Space Sci. **108**, 98–107 (2015)
P.N. Peplowski, R.L. Klima, D.J. Lawrence, C.M. Ernst, B.W. Denevi, E.A. Frank, J.O. Goldsten, S.L. Murchie, L.R. Nittler, S.C. Solomon, Remote sensing evidence for an ancient carbon-bearing crust on Mercury. Nat. Geosci. **9**, 273–276 (2016)
A.H. Peslier, M. Schönbächler, H. Busemann, S.I. Karato, Water in the Earth's interior: distribution and origin. Space Sci. Rev. **212**, 743–810 (2017) (this book). https://doi.org/10.1007/s11214-017-0387-z
A.E. Potter, T.H. Morgan, Discovery of sodium in the atmosphere of Mercury. Science **229**, 651–653 (1985)
A.E. Potter, T.H. Morgan, Potassium in the atmosphere of Mercury. Icarus **67**, 336–340 (1986)
B. Rava, B. Hapke, An analysis of the Mariner 10 color ratio map of Mercury. Icarus **71**, 397–429 (1987)
R.C. Reedy, Cosmic-ray produced stable nuclides: various production rates and their implications. Lunar Planet. Sci. Conf. **XII**, 871 (1981)
J. Riker, M.C.S. Humphreys, R.A. Brooker, J.C.M. De Hoog (EIMF), First measurements of OH-C exchange and temperature-dependent partitioning of OH and halogens in the system apatite–silicate melt. Am. Mineral. **103**, 260–270 (2018)
M.A. Riner, P.G. Lucey, S.J. Desch, F.M. McCubbin, Nature of opaque components on Mercury: insights into a Mercurian magma ocean. Geophys. Res. Lett. **36**, L02201 (2009)
K.L. Robinson, G.J. Taylor, Heterogeneous distribution of water in the Moon. Nat. Geosci. **7**, 401–408 (2014)
M.S. Robinson, S.L. Murchie, D.T. Blewett, D.L. Domingue, S.E. Hawkins, J.W. Head III, G.M. Holsclaw, W.E. McClintock, T.J. McCoy, R.L. McNutt Jr., L.M. Prockter, S.C. Solomon, T.R. Watters, Reflectance and color variations on Mercury: regolith processes and compositional heterogeneity. Science **321**, 66–69 (2008)
K.L. Robinson, J.J. Barnes, K. Nagashima, A. Thomen, I.A. Franchi, G.R. Huss, M. Anand, G.J. Taylor, Water in evolved lunar rocks: evidence for multiple reservoirs. Geochim. Cosmochim. Acta **188**, 244–260 (2016)
D.C. Rubie, D.J. Frost, U. Mann, Y. Asahara, F. Nimmo, K. Tsuno, P. Kegler, A. Holzheid, H. Palme, Heterogeneous accretion, composition and core-mantle differentiation of the Earth. Earth Planet. Sci. Lett. **301**, 31–42 (2011). https://doi.org/10.1016/j.epsl.2010.11.030

A.E. Saal, E.H. Hauri, M. Lo Cascio, J.A. Van Orman, M.C. Rutherford, R.F. Cooper, Volatile content of lunar volcanic glasses and the presence of water in the Moon's interior. Nature **454**, 192–196 (2008)

A.E. Saal, E.H. Hauri, J.A. Van Orman, M.C. Rutherford, Hydrogen isotopes in lunar volcanic glasses and melt inclusions reveal a carbonaceous chondrite heritage. Science **340**, 1317–1320 (2013)

J. Salmon, R.M. Canup, Lunar accretion from a Roche-interior fluid disk. Astrophys. J. **760**, 83 (2012). https://doi.org/10.1088/0004-637X/760/1/83

C. Sanloup, A. Jambon, P. Gillet, A simple chondritic model of Mars. Phys. Earth Planet. Inter. **112**, 43–54 (1999)

A.R. Sarafian, S.G. Nielsen, H.R. Marschall, F.M. McCubbin, B.D. Monteleone, Early accretion of water in the inner solar system from a carbonaceous chondrite-like source. Science **346**, 623–626 (2014)

A.R. Sarafian, E.H. Hauri, F.M. McCubbin, T.J. Lapen, E.L. Berger, S.G. Nielsen, H.R. Marschall, G.A. Gaetani, K. Righter, E. Sarafian, Early accretion of water and volatile elements to the inner Solar System: evidence from angrites. Philos. Trans. R. Soc. Lond. A **375**, 1–27 (2017)

G.G. Schaber, R.G. Strom, H.J. Moore, L.A. Soderblom, R.L. Kirk, D.J. Chadwick, D.D. Dawson, L.R. Gaddis, J.M. Boyce, J. Russell, Geology and distribution of impact craters on Venus: what are they telling us? J. Geophys. Res. **97**, 13257–13301 (1992)

Z.D. Sharp, C.K. Shearer, K.D. McKeegan, J.D. Barnes, Y.Q. Wang, The chlorine isotope composition of the Moon and implications for an anhydrous mantle. Science **329**, 1050–1053 (2010)

Z.D. Sharp, F.M. McCubbin, C.K. Shearer, A hydrogen-based oxidation mechanism relevant to planetary formation. Earth Planet. Sci. Lett. **380**, 88–97 (2013)

J.A. Singer, J.P. Greenwood, S. Itoh, N. Sakamoto, H. Yurimoto, Evidence for the solar wind in lunar magmas: a study of slowly-cooled samples of the Apollo 12 olivine basalt suite. Geochem. J. **51**, 95–104 (2017)

M.A. Slade, B.J. Butler, D.O. Muhleman, Mercury radar imaging: evidence for polar ice. Science **258**, 635–640 (1992)

G. Slodzian, T.D. Wu, N. Bardin, J. Duprat, C. Engrand, J.L. Guerquin-Kern, Simultaneous hydrogen and heavier element isotopic ratio images with a scanning submicron ion probe and mass resolved polyatomic ions. Microsc. Microanal. **20**(2), 577–581 (2014)

P.H. Smith, L.K. Tamppari, R.E. Arvidson, D. Bass, D. Blaney, W.V. Boynton, A. Carswell, D.C. Catling, B.C. Clark, T. Duck, E. DeJong, D. Fisher, W. Goetz, H.P. Gunnlaugsson, M.H. Hecht, V. Hipkin, J. Hoffman, S.F. Hviid, H.U. Keller, S.P. Kounaves, C.F. Lange, M.T. Lemmon, M.B. Madsen, W.J. Markiewicz, J. Marshall, C.P. McKay, M.T. Mellon, D.W. Ming, R.V. Morris, W.T. Pike, N. Renno, U. Staufer, C. Stoker, P. Taylor, J.A. Whiteway, A.P. Zent, H_2O at the Phoenix landing site. Science **325**, 58–61 (2009). https://doi.org/10.1126/science.1172339

D.E. Smith, M.T. Zuber, R.J. Phillips, S.C. Solomon, S.A. Hauck, F.G. Lemoine, E. Mazarico, G.A. Neumann, S.J. Peale, J.-L. Margot, C.L. Johnson, M.H. Torrence, M.E. Perry, D.D. Rowlands, S. Goossens, J.W. Head, A.H. Taylor, Gravity field and internal structure of Mercury from MESSENGER. Science **336**, 214–217 (2012)

S.C. Solomon, R.L. McNutt Jr., R.E. Gold, M.H. Acuna, D.N. Baker, W.V. Boynton, C.R. Chapman, A.F. Cheng, G. Gloeckler, J.W. Head, S.M. Krimigis, W.E. McClintock, S.L. Murchie, S.J. Peale, R.J. Phillips, M.S. Robinson, J.A. Slavin, D.E. Smith, R.G. Strom, J.I. Trombka, M.T. Zuber, The MESSENGER mission to Mercury: scientific objectives and implementation. Planet. Space Sci. **49**, 1445–1465 (2001)

C.P. Sonett, D.S. Colburn, P. Dyal, C.W. Parkin, B.F. Smith, G. Schubert, K. Schwartz, Lunar electrical conductivity profile. Nature **230**, 359–362 (1971)

D.J. Stevenson, Origin of the Moon—the collision hypothesis. Annu. Rev. Earth Planet. Sci. **15**, 271–315 (1987)

K.R. Stockstill-Cahill, T.J. McCoy, L.R. Nittler, S.Z. Weider, S.A. Hauck II, Magnesium-rich crustal compositions on Mercury: implications for magmatism from petrologic modeling. J. Geophys. Res. **117**, E00L15 (2012)

N. Sugiura, H. Hoshino, Hydrogen-isotopic compositions in Allan Hills 84001 and the evolution of the martian atmosphere. Meteorit. Planet. Sci. **35**, 373–380 (2000)

R. Tartèse, M. Anand, Late delivery of chondritic hydrogen into the lunar mantle: insights from mare basalts. Earth Planet. Sci. Lett. **361**, 480–486 (2013)

R. Tartèse, M. Anand, J.J. Barnes, N.A. Starkey, I.A. Franchi, Y. Sano, The abundance, distribution, and isotopic composition of hydrogen in the Moon as revealed by basaltic lunar samples: implications for the volatile inventory of the Moon. Geochim. Cosmochim. Acta **122**, 58–74 (2013)

R. Tartèse, M. Anand, F.M. McCubbin, S.M. Elardo, C.K. Shearer, I.A. Franchi, Apatite in lunar KREEP basalts: the missing link to understanding the H isotope systematics of the Moon. Geology **42**, 363–366 (2014a)

R. Tartèse, M. Anand, K.H. Joy, I.A. Franchi, H and Cl isotope systematics of apatite in brecciated lunar meteorites Northwest Africa 4472, Northwest Africa 773, Sayh al Uhaymir 169, and Kalahari 009. Meteorit. Planet. Sci. **49**, 2266–22894 (2014b). https://doi.org/10.1111/maps.12398

B.E. Taylor, Magmatic volatiles: isotopic variation of C, H, and S. Rev. Mineral. Geochem. **16**(1), 185–225 (1986). *Stable Isotopes in High Temperature Geological Processes*, ed. by J.W. Valley, H.P. Taylor, Jr., J.R. O'Neil (MSA, Blacksburg, 1986)
G.J. Taylor, The bulk composition of Mars. Chem. Erde **73**, 401–420 (2013)
R.J. Thomas, D.A. Rothery, S.J. Conway, M. Anand, Long-lived explosive volcanism on Mercury. Geophys. Res. Lett. **41**, 6084–6092 (2014)
R.J. Thomas, D.A. Rothery, S.J. Conway, M. Anand, Explosive volcanism in complex impact craters on Mercury and the Moon: influence of tectonic regime on depth of magmatic intrusion. Earth Planet. Sci. Lett. **431**, 164–172 (2015)
A. Treiman, E. Harrington, V. Sharpton, Venus' radar bright highlands: different signatures and materials on Ovda Regio and on Maxwell Montes. Icarus **280**, 172–182 (2016a)
A.H. Treiman, J.W. Boyce, J.P. Greenwood, J.M. Eiler, J. Gross, Y. Guan, C. Ma, E.M. Stolper, D-poor hydrogen in lunar mare basalts assimilated from lunar regolith. Am. Mineral. **101**, 1596–1603 (2016b)
J. Tuff, J. Wade, B. Wood, Volcanism on Mars controlled by early oxidation of the upper mantle. Nature **498**, 342 (2013)
G. Ustunisik, H. Nekvasil, D.H. Lindsley, F.M. McCubbin, Degassing pathways of Cl-, F-, H-, and S-bearing magmas near the lunar surface: implications for the composition and Cl isotopic values of lunar apatite. Am. Mineral. **100**(8–9), 1717–1727 (2015)
T. Usui, Hydrogen reservoirs in Mars as revealed by martian meteorites, in *Volatiles in the Martian Crust*, ed. by J. Filiberto, S.S. Schwenzer (Elsevier, Amsterdam, 2018)
T. Usui, C.M.O'D. Alexander, J. Wang, J.I. Simon, J.H. Jones, Origin of water and mantle-crust interactions on Mars inferred from hydrogen isotopes and volatile element abundances of olivine-hosted melt inclusions of primitive shergottites. Earth Planet. Sci. Lett. **357–358**, 119–129 (2012)
T. Usui, C.M.O'D. Alexander, J. Wang, J.I. Simon, J.H. Jones, Meteoritic evidence for a previously unrecognized hydrogen reservoir on Mars. Earth Planet. Sci. Lett. **410**, 140–151 (2015)
K.E. Vander Kaaden, F.M. McCubbin, Exotic crust formation on Mercury: consequences of a shallow, FeO-poor mantle. J. Geophys. Res. **120**, 195–209 (2015)
K.E. Vander Kaaden, F.M. McCubbin, The origin of boninites on Mercury: an experimental study of the northern volcanic plains lavas. Geochim. Cosmochim. Acta **173**, 246–263 (2016)
K.E. Vander Kaaden, F.M. McCubbin, D.K. Ross, J.F. Rapp, L.R. Danielson, L.P. Keller, K. Righter, Carbon solubility in Si-Fe-bearing metals during core formation on Mercury, in *Lunar Plan. Sci. Conf. XLVII*, abstract #1474 (2016)
K.E. Vander Kaaden, F.M. McCubbin, L.R. Nittler, P.N. Peplowski, S.Z. Weider, E.A. Frank, T.J. McCoy, Geochemistry, mineralogy, and petrology of boninitic and komatiitic rocks on the mercurian surface: insights into the mercurian mantle. Icarus **285**, 155–168 (2017)
O. Verhoeven, P. Vasher, Laboratory-based electrical conductivity at Martian mantle conditions. Planet. Space Sci. **134**, 29–35 (2016)
G.L. Villanueva, M.J. Mumma, R.E. Novak, H.U. Kaufl, P. Hartogh, T. Encrenaz, A. Tokunaga, A. Khayat, M.D. Smith, Strong water isotopic anomalies in the martian atmosphere: probing current and ancient reservoirs. Science **348**, 218–221 (2015)
M. Wadhwa, Redox conditions on small bodies, the Moon, and Mars. Rev. Mineral. Geochem. **68**, 493–510 (2008). *Oxygen in the Solar System*, ed. by G.J. MacPherson
D. Walker, J. Longhi, J. Kirkpatrick, J.F. Hays, Differentiation of an Apollo 12 picrite magma. Proc. Lunar Planet. Sci. Conf. **7**, 1365–1389 (1976)
K.J. Walsh, A. Morbidelli, S.N. Raymond, D.P. O'Brien, A.M. Mandell, A low mass for Mars from Jupiter's early gas-driven migration. Nature **475**, 206–209 (2011). https://doi.org/10.1038/nature10201
K. Wang, S.B. Jacobsen, Potassium isotopic evidence for a high-energy giant impact origin of the Moon. Nature **538**, 487 (2016)
H. Wänke, G. Dreibus, Chemical composition and accretion history of terrestrial planets. Philos. Trans. R. Soc. Lond. Ser. A, Math. Phys. Sci. **325**, 545–557 (1988)
P.H. Warren, The magma ocean concept and lunar evolution. Annu. Rev. Earth Planet. Sci. **13**, 201–240 (1985)
P.H. Warren, "New" lunar meteorites: implications for composition of the global lunar surface, of the lunar crust, and of the bulk Moon. Meteorit. Planet. Sci. **40**, 477–506 (2005)
P.H. Warren, G.W. Kallemeyn, Geochemistry of lunar meteorite Yamato-791197: comparison with ALHA81005 and other lunar samples. Mem. Natl. Inst. Polar Res., Spec. Issue **41**, 3–16 (1986). *Proceedings of the NIPR Symposium on Antarctic Meteorites (Tokyo)*
P.H. Warren, G.J. Taylor, The Moon, in *Planets, Asteroids, Comets, and the Solar System*, ed. by A.M. Davis. Treatise on Geochemistry, 2nd edn. (2014), pp. 213–250
P.H. Warren, J.T. Wasson, The origin of KREEP. Rev. Geophys. Space Phys. **17**, 73–88 (1979)

J.T. Wasson, G.W. Kallemeyn, Compositions of chondrites. Philos. Trans. R. Soc. Lond. A **325**, 535–544 (1988)

L.L. Watson, I.D. Hutcheon, S. Epstein, E.M. Stolper, Water on Mars: clues from deuterium/hydrogen and water contents of hydrous phases in SNC meteorites. Science **265**, 86–90 (1994)

C.R. Webster, P.R. Mahaffy, G.J. Flesch, P.B. Niles, J.H. Jones, L.A. Leshin, S.K. Atreya, J.C. Stern, L.E. Christensen, T. Owen, H. Franz, R.O. Pepin, A. Steele (MSL Science Team), Isotope ratios of H, C, and O in CO_2 and H_2O of the martian atmosphere. Science **341**, 260–263 (2013)

S.Z. Weider, L.R. Nittler, R.D. Starr, T.J. McCoy, K.R. Stockstill-Cahill, P.K. Byrne, B.W. Denevi, J.W. Head, S.C. Solomon, Chemical heterogeneity on Mercury's surface revealed by the MESSENGER X-Ray Spectrometer. J. Geophys. Res. **117**, E00L05 (2012). https://doi.org/10.1029/2012JE004153

J.G. Williams, D.H. Boggs, C.F. Yoder, J.T. Ratcliff, J.O. Dickey, Lunar rotational dissipation in solid body and molten core. J. Geophys. Res. **106**, 27933–27968 (2001)

B.J. Wood, A.N. Halliday, M. Rehkamper, Volatile accretion history of the Earth. Nature **467**, E6–E7 (2010)

R.D. Wordsworth, The climate of early Mars. Annu. Rev. Earth Planet. Sci. **44**, 381–408 (2016)

Y. Xu, T.J. Shankland, A.G. Duba, Pressure effect on electrical conductivity of mantle olivine. Phys. Earth Planet. Inter. **118**, 149–161 (2000)

T. Yoshino, A. Shimojuku, S. Shan, X. Guo, D. Yamazaki, E. Ito, Y. Higo, K. Funakoshi, Effect of temperature, pressure and iron content on the electrical conductivity of olivine and its high-pressure polymorphs. J. Geophys. Res. **117**, B08205 (2012). https://doi.org/10.1029/2011JB008774

E.D. Young, Evaporating planetesimals. Nature **549**, 461–462 (2017)

H. Yurimoto, K. Nagashima, T. Kunihiro, High precision isotope micro-imaging of materials. Appl. Surf. Sci. **203–204**, 793–797 (2003)

H. Yurimoto, S. Itoh, M. Zolensky, M. Kusakabe, A. Karen, R. Bodnar, Isotopic compositions of asteroidal liquid water trapped in fluid inclusions of chondrites. Geochem. J. **48**, 549–560 (2014)

M.Y. Zolotov, A.L. Sprague, L.R. Nittler, S.Z. Weider, R.D. Starr, L.G. Evans, W.V. Boynton, J.O. Goldstein, S.A. Hauck, S.C. Solomon, Implications of the MESSENGER discovery of high sulfur abundance on the surface of Mercury. EOS (Trans., Am. Geophys. Union) Abstract #P41A-1584 (2011)

M.Y. Zolotov, A.L. Sprague, S.A. Hauck, L.R. Nittler, S.C. Solomon, S.Z. Weider, The redox state, FeO content, and origin of sulfur-rich magmas on Mercury. J. Geophys. Res. **118**, 138–146 (2013). https://doi.org/10.1029/2012JE004274

Space Sci Rev (2017) 212:835–875
DOI 10.1007/s11214-017-0407-z

Water and Volatiles in the Outer Solar System

O. Grasset[1] · J. Castillo-Rogez[2] · T. Guillot[3] · L.N. Fletcher[4] · F. Tosi[5]

Received: 17 November 2016 / Accepted: 4 August 2017 / Published online: 29 August 2017

Abstract Space exploration and ground-based observations have provided outstanding evidence of the diversity and the complexity of the outer solar system. This work presents our current understanding of the nature and distribution of water and water-rich materials from the water snow line to the Kuiper Belt. This synthesis is timely, since a thorough exploration of at least one object in each region of the outer solar system has now been achieved. Next steps, starting with the Juno mission now in orbit around Jupiter, will be more focused on understanding the processes at work than on describing the general characteristics of each giant planet systems.

The Delivery of Water to Protoplanets, Planets and Satellites
Edited by Michel Blanc, Alessandro Morbidelli, Yann Alibert, Lindy Elkins-Tanton, Paul Estrada, Keiko Hamano, Helmut Lammer, Sean Raymond and Maria Schönbächler

✉ O. Grasset
Olivier.Grasset@univ-nantes.fr

J. Castillo-Rogez
julie.c.castillo@jpl.nasa.gov

T. Guillot
tristan.guillot@oca.eu

L.N. Fletcher
leigh.fletcher@leicester.ac.uk

F. Tosi
federico.tosi@iaps.inaf.it

1 Laboratoire de Planétologie et Géodynamique, LPG-Nantes UMR 6112, 2 rue de la Houssinière, 44322 Nantes Cedex 03, France

2 Jet Propulsion Laboratory, California Institute of Technology, 4800 Oak Grove Drive, Pasadena, CA 91109, USA

3 Observatoire de la Côte d'Azur, Bd de l'Observatoire, 06304 Nice Cedex 4, France

4 Department of Physics and Astronomy, University of Leicester, University Road, Leicester, LE1 7RH, UK

5 INAF-IAPS Istituto di Astrofisica e Planetologia Spaziali, Via del Fosso del Cavaliere, 100, Roma 00133, Italy

This review is organized in three parts. First, the nature and the distribution of water and volatiles in giant and intermediary planets are described from their inner core to their outer envelopes. A special focus is given to Jupiter and Saturn, which are much better understood than the two ice giants (Uranus and Neptune) thanks to the Galileo and Cassini missions. Second, the icy moons will be discussed. Space missions and ground-based observations have revealed the variety of icy surfaces in the outer system. While Europa, Enceladus, and maybe Titan present past or even active tectonic and volcanic activities, many other moons have been dead worlds for more than 3 billion years. Ice compositions found at these bodies are also complex and it is now commonly admitted that icy surfaces are never composed of pure ices. A detailed review of the distribution of non-ice materials on the surfaces and in the tenuous atmospheres of the moons is proposed, followed by a more focused discussion on the nature and the characteristics of the liquid layers trapped below the cold icy crusts that have been suggested in the icy Galilean moons, and in Enceladus, Dione, and Titan at Saturn. Finally, the recent observations collected by Dawn at Ceres and New Horizons at Pluto, as well as the state of knowledge of other transneptunian objects, are summarized, and complete this overview of the nature and distribution of ice-rich material in the outer solar system.

Keywords Water and volatiles · Giant and Intermediate planets · Giant moons and small bodies · Chemistry and Physics

1 Introduction

The outer solar system hosts a significant amount of water and volatiles, about half of the total mass of material once hydrogen and helium have been excluded, mostly located in the four outer planets. Beyond the “snow line” or “frost line” located at roughly 2.7 AU from the Sun, where radiation equilibrium temperature drops below $\sim$150 K, water ice is stable over geologic timescales on the surface of airless bodies. The dwarf planet Ceres, the most massive object in the main asteroid belt, is located at $\sim$2.77 AU from the Sun and has a low bulk density of 2.16 $\mathrm{g\,cm^{-3}}$ that reveals an icy mantle, while water ice was discovered only in a few specific places at the surface (Combe et al. 2016, 2017). At larger heliocentric distances, water ice is a major constituent of the surface and interior of the satellites revolving around the outer planets, and even more it dominates the bulk composition of the Kuiper Belt Objects (KBO). The abundance of water in the deep interior of both giant and intermediary planets is also ascertained, even if its amount is still poorly constrained. In the following sections, it is not only the water molecule but also the volatiles and the organic matter that will be studied for each planetary object in the outer solar system. Water ices can exist in several solid phases depending on temperature and pressure conditions (crystalline versus amorphous, low or high pressure polymorphs from the surface to the deep interiors of planetary objects), but also in the liquid state at least in a few icy moons of Jupiter and Saturn, and finally in the gaseous state especially in the exospheres of the small bodies and the deep atmospheres of the giant planets. In addition, the water ice is rarely found in a pure state. Water is indeed easily mixed with other species, either in a hydrated form or in a clathrate-hydrate structure. The pressure and temperature conditions, the nature of the volatiles as well as their abundance with respect to the water compound, influence the nature of the water phases that can be found. This diversity of components and phases will be explored from the asteroid belt to the Kuiper belt. For each domain, a discussion based on the most recent discoveries issued from recent space missions has been chosen.

Even if our current knowledge of giant and intermediary planets is subject to numerous uncertainties, especially in term of atmospheric/interior structure and compositions, previous space missions and ground-based observations have provided sufficient information to understand their main characteristics. This will be discussed in Sect. 2 first by looking at the deep interior of these planets (Sect. 2.1), and second by exploring the composition and the dynamics of their outer envelopes (Sect. 2.2). Although volatile species are only found in trace quantities within the gaseous atmospheres of the four outer planets, their abundance and spatial distribution play a vital role in shaping the dynamics, clouds, and general circulation of the visible atmosphere, from the churning cloud decks of the troposphere through to the stable middle atmosphere. Spectroscopic measurements, combined with in situ sampling by the Galileo probe at Jupiter, have revealed that the most cosmogenically-abundant elements (carbon, oxygen, nitrogen, sulphur) are present within the gas giants Jupiter and Saturn in their reduced, hydrogenated forms (CH_4, H_2O, NH_3, H_2S). To date, the colder atmospheric temperatures of Uranus and Neptune have prevented the direct detection of all but CH_4 gas on the ice giants, although this is found to be greatly enhanced (by 30–50 times) over the solar composition. By comparison to their gas giant cousins, we might expect the other chemical species to be similarly enriched in the ice giants (Lunine 1993), and this is borne out by microwave and radio observations.

In Sect. 3, the icy moons of the outer planetary systems will be explored. The flybys of the Pioneer and Voyager space missions have revealed the diversity of icy moon surfaces, suggesting that each moon may have had a unique history. This complexity and variety of features have been highlighted since then by the Galileo mission at Jupiter, the Cassini-Huygens mission at Saturn, Dawn at Ceres, and finally New Horizons at Pluto and beyond. It is beyond the scope of this paper to describe the numerous patterns of all the icy moons in detail—the giant planets possess a very large number of moons (> 67, > 52, > 27, > 14 for Jupiter, Saturn, Uranus, and Neptune, respectively), each of them being unique. Detailed descriptions can be found in the literature if not for all but for most of them, e.g., in Bagenal et al. (2006) for the jovian moons and in Dougherty et al. (2009) for the saturnian moons. In this paper, the focus will be first on the nature, the origin, and the distribution of the volatiles detected both in the tenuous atmospheres and on the surfaces of the moons. Second, we will be focused on the deep liquid reservoirs existing in a few moons only. The large amount of water available in the outer solar system implies that the hydrosphere can be extremely thick in differentiated bodies, namely several hundreds of kilometers in the case of the largest moons. In the Pioneer and Voyager era, it was commonly admitted that due to the very cold environment and to the vigorous convection within icy layers, icy satellites had to be entirely solid (Consolmagno and Lewis 1978; Reynolds and Cassen 1979). But it was also suggested that these bodies could host a deep ocean (Cassen et al. 1982). It turned out, thanks to the observations returned by the Galileo and Cassini missions, that some moons could indeed preserve deep liquid reservoirs until present. Over the past two decades many studies have explored the properties and evolution of icy satellite shells. These are driven on one side by the dynamical processes (especially tidal forcing—Hussmann et al. 2010) and on the other side by the amount and nature of the volatiles compounds trapped in the moons (Fortes and Choukroun 2010). One may have the feeling that deep liquid reservoirs are indeed very common and expected in all midsized and large icy moons. This idea was explored in the past (see for example Hussmann et al. 2006), especially knowing that it is indeed the case for at least one of the midsized moons, Enceladus (Cadek et al. 2016) and suspected in Dione (Beuthe et al. 2016). Nonetheless, there is no way to ascertain such a hypothesis at the present time in absence of geophysical observations at most of these bodies. The only moons, for which a liquid layer is envisaged (in terms of expected composition, depth and thickness, and habitability potential), are the following: Europa, Ganymede, and Callisto at Jupiter; Titan and Enceladus at Saturn. To our knowledge, those are the moons for which significant evidence of the existence of liquid layers has been ascertained by geophysical measurements.

The last section will be devoted to the special cases of dwarf planets found in or believed to have originated from the outer solar system. This paper will be focused on the cases where significant and recent progresses have been made thanks to the Dawn and New Horizon missions.

2 Giant and Intermediate planets

First, the deeper parts of the planets will be described. Second, the outer envelopes, on which many more constraints and measurements are available, will be characterized. Our goal is not to provide an exhaustive and complete description of all the layers of each planet, but rather to provide a state-of-the-art discussion focused on the distribution, the origins, and the transfer processes of water and volatiles trapped in these planets. Two space missions, Cassini orbiting Saturn since 2004, and Juno orbiting Jupiter since mid-2016, provide further important data to constrain the deep interiors of the giant planets. Both use polar eccentric orbits with periapses located extremely close to the planets' atmosphere in order to make high accuracy determination of their gravity fields. In addition, Juno will probe the deep atmosphere down to about 100 bars using microwave antennas, measuring ammonia abundances and providing constraints on the deep abundance of water in the planet. This will be extremely important to link measurements made in the atmosphere (see Sect. 2.2) to those inferred for the deep interior (see Sect. 2.1) and to really constrain the amount of heavy elements and ices present in the planets, both in their central cores and in their envelopes.

2.1 Ices and Volatiles in Giant Planet Interiors

In this section, the composition and the deep internal structure of the four outer planets are described, with a special focus on the amount and distribution of "ices"[1] within the cores and the inner envelopes. A tentative description of the origin and evolution of these deep layers, especially regarding the distribution of the ices, is also provided.

2.1.1 Jupiter and Saturn

Jupiter and Saturn are composed primarily of hydrogen and helium but they most likely contain more volatiles than all the other objects in the solar system added together, the Sun excluded. Jupiter is thought to contain between 10 and 40 Earth masses in heavy elements (i.e., all elements except hydrogen and helium) and Saturn between 20 and 30 Earth masses (Guillot 2005; Nettelmann et al. 2012; Helled and Guillot 2013; Hubbard and Militzer 2016; Miguel et al. 2016). Even larger quantities of heavy elements may be present in the interior if the structure departs significantly from an adiabat (Leconte and Chabrier 2012). We do not know what fraction of these heavy elements are in the form of "ices": The problem is highly degenerate and because all heavy elements are significantly denser than hydrogen and helium, their contribution in terms of global density is very similar. It is natural and convenient to assume that the respective abundance ratio between heavy elements remain solar. The standard mix of "ices" (including H_2O, CH_4 and NH_3 ices) and "rocky material"

[1] In this section, the term "ices" stands for H_2O, CO, NH_3 (and possibly CO and CO_2) because of their condensation in the form of solid ices at low temperature in the protosolar disk. However, these compounds become fluid in the high-pressure and high-temperature interiors of giant planets.

(including silicates and iron, called silicate-rich hereafter) obtained from the condensation of a solar composition gas at the distance of Jupiter is 2/3 ices and 1/3 silicate-rich (Lodders 2003). This implies that between 20 and 50 Earth masses of ices could be hidden inside Jupiter and Saturn.

These heavy elements (and ices) are believed to be partly hidden as a central core and partly distributed in the envelopes of the planets. Constraints from gravitational moments indicate that the mass of the central core is less than about 20 M• in both Jupiter and Saturn with values which depend significantly on the equations of state of hydrogen used (Nettelmann et al. 2012; Militzer and Hubbard 2013; Hubbard and Militzer 2016; Miguel et al. 2016) and on the location at which helium separates out from hydrogen (Helled and Guillot 2013).

This central core is an important ingredient for the formation of a giant planet: The accretion scenario shows that a giant planet can form by the collapse of a hydrogen-helium envelope around a high-density core estimated to be around 10 $M_\oplus$ (e.g., Mizuno 1980; Pollack et al. 1996; Wuchterl et al. 2000; Alibert et al. 2005). Another class of solution invokes a direct instability of the gas disk followed by the capture of heavy elements (e.g., Boley et al. 2010; Nayakshin et al. 2014). In all cases, the presence of an abundant source of solids in the outer solar system in the form of dust grains and pebbles appears to be essential for the formation of giant planets as it helps building a core quickly before the removal of the gas disk (e.g., Lambrechts and Johansen 2012; Guillot and Gautier 2015; Levison et al. 2015).

It is presently unclear how to account for the large masses in heavy elements in the envelopes of Jupiter and Saturn and for the enrichment of their atmospheres (see section hereafter). This must be done early, the giant planets becoming rapidly very efficient scatterers of planetesimals (Matter et al. 2009). One possibility is to invoke an efficient capture of planetesimals during the growth of the envelope phase (e.g., Alibert et al. 2005). Another option is that the envelopes may have been enriched by the progressive erosion of the core (Guillot et al. 2004) or generally by upward mixing of material present in the lower envelope (Vazan et al. 2016). Finally the protosolar disk itself may have been enriched in heavy elements by selective photo-evaporation at the time of the formation of the giant planets (Guillot and Hueso 2006; Monga and Desch 2015).

2.1.2 Uranus and Neptune

With masses of 14 $M_\oplus$ and 17 $M_\oplus$ for densities of 1.2 g/cm^3 and 1.7 g/cm^3, respectively, Uranus and Neptune look like failed giant planets. While they seem relatively similar, Neptune's larger mean density compared to Uranus indicates a slightly different composition: either more heavy elements compared to hydrogen and helium or a larger rocky to icy materials ratio. The gravitational moments impose that the density profiles lie close to that of 'ices' (a mixture initially composed of, for example, H2O, CH4, and NH3 but that rapidly becomes an ionic fluid of uncertain chemical composition in the planetary interior), except in the outermost layers, which have a density closer to that of hydrogen and helium (Marley et al. 1995; Podolak et al. 2000). Most models of these planets assume the presence of three layers, a central core (magnesium-rich silicate and iron material), an ice layer, and a hydrogen–helium gas envelope (e.g., Helled et al. 2011; Nettelmann et al. 2013).

According to the models of Nettelmann et al. (2013), Uranus contains a minimum of 1.8–2.2 $M_\oplus$ of hydrogen and helium and Neptune 2.7–3.3 $M_\oplus$. The global ice to silicate-rich ratio that is derived is very high (19–36) in Uranus, while Neptune has a wide range of solutions from 3.6 to 14. These values are much larger than the canonical ice to silicate-rich ratio of 2–3 for the protosun that accounts for the abundances of all elements condensing at low temperatures ('ices') versus that of more refractory elements ('silicate-rich'). The fact that either planet would have accreted much less silicate-rich phases than ices is puzzling and unexplained by formation models. It is probably an artefact from assuming that ices are confined to the envelope and silicates and iron to the core.

In fact, it is likely that all present models of Uranus and Neptune are inadequate because of the assumption of an adiabatic temperature structure across interfaces with different compositions. The high abundance of methane is shown to inhibit convection in its cloud formation region (between 1 and 2 bar) and leads to a superadiabatic temperature gradient that was observed in Voyager radio-occultation data (Guillot 1995). If water is more than ten times solar, a similar but quantitatively stronger effect may occur, characterized by the presence of a deep radiative zone with a strong increase of the temperature with depth (Guillot 1995; Leconte et al. 2017). Still deeper, diffusive convection should occur in regions with compositional changes, also leading to a superadiabatic temperature gradient (e.g., Rosenblum et al. 2011). As in the case of Jupiter and Saturn (see Leconte and Chabrier 2012), this would lead to higher temperatures in the interior and very different constraints on the interior composition. The amount of rocky material required to fit the mean density and gravitational moments would certainly rise, potentially solving the ice to rocky material ratio problem.

2.2 Water and Volatiles in the Gaseous Envelopes

In this section, we review the present state of knowledge of the abundances of the simple molecules (CH_4, H_2O, NH_3, H_2S) and their isotopologues in the gaseous envelopes of the outer planets. We will describe their importance for cloud formation, their role in atmospheric dynamics, and isotopic ratios within these species as a constraint on planetary origins. We will also briefly review the present-day supply of oxygenated species to their upper atmospheres, via cometary impacts and influx from meteorites, satellites and rings.

2.2.1 Volatiles and Clouds

The vertical distribution of volatile species (methane, ammonia, hydrogen sulphide and water) is governed by a variety of processes, as depicted in Fig. 1. Thermochemical equilibrium modelling (e.g., Weidenschilling and Lewis 1973) provides a first estimate of the importance of condensation, whereby a cloud deck condenses to separate a well-mixed internal reservoir of the species at depth from a higher-altitude region where the volatile follows its own relative humidity curve. The altitudes of the cloud decks are governed by the abundance of the species in question, their volatility (i.e., condensation point) and the temperature structure of the environment.

Methane Methane is the most volatile species, condensing only on the ice giants, Uranus and Neptune, between the $\sim$1 bar pressure level and the tropopause, meaning that the topmost clouds are expected to consist of methane ices. On the gas giants, where temperatures never get this cold, methane is expected to be well-mixed everywhere, and only begins to decrease in abundance at high stratospheric altitudes due to the effects of photochemistry and diffusion (Moses et al. 2005). Indeed, the presence of methane in the upper tropospheres and stratospheres of all four giants has two notable effects. Firstly, the stratospheres are replete with hydrocarbon species derived from the photolysis of CH_4 (see Moses et al. 2005 for a review). Secondly, the abundance of methane has been used as an estimate of the bulk carbon enrichment of all four giants, increasing from $\sim 5\times$ enrichment for Jupiter

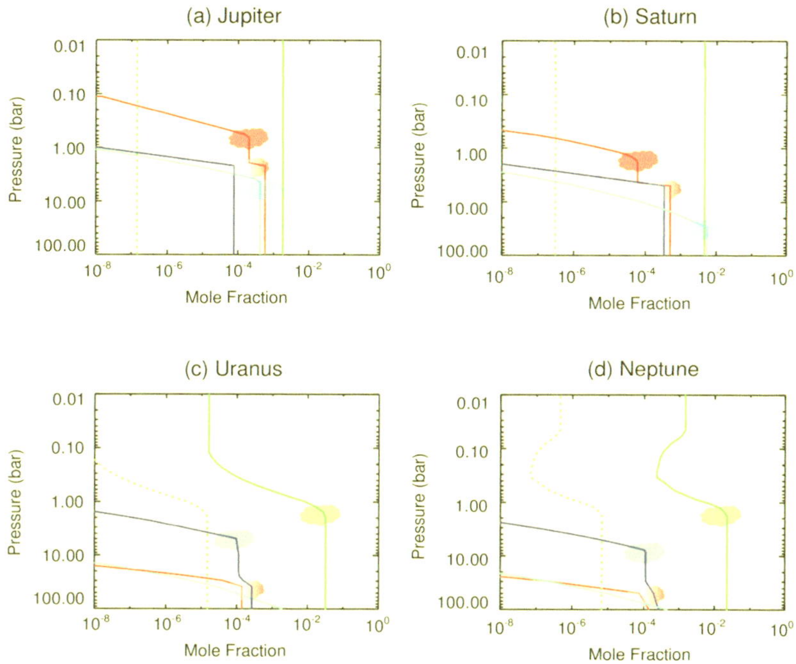

Fig. 1 Observed and modeled vertical distributions of volatiles on all four giant planets, including CH_4 (*green*), NH_3 (*red*), H_2S (*blue*), H_2O (*cyan*) and CH_3D (*dotted green*). Approximate locations of condensate clouds are also shown (CH_4 ice in *green*, NH_3 ice in *red*, H_2 S in *blue*, NH_4 SH as a hybrid of *red/blue*, and H_2O ice in *cyan*). CH_4 distributions are from Galileo (Jupiter), Cassini (Saturn), Spitzer (Uranus) and AKARI (Neptune) analyses. NH_3, H_2S and H_2O are from combinations of direct observation (Galileo), remote sensing of the $p < 2$ bar range (Jupiter and Saturn), and cloud condensation modelling following Irwin et al. (2009)—this allows us to see the interplay between NH_3 and H_2S in the top two cloud layers, where cloud condensation causes discrete steps in the profiles. The true vertical distributions of these species for $p > 2$ bar remain to be properly quantified by measurements (see main text)

(Wong et al. 2004a, 2004b), to ∼10× enrichment for Saturn (Fletcher et al. 2009), to ∼90× enrichment for Uranus (Sromovsky et al. 2011) and Neptune (Karkoschka and Tomasko 2011). Given that Jupiter showed uniform enrichments of all elements by 2–5 over protosolar composition (excluding oxygen and neon), we might expect similar enrichments on all the giants, with a greater proportion of heavy elements in the ice giants than the gas giants due to their increased core-to-envelope ratios (Owen and Encrenaz 2003). We note, however, that methane gas appears to vary with latitude on the ice giants (Sromovsky et al. 2011; Karkoschka and Tomasko 2011), being more abundant at low latitudes than the poles, highlighting the importance of atmospheric dynamics in governing the abundances of these species.

Ammonia and Hydrogen Sulphide Below the methane clouds, the condensation levels of ammonia (NH_3) and hydrogen sulphide (H_2S) is made more complex by a two-component reaction that produces a solid-phase cloud of NH_4SH (ammonium hydrosulphide), which either serves to consume all of the NH_3 or all of the H_2S, depending on which species is more abundant (i.e., depending on the N/S ratio). If the giant planets showed a steady enrichment of the solar composition, then NH_3 should always be more abundant than H_2S (N/S > 1), so that there would be NH_3 left over to form the topmost condensation cloud. However, this has not been observationally verified on all four giants. On Jupiter, the in situ measurements of Galileo demonstrated that the NH_3 and H_2S abundances were both enriched over the solar composition in the 8–12 bar range, at least within the descent site in the Northern Equatorial Belt. NH_3 was found to be more enriched than H_2S (N/S $>$ solar), implying that all of the H_2S would be consumed by the combination reaction. Recent microwave mapping (de Pater et al. 2016) has confirmed that NH_3 enrichments are present in some locations on Jupiter, but that this is not the case over the whole planet. Jupiter's topmost cloud, near 800 mbar and $\sim$150 K, should be composed of NH_3 ice, with a deeper cloud of NH_4SH solid near $\sim$2 bar and 210 K (West et al. 2004). Indeed, NH_3 ice has been spectroscopically identified in regions of powerful cloud activity on Jupiter (Baines et al. 2002; Wong et al. 2004b)

Unfortunately, the bulk abundances of NH_3 and H_2S remain poorly understood on the other giants for two reasons—the lack of in situ measurement, and the colder temperatures pushing these species to deeper and more inaccessible pressures. Infrared (IR) observations, such as those from Cassini, ISO, Voyager and Spitzer, are sensitive to ammonia absorption in the upper tropospheres, above the main cloud decks in the region where NH_3 is declining with height. NH_3 dominates the 5-, 10- and 100-μm regions in Jupiter's spectrum (Fouchet et al. 2000; Achterberg et al. 2006) and has a clear influence on Saturn (de Graauw et al. 1997), but NH_3 estimates from these observations are only lower limits on the bulk NH_3 content. No NH_3 signatures have ever been identified on Uranus and Neptune in this range, and no infrared H_2S gas signatures have ever been observed. Microwave observations hold more promise, sensing the deeper troposphere by utilizing a broad NH_3 absorption band near 1-cm wavelength to sense below the clouds. Juno's Microwave Radiometer (MWR), which successfully began operations at Jupiter in 2016, will be the first to map this absorption from a planetary spacecraft, but all other inferences come from challenging ground-based microwave observations. H_2S, in particular, has only a broad-band influence over the microwave spectrum, so the abundance remains highly uncertain. On Saturn, the only reported measurement (Briggs and Sackett 1989) suggests a large enrichment of H_2S compared to NH_3 (de Pater and Massie 1985; Janssen et al. 2013), suggesting N/S $<$ solar. This controversial result is rarely incorporated into cloud-equilibrium modelling—either Saturn's abundance of NH_3 is expected to be higher, or the abundance of H_2S is expected to be lower. Most authors assume a Jupiter-like vertical structure, with NH_3 ice overlaying a cloud of solid NH_4SH. Recent spectroscopic observations during Saturn's springtime storm (Sromovsky et al. 2013) were consistent with the presence of NH_3 ice.

The situation is even more challenging for Uranus and Neptune, where no NH_3 or H_2S signatures have been detected in the infrared. Microwave observations have suggested that Uranus' ammonia is strongly depleted in the 10–20 bar range with respect to solar (de Pater and Massie 1985), whereas H_2S is much more enriched (by factors of 10–35, de Pater et al. 1991). Microwave observations of Neptune tell the same story (de Pater and Massie 1985; de Pater et al. 1991). The ammonia measurements on both planets could be upper limits, if all the NH_3 were consumed by the formation of the NH_4SH cloud near 40–50 bar, or if substantial amounts of NH_3 were incorporated into an aqueous NH_3-H_2O cloud at greater depths. Alternatively, the ice giants could genuinely be nitrogen-poor when compared to the gas giants, due to the different source regions for the icy materials from which they accreted. Or maybe ice giant nitrogen exists mainly as N_2, a disequilibrium species in the observable atmospheres that is difficult to detect spectroscopically. In any case, the N/S < 1 implied by these measurements suggests that the ice giants differ from the gas giants in that an H_2S cloud forms near 4–8 bar above the NH_4SH solid at 40–50 bar, rather than having NH_3 ice as the topmost cloud (as shown in Fig. 1).

Water Below the carbon, nitrogen and sulphur-based clouds, the water cloud is the most challenging to observe, and yet the most important for constraining theories of atmospheric dynamics and planetary origins. The condensation of the water ice cloud, near 5 bar on Jupiter, 10 bar on Saturn, and hundreds of bars on the ice giants, loosely defines the base of the weather layer and separates it from the deep interior. Identifying this cloud in remote sensing data is challenging, given the wide range of gaseous and aerosol absorptions overlying this cloud deck, and only on Jupiter has a putative cloud been identified at the right altitude to be composed of H_2O ice (Nixon et al. 2001; Banfield et al. 1998; Bjoraker et al. 2015). Lighting activity, related to charge-separation in the water ice cloud, has also been observed directly on Jupiter (Little et al. 1999; Gierasch et al. 2000). Spectroscopically-identifiable water ice has only ever observed in Voyager data near 44 μm (Simon-Miller et al. 2000), and is not widespread. Gaseous water, residing above the cloud deck and following a saturated vapor pressure curve, influences the 5-μm spectrum of Jupiter (measured from Galileo and Cassini data by Roos-Serote et al. 2000; Giles et al. 2015). However, precise estimates of its abundance and vertical distribution are extremely challenging. Given that the Galileo probe measurements showed that the water abundance was still increasing, even when they ceased near 22 bar (Niemann et al. 1998; Wong et al. 2004a, 2004b), the question of Jupiter's deep water abundance remains an unresolved problem. Juno's Microwave Radiometer aims to determine both the deep H_2O and NH_3 abundances to 100 bar of pressure.

For Saturn, the powerful 2010 storm system featured spectroscopic signatures that could be reproduced by a mixture of water ice, ammonia ice and NH_4SH (Sromovsky et al. 2013), although this depends on the assumed optical properties of the particles. Upper tropospheric signatures of gaseous water have been observed by ISO at 5 μm (de Graauw et al. 1997), but Cassini lacked the spectral resolution to map these across the planet, and ground-based observations must deal with considerable contamination from terrestrial water. Given the challenge of direct measurements of Saturn's water (Mousis et al. 2014), the abundances of disequilibrium species (CO, PH_3, etc.) that react chemically with water can be used to infer the abundance indirectly (Visscher et al. 2010). Although these typically predict a supersolar water enrichment, there remains huge uncertainties in the water abundance due to the need to assume a vertical mixing rate in the calculations and the criticality of the choice of reaction network.

Despite being known as ice giants, the water on Uranus and Neptune is even more inaccessible, with liquid and ice clouds forming at hundreds of bars of pressure, and no detections of gaseous or solid water to date (with the exception of exogenic sources, see below). Indirect estimates based on the CO abundance (Lodders and Fegley 1994; Marten et al. 1993; Cavalié et al. 2014) suggest that the O/H enrichment of the ice giants may reach factors of several hundred. Microwave observations of Uranus suggested a deep absorbing layer that could have been related to a water cloud (Hofstadter and Muhleman 1989; Hofstadter and Butler 2003), whereas analysis of Neptune microwave data by de Pater et al. (1991) simply assumed a water abundance. Indeed, even the microwave spectra have a limited sensitivity to this deep water, but assessing its abundance should be a primary goal of any future ice giant mission.

2.2.2 *Spatial Distribution and Atmospheric Dynamics*

Although a detailed review of giant planet dynamics is beyond the scope of this book (see Ingersoll et al. 2004; Del Genio et al. 2009 for comprehensive reviews), we note that the distribution and condensation of volatiles plays a crucial role in shaping the general circulation and 'weather' of these giant planet atmospheres. Each condensate cloud forms a molecular weight gradient in the planet's density structure, serving to inhibit convective motions through that layer (e.g., Stoker 1986; Sugiyama et al. 2014). They can thus serve to isolate and stabilize different vertical regions, permitting circulation cells with differing directions between the cloud layers. However, strong vertical motions can overcome these barriers, resulting in powerful convective activity being witnessed in the topmost cloud decks, from billowing storm activity and their associated lightning on the gas giants (e.g., Ingersoll and Vasavada 2000; Gierasch et al. 2000; Sanchez-Lavega et al. 2011; Fischer et al. 2011; Fletcher et al. 2011a, 2011b) to large storm eruptions on the ice giants (e.g., de Pater et al. 2015). These storms are superimposed onto an intermediate-scale banded structure (with convective storm activity preferentially occurring in cyclonic, cloud-free belts) and, possibly, a large scale circulation pattern with air moving from the seasonally-dependant sub-solar point to higher latitudes (e.g., Conrath et al. 1990). Vertical and horizontal motions associated with these circulation patterns will affect atmospheric temperatures, cloud formation, and the abundances of these species, potentially explaining why ices of ammonia and water have only been identified in regions of powerful storm activity, and why methane exhibits strong equator-to-pole contrasts on the ice giants.

This energy cascade from the smallest to the largest scales (or vice versa) is a topic of active research, but the condensation of volatiles—primarily water—provides the energy source through latent heating to drive the convective motions. Once a parcel of air is forced upwards, the water condensation will provide energy that makes the parcel buoyant, able to penetrate through the overlying stable layers until it reaches the stably-stratified upper troposphere. Here the divergent air cools, condensates begin to precipitate out as rain or snow, to re-evaporate again in the deep atmosphere. Energy can penetrate higher into the stratosphere in the form of waves spawned by this powerful convection (Fletcher et al. 2012), serving to modify the circulation of the middle-atmosphere. Giant planet volatiles are therefore directly connected to the dynamic motions shaping the ever-changing appearance of their atmospheres.

2.2.3 Bulk Abundances and Isotopic Composition

Recent reviews of the bulk elemental and isotopic compositions of the giant planets can be found in Mousis et al. (2014) and Guillot and Gautier (2015), here we summarize the key findings for the four key elements (C, N, S, and O). However, we note that the atmospheric and interior compositions may be decoupled by phase transitions, cloud formation, chemical reactions and phase separation, implying that the atmospheric composition may not be a good representation of the bulk (Guillot and Gautier 2015). Furthermore, the fact that most remote sensing measures volatile abundances above their primary condensation altitudes means that large uncertainties are involved. As described above, C/H appears to be enriched over protosolar composition by $5\times$ on Jupiter, $10\times$ on Saturn, and $90\times$ on Uranus and Neptune. N/H appears to be more enriched on Jupiter ($4\times$, as measured by the Galileo probe, Wong et al. 2004a, 2004b) than on Saturn ($3\times$, determined from Cassini remote sensing, Fletcher et al. 2009, 2011a, 2011b), although this could be a result of an inaccessible reservoir of nitrogen below the clouds. Conversely, microwave modelling by de Pater and Massie (1985) suggested a substantial sub-solar N/H on the ice giants, which may also be associated with sequestration in deeper cloud decks. Like carbon, S/H is higher on Saturn ($8\times$, based on the radio-wavelength measurements of Briggs and Sackett 1989) than on Jupiter ($3\times$, based on the Galileo probe, Wong et al. 2004a, 2004b), and substantially enriched ($10–35\times$) on both ice giants (de Pater et al. 1991). Finally, the deep abundance of oxygen is a key unknown on all four giant planets, with indirect estimates of supersolar enrichments varying by an order of magnitude, from tens to hundreds of times solar. Given that trapping of volatile species within ices may have formed the building blocks of the protoplanetary cores, this uncertainty in the O/H is a crucial problem to solve.

Measurements of isotopic ratios within these molecules, however, can provide further insights even if the absolute abundances remain unknown. This is because the isotopic fractionation is not expected to have changed over time and should be the same irrespective of the phase transitions and cloud condensation processes that may be occurring. The isotopic composition in the volatiles that are accessible to remote sensing include those in the primary atom ($^{13}C/^{12}C$, $^{15}N/^{14}N$) and in hydrogen (D/H). There have been no measurements of the isotopic composition in sulphur or deep oxygen (see below for exogenic sources). Measurements of the $^{13}C/^{12}C$ ratio are available for Jupiter (Galileo probe, Atreya et al. 2003), Saturn (Cassini infrared measurements, Fletcher et al. 2009) and for Neptune (ground-based observations, Orton et al. 1992), and all show consistency with the terrestrial value of $89.9^{+2.6}_{-2.4}$, and therefore imply that the giants accreted from the same material as the Sun and that no further isotopic fractionation has taken place.

Deuterium has been more widely measured on the giant planets, in both pure H_2 and in CH_3D, using infrared and sub-millimeter spectroscopy. The most recent gas giant values come from ISO (e.g., Lellouch et al. 2001) and Cassini (Fletcher et al. 2009), whereas the most recent ice giant values come from Herschel (Feuchtgruber et al. 2013), Spitzer (Orton et al. 2014a, 2014b), AKARI (Fletcher et al. 2010) and ground-based observations (Irwin et al. 2014). The results fall into two natural classes, with Jupiter and Saturn exhibiting approximately protosun values of D/H (1.94×10^{-5}), and Uranus and Neptune showing an enrichment of 2–3 compared to protosun values. In the case of Uranus and Neptune, this is considerably lower than what might be expected for an ice-dominated planet (assuming D/H ratios in known ice sources in the solar system, such as cometary ices), prompting Feuchtgruber et al. (2013) to postulate that these worlds are in fact dominated by rocky material, rather than ices. However, alternative sources of protoplanetary ices with lower D/H ratios than in comets, still awaiting identification, might explain this result. It is also possible, and even likely, that the ices in the interiors of Uranus and Neptune were only incompletely mixed, therefore preventing any such inference on the ice-to-rock ratio in the planetary interior. Furthermore, there are tantalizing hints that Saturn's D/H ratio is actually smaller than that of Jupiter (Fletcher et al. 2009), suggesting that the volatile reservoirs for Jupiter and Saturn were different, although this remains a topic of active study.

Further constraints on the volatile origins arise from the $^{15}N/^{14}N$ ratio measured in ammonia on Jupiter and Saturn. Molecular N_2 is the dominant form of nitrogen in molecular clouds, but requires low environmental temperatures to trap within icy cages to form protoplanets. NH_3, conversely, is easier to trap at warmer temperatures. Making use of the fact that ^{15}N is enriched in nitrogen molecules compared to N_2, the bulk $^{15}N/^{14}N$ ratio therefore reveals whether N_2 or NH_3 was the dominant carrier of nitrogen to the giant planets. Jupiter's $^{15}N/^{14}N$ has been measured in situ and remotely (Niemann et al. 1998; Owen et al. 2001; Fouchet et al. 2000, 2004; Fletcher et al. 2014) and was found to be considerably smaller than the terrestrial value, and is thought to be the same as that found in the protosun. Upper limits on Saturn's $^{15}N/^{14}N$ have been reported via ground-based spectroscopy (Fletcher et al. 2014), and found to be consistent with that of Jupiter. Both of these values are considerably smaller than the $^{15}N/^{14}N$ found in cometary molecules in the outer solar system. This suggests that molecular N_2 dominates the nitrogen inventory of both gas giants, rather than ice-trapped NH_3. This delivery could have been in the gas phase (i.e., along with the H_2 and He, although some method for enriching nitrogen must then be found) or trapped in icy planetesimals formed at cold, distant locations (and then transported inwards without vaporisation into the feeding zones of the giants). Furthermore, the Saturnian value is smaller than that of Titan (Niemann et al. 2010), suggesting different accretion reservoirs for the planet and its satellites.

2.2.4 Exogenic Sources

So far, we have discussed the vertical and horizontal distribution of volatiles that were incorporated into the planets during their formation. However, high-resolution spectroscopy over the past two decades have revealed a continuing source of volatile delivery—particularly oxygenated species—to their upper atmospheres (Feuchtgruber et al. 1997; Moses et al. 2000). This oxygen arises from a range of processes, including cometary impacts, diffusion of ring particles, debris from satellites, and the ablation of micrometeroids in the upper atmosphere. The subsequent chemistry gives rise to water, CO_2 and CO detected as fine emission lines in the infrared and sub-millimeter spectrum, and measurements of their abundances and isotopic composition (including $^{18}O/^{16}O$) tell us about the present-day materials falling into the giants, rather than the origins of volatiles during planetary formation.

These stratospheric oxygen compounds were first studied by ISO (Feuchtgruber et al. 1997). The Shoemaker-Levy 9 cometary impact on Jupiter (1994) was a source of water and CO_2 (Lellouch et al. 2002); the geysers of Enceladus may well provide water to Saturn's upper atmosphere (Hartogh et al. 2011). Spitzer has measured Uranus' CO_2 (Orton et al. 2014b), and Uranus' CO was detected in fluorescence, its vertical distribution suggesting an external source (Encrenaz et al. 2004) which was more comprehensively studied by Herschel (Cavalié et al. 2014). Neptune's CO has also been revealed to have an external source (Hesman et al. 2007; Lellouch et al. 2010; Fletcher et al. 2010), possibly related to an ancient cometary strike, or to the slow rain of dust and micrometeorites.

These oxygen compounds can affect the stratospheric and ionospheric chemistry (particularly via the photolysis of H_2O, which then leads to CO and CO_2 formation), and could lead to the formation of thin water ice clouds in the upper atmosphere (Moses et al. 2000). Future missions to Jupiter (e.g., JUICE) will make use of the distribution of this infalling watery debris to understand the properties of the stratosphere and the nature of the materials being accreted today.

2.2.5 Summary

Carbon, nitrogen, sulphur and oxygen are found in their reduced, hydrogenated forms in giant planet atmospheres, and condense (or react) to form discrete cloud decks in planetary atmospheres. The latent heat of water condensation plays a vital role in powering the dynamic activity which shapes the meteorology and climate of these worlds. Ammonia and methane ices dominate the upper clouds of the gas and ice giants, respectively, whereas the water clouds are deep and hidden from the view of remote sensing. This presents a significant challenge for assessing the chemical inventories of these worlds, as neither in situ probes nor remote sensing have penetrated the well-mixed, deep reservoirs for these species. A comparison of the bulk elemental composition and isotopic ratios on the four giant planets can be instructive in understanding the different volatile reservoirs in the proto-solar system from which the planets accreted. The bulk oxygen content is crucial, given the potential importance of volatile trapping within water ice in for formation of protoplanets in the outer solar system. Microwave spectroscopy, coupled with in situ sampling, offer some hope of determining these values in the four giant planets.

3 Giant Icy Moons

In this section, we describe our current knowledge of the surface and subsurface characteristics of most of these objects, thus providing a general overview of the context (Fig. 2). Then, a second part will focus on the largest moons for which a thick liquid layer is suspected below the icy crust (Europa, Ganymede, and Callisto at Jupiter; Titan and Enceladus at Saturn). Those are the objects strongly motivating future space missions, partly because of their high astrobiological interest. The deep layers of these moons, and more specifically the main characteristics of the liquid layers, will be discussed.

3.1 Surface and Space Environments

The goal of this section is to describe surface compositions and volatile contents in the environments of Jupiter and Saturn for which most of the data have been acquired, but also to a lesser degree around Uranus and Neptune. A special focus is first given on water ice, which is by far the most abundant volatile compound making up the composition of the outer planets' moons. The nature of the water ice phases (amorphous versus crystalline) is discussed using the space-based hyperspectral evidences. Secondly, we will review non-water-ice materials that have been detected on the surfaces of the icy moons. Finally, a third section will describe our current understanding of the composition of the atmospheres that surround the major moons.

3.1.1 Water Signatures on Solid Surfaces

Water ice, which formally is a mineral, shows diagnostic spectral signatures in the infrared range. Overall, the reflectance spectrum of water ice is temperature-dependent, and very sensitive to the average size of the grains composing the surface regolith (Clark 1981). The spectrum of water ice is dominated by a strong absorption band centered at about 3 μm ($3250\ cm^{-1}$), resulting from the spectral counterparts of two fundamental vibration modes of the water molecule (H_2O): ν_1 symmetric stretch, occurring at 2.941 μm or $3400\ cm^{-1}$, and ν_3 asymmetric stretch, occurring at 3.105 μm or $3221\ cm^{-1}$, while the fundamental bending mode ν_2 is responsible for a less intense absorption located just beyond 6 μm ($1660\ cm^{-1}$). Combinations and overtones of these fundamental modes are represented by absorption bands falling in the near-infrared domain, centered at about 1.04, 1.25, 1.50, and 2.02 μm, and whose strength increases with increasing wavelength (Fig. 3, bottom).

Amorphous and crystalline ices are spectrally distinct, and as such they can be identified by comparing their spectra. Amorphous ice forms generally at temperatures lower than ~130 K. It can also be formed from crystalline ice by radiation of high-energy particles, such as ultraviolet photons, helium ions and high-energy electrons. Conversely, crystalline ice can be formed from amorphous ice by heating above 110–120 K (e.g. Newman et al. 2008). In crystalline water ice, one additional band is found at 1.65 μm ($6061\ cm^{-1}$), on the edge of the 1.5-μm band (Fink and Larson 1975). This feature is temperature-sensitive, such that it is deeper for colder crystalline ice. In spectroscopic measurements, this makes it a suitable marker of ice temperature, although this signature is also strongly affected by the grain size (Clark 1981). A reflectance peak at 3.1 μm comes from the Fresnel reflection off the facets of the water ice grains in the surface. For this reason, the strength and shape of the 3.1 μm peak does not vary significantly with grain size, as long as the grains are larger than a few microns in size. Amorphous ice can be qualitatively identified from a generally broad and featureless 3.1 μm reflection peak, compared to a more pronounced and three-peaked

Fig. 2 Moons of the outer solar systems which are discussed in this paper. Except for Io, all these moons possess a significant amount of water and also volatiles and organics (see text for details). The *horizontal axis* indicates their distance to the planet (in million kilometers). All moons are at scale, except for the four giant moons of Jupiter, and for Titan at Saturn, which are scaled to 1/4 of their real size relative to the other moons

structure in the crystalline spectrum (Grundy and Schmitt 1998; Hansen and McCord 2004; Mastrapa et al. 2009).

Long-period, Centaur, and Jupiter Family comets at heliocentric distances beyond ~6 AU are expected to host substantial amounts of amorphous ice. As for the icy moons, the distribution of crystalline vs. amorphous water ice on the surfaces is mapped by using a combination of some spectral markers diagnostic of ice crystallinity: the strength of the (temperature-sensitive) 1.65-μm band, the position of the minimum of the 1.5-μm and 2.0-μm bands, and the intensity and shape of the 3.1-μm Fresnel reflectance peak (e.g., Mastrapa et al. 2009). In general, amorphous ice spectra are characterized by wide and relatively unstructured bands. Upon crystallization, ice bands shift to longer wavelengths, become narrower, increase in intensity, and generally reveal new structures (Schmitt et al. 1998; Cruikshank et al. 1998a, 1998b, 2000a, 2000b; Hansen and McCord 2004; Clark et al. 2012). Such maps may reveal longitudinal variations due to weathering effects as well as latitudinal gradients due to radiation and local conditions (surface albedo). The average grain sizes can be inferred from the spectral continuum profile in the range between 3 and 4 μm, with finer grains showing a higher reflectance peak in this region and vice versa.

Based on hyperspectral data returned by the *Near Infrared Mapping Spectrometer* (NIMS) on-board the Galileo spacecraft, operating in the range 0.7–5.2 μm (Carlson et al. 1992), the icy moons of Jupiter mostly show evidence of crystalline ice at the surface, with some exceptions. On Callisto, NIMS spectra revealed that ice appears crystalline everywhere (where it is abundant enough to see), as expected from the average temperature of the satellite and from the low radiolytic fluxes. On Ganymede, sputtered ice particles easily reach the cold polar regions along Ganymede's magnetic field lines and form a thin layer of frost; the surface boundary between the open and closed field lines is clearly visible in color images (Khurana et al. 2007), with material mantling the polar regions that shows a higher albedo. In addition, the interaction with the Jovian magnetosphere was found to cause spectral differences with respect to the crystallinity and particle size of H_2O ice between the equatorial and the polar regions of Ganymede, with a general trend from more crystalline ice at low latitudes to more amorphous ice in the polar regions (Hansen and McCord 2000b, 2004; Stephan 2006; Stephan et al. 2009). This is caused by the configuration of Ganymede's own magnetic field that protects at least partly surface regions between ±40° latitude from impacting magnetospheric particles (however, local conditions produce large variations in this trend). On Europa, where cold temperatures and high radiation fluxes would favor amorphous ice, the resemblance to the crystalline ice model is poor (Hansen and McCord 2000a), and although the amorphous ice model is not an exact fit, the spectrum is very sensitive to the exact morphological state of the amorphous deposit. Spectral signatures reveal an average fine grain size on Europa (< 100 μm), while larger and rough grains are found on Callisto (Roush et al. 1990; Calvin and Clark 1991). In all satellites of the Jovian system, finer grains tend to concentrate in the leading hemispheres, which is probably the result of micrometeorite gardening, combined with sputtering effects on the trailing side (Calvin and Clark 1993).

Hyperspectral images returned by the *Visual and Infrared Mapping Spectrometer* (VIMS) on-board the Cassini orbiter, covering the overall spectral range 0.35–5.1 μm (Brown et al. 2004), enabled a thorough comparative analysis of Saturn's icy moons. Disk-integrated data revealed that crystalline water ice is also dominant on those surfaces (Filacchione et al. 2010). On Enceladus, whose very high albedo drives low daytime surface temperatures, spatially-resolved maps show water ice to be mostly crystalline, with a higher degree of crystallinity at the "tiger-stripe" cracks and a larger amorphous signature between these stripes (Newman et al. 2008). These results suggest an intriguing atmospheric environment over the south pole where amorphous ice is produced either through intense radiative bombardment, flash-freezing of cryovolcanic liquid, or rapid condensation of water vapor particles on icy microspherules or on the surface of Enceladus (Newman et al. 2008). Furthermore, by considering six spectral indices diagnostic of sub-micron particles, those terrains showing the deepest water ice absorption bands were found to correlate with the lowest amount of sub-micron particles (Scipioni et al. 2017). For a surface depleted in visually dark, non-ice contaminants, like the average surface of Enceladus, this leads to the interpretation that the variation measured in the depth of the water ice absorption bands is directly related to a variation in grain size, rather than to a compositional gradient (Scipioni et al. 2017). On Rhea, both amorphous and crystalline ice are present in different amounts across the craters. While neither pure amorphous nor pure crystalline H_2O ice are widespread, crystalline ice is predominant in the center of the craters (Dalle Ore et al. 2015). A phase change from amorphous to crystalline may also occur as a result of localized heating of the ice by impacts of meteoroids in some circumstances (Stewart et al. 2008), and probably by micrometeoroids, which would explain the crystalline signature at 1.65 μm on most outer Solar System bodies (Dalle Ore et al. 2015). Typical regolith grain sizes are smaller than on Jupiter's icy Galilean satellites, ranging between 38 μm (Rhea) and 69 μm (Tethys) (Filacchione et al. 2012).

3.1.2 Nature and Distribution of Non-Water-Ice Surface Materials

The nature and distribution of non-water-ice materials, and their relation with water ice, is a crucial point to understand the origin and evolution of the surfaces of the icy satellites, as surface material can be a link to the interior of the moons and provides constraints about the environment in which these bodies were formed and currently exist.

Distribution of Oxides In the icy Galilean satellites around Jupiter, the strength of the CO_2 signature increases with increasing radial distance, moving away from Europa to Callisto. On Callisto, CO_2 is more abundant on the trailing hemisphere, and in the interior, the rim and the ejecta of the major icy impact basins and many impact craters, with the youngest craters showing the largest abundance of this compound (Hibbitts et al. 2002). Since the impactor bodies cannot be the source of CO_2, as this compound would rapidly sublimate at the temperature typical of the satellite's subsolar point ($\sim$170 K), some kind of trapping structures (e.g., ice clathrates, physisorption) is invoked as a stable underground reservoir of CO_2. On Ganymede, CO_2 appears to be mostly correlated with the moderately hydrated non-ice material primarily associated with the dark *regiones* (Hibbitts et al. 2009) and, unlike Callisto, there does not appear to be any leading/trailing hemisphere asymmetry in the distribution of CO_2 nor do impact craters tend to be CO_2-rich (Hibbitts et al. 2003), while CO_2 is occasionally enriched in terrain containing larger grained ice in comparison with adjacent terrain of similar morphology and ice abundance. On Europa, the CO_2 band is present in the equatorial region of the leading hemisphere but is absent on the trailing hemisphere (Smythe et al. 1998; McCord et al. 1998a; Hansen and McCord 2009). Since impacts and meteoritic fall are greatest on the leading hemisphere, Europa's CO_2 may be radiolytically produced in dark carbonaceous meteoritic deposits.

Ganymede also shows evidences of oxygen species, particularly solid O_2 (Spencer et al. 1995) and O_3 (Noll et al. 1996), in its trailing hemisphere, consistent with the preferential orientation of that side of the satellite with Jupiter's magnetosphere. Both of these species appear to be trapped within the ice matrix, and probably originate from ionic bombardment of the icy surface. Ultraviolet (UV) measurements of Ganymede carried out by the Galileo spacecraft confirmed the presence of ozone, whose abundance varies with latitude, with the strongest absorption measured at higher latitudes (Hendrix et al. 1999). This was interpreted as being the result of plasma bombardment creating O_3 in the ice matrix and photodissociation destroying it, on a continual basis.

Io's colorful appearance is the result of various materials produced by its extensive volcanism. Io contains little to no water, though small pockets of water ice or hydrated minerals have been tentatively identified (Douté et al. 2004). Io's surface is largely dominated by sulphur species: in particular, SO_2 frost is omnipresent (e.g. Douté et al. 2001), but there is also evidence for S_2, SO, SO_2 gas and NaCl erupted from plumes (Lopes and Spencer 2007), as well as indications of Fe-bearing salts, silicates (feldspars and pyroxenes) consistent with high-temperature lava flows, FeS_2, and iron sulphide minerals. Concerning SO_2, it is also distributed on Callisto where it is generally mottled, with some areas of high concentrations correlated with ice-rich impact craters (Hibbitts et al. 2000). Large-scale patterns include the depletion of SO_2 in the polar regions; and a depletion of SO_2 on the trailing side relative to the leading side is observed. High-resolution data from Galileo/UVS (Hord et al. 1992) allowed the SO_2 absorption to be mapped out across the surface of Europa (Hendrix et al. 1998, 2011), with a peak concentration near the apex of the trailing hemisphere. NIMS-measured SO_2 shows that it is not strongly correlated with CO_2 but has a similar sparse distribution (Hansen and McCord 2008).

Volatiles such as CO_2 were also found to be common on Saturn's larger icy satellites like Dione (Clark et al. 2008; Cruikshank et al. 2010; Stephan et al. 2010), Iapetus (Buratti et al. 2005), Hyperion (Cruikshank et al. 2008) and Phoebe (Clark et al. 2005), while it is less abundant on Rhea (Clark et al. 2008; Cruikshank et al. 2010; Stephan et al. 2012). Results from VIMS actually show that CO_2's most diagnostic feature in the 3–5 μm range is slightly shifted to shorter wavelengths and broadened on Iapetus, Hyperion, and Dione compared to Phoebe (Cruikshank et al. 2010; Filacchione et al. 2010), thus revealing a different nature of the environment in which the CO_2 molecule is trapped (Chaban et al. 2007). In the Saturnian system, CO_2 is usually associated with dark material, suggesting that it is related to the interaction with incoming energetic particles from Saturn's magnetosphere or UV radiation and the emplacement or the formation of the rocky non-ice material (Stephan et al. 2010, 2012; Scipioni et al. 2013, 2014). However, Phoebe's CO_2 is native to the body as part of the initial inventory of condensates and now exposed on the surface, strongly mixed with water ice and hydrocarbons (Coradini et al. 2008).

The albedos of the major Uranian moons, considerably lower than those of Saturn's moons (except Phoebe and the dark hemisphere of Iapetus), reveal that water ice is mixed in varying proportions to non-ice, visually dark and spectrally bland, material that is possibly carbonaceous in origin (Brown and Cruikshank 1983). CO_2 has been detected from telescopic observations on Ariel, Umbriel and Titania, but has not been detected on the farthest regular Uranian satellite, Oberon (Grundy 2006). The detected CO_2 ice appears to be concentrated on the trailing hemispheres of these satellites, and it decreases in abundance with increasing semi-major axis (Grundy 2006), as opposed to what is observed in the Saturnian system. The surface of Neptune's largest moon Triton is composed of a range of volatile-rich ices, including N_2, H_2O and CO_2 (Quirico et al. 1999; Cruikshank et al. 2000b).

Nature and Distribution of Hydrated Minerals Hydrated, non-ice materials were identified by Galileo/NIMS on the icy Galilean moons Europa and Ganymede. These materials are concentrated in the visually dark regions of these satellites, showing low reflectance at wavelengths shortward of 3 μm plus subdued and distorted water ice bands at 1.5 and 2.0 μm, as well as relatively high reflectance in the 3–5 μm range compared to water ice

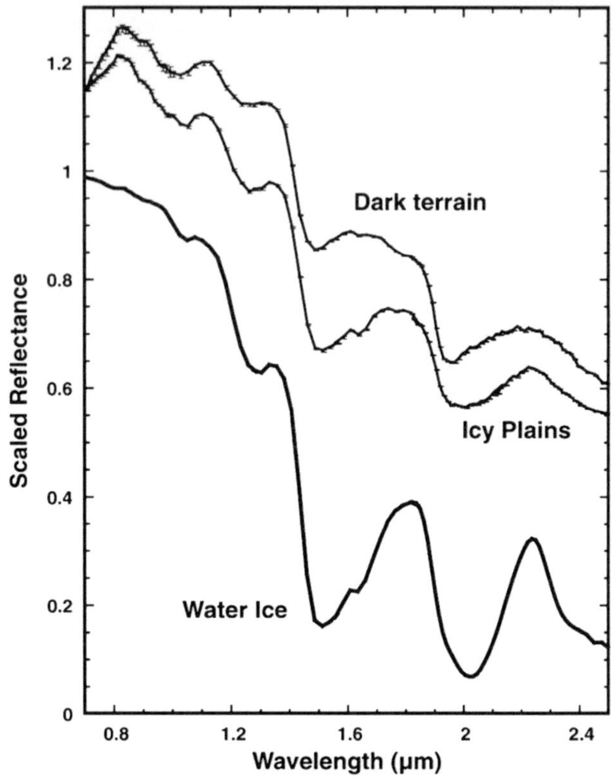

Fig. 3 Galileo Near-Infrared Mapping Spectrometer (NIMS) spectra of Europa. The icy plains (*middle*) and dark, non-icy terrain (*top*) display absorption features arising from vibrational modes of the water molecule. The spectrum of crystalline water ice at 100 K (*bottom*) is shown for comparison. Note the appearance of the broad absorption features centered at 1.5 and 2.0 μm, and how their shapes differ in the Europa spectra. The asymmetric, distorted shapes seen in the less-icy terrain are attributed to a bound state such as water of hydration (from: Dalton et al. 2005)

(McCord et al. 1998b, 1999) (Fig. 3). Compositional end-members invoked for this non-ice material are hydrated salt minerals enriched in Mg and Na, which may form from the crystallization of brines erupted from the subsurface—these compounds were actually expected by the formation models of Europa (McCord et al. 1998b, 1999, 2001, 2002; Dalton 2003; Dalton et al. 2005; Dalton et al. 2012)—and hydrated sulphuric acid, which may result from the radiolysis of water and of the sulphur-bearing species, or from the decomposition of sulphate salts (Carlson et al. 1999). Recently, an alternative based on Mg-bearing chlorinated salts has been proposed for Europa (Ligier et al. 2016). Similar to water ice, the reflectance spectrum of hydrated salt minerals is very sensitive to grain size and temperature (e.g., Dalton 2003; Dalton et al. 2005; De Angelis et al. 2017).

Spectra of Europa's non-ice material are actually best matched by mixtures of these two compositional classes, i.e. sulphuric acid hydrate together with hydrated salts (McCord et al. 2001, 2002; Dalton 2007). One possible mechanism is that Na associated with some salts could be easily sputtered away and abundant H^+ could take its place, forming sulphuric acid (McCord et al. 2001, 2002). On Europa, the abundance of sulphuric acid hydrate is linked to the magnetospheric charged particle energy flux, and does not seem to strongly correlate with the underlying geologic strata. This could result from radiolytic processing of implanted sulphur from Io, or of sulphur emplaced as part of the surface deposits that came from the interior (Shirley et al. 2010; Dalton et al. 2013). The signature of the hydrated material, or the dark terrain, also varies according to geologic unit types (Shirley et al. 2010; Dalton et al. 2012). In particular, the hydrate absorption, perhaps salts and reformed sulphuric acid, are closely associated with at least some lineaments and is especially concentrated near their centers, suggesting a source associated with the lineament formation mechanism (McCord et al. 2010; Dalton et al. 2012). The impact of particles trapped in Jupiter's magnetic field is very different on Ganymede than on Europa and no association with the Ganymede impact pattern is seen. Another possible source could be from exogenic material falling onto Ganymede's surface, but again the distribution only at some dark regions that are clearly exogenic from geologic arguments seems to argue for an exogenic origin.

The surfaces of Callisto and Ganymede also bear evidence of the presence of organic compounds, possibly tholins, i.e. nitrogen-rich organics that are found in cometary nuclei or that form by the action of charged particles impacting the icy surfaces (McCord et al. 1997, 1998a). The abundance of these compounds is tightly related to dark material, being weaker on Ganymede than on Callisto.

In the Saturn system, there is a trend of decreasing water ice band strength as a function of the radial distance, marking an increasing contribution of non-ice contaminants moving away from the planet (Filacchione et al. 2007). Using diagnostic spectral indices, such as spectral slopes and band depths in the VIS-IR range, a comparative analysis of the icy Saturnian moons carried out on the basis of disk-integrated VIMS data revealed striking differences among ring regions and satellites, ranging from Enceladus and Calypso's bluish surfaces, which appear very bright, water-ice-rich, and almost uncontaminated, to the more distant Hyperion, Iapetus, and Phoebe where metals, organics, and carbon dioxide are mixed within water ice at hemispheric scale, resulting in overall lower albedos and—for Hyperion and the leading side of Iapetus—redder spectra (Cruikshank et al. 2008; Tosi et al. 2010; Filacchione et al. 2010, 2012, 2013). Phoebe appears to be compositionally different from the intermediate-sized regular satellites of Saturn: its organic-rich composition as revealed by VIMS is unlike any surface yet observed in the inner Solar System, which supports the hypothesis that this satellite is coated by material of cometary or outer Solar System origin (Clark et al. 2005), ultimately confirming that it is a captured body (Johnson and Lunine 2005).

Finally, organic material has been detected in Saturn's ring particles. In general, ring particles appear to have peculiar properties, being the reddest objects in Saturn's system at visible wavelengths while maintaining sharp and intense water ice bands in the infrared range (Nicholson et al. 2008; Filacchione et al. 2012; Cuzzi et al. 2010; Hedman et al. 2013). This spectral behavior is compatible with crystalline water ice polluted by chromophores, e.g., organic material resulting from the irradiation of simple hydrocarbons (Johnson et al. 1983; Moore et al. 1983), nanophase iron or hematite (Clark et al. 2012, 2008), tholins in intimate mixing (Ciarniello et al. 2011), amorphous silicates (Poulet et al. 2003), carbonaceous particles (Cuzzi and Estrada 1998), or different combinations of these end-members.

3.1.3 Atmospheres, Plumes and Jets in the Icy Moons of the Outer Planets

All of the four Galilean satellites are known to have thin atmospheres (McGrath et al. 2004), whose properties are indicative of surface processes and composition. Io's SO_2 atmosphere, arising from volcanic emissions and sublimation, is the densest with an estimated surface pressure of the order of 10^{-9} bar. Additional species, including SO, S_2, O, S, Na, K and Cl, have also been detected as constituents of Io's atmosphere. The presence of an O_2 atmosphere at Europa has been inferred from measurements of UV emissions (Hall et al. 1995, 1998); Na and K have also been measured at Europa, in ground-based observations (Brown and Hill 1996; Brown 2001). Ganymede also has a thin O_2 atmosphere, inferred from measurements of UV emissions, and a hydrogen exosphere, measured by Galileo/UVS in a limb scan (Barth et al. 1997). The exosphere of Callisto is dominated by O_2, detected on the basis of Hubble Space Telescope (HST) spectra acquired in the far-ultraviolet (Cunningham et al. 2015). Callisto's atmosphere is collisional and is the fourth-densest satellite atmosphere in the Solar System, in addition to being the second-densest O_2-rich collisional atmosphere in the Solar System, after Earth (ibid). The Galileo/NIMS spectrometer also discovered exospheric CO_2 (Carlson 1999). In 2012, direct images obtained by HST in the far-UV enabled the first detection of water vapor plumes at Europa, possibly as high as 200 km (Roth et al. 2014). Absorption from gas or aerosols beyond the Europa limb was then confirmed on the basis of further far-UV images obtained by HST in 2014, as the satellite transited the smooth face of Jupiter (Sparks et al. 2016).

Jets of water ice from surface fractures near the south pole of Saturn's icy moon Enceladus produce a plume of gas and particles (Dougherty et al. 2006; Waite et al. 2006). Enceladus' gaseous plumes were sampled with the Cassini Saturn Orbiter's INMS (Waite et al. 2004) during several fly-bys of the moon and found to be composed predominantly of H_2O vapor, with CO_2 as the second most abundant species. Ammonia is also present in the plumes, along with various organic compounds, deuterium and ^{40}Ar (Waite et al. 2009). The Cosmic Dust Analyzer (CDA) on-board the Cassini Orbiter (Srama 2004) also discovered salts in E-ring particles derived from the plume. These E-ring particles were determined to be rich in sodium salts ($\sim$0.5–2% by mass) and were said to arise only if Enceladus' plume originates from liquid water that had been in contact with rocky interior material (Postberg et al. 2009).

Titan is the only satellite in the Solar System to have a thick, dense atmosphere, mainly composed of N_2 (98.4%), CH_4 (1.4%) and H_2 (0.1%), with trace amounts of several other gases and hydrocarbons. Thanks to the Cassini-Huygens mission, the atmosphere of Titan has now been extensively studied (e.g. Niemann et al. 2010) and presents a very complex chemistry. In particular, Titan's lower atmosphere harbor organic aerosols (tholins) presumed to be formed from simple molecules, such as methane and nitrogen. Cassini particle spectrometers indicate that a complex ion-neutral chemistry in Titan's upper atmosphere near 1000 km plays a major role in the formation of progressively more complex hydrocarbon molecules, from benzene to polycyclic aromatic hydrocarbons (PAH) and ultimately to aerosol particles with $\sim$260-nm radii (Waite et al. 2007).

In situ measurements obtained by the Cassini spacecraft at Saturn's moons Dione and Rhea revealed tenuous atmospheres composed of O_2 and CO_2 (Teolis et al. 2010; Teolis and Waite 2016). Cassini magnetic field and plasma observations from the close flybys of Saturn's icy moon Dione are also consistent with the presence of an exosphere rich in molecular oxygen ions (O_2^+) (Simon et al. 2011; Tokar et al. 2012). The atmospheres of Dione and Rhea appear to be sustained by chemical decomposition of the surface water ice under irradiation from Saturn's magnetospheric plasma. The presence of CO_2 suggests radiolysis reactions between surface oxidants and organics or sputtering and/or outgassing of CO_2 endogenic to ice. O_2:CO_2 source rates at Dione (Rhea) are estimated to be $\sim$30:1 (2:1) relative, consistent with a stoichiometric bulk composition below the lag layer of 0.01 (0.13) C atoms per water molecule, deriving from endogenic constituents, implanted micrometeoritic organics, and (especially at Dione) exogenous H_2O delivery by E-ring grains (Teolis and Waite 2016).

Finally, plumes consisting of nitrogen gas and dust were observed emanating from the southern polar cap of Triton (Soderblom et al. 1990). It is also known that Triton possesses a tenuous global atmosphere composed mainly of molecular nitrogen, with a trace of methane near the surface (Broadfoot et al. 1989).

3.2 Deep Water Reservoirs in Giant Satellites

In this section, we summarize the main evidence from the Galileo era that have been driving the science community towards the hypothesis that liquid layers are indeed present in several moons. For each moon that may possess an ocean, it reviews briefly what are the main characteristics from the subsurface to the core, with a special focus on the evidence that suggests the presence of deep liquid reservoirs below these crusts. The main characteristics of the crustal material inferred from the surface observations are briefly summarized. Europa at Jupiter and Enceladus at Saturn will be looked at in a comparative manner, and a similar approach will be used for discussing Callisto and Ganymede at Jupiter, and Titan at Saturn. In the second subsection, the main characteristics of these liquid reservoirs both in term of composition and dynamics will be looked at.

3.2.1 How Deep Are the Oceans?

Recent or present geological activity, involving water rising from the shallow subsurface, has been demonstrated on Enceladus and Europa. Enceladus is indeed famous because of the geysers activity on its south pole. Globally, Enceladus exhibits four broad terrain regions: cratered plains, western hemisphere fractured plains, eastern hemisphere fractured plains, and the only active area called South Polar Terrain (Spencer et al. 2009). This is the area where Cassini found the tiger stripes, linear cracks about 130 km in length and 2 km wide. The tiger stripes are extremely active, producing a huge thermal emission, and outstanding plume jets, which composition and nature have been extensively studied (see Sect. 3.1.3). Even if these geysers cannot be considered as a proof of the existence of a global ocean, they represent at least a clear indication of the fact that a liquid reservoir must be present below the icy crust. The exact nature of this reservoir (locally distributed below the south pole, or global ocean) has been debated for about 10 years and is not yet entirely solved. First interpretations of the Cassini data favored an icy crust of about 40 km thick at the South pole and a thickening of the shell when approaching the equator of the moon. Global ocean or deep reservoir at the South pole were both envisaged. But in 2015, libration measurements were obtained, and favor the existence of a global ocean (Cadek et al. 2016). Cadek et al. (2016) also suggest that the ice shell might be much thinner than previously thought (< 20 km), a prediction difficult to reconcile with other geophysical data. Nonetheless, the authors propose an original concept which might be consistent with all available data. Enceladus could be layered as follows: a rocky core, an internal global ocean approximately 45 km deep, and a 20-km thick ice shell averaged. The South pole should be thinner (< 5 km thick). In this model, the ocean beneath the ice is about 40% of the total volume of the moon, and in direct contact with the silicate core.

Europa's surface is characterized by an extremely young surface, as indicated by its very low impact crater density (only 16 craters with diameters of 3–27 km). Europa's surface can be subdivided into bright (bluish color) plains, featuring numerous parallel ridges in a wide range of orientations, and darker, brownish mottled terrains. Linear ridges are the most widespread landforms on Europa; the most common of which are double ridges, consisting of a pair of ridges with a medial trough. The latter are thought to have originated through a variety of mechanisms, including, e.g., tectonism, cryovolcanism, or diapirism, and require either the presence of liquid water in the shallow subsurface, or warm mobile ice underlain by a deep ocean (Greeley et al. 2004; Schmidt et al. 2011). More recently, evidence for convergent areas were also suggested, thus providing for the first time clear indications of the possible recycling of the crust through a process similar to plate tectonics on Earth (Kattenhorn and Prockter 2014). The evidence for a large

(~200-km high) plume of water vapor provided by Hubble Space Telescope (HST) imaging of far-UV emissions from excited H and O water dissociation products (Roth et al. 2014; Sparks et al. 2016) suggest that active processes may currently be at work at Europa. In any case, the surface of Europa suggests recently active processes operating in the ice shell. Although relatively little is known about the internal structure, most models include an outer ice shell underlain by liquid water, a silicate mantle, and an iron-rich core (Schubert et al. 2009). The existence of the liquid ocean, globally distributed below the icy crust, is still not proven without any doubt, but it corresponds to the best model that explains all geophysical data of the Galileo mission. In particular, it is now considered to be the only way to produce a shallow electrical current interacting with the jovian field to produce the induced magnetic field that has been detected by Galileo (Kivelson et al. 2009). The thickness of the crust is still highly uncertain. Two categories of models have been proposed suggesting either a very shallow liquid ocean only a few kilometers deep or an ocean trapped below a few tens of kms (Billings and Kattenhorn 2005). In Nimmo et al. (2003), a synthesis and analysis of all the data available at that time were provided suggesting a rather strong preference for the second hypothesis with an averaged value of the crustal thickness around 25 km.

Ganymede, the largest moon of the Jovian system, presents a mix of very old and younger terrains revealing a geological record of several billions of years, ancient impact basins and fresh craters, and landscapes formed via tectonism, possibly icy volcanism, and slow-rate degradation by space weathering. In the Jovian system, Ganymede holds a key position because it features old, densely-cratered terrain, similar to most of Callisto's surface, but also widespread tectonically resurfaced regions, resembling a large part of Europa's surface. Ganymede does not show any evidence of activity at present, or even recent features. In its past history, the locally restricted depressions called *paterae* are the only features, which, to our knowledge, are seriously interpreted as being formed from cryovolcanic processes. At least one of them was interpreted as an icy flow that produced smooth bright lanes within the grooved terrain (Giese et al. 1998). This suggests that cryovolcanism in combination with tectonic processes may have played a role in the past for the formation of the bright terrain. But geologic evidences do not support the existence of shallow liquid reservoirs, similar to what has been suspected on Europa (Schmidt et al. 2011). On the other hand, strong evidences of the presence of deep and global liquid layers below the icy crust have been provided by the Galileo mission, with the detection of an induced magnetic field generated in response to the time-variable rotating magnetosphere of Jupiter (Kivelson et al. 2002; Kivelson 2004). The available data are still inconclusive but it is worth noting that the presence of the ocean is in agreement with geophysical models which predict that tidal dissipation and radiogenic energy keep the water liquid (e.g., Spohn and Schubert 2003; Hussmann et al. 2006), and also with the very recent observation from the Hubble Space Telescope of the oscillations of Ganymede's auroral ovals (Saur et al. 2015). Ganymede possesses a high-level of differentiation, meaning that there must be a clear separation between an iron metallic core (which existence is proven by the intrinsic magnetic field of the moon), a silicate mantle, and a hydrosphere about 500 km thick. Grasset et al. (2013) showed that all models converge towards an ocean thickness that varies from 50 km at 130 km in depth in the case of pure water, up to 230 km at about 60 km depth if ammonia were added to the liquid.

Callisto is characterized by globally abundant dark, densely cratered plains, indicating a very old surface, almost unchanged for billions of years. Callisto should be considered as a window into the early history of the outer solar system. Its surface is dominated by various impact features, similar in morphology to those observed on Ganymede, and by landforms indicative of intense surface erosion and degradation. Callisto is also characterized by a dark blanket covering most of its surface. This is due to a process called sublimation degradation, most certainly triggered by the presence of CO_2, that caused the degradation of bright high-standing terrain (e.g., crater rims) and the formation of this globally abundant dark, smooth blanket. Unfortunately, the time-scale of this dark lag formation is not known (Moore et al. 2004). Crater forms suggest that rheological properties and subsurface layering may be similar to Ganymede, but degradation states and ages of the palimpsests (craters with a specific morphology) infer different rates of change of these properties with time. Tectonism is not widespread but a few systems (furrows and lineaments) indicate that a small geologic activity occurred in the past (Moore et al. 2004; Prockter et al. 2010). Galileo data suggest that Callisto is a partially differentiated body. The very old surface of Callisto indicating an absence of activity for billions of years, may imply a total lack of dynamics in its deep interior. But this is not consistent with Galileo detection of an induced magnetic field that could possibly be due to the presence of a large liquid reservoir below the outer icy shell, or an induced magnetosphere produced by a substantial and conducting ionosphere. This moon being not in the Laplace resonance (resonance that locks the three inner moons into a state where tidal heating is intense, especially for Io and Europa), there is no other sources of heat than the radiogenic heating from the silicates. That fact, plus the fact that the moon may not be fully differentiated indicate that Callisto remains enigmatic. The discovery by Galileo of the induced magnetic field is still hard to reconcile with the geophysical state of the moon. In any case, the putative ocean should be located under a thick crust, most certainly thicker than all other moons since this crust has been there almost from the origins of Callisto.

Titan is very similar to Ganymede in terms of size and bulk composition. Its internal structure is still mostly unknown but the moment of inertia indicates a moderately differentiated body not possessing an inner metallic core but an hydrous silicate core, overlain by an hydrosphere which must be about the same thickness as on Ganymede. Cassini gravity science observations also revealed that Titan hosts a deep ocean (Iess et al. 2012). But Titan is unique in the solar system because of its thick atmosphere, and because of the exchange processes between this atmosphere, the surface, and possibly the icy crust. The massive atmosphere is mainly composed of N_2 ($\sim$98%) and CH_4 ($\sim$2%) (e.g. Griffith et al. 2013). It is beyond the scope of this paper to elaborate on the extreme complexity of Titan's atmospheric dynamics, but it is worth noting that Titan is an excellent laboratory for exploring the diversity of volatiles that can be found with the water-rich material of the outer solar system, and may still be present inside the large satellites. First, the low $^{36}Ar/N_2$ abundance measured by the Gas Chromatograph Mass Spectrometer onboard Huygens indicates that nitrogen has been brought on Titan in the form of N-bearing compounds, most likely NH_3 (Atreya et al. 2009). The $^{14}N/^{15}N$ ratio measured in N_2 is very close to the value recently inferred in NH_2 radicals in comets (Rousselot et al. 2014; Shinnaka et al. 2014), suggesting a similar origin. Huygens also detected a small amount of CO_2 in the upper atmosphere, that could have originated from a recombination of CO and OH^- (Hörst et al. 2008). CO_2 has also been observed at different locations on Titan's surface from IR spectral absorption (McCord et al. 2008), and possibly in the upper crust as suggested by the CO_2 evaporation subsequent to the landing of the Huygens probe (Niemann et al. 2010). Mandt et al. (2014) suggested that Titan's building blocks might have a composition close to comets. Unfortunately, even if comets were representative of the initial composition of planetesimals in the vicinity of Saturn's forming region, the fraction of volatile compounds that was lost or modified during the accretion processes is still unknown (e.g. Estrada et al. 2009). Nonetheless, it is worth noting that the major species which can

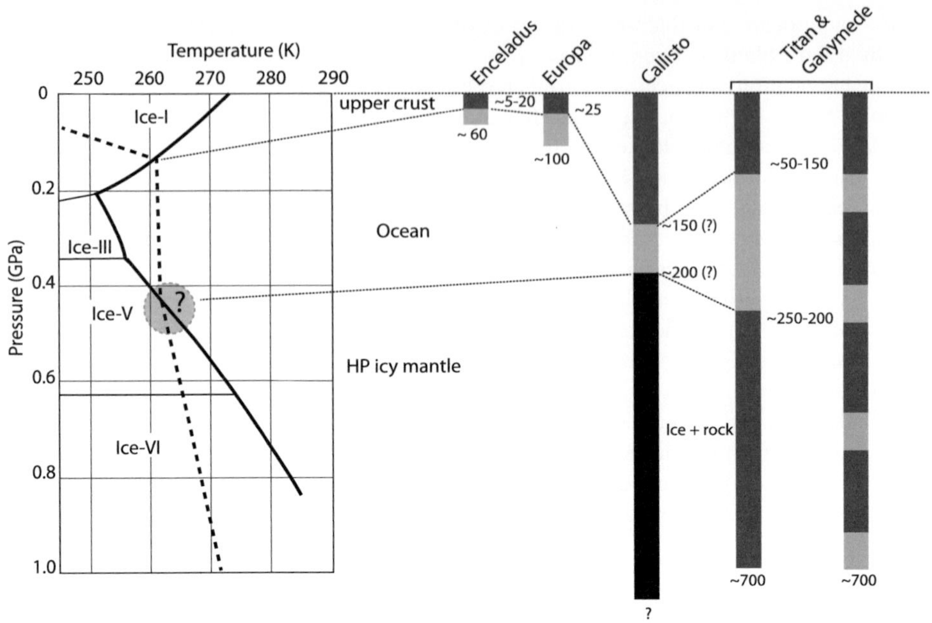

Fig. 4 *Left*: Phase diagram of the pure water system in the pressure–temperature range relevant for icy moons (*bold lines*: melting curves; *thin lines*: solid-solid transitions; *dashed line*: a possible thermal profile—see text for details) *Right*: hydrosphere layering of the giant moons. Positions of the interfaces are indicative of the pressure range only, but their precise location depends on temperature profiles within the hydrospheres and on the composition of the icy material. Numbers are depth ranges in km (not at scale) (Enceladus; Cadek et al. 2016; Europa: Nimmo et al. 2003; Callisto: Kuskov and Kronrod 2005; Titan & Ganymede—*left*: Tobie et al. 2005; Hussmann et al. 2015; *right*: from Vance et al. 2014)

be easily mixed with liquid water in small amounts, or combined with water ice through hydration or clathration processes must have been incorporated into Titan's interior in the primordial stages (Tobie et al. 2006). Further observations are needed to better describe the exchange processes within this moon from its primordial stage to its present state, which calls for future missions. As for Ganymede, there is no way to derive a precise estimate of the crustal thickness and of the size of the liquid reservoir. Too many uncertainties remain, especially when considering the possible compositions of the liquid and the icy crust, key parameters in determining the position of the liquid-ice interfaces. Nonetheless, it is commonly agreed that Titan's liquid reservoir should be about 100 km deep (e.g., Tobie et al. 2005).

3.2.2 Structure and Dynamics of Deep Hydrospheres

Deep liquid layers trapped below a thick icy crust seem to be a common feature in the giant moons of the outer solar system. At this point, it is necessary to make a distinction between the two, and possibly more, different natures of oceans that could exist. At first order this is related to the amount of water available, i.e., to the pressure range that is imposed on it. Coming back to the pressure-temperature phase diagram of water (Fig. 4) first established by Bridgman (1937), and just slightly modified since then (see Choukroun and Grasset 2010 for a review), one can see that the pressure domain is a key issue for understanding the nature of deep reservoirs. Without considering any chemical complexity (pure water only), many ice polymorphs already exist in the moderate pressure range 0–2 GPa, relevant to icy moons. But the main characteristics of the diagram that is worth noticing is that the melting curve of the low-pressure ice Ih decreases with pressure, which is related thermodynamically to the fact that ice Ih is less dense than its liquid. The other ices, existing above 0.209 GPa in the case of the pure water system, are denser than the liquid and display melting curves with positive slopes. This is why two kinds of liquid reservoirs can already be imagined. When the hydrosphere is thin (such as for Enceladus or Europa), the liquid reservoir is most certainly trapped in between an icy crust and a silicate mantle, because the pressure at the bottom of the hydrosphere is below the critical pressure above which high pressure polymorphs exist. In that case, the liquid ocean is most certainly in contact with the rocky core, thus providing interesting possibilities for chemical exchanges through geothermal processes, extremely relevant for the astrobiological potential of these moons. Hydrothermal vents are indeed extremely favorable places for complex chemical evolution, thanks to the heterogeneous processes occurring between the hot gases and liquid and solid phases (Raulin 2005). On the other hand, for the largest moons and most probably all water-rich exoplanets with cold surfaces, the only way for a liquid layer to exist is to be trapped between the icy crust and a lower layer of high-pressure (HP) ices. Internal pressures are sufficient to allow for the formation of high-pressure ice phases below the ocean, thus splitting the hydrosphere into a high-pressure ice layer consisting of various water-rich ices denser than liquid water, the subsurface water ocean, and an ice-I layer forming the outer crust of the satellites.

The thermal structure of the hydrospheres is commonly described from top to bottom using the scaling laws derived from numerical modelling of convective processes within thick icy mantles. In fact, they were first modeled in the early 80s after the Voyager missions returned evidence for endogenic activity (Cassen et al. 1982; Friedson and Stevenson 1983) and the field has been expanding ever since. Following the pioneering models of Lewis (1971) or Consolmagno and Lewis (1978), early thermal evolutions for large icy moons (Reynolds and Cassen 1979; Kirk and Stevenson 1987) either did not consider internal oceans or put enormous emphasis on the freezing of transient oceans in terms of convective and tectonic implications—see for example the work of Kirk and Stevenson (1987) in the case of Ganymede. More recent studies have taken into account the possibility/likelihood of present-day internal oceans above high-pressure icy mantles (e.g. Showman et al. 1997; Tobie et al. 2005; Hussmann et al. 2015). The inner structures displayed in Fig. 4 were calculated assuming hydrostatic and thermal equilibrium and matching available moments of inertia and mass balance constraints. All these models assume thermal equilibrium, i.e., that the heat produced both from radiogenic decay in the silicates and tidal heating is entirely transferred through the thick high-pressure layer, if present, the ocean, and finally the upper icy mantle. The interface depths vary between studies because of the chosen physical parameters and scaling laws. For example, the ocean thickness depends on assumptions made on thermal gradients in the icy layer. Since thermal profiles cannot be measured, our understanding of the deep interior structure remains model-dependent.

Features commonly accepted for the two giant moons Ganymede and Titan include: a high-pressure ice mantle about 400 km thick; a deep ocean about 250 km thick; an icy upper layer ranging from a few tens to about one hundred km thick. For Callisto, the upper icy layer and the ocean may have the same characteristics than for the two other large satellites, but the limited differentiation inferred from available gravity data implies that there is no high-pressure icy mantle. The interior model of Europa displayed in Fig. 4 has an upper crust a few tens of km thick in average (Nimmo et al. 2003). An alternative scenario suggests that the icy crust could not be more than a few km thick (see Billings and Kattenhorn 2005). Future missions such as the Europa Orbiter mission planned by NASA for the next decade will help retire this uncertainty via a combination of ground penetrating radar and gravity science. If Europa's crust is thinner than a few kilometers, the ESA JUICE mission will also be capable to detect the ocean below (Grasset et al. 2013). Finally, the interior structure of Enceladus displayed in Fig. 4 is based on the latest model of Cadek et al. (2016).

Icy and liquid layers are composed of pure water. Icy compositions found in cometary material and on moon surfaces indicate that hydrospheres must host a complex chemistry. Unfortunately, the nature of this complex chemistry is still mostly unknown. As already mentioned in Sect. 3.1, many volatiles and organic compounds have been detected on icy surfaces and may originate from the deep interior of these bodies. Ammonia is also suspected, especially on Titan. These compounds add complexity to the simple model proposed above, for two mains reasons: volatiles reduce the melting temperature of the ice (especially ammonia); clathrate and hydrate structures may form, hence increasing the fraction of solid materials present in the hydrosphere. Regarding the first point, the minimum temperature allowing for deep oceans to exist is about 250 K in case of pure water ice, but could be substantially smaller with the presence of antifreeze compounds such as ammonia (down to 180 K). This property influences significantly the size of the liquid reservoir for a given thermal profile. As for the second point, recent studies of Ganymede (Vance and Brown 2013; Vance et al. 2014) illustrate the importance of hydrated materials. An original internal structure has been proposed based on the sandwiched ocean scenario, but with more than one liquid layer because of the density contrasts between the liquid, the pure ices, and the hydrated and/or clathrated ices through depth. It is rather hard to reconcile this alternative model with dynamical constraints based on thermochemical models of the moons, but it is a very attractive and interesting alternative that shows clearly how imprecise is our current understanding of the deep interior of the moons.

In Fig. 4, the question mark at the interface at the bottom of the ocean illustrates a singularity, which to our knowledge, has not yet been resolved for the largest moons and possibly water-rich exoplanets. While the thermal profile postulates a simple temperature increase in the HP ice mantle, either corresponding to an isentropic increase (ascribed to extremely efficient convection) or to an increase in conductive transfer, the actual temperature profile most probably differs from this simple picture. As the HP ice mantle thickness probably amounts to several hundreds of kilometers, and knowing the orders of magnitudes for the viscosity values of high pressure polymorphs, vigorous subsolidus convective instabilities must occur. For this reason, a conductive profile is not realistic. Subsolidus thermal convection is, notwithstanding, much less efficient than in the ocean so that an isentropic profile will not extend up to the liquid ocean/HP mantle interface and a cold boundary layer will build up. Estimates (Choblet and Tobie 2015) show that the slope of this cold thermal boundary layer should exceed that of the ice melting curve which suggests that the upper portion of the HP mantle is partially molten. This state is not stable over geologic time scales since the entire system is out of thermal equilibrium. Preliminary simulations also show that significant melting should occur in the warmest regions of the HP mantle with two major consequences: (1) melt production and melt delivery to the ocean above is a major heat sink; (2) in most models, melt is predicted to occur in the deeper-most region of the HP ice layer in contact with the rock core. These calculations suggests that chemical/mass transfer is probably not prohibited by the HP mantle and point to the need for a better assessment of the dynamics of the HP mantle in the context of an evolving ocean/HP mantle interface. As noted above, planetary models lack both a proper description of the high-pressure ice mantle dynamics and an adequate description of the liquid-solid interface. While the configuration of an evolving melting/crystallizing front is encountered in many different settings in Earth and planetary sciences (e.g. magma oceans evolution, growth of a solid inner core within metallic cores), it has mostly been addressed with a focus on convection in the low viscosity liquid. A proper description at longer time-scales, involving solid-state convection, remains to be developed.

4 Transneptunian Objects and Ceres

This section focuses on large planetesimals found beyond Neptune's orbit. The past decade has seen the discovery of a large number of large bodies in the Kuiper Belt at a large pace (see Brown 2013, for a review). Larger objects are suspected beyond the detection capability of the current telescopes via their influence on the dynamical behavior of the Kuiper Belt and scattered disk (e.g., Batygin and Brown 2016). These objects are expected to be dominated in volume by water ice and other volatiles such as methane, methanol, nitrogen, etc. McKinnon et al. (2016) pointed out that the activity inferred at Sputnik Planitia based on nitrogen ice is naturally expected at other Kuiper belt objects. More generally, Schenk and Nimmo (2016) demonstrated that similar geological processes are found across the outer solar system driven by whichever volatiles condensed in abundance, or even second-phase impurities as a function of the distance to the Sun. This section also includes Ceres, a dwarf planet located in the main belt of asteroids, but believed to have acquired its volatiles from outer solar system planetesimals (De Sanctis et al. 2015).

4.1 Dwarf Planet Ceres

Evidence for a large fraction of volatiles, about 50 vol.%, was inferred from Ceres' density of about 2.16 g/cm^3 (McCord and Sotin 2005). Water may be partitioned between phyllosilicates and other hydrated materials (e.g., salts) and ice. Early models of Ceres' internal evolution suggested that Ceres would have differentiated a thick ice shell (McCord and Sotin 2005; Castillo-Rogez and McCord 2010). However, Ceres' geomorphology turned out to tell a different story (Fig. 5a). While its shape is globally relaxed to hydrostatic equilibrium (Park et al. 2016), Ceres has preserved topographic features of scale 130 km and smaller, including unrelaxed craters that require a mechanically strong shell (Bland et al. 2017). Bland et al. (2017) inferred that the shell should contain at least 50vol.% of non-ice, material to explain the observed geomorphology, which may point to an enrichment in silicates and hydrated salts, possibly also clathrates (Fu et al. 2017; Castillo-Rogez et al. 2017). While water ice might be a minor component of Ceres' shell, its presence is expressed in a variety of features: surface exposure (Combe et al. 2016, 2017), ground ice flows (Schmidt et al. 2017), permanently shadowed regions (Schorghofer et al. 2016). The gamma ray and neutron detector (GRaND) on-board the Dawn mission also detected a significant increase in near-surface (1 m) ice at latitudes greater than 45 deg. (Prettyman et al. 2017) consistently with the distribution of ground ice features (Schmidt et al. 2017). Indirect evidence for volatiles in Ceres' shell is expressed via large edifices such as Ahuna mons and the bright deposits in Occator's crater, whose emplacement was facilitated by the presence of liquid. Ahuna mons is a 4-km mean height mountain that requires the presence of a few percent melt, likely brines expected at depth by thermal models (Ruesch et al. 2016; Castillo-Rogez and McCord 2010). The Occator bright spots are dominated by sodium carbonates and ammonium salts (carbonates or chlorides, De Sanctis et al. 2016). The concentration of these species is explained by the creation of a local reservoir of melt upon impact-induced heat (Bowling et al. 2016). In this context the deposits are evaporites that concentrated as water vaporized from Ceres' airless surface.

Fig. 5 Enhanced color images of (**a**) Ceres as seen by the Dawn mission (Credit NASA/JPL/DLR); (**b**) Pluton and (**c**) Charon as seen by the New Horizons mission (Credit: NASA/SwRI/APL). Color variations across the surface of these bodies reveal lithological variations that may be a combination of composition and physical processing (e.g., fresh vs. altered material). The prominent bright spot in Ceres' Occator crater has been characterized as mixture of salts but the very dark ejecta surrounding that crater remain to be interpreted. Reddish colors of Charon's and Pluto's surfaces reveal exotic ices and processed organics, in particular tholins

Besides water, Ceres accreted non-negligible amounts of ammonia and CO/CO_2. This is evidenced by the widespread occurrence of ammoniated materials and carbonates across Ceres' surface detected from ground-based observations (King et al. 1992; Rivkin et al. 2006) and confirmed by the Dawn mission (De Sanctis et al. 2015). Ammonia may have been accreted in the form of ammonia hydrates that turned into ammonium in solution during a phase when Ceres held a deep ocean (Castillo-Rogez et al. 2017). An alternative is that ammonia was accreted in the form of organics present in the early solar nebula and desorbed as a result of thermal metamorphism (McSween et al. 2017).

Several other main belt objects share similar evidence for the accretion of volatiles found in abundance in outer solar system planetesimals: the large 10 Hygiea and the smaller (< 200 km) asteroids part of the Europa group of asteroids that display a 3.15 μm absorption band associated with ammonia compounds by Takir et al. (2015). An origin of these objects in the outer solar system is likely and supports recent solar system dynamical models that simulate a large migration events from planetesimals in the 8 to 15 AU region in the early solar system history (e.g., Walsh et al. 2011). An alternative scenario is these asteroids formed in situ and accreted planetesimals migrated from the outer solar system (Mousis et al. 2008; Grazier et al. 2014). However, migration across the nebula gap opened by Jupiter would have led to the depletion of these planetesimals of their most volatile species (especially ammonia) (Turner et al. 2012). Hence Ceres and other large asteroids are likely to be related to outer solar system bodies, and potentially giant planet satellites.

4.2 Pluto and Charon

Pluto and its satellite system were revealed in great detail during the approach and flyby of the New Horizons mission, between January 2014 and July 2015 (Figs. 5b and 5c). New Horizons confirmed the abundance of water ice as well as other volatiles detected from ground-based and space telescopes (Cruikshank et al. 2015). The dwarf planet's surface also presents a variety of additional volatiles, including nitrogen (N_2), methane (CH_4), and carbon monoxide (CO) ices (Grundy et al. 2016). Pluto's landscape has been shaped by glacial flow where water plays the role of bedrock (Moore et al. 2016). It experienced major extensional tectonism and resurfacing, probably of cryovolcanic origin, nearly 4 billion years ago. It also shows evidence for cryovolcanism in the form of mounds and a possible supervolcano in the Southern hemisphere (http://www.nasa.gov/feature/possible-ice-volcano-on-pluto-has-the-wright-stuff).

The impact of volatile materials on Pluto's geology is expressed in the diversity of the landscapes found on this dwarf planet. Like Ceres and Mars, Pluto displays pitted terrains that express the loss of volatiles by sublimation and other volatile-driven processes. Furthermore, several features suggest that Pluto could be endogeneously active today. One is the presence of a large basin, called Sputnik Planitia, that was formed less than 10 My ago. Sputnik Planitia is dominated by nitrogen ice. McKinnon et al. (2016) suggested that the peculiar morphology of this region, which displays an assemblage of polygons, is evidence for sluggish convection (a special style of convection in which deep material is upwelled up to the surface) over a 1-km deep layer. Nimmo et al. (2016) pointed out that Sputnik Planitia had to reorient, which implies the presence of a subsurface ocean in Pluto, such as those already described in Sect. 3.2. Furthermore, Singer and Stern (2016) pointed out that Pluto's nitrogen atmosphere is not stable over long periods and cannot be replenished solely from cometary supply, hence it may come from Pluto's interior.

Less information is available for Pluto's satellites. However, water ice has been found to dominate their surfaces and geology (Weaver et al. 2016). Pluto's largest satellite, Charon is relatively simpler than Pluto and more akin to icy satellites in its geological properties. Grundy et al. (2016) found a surface dominated by water ice as well as ammonia in some fresh craters. Reddish material at its poles appears due to tholins. Large, hemispheric-wide scarps observed on Charon may be ascribed to stresses generated by the cooling of a global ocean.

4.3 Transneptunian Objects (KBOs, Centaurs, Scattered Disk Objects)

This subsection reviews the state of knowledge of icy composition found at transneptunian objects. These are known to be volatile rich based on their relatively low densities, in general lower than 2 g/cm^3. Ice has been commonly found on these bodies (see, e.g., Barucci et al. 2006, 2011; Schaller and Brown 2008). Brown et al. (2012) found that members in the Haumea family exhibit almost pure water ice surfaces, which they interpreted as evidence for the differentiation of a thick icy mantle in their parent. These authors also found a possible relationship between the occurrence of ice and ammonia and large KBOs that could point to recent or ongoing internal activity. Terai et al. (2016) showed that water ice on at least three large KBOs is a mixture of crystalline and amorphous phases. These authors suggested

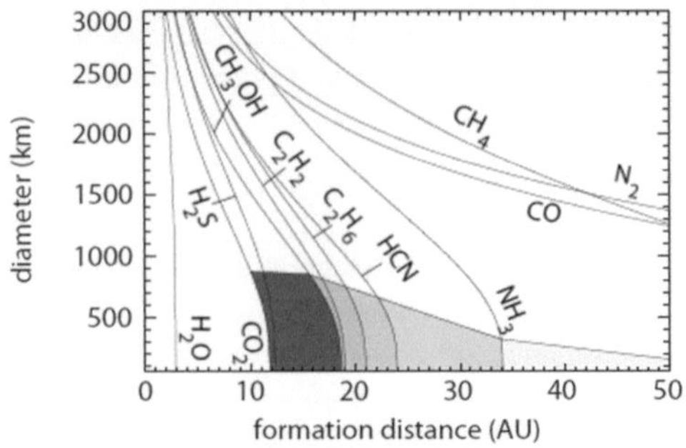

Fig. 6 Evaporation gradients in the early Solar system as a function of object size, which drives the evaporation timescale. The *shaded regions* show the formation regions of the three main surface types of KBOs: in the *inner part* of the primordial Kuiper Belt only water ice and carbon dioxide are stable over long timescales. In the *middle part* of the Kuiper Belt methanol is retained, in the *outer part* ammonia is preserved (from Brown et al. 2011, permission pending)

a possible correlation between the abundance of crystalline ice and the size of the objects, although more observations are required for robust linkage.

Kuiper Belt objects are frequently featureless (see for example Fornasier et al. 2009). However, ammonia, methane, and nitrogen have been frequently found on the surfaces of KBOs and TNOs, for example (50000) Quaoar, and the dwarf planet (13699) Eris (e.g., Merlin et al. 2009). Carry et al. (2011) confirmed the presence of ammonia at a level of a few percent on the surface of Orcus and further identified the presence of methane. Barucci et al. (2005) found a mixture of nitrogen, methane, and methanol ices, besides tholins. Methanol has been suggested to be responsible for the red colors of some Centaurs and KBOs believed to have formed beyond the methanol evaporation line (Brown et al. 2012). Specifically, the very red crustal material observed on some of these bodies has been suggested as the product of irradiation of a mixture of ice, carbon dioxide, and methanol (e.g., Brunetto et al. 2006). However, other explanations are possible, for example mixtures of organic tholins (e.g., Cruiskshank and Dalle Ore 2013).

Schaller and Brown (2007) showed that the escape of volatile compounds can explain the diversity in surface volatile content, and the frequent absence of volatile altogether, at these bodies. These authors also suggested that KBOs smaller than 1000 km are too small to retain volatiles on their surface for extended periods of time. They mapped out the stability of various volatiles as a function of distance to the Sun (Fig. 6). Ammonia is not expected to be stable over extended periods of time because it is rapidly destroyed by irradiation (e.g., Cooper et al. 2004), less than 0.1 Myr after Cook et al. (2007). Hence the occurrence of ammonia implies a relatively recent emplacement either via cryovolcanism acting on a regional scale or a large impact that excavated part of the deep interior. DeMeo et al. (2008) pointed out that the widespread presence of crystalline ice in the TNO population also suggests recent emplacement since ice tends to amorphize upon long exposure to space weathering.

Centaurs are very interesting objects in that some can be outgassing and reveal their primitive composition. Observations so far have detected ice and frozen methanol or light hydrocarbons, for example at 5145 Pholus (Cruiskshank et al. 1998). Carbon monoxide has also been found on a few Centaurs including 174P/Echelus and 2060 Chiron (e.g., Wierzchos et al. 2017). Many observations of Centaurs have revealed surface variations, which Merlin et al. (2005) attributed to impacts (e.g., Licandro and Pinilla-Alonso 2005) or as signature of activity (outgassing), as expected for these bodies when they start approaching their perihelia.

A major open question for that class of objects is whether some of their volatiles (e.g., methane, carbon monoxide and dioxide) were trapped as ice, in amorphous ice, and/or acquired via clathrate hydrates (e.g., McKinnon 2002). A similar question applies to comets. This could a priori be addressed by tracking the type of ice in Centaurs outgassed as a function of distance to the Sun (e.g., Leliwa-Kopystynski et al. 2013), since the release temperature would be different for these materials. However, modelling this is complicated while observations are scarce, so this questions remains open at present.

Abundant ice can be a vector of long-term activity in dwarf planets and other large TNOs provided that they also contain enough rock as a source of radioisotope-based heating. Hydrated silicates have been suggested at the surface of some of these bodies (e.g., Busarev et al. 2003) that would testify of a protracted phase of aqueous alteration involving at least partial melting of their interiors (Desch et al. 2009). The Plutino (90482) Orcus displays potential spectral evidence for ammonium (Delsanti et al. 2010), a marker of aqueous alteration, like in the case of Ceres (De Sanctis et al. 2015), although other types of nitrogen compounds can explain the Orcus spectra (e.g., NH_3, Barucci et al. 2008). Our current understanding of the evolution of large icy satellites and Pluto is relevant to that class of bodies, but key unknowns are the conditions under which cryovolcanism would develop and the role of second-phase, low-eutectic species in promoting ice mobility, as explored by Kargel et al. (1991).

5 Conclusion

In this paper, a broad overview of the nature and distribution of water and volatiles in the outer solar system has been proposed. In the first part, the focus was placed on the giant and intermediary planets. The moons, and especially those in which a liquid ocean is expected, were described in the second part. Finally, a third section was dedicated to Ceres and the dwarf planets in the Kuiper Belt.

To summarize, Jupiter and Saturn most likely contain more volatiles than all the other objects in the solar system added together. But the volatile abundances remain poorly known as they are hidden away at great depths within the planets. On Uranus and Neptune, the amount of volatiles and ices is even more uncertain due to the lack of constraints. The exact nature and distribution of the elements in the deep interior of giant and intermediary planets are still highly unknown. In the gaseous atmospheres, volatile species are in trace quantities but they play a crucial role in shaping the dynamics, clouds, and general circulation of the atmospheres. On that specific matter, a detailed analysis of the current knowledge on the four planets has been proposed. Most of the water and volatiles not trapped in the planets are in the moons, and to a much lesser extent in their tenuous atmospheres (Titan excepted). The exact nature and distribution of volatiles and ices differ significantly from one planetary system to the other, and among moons within a system. But there is a general consensus that volatiles and possibly organic compounds can be found on all surfaces, even if it is sometimes marginal. For most of the moons, the relationships between the surface characteristics and the internal structure and dynamics is highly uncertain, except in a few cases where evidence for deep oceans yields insights into the coupling processes at work. Finally, the two best known examples of dwarf planet systems (Ceres, Pluto and its moons), and other KBOs have been addressed. It turns out that their bulk characteristics in terms of nature and distribution of water and ices at their surfaces do not differ significantly from the icy moons.

Significant progress has been made these last two decades thanks to the two flagship missions Cassini-Huygens at Saturn, and Galileo at Jupiter. Additional datasets recently returned by New Horizons in the Pluto system, Dawn at Ceres, and also from ground-based observations at all wavelengths and for all planetary objects have provided a rather complete overview of the nature and the diversity of the water-rich bodies in the outer system. A next and significant step will be achieved by the Juno mission at Jupiter, especially in probing the deep atmosphere to better constrain the coupling processes from the atmosphere to the deep interior of Jupiter. After the completion of the Cassini mission in 2017, there will not be any further exploration of the icy moons for at least one decade. Then, the planned NASA mission to Europa and the ESA JUICE mission to Ganymede will provide new constraints on the Jovian system. These two missions will carry the instrumentation needed to accurately determine the nature and distribution of the volatile and organic compounds on the two moons, and to better characterize their hydrospheric structures and thermal states. In 2016 NASA and ESA kicked off a joint study about the exploration of the icy giant systems. These missions are envisioned for the 2020–2030 timeframe. Other advanced concepts for the exploration of the outer solar system are being devised, including a probe into Saturn's atmospheric composition (e.g., Mousis et al. 2016) and Titan Lake Lander (Mitri et al. 2014).

References

R.K. Achterberg, B.J. Conrath, P.J. Gierasch, Icarus **182**, 169–180 (2006)
Y. Alibert, C. Mordasini, W. Benz, C. Winisdoerffer, Astron. Astrophys. **434**, 343–353 (2005). doi:10.1051/0004-6361:20042032
S.K. Atreya, P.R. Mahaffy, H.B. Niemann, M.H. Wong, T.C. Owen, Planet. Space Sci. **51**, 105–112 (2003)
S.K. Atreya, R.D. Lorenz, J.H. Waite, in *Titan from Cassini-Huygens* (2009), pp. 177–199
F. Bagenal, T.E. Dowling, W.B. McKinnon, *Jupiter: The Planet, Satellites and Magnetosphere*, 2nd edn. (Cambridge University Press, Cambridge, 2006)
K.H. Baines, R.W. Carlson, L.W. Kamp, Icarus **159**, 74 (2002)
D. Banfield, P. Gierasch, M. Bell, E. Ustinov, A. Ingersoll, A. Vasavada, R. West, M. Belton, Icarus **135**(1), 230–250 (1998)
C.A. Barth et al., Geophys. Res. Lett. **24**, 2147–2150 (1997)
M.A. Barucci, D.P. Cruikshank, E. Dotto, F. Merlin, F. Poulet, C. Dalle Ore, S. Fornasier, C. de Bergh, Astron. Astrophys. **439**, L1–L4 (2005)
M.A. Barucci, F. Merlin, E. Dotto, A. Doressoundiram, C. de Bergh, Observations of the TNO 55638 (2002 VE_{95}) and analysis of the population's spectral properties. Astron. Astrophys. **455**, 725–730 (2006)
M.A. Barucci, F. Merlin, A. Guilbert, C. de Bergh, A. Alvarez-Candal, O. Hainaut, A. Doressoundiram, C. Dumas, T. Owen, A. Coradini, Astron. Astrophys. **479**, L13–L16 (2008)
M.A. Barucci, A. Alvarez-Candal, F. Merlin, I.N. Belskaya, C. de Bergh, D. Perna, F. DeMeo, S. Fornasier, Icarus **214**, 297–307 (2011)
K. Batygin, M.E. Brown, Astron. J. **151**(2), 22 (2016)
M. Beuthe, A. Rivoldini, A. Trinh, Geophys. Res. Lett. (2016). doi:10.1002/2016GL070650
S.E. Billings, S.A. Kattenhorn, Icarus **177**, 397–412 (2005)
G.L. Bjoraker, M.H. Wong, I. de Pater, M. Adamkovics, Astrophys. J. **810**, 122 (2015)
M.T. Bland, C.A. Raymond, R.R. Fu, P. Schenk, T. Kneissl, J.H. Pasckert, H. Hiesinger, F. Preusker, R. Park, S. Marchi, S. King, J.C. Castillo-Rogez, C.T. Russell, Nat. Geosci. **10**, 338–343 (2017). doi:10.1038/ngeo2936
A.C. Boley, T. Hayfield, L. Mayer, R.H. Durisen, Icarus **207**, 509–516 (2010). doi:10.1016/j.icarus.2010.01.015
T. Bowling, F.J. Ciesla, S. Marchi, B.C. Johnson, T.M. Davison, J.C. Castillo-Rogez, M.C. De Sanctis, C.A. Raymond, C.T. Russell, *Lunar and Planetary Science Conference* (2016), p. 47
P.W. Bridgman, J. Chem. Phys. **5**, 964 (1937)
F.H. Briggs, P.D. Sackett, Icarus **80**, 77–103 (1989)
A.L. Broadfoot et al., Science **246**, 1459–1466 (1989)
M.E. Brown, Icarus **151**(2), 190–195 (2001)
M.E. Brown, Astrophys. J. Lett. **778**(2), L34 (2013)

R.H. Brown, D.P. Cruikshank, Icarus **55**, 83–92 (1983)
M.E. Brown, R.E. Hill, Nature **380**(6571), 229–231 (1996)
R.H. Brown et al., Space Sci. Rev. **115**(1–4), 111–168 (2004)
M.E. Brown, E.L. Schaller, W.C. Fraser et al., Astrophys. J. Lett. **739**, L60 (2011)
M.E. Brown, E.L. Schaller, W.C. Fraser, Astrophys. J. **143**, 146 (2012)
R. Brunetto, M.A. Barucci, E. Dotto, G. Strazzulla, Ion irradiation of frozen methanol, methane, and benzene: linking to the colors of Centaurs and trans-Neptunian objects. Astrophys. J. **644**, 646 (2006)
B.J. Buratti et al., Astrophys. J. **622**(2), L149–L152 (2005)
V.V. Busarev, V.A. Dorofeeva, A.B. Makalkin, Earth Moon Planets **92**, 345–357 (2003)
O. Cadek, G. Tobie, T. Van Hoolst, M. Masse, G. Choblet, L. Lefevre, G. Mitri, R.-M. Baland, M. Behounkova, O. Bourgeois, A. Trinh, Geophys. Res. Lett. (2016). doi:10.1002/2016GL068634
W.M. Calvin, R.N. Clark, Icarus **89**, 305–317 (1991)
W.M. Calvin, R.N. Clark, Icarus **104**(1), 69–78 (1993)
R.W. Carlson, Science **283**(5403), 820 (1999)
R.W. Carlson et al., Space Sci. Rev. **60**(1–4), 457–502 (1992)
R.W. Carlson et al., Science **286**(5437), 97–99 (1999)
B. Carry, D. Hestroffer, F.E. DeMeo, A. Thirouin, J. Berthier, P. Lacerda, B. Sicardy, A. Doressoundiram, C. Dumas, D. Farrelly, T.G. Muller, Astron. Astrophys. **534**, A115 (2011)
P.M. Cassen, S.J. Peale, R.T. Reynolds, in *Satellites of Jupiter* (University of Arizona Press, Tucson, 1982), pp. 93–128 (A83-16226 04-91)
J.C. Castillo-Rogez et al. Meteorit. Planet. Sci. (2017, submitted)
J.C. Castillo-Rogez, T.B. McCord, Icarus **205**, 443–459 (2010). doi:10.1016/j.icarus.2009.04.008
T. Cavalié, R. Moreno, E. Lellouch, P. Hartogh, O. Venot, G.S. Orton, C. Jarchow, T. Encrenaz, F. Selsis, F. Hersant, L.N. Fletcher, Astron. Astrophys. **562**, A33 (2014). doi:10.1051/0004-6361/201322297c
G.M. Chaban et al., Icarus **187**(2), 592–599 (2007)
G. Choblet, G. Tobie, AGU abstract P23D-06. AGU Fall Meeting (2015)
M. Choukroun, O. Grasset, J. Chem. Phys. **133**, 144502 (2010)
M. Ciarniello et al., Icarus **214**(2), 541–555 (2011)
R.N. Clark, J. Geophys. Res. **86**, 3087–3096 (1981)
R.N. Clark et al., Nature **435**(7038), 66–69 (2005)
R.N. Clark et al., Icarus **193**(2), 372–386 (2008)
R.N. Clark et al., Icarus **218**(2), 831–860 (2012)
J.-Ph. Combe, T.B. McCord, F. Tosi, E. Ammannito, F. Carrozzo, M.-C. De Sanctis A. Raponi, S. Byrne, M.E. Landis, K. Hughson, C. Raymond, C. Russell, Science **353**(6303), id.aaf3010 (2016)
J.-Ph. Combe, A. Raponi, F. Tosi, M.C. De Sanctis, F.G. Carrozzo, F. Zambon, E. Ammannito, K.H.G. Hughson, A. Nathues, M. Hoffmann, T. Platz, G. Thangjam, N. Schorghofer, S. Schröder, S. Byrne, M.E. Landis, O. Ruesch, T.B. McCord, K.E. Johnson, S.M. Singh, C.A. Raymond, C.T. Russell, Exposed H_2O-rich areas detected on Ceres with the Dawn Visible and InfraRed mapping spectrometer. Icarus (2017, submitted)
B.J. Conrath, P.J. Gierasch, S.S. Leroy, Icarus **83**, 255–281 (1990)
G.J. Consolmagno, J.S. Lewis, Icarus **34**, 280–293 (1978)
J.C. Cook, S.J. Desch, T.L. Roush, C.A. Trujillo, T.R. Geballe, Astrophys. J. **663**, 1406 (2007)
J.F. Cooper, E.R. Christian, J.D. Richardson, C. Wang, in *The First Decadal Review of the Edgeworth-Kuiper Belt* (2004), pp. 261–277
A. Coradini et al., Identification of spectral units on Phoebe. Icarus **193**(1), 233–251 (2008)
D.P. Cruikshank et al., Ices on the satellites of Jupiter, Saturn, and Uranus, in *Solar System Ices*, ed. by B. Schmitt, C. deBergh, M. Festou (Kluwer Academic, Dordrecht, 1998a), pp. 579–606
D.P. Cruikshank, T.L. Roush, T.C. Owen, E. Quirico, C. deBergh, The surface compositions of Triton, Pluto, and Charon, in *Solar System Ices*, ed. by B. Schmitt, C. deBergh, M. Festou (Kluwer Academic, Dordrecht, 1998b), pp. 655–684
D.P. Cruikshank et al., Surface composition of Hyperion. Nature **448**(7149), 54–56 (2000a)
D.P. Cruikshank et al., Water ice on Triton. Icarus **147**(1), 309–316 (2000b)
D.P. Cruikshank et al., Hydrocarbons on Saturn's satellites Iapetus and Phoebe. Icarus **193**(2), 334–343 (2008)
D.P. Cruikshank et al., Carbon dioxide on the satellites of Saturn: results from the Cassini VIMS investigation and revisions to the VIMS wavelength scale. Icarus **206**(2), 561–572 (2010)
D.P. Cruikshank, W.M. Grundy, F.E. DeMeo, M.W. Buie, R.P. Binzel, D.E. Jennings, C.B. Olkin, J.W. Parker, D.C. Reuter, J.R. Spencer, S.A. Stern, L.A. Young, H.A. Weaver, The surface compositions of Pluto and Charon. Icarus **246**, 82–92 (2015)
D.P. Cruiskshank, C.M. Dalle Ore, Earth Moon Planets **92**, 315–330 (2013)

D.P. Cruiskshank, T.L. Roush, M.J. Bartholomew, T.R. Geballe, Y.J. Pendleton, S.M. White, J.F. Bell III, J.K. Davies, T.C. Owen, C. de Bergh, D.J. Tholen, Icarus **135**, 389–407 (1998)
N.J. Cunningham, J.R. Spencer, P.D. Feldman, D.F. Strobel, K. France, S.N. Osterman, Icarus **254**, 178–189 (2015)
J.N. Cuzzi, P.R. Estrada, Icarus **132**(1), 1–35 (1998)
J.N. Cuzzi et al., An evolving view of Saturn's dynamic rings. Science **327**(5972), 1470–1475 (2010)
C.M. Dalle Ore, D.P. Cruikshank, R.M.E. Mastrapa, E. Lewis, O.L. White, Impact craters: an ice study on Rhea. Icarus **261**, 80–90 (2015)
J.B. Dalton, Astrobiology **3**(4), 771–784 (2003)
J.B. Dalton, Geophys. Res. Lett. **34**(21), L21205 (2007)
J.B. Dalton et al., Spectral comparison of heavily hydrated salts with disrupted terrains on Europa. Icarus **177**(2), 472–490 (2005)
J.B. Dalton et al., Europa's icy bright plains and dark linea: exogenic and endogenic contributions to composition and surface properties. J. Geophys. Res. **117**(E3), E03003 (2012)
J.B. Dalton et al., Exogenic controls on sulfuric acid hydrate production at the surface of Europa. Planet. Space Sci. **77**, 45–63 (2013)
S. De Angelis et al., Temperature-dependent VNIR spectroscopy of hydrated Mg-sulfates. Icarus **281**, 444–458 (2017)
T. de Graauw, H. Feuchtgruber, B. Bezard, P. Drossart, T. Encrenaz, D.A. Beintema, M. Griffin, A. Heras, M. Kessler, K. Leech, E. Lellouch, P. Morris, P.R. Roelfsema, M. Roos-Serote, A. Salama, B. Vandenbussche, E.A. Valentijn, G.R. Davis, D.A. Naylor, Astron. Astrophys. **321**, L13–L16 (1997)
I. de Pater, S.T. Massie, Icarus **62**, 143–171 (1985)
I. de Pater, P.N. Romani, S.K. Atreya, Icarus **91**, 220–233 (1991)
I. de Pater, L.A. Sromovsky, P.M. Fry, H.B. Hammel, C. Baranec, K.M. Sayanagi, Icarus **252**, 121–128 (2015)
I. de Pater, R.J. Sault, B. Butler, D. DeBoer, M.H. Wong, Science **352**, 1198–1201 (2016)
M.C. De Sanctis et al., Ammoniated phyllosilicates with a likely outer Solar System origin on (1) Ceres. Nature **528**, 241–244 (2015)
M.C. De Sanctis, A. Raponi, E. Ammannito, M. Ciarniello, M.J. Toplis, H.Y. McSween, J.C. Castillo-Rogez, B.L. Ehlmann, F.G. Carrozzo, S. Marchi, F. Tosi, F. Zambon, F. Capaccioni, M.T. Capria, S. Fonte, M. Formisano, A. Frigeri, M. Giardino, A. Longobardo, G. Magni, E. Palomba, L.A. McFadden, C.M. Pieters, R. Jaumann, P. Schenk, R. Mugnuolo, C.A. Raymond, C.T. Russell, Nature **536**, 54–57.12 (2016)
A.D. Del Genio, R.K. Achterberg, K.H. Baines, F.M. Flasar, P.L. Read, A. Sanchez-Lavega, A.P. Showman, in *Saturn from Cassini-Huygens*, vol. 6 (Springer, Berlin, 2009), pp. 113–159
A. Delsanti, F. Merlin, A. Guilbert-Lepoutre, J. Bauer, B. Yang, K.J. Meech, Astron. Astrophys. **520**, A40 (2010)
F. DeMeo, M.A. Barucci, A. Alvarez-Candal, C. de Bergh, S. Fornasier, F. Merlin, D. Perna, I. Belskaya, Icy bodies of the solar system, in *Proceedings of the 263th of the International Astronomical Union*, ed. by J.A. Fernandez, D. Lazzaro, D. Prialnik, R. Schulz (2008), pp. 186–191
S.J. Desch, J.C. Cook, T.C. Doggett, S.B. Porter, Icarus **202**, 694–714 (2009)
M.K. Dougherty et al., Identification of a dynamic atmosphere at Enceladus with the Cassini magnetometer. Science **311**(5766), 1406–1409 (2006)
M.K. Dougherty, L.W. Esposito, S.M. Krimigis, in *Saturn from Cassini-Huygens*, ed. by M.K. Dougherty, L.W. Esposito, S.M. Krimigis (Springer, Berlin, 2009). doi:10.1007/978-1-4020-9217-6
S. Douté et al., Mapping SO_2 frost on Io by the modeling of NIMS hyperspectral images. Icarus **149**(1), 107–132 (2001)
S. Douté et al., Geology and activity around volcanoes on Io from the analysis of NIMS spectral images. Icarus **169**(1), 175–196 (2004)
T. Encrenaz, E. Lellouch, P. Drossart, H. Feuchtgruber, G.S. Orton, S.K. Atreya, Astron. Astrophys. **413**, L5–L9 (2004)
P.R. Estrada, I. Mosqueira, J.J. Lissauer, G. D'Angelo, D.P. Cruikshank, Formation of Jupiter and conditions for accretion of the Galilean satellites, in *Europa*, ed. by R.T. Pappalardo (University of Arizona Press, Tucson, 2009), pp. 27–58
H. Feuchtgruber, E. Lellouch, T. de Graauw, B. Bezard, T. Encrenaz, M. Griffin, Nature **389**, 159–162 (1997)
H. Feuchtgruber, E. Lellouch, G. Orton, T. de Graauw, B. Vandenbussche, B. Swinyard, R. Moreno, C. Jarchow, F. Billebaud, T. Cavalie, S. Sidher, P. Hartogh, Astron. Astrophys. **551**, A126 (2013)
G. Filacchione et al., Saturn's icy satellites investigated by Cassini–VIMS: I. Full-disk properties: 350–5100 nm reflectance spectra and phase curves. Icarus **186**(1), 259–290 (2007)
G. Filacchione et al., Saturn's icy satellites investigated by Cassini–VIMS: II. Results at the end of nominal mission. Icarus **206**(2), 507–523 (2010)

G. Filacchione et al., Saturn's icy satellites and rings investigated by Cassini–VIMS: III—radial compositional variability. Icarus **220**(2), 1064–1096 (2012)

G. Filacchione et al., The radial distribution of water ice and chromophores across Saturn's system. Astrophys. J. **766**(2), 76 (2013)

U. Fink, H.P. Larson, Icarus **24**, 411–420 (1975)

G. Fischer, W.S. Kurth, D.A. Gurnett, P. Zarka, U.A. Dyudina, A.P. Ingersoll, S.P. Ewald, C.C. Porco, A. Wesley, C. Go, M. Delcroix, Nature **475**, 75–77 (2011)

L.N. Fletcher, G.S. Orton, N.A. Teanby, P.G.J. Irwin, G.L. Bjoraker, Icarus **199**, 351–367 (2009)

L.N. Fletcher, P. Drossart, M. Burgdorf, G.S. Orton, T. Encrenaz, Astron. Astrophys. **514**, A17+ (2010)

L.N. Fletcher, K.H. Baines, T.W. Momary, A.P. Showman, P.G.J. Irwin, G.S. Orton, M. Roos-Serote, C. Merlet, Icarus **214**, 510–533 (2011a)

L.N. Fletcher, B.E. Hesman, P.G.J. Irwin, K.H. Baines, T.W. Momary, A. Sanchez-Lavega, F.M. Flasar, P.L. Read, G.S. Orton, A. Simon-Miller, R. Hueso, G.L. Bjoraker, A. Mamoutkine, T. del Rio-Gaztelurrutia, J.M. Gomez, B. Buratti, R.N. Clark, P.D. Nicholson, C. Sotin, Science **332**, 1413–1417 (2011b)

L.N. Fletcher, B.E. Hesman, R.K. Achterberg, P.G.J. Irwin, G. Bjoraker, N. Gorius, J. Hurley, J. Sinclair, G.S. Orton, J. Legarreta, E. García-Melendo, A. Sańchez-Lavega, P.L. Read, A.A. Simon-Miller, F.M. Flasar, Icarus **221**, 560–586 (2012)

L.N. Fletcher, T.K. Greathouse, G.S. Orton, P.G.J. Irwin, O. Mousis, J.A. Sin- clair, R.S. Giles, Icarus **238**, 170–190 (2014)

S. Fornasier, M.A. Barucci, C. de Bergh, A. Alvarez-Candal, F. DeMeo, F. Merlin, D. Perna, A. Guilbert, A. Delsanti, E. Dotto, A. Doressoundiram, Astron. Astrophys. **508**, 457–465 (2009)

A.D. Fortes, M. Choukroun, Space Sci. Rev. **153**, 185–218 (2010). doi:10.1007/s11214-010-9633-3

T. Fouchet, E. Lellouch, B. Bezard, T. Encrenaz, P. Drossart, H. Feuchtgruber, T. de Graauw, Icarus **143**, 223–243 (2000)

T. Fouchet, P.G.J. Irwin, P. Parrish, S.B. Calcutt, F.W. Taylor, C.A. Nixon, T. Owen, Icarus **172**, 50–58 (2004)

A.J. Friedson, D.J. Stevenson, Icarus **56**(1), 1–14 (1983)

R. Fu, E. Ermakov, S. Marchi, J.C. Castillo-Rogez, C.A. Raymond, B.H. Hager, M.T. Zuber, S.D. King, M.T. Bland, M.C. De Sanctis, F. Preusker, R.S. Park, C.T. Russell, Earth Planet. Sci. Lett. (2017, submitted)

P.J. Gierasch, A.P. Ingersoll, D. Banfield, S.P. Ewald, P. Helfenstein, A. Simon- Miller, A. Vasavada, H.H. Breneman, D.A. Senske (Galileo Imaging Team), Nature **403**, 628–630 (2000)

B. Giese, J. Oberst, T. Roatsch, G. Neukum, J.W. Head, R.T. Pappalardo, Icarus **135**, 303–316 (1998)

R.S. Giles, L.N. Fletcher, P.G.J. Irwin, Icarus **257**, 457–470 (2015)

O. Grasset, E.J. Bunce, A. Coustenis, M.K. Dougherty, C. Erd, H. Hussmann, R. Jaumann, O. Prieto-Ballesteros, Astrobiology **13**(10), 991–1004 (2013). doi:10.1089/ast.2013.1013

K.R. Grazier, J.C. Castillo-Rogez, P.W. Sharp, Icarus **232**, 13–21 (2014)

R. Greeley, C.E. Chyba, J.W. Head, T.B. McCord, W. McKinnon, R.T. Pappalardo, P.H. Figueredo, Geology of Europa, in *Jupiter. The Planet, Satellites and Magnetosphere*, ed. by F. Bagenal, T.E. Dowling, W.B. McKinnon. Cambridge Planetary Science, vol. 1 (Cambridge University Press, Cambridge, 2004), pp. 329–362. ISBN 0-521-81808-7

C.A. Griffith, J.L. Mitchell, P. Lavvas, G. Tobie, Titan's evolving climate, in *Comparative Climatology of the Terrestrial Planets*, ed. by S. Mackwell, M. Bullock, J. Harder (University of Arizona Press, Tucson, 2013)

W.M. Grundy, Icarus **184**(2), 543–555 (2006)

W.M. Grundy, B. Schmitt, The temperature-dependent near-infrared absorption spectrum of hexagonal H_2O ice. J. Geophys. Res. **103**(E11), 25809–25822 (1998)

W.M. Grundy, R.P. Binzel, B.J. Buratti, J.C. Cook, D.P. Cruikshank, C.M. Dalle Ore, A.M. Earle, K. Ennico, C.J.A. Howett, A.W. Lunsford, C.B. Olkin, A.H. Parker, S. Philippe, S. Protopapa, E. Quirico, D.C. Reuter, B. Schmitt, K.N. Singer, A.J. Verbiscer, R.A. Beyer, M.W. Buie, A.F. Cheng, D.E. Jennings, I.R. Linscott, J.W. Parker, P.M. Schenk, J.R. Spencer, J.A. Stansberry, S.A. Stern, H.B. Throop, C.C.C. Tsang, H.A. Weaver, G.E. Weigle, L.A. Young, Science **351**, id.aad9189 (2016)

T. Guillot, Science **269**, 1697–1699 (1995). doi:10.1126/science.7569896

T. Guillot, Annu. Rev. Earth Planet. Sci. **33**, 493–530 (2005)

T. Guillot, D. Gautier, Giant planets, in *Treatise on Geophysics*, vol. 10, ed. by G. Schubert 2nd edn. (Elsevier, Oxford, 2015), pp. 529–557

T. Guillot, R. Hueso, Mon. Not. R. Astron. Soc. **367**, L47–L51 (2006). doi:10.1111/j.1745-3933.2006.00137.x

T. Guillot, D.J. Stevenson, W.B. Hubbard, D. Saumon, in *Jupiter. The Planet, Satellites and Magnetosphere*, ed. by F. Bagenal, T.E. Dowling, W.B. McKinnon (2004), pp. 35–57

D.T. Hall, D.F. Strobel, P.D. Feldman, M.A. McGrath, H.A. Weaver, Nature **373**(6516), 677–679 (1995)

D.T. Hall, P.D. Feldman, M.A. McGrath, D.F. Strobel, Astrophys. J. **499**(1), 475–481 (1998)

G.B. Hansen, T.B. McCord, in *31st Annual Lunar and Planetary Science Conference*, March 13–17, 2000, Houston, TX (2000a), abstract no. 1630
G.B. Hansen, T.B. McCord, Bull. Am. Astron. Soc. **32**, 1068 (2000b)
G.B. Hansen, T.B. McCord, J. Geophys. Res. **109**(E1), E01012 (2004)
G.B. Hansen, T.B. McCord, Geophys. Res. Lett. **35**, L01202 (2008)
G.B. Hansen, T.B. McCord, AGU Fall Meeting 2009, abstract #P51E-1163 (2009)
P. Hartogh, E. Lellouch, R. Moreno, D. Bockelee-Morvan, N. Biver, T. Cassidy, M. Rengel, C. Jarchow, T. Cavalie, J. Crovisier, F.P. Helmich, M. Kidger, Astron. Astrophys. **532**, L2+ (2011)
M.M. Hedman et al., Icarus **223**(1), 105–130 (2013)
R. Helled, T. Guillot, Astrophys. J. **767**, 113 (2013). doi:10.1088/0004-637X/767/2/113
R. Helled, J.D. Anderson, M. Podolak, G. Schubert, Astrophys. J. **726**, 15 (2011). doi:10.1088/0004-637X/726/1/15
A.R. Hendrix et al., Icarus **135**(1), 79–94 (1998)
A.R. Hendrix et al., J. Geophys. Res. **104**(E6), 14169–14178 (1999)
A.R. Hendrix et al., Europa's disk-resolved ultraviolet spectra: relationships with plasma flux and surface terrains. Icarus **212**(2), 736–743 (2011)
B.E. Hesman, G.R. Davis, H.E. Matthews, G.S. Orton, Icarus **186**, 342–353 (2007)
C.A. Hibbitts et al., Composition and distribution of nonice and trace materials on Ganymede as derived from Galileo observations, abstract n. 632. EPSC (2009)
C.A. Hibbitts et al., Distributions of CO_2 and SO_2 on the surface of Callisto. J. Geophys. Res. **105**(E9), 22541–22558 (2000)
C.A. Hibbitts et al., CO_2-rich impact craters on Callisto. J. Geophys. Res. **107**(E10), 5084–5100 (2002)
C.A. Hibbitts et al., Carbon dioxide on Ganymede. J. Geophys. Res. **108**(E5), 5036 (2003)
M.D. Hofstadter, B.J. Butler, Icarus **165**, 168–180 (2003)
M.D. Hofstadter, D.O. Muhleman, Icarus **81**, 396–412 (1989)
C.W. Hord et al., Galileo Ultraviolet Spectrometer experiment. Space Sci. Rev. **60**(1–4), 503–530 (1992)
S.M. Hörst, V. Vuitton, R.V. Yelle, J. Geophys. Res. **113**(E10), E10006 (2008)
W.-B. Hubbard, B. Militzer, Astrophys. J. **820**, 80 (2016). doi:10.3847/0004-637X/820/1/80
H. Hussmann, F. Sohl, T. Spohn, Icarus **185**, 258–273 (2006)
H. Hussmann, G. Choblet, V. Lainey et al., Space Sci. Rev. **153**, 317–334 (2010)
H. Hussmann, C. Sotin, J. Lunine, in *Treatise of Geophysics* 2nd edn. (Elsevier, Amsterdam, 2015), pp. 605–635. doi:10.1016/B978-0-444-53802-4.00178-0
L. Iess, R.A. Jacobson, M. Ducci, D.J. Stevenson, J.I. Lunine, J.W. Armstrong, S.W. Asmar, P. Racioppa, N.J. Rappaport, P. Tortora, Science **337**(6093), 457–459 (2012). doi:10.1126/science.1219631
A. Ingersoll, A.R. Vasavada (Galileo Imaging Team), Nature **403**, 630–632 (2000)
A.P. Ingersoll, T.E. Dowling, P.J. Gierasch, G.S. Orton, P.L. Read, A. Sanchez-Lavega, A.P. Showman, A.A. Simon-Miller, A.R. Vasavada, in *Jupiter. The Planet, Satellites and Magnetosphere* (2004), pp. 105–128
P. Irwin, N.A. Teanby, G.R. Davis, Icarus **203**, 287–302 (2009)
P.G.J. Irwin, E. Lellouch, C. de Bergh, R. Courtin, B. Bezard, L.N. Fletcher, G.S. Orton, N.A. Teanby, S.B. Calcutt, D. Tice, J. Hurley, G.R. Davis, Icarus **227**, 37–48 (2014)
M.A. Janssen, A.P. Ingersoll, M.D. Allison, S. Gulkis, A.L. Laraia, K.H. Baines, S.G. Edgington, Y.Z. Anderson, K. Kelleher, F.A. Oyafuso, Icarus **226**, 522–535 (2013)
T.V. Johnson, J.I. Lunine, Saturn's moon Phoebe as a captured body from the outer solar system. Nature **435**, 69–71 (2005)
R.E. Johnson et al., Astron. Astrophys. **123**(2), 343–346 (1983)
J.S. Kargel, S.K. Croft, J.I. Lunine, J.S. Lewis, Icarus **89**, 93–112 (1991)
E. Karkoschka, M.G. Tomasko, Icarus **211**, 780–797 (2011)
S. Kattenhorn, L.M. Prockter, Nat. Geosci. **7**, 762–767 (2014). doi:10.1038/ngeo2245
K.K. Khurana et al., Icarus **191**(1), 193–202 (2007)
T.V.V. King, R.N. Clark, W.M. Calvin, D.M. Sherman, R.H. Brown, Science **255**, 1551–1553 (1992)
R.L. Kirk, D.J. Stevenson, Icarus **69**, 91–134 (1987)
M.G. Kivelson, Adv. Space Res. **33**(11), 2061–2077 (2004)
M.G. Kivelson, K.K. Khurana, M. Volwerk, Icarus **157**(2), 507–522 (2002)
M.G. Kivelson, K.K. Khurana, M. Volwerk, in *Europa*, ed. by R.T. Pappalardo, W.B. McKinnon, K.K. Khurana (University of Arizona Press, Tucson, 2009), pp. 545–570
O.L. Kuskov, V.A. Kronrod, Sol. Syst. Res. **39**, 283 (2005). doi:10.1007/s11208-005-0043-0
M. Lambrechts, A. Johansen, Astron. Astrophys. **544**, A32 (2012). doi:10.1051/0004-6361/201219127
J. Leconte, G. Chabrier, Astron. Astrophys. Rev. **540**, A20 (2012). doi:10.1051/0004-6361/201117595
J. Leconte, F. Selsis, F. Hersant, T. Guillot, Astron. Astrophys. (2017). doi:10.1051/0004-6361/201629140
J. Leliwa-Kopystynski, M. Satorre, C. Santonja, Acta Geophys. **61**, 1304–1321 (2013)

E. Lellouch, B. Bezard, T. Fouchet, H. Feuchtgruber, T. Encrenaz, T. de Graauw, Astron. Astrophys. **370**, 610–622 (2001)
E. Lellouch, B. Bezard, J.I. Moses, G.R. Davis, P. Drossart, H. Feuchtgruber, E.A. Bergin, R. Moreno, T. Encrenaz, Icarus **159**, 112–131 (2002)
E. Lellouch, P. Hartogh, H. Feuchtgruber, B. Vandenbussche, T. de Graauw, R. Moreno, C. Jarchow, T. Cavalié, G. Orton, M. Banaszkiewicz, M.I. Blecka, D. Bockelée-Morvan, J. Crovisier, T. Encrenaz, T. Fulton, M. Küppers, L.M. Lara, D.C. Lis, A.S. Medvedev, M. Rengel, H. Sagawa, B. Swinyard, S. Szutowicz, F. Bensch, E. Bergin, F. Billebaud, N. Biver, G.A. Blake, J.A.D.L. Blommaert, J. Cernicharo, R. Courtin, G.R. Davis, L. Decin, P. Encrenaz, A. Gonzalez, E. Jehin, M. Kidger, D. Naylor, G. Portyankina, R. Schieder, S. Sidher, N. Thomas, M. de Val-Borro, E. Verdugo, C. Waelkens, H. Walker, H. Aarts, C. Comito, J.H. Kawamura, A. Maestrini, T. Peacocke, R. Teipen, T. Tils, K. Wildeman, Astron. Astrophys. **518**, L152 (2010)
H.F. Levison, K.A. Kretke, M.J. Duncan, Nature **524**, 322–324 (2015). doi:10.1038/nature14675
J.S. Lewis, Icarus **15**, 174–185 (1971)
J. Licandro, N. Pinilla-Alonso, Astrophys. J. Lett. **630**, L93 (2005)
N. Ligier, F. Poulet, J. Carter, R. Brunetto, F. Gourgeot, Astron. J. **151**, 163 (2016). doi:10.3847/0004-6256/151/6/163
B. Little, C.D. Anger, A.P. Ingersoll, A.R. Vasavada, D.A. Senske, H.H. Breneman, W.J. Borucki (The Galileo SSI Team), Icarus **142**, 306–323 (1999)
K. Lodders, Astrophys. J. **591**, 1220–1247 (2003). doi:10.1086/375492
K. Lodders, B. Fegley Jr., Icarus **112**, 368–375 (1994)
R.M. Lopes, J.R. Spencer, *Io After Galileo* (Springer, Berlin, 2007). ISBN 978-3-540-34681-4
J.I. Lunine, Annu. Rev. Astron. Astrophys. **31**, 217–263 (1993)
K.E. Mandt, O. Mousis, J. Lunine, D. Gautier, Astrophys. J. **788**(2), L24 (2014)
M.S. Marley, P. Gomez, M. Podolak, J. Geophys. Res. **100**, 23349–23354 (1995). doi:10.1029/95JE02362
A. Marten, D. Gautier, T. Owen, D.B. Sanders, H.E. Matthews, S.K. Atreya, R.P.J. Tilanus, J.R. Deane, Astrophys. J. **406**, 285–297 (1993)
R.M. Mastrapa, S.A. Sandford, T.L. Roush, D.P. Cruikshank, C.M. Dalle Ore, Astrophys. J. **701**(2), 1347–1356 (2009)
A. Matter, T. Guillot, A. Morbidelli, Planet. Space Sci. **57**, 816–821 (2009). doi:10.1016/j.pss.2009.01.010
T.B. McCord, C. Sotin, J. Geophys. Res. **110**(E5), E05009 (2005)
T.B. McCord et al., Organics and other molecules in the surfaces of Callisto and Ganymede. Science **278**, 271–275 (1997)
T.B. McCord et al., Non-water ice constituents in the surface material of the icy Galilean satellites from the Galileo near-infrared mapping spectrometer investigation. J. Geophys. Res. **103**(E4), 8603–8626 (1998a)
McCord et al., Salts on Europa's surface detected by Galileo's Near Infrared Mapping Spectrometer. Science **280**(5367), 1242 (1998b)
T.B. McCord et al., Hydrated salt minerals on Europa's surface from the Galileo Near-Infrared Mapping Spectrometer (NIMS) investigation. J. Geophys. Res. **104**(E5), 11827–11852 (1999)
T.B. McCord et al., Thermal and radiation stability of the hydrated salt minerals epsomite, mirabilite, and natron under Europa enviromental conditions. J. Geophys. Res. **106**(E2), 3311–3319 (2001)
T.B. McCord et al., Brines exposed to Europa surface conditions. J. Geophys. Res. **107**(E1), 5004 (2002)
T.B. McCord, P. Hayne, J.-P. Combe (The Cassini VIMS Team), Icarus **194**, 212–242 (2008). doi:10.1016/j.icarus.2007.08.039
McCord et al., Hydrated minerals on Europa's surface: an improved look from the Galileo NIMS investigation. Icarus **209**(2), 639–650 (2010)
M.A. McGrath, E. Lellouch, D.F. Strobel, P.D. Feldman, R.E. Johnson, in *Jupiter*, ed. by F. Bagenal, T. Dowling, W. McKinnon (Cambridge University Press, Cambridge, 2004)
W.B. McKinnon, in *Proceedings of International Conference on Asteroids, Comets, Meteors*, ACM 2002, 29 July–2 August 2002, Berlin, Germany, ed. by B. Warmbein. ESA, vol. SP-500 (ESA Publications Division, Noordwijk, 2002), pp. 29–38. ISBN 92-9092-810-7
W.B. McKinnon, F. Nimmo, T. Wong, P.M. Schenk, O.L. White, J.H. Roberts, J.M. Moore, J.R. Spencer, A.D. Howard, O.M. Umurhan, S.A. Stern, H.A. Weaver, C.B. Olkin, L.A. Young, K.E. Smith, R. Beyer, M. Buie, B. Buratti, A. Cheng, D. Cruikshank, C. Dalle Ore, R. Gladstone, W. Grundy, T. Lauer, I. Linscott, J. Parker, S. Porter, H. Reitsema, D. Reuter, S. Robbins, M. Showalter, K. Singer, D. Strobel, M. Summers, L. Tyler, M. Banks, O. Barnouin, V. Bray, B. Carcich, A. Chaikin, C. Chavez, C. Conrad, D. Hamilton, C. Howett, J. Hofgartner, J. Kammer, C. Lisse, A. Marcotte, A. Parker, K. Retherford, M. Saina, K. Runyon, E. Schindhelm, J. Stansberry, A. Steffl, T. Stryk, H. Throop, C. Tsang, A. Verbiscer, H. Winters, A. Zangari, New horizons geology, geophysics, and imaging theme team. Nature **534**, 82–85 (2016)

H.Y. McSween, J.P. Emery, A.S. Rivkin, M.J. Toplis, J. Castillo-Rogez, T.H. Prettyman, M.C. De Sanctis, C.M. Pieters, C.A. Raymond, C.T. Russell, Carbonaceous chondrites as analogs for the composition and alteration of Ceres. Meteorit. Planet. Sci. (2017, in press). doi:10.1111/maps.12947

F. Merlin, M.A. Barucci, E. Dotto, C. de Bergh, G. Lo Curto, Astron. Astrophys. **444**, 977–982 (2005)

F. Merlin, A. Alvarez-Candal, A. Delsanti, S. Fornasier, M.A. Barucci, F.E. DeMeo, C. de Bergh, A. Doressoundiram, E. Quirico, B. Schmitt, Astron. J. **137**, 315 (2009)

Y. Miguel, T. Guillot, L. Fayon, Astron. Astroph. **596**, A114 (2016). doi:10.1051/0004-6361/201629732

B. Militzer, W.B. Hubbard, Astrophys. J. **774**, 148 (2013). doi:10.1088/0004-637X/774/2/148

G. Mitri, A. Coustenis, G. Fanchini, A.G. Hayes, L. Iess, K. Khurana, J.-P. Lebreton, R.M. Lopes, R.D. Lorenz, R. Meriggiola, M.L. Moriconi, R. Orosei, C. Sotin, E. Stofan, G. Tobie, T. Tokano, F. Tosi, Planet. Space Sci. **104**, 78–92 (2014)

H. Mizuno, Prog. Theor. Phys. **64**, 544–557 (1980). doi:10.1143/PTP.64.544

N. Monga, S. Desch, Astrophys. J. **798**, 9 (2015). doi:10.1088/0004-637X/798/1/9

M.H. Moore et al., Studies of proton-irradiated cometary-type ice mixtures. Icarus **54**, 388–405 (1983)

J.M. Moore, C.R. Chapman, E.B. Bierhaus, R. Greeley, F.C. Chuang, J. Klemaszewski, R.N. Clark, J.B. Dalton, C.A. Hibbitts, P.M. Schenk, J.R. Spencer, R. Wagner, in *Jupiter—The Planet, Satellites, and Magnetosphere*, ed. by F. Bagenal, T. Dowling, W.B. McKinnon (Cambridge University Press, Cambridge, 2004), pp. 397–426. ISBN 0-521-81808-7

J.M. Moore, W.B. McKinnon, J.R. Spencer, A.D. Howard, P.M. Schenk, R.A. Beyer, F. Nimmo, K.N. Singer, O.M. Umurhan, O.L. White, S.A. Stern, K. Ennico, C.B. Olkin, H.A. Weaver, L.A. Young, R.P. Binzel, M.W. Buie, B.J. Buratti, A.F. Cheng, D.P. Cruikshank, W.M. Grundy, I.R. Linscott, H.J. Reitsema, C. Reuter Dennis, R. Showalter Mark, V.J. Bray, C.L. Chavez, C.J.A. Howett, T.R. Lauer, C.M. Lisse, A.H. Parker, S.B. Porter, S.J. Robbins, K. Runyon, T. Stryk, H.B. Throop, C.C. Tsang Constantine, J. Verbiscer Anne, A.M. Zangari, A.L. Chaikin, D.E. Wilhelms, F. Bagenal, G.R. Gladstone, T. Andert, J. Andrews, M. Banks, B. Bauer, J. Bauman, O.S. Barnouin, P. Bedini, K. Beisser, S. Bhaskaran, E. Birath, M. Bird, D.J. Bogan, A. Bowman, M. Brozovic, C. Bryan, M.R. Buckley, S.S. Bushman, A. Calloway, B. Carcich, S. Conard, C.A. Conrad, J.C. Cook, O.S. Custodio, C.M. Dalle Ore, C. Deboy, Z.J.B. Dischner, P. Dumont, A.M. Earle, H.A. Elliott, J. Ercol, C.M. Ernst, T. Finley, S.H. Flanigan, G. Fountain, M.J. Freeze, T. Greathouse, J.L. Green, Y. Guo, M. Hahn, D.P. Hamilton, S.A. Hamilton, J. Hanley, A. Harch, H.M. Hart, C.B. Hersman, A. Hill, M.E. Hill, D.P. Hinson, M.E. Holdridge, M. Horanyi, C. Jackman, R.A. Jacobson, D.E. Jennings, J.A. Kammer, H.K. Kang, D.E. Kaufmann, P. Kollmann, S.M. Krimigis, D. Kusnierkiewicz, J.E. Lee, K.L. Lindstrom, A.W. Lunsford, V.A. Mallder, N. Martin, D.J. McComas, R.L. McNutt, D. Mehoke, T. Mehoke, E.D. Melin, M. Mutchler, D. Nelson, J.I. Nunez, A. Ocampo, W.M. Owen, M. Paetzold, B. Page, J.W. Parker, F. Pelletier, J. Peterson, N. Pinkine, M. Piquette, S. Protopapa, J. Redfern, J.H. Roberts, G. Rogers, D. Rose, K.D. Retherford, M.G. Ryschkewitsch, E. Schindhelm, B. Sepan, M. Soluri, D. Stanbridge, A.J. Steffl, D.F. Strobel, M.E. Summers, J.R. Szalay, M. Tapley, A. Taylor, H. Taylor, G.L. Tyler, M.H. Versteeg, M. Vincent, R. Webbert, S. Weidner, G.E. Weigle, K. Whittenburg, B.G. Williams, K. Williams, S. Williams, W.W. Woods, E. Zirnstein, Science **351**, 1284–1293 (2016)

J.I. Moses, E. Lellouch, B. Bezard, G.R. Gladstone, H. Feuchtgruber, M. Allen, Icarus **145**, 166–202 (2000)

J.I. Moses, T. Fouchet, B. Bézard, G.R. Gladstone, E. Lellouch, H. Feucht- gruber, J. Geophys. Res., Planets **110**, E08001 (2005)

O. Mousis, Y. Alibert, D. Hestroffer, U. Marboeuf, C. Dumas, B. Carry, J. Horner, F. Selsis, Mon. Not. R. Astron. Soc. **383**, 1269–1280 (2008)

O. Mousis, L.N. Fletcher, J.-P. Lebreton et al., Planet. Space Sci. **104**, 29–47 (2014)

O. Mousis, D.H. Atkinson, T. Spilker, E. Venkatapathy, J. Poncy, R. Frampton, A. Coustenis, K. Reh, J.-P. Lebreton, L.N. Fletcher, R. Hueso, M.J. Amato, A. Colaprete, F. Ferri, D. Stam, P. Wurz, S. Atreya, S. Aslam, D.J. Banfield, S. Calcutt, G. Fischer, A. Holland, C. Keller, E. Kessler, M. Leese, P. Levacher, A. Morse, O. Muñoz, J.-B. Renard, S. Sheridan, F.-X. Schmider, F. Snik, J.H. Waite, M. Bird, T. Cavalié, M. Deleuil, J. Fortney, D. Gautier, T. Guillot, J.I. Lunine, B. Marty, C. Nixon, G.S. Orton, A. Sánchez-Lavega, Planet. Space Sci. **130**, 80–103 (2016)

S. Nayakshin, R. Helled, A.C. Boley, Mon. Not. R. Astron. Soc. **440**, 3797–3808 (2014). doi:10.1093/mnras/stu473

N. Nettelmann, A. Becker, B. Holst, R. Redmer, Astrophys. J. **750**, 52 (2012). doi:10.1088/0004-637X/750/1/52

N. Nettelmann, R. Helled, J.J. Fortney, R. Redmer, Planet. Space Sci. **77**, 143–151 (2013). doi:10.1016/j.pss.2012.06.019

S. Newman et al., Photometric and spectral analysis of the distribution of crystalline and amorphous ices on Enceladus as seen by Cassini. Icarus **193**, 397–406 (2008)

P.D. Nicholson et al., A close look at Saturn's rings with Cassini VIMS. Icarus **193**(1), 182–212 (2008)

H.B. Niemann, S.K. Atreya, G.R. Carignan, T.M. Donahue, J.A. Haberman, D.N. Harpold, R.E. Hartle, D.M. Hunten, W.T. Kasprzak, P.R. Mahaffy, T.C. Owen, S.H. Way, J. Geophys. Res. **103**(12), 22831–22846 (1998)
H.B. Niemann, S.K. Atreya, J.E. Demick, D. Gautier, J.A. Haberman, D.N. Har- pold, W.T. Kasprzak, J.I. Lunine, T.C. Owen, F. Raulin, J. Geophys. Res., Planets **115**, 12006 (2010)
F. Nimmo, B. Giese, R.T. Pappalardo, Geophys. Res. Lett. **30**(5), 1233 (2003). doi:10.1029/2002GL016660
F. Nimmo, D.P. Hamilton, W.B. McKinnon, P.M. Schenk, R.P. Binzel et al., Nature **540**, 94–96 (2016). doi:10.1038/nature20148
C.A. Nixon, P.G.J. Irwin, S.B. Calcutt, F.W. Taylor, R.W. Carlson, Icarus **150**, 48–68 (2001)
K.S. Noll et al., Detection of ozone on Ganymede. Science **273**(5273), 341–343 (1996)
G.S. Orton, J.H. Lacy, J.M. Achtermann, P. Parmar, W.E. Blass, Icarus **100**, 541–555 (1992)
G.S. Orton, L.N. Fletcher, J.I. Moses, A.K. Mainzer, D. Hines, H.B. Hammel, F.J. Martin-Torres, M. Burgdorf, C. Merlet, M.R. Line, Icarus **243**, 494–513 (2014a)
G.S. Orton, J.I. Moses, L.N. Fletcher, A.K. Mainzer, D. Hines, H.B. Hammel, J. Martin-Torres, M. Burgdorf, C. Merlet, M.R. Line, Icarus **243**, 471–493 (2014b)
T. Owen, T. Encrenaz, Space Sci. Rev. **106**, 121–138 (2003)
T. Owen, P.R. Mahaffy, H.B. Niemann, S. Atreya, M. Wong, Astrophys. J. Lett. **553**, L77–L79 (2001)
R. Park, A.S. Konopliv, B.G. Bills, N. Rambaux, J.C. Castillo-Rogez, C.A. Raymond, A.T. Vaughan, A.I. Ermakov, M.T. Zuber, R.R. Fu, M.J. Toplis, C.T. Russell, A. Nathues, Nature **537**, 515–517 (2016). doi:10.1038/nature18955
M. Podolak, J.I. Podolak, M.S. Marley, Planet. Space Sci. **48**, 143–151 (2000). doi:10.1016/S0032-0633(99)00088-4
J.B. Pollack, O. Hubickyj, P. Bodenheimer, J.J. Lissauer, M. Podolak, Y. Greenzweig, Icarus **124**, 62–85 (1996). doi:10.1006/icar.1996.0190
F. Postberg et al., Sodium salts in E-ring ice grains from an ocean below the surface of Enceladus. Nature **459**(7250), 1098–1101 (2009)
F. Poulet et al., Compositions of Saturn's rings A, B, and C from high resolution near-infrared spectroscopic observations. Astron. Astrophys. **412**, 305–316 (2003)
T.H. Prettyman, N. Yamashita, M.J. Toplis, H.Y. McSween, N. Schorghofer, S. Marchi, W.C. Feldman, J. Castillo-Rogez, O. Forni, D.J. Lawrence, E. Ammannito, B.L. Ehlmann, H.G. Sizemore, S.P. Joy, C.A. Polanskey, M.D. Rayman, C.A. Raymond, C.T. Russell, Science **355**(6320), 55–59 (2017). doi:10.1126/science.aah6765
L.M. Prockter, R.M.C. Lopes, B. Giese, R. Jaumann, R.D. Lorenz, R.T. Pappalardo, G.W. Patterson, P.C. Thomas, E.P. Turtle, R.J. Wagner, Space Sci. Rev. **153**, 63–111 (2010). doi:10.1007/s11214-010-9649-8
E. Quirico et al., Composition, physical state, and distribution of ices at the surface of Triton. Icarus **139**(2), 159–178 (1999)
F. Raulin, Space Sci. Rev. **116**, 471–487 (2005). doi:10.1007/s11214-005-1967-x
R.T. Reynolds, P.M. Cassen, Geophys. Res. Lett. **6**(2), 121–124 (1979)
A.S. Rivkin, E.L. Volquardsen, B.E. Clark, Icarus **185**, 563–567 (2006)
M. Roos-Serote, A.R. Vasavada, L. Kamp, P. Drossart, P. Irwin, C. Nixon, R.W. Carlson, Nature **405**, 158–160 (2000)
E. Rosenblum, P. Garaud, A. Traxler, S. Stellmach, Astrophys. J. **731**, 66 (2011). doi:10.1088/0004-637X/731/1/66
L. Roth et al., Transient water vapor at Europa's south pole. Science **343**(6167), 171–174 (2014)
T.L. Roush et al., Ice and minerals on Callisto—a reassessment of the reflectance spectra. Icarus **86**, 355–382 (1990)
P. Rousselot, O. Pirali, E. Jehin, M. Vervloet, D. Hutsemekers, J. Manfroid, D. Cordier, M.-A. Martin-Drumel, S. Gruet, C. Arpigny, A. Decock, O. Mousis, Astrophys. J. **780**(2), L17 (2014)
O. Ruesch, T. Platz, P. Schenk, L.A. McFadden, J.C. Castillo-Rogez et al., Science **353**(6303), aaf4286 (2016). doi:10.1126/science.aaf4286
A. Sanchez-Lavega, T. del Rio-Gaztelurrutia, R. Hueso, J.M. Gomez-Forrellad, J.F. Sanz-Requena, J. Legarreta, E. Garcia-Melendo, F. Colas, J. Lecacheux, L.N. Fletcher, D. Barrado y Navascués, D. Parker, T. Akutsu, T. Barry, J. Beltran, S. Buda, B. Combs, F. Carvalho, P. Casquinha, M. Delcroix, S. Ghomizadeh, C. Go, J. Hotershall, T. Ikemura, G. Jolly, A. Kazemoto, T. Kumamori, M. Lecompte, P. Maxson, F.J. Melillo, D.P. Milika, E. Morales, D. Peach, J. Phillips, J.J. Poupeau, J. Sussenbach, G. Walker, S. Walker, T. Tranter, A. Wesley, T. Wilson, K. Yunoki (International Outer Planet Watch Team), Nature **475**, 71–74 (2011)
J. Saur, S. Duling, L. Roth et al., J. Geophys. Res. Space Phys. **120**(3), 1715–1737 (2015)
E.L. Schaller, M.E. Brown, Astrophys. J. Lett. **659**, L61 (2007)
E.L. Schaller, M.E. Brown, Astrophys. J. Lett. **684**, L107 (2008)
P. Schenk, F. Nimmo, Nat. Geosci. **9**, 411–412 (2016)

B.E. Schmidt, D.D. Blankenship, G.W. Patterson, P.M. Schenk, Nature (2011). doi:10.1038/nature10608
B.E. Schmidt et al., Nat. Geosci. **10**, 338–343 (2017)
B. Schmitt, E. Quirico, F. Trotta, W.M. Grundy, in *Solar System Ices*, ed. by B. Schmitt, C. De Bergh, M. Festou (Kluwer Academic, Boston, 1998), pp. 199–240
N. Schorghofer, E. Mazarico, T. Platz, F. Preusker, S.E. Schröder, C.A. Raymond, C.T. Russell, Geophys. Res. Lett. **43**, 6783–6789 (2016)
G. Schubert, F. Sohl, H. Hussmann, in *Europa*, ed. by R.T. Pappalardo, W.B. McKinnon, K.K. Khurana (University of Arizona Press, Tucson, 2009), pp. 353–368
F. Scipioni, F. Tosi, K. Stephan, G. Filacchione, M. Ciarniello, F. Capaccioni, P. Cerroni (the VIMS Team), Spectroscopic classification of icy satellites of Saturn I: identification of terrain units on Dione. Icarus **226**, 1331–1349 (2013)
F. Scipioni, F. Tosi, K. Stephan, G. Filacchione, M. Ciarniello, F. Capaccioni, P. Cerroni (the VIMS Team), Spectroscopic classification of icy satellites of Saturn II: identification of terrain units on Rhea. Icarus **234**, 1–16 (2014)
F. Scipioni, P. Schenk, F. Tosi, E. D'Aversa, R. Clark, J.-P. Combe, C.M. Dalle Ore, Deciphering sub-micron ice particles on Enceladus surface. Icarus **290**, 183–200 (2017)
Y. Shinnaka, H. Kawakita, H. Kobayashi, M. Nagashima, D.C. Boice, Astrophys. J. **782**, L16 (2014)
J.H. Shirley et al., Icarus **210**(1), 358–384 (2010)
A. Showman, D.J. Stevenson, R. Mahlotra, Icarus **129**, 367–383 (1997)
S. Simon et al., Magnetic signatures of a tenuous atmosphere at Dione. Geophys. Res. Lett. **38**(15), L15102 (2011)
A.A. Simon-Miller, B. Conrath, P.J. Gierasch, R.F. Beebe, Icarus **145**, 454–461 (2000)
K.N. Singer, S.A. Stern, Astrophys. J. Lett. **808**, L50 (2016)
W.D. Smythe, T.B. McCord, A. Coradini, C.A. Hibbitts, F. Capaccioni, G.B. Hansen, G. Filacchione, R.N. Clark, P. Cerroni, R.H. Brown, K.H. Baines, G. Bellucci, J.-P. Bibring, B.J. Buratti, E. Bussoletti, M. Combes, D.P. Cruikshank, P. Drossart, V. Formisano, R. Jaumann, Y. Langevin, D.L. Matson, R.M. Nelson, P.D. Nicholson, B. Sicardy, C. Sotin, *29th Annual Lunar and Planetary Science Conference* (1998), p. 1532
L.A. Soderblom, S.W. Kieffer, T.L. Becker, R.H. Brown, A.F. Cook, C.J. Hansen, T.V. Johnson, R.L. Kirk, E.M. Shoemaker, Science **250**, 410–415 (1990)
W.B. Sparks, K.P. Hand, M.A. McGrath, E. Bergeron, M. Cracraft, S.E. Deustua, Astrophys. J. **829**(2), 121 (2016)
J.R. Spencer, W.M. Calvin, M.J. Person, J. Geophys. Res. **100**(E9), 19049–19056 (1995)
J.R. Spencer, A.C. Barr, L.W. Esposito, P. Helfenstein, A.P. Ingersoll, R. Jaumann, C.P. McKay, F. Nimmo, J.H. Waite, in *Saturn from Cassini-Huygens*, ed. by M.K. Dougherty, L.W. Esposito, S.M. Krimigis (Springer, Dordrecht, 2009), pp. 683–724. doi:10.1007/978-1-4020-9217-6_21
T. Spohn, G. Schubert, Icarus **161**, 456–467 (2003)
R. Srama, Space Sci. Rev. **114**(1–4), 465–518 (2004)
L.A. Sromovsky, P.M. Fry, J.H. Kim, Icarus **215**, 292–312 (2011)
L.A. Sromovsky, K.H. Baines, P.M. Fry, Icarus **226**, 402–418 (2013)
K. Stephan, Ph.D. Thesis, Freie Universität Berlin (2006). http://diss.fu-berlin.de/2007/806
K. Stephan, H. Hoffmann, K. Hibbitts, R. Wagner, R. Jaumann, EPSC 2009, abstract n. 633 (2009)
K. Stephan, R. Jaumann, R. Wagner, R.N. Clark, D.P. Cruikshank, C.A. Hibbitts, T. Roatsch, H. Hoffmann, R.H. Brown, G. Filiacchione, B.J. Buratti, G.B. Hansen, T.B. McCord, P.D. Nicholson, K.H. Baines, Icarus **206**(2), 631–652 (2010)
K. Stephan, R. Jaumann, R. Wagner, R.N. Clark, D.P. Cruikshank, B. Giese, C.A. Hibbitts, T. Roatsch, K. Matz, R.H. Brown, G. Filacchione, F. Cappacioni, F. Scholten, B.J. Buratti, G.B. Hansen, P.D. Nicholson, K.H. Baines, R. Nelson, D.L. Matson, Planet. Space Sci. **61**(1), 142–160 (2012)
S.T. Stewart, A. Seifter, A.W. Obst, Shocked H_2O ice: thermal emission measurements and the criteria for phase changes during impact events. Geophys. Res. Lett. **35**(23), L23203 (2008)
C.R. Stoker, Icarus **67**, 106–125 (1986)
K. Sugiyama, K. Nakajima, M. Odaka, K. Kuramoto, Y.-Y. Hayashi, Icarus **229**, 71–91 (2014)
D. Takir, J.P. Emery, H.Y. McSween, Icarus **257**, 185–193 (2015)
B.D. Teolis, J.H. Waite, Icarus **272**, 277–289 (2016)
B.D. Teolis, G.H. Joneas, P.F. Miles, R.L. Tokar, B.A. Magee, J.H. Waite et al., Science **330**(6012), 1813–1815 (2010)
T. Terai, Y. Itoh, Y. Oasa, R. Furusho, J. Watanabe, Astrophys. J. **827**, 65 (2016)
G. Tobie, O. Grasset, J.I. Lunine, A. Mocquet, C. Sotin, Icarus (2005). doi:10.1016/j.icarus.2004.12.007
G. Tobie, J.I. Lunine, C. Sotin, Nature **440**(7080), 61–64 (2006)
R.L. Tokar, R.E. Johnson, M.F. Thomsen, E.C. Sittler, A.J. Coates, R.J. Wilson, D.T. Young, G.H. Jones, Geophys. Res. Lett. **39**(3), L03105 (2012)

F. Tosi, D. Turrini, A. Coradini, G. Filacchione (the VIMS Team), Mon. Not. R. Astron. Soc. **403**(3), 1113–1130 (2010)
N. Turner, M. Choukroun, J. Castillo-Rogez, G. Bryden, Astrophys. J. Lett. **748**, 92 (2012)
S. Vance, J.M. Brown, Geochim. Cosmochim. Acta **110**, 176–189 (2013)
S. Vance, M. Bouffard, M. Choukroun, C. Sotin, Planet. Space Sci. **96**, 62–70 (2014). doi:10.1016/j.pss.2014.03.011
A. Vazan, R. Helled, M. Podolak, A. Kovetz, Astrophys. J. **829**, 118 (2016). doi:10.3847/0004-637X/829/2/118
C. Visscher, J.I. Moses, S.A. Saslow, Icarus **209**, 602–615 (2010)
J.H. Waite, W.S. Lewis, W.T. Kasprzak, V.G. Anicich, B.P. Block, T.E. Cravens, G.G. Fletcher, W.-H. Ip, J.G. Luhmann, R.L. Mcnutt, H.B. Niemann, J.K. Parejko, J.E. Richards, R.L. Thorpe, E.M. Walter, R.V. Yelle, Space Sci. Rev. **114**(1–4), 113–231 (2004)
J.H. Waite, M.R. Combi, W.H. Ip, T.E. Cravens, R.L. McNutt, W. Kasprzak, R. Yelle, J. Luhmann, H. Niemann, D. Gell, B. Magee, G. Fletcher, J. Lunine, W.L. Tseng, Science **311**(5766), 1419–1422 (2006)
J.H. Waite, D.T. Young, T.E. Cravens, A.J. Coates, F.J. Crary, B. Magee, J. Westlake, Science **316**(5826), 870 (2007)
J.H. Waite, W.S. Lewis, B.A. Magee, J.I. Lunine, W.B. McKinnon, C.R. Glein, O. Mousis, D.T. Young, T. Brockwell, J. Westlake, M.-J. Nguyen, B.D. Teolis, H.B. Niemann, R.L. McNutt Jr., M. Perry, W.-H. Ip, Nature **460**(7254), 487–490 (2009)
K.J. Walsh, A. Morbidelli, S.N. Raymond, D.P. O'Brien, A.M. Mandell, Nature **475**, 206–209 (2011)
H.A. Weaver, M.W. Buie, B.J. Buratti, W.M. Grundy, T.R. Lauer, C.B. Olkin, A.H. Parker, S.B. Porter, M.R. Showalter, J.R. Spencer, S.A. Stern, A.J. Verbiscer, W.B. McKinnon, J.M. Moore, S.J. Robbins, P. Schenk, K.N. Singer, O.S. Barnouin, A.F. Cheng, C.M. Ernst, C.M. Lisse, D.E. Jennings, A.W. Lunsford, D.C. Reuter, D.P. Hamilton, D.E. Kaufmann, K. Ennico, L.A. Young, R.A. Beyer, R.P. Binzel, V.J. Bray, A.L. Chaikin, J.C. Cook, D.P. Cruikshank, C.M. Dalle Ore, A.M. Earle, G.R. Gladstone, C.J.A. Howett, I.R. Linscott, F. Nimmo, J.W. Parker, S. Philippe, S. Protopapa, H.J. Reitsema, B. Schmitt, T. Stryk, M.E. Summers, C.C.C. Tsang, H.H.B. Throop, O.L. White, A.M. Zangari, Science **351**, aae0030 (2016)
S.J. Weidenschilling, J.S. Lewis, Icarus **20**, 465–476 (1973)
R.A. West, K.H. Baines, A.J. Friedson et al., in *Jovian Clouds and Haze*, ed. by F. Bagenal, T.E. Dowling, W.B. McKinnon (2004), pp. 79–104
K. Wierzchos, M. Womack, G. Sarid, Astron. J. **153**, 230 (2017)
M.H. Wong, G.L. Bjoraker, M.D. Smith, F.M. Flasar, C.A. Nixon, Planet. Space Sci. **52**, 385–395 (2004a)
M. Wong, P. Mahaffy, S. Atreya, H. Niemann, T. Owen, Icarus **171**(1), 153–170 (2004b)
G. Wuchterl, T. Guillot, J.J. Lissauer, in *Giant Planet Formation, Protostars and Planets IV* (2000), p. 1081

Space Sci Rev (2017) 212:813–834
DOI 10.1007/s11214-017-0406-0

Disk Evolution and the Fate of Water

Lee Hartmann[1] · **Fred Ciesla**[2] · **Oliver Gressel**[3] · **Richard Alexander**[4]

Received: 18 July 2016 / Accepted: 2 August 2017 / Published online: 21 August 2017

Abstract We review the general theoretical concepts and observational constraints on the distribution and evolution of water vapor and ice in protoplanetary disks, with a focus on the Solar System. Water is expected to freeze out at distances greater than 1–3 AU from solar-type central stars; more precise estimates are difficult to obtain due to uncertainties in the complex processes involved in disk evolution, including dust growth, settling, and radial drift, and the level of turbulence and viscous dissipation within disks. Interferometric observations are now providing constraints on the positions of CO snow lines, but extrapolation to the unresolved regions where water ice sublimates will require much better theoretical understanding of mass and angular momentum transport in disks as well as more refined comparison of observations with sophisticated disk models.

Keywords Accretion and accretion disks · Protoplanetary disks · Atomic, molecular, chemical, and grain processes

The Delivery of Water to Protoplanets, Planets and Satellites
Edited by Michel Blanc, Alessandro Morbidelli, Yann Alibert, Lindy Elkins-Tanton, Paul Estrada, Keiko Hamano, Helmut Lammer, Sean Raymond and Maria Schönbächler

✉ L. Hartmann
lhartm@umich.edu

F. Ciesla
fciesla@uchicago.edu

O. Gressel
gressel@nbi.ku.dk

R. Alexander
richard.alexander@leicester.ac.uk

1 Dept. of Astronomy, University of Michigan, Ann Arbor, MI, USA

2 Dept. Geophysical Sciences, University of Chicago, Chicago, IL, USA

3 The Niels Bohr Institute, Blegdamsvej 17, 2100 Copenhagen Ø, Denmark

4 Dept. of Physics & Astronomy, University of Leicester, University Road, Leicester, LE1 7RH, UK

1 Introduction

The first stages of planet formation begin within protoplanetary disks—the cloud of dust and gas that circle young stars during their first few million years of evolution. These disks form at the same time as their central stars, as a molecular cloud core collapses under the force of gravity. While much of the mass in the core eventually coalesces at the center to form a young star, the large amount of angular momentum in the core dictates that some of the mass settles into orbit around that star. It is within these disks that solid particles aggregate to form the seeds of planets and satellites, with the chemical compositions of these solids determined by the different environments that they are exposed to within the disk.

The physical and chemical conditions within protoplanetary disks evolve over millions of years as mass and angular momentum are transported during the final stages of their star's growth. Their structures are determined by a complex interplay between dust evolution, which affects the disk temperature, density and ionization distributions through absorption of light from the central star and any internal viscous heating, and transport processes which cause accretion, diffusion, and possibly turbulence (each of which affect the dust distribution). In particular, the specific locations of water ice and vapor within the disk, and therefore the implications for the possible delivery of water to terrestrial planets, remain uncertain due to uncertainties in transport, viscous heating, and dust growth and settling, as we discuss in this review.

1.1 Protoplanetary Disk Basics

The standard reference point in discussing the structure of protoplanetary disks is the Minimum Mass Solar Nebula (Weidenschilling 1977; Hayashi 1981). The idea for this hypothetical disk came from trying to reconstruct the solar nebula by taking the mass of planets in our Solar System and distributing it in a series of "feeding zones" for each planet to recover a smooth distribution. Hydrogen gas was then added to recover a solar mix of elements (dust-to-gas ratio of ~ 0.01 when water ice was condensed and ~ 0.005 when water remained as a vapor). The total surface density (mass per unit area in a given annulus) of such a disk is given by:

$$\Sigma(r) = 1700\left(\frac{r}{1\ \text{AU}}\right)^{-\frac{3}{2}}\ \text{g}\,\text{cm}^{-2}. \tag{1}$$

This gives a total disk mass out to 100 AU of $\sim 0.01 M_\odot$. The temperature structure was found by determining the directly-irradiated blackbody temperature as a function of distance from the Sun (Hayashi 1981):

$$T(r) = 280\left(\frac{r}{1\ \text{AU}}\right)^{-\frac{1}{2}}\ \text{K}. \tag{2}$$

It is not clear how accurately the above structure reflects the properties of the solar nebula. For example, there are significant uncertainties in the temperature structures of protoplanetary disks because they are optically thick, at least to the star's radiation and in parts to their own internal radiation field, and viscous dissipation may serve as an additional source of heating, so that the application of Eq. (2) is questionable (see discussion in Sect. 1.4). As we discuss further below, disks are also evolving due to a variety of processes, transporting mass inward to be accreted by the central star or losing mass due to photoevaporative radiation; thus the surface density and temperature are unlikely to be constant in time. Moreover,

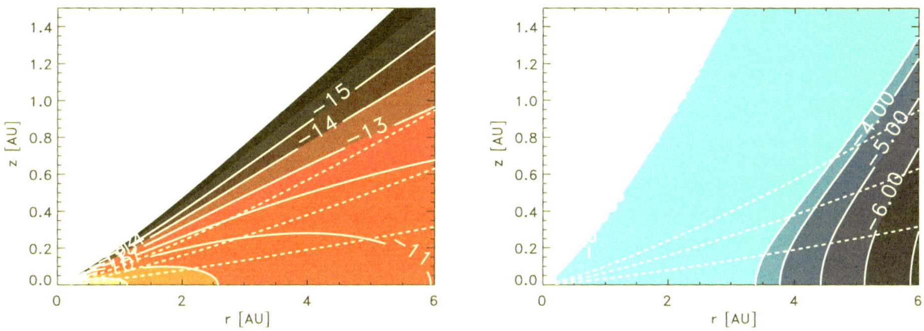

Fig. 1 *Left*: Two-dimensional density structure of the inner regions of the Minimum Mass Solar Nebula as described in the text. *Contours* represent $\log_{10}$ of the total gas mass density. *The dashed lines* represent the heights of 1, 2, and 3 scale heights within the disk. *Right*: Two-dimensional distribution of water vapor in the disk. Canonical H_2O/H_2 number ratios are 5×10^{-4}, which are found inwards of the midplane snow line (just inward of 3.5 AU here). *Contours* are $\log_{10}$ of the H_2O/H_2 ratio

the mass of the MMSN is a *minimum* by definition, and other models suggest higher surface densities (Desch 2007).

Despite these issues, the MMSN serves as a reference model to illustrate key properties and structures by which we can understand real protoplanetary disks. Proceeding further, one must consider the vertical structure of the disk, assumed to be set by hydrostatic equilibrium as follows:

$$\rho_g(z) = \frac{\Sigma}{(2\pi)^{1/2} h_g} \exp\left(-\frac{z^2}{2h_g^2}\right), \tag{3}$$

where ρ_g is the z is the height above/below the disk midplane and h_g is the disk scale-height, with $h_g = c_s/\Omega$, where c_s is the local speed of sound and Ω the local Keplerian frequency. For simplicity, the disk is assumed to be vertically isothermal, allowing c_s and h_g to be constant with height. Figure 1 shows the plot of the gas density structure for the MMSN around the terrestrial planet region of the Solar System. In the absence of any dust dynamics (see below), small dust grains will be suspended within the disk, at least initially, determined largely by whether water is condensed or not as described above.

1.2 Snow Line

A key element in the discussion of water is the location of the snow line (SL) in the disk; that is the distance from the star where water begins to freeze-out and become a solid. Any water vapor that moves outward from this location condenses, while any water ice that migrates inward evaporates, and thus this location is also referred to as the water ice condensation/evaporation front. This location can be determined from both equilibrium and kinetic arguments. In the equilibrium approach, the snow line is set based on thermodynamic data; that is, it is the location in the disk where, if all water was present as a vapor in the disk, its partial pressure would exceed the equilibrium vapor pressure at the local temperature.

An alternative approach defines the location of the snow line based on chemical kinetics, which is useful as thermodynamic equilibrium is not always achieved in a protoplanetary disk. In this manner, the snow line can be determined by finding where the freeze-out (adsorption) flux of water molecules onto grains equals the desorption flux of molecules off of grain surfaces. Following Bergin et al. (2013), the two fluxes can be expressed as:

$$F_{\rm ads} = 0.25 n_{\rm H_2O} v_{th}, \tag{4}$$

$$F_{\rm des} = \nu_0 \exp(-E_b/kT) N_s, \tag{5}$$

where $n_{\rm H_2O}$ is the local number density of gaseous water molecules, v_{th} is the thermal velocity of those water molecules, E_b is the experimentally determined binding energy of water onto a grain surface, N_s is the density of adsorption sites on the surface of a grain ($\sim 10^{15}$ cm^{-2}), and ν_0 is a vibrational frequency of order 10^{12} s^{-1}. In instances where $F_{\rm des} > F_{\rm ads}$, water remains completely in the vapor phase. When $F_{\rm ads} > F_{\rm des}$, water molecules begin to freeze-out onto dust grains until $n_{\rm H_2O}$ drops to a value such that the two fluxes are equal. Taking the water (vapor + ice) to hydrogen ratio throughout the disk as a constant $n^{Tot}_{\rm H_2O}/n_{H_2} \sim 5 \times 10^{-4}$ (Lodders 2003; Cleeves et al. 2014) and typical protoplanetary disk midplane ($z = 0$) conditions, the snow line will generally occur at temperatures of $T \sim$ 120–170 K. These temperatures are in general agreement with those predicted from equilibrium studies (e.g. Lodders 2003).

The snow line is often pictured as being a singular radial location in the disk, $R_{\rm SL}$, the point at the disk midplane where $F_{\rm ads} = F_{\rm des}$ can only be maintained if some of the water in the vapor phase is removed and put onto solids. However, due to the decrease in gas density with height above the disk midplane, the snow line actually has a vertical structure. If the disk is vertically isothermal, $F_{\rm ads}$ will decrease exponentially with height (following the background gas density as assumed in Eq. (3)), while the desorption flux will remain constant. Thus, the snow line will begin at the disk midplane at $R_{\rm SL}$, with only a small fraction of water being incorporated into solids and the rest remaining as vapor with height. As seen in Fig. 1. for $r > R_{\rm SL}$, the amount of water in the disk present as ice increases; this is due both to the increasing height of the snow line and the drop of temperatures leading to more water being removed from the gas phase.

1.3 Dust and Gas Dynamics in Protoplanetary Disks

Upon formation of a protoplanetary disk, the dust grains and gas are generally thought to be uniformly mixed, with $\sim 99\%$ of the mass found in gas and $\sim 1\%$ as condensed solid grains ranging from submicron to microns in size (Lodders 2003). Over time, the solids and gas will be subjected to a series of dynamic processes which will lead to them evolving differently from one another and aid in the formation of planetary building blocks.

While the vertical structure of the gas in a protoplanetary disk is determined by the hydrostatic balance between the component of stellar gravity that pulls material towards the disk midplane with the pressure of the gas, solids feel less pressure support and will settle through the gas toward the midplane. Larger solids settle more rapidly than fine dust. Turbulence, if present, will work to offset the growing concentration of solids at the disk midplane, lofting materials to higher altitudes again. The extent to which solids of a given radius, a, settle in a disk is determined by the dimensionless stopping time of the particle:

$$\tau_s = \frac{\rho a}{\rho_g c_s} \Omega \tag{6}$$

where ρ is the material density of the solid. (This quantity is also referred to as the Stokes number in the literature. e.g. Cuzzi et al. 1993). The amount of turbulence in a disk is uncertain, as described below. Models typically quantify it by defining a dimensionless parameter,

α which produces an effective viscosity, $\nu = \alpha c_s h_g$. This turbulence would also produce a diffusivity within the gas which is often taken as D as being of order ν, though differences in magnitude are possible. If we assume $D = \nu$ for the particles we are interested in, then those with $\tau_s/\alpha \ll 1$ will remain suspended throughout the height of the disk (giving them a dust layer thickness of $h_d \sim h_g$), while larger particles will settle into a dust layer with thickness (Youdin and Lithwick 2007):

$$h_d = \left[1 + \frac{\tau_s}{\alpha}\left(\frac{1+2\tau_s}{1+\tau_s}\right)\right]^{-\frac{1}{2}} h_g. \tag{7}$$

It is worth noting that larger particles, as they decouple from the gas, will have effective diffusivities less than that of the gas. Turbulence may also redistribute materials radially in the disk, but this is only important for gases and small solids (those small enough to remain coupled to the gas with $\tau_s < \alpha$).

Larger particles will also begin to decouple from the gas through their radial motions in the disk. While the gas in the disk orbits its central star, its velocity is usually slightly less than Keplerian due to the generally outward pressure gradient (hot, dense gas near the star; cool, sparse gas further out). Thus, while the star exerts a gravitational force on the gas, the net central force is somewhat less due to the pressure support it experiences. Solid particles get negligible support from this pressure gradient, and thus would otherwise orbit the star at Keplerian rates. As a result, the solids experience a headwind in their orbits around the star. Solids with $\tau_s \ll 1$ continue to remain coupled to the gas and are minimally affected by aerodynamic drag, while those with $\tau_s \gg 1$ have so much inertia that the headwind can largely be ignored. However, those particles with $\tau_s \sim 0.01$–10 (typically mm to meter sizes) drift inwards as a result of the loss of energy and angular momentum arising due to the headwind.

The collective effect of these dynamical mechanisms results in a complex evolution of the dust-to-gas ratio as a function of time, with important but uncertain consequences for the building of large bodies and the global water distribution, as discussed next.

1.4 Viscous Heating and Radiative Transfer Effects

The simple optically-thin equation (2) for the disk temperature is unrealistic because protoplanetary disks are optically thick to radiation from the central star. Beyond a few AU or so, temperatures are dominated by irradiation from the central star. The precise heating rate is dominated by the "flaring" of the disk (i.e., the layers which absorb the starlight lie at heights $z/R \propto R^{\eta}$, $\eta > 0$; Kenyon and Hartmann (1987)). In the case of dust particles with properties similar to that of the interstellar medium (ISM) that are well-mixed with the gas, the simple model of Chiang and Goldreich (1997) for optically-thick disk absorption of starlight yields an effective (disk surface) temperature distribution $T_{e,r} \propto R^{-3/7}$, while more detailed models yield a slightly steeper dependence on radius,

$$T_{e,r} \approx 220 L_1^{1/4} R_{AU}^{-1/2}\ \mathrm{K} \tag{8}$$

(D'Alessio et al. 1998, 2001), where the stellar luminosity is in units of solar. However, the *midplane* temperature for such optically-thick disks is considerably lower, often by a factor of two at a few AU. It appears that irradiation heating by itself is not able to move the water snow line significantly out beyond 1 AU, which poses problems for understanding how chondritic meteorite parent bodies, thought to form outside of ~ 1 AU late in solar nebula evolution, are water-poor (e.g. Abe et al. 2000).

The apparent limitations of stellar irradiation for explaining the variation of water abundances in the solar system has led to proposals that accretion heating is the dominant factor setting the disk central temperature (e.g., Garaud and Lin 2007; Kennedy and Kenyon 2008; Bitsch et al. 2015). To fix ideas, consider a toy model of a optically-thick, steady, viscously accreting disk, with effective temperature of

$$T_{e,a} = \left(\frac{3GM\dot{M}}{8\pi\sigma R^3}\right)^{\frac{1}{4}} \sim 85 M_1 \dot{M}_{-8} R_{AU}^{-3/4}\ \mathrm{K}, \tag{9}$$

where we have assumed fiducial values of a central star of one solar mass and an accretion rate of $10^{-8} M_\odot\,\mathrm{yr}^{-1}$ as typically observed. If the disk is very optically thick and the viscous dissipation occurs near the midplane, radiative trapping can increase the central temperature $T_{c,a}$ over $T_{e,a}$ by a factor of $\sim (3\tau/8)^{1/4}$ (Hubeny 1990), where τ is the vertical Rosseland mean optical depth. The optical depth is $\tau = k_R \Sigma/2$, where k_R is the Rosseland mean opacity and Σ is the surface density of the disk, which is related to the mass accretion rate $\dot{M}$ in the standard (vertically-isothermal) viscous disk as

$$\dot{M} = 3\pi\alpha c_s^2 \Sigma \Omega^{-1}, \tag{10}$$

where α is the usual parameterization of the viscosity (see Sect. 3) and c_s is the sound speed. The result is

$$T_{c,a} = \left[\frac{3}{64\pi^2}\frac{GM\dot{M}^2}{\sigma R^3}\frac{\mu\Omega k_R}{\alpha R_g}\right]^{1/5} = 330 M_1^{3/10} \dot{M}_{-8}^{2/5} k_R^{1/5} \alpha_{-3}^{-1/5} R_{AU}^{-9/10}\ \mathrm{K}, \tag{11}$$

where α_{-3} is in units of 10^{-3} and R_g is the gas constant, and for simplicity we have ignored any temperature or density dependence of k_R.

Assuming ice condenses at temperatures 170–120 K, and setting $k_R \sim 1\ \mathrm{cm}^2\,\mathrm{g}^{-1}$, the crude estimate of Eq. (11) indicates a snow line at ~ 2.1–3.1 AU for the fiducial parameters, reasonably consistent with more detailed calculations (e.g., Bitsch et al. 2015, see Fig. 2). However, this estimate may well be an upper limit to the increase in temperature due to viscous heating. As discussed in Sect. 1.3, dust evolution is likely to reduce the optical depth, as growth beyond the ISM grain size distribution will have lower opacities. Empirical estimates suggest depletion of small dust in the upper disk layers by factors of 10^{-1} to 10^{-3} (Sect. 2.1, Furlan et al. 2006); presumably this is the result of growth as the large particles settle to the midplane (as needed to explain mm emission; D'Alessio et al. 2001). The lower panels of Fig. 2 show that the snow line moves in to $R \sim 1$ AU for a depletion of 10^{-2} of small dust in the upper layers. Theory suggests that such depletion should occur rapidly (depending upon the level of turbulence present), so the assumption of no depletion seems problematic.

Higher accretion rates and fragmentation to replenish small dust grains can help compensate for reduced trapping of heat by dust depletion (e.g., Bitsch et al. 2015). However, mass accretion rates drop significantly with time, on timescales of 3 Myr or less (Sect. 2.2; Hartmann et al. (2016)) and dust growth even with fragmentation is rapid (Sect. 2.1). Moreover, it is not clear that viscous dissipation occurs mostly near the midplane, or even that viscosity is important in transporting angular momentum in large regions of the disk, making its importance as a heat source questionable (see Sect. 3).

Major outbursts of rapid accretion—up to $10^{-4} M_\odot\,\mathrm{yr}^{-1}$—are known to occur in some young objects (the FU Ori variables; Hartmann and Kenyon 1996), whose disks are demonstrably internally-heated and thus viscous-dissipation dominated. Such high accretion rates

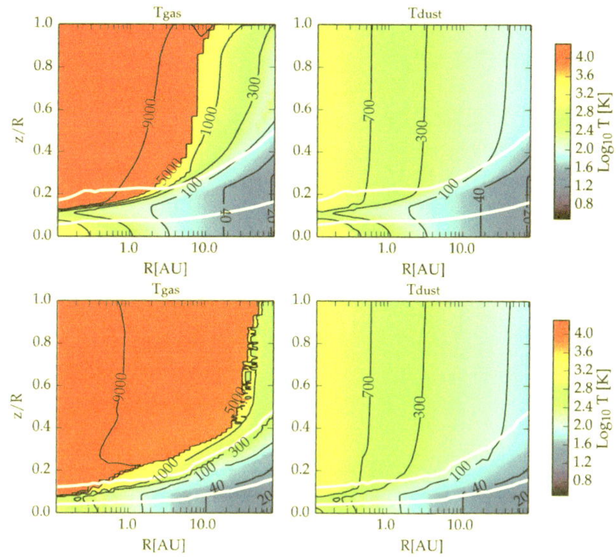

Fig. 2 *Upper panels*: gas (*left*) and dust (*right*) temperatures for a detailed radiative transfer model with viscous heating with $\dot{M} = 10^{-8} M_\odot \, yr^{-1}$, $\alpha = 10^{-3}$. The SL at the midplane lies at $R \sim 2$–3 AU, in agreement with Eq. (11) for the toy model. *Lower panels*: the same global model properties, but now assuming that small dust is depleted by a factor of 100 in the *upper layers*. Now the SL moves inward to ~ 1 AU, due to the decreased trapping of the viscously-generated radiation. Courtesy Nuria Calvet and Ramiro Franco Hernandez

can move snow lines farther out (Cieza et al. 2016), but these large outbursts are confined to early evolutionary phases (ages < 1 Myr); any remnants of these episodes may well be accreted into the star by the time of planetesimal formation.

2 Observational Constraints on Disk Structure and Evolution

The density and internal temperature structure of protoplanetary disks are not well constrained. Most of our direct information is limited to scales of several to hundreds of AU. The inner disk structure near the expected snow line at a few AU (for a solar-luminosity star) is not easily resolvable with current techniques, and there are indications that inner disks are optically thick (due to dust) at mm wavelengths. One therefore must work to connect models and observations on larger scales to attempt any extrapolation to the water SL.

2.1 Dust

Most of what we infer about the radial and vertical temperature and density distributions of disks comes from observations of dust emission, as cold molecular hydrogen is not detectable and most other molecular tracers are either optically-thick, are activated only in disk upper layers, or are subject to uncertainties in chemical reaction rates. As shown in Fig. 3, we infer that the dust in disks around solar-mass pre-main sequence (T Tauri) stars exhibit a sort of "two layer" structure, with large dust concentrated near the disk midplane and small dust suspended in upper layers to a few scale heights. The 10–20 μm silicate emission features seen in most T Tauri disks can be explained by dust with sizes smaller than a few μm, while the amount of mm-wave emission is only explainable with reasonable dust masses if the grains have sizes of order 1 mm or larger (e.g., D'Alessio et al. 2001). Placing the small dust in upper layers where the vertical temperature inversion (caused by irradiation due to the central star) results in emission features, while distributing the large dust closer to or in the midplane avoids the elimination of silicate features.

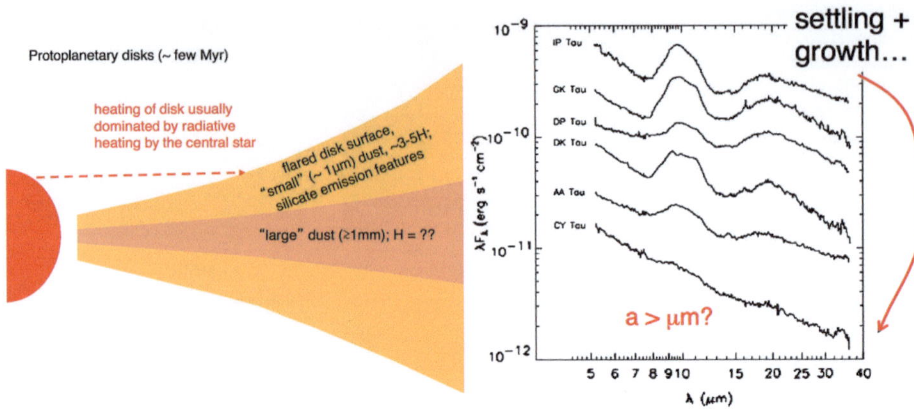

Fig. 3 The *left panel* schematically indicates structure of solar-mass pre-main sequence (T Tauri) disks. Small dust must be mixed with the gas to at least a few scale heights to explain the observed infrared dust emission (*right panel*), both in terms of a continuum and also to explain the silicate emission feature at $\lambda \sim 10$ μm. Large (mm) dust must be present to explain the long-wavelength emission; presumably this dust resides in a relatively thin layer at the disk midplane. The set of mid-infrared spectral energy distributions in the right panel, taken from Furlan et al. (2006), is suggestive of a sequence of increased settling of dust, making the infrared excess fall off faster to longer wavelengths, along with a decrease in the strength of the silicate emission feature which may indicate grain growth. Note that while some stars exhibit emission consistent with amorphous silicates (e.g., IP Tau), others clearly show the presence of crystalline silicate particles (e.g., DK Tau)

As dust grows and settles to the midplane (see Sect. 1.3), small dust becomes depleted in the upper layers. Indeed, theoretical models of this process in the absence of significant turbulence suggest this should happen quickly, posing challenges to explain the observed existence of suspended small dust in disks of ages as small as ~ 1 Myr (Dullemond and Dominik 2005; Birnstiel et al. 2011). Empirical estimates of small dust depletion in upper disk layers are difficult to make with any accuracy, but current models to explain observations suggest depletions by factors of 10 to 1000 (D'Alessio et al. 2001; Furlan et al. 2006), with significant implications for radiative trapping of viscous heat (Sect. 1.4).

Efforts to demonstrate grain growth generally are based on observations of spectral indices in the mm-wave region. Small ISM dust has a steep dependence of opacity on frequency $k_\nu \propto \nu^{-\beta}$, with $\beta \sim 1.7$, as a consequence of the limited sizes of the largest grains ($\sim 0.3\mu$). Observations of mm-wave and cm-wave emission from disks at distances of $\sim$ 20–100 AU indicate much smaller values of $\beta \sim 1$ or even 0, indicating growth to at least mm or even cm sizes (Pérez et al. 2012, 2015; Tazzari et al. 2015). Unfortunately, maximum particle sizes cannot be determined once dust grows in size beyond the wavelengths of observation; in this case β is determined by the shape of the particle size distribution. As emission at cm wavelengths is generally quite weak in outer disks (50–100 AU), currently it is not possible say anything about $\gtrsim$ cm-sized dust at large radii.

Given the uncertainties in dust (and solid) size distributions, dust disk masses must be considered to be lower limits; that is, what is being measured at mm wavelengths is only the amount of mass in solids of sizes ~ 100 μm to $\sim$ a few mm. The survey of (Andrews and Williams 2005) yielded a median "disk mass" (100× the solid mass) of $\sim 0.005 M_\odot$. Statistical estimates from gas accretion yield somewhat larger values (Sect. 2.2).

Another complicating effect in addition to vertical settling is radial drift. This effect has been seen in a number of objects, most clearly in the nearby T Tauri star TW Hya with

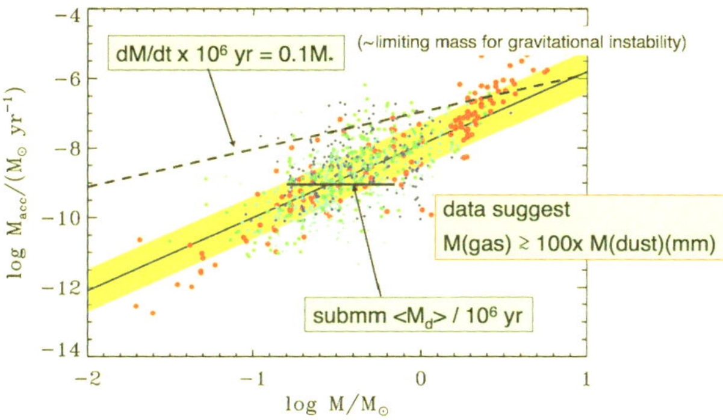

Fig. 4 Observationally estimated accretion rates of T Tauri stars at typical ages $\sim$ 1–2 Myr as a function of stellar mass. There is a clear trend of increased mass accretion with increasing stellar mass, though there is substantial scatter. The *dashed line* indicates the rate that would result in accreting 10% of the stellar mass from the disk; it seems unlikely that disks can be considerably more massive than this, as they would be gravitationally unstable. (Stars with masses $> 1 M_{\odot}$ appear to have shorter average disk lifetimes.) The data suggest that T Tauri stars typically accrete a mass equal to the inferred dust masses, suggesting that the mm-wave emission analysis generally underestimates the gas mass. Modified from Hartmann et al. (2016)

mm-dust grains having a sharp edge at $\sim$ 60 AU, while CO gas emission is seen out to $\sim$ 200 AU (Andrews et al. 2012). As drift of mm–cm sized particles is quite rapid, this causes some problems in understanding the existence of dust in this size range out at $\gtrsim$ 50 AU, particularly in this relatively old object (age $\sim$ 10 Myr).

2.2 Gas

The only current constraint on the mass of molecular hydrogen in disks is the estimate of $M_d \sim 0.06 M_{\odot}$ from analysis of HD in TW Hya (Bergin et al. 2013). Even this result is model-dependent, as it relies both on the assumed isotopic ratio and, more importantly, the disk temperature structure, as HD does not probe the coldest regions where most of the mass resides.

Williams and Best (2014) attempted to use rotational emission lines of ^{12}CO, ^{13}CO, and $C^{18}O$ isotopologues to correct for optical depth effects and variations in gas temperatures as a function of radius and height. Assuming a conversion of CO to H_2, they derived gas masses of order $10^{-3} M_{\odot}$, an order of magnitude smaller than expected from dust masses (times 100). More importantly, such gas masses are also an order of magnitude too small to sustain typical accretion rates of $10^{-8} M_{\odot}\ \mathrm{yr}^{-1}$ over timescales of 1–3 Myr (see below). One possible resolution is that CO freezes out onto grains, removing it from the gas phase and thus rendering it undetectable in mm-wave rotational line emission.

Gas mass accretion rates ($\dot{M}$) are important in providing statistical upper limits to the possible viscous heating in disks. $\dot{M}$ increases with increased stellar mass and decay with age. The best fit for T Tauri stars with a median age of about 1–2 Myr derived by G. Herczeg is

$$\log(\dot{M}/M_{\odot}\ \mathrm{yr}^{-1}) \approx -7.9 + 2.1 \times \log(M_*/M_{\odot}) \tag{12}$$

(see Hartmann et al. 2016). As one can see from Fig. 4, there is a large scatter at any mass, more than the factor $\sim$ 0.5 dex expected for observational errors. Accretion rates also generally decrease with increasing age (Fig. 5), although the rates of decay are uncertain, due

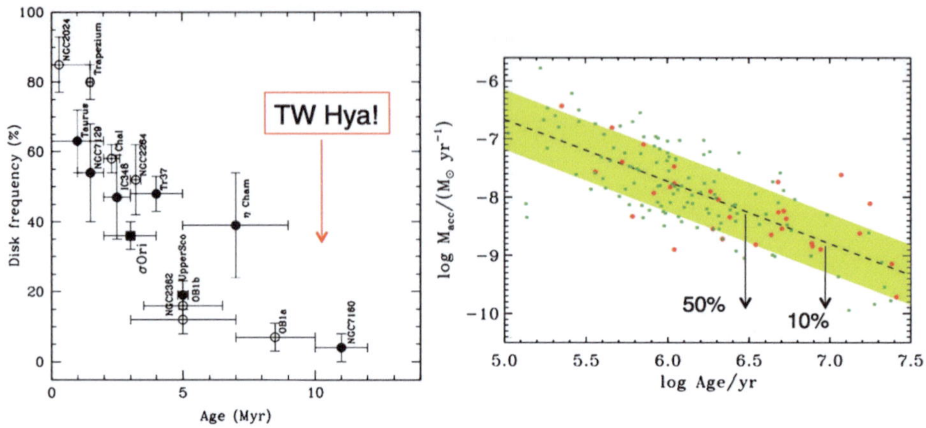

Fig. 5 *Left*: frequency of detectable infrared dust disk emission in the IRAC bands 3.5–8 μm as a function of group or cluster age. From Hernández et al. (2008). *Right*: dependence of mass accretion rates on age, scaled to account for the mass dependence of Eq. (12) for stars with masses in the range $0.3–1.0 M_\odot$ (see also Fig. 4). The mass-accretion rate relationship is calculated for objects with mass between $0.05–1 M_\odot$. The resulting fit is uncertain due to correlated errors (see text). The *arrows* emphasize that significant numbers of accreting disks "drop out" with increasing age. From Hartmann et al. (2016)

in part to difficulties in assigning ages to individual young stars (Soderblom et al. 2014). This decrease in $\dot{M}$ with time, along with the likely depletion of small dust with increasing age, indicates that the snow line should move inward during the main optically thick disk lifetime (Sect. 1.4).

The "lifetimes" of disks as judged from either infrared dust emission (left panel of Fig. 5) or from the ending of detectable accretion onto the central star (Fedele et al. 2010), exhibit a significant spread. The "usual" disk lifetime is often given as 3 Myr, but a few (optically-thick) disks survive to ~ 10 Myr, while others last for $\lesssim 2$ Myr. It appears that both dust emission and accretion disappear fairly rapidly (see below), so there must be something like a two-phase behavior, with a moderate decay lasting for 1–10 Myr and a rapid decrease over < 1 Myr. This rapid clearing could be produced by photoevaporative mass loss from the disk driven by X-ray or other high-energy radiation (Alexander et al. 2006; Owen et al. 2011), or by giant planet formation.

2.3 Gaps and Holes

The advent of sensitive, high-spatial resolution mm-wave interferometry, especially using the Atacama Large Millimeter and Submillimeter Array (ALMA), has revealed that major radial structuring of protoplanetary disks is far more common than previously realized. Disks of substantial masses but with inner holes strongly depleted in small dust have been increasingly recognized (the "transitional disks"; Calvet et al. 2002; Espaillat et al. 2014; Fig. 6). In addition, there are also disks with inner disk dust as well as a large gap (the "pre-transitional disks"; Espaillat et al. 2007). Recent observations suggest that the gas within these holes and gaps (Bergin et al. 2003; Najita et al. 2007; Ingleby et al. 2009) is also depleted, although by smaller factors than of the dust (van der Marel et al. 2016). Because observations still generally detect only the largest holes and gaps, the new results suggest that there may not be any truly "primordial" or "protoplanetary" disks once the infall of envelope material to the disk during the protostellar phase has ended. Significant formation

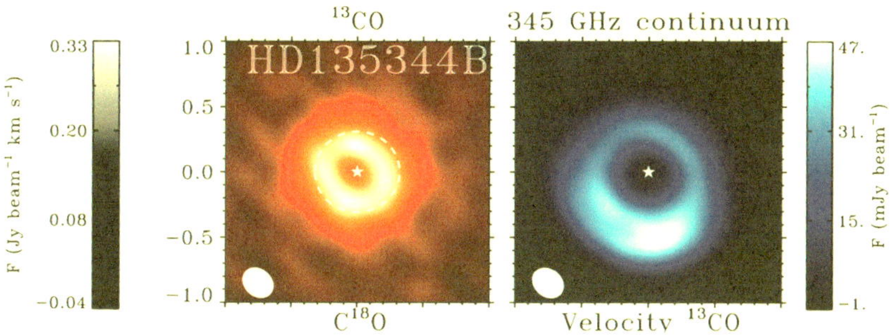

Fig. 6 ALMA imaging of the $1.6 M_\odot$, $8 L_\odot$ young star HD 135344B, showing that gas emission ($C^{18}O$) is present inside the dust disk as seen in mm-wave emission. Similar structures are seen in several other systems, suggesting that planet formation has trapped large dust particles in the outer disk while allowing some gas to move inward, helping to explain the observed gas accretion onto central stars (e.g., Ingleby et al. 2014). The non-axisymmetric structure in the dust may indicate a vortex, suggesting perturbations by planets. From van der Marel et al. (2016)

of planetesimals and embryos, if not planetary cores, probably begins very early, $\lesssim 1$ Myr, especially if giant planets are responsible for creating large disk holes and gaps. This is consistent with evidence from the Solar System; rapid formation of planetesimals is supported by radiometric dating of iron meteorites, which suggest planetesimals formed in our solar nebula concurrently or within 10^5 years of the formation of CAIs, the oldest objects in our Solar System, and which define $t = 0$ in cosmochemical clocks (Kruijer et al. 2014). Such bodies would predate the accretion of chondritic meteorite parent bodies by 2–3 Myr (Connelly et al. 2012).

An interesting sidelight to the presence of gaps and holes strongly cleared of small dust in young disks is the possibility that even rapid planetesimal and embryo formation might occur in optically-thin regions (see Sect. 5).

2.4 Observations of Water in Protoplanetary Disks

Water vapor has been detected in several T Tauri and Herbig Ae/Be stars (Carr and Najita 2011; Brittain et al. 2015; Blevins et al. 2015). H_2O emission lines are seen superposed on the strong dust continuum, showing that they arise in upper layers of the disk, with typical gas temperatures ~ 600 K (Carr and Najita 2011). Unfortunately, translating these observations of surface water vapor, estimated to originate at radii ~ 0.5–5 AU, into midplane snowline positions is highly model-dependent (Blevins et al. 2015), particularly because the high temperatures of the upper disk gas layers tend to reflect heating by stellar X-rays and ultraviolet radiation (Glassgold et al. 2009).

Detection of where water starts to freeze out onto grains has proven to be more challenging. Even with the vast improvements made by the ALMA interferometry, spatial resolution is limited to a few AU in the nearest star-forming regions, and to the brightest objects. As a check on our general understanding of disk temperature structures, CO snow lines have been detected indirectly at scales of tens to nearly 100 AU (Qi et al. 2013), depending upon the luminosity of the central star. CO was not directly observed; instead, the inner edge of emission from N_2H^+ was taken to be the position of the CO snow line because this ion is only abundant in regions where CO is depleted from the gas. Similar tracers do not exist for H_2O, making detection of the water snow line a challenge. Further, the likely positions of water snow lines in the nearest protoplanetary disks are at the edge or below current imaging capabilities. Banzatti et al. (2015) suggested that the water snow line might be inferred

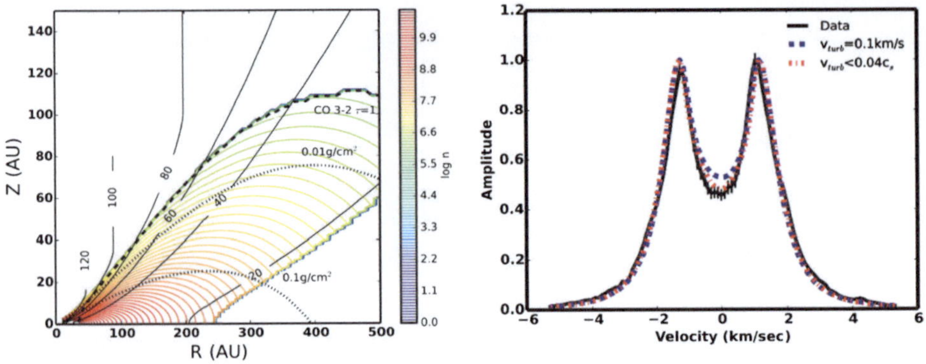

Fig. 7 *Left*: disk model constructed to explain the CO 3-2 emission from the disk of HD 163296. The *black solid curves* denote isotherms in K, the *dotted curves* show contours of constant vertical column density, and the *colored curves* show isodensity contours for the CO lower level, with the *upper heavy dashed curve* showing the layer where the line becomes optically thick. The *lower right* region shows where CO freezes out. This model produces the emission line profile shown in the *right panel*; the sharpness of the peaks and overall profile shape (modeled spatially) provides an indication that any turbulent broadening must be small. From Flaherty et al. (2015)

from a change in the spectral index due to the decrease in dust size and optical properties as water ice is sublimated, but the potentially optically-thick nature of the inner disks may make these observations difficult. Crystalline water ice has been detected in a few disks from spatially-unresolved observations (McClure et al. 2015), but these observations undoubtedly trace outer regions at tens of AU or more.

2.5 Observational Limits on Turbulence

The magnitude of turbulence in planet-forming disks plays an important role in allowing (or preventing) coagulation/growth of solids, as well as affecting both mass and angular momentum transport. Unfortunately observational constraints on turbulence are extremely limited. Observations of the first-overtone lines of CO have yielded estimates of transsonic turbulence in the upper layers of disks at radii $\lesssim 1$ AU in a few systems, but these objects are either relatively luminous (Najita et al. 1996, 2009) or very rapidly accreting (Hartmann et al. 2004), such that temperatures are likely high enough to support the magnetorotational instability (MRI) via thermal ionization (Sect. 3). The most detailed limits so far for colder disk regions come from the recent ALMA study of the large disk surrounding the $L \sim 35 L_{\odot}$ A-type star HD 163296 (Flaherty et al. 2015). Using a sequence of CO isotopologues to deal with issues of temperature structure, excitation, and optical depth, Flaherty et al. conclude that non-thermal motions in the upper disk layers are no more than about 3–6% of the local sound speed at $R \gtrsim 30$ AU (see Fig. 7), an order of magnitude less than expected for full MRI turbulence, though still significant enough to play a role in mixing and transport of disk materials.

3 Turbulence, Winds, and Mass and Angular Momentum Transport

Turbulence can play an important role in disk density and temperature structures via its effects on viscous dissipation and mass transport. As discussed in Sect. 2.2, to maintain observed accretion rates over evolutionary timescales seems to imply transport of a significant fraction of initial disk masses into the central star. Turbulence can also affect the amount of heating by light from the central star if it stirrs up small dust grains (Dullemond and Dominik 2005), enhancing their vertical scale height relative to the gas (Fromang and Papaloizou 2006).

In the case of fully-ionized disks, the magneto-rotational instability (MRI: Balbus and Hawley 1998) produces efficient turbulent angular momentum transport, with viscosities parameterized by the parameter $\alpha \sim 10^{-1}$–10^{-2} (Sect. 1.4). However, as they are comparatively dense and cold, protoplanetary disks (PPDs) are generally expected to have extremely low levels of ionization deep in their interiors (see Turner et al. 2014, for a recent review). Only inner ($\lesssim 0.3$ AU) regions of protoplanetary disks with higher temperatures and surface layers ionized by hard radiation (such as FUV photons, X-rays, cosmic rays) from the magnetically active central star (see Sect. 2.5 in Turner et al. 2014), or potentially from an actively star-forming nearby cluster environment (Kalyaan et al. 2015), can result in effective coupling of the ions with the neutral gas. Cosmic rays could be especially important in ionizing gas, particularly in the outer disk where column densities are low, but deflection by the stellar wind or magnetic fields might prevent them from becoming important (Cleeves et al. 2013).

The ionizing events from external sources are balanced by recombination with gas-phase ions and, in the presence of sub-micron-sized dust, by absorption on the grain surface. Based on estimates of the abundance of free electrons and the collision regimes characteristic of various regions within the disc, it is now commonly accepted that all three non-ideal MHD effects, namely, Ohmic resistivity, ambipolar diffusion (AD), and the Hall term have to be considered in the generalized Ohm's law to arrive at sound conclusions for the overall dynamical state of PPDs (e.g. Salmeron and Wardle 2003; Bai 2011).

3.1 Ambipolar Diffusion Affected Laminar Disks

The inclusion of ambipolar diffusion has strongly modified the original layered model assuming only Ohmic resistivity, with an MRI-active upper disk layers providing the angular momentum transport and viscous heating (Gammie 1996). Studies by Bai and Stone (2013), Bai (2013), and Simon et al. (2013a,b) show that in the absence of net vertical flux, the MRI is typically suppressed in wide radial stretches of typical T Tauri disks and standard solar-nebular models, and the resulting state is a laminar flow with negligible angular momentum transport. Moreover, otherwise laminar regions in the inner disk (roughly speaking $R \simeq 1$–20 au) can launch a magnetocentrifugal wind within the FUV-ionized surface layers. Such AD-assisted magnetic winds were actually explored in prescient work by Königl (1989) and Wardle and Königl (1993), who developed self-similar steady disk solutions matching to the wind solutions of Blandford and Payne (1982), and illustrate many of the main features of subsequent numerical simulations Bai and Stone (2013), Simon et al. (2013a), Lesur et al. (2014). However, due to technical limitations (such as finite computational volumes), mass loading and mass loss rates are uncertain. Since the role of the far-UV ionized upper disk surface is crucial (Perez-Becker and Chiang 2011), the shielding of the required energetic photons needs to be assessed to corroborate the foundation of the mechanism.

A noteworthy feature of AD laminar accretion disks is the existence of thin current layers, which have both been found in the local models of Bai and Stone (2013), and the global ones of Gressel et al. (2015). These current layers can produce strong radial streams of material (at ~ 30–40% of the sound speed), and can potentially provide non-negligible amounts of radial mass transport. As has recently been argued by Bai et al. (2016), laminar MHD winds can in principle team up with photoevaporative acceleration (e.g. Alexander et al. 2006) to remove mass and limit disk lifetimes (see Sect. 3.3).

3.2 Magnetic Self-Organization and the Hall Effect

Although suggestions of the importance of the Hall effect in PPDs have been made for some time (e.g., Wardle and Königl 1993), it has only recently been included in detailed numerical simulations. Complicating matters, the dynamical outcome of the Hall effect depends upon the mutual sign of the rotation vector (represented by $\boldsymbol{\Omega}$) and large-scale magnetic flux (represented by $\mathbf{B}$). The Hall drift has the effect to either boost (in the case of $\boldsymbol{\Omega} \cdot \mathbf{B} > 0$) or impede (conversely for $\boldsymbol{\Omega} \cdot \mathbf{B} < 0$) linear growth (Lesur et al. 2014; Bai 2015) of the MRI. Notably, with a somewhat modified ionization model, Simon et al. (2015) found bursty behavior in the latter case.

While the MRI relies on the combination of field induction and plasma inertia (described by perturbations $\delta\mathbf{v}$ and $\delta\mathbf{B}$), there is an even simpler instability that strictly only relies on the induction equation (and perturbations $\delta\mathbf{B}$)—this has been termed the Hall-shear instability (HSI, Kunz 2008). As studied for the case of a simplified Hall coefficient that results in a fixed length scale, $l_{\rm H} \equiv \eta_{\rm H}/v_{\rm A}$, the HSI can lead to self-organization behavior (for $l_{\rm H} \simeq H$) and, depending on the strength of the effect, creates azimuthally confined zonal bands and/or vortices.

MRI in the context of Hall-MHD has so far been studied in unstratified (Sano and Stone 2002; Kunz and Lesur 2013) and stratified (Lesur et al. 2014; Bai 2014) shearing-box models, and in unstratified cylindrical disk models (O'Keeffe and Downes 2014; Béthune et al. 2016). The microphysics of the Hall effect is complicated by many uncertainties, such as the population of small grains and their charge.

The presence of strong self-organizing azimuthal field belts naturally begs the question of buoyant stability of these configurations, the study of which will demand vertically-stratified global disk models. As emphasized by Xu and Bai (2016), and demonstrated by Gressel et al. (2015) in the context of AD, the field-strength dependence of diffusion coefficients, that is $\eta_{\rm H} \sim B$ and $\eta_{\rm A} \sim B^2$, opens the possibility of highly non-trivial evolution of the system once it reaches non-linear field amplitudes. Finally, a spatial gradient in the coefficients (for instance at the FUV interface as shown by Gressel et al. 2015) can produce non-negligible electromotive forces (potentially leading to significant dynamical effects) all by itself.

3.3 Photoevaporative Mass-Loss

In addition to magnetically driven winds and jets, protoplanetary disks can also lose mass through photoevaporation. Stellar UV and/or X-ray irradiation heats the disk surface to $\sim 10^3$–10^4 K, and beyond some critical radius (typically 1–10AU) the hot gas is unbound and escapes as a wind (e.g. Alexander et al. 2014; Gorti et al. 2016). Photoevaporative disk winds have now been observed directly in a number of objects (e.g., Pascucci and Sterzik 2009; Sacco et al. 2012), and are probably responsible for clearing the outer disk. If a magnetic field is present, the thermal pressure can complement the magnetic acceleration to enhance mass loss and remove angular momentum (Sect. 3.1).

The impact of photoevaporative mass-loss on volatiles, and water in particular, is less clear. The wind is typically launched from the upper layers of the disk, and can entrain particles at least as large as μm-size grains (Takeuchi et al. 2005). Moreover, the mass-loss is concentrated at a radius ($\sim$ 1–10 AU) comparable to the snow line, and could therefore carry

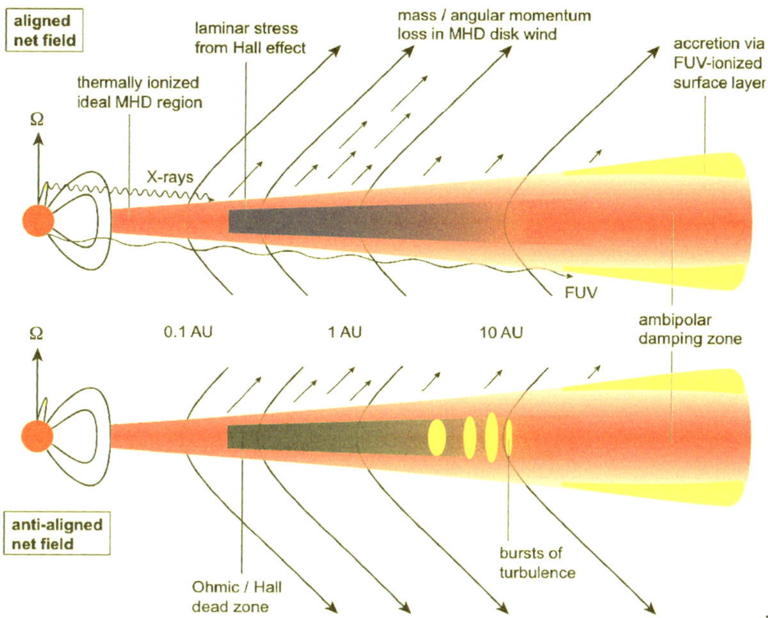

Fig. 8 Schematic disk structure suggested for protoplanetary disks with non-ideal MHD transport, including the Hall effect. Modified from Simon et al. (2015)

away water in both gaseous and solid phases. The impact on the disk's chemical composition therefore depends critically on its radial and (especially) vertical structure, but there has been relatively little exploration of this issue in the literature. Most relevant is the work of Guillot and Hueso (2006), who showed that photoevaporation can play a role in enriching the solar system gas giants in noble gases. However, in light of our improved understanding of both mass-loss and turbulence, more detailed modelling is required if we are to assess the importance of this process more critically.

3.4 Implications for Disk Structure, Turbulence and Heating

Figure 8 summarizes the recent ideas about PPD structure, turbulence, and transport. In the inner disk, thermal ionization results in MRI transport. Beyond about 0.3 AU (for a low-mass central star), ionization levels are high only in surface layers due to the limited penetration of X-rays (and maybe cosmic rays). Magnetic fields threading the disk can produce outflows (probably enhanced by photoevaporative heating), resulting in angular momentum transport and accretion in thin layers where AD begins to become important. The Hall effect can produce strong azimuthal fields, and may result in qualitatively different behavior depending upon the orientation of the magnetic field relative to the angular momentum vector.

It must be emphasized that the operation of all these non-ideal MHD processes depend upon the amount to which the disk is able to keep magnetic flux, which is poorly constrained (Okuzumi et al. 2014; Guilet and Ogilvie 2014), and will likely depend on the level of disk turbulence, as well as the effective magnetic Prandtl number (Fromang and Stone 2009) of the turbulent flow. This has helped stimulate interest in hydrodynamic instabilities which can operate in the absence of magnetic stresses, such as the vertical shear instability (Nelson et al. 2013; Stoll and Kley 2016), the convective overstability instability (Lyra 2014; Klahr and Hubbard 2014), the "zombie vortex" instability (Marcus et al. 2015; but see Lesur and Latter 2016). These instabilities (which depend critically on rates of radiative relaxation; Malygin et al. 2017) generally result in low viscosities and/or transport, with $\alpha \sim 10^{-4}$, but this could be enough to produce important effects on mass transport and dust coagulation. Finally, it has even been suggested that, in very low viscosity disks, planets (super-Earths) can actually produce strong enough spiral waves which reduce their inward migration while producing observable accretion rates (Fung and Chiang 2017). Obviously our understanding of PPDs is highly uncertain; tentatively, one may conclude that both theory and observations are driving us to a picture in which most of the disks are subject to low levels of turbulence.

4 A Framework for Water Evolution in Protoplanetary Disks

As discussed above, there are still significant uncertainties in the extent to which the different processes described above impact how protoplanetary disks are structured and evolve. A greater complication arises as we work to understand how all of these processes operate *simultaneously* to shape the distribution of water in the disk. That is, as solids grow to different sizes, with water being incorporated into bodies with differing dynamical properties, the distribution of water ice and vapor will continuously evolve over the lifetime of the protoplanetary disk. Here we outline a framework for understanding how this evolution may proceed, and the impact it may have on different aspects of planet formation.

Early on in disk history, dust grains will begin to coagulate and form larger and larger solids. Outside the snow line ($r > R_{SL}$), water ice-bearing solids will concentrate at the disk midplane as a result of settling, resulting in enhanced H_2O/H_2 values. Water vapor at high altitudes can diffuse downward and freeze-out on these solids, only returning again as the fine-dust that remains strongly coupled to the gas is lofted to high altitudes allowing desorption to occur again. As the abundance of fine grained dust becomes depleted, the amount of vapor present at high altitudes in the disk diminishes, effectively lowering the height of the snow line. This is a result of the water vapor concentration decreasing, meaning the desorption flux will exceed the adsorption flux through more of the disk (Krijt et al. 2016).

As radial drift becomes important, with solids millimeters and centimeters in size marching inwards, the location of the snow line will become blurred. The most rapidly drifting solids, those millimeters to meters in scale, have drift timescales that are short compared to their sublimation timescale (Cyr et al. 1998; Piso et al. 2015). As a result, icy solids may be found inside of R_{SL}. The exact distance to which icy solids would survive drifting inside the snow line depends on the details of the temperature gradient in the disk, but may be as much as 1–2 AU.

As these icy solids sublimate, the water will return to the gas phase. Because the inward drift of solids can be rapid (inward drift rates will be of order cm/s to m/s relative to the gas), the delivery of water vapor to the inner disk can be much faster than the rate at which it is redistributed by radial diffusion. This imbalance will lead to water vapor becoming enhanced in the inner disk, with concentrations increasing either until the outward flux of water vapor via diffusion equals the inward flux from drifting pebbles, or until inward drift ceases either because the pebbles are accreted by growing planets or are no longer able to form due to the depleted reservoir of solids in the outer disk (Cuzzi and Zahnle 2004; Ciesla and Cuzzi 2006; Estrada et al. 2016). Detailed models suggest that water vapor enhancements of a

3–5× over the "canonical" value could be seen inside the snow line (Ciesla and Cuzzi 2006; Estrada et al. 2016).[1]

Once the inward drift of icy solids slows, vapor diffusion will remove water vapor by carrying it beyond the snow line to freeze-out again. If the ice is locked up efficiently beyond the snow line so that it cannot drift inward again, then the inner disk will become continuously depleted in water vapor throughout the rest of its lifetime (Stevenson and Lunine 1988; Cyr et al. 1998; Cuzzi and Zahnle 2004; Ciesla and Cuzzi 2006). This diffusion serves to increase the ice concentration just beyond the snow line, meaning that the region just beyond the snow line is expected to become enhanced in solids. The migration of the snow line over time, as a result of the evolution of the disk, would set exactly where such enhancements would occur. The particulars of disk evolution may change things, as if the net flows of gas around the snow line, depending on their orientation, could carry the water vapor enhanced/depleted gas inside the snow line outward/inward, or the ice enhanced/depleted gas inward, which would impact the overall distribution of water in the disk over time.

4.1 Implications of Water Evolution for Planet Formation

The varying dynamic properties of water-bearing species will lead to continuous evolution in the distribution of water within the disk, both in the gas and solid phases. The magnitude and timescale for this evolution will depend on many uncertain parameters, including the efficiency of particle growth and the level of turbulence (diffusion) within the disk. There is, however, a number of features in full-born and proto-planetary systems that may be linked to this evolution which we discuss here. By understanding these relationships, perhaps we can begin to evaluate the uncertainties in the parameters which describe protoplanetary disk processes.

The concentration of solids that would develop just outside of the snow line due to the inward drift of ices from greater heliocentric distances and the outward diffusion of ices may serve to seed the early stages of giant planet formation. Stevenson and Lunine (1988) originally suggested that the outward diffusion of vapor from the inner Solar System to beyond the assumed snow line location of 5 AU could have been a catalyst for the formation of Jupiter, as a massive core could grow more readily from the higher concentration of mass then expected. While disk temperatures are predicted to be such for the snow line around a solar mass star to have been located closer to 1–3 AU, it is also possible that Jupiter did not form at its current orbit, and may have migrated outwards to its current location at some later point (e.g. Walsh et al. 2011). Further, a general view for many large exoplanets close to their host stars is that they formed further out in their protoplanetary disk then migrated inwards to where we observe them today; perhaps this was due to a very distant snow line early in disk evolution when mass accretion rates (viscous dissipation) were high (assuming there is significant viscosity; Sect. 3).

The fluctuating water conditions in the inner protoplanetary disk may play a role in shaping the chemistry and mineralogy of planetary building blocks that form there. Cuzzi and Zahnle (2004) originally suggested that water vapor levels could be enhanced by many orders of magnitude by the inward drift of solids, which may have explained how iron-bearing silicates were able to form (in the absence of extreme oxygen enhancements, Fe is expected

[1] Ciesla and Cuzzi (2006) used an incorrect form of the equation describing the diffusive redistribution of materials; correction of this term leads to changes in the predicted concentrations of a factor of a few. Estrada et al. (2016) provided a more rigorous treatment of growth and transport and found that the qualitative relationships and behaviors hold true.

only to exist in metallic form (e.g. Fedkin and Grossman 2006)). However, it appears that such enhancements are limited to just a factor of a few, due to finite reservoirs of ices and the difficulty in growing large populations of very rapidly drifting solids (Ciesla and Cuzzi 2006; Estrada et al. 2016). Later water vapor depletions would aid in the development of very reducing conditions, and may be required to explain the mineralogy observed in enstatite chondrites (Cyr et al. 1999; Hutson and Ruzicka 2000).

It is worth noting that the evolution described here would not be limited to water alone. CO will freeze-out onto solids in protoplanetary disks at temperatures below ~ 20 K (Qi et al. 2013), and will be subjected to the same settle/drift-adsorb/desorb cycling described above. This will also lead to fluctuating distributions of CO in protoplanetary disks, something that must be remembered as this molecule is readily observed and used to trace gas evolution and mass (Williams and Best 2014).

5 Final Remarks

Theoretical expectations of the positions of water snow lines in protoplanetary disks are unclear, due to the large uncertainties in dust opacities and the amount of viscous dissipation present. Unfortunately, observations so far do not provide meaningful constraints because of limited angular resolution and the significant dust optical depths in inner disks. Further progress might result from longer-wavelength interferometry to detect opacity transitions due to ice that arise from the fact that water ice on the surface of grains makes them more sticky, and thus able to grow to larger sizes than the rocky grains inside of the snow line (Banzatti et al. 2015). Additionally, more sophisticated disk models with advanced chemistry and radiative transfer is needed to take advantage of the current improvements in spatially-resolved dust and molecular emission in disks to provide further insight into the distribution of water. For now, all that can be said with reasonable confidence is that snow lines in protoplanetary disks around solar-type stars should lie in the range of a few to one AU, and move in with time as accretion slows and dust opacities decrease.

The existence of transitional disks, with large holes cleared of small dust at early ages, suggests that some planetesimal, embryo, or planet formation stages might occur during optically-thin phases of inner disks. If the formation of Jupiter could have produced such a transitional disk structure by preventing small dust from entering the inner Solar System, the effectively optically-thin temperature estimate of Eq. (2) might be relevant after all.

Acknowledgements The research of L. Hartmann was supported by the University of Michigan and in part by NASA grant NNX16AB46G. F. Ciesla acknowledges support from NASA's Exobiology and Outer Planets Research Programs (NNX12AD59G and NNX14AQ17G). O. Gressel has received funding from the European Research Council (ERC) under the European Union's Horizon 2020 research and innovation programme (grant agreement No. 638596). R. Alexander has received funding from the European Research Council (ERC) under the European Union's Horizon 2020 research and innovation programme (grant agreement No. 681601), and also acknowledges support from the Leverhulme Trust through a Philip Leverhulme Prize.

References

Y. Abe, E. Ohtani, T. Okuchi, K. Righter, M. Drake, in *Water in the Early Earth*, ed. by R.M. Canup, K. Righter et al.(2000), pp. 413–433

R.D. Alexander, C.J. Clarke, J.E. Pringle, Photoevaporation of protoplanetary discs—I. Hydrodynamic models. Mon. Not. R. Astron. Soc. **369**, 216–228 (2006)

R. Alexander, I. Pascucci, S. Andrews, P. Armitage, L. Cieza, The dispersal of protoplanetary disks, in *Protostars and Planets VI* (2014), pp. 475–496

S.M. Andrews, J.P. Williams, Circumstellar dust disks in Taurus-Auriga: the submillimeter perspective. Astrophys. J. **631**, 1134–1160 (2005)

S.M. Andrews, D.J. Wilner, A.M. Hughes, C. Qi, K.A. Rosenfeld, K.I. Öberg, T. Birnstiel, C. Espaillat, L.A. Cieza, J.P. Williams, S.-Y. Lin, P.T.P. Ho, The TW Hya disk at 870 μm: comparison of CO and dust radial structures. Astrophys. J. **744**, 162 (2012)

X.-N. Bai, Magnetorotational-instability-driven accretion in protoplanetary disks. Astrophys. J. **739**, 50 (2011)

X.-N. Bai, Wind-driven accretion in protoplanetary disks. II. Radial dependence and global picture. Astrophys. J. **772**, 96 (2013)

X.-N. Bai, Hall-effect-controlled gas dynamics in protoplanetary disks. I. Wind solutions at the inner disk. Astrophys. J. **791**, 137 (2014)

X.-N. Bai, Hall effect controlled gas dynamics in protoplanetary disks. II. Full 3D simulations toward the outer disk. Astrophys. J. **798**, 84 (2015)

X.-N. Bai, J.M. Stone, Wind-driven accretion in protoplanetary disks. I. Suppression of the magnetorotational instability and launching of the magnetocentrifugal wind. Astrophys. J. **769**, 76 (2013)

X.-N. Bai, J. Ye, J. Goodman, F. Yuan, Magneto-thermal disk winds from protoplanetary disks. Astrophys. J. **818**, 152 (2016)

S.A. Balbus, J.F. Hawley, Instability, turbulence, and enhanced transport in accretion disks. Rev. Mod. Phys. **70**, 1–53 (1998)

A. Banzatti, P. Pinilla, L. Ricci, K.M. Pontoppidan, T. Birnstiel, F. Ciesla, Direct imaging of the water snow line at the time of planet formation using two ALMA continuum bands. Astrophys. J. Lett. **815**, 15 (2015)

E.A. Bergin, L.I. Cleeves, U. Gorti, K. Zhang, G.A. Blake, J.D. Green, S.M. Andrews, N.J. Evans II, T. Henning, K. Öberg, K. Pontoppidan, C. Qi, C. Salyk, E.F. van Dishoeck, An old disk still capable of forming a planetary system. Nature **493**, 644–646 (2013)

E. Bergin, N. Calvet, P. D'Alessio, G.J. Herczeg, The effects of UV continuum and Lyα radiation on the chemical equilibrium of T Tauri disks. Astrophys. J. Lett. **591**, 159–162 (2003)

W. Béthune, G. Lesur, J. Ferreira, Self-organization in protoplanetary discs. Global, non-stratified Hall-MHD simulations. Astron. Astrophys. **589**, 87 (2016)

T. Birnstiel, C.W. Ormel, C.P. Dullemond, Dust size distributions in coagulation/fragmentation equilibrium: numerical solutions and analytical fits. Astron. Astrophys. **525**, 11 (2011)

B. Bitsch, A. Johansen, M. Lambrechts, A. Morbidelli, The structure of protoplanetary discs around evolving young stars. Astron. Astrophys. **575**, 28 (2015)

R.D. Blandford, D.G. Payne, Hydromagnetic flows from accretion discs and the production of radio jets. Mon. Not. R. Astron. Soc. **199**, 883–903 (1982)

S.M. Blevins, K.M. Pontoppidan, A. Banzatti, K. Zhang, J.R. Najita, J.S. Carr, C. Salyk, G.A. Blake, Measurements of water surface snow lines in classical protoplanetary disks. ArXiv e-prints (2015)

S.D. Brittain, J.R. Najita, J.S. Carr, Near infrared high resolution spectroscopy and spectro-astrometry of gas in disks around Herbig Ae/Be stars. Astrophys. Space Sci. **357**, 54 (2015)

N. Calvet, P. D'Alessio, L. Hartmann, D. Wilner, A. Walsh, M. Sitko, Evidence for a developing gap in a 10 Myr old protoplanetary disk. Astrophys. J. **568**, 1008–1016 (2002)

J.S. Carr, J.R. Najita, Organic molecules and water in the inner disks of T Tauri stars. Astrophys. J. **733**, 102 (2011)

E.I. Chiang, P. Goldreich, Spectral energy distributions of T Tauri stars with passive circumstellar disks. Astrophys. J. **490**, 368–376 (1997)

F.J. Ciesla, J.N. Cuzzi, The evolution of the water distribution in a viscous protoplanetary disk. Icarus **181**, 178–204 (2006)

L.A. Cieza, S. Casassus, J. Tobin, S.P. Bos, J.P. Williams, S. Perez, Z. Zhu, C. Caceres, H. Canovas, M.M. Dunham, A. Hales, J.L. Prieto, D.A. Principe, M.R. Schreiber, D. Ruiz-Rodriguez, A. Zurlo, Imaging the water snow-line during a protostellar outburst. Nature **535**, 258–261 (2016). doi:10.1038/nature18612

L.I. Cleeves, F.C. Adams, E.A. Bergin, Exclusion of cosmic rays in protoplanetary disks: stellar and magnetic effects. Astrophys. J. **772**, 5 (2013)

L.I. Cleeves, E.A. Bergin, C.M.O. Alexander, F. Du, D. Graninger, K.I. Öberg, T.J. Harries, The ancient heritage of water ice in the solar system. Science **345**, 1590–1593 (2014)

J.N. Connelly, M. Bizzarro, A.N. Krot, Å. Nordlund, D. Wielandt, M.A. Ivanova, The absolute chronology and thermal processing of solids in the solar protoplanetary disk. Science **338**, 651 (2012)

J.N. Cuzzi, K.J. Zahnle, Material enhancement in protoplanetary nebulae by particle drift through evaporation fronts. Astrophys. J. **614**, 490–496 (2004)

J.N. Cuzzi, A.R. Dobrovolskis, J.M. Champney, Particle-gas dynamics in the midplane of a protoplanetary nebula. Icarus **106**, 102 (1993). doi:10.1006/icar.1993.1161
K.E. Cyr, W.D. Sears, J.I. Lunine, Distribution and evolution of water ice in the solar nebula: implications for solar system body formation. Icarus **135**, 537–548 (1998)
K.E. Cyr, C.M. Sharp, J.I. Lunine, Effects of the redistribution of water in the solar nebula on nebular chemistry. J. Geophys. Res. **104**, 19003–19014 (1999)
P. D'Alessio, N. Calvet, L. Hartmann, Accretion disks around young objects. III. Grain growth. Astrophys. J. **553**, 321–334 (2001)
P. D'Alessio, J. Cantö, N. Calvet, S. Lizano, Accretion disks around young objects. I. The detailed vertical structure. Astrophys. J. **500**, 411–427 (1998)
S.J. Desch, Mass distribution and planet formation in the solar nebula. Astrophys. J. **671**, 878–893 (2007)
C.P. Dullemond, C. Dominik, Dust coagulation in protoplanetary disks: a rapid depletion of small grains. Astron. Astrophys. **434**, 971–986 (2005)
C. Espaillat, N. Calvet, P. D'Alessio, J. Hernández, C. Qi, L. Hartmann, E. Furlan, D.M. Watson, On the diversity of the Taurus transitional disks: UX Tauri A and LkCa 15. Astrophys. J. Lett. **670**, 135–138 (2007)
C. Espaillat, J. Muzerolle, J. Najita, S. Andrews, Z. Zhu, N. Calvet, S. Kraus, J. Hashimoto, A. Kraus, P. D'Alessio, An observational perspective of transitional disks, in *Protostars and Planets VI* (2014), pp. 497–520
P.R. Estrada, J.N. Cuzzi, D.A. Morgan, Global modeling of nebulae with particle growth, drift, and evaporation fronts. I. Methodology and typical results. Astrophys. J. **818**, 200 (2016)
D. Fedele, M.E. van den Ancker, T. Henning, R. Jayawardhana, J.M. Oliveira, Timescale of mass accretion in pre-main-sequence stars. Astron. Astrophys. **510**, 72 (2010)
A.V. Fedkin, L. Grossman, in *The Fayalite Content of Chondritic Olivine: Obstacle to Understanding the Condensation of Rocky Material*, ed. by D.S. Lauretta, H.Y. McSween (2006), pp. 279–294
K.M. Flaherty, A.M. Hughes, K.A. Rosenfeld, S.M. Andrews, E. Chiang, J.B. Simon, S. Kerzner, D.J. Wilner, Weak turbulence in the HD 163296 protoplanetary disk revealed by ALMA CO observations. Astrophys. J. **813**, 99 (2015)
S. Fromang, J. Papaloizou, Dust settling in local simulations of turbulent protoplanetary disks. Astron. Astrophys. **452**, 751–762 (2006)
S. Fromang, J.M. Stone, Turbulent resistivity driven by the magnetorotational instability. Astron. Astrophys. **507**, 19–28 (2009)
J. Fung, E. Chiang, Save the planet, feed the star: how super-earths survive migration and drive disk accretion. Astrophys. J. **839**, 100 (2017). doi:10.3847/1538-4357/aa6934
E. Furlan, L. Hartmann, N. Calvet, P. D'Alessio, R. Franco-Hernández, W.J. Forrest, D.M. Watson, K.I. Uchida, B. Sargent, J.D. Green, L.D. Keller, T.L. Herter, A survey and analysis of spitzer infrared spectrograph spectra of T Tauri stars in Taurus. Astrophys. J. Suppl. Ser. **165**, 568–605 (2006)
C.F. Gammie, Layered accretion in T Tauri disks. Astrophys. J. **457**, 355 (1996)
P. Garaud, D.N.C. Lin, The effect of internal dissipation and surface irradiation on the structure of disks and the location of the snow line around Sun-like stars. Astrophys. J. **654**, 606–624 (2007)
A.E. Glassgold, R. Meijerink, J.R. Najita, Formation of water in the warm atmospheres of protoplanetary disks. Astrophys. J. **701**, 142–153 (2009)
U. Gorti, R. Liseau, Z. Sándor, C. Clarke, Disk dispersal: theoretical understanding and observational constraints. Space Sci. Rev. **205**(1–4), 125–152 (2016). doi:10.1007/s11214-015-0228-x
O. Gressel, N.J. Turner, R.P. Nelson, C.P. McNally, Global simulations of protoplanetary disks with ohmic resistivity and ambipolar diffusion. Astrophys. J. **801**, 84 (2015)
J. Guilet, G.I. Ogilvie, Global evolution of the magnetic field in a thin disc and its consequences for protoplanetary systems. Mon. Not. R. Astron. Soc. **441**, 852–868 (2014)
T. Guillot, R. Hueso, The composition of Jupiter: sign of a (relatively) late formation in a chemically evolved protosolar disc. Mon. Not. R. Astron. Soc. **367**, 47–51 (2006)
L. Hartmann, S.J. Kenyon, The FU Orionis phenomenon. Annu. Rev. Astron. Astrophys. **34**, 207–240 (1996)
L. Hartmann, G. Herczeg, N. Calvet, Accretion onto pre-main-sequence stars. Annu. Rev. Astron. Astrophys. **54**, 135–180 (2016). doi:10.1146/annurev-astro-081915-023347
L. Hartmann, K. Hinkle, N. Calvet, High-resolution near-infrared spectroscopy of FU Orionis objects. Astrophys. J. **609**, 906–916 (2004)
C. Hayashi, Structure of the solar nebula, growth and decay of magnetic fields and effects of magnetic and turbulent viscosities on the nebula. Prog. Theor. Phys. Suppl. **70**, 35–53 (1981)
J. Hernández, L. Hartmann, N. Calvet, R.D. Jeffries, R. Gutermuth, J. Muzerolle, J. Stauffer, A Spitzer view of protoplanetary disks in the γ Velorum cluster. Astrophys. J. **686**, 1195–1208 (2008)
I. Hubeny, Vertical structure of accretion disks—a simplified analytical model. Astrophys. J. **351**, 632–641 (1990)

M. Hutson, A. Ruzicka, A multi-step model for the origin of E3 (enstatite) chondrites. Meteorit. Planet. Sci. **35**, 601–608 (2000)
L. Ingleby, N. Calvet, E. Bergin, A. Yerasi, C. Espaillat, G. Herczeg, E. Roueff, H. Abgrall, J. Hernández, C. Briceño, I. Pascucci, J. Miller, J. Fogel, L. Hartmann, M. Meyer, J. Carpenter, N. Crockett, M. McClure, Far-ultraviolet H_2 emission from circumstellar disks. Astrophys. J. Lett. **703**, 137–141 (2009)
L. Ingleby, N. Calvet, J. Hernández, L. Hartmann, C. Briceno, J. Miller, C. Espaillat, M. McClure, The evolution of accretion in young stellar objects: strong accretors at 3–10 Myr. Astrophys. J. **790**, 47 (2014)
A. Kalyaan, S.J. Desch, N. Monga, External photoevaporation of the solar nebula. II. Effects on disk structure and evolution with non-uniform turbulent viscosity due to the magnetorotational instability. Astrophys. J. **815**, 112 (2015)
G.M. Kennedy, S.J. Kenyon, Planet formation around stars of various masses: the snow line and the frequency of giant planets. Astrophys. J. **673**, 502–512 (2008)
S.J. Kenyon, L. Hartmann, Spectral energy distributions of T Tauri stars—disk flaring and limits on accretion. Astrophys. J. **323**, 714–733 (1987)
H. Klahr, A. Hubbard, Convective overstability in radially stratified accretion disks under thermal relaxation. Astrophys. J. **788**, 21 (2014)
A. Königl, Self-similar models of magnetized accretion disks. Astrophys. J. **342**, 208–223 (1989)
S. Krijt, F.J. Ciesla, E.A. Bergin, Tracing water vapor and ice during dust growth. Astrophys. J. **833**, 285 (2016)
T.S. Kruijer, M. Touboul, M. Fischer-Gödde, K.R. Bermingham, R.J. Walker, T. Kleine, Protracted core formation and rapid accretion of protoplanets. Science **344**, 1150–1154 (2014)
M.W. Kunz, On the linear stability of weakly ionized, magnetized planar shear flows. Mon. Not. R. Astron. Soc. **385**, 1494–1510 (2008)
M.W. Kunz, G. Lesur, Magnetic self-organization in Hall-dominated magnetorotational turbulence. Mon. Not. R. Astron. Soc. **434**, 2295–2312 (2013)
G. Lesur, M.W. Kunz, S. Fromang, Thanatology in protoplanetary discs. The combined influence of Ohmic, Hall, and ambipolar diffusion on dead zones. Astron. Astrophys. **566**, 56 (2014)
G.R.J. Lesur, H. Latter, On the survival of zombie vortices in protoplanetary discs. Mon. Not. R. Astron. Soc. **462**, 4549–4554 (2016). doi:10.1093/mnras/stw2172
K. Lodders, Solar system abundances and condensation temperatures of the elements. Astrophys. J. **591**, 1220–1247 (2003)
W. Lyra, Convective overstability in accretion disks: three-dimensional linear analysis and nonlinear saturation. Astrophys. J. **789**, 77 (2014)
M.G. Malygin, H. Klahr, D. Semenov, T. Henning, C.P. Dullemond, Efficiency of thermal relaxation by radiative processes in protoplanetary discs: constraints on hydrodynamic turbulence. ArXiv e-prints (2017)
P.S. Marcus, S. Pei, C.-H. Jiang, J.A. Barranco, P. Hassanzadeh, D. Lecoanet, Zombie vortex instability. I. A purely hydrodynamic instability to resurrect the dead zones of protoplanetary disks. Astrophys. J. **808**, 87 (2015). doi:10.1088/0004-637X/808/1/87
M.K. McClure, C. Espaillat, N. Calvet, E. Bergin, P. D'Alessio, D.M. Watson, P. Manoj, B. Sargent, L.I. Cleeves, Detections of trans-Neptunian ice in protoplanetary disks. Astrophys. J. **799**, 162 (2015)
J.R. Najita, S.E. Strom, J. Muzerolle, Demographics of transition objects. Mon. Not. R. Astron. Soc. **378**, 369–378 (2007)
J.R. Najita, G.W. Doppmann, J.S. Carr, J.R. Graham, J.A. Eisner, High-resolution K-band spectroscopy of MWC 480 and V1331 Cyg. Astrophys. J. **691**, 738–748 (2009)
J. Najita, J.S. Carr, A.E. Glassgold, F.H. Shu, A.T. Tokunaga, Kinematic diagnostics of disks around young stars: CO overtone emission from WL 16 and 1548C27. Astrophys. J. **462**, 919 (1996)
R.P. Nelson, O. Gressel, O.M. Umurhan, Linear and non-linear evolution of the vertical shear instability in accretion discs. Mon. Not. R. Astron. Soc. **435**, 2610–2632 (2013)
W. O'Keeffe, T.P. Downes, Multifluid simulations of the magnetorotational instability in protostellar discs. Mon. Not. R. Astron. Soc. **441**, 571–581 (2014)
S. Okuzumi, T. Takeuchi, T. Muto, Radial transport of large-scale magnetic fields in accretion disks. I. Steady solutions and an upper limit on the vertical field strength. Astrophys. J. **785**, 127 (2014)
J.E. Owen, B. Ercolano, C.J. Clarke, Protoplanetary disc evolution and dispersal: the implications of X-ray photoevaporation. Mon. Not. R. Astron. Soc. **412**, 13–25 (2011)
I. Pascucci, M. Sterzik, Evidence for disk photoevaporation driven by the Central Star. Astrophys. J. **702**, 724–732 (2009)
L.M. Pérez, J.M. Carpenter, C.J. Chandler, A. Isella, S.M. Andrews, L. Ricci, N. Calvet, S.A. Corder, A.T. Deller, C.P. Dullemond, J.S. Greaves, R.J. Harris, T. Henning, W. Kwon, J. Lazio, H. Linz, L.G. Mundy,

A.I. Sargent, S. Storm, L. Testi, D.J. Wilner, Constraints on the radial variation of grain growth in the AS 209 circumstellar disk. Astrophys. J. Lett. **760**, 17 (2012)

L.M. Pérez, C.J. Chandler, A. Isella, J.M. Carpenter, S.M. Andrews, N. Calvet, S.A. Corder, A.T. Deller, C.P. Dullemond, J.S. Greaves, R.J. Harris, T. Henning, W. Kwon, J. Lazio, H. Linz, L.G. Mundy, L. Ricci, A.I. Sargent, S. Storm, M. Tazzari, L. Testi, D.J. Wilner, Grain growth in the circumstellar disks of the young stars CY Tau and DoAr 25. Astrophys. J. **813**, 41 (2015)

D. Perez-Becker, E. Chiang, Surface layer accretion in conventional and transitional disks driven by far-ultraviolet ionization. Astrophys. J. **735**, 8 (2011)

A.-M.A. Piso, K.I. Öberg, T. Birnstiel, R.A. Murray-Clay, C/O and snowline locations in protoplanetary disks: the effect of radial drift and viscous gas accretion. Astrophys. J. **815**, 109 (2015)

C. Qi, K.I. Öberg, D.J. Wilner, P. D'Alessio, E. Bergin, S.M. Andrews, G.A. Blake, M.R. Hogerheijde, E.F. van Dishoeck, Imaging of the CO snow line in a solar nebula analog. Science **341**, 630–632 (2013)

G.G. Sacco, E. Flaccomio, I. Pascucci, F. Lahuis, B. Ercolano, J.H. Kastner, G. Micela, B. Stelzer, M. Sterzik, High-resolution Spectroscopy of Ne II emission from young stellar objects. Astrophys. J. **747**, 142 (2012)

R. Salmeron, M. Wardle, Magnetorotational instability in stratified, weakly ionized accretion discs. Mon. Not. R. Astron. Soc. **345**, 992–1008 (2003)

T. Sano, J.M. Stone, The effect of the Hall term on the nonlinear evolution of the magnetorotational instability. II. Saturation level and critical magnetic Reynolds number. Astrophys. J. **577**, 534–553 (2002)

J.B. Simon, X.-N. Bai, J.M. Stone, P.J. Armitage, K. Beckwith, Turbulence in the outer regions of protoplanetary disks. I. Weak accretion with no vertical magnetic flux. Astrophys. J. **764**, 66 (2013a)

J.B. Simon, X.-N. Bai, P.J. Armitage, J.M. Stone, K. Beckwith, Turbulence in the outer regions of protoplanetary disks. II. Strong accretion driven by a vertical magnetic field. Astrophys. J. **775**, 73 (2013b)

J.B. Simon, G. Lesur, M.W. Kunz, P.J. Armitage, Magnetically driven accretion in protoplanetary discs. Mon. Not. R. Astron. Soc. **454**, 1117–1131 (2015)

D.R. Soderblom, L.A. Hillenbrand, R.D. Jeffries, E.E. Mamajek, T. Naylor, Ages of young stars, in *Protostars and Planets VI* (2014), pp. 219–241

D.J. Stevenson, J.I. Lunine, Rapid formation of Jupiter by diffuse redistribution of water vapor in the solar nebula. Icarus **75**, 146–155 (1988)

M.H.R. Stoll, W. Kley, Particle dynamics in discs with turbulence generated by the vertical shear instability. Astron. Astrophys. **594**, 57 (2016). doi:10.1051/0004-6361/201527716

T. Takeuchi, C.J. Clarke, D.N.C. Lin, The differential lifetimes of protostellar gas and dust disks. Astrophys. J. **627**, 286–292 (2005)

M. Tazzari, L. Testi, B. Ercolano, A. Natta, A. Isella, C.J. Chandler, L.M. Pérez, S. Andrews, D.J. Wilner, L. Ricci, T. Henning, H. Linz, W. Kwon, S.A. Corder, C.P. Dullemond, J.M. Carpenter, A.I. Sargent, L. Mundy, S. Storm, N. Calvet, J.A. Greaves, J. Lazio, A.T. Deller, A multi-wavelength analysis for interferometric (sub-)mm observations of protoplanetary disks: radial constraints on the dust properties and the disk structure. ArXiv e-prints (2015)

N.J. Turner, S. Fromang, C. Gammie, H. Klahr, G. Lesur, M. Wardle, X.-N. Bai, Transport and accretion in planet-forming disks, in *Protostars and Planets VI* (2014), pp. 411–432

N. van der Marel, E.F. van Dishoeck, S. Bruderer, S.M. Andrews, K.M. Pontoppidan, G.J. Herczeg, T. van Kempen, A. Miotello, Resolved gas cavities in transitional disks inferred from CO isotopologs with ALMA. Astron. Astrophys. **585**, 58 (2016)

K.J. Walsh, A. Morbidelli, S.N. Raymond, D.P. O'Brien, A.M. Mandell, A low mass for Mars from Jupiter's early gas-driven migration. Nature **475**, 206–209 (2011)

M. Wardle, A. Königl, The structure of protostellar accretion disks and the origin of bipolar flows. Astrophys. J. **410**, 218–238 (1993)

S.J. Weidenschilling, The distribution of mass in the planetary system and solar nebula. Astrophys. Space Sci. **51**, 153–158 (1977)

J.P. Williams, W.M.J. Best, A parametric modeling approach to measuring the gas masses of circumstellar disks. Astrophys. J. **788**, 59 (2014)

R. Xu, X.-N. Bai, On the grain-modified magnetic diffusivities in protoplanetary disks. Astrophys. J. **819**, 68 (2016)

A.N. Youdin, Y. Lithwick, Particle stirring in turbulent gas disks: including orbital oscillations. Icarus **192**, 588–604 (2007)

Space Sci Rev (2018) 214:38
https://doi.org/10.1007/s11214-018-0472-y

Giant Planet Formation and Migration

Sijme-Jan Paardekooper[1] · **Anders Johansen**[2]

Received: 27 July 2016 / Accepted: 11 January 2018 / Published online: 24 January 2018

Abstract Planets form in circumstellar discs around young stars. Starting with sub-micron sized dust particles, giant planet formation is all about growing 14 orders of magnitude in size. It has become increasingly clear over the past decades that during all stages of giant planet formation, the building blocks are extremely mobile and can change their semimajor axis by substantial amounts. In this chapter, we aim to give a basic overview of the physical processes thought to govern giant planet formation and migration, and to highlight possible links to water delivery.

Keywords Planet-disc interactions · Planets and satellites: formation · Planets and satellites: gaseous planets · Protoplanetary discs

1 Introduction

For many centuries, ideas of how planets come into being were necessarily aimed at explaining the Solar system. The observation that the orbits of the planets in the Solar system lie more or less in a single plane led to the Nebular hypothesis, developed originally by Emanuel Swedenborg and Immanuel Kant and independently by Pierre-Simon Laplace in the 18th century: the planets were born in a cloud of gas around the Sun that we now identify with protoplanetary discs observed around young stars. The wealth of data available

The Delivery of Water to Protoplanets, Planets and Satellites
Edited by Michel Blanc, Alessandro Morbidelli, Yann Alibert, Lindy Elkins-Tanton, Paul Estrada, Keiko Hamano, Helmut Lammer, Sean Raymond and Maria Schönbächler

✉ S.-J. Paardekooper
s.j.paardekooper@qmul.ac.uk

A. Johansen
anders@astro.lu.se

1 Astronomy Unit, School of Physics and Astronomy, Queen Mary, University of London, 327 Mile End Road, London E1 4NS, UK

2 Lund Observatory, Department of Astronomy and Theoretical Physics, Lund University, Box 43, 22100 Lund, Sweden

on bodies in the Solar system and their histories shaped the ideas of planet formation into very detailed scenarios and explanation as to why the architecture of our planetary system looks the way it does. One example is the separation between the inner Solar system with the terrestrial planets close to the Sun, while gas giant planets can be found further out. The classical explanation for this dichotomy has been that further out, especially beyond the water ice line at a few Astronomical Units (AU) from the Sun, much more material was available in the form of solids to rapidly build big objects.

This made the discovery of the first extrasolar planet around a Solar-type star, the hot Jupiter 51 Peg b (Mayor and Queloz 1995), all the more a surprise. Here we have a gas giant planet like Jupiter, but not in an orbit of 12 years, but 5 days. It was soon realised away that the most straightforward explanation for this class of planets was that they were formed further out in colder regions and subsequently migrated inward due to interaction with the disc (Lin et al. 1996). Since then, migration has become an essential ingredient in planet formation theory. Interestingly, the ideas that are now invoked for the migration of hot Jupiters long predate the discovery of 51 Peg b (Lin and Papaloizou 1979; Goldreich and Tremaine 1980), but they never made it into the standard picture because there was no evidence for large-scale inward migration in the Solar system.

However, while simple models of migration can explain the existence of hot Jupiters, from the population of extrasolar planets it is clear that there is a wide possible range of outcomes in terms of final orbital periods. Most giant planets do not become hot Jupiters, and in general planets are found in orbits that range from less than a day all the way up to several 100s of years. Clearly, some planets have migrated inward over vast distances, while others appear to have experienced little or no migration. Over the past decades, various models of disc migration have become more and more sophisticated in an attempt to account for this diversity (for a recent review see Baruteau et al. 2014).

It is now becoming more and more clear that growth and migration are intimately linked. This is why in this chapter, rather than treating formation and migration separately, we go through the story of planet formation roughly in chronological order, and for each stage highlight the migration processes that are important and how they may affect further growth. We will discuss the two main scenarios for giant planet formation: core accretion and disc fragmentation. At all times, we focus on the basic physical concepts only, in the hope to keep the text accessible for those not familiar with planet formation and migration.

This chapter is organised as follows: we start in Sect. 2 with an introduction to the birth environment of planets, namely protoplanetary discs. We focus only on the ingredients we need in order to discuss planet formation and migration; a more detailed introduction on the subject can be found elsewhere in this volume. We go on in Sect. 3 to describe the first steps in the two competing models of core accretion and disc fragmentation. As it turns out, the first step in disc fragmentation leaves us already with massive planets, while core accretion needs a few more steps to catch up, building large objects from initially small solid particles. These steps are discussed in Sects. 4–8, starting with a discussion on the migration of small solid bodies in Sect. 4, followed by an overview of how massive solid cores could form in Sect. 5. Section 6 considers the migration of such cores. Migration of massive planets, formed by either core accretion or disc fragmentation, is dealt with in Sect. 7. In Sect. 8 we discuss gas accretion and show maps for the formation of planets that include pebble accretion and planetary migration. We finally discuss in Sect. 9 how all these processes affect possible water delivery to terrestrial planets, before we end with concluding remarks in Sect. 10.

2 Protoplanetary Discs

The discs around young stars that are the birthplace of planetary systems are discussed in detail elsewhere in this volume. Here, we introduce a very simple, axisymmetric toy model of a disc that we will use to introduce giant planet formation and migration.

2.1 Global Disc Properties

First of all, consider the mid plane ($z = 0$) of the disc. Discs are known to survive for many dynamical time scales (one dynamical time scale is basically a local orbit), which means they must be close to equilibrium with respect to gravitational and centrifugal forces,

$$r\Omega^2 \approx \frac{GM_*}{r^2}, \tag{1}$$

where r is the radial distance to the central star, Ω is the angular velocity and M_* is the mass of the central star. In other words, the angular velocity Ω must be close to its Keplerian value

$$\Omega_{\rm K} = \sqrt{\frac{GM_*}{r^3}}. \tag{2}$$

This immediately leads to the very important observation that Keplerian discs rotate *differentially*, and one can define a shear rate

$$S = -r\frac{d\Omega}{dr} \approx \frac{3}{2}\Omega, \tag{3}$$

the last approximation holding for discs that are almost Keplerian.

The Keplerian angular velocity, as given by (2), is an exact equilibrium solution if no other forces besides gravity act on the disc. It could represent a circular disc of non-interacting particles (any non-circular motion is excluded by the assumption of axisymmetry). Real protoplanetary discs consist of gas and dust, where both components interact with themselves and with each other, and the simple picture above has to be modified accordingly.

During most of the lifetime of the disc, the gaseous component is dynamically the most important since it contains most of the mass. The most obvious modification of the Keplerian picture concerns gas pressure since it can affect the radial force balance. For gas of density ρ and pressure p, the equilibrium angular velocity is found from

$$r\Omega^2 = r\Omega_{\rm K}^2 + \frac{1}{\rho}\frac{\partial p}{\partial r}. \tag{4}$$

Therefore, any radial pressure gradient modifies the local angular velocity.

For a gaseous disc to be in equilibrium in the vertical direction, the vertical component of the stellar gravity has to be balanced by a pressure gradient,

$$\frac{1}{\rho}\frac{\partial p}{\partial z} = -\frac{\partial \Phi}{\partial z}, \tag{5}$$

where Φ indicates the gravitational potential. If we take the disc to be vertically isothermal, which is usually a good approximation in the bulk of the disc where most of the mass resides

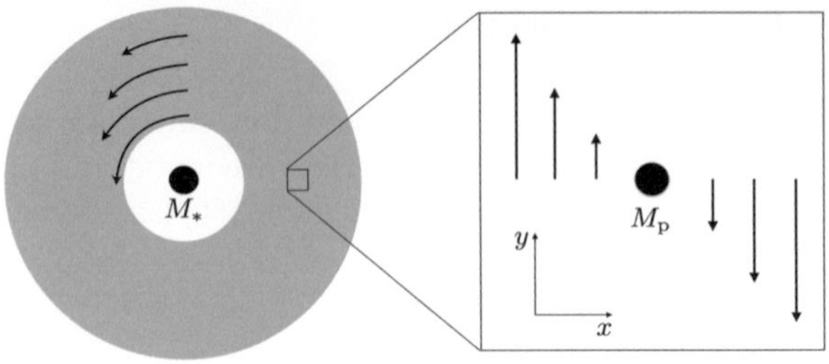

Fig. 1 Local model of a protoplanetary disc. Consider a small patch rotating around the central star at the local orbital velocity Ω. Within the patch, define a Cartesian coordinate frame so that circular motion in the full disc corresponds to a velocity in the y-direction in the patch

(e.g. Woitke et al. 2009), so that $p = c_s^2 \rho$ with c_s the isothermal sound speed, we can easily solve for the density, yielding

$$\rho(z) = \rho(z=0)\exp\left(\frac{\Phi(z) - \Phi(z=0)}{c_s^2}\right) \approx \rho(z=0)\exp\left(-\frac{\Omega_K^2 z^2}{2c_s^2}\right), \tag{6}$$

the last approximation being valid in the limit $z \ll r$. Therefore, the vertical profile of the disc is roughly a Gaussian, and one can define a disc thickness H,

$$H = \frac{c_s}{\Omega_K} \tag{7}$$

and an aspect ratio $h = H/r$. The disc thickness is determined by the local temperature through c_s, and for protoplanetary discs we usually expect $h \sim 0.03$–0.1, depending on the radial location and the evolutionary stage of the disc. In all cases, the disc can be assumed to be thin, i.e. $h \ll 1$, which means that the angular velocity is close to Keplerian, see (4).

Protoplanetary discs are also observed to accrete, or, more accurately, a signature of ongoing accretion is regularly observed on the central star (Ménard and Bertout 1999). If disc material is to flow onto the central star, it has to somehow lose its angular momentum. The exact mechanisms of angular momentum transport in protoplanetary discs are subject of very active research, but here we restrict ourselves to the standard α-prescription: we assume that angular momentum transport can be modelled as a viscous process, with a kinematic viscosity that is set by a parameter α,

$$\nu = \alpha c_s H, \tag{8}$$

where α is usually taken to be 0.001–0.01 to explain observed accretion rates.

2.2 Local Disc Model

It is often useful to consider a local model of a protoplanetary disc, often called the shearing sheet or the shearing box, see Fig. 1. Consider a small patch of the disc, rotating at $r = r_0$ at the local angular velocity Ω_0. Now define a local Cartesian coordinate system

$$x = r - r_0, \tag{9}$$

$$y = r_0(\varphi - \varphi_0 - \Omega_0 t), \tag{10}$$

$$z = z. \tag{11}$$

The effective potential, with contributions from gravity due to the central star and a centrifugal potential from orbital motion, acting on the patch is

$$\Phi_{\rm eff} = \Phi_*(r, z) - \frac{1}{2}\Omega_0 r^2 \approx -\Omega_0 S_0 x^2 + \frac{1}{2}\Omega_0^2 z^2, \tag{12}$$

where the approximation is obtained by expanding the effective potential up to second order in x and z.

Particle motion in the rotating patch under the influence of this effective potential plus the contribution from a massive perturber takes the form of Hill's lunar problem,

$$\ddot{x} - 2\Omega_0 \dot{y} = 2\Omega_0 S_0 x - \frac{\partial \Phi_{\rm p}}{\partial x}, \tag{13}$$

$$\ddot{y} + 2\Omega_0 \dot{x} = -\frac{\partial \Phi_{\rm p}}{\partial y}, \tag{14}$$

$$\ddot{z} = -\Omega_0^2 z - \frac{\partial \Phi_{\rm p}}{\partial z}, \tag{15}$$

where $\Phi_{\rm p}$ is the gravitational potential due to a planet. Unperturbed motion in the patch, corresponding to Keplerian rotation in the full disc, is given by $\dot{x} = \dot{z} = 0$ and $\dot{y} = -S_0 x$ (setting $\Phi_{\rm p} = 0$). Therefore, the Keplerian shear of the full disc is replaced by a linear shear in the local patch. One can either take the sheet to be unbounded in the horizontal directions or apply modified periodic boundary conditions.[1]

While the shearing sheet or box makes the analysis of several problems much easier, the simplicity comes at a price: for example, it is impossible to determine the direction to the central star (it is located on the x-axis, but both directions are equivalent). Therefore, there can be no accretion flow through the patch. Despite such limitations, the shearing sheet/box is a very powerful tool to analyse and simulate the dynamics of protoplanetary discs.

3 The First Steps of Growing Giant Planets

There are two competing scenarios of how giant planets form. The standard model, referred to as the *core accretion* model, builds giant planets from the bottom up: first, a core of typically a few Earth masses is built out of the solids present in the disc, in a similar way as terrestrial planets are formed, after which large amounts of gas are accreted on top of this core. In an alternative model, referred to as the *disc instability* model, giant planets collapse directly from the disc through a gravitational instability.

3.1 Gravitational Instabilities

The two competing scenarios of how giant planets are formed therefore diverge from the very start. In the disc instability model, we start from a massive protoplanetary disc, relatively early in the lifetime of the disc. The simplest possible setup in which to study gravitational instabilities is the periodic shearing sheet, where we neglect any variation in y, so

[1] While in the y-direction ordinary periodic boundary conditions can be applied, in the x direction we have to correct for the shear.

that the governing equations become

$$\frac{\partial \Sigma}{\partial t} + \frac{\partial}{\partial x}(\Sigma u) = 0, \tag{16}$$

$$\frac{\partial u}{\partial t} + u\frac{\partial u}{\partial x} - 2\Omega v = 2\Omega S x - \frac{\partial \Phi_{\rm d}}{\partial x} - \frac{1}{\Sigma}\frac{\partial p}{\partial x}, \tag{17}$$

$$\frac{\partial v}{\partial t} + u\frac{\partial v}{\partial x} + 2\Omega u = 0, \tag{18}$$

where the velocity $\mathbf{v} = (u, v)^T$ and the gravitational potential of the disc $\Phi_{\rm d}$ is calculated from Poisson's equation, assuming all mass is concentrated in the plane $z = 0$,

$$\frac{\partial^2 \Phi_{\rm d}}{\partial x^2} + \frac{\partial^2 \Phi_{\rm d}}{\partial z^2} = 4\pi G \Sigma \delta(z). \tag{19}$$

For simplicity, we take the equation of state to be barotropic, i.e. $p = p(\Sigma)$. As the basic state, take a constant surface density, $u = 0$ and $v = -Sx$, the latter representing Keplerian shear.

Assume small perturbations Σ', u' and v', all $\propto \exp(\mathrm{i}(kx - \omega t))$. The gravitational potential perturbation can be found from

$$\left(\frac{\partial^2}{\partial z^2} - k^2\right)\Phi'_{\rm d} = 4\pi G \Sigma' \delta(z), \tag{20}$$

using a Gaussian surface of vanishing height,

$$\Phi'_{\rm d} = -\frac{2\pi G \Sigma'}{|k|}\exp(-|kz|). \tag{21}$$

Using the above expression, evaluated at $z = 0$, together with only linear terms in Σ', u' and v', and $p' = c_{\rm s}^2 \Sigma'$, we find the following relations for the amplitudes,

$$-\mathrm{i}\omega \Sigma' + \Sigma \mathrm{i} k u' = 0, \tag{22}$$

$$-\mathrm{i}\omega u' - 2\Omega v' = -\mathrm{i}k\left(c_{\rm s}^2 - \frac{2\pi G \Sigma}{|k|}\right)\frac{\Sigma'}{\Sigma}, \tag{23}$$

$$-\mathrm{i}\omega v' + (2\Omega - S)u' = 0, \tag{24}$$

from which we can deduce the dispersion relation for *density waves*,

$$\omega^2 = \kappa^2 - 2\pi G \Sigma |k| + c_{\rm s}^2 k^2, \tag{25}$$

where $\kappa = 2\Omega(2\Omega - S)$ is the epicyclic frequency. In the dispersion relation, we see the stabilising effects of rotation (κ^2) and pressure ($c_{\rm s}^2$) and the destabilising effect of self-gravity. The most unstable wavelength is given by

$$|k_{\rm min}| = \pi G \Sigma / c_{\rm s}^2, \tag{26}$$

for which

$$\omega_{\rm min}^2 = \kappa^2\left(1 - \frac{1}{Q^2}\right), \tag{27}$$

where

$$Q = \frac{\kappa c_s}{\pi G \Sigma} \tag{28}$$

is the Toomre Q parameter (Safronov 1960; Toomre 1964). The disc is unstable to axisymmetric (i.e., no y dependence) perturbations for $Q < 1$. Therefore, in order for density perturbations to grow large, we need a disc that is massive (large Σ) and cold (low c_s). Note that this Q-criterion only applies to axisymmetric, razor-thin discs. Discs of finite thickness tend to be more stable for a given Σ and c_s (Mamatsashvili and Rice 2010).

Non-axisymmetric razor-thin discs show instability at larger values of Q. This can be understood qualitatively from the fact that an axisymmetric setup necessarily involves conservation of angular momentum, which substantially limits the ability of material to move radially. This may seem good news if we want to build giant planets, in the sense that it is easier to get large density perturbations, but at the same time non-axisymmetry unlocks a defence mechanism of the disc against collapse: spiral density waves. These waves dissipate energy, increasing the temperature and hence Q.

It is therefore not straightforward to make a disc unstable. If we for example extract energy from the sheet at a specified rate

$$\frac{dU}{dt} = -\beta \Omega U, \tag{29}$$

where U is the internal energy of the gas and β is a constant, and thereby lowering Q, taking the disc towards instability, this cooling can in principle be balanced by heating through wave dissipation, leading to a quasi-steady state often referred to as *gravito-turbulence*. In order to beat the disc's defences so that it fragments into bound objects, cooling needs to be rapid enough so that the heating can not keep up (Gammie 2001). The exact limit is difficult to establish because the problem is fraught with numerical difficulties (Meru and Bate 2011; Paardekooper et al. 2011b), and, at least for razor-thin discs, there is a stochastic element to disc fragmentation (Paardekooper 2012). However, since realistic calculations make the cooling time fall off very rapidly with distance to the central star, it is clear that disc fragmentation will only occur in the outer parts of protoplanetary discs, typically beyond 50 AU (Rafikov 2005). Examples of local simulations of gravitationally unstable discs are shown in Fig. 2, for two values of the dimensionless cooling rate β. The left panel ($\beta = 10$) shows steady gravito-turbulence, while the right panel ($\beta = 3$) shows a fragmented disc.

From (26), we can deduce that for $Q \sim 1$, the most unstable wavelength is H. Therefore, fragment masses will be

$$M_{\rm frag} \sim \Sigma H^2. \tag{30}$$

For $Q \sim 1$, and approximating $\kappa \approx \Omega$, we get

$$\frac{\Sigma H^2}{M_*} \sim h^3. \tag{31}$$

In the outer regions of protoplanetary discs, h is probably at least 0.1, so that fragments will initially have at least the mass of Jupiter. A more realistic calculation by Rafikov (2005) results in fragment masses of at least 5 times the mass of Jupiter. And this is only the initial mass of the fragment: it is embedded in a massive disc (by construction), which may still be experiencing infall from the surrounding molecular cloud. It is very difficult to stop any of this material from accreting onto the fragment, and for example Kratter et al. (2010) conclude that if discs fragment, they are more likely to produce brown dwarfs rather than

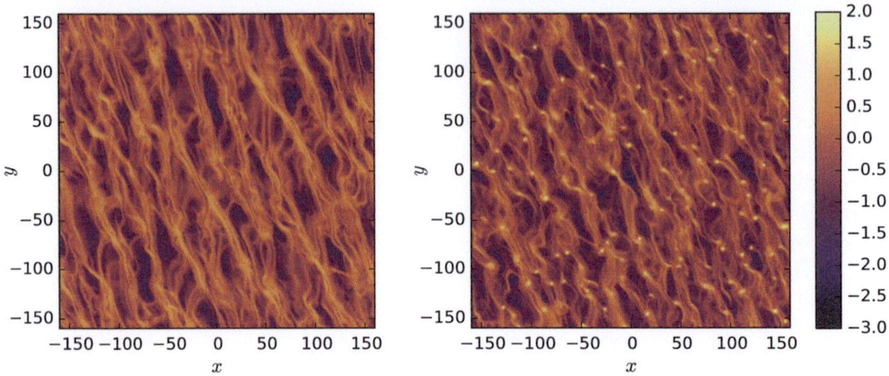

Fig. 2 Local simulations of a gravitationally unstable disc for two different cooling rates. Shown is $\log_{10} \Sigma$. *Left panel*: $\beta = 10$ leads to quasi-steady gravito-turbulence. *Right panel*: at $\beta = 3$, the disc fragments into bound objects

giant planets. A possible solution is to move these planets inward so that they overflow their Roche lobe and lose mass (Boley et al. 2010; Nayakshin 2010).

Despite these problems, and in addition the many uncertainties that haunt the thermal evolution of fragments (for a recent review see Kratter and Lodato 2016), disc fragmentation is still on the table as a possible way to form giant planets; its appeal coming mainly from the short time scale on which it can generate planets: since it is a dynamical instability, planets form roughly on a dynamical time scale. This is of particular interest for giant planets observed at very large distances from their central star (Marois et al. 2008), where dynamical time scales are long and core accretion would take much longer than the disc life time. This rapid formation mechanism is also appealing because it appears to be able to jump over many difficult problems that the core accretion model has (see however Sect. 7.4).

3.2 Dust Coagulation

The core accretion model starts with the smallest solid building blocks that are present in the disc, namely sub-micron dust particles that are present in the interstellar medium as well. In this scenario, the first step is to grow these particles by sticking them together whenever they meet. Modelling this growth process is complicated by the fact that a particle of given size can collide with another particle of *any* size: in other words, the process in non-local in size space. Some insight can be gained though by considering *monodisperse growth*: at all times, all particles are assumed to have the same size.

If the particles of size a have number density n and geometrical cross-section σ, then the average time between collisions is given by

$$\tau_{\rm col} = \frac{1}{n\sigma\Delta v}, \tag{32}$$

where Δv is the relative velocity between particles. This relative velocity can for example come from Brownian motion

$$\Delta v_{\rm BM} = \sqrt{\frac{16 k_{\rm B} T}{\pi m}}, \tag{33}$$

where m is the mass of a single particle. Other sources of relative velocities include turbulence and dust drift (see below).

Assuming perfect sticking, every collision leads to a doubling in the mass, so

$$\frac{dm}{dt} \approx \frac{m}{\tau_{\mathrm{col}}} = \rho_{\mathrm{d}}\sigma\,\Delta v, \tag{34}$$

where $\rho_{\mathrm{d}} = mn$ is the (constant) mass density of dust particles. Focusing on Brownian motion only, which is typically the dominant contribution to relative velocities up to particle sizes of $\sim 1\ \mu\mathrm{m}$ (Dullemond and Dominik 2005), we find that

$$\frac{dm}{dt} = K m^{1/6}, \tag{35}$$

for some constant K, which can be integrated to

$$m(t) = \left(m_0^{5/6} + \frac{5}{6}Kt\right)^{6/5}, \tag{36}$$

where m_0 is the mass of the particles at $t = 0$. Similar calculations can be done for other sources of relative velocities, and they tend to reproduce the early phases of full numerical simulations quite well (e.g. Ormel et al. 2009; Birnstiel et al. 2010).

While perfect sticking is quite a good approximation early on, when particles grow and relative velocities increase at some point collisions start to have a different outcome, from bouncing (Zsom et al. 2010), where particles just bounce off each other without any mass transfer, or even fragmentation, where a collision results in particles that are all smaller in size than the original particles. For silicate grains, the "bouncing barrier" occurs at sizes of roughly one millimetre. However, at this stage the composition of the particles matters a lot; for example, for particles made of water ice the velocity at which they fragment is an order of magnitude larger than for silicate grains (Gundlach and Blum 2015). It therefore helps to have water ice around to grow beyond one millimetre, as anyone who has seen golf-ball-sized hail stones can appreciate.

There is much more to say about these early stages of planet formation, see for example the recent reviews Johansen et al. (2014), Birnstiel et al. (2016). Growing to larger sizes brings new problems as well, which we discuss next.

4 Dust and Pebble Migration (Type 0)

Even the smallest dust particle interacts with the surrounding gas through frictional forces. The nature of the interaction depends on the size of the particle with respect to the mean free path of the gas molecules. Small particles (smaller than 1 cm in typical protoplanetary discs) feel free-streaming gas molecules and find themselves in the Epstein regime, where the drag force is given by (assuming subsonic relative velocities between gas and dust),

$$\mathbf{F}_{\mathrm{fric}} = -\frac{4}{3}\rho c_{\mathrm{s}}\sigma\,\mathbf{\Delta v}, \tag{37}$$

where σ is the collisional cross-section of the dust particle and $\mathbf{\Delta v}$ is the relative velocity between gas and dust (Dullemond and Dominik 2004). One can define a friction time

$$t_{\mathrm{fric}} = \frac{m|\mathbf{\Delta v}|}{|\mathbf{F}_{\mathrm{fric}}|} \propto \frac{m}{\sigma}, \tag{38}$$

where m is the mass of the dust particle. This friction time is basically the time it takes for the dust particle to adjust to the motion of the gas. Since it is proportional to m/σ, which for spherical particles is proportional to the size of the particle,[2] the smallest particles are tightly coupled to the gas, while larger boulders can move relatively independently of the gas. Differently shaped particles will have different values of m/σ and will therefore the magnitude of the drag force will be different for a given size. For example, for fractal aggregates of fractal dimension $\lesssim 2$, m/σ approaches a constant, and therefore the friction time becomes independent of size, which can have profound implications for coagulation (e.g. Okuzumi et al. 2012). By writing the equations in terms of $t_{\rm fric}$ they become valid for any particle shape (Testi et al. 2014).

For dust particles that are large compared to the mean free path of the gas, drag occurs via gas molecular viscosity as a boundary layer forms around the particle and experimental results are available (e.g. Lain et al. 1999). The exact form of the drag force depends on whether the flow around the dust particle is laminar (Stokes friction) or turbulent (Newtonian friction). We refer the reader to the paper by Whipple (1972) for a discussion of these friction regimes.

The equations of motion of a dust particle with radial velocity $v_{\rm d}$ and angular velocity $\Omega_{\rm d}$, including friction, are

$$\frac{dv_{\rm d}}{dt} = r\Omega_{\rm d}^2 - r\Omega_{\rm K}^2 - \frac{\Omega_{\rm K} v_{\rm d}}{\rm St}, \tag{39}$$

$$\frac{d}{dt}\left(r^2\Omega_{\rm d}\right) = -\frac{r^2\Omega_{\rm K}(\Omega_{\rm d} - \Omega)}{\rm St}, \tag{40}$$

where we have defined a Stokes number $\mathrm{St} = \Omega_{\rm K} t_{\rm fric}$ and ignored any radial velocity of the gas. Two interesting limits are $\mathrm{St} \ll 1$ (tightly coupled particles) and $\mathrm{St} \gg 1$ (loosely coupled particles).

Consider first the case $\mathrm{St} \ll 1$. From (40) we see that $\Omega_{\rm d} \approx \Omega$. Inserting this in (39) and solving for the terminal radial velocity (i.e. $dv_{\rm d}/dt = 0$), we get,

$$v_{\rm d} = \frac{\rm St}{\Omega_{\rm K}}\left(r\Omega^2 - r\Omega_{\rm K}^2\right) = \frac{\rm St}{\Omega_{\rm K}}\frac{1}{\rho}\frac{dp}{dr}, \tag{41}$$

where in the last step we have used (4). Note that dust particles always move towards regions of highest gas pressure, and that for moderate gradients in pressure ($|d\log p/d\log r| \approx 2\ldots3$), $v_{\rm d} \sim \mathrm{St} h^2 r\Omega_{\rm K}$.

In the opposite limit of $\mathrm{St} \gg 1$, we have that dust particles stay close to their equilibrium orbit $\Omega_{\rm d} \approx \Omega_{\rm K}$, but with a torque applied as given by (40). Assuming the orbit stays circular, this torque translates into a radial velocity

$$v_{\rm d} = -\frac{2r(\Omega_{\rm K} - \Omega)}{\rm St} \approx \frac{1}{\mathrm{St}\Omega_{\rm K}}\frac{1}{\rho}\frac{dp}{dr}, \tag{42}$$

where the last approximation is valid in the limit $h \ll 1$. Again, dust particles drift towards regions of high pressure, but now $v_{\rm d} \sim h^2 r\Omega_{\rm K}/\mathrm{St}$.

[2] Note that any contribution from the gravitational field of the particle can be safely ignored in the cross-section. When gravity starts to play a role, we are in the regime of Type I migration (see Sect. 6).

By ignoring $dv_{\rm d}/dt$ and writing $\Omega_{\rm d} = \Omega_{\rm K} + \delta\Omega$ with $\delta\Omega \ll \Omega_{\rm K}$ one can solve for the radial drift speed for arbitrary values of St (Weidenschilling 1977),

$$v_{\rm d} = \frac{\rm St}{1+{\rm St}^2}\frac{1}{\Omega_{\rm K}\rho}\frac{dp}{dr}. \tag{43}$$

Therefore, we expect inward drift of dust, since in protoplanetary discs the pressures is generally higher in the inner regions, on time scales that can be as short as a few 100 dynamical time scales for particles that have St $\sim$ 1. This is an enormous challenge for planet formation theory, since as soon as particles grow to St $\sim$ 1 (1 cm–1 m in typical protoplanetary discs), you lose them through very efficient drift (Weidenschilling 1977).

Dust drift occurs in the vertical direction as well, in which case it is usually called *settling*. Since the gas is completely pressure supported in the vertical direction, settling speeds can be quite fast (typically, the settling time scale is 1/St times the dynamical time scale for small particles). This allows for particles to grow by sweeping up smaller particles on their way to the mid plane of the disc, similar to what happens on Earth when raindrops form in clouds (Dullemond and Dominik 2005). If the gas disc is turbulent, dust settling will be reduced (e.g. Johansen and Klahr 2005; Fromang and Papaloizou 2006).

5 Forming Giant Cores

In this section we review the physics of forming the cores of giant planets by planetesimal accretion and pebble accretion. We also discuss the formation of planetesimals, as this stage has important implications for the growth rates of protoplanets.

5.1 Planetesimal Formation

Radial drift limits the particle sizes that can be obtained by dust coagulation. Very fluffy ice particles can in principle break through the drift barrier (Okuzumi et al. 2012), due to their low internal density of down to 10^{-5} times the material density. The material density actually does not play a role in the growth time-scale in the Epstein drag force regime (see discussion in Johansen et al. 2014), but fluffy grains enter the Stokes regime and increase their friction time, and hence the drift time-scale if the particle is already large, proportional to the squared size. However, this fluffy growth requires very small ice monomers, of 0.1 μm in size, in order to avoid compactification and allow sticking beyond 10 m/s. Another threat to fluffy ice growth is the erosion of the aggregates by small dust grains. Krijt et al. (2015) demonstrated that the fluffy ice particles do not grow beyond a Stokes number of unity, unless erosion by small dust grains is ineffective.

The streaming instability provides another avenue for making planetesimals (Youdin and Goodman 2005; Youdin and Johansen 2007; Johansen and Youdin 2007). This instability feeds of the relative velocity between particles (moving sub-Keplerian in the orbit and radially inwards) and gas (moving strongly sub-Keplerian in the orbit and radially outwards).[3] Above a threshold metallicity the particles in the mid-plane layer form dense nearly axisymmetric filaments due to particle pile-ups (Johansen et al. 2009b; Bai and Stone 2010). The threshold metallicity is around 1.5% at a Stokes number of St $\sim$ 0.1, but increases towards both smaller and larger particles (Carrera et al. 2015). The case of marginally coupled

[3]The streaming instability analysis ignores the possible presence of a radial accretion flow of the gas.

particles (St ~ 0.3) has been studied most extensively. These particles reach densities up to 10,000 times the local gas density in the absence of self-gravity (Johansen et al. 2012, 2015). Including self-gravity leads to the formation of planetesimals of a wide range of sizes, following the approximate differential mass distribution $\mathrm{d}N/\mathrm{d}M \propto M^{-1.6}$ (Johansen et al. 2015; Simon et al. 2016). This distribution function is dominated in number by the smallest planetesimals and in mass by the largest. The characteristic sizes of the planetesimals that form following gravitational collapse are around 100 km. This planetesimal birth size is in good qualitative agreement with the observed steepening of the size distribution above 100 km sizes of both asteroids and Kuiper belt objects (Bottke et al. 2005; Morbidelli et al. 2009; Sheppard and Trujillo 2010).

The streaming instability scenario has mainly been tested under the assumption that there are no other sources of turbulence in the disc region under consideration. The streaming instability can actually interact constructively with turbulence driven by the magnetorotational instability. Some of the energy in the turbulence cascades to large scales where it organises into long-lived zonal flows, alternating axisymmetric regions of faster and slower orbital speed (Johansen et al. 2009a; Simon et al. 2012). The zonal flow forms a geostrophic structure in force balance with the pressure gradient from a pressure bump that peaks between the points of fastest and slowest orbital speed. These pressure bumps in turn can collect very large pebbles (boulders) of meter sizes (Johansen et al. 2006; Dittrich et al. 2013). The concentrations of boulders can in turn trigger the streaming instability by accelerating the gas towards the Keplerian speed inside of the pressure bump, leading to the formation of planetesimals (Johansen et al. 2007).

The dead zone extending out to several 10 AU has too high ohmic resistivity for the magnetorotational instability to grow. Exterior of this dead zone the development of the magnetorotational instability will be influenced by ambipolar diffusion. Bai and Stone (2014) demonstrated that turbulence can develop very strong pressure bumps and zonal flows in these regions. In the dead zone itself several hydrodynamical instabilities are currently under investigation to drive turbulence. The vertical shear instability arises as a consequence of radial temperature gradients, which induce vertical shear, but the instability furthermore requires a very low cooling time to suppress buoyancy oscillations (Nelson et al. 2013; Lin and Youdin 2015). In order to obtain the most realistic cooling times, Stoll and Kley (2014) performed radiative transfer simulations of turbulence driven by the vertical shear instability and found turbulent stresses on the order of 10^{-4} in the mid-plane of the disc. This low level of turbulence is promising for a constructive interaction with the streaming instability, but simulations of the vertical shear instability including particle drag on the gas have not been reported in the literature yet. Richard et al. (2016) showed in simulations of relatively thick protoplanetary discs that the vertical shear instability can lead to the formation of long-lived vortices, likely stabilised against destruction by the sub-critical baroclinic instability (Klahr and Bodenheimer 2003; Lesur and Papaloizou 2010; Raettig et al. 2013). Raettig et al. (2015) included the back-reaction of the pebbles in simulations of the baroclinic instability and found that the vortices are actually destroyed by the back-reaction, a result that is also known for single vortices that are set as an initial condition (Johansen et al. 2004; Crnkovic-Rubsamen et al. 2015). The relative roles of the streaming instability and the vertical shear and baroclinic instabilities in concentrating pebbles is still poorly understood and should be an important priority for future research.

5.2 Planetesimal and Pebble Accretion

The largest planetesimals that form at the planetesimal formation stage will continue to grow towards planetary masses by accreting other planetesimals and the remaining pebbles. The accretion rate of a protoplanet is given by (Safronov 1969)

$$\frac{\mathrm{d}M}{\mathrm{d}t} = \pi R^2 \rho_{\mathrm{pla}} v_{\mathrm{pla}} \left[1 + \left(\frac{v_{\mathrm{e}}}{v_{\mathrm{pla}}} \right)^2 \right]. \tag{44}$$

Here M and R are the mass and the radius of the planetesimals, ρ_{pla} is the spatial density of planetesimals, v_{pla} is the approach speed of the planetesimals and v_{e} is the escape speed from the surface of the core. The gravitational focusing term in the hard brackets can boost planetesimal accretion rates significantly if the random planetesimal speed is slower than the escape speed of the protoplanet. In the limit of strong gravitational focusing, we get the expression

$$\frac{\mathrm{d}M}{\mathrm{d}t} = \pi R^2 \Sigma_{\mathrm{p}} \Omega \frac{6p^{-1}}{\zeta^2}. \tag{45}$$

Here the parameter $\zeta \equiv v_{\mathrm{r}}/v_{\mathrm{H}}$ defines the random planetesimal speed relative to the Hill speed $v_{\mathrm{H}} = \Omega R_{\mathrm{H}}$ and $p \equiv R/R_{\mathrm{H}}$ is the size of the protoplanet relative to its Hill radius. The expression is only valid when $\zeta > 1$. The parameter p can be written as

$$p = \frac{R}{R_{\mathrm{H}}} = \left(\frac{4\pi G \rho_{\bullet}}{9\Omega^2} \right)^{-1/3} \approx 0.001 \left(\frac{r}{5\ \mathrm{AU}} \right)^{-1} \left(\frac{\rho_{\bullet}}{2.0\ \mathrm{g\,cm^{-3}}} \right)^{-1/3}. \tag{46}$$

This allow us to write the growth rate as

$$\frac{\mathrm{d}M}{\mathrm{d}t} \approx 3.8 M_{\mathrm{E}}\ \mathrm{Myr}^{-1}\ f_{\mathrm{pla}} \left(\frac{M}{M_{\mathrm{E}}} \right)^{2/3} \left(\frac{r}{5\ \mathrm{AU}} \right)^{-2} \zeta^{-2}. \tag{47}$$

We have here scaled the planetesimal column density according to $\Sigma_{\mathrm{pla}} = 10\ \mathrm{g\,cm^{-2}} f_{\mathrm{pla}}\ (r/\mathrm{AU})^{-1}$. The value of ζ makes it tricky to evaluate the actual growth rate of planetesimal accretion. In principle, ζ can be less than unity if gas drag or mutual collisions damp the scale-height of the planetesimals (Rafikov 2004). However, planetesimals must be small (< 1 km) for this process to be efficient. Greenzweig and Lissauer (1990) found that a single scattering of a planetesimal increases its random motion to a few times the Hill speed. Realistic values of the planetesimal accretion rates can be obtained from simulations of a large number of protoplanets growing simultaneously. Levison et al. (2010) showed that the protoplanets excite the planetesimal orbits to very high random speeds and that planetesimals are transported rapidly away from the formation region of the giant planets.

The pebbles left over from the planetesimal formation stage provide another source of planetary growth. Pebbles can be accreted very efficiently due to the friction with the gas. While most planetesimals are scattered by a growing protoplanet, pebbles of optimal sizes are accreted from the entire Hill radius (Johansen and Lacerda 2010; Ormel and Klahr 2010; Lambrechts and Johansen 2012). The optimally-accreted pebbles have a Stokes number of approximately ~ 0.1, in good agreement with the results of coagulation models (Brauer et al. 2007; Birnstiel et al. 2012; Lambrechts and Johansen 2014).

The growth rate by pebble accretion in the Hill accretion regime can be written as

$$\dot{M} = 2 R_{\mathrm{H}} v_{\mathrm{H}} \Sigma_{\mathrm{p}}, \tag{48}$$

where Σ_{p} is the pebble column density. This expression is valid above the transition mass from Bondi accretion to Hill accretion, which happens at around $10^{-3}\ M_{\mathrm{E}}$ in the inner regions of the protoplanetary disc. We furthermore assumed that the accretion is in the 2-D regime, i.e. that the Hill radius is larger than the scale-height of the particle mid-plane layer. These two requirements are typically fulfilled at the same time (Lambrechts and Johansen 2014; Bitsch et al. 2015). Scaling the pebble accretion rate with typical quantities gives

$$\dot{M}_{\rm H,2D} = 210 M_{\rm E}\,{\rm Myr}^{-1}\, f_{\rm p} \left(\frac{\rm St}{0.1}\right)^{2/3} \left(\frac{M}{M_{\rm E}}\right)^{2/3} \left(\frac{r}{5\,{\rm AU}}\right)^{-0.5}. \tag{49}$$

Here we incorporated the Stokes number dependence in the expression (Lambrechts and Johansen 2014; Morbidelli et al. 2015) and introduced a parameter $f_{\rm p}$ that gives the pebble column density relative to a nominal value of $\Sigma_{\rm p} = 10\,{\rm g\,cm^{-2}}\, f_{\rm p}\,(r/{\rm AU})^{-1}$. The growth rates by pebble accretion is approximately two orders of magnitude higher than the planetesimal accretion rate (which is in itself in upper limit, given the uncertain value of ζ).

An advantage of the very high accretion rates obtainable with pebble accretion is that planets can outgrow Type I migration. We discuss this in Sect. 8. But first we will review the migration of planets.

6 Low Mass Planet Migration (Type I)

In this section we consider the migration of low-mass planets. We will give a precise definition of what constitutes a low-mass planet later on, but as a rough guide consider planets less massive than Neptune. While planets have a surface and are therefore subject to aerodynamic drag, when planets reach a mass of roughly 1% of that of the Earth, gravitational interactions will start to dominate. The resulting migration is usually referred to as Type I migration (Ward 1997). Below, we will introduce a simple toy model that highlights the physics of these gravitational interactions. For readers who want more details, Baruteau et al. (2014) provide a recent more in-depth review.

Consider the local, two dimensional shearing sheet introduced in Sect. 2.2, and put a planet of mass $M_{\rm p}$ at the origin of the local coordinate system by adding a gravitational potential to the equations,

$$\Phi_{\rm p} = -\frac{GM_{\rm p}}{\sqrt{x^2+y^2+\epsilon^2}}, \tag{50}$$

where ϵ is a smoothing length which is at the moment added to keep all gravitational forces finite everywhere, but for which we will give a physical interpretation later on.

6.1 Scattering Torque

Particles streaming past the planet, on initially unperturbed orbits, feel the gravitational force from the planet and therefore change their trajectory. A simple model capturing this is the so-called impulse approximation (Lin and Papaloizou 1979), where it is assumed that the particle motion remains unperturbed until it reaches the planet, at which point it is scattered, see the left panel of Fig. 3. Throughout we assume that the interaction of the particle with the planet is weak. The total change in the x-component of the velocity is given by

$$\Delta v_x \approx -\int_0^\infty \frac{GM_{\rm p}x_0}{(x_0^2+y(t)^2+\epsilon^2)^{3/2}}dt. \tag{51}$$

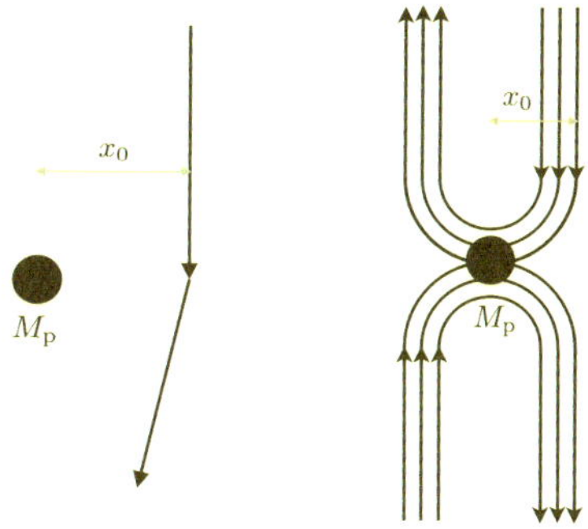

Fig. 3 *Left*: gravitational scattering of a particle by a planet in the local patch. The particle comes from the top, and its unperturbed Keplerian motion takes it past the planet, where it is scattered and put on a different trajectory. *Right*: horseshoe trajectories near the planet

Taking $y(t) = -S_p x_0 t$, where a subscript p means evaluation at the location of the planet, the integral evaluates in a straightforward way to,

$$\Delta v_x \approx -\frac{GM_p}{(x_0^2 + \epsilon^2)S_p}. \tag{52}$$

From Eqs. (13) and (14) we find that the total energy

$$E = \frac{1}{2}v_x^2 + \frac{1}{2}v_y^2 - \Omega_p S_p x^2 + \Phi_p \tag{53}$$

of the particle is conserved. Neglecting any changes in x coordinate, and evaluating E far away from the planet so that the contribution of Φ_p can be neglected, energy conservation dictates that $\Delta v_x^2 + \Delta v_y^2 = 0$. Assuming the change in v_y to be small compared to the unperturbed orbital motion,

$$\left(\frac{GM_p}{(x_0^2 + \epsilon^2)S_p}\right)^2 + 2S_p x_0 \Delta v_y \approx 0, \tag{54}$$

which leads to

$$\Delta v_y(x_0) \approx -\frac{(GM_p)^2}{2S_p^3 x_0 (x_0^2 + \epsilon^2)^2}. \tag{55}$$

As a consequence of Newton's third law, the planet feels a force when it scatters a particle. The encounter rate of particles with impact parameter x is $|S_p x|$, and if the disc has surface density $\Sigma(x)$, the force per unit radius in the y-direction is

$$f_y(x) = \Delta v_y(x)\Sigma(x)|S_p x|, \tag{56}$$

and the total force on the planet can be found by integrating over x. Note that for constant Σ, f_y is an odd function of x and therefore the total force evaluates to zero. This is a consequence of the simplification that is the shearing sheet compared to the full disc. We can restore some of the properties of the full disc by letting S be a function of x,[4] and approximate

$$\frac{1}{S^2} \approx \frac{1 + 3x/r_p}{S_p^2}, \tag{57}$$

[4]The same effect can be achieved by choosing an appropriate surface density profile.

which is consistent with (3). Then, for constant surface density the total force is

$$F_y = \int_{-\infty}^{\infty} f_y(x)dx \approx -\frac{3(GM_\mathrm{p})^2\Sigma}{2r_\mathrm{p}S_\mathrm{p}^2\epsilon^2} \tag{58}$$

The torque due to such scattering events on the planet is given by $\Gamma = r_\mathrm{p}F_y$, which then takes the form

$$\Gamma_\mathrm{s} = \gamma_\mathrm{s}\frac{q^2}{\epsilon^2}\Sigma_\mathrm{p}r_\mathrm{p}^6\Omega_\mathrm{p}^2, \tag{59}$$

where γ_s is a constant of order unity, and q is the planet-to-star mass ratio. It should be noted that although for simplicity we have focused on a simple shearing sheet disc model, the same functional form of the scattering torque is found in global cylindrical geometry, in which γ_s depends on the radial gradients of the gas density and temperature (e.g. Tanaka et al. 2002).

6.2 Horseshoe Drag

The scattering analysis presented above is only valid for impact parameters that are large enough. Closer to the planet, the change in v_y is large enough to change the sign of v_y, and the result is material executing horseshoe orbits. The right panel of Fig. 3 shows particle trajectories close to the planet. These horseshoe turns are responsible for a different kind of torque: the horseshoe drag (Ward 1991).

We can play the same game as in the previous section, by considering a particle with impact parameter x_0 (see the right panel of Fig. 3). Assuming the particle makes a symmetric turn, the change in y-velocity is

$$\Delta v_y(x_0) = 2S_\mathrm{p}x_0, \tag{60}$$

and therefore the total force can be found from

$$F_y = \int_{-w}^{w} \Delta v_y(x)\Sigma(x)|S_\mathrm{p}x|dx, \tag{61}$$

where w is the maximum impact parameter for which a horseshoe turn occurs, or, in other words, the half-width of the horseshoe region.

As in the scattering calculation, the shearing sheet is too symmetric to yield a torque: we either have to allow S^2 to vary with x, or Σ. Expanding any of these quantities to first order in x, we get

$$F_y = f_h\frac{S_\mathrm{p}^2\Sigma_\mathrm{p}w^4}{r_\mathrm{p}}, \tag{62}$$

where f_h is a constant of order unity and depends on the gradients in S and Σ.

Interestingly, it is relatively straightforward to calculate w (Masset et al. 2006; Paardekooper and Papaloizou 2009b). Equations (13) and (14) admit a Jacobi integral,

$$J = \Omega_0 S_0 x^2 - \Phi_\mathrm{p} - \frac{1}{2}(\dot{x}^2 + \dot{y}^2). \tag{63}$$

Consider the particle orbit that gets closest to the planet. Clearly, the impact parameter of this orbit is the maximum impact parameter for which a horseshoe turn will occur. At $x = y = 0$, there is an X-point in the particle stream lines, and therefore all velocities must vanish. Therefore, the outermost horseshoe orbit has $J = -\Phi_{\rm p}(x = 0, y = 0)$. Far away from the planet, motion is purely Keplerian again, with $\dot{x} = \Phi_{\rm p} = 0$ and $\dot{y} = -S_{\rm p}x$. Equating J for both locations gives

$$\frac{GM_{\rm p}}{\epsilon} = S_{\rm p}\left(\Omega_{\rm p} - \frac{S_{\rm p}}{2}\right)x^2. \tag{64}$$

Solving for x gives the half-width of the horseshoe region w,

$$w = \sqrt{\frac{8GM_{\rm p}}{3\epsilon\Omega_{\rm p}^2}}. \tag{65}$$

The final form of the horseshoe drag is then

$$\Gamma_{\rm h} = r_{\rm p}F_y = \gamma_{\rm h}\frac{q^2}{\epsilon^2}\Sigma_{\rm p}r_{\rm p}^6\Omega_{\rm p}^2, \tag{66}$$

where $\gamma_{\rm h}$ is a constant of order unity. Remarkably, the two types of torque (59) and (66) have exactly the same form and only differ by a numerical prefactor of order unity. As in the case of the scattering torque, it should be noted that although for simplicity we have focused on a simple shearing sheet disc model, the same functional form of horseshoe drag is found in global cylindrical geometry, in which $\gamma_{\rm h}$ depends on the radial gradients of the gas density and temperature (e.g. Ward 1991; Paardekooper and Papaloizou 2008).

It should be noted that in a full disc, the horseshoe region is a closed system. Particles leaving the domain after a horseshoe turn at the bottom of the right panel of Fig. 3 come back at the top and make another horseshoe turn. Unless the particle loses its identity before returning, the total torque averaged over many encounters will be zero. This is a process that is called *saturation*. Another way of thinking about this is that the total angular momentum content of the closed system that is the horseshoe region is finite, and it therefore only has a finite amount of angular momentum to give to the planet. Unless 'fresh' angular momentum can be supplied to the horseshoe region, perhaps through viscous diffusion (e.g. Masset 2001), the horseshoe drag will only be a transient phenomenon. Fortunately, the conditions in typical protoplanetary discs allow for unsaturated horseshoe drag in large regions (Paardekooper et al. 2011a; Bitsch et al. 2013).

6.3 Dynamic Torques (Type III Migration)

From the discussion above, it is clear that for the horseshoe drag to keep going, some form of asymmetry between the two horseshoe legs is needed. In the local model, a radial surface density gradient will do the trick; in a global isothermal gaseous disc, a gradient in specific vorticity $\nabla \times \mathbf{v}/\Sigma$ is needed (Ward 1991; Paardekooper and Papaloizou 2009a; Casoli and Masset 2009), while in a global adiabatic disc a gradient in entropy is important (Baruteau and Masset 2008; Paardekooper and Papaloizou 2008; Masset and Casoli 2009). These results were all obtained by taking the orbit of the planet as fixed, and measuring or calculating the torque on the planet.

Interestingly, a migrating planet naturally introduces asymmetry between the two horseshoe legs. This is illustrated schematically in Fig. 4.The left panel shows the horseshoe (or corotation) region around a non-migrating planet in red. Note that the planet is shifted to the top (and bottom because the domain is periodic in φ) to show the full extent of the horseshoe

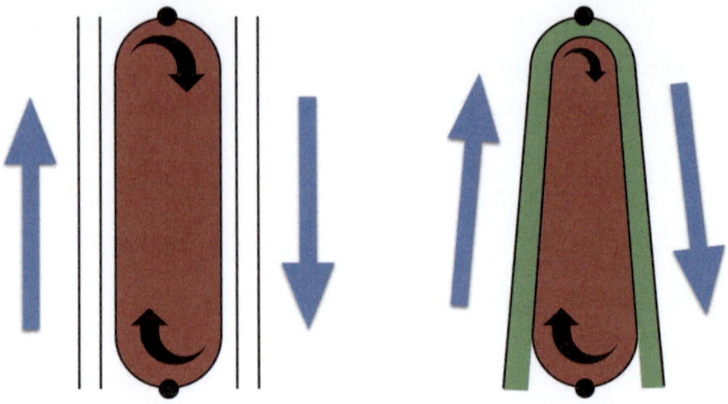

Fig. 4 Schematic stream lines for a non-migrating (*left panel*) and a migrating (*right panel*) planet. The *vertical coordinate* is φ, for which the full 2π is shown, and the *horizontal coordinate* is r. The planet is denoted by the *black circles*, the *blue arrows* indicate the disc rotation in the frame of the planet, and the *black arrows* indicate the velocity of material in the corotation region, which is marked *red*. In the *right panel*, the planet is migrating to the left, which means that in the rest frame of the planet, unperturbed disc material moves outward. Material in the *green region* gets transported from one side of the planet to the other in a single horseshoe turn

region. In the rest frame of the planet, material in the red region executes horseshoe turns all the time, while material outside the red region is on circular orbits (note that we neglect any effect of the scattering torque here), in the direction of the blue arrows.

The right panel of Fig. 4 shows the case of a migrating planet, with the planet moving towards the left. For the sake of definiteness, let us assume the planet is migrating inwards, so that the star is located far away to the left. In the rest frame of the planet, gas on circular orbits now moves to the right, again in the direction of the blue arrows. Note that the red corotation region gets deformed, one side getting thinner. This region in fact migrates inward together with the planet. On the thin side of the corotation region, material in the green region executes a single horseshoe turn. This material was on circular orbits before in the inner disc (left of the planet, *outside* the corotation region), and gets transferred to the outer disc (right of the planet, again *outside* the corotation region).

In this single horseshoe turn, the material exerts a torque on the planet as usual. However, note that there is no corresponding turn at the bottom of the figure, where only red material executes horseshoe turns. Therefore, all we need for the horseshoe legs to be asymmetric is a difference between the red and the green material. Such a torque can be called *dynamic* because it requires the planet to move through the disc. Suppose for example that the red region is underdense compared to the rest of the disc, perhaps through partial gap opening (see Sect. 7.1). Then the torque on the planet is negative, because of the green material getting pushed outward. Moreover, this torque is proportional to the migration rate: the faster the migration, the more green material, the stronger the torque. Hence there exists a positive feedback loop (e.g. Masset 2008) with the possibility of an instability, termed *runaway migration* by Masset and Papaloizou (2003). Since there exist steady forms of very fast migration, Artymowicz (2004) introduced the term Type III migration, since Type II migration is reserved for massive, gap-opening planets (see Sect. 7).

Originally, Type III migration was studied for intermediate-mass planets (roughly Saturn's mass) with a slight density depression in their corotation region (Masset and Papaloizou 2003), which naturally leads to an asymmetry between the red and green region in the right panel of Fig. 4. Such an asymmetry can also be imposed by placing the planet on a sharp density gradient initially, in which case even massive planets like Jupiter undergo Type III migration (Pepliński et al. 2008). Yet another way to introduce an asymmetry is through migration itself: since the planet takes with it the red region, if the planet migrates into a disc region with a different density this will again lead to a difference between the red and the green material and to a dynamic corotation torque (Paardekooper 2014). This has been studied both isothermal (Paardekooper 2014) and non-isothermal (Pierens 2015) discs.

6.4 Three Dimensions

So far, we have discussed only test particles in the two-dimensional shearing sheet. Real discs are of course three-dimensional and made of gas, but fortunately we can incorporate three-dimensional effects in an approximate way while at the same time fixing the free parameter ϵ, thereby killing two birds with one stone.

The idea is to view the two-dimensional governing equations as vertically integrated versions of their three-dimensional counterparts. For example, the continuity (or mass conservation) equation reads

$$\frac{\partial \rho}{\partial t} + \frac{\partial}{\partial x}(\rho v_x) + \frac{\partial}{\partial y}(\rho v_y) + \frac{\partial}{\partial z}(\rho v_z) = 0. \tag{67}$$

Integrating this equation from $z = -\infty$ to $z = \infty$, assuming ρv_z vanishes for large values of $|z|$, we get

$$\frac{\partial}{\partial t}\left(\int_{-\infty}^{\infty} \rho dz\right) + \frac{\partial}{\partial x}\left(\int_{-\infty}^{\infty} \rho v_x dz\right) + \frac{\partial}{\partial y}\left(\int_{-\infty}^{\infty} \rho v_y dz\right) = 0. \tag{68}$$

Defining a vertically averaged velocity

$$\bar{\mathbf{v}} = \frac{1}{\Sigma}\int_{-\infty}^{\infty} \rho \mathbf{v} dz, \tag{69}$$

we get

$$\frac{\partial \Sigma}{\partial t} + \frac{\partial}{\partial x}(\Sigma \bar{v}_x) + \frac{\partial}{\partial y}(\Sigma \bar{v}_y) = 0, \tag{70}$$

which is just the two-dimensional continuity equation. Things do not work out as nicely for the momentum equations, for example the x-component

$$\frac{\partial}{\partial t}(\rho v_x) + \frac{\partial}{\partial x}\left(\rho v_x^2 + p\right) + \frac{\partial}{\partial y}(\rho v_x v_y) + \frac{\partial}{\partial z}(\rho v_x v_z)$$
$$= 2\Omega_{\mathrm{p}}\rho(v_y + S_{\mathrm{p}}x) - \rho\frac{\partial \Phi_{\mathrm{p}}}{\partial x}, \tag{71}$$

but following this idea of vertical integration suggests that in two dimensions, we should replace the last term on the right-hand side with (for a more detailed discussion see Müller et al. 2012),

$$-\int_{-\infty}^{\infty} \rho \frac{\partial \Phi_{\mathrm{p}}}{\partial x} dz. \tag{72}$$

Recall from (6) that the vertical density profile is essentially a Gaussian of width H. This means that integrating vertically is the same as applying Gaussian smoothing with smoothing length H. Therefore, for consistency we must choose $\epsilon \sim H$, and our final expression for the total torque, adding the scattering torque and the horseshoe drag, becomes

$$\Gamma = \gamma \frac{q^2}{h^2} \Sigma_\mathrm{p} r_\mathrm{p}^4 \Omega_\mathrm{p}^2, \tag{73}$$

where γ is a constant of order unity, which, in the case of global cylindrical geometry, depends on local gradients in density and temperature.

6.5 Gas Pressure

While a full analysis including effects of gas pressure on the dynamics is much more complicated than our simple test particle approach, the resulting form of the torque is exactly that of (73). Most of the work goes into getting the numerical constant γ correct, but three important other effects of gas pressure must be mentioned first. First of all, in the scattering calculation, it may seem that the sign of the torque can be altered by choosing an appropriate surface density profile. In practice, this is not the case, since changing the density usually leads to a change in pressure, which in turn affects the rotation profile through (4). It turns out that any attempt to change the torque this way is doomed to fail, since changing the density in one side of the disc to affect the torque leads to a change in rotation profile that almost completely cancels any effect on the torque. This is called the pressure buffer effect (Ward 1986), and makes the scattering torque rather independent of local density gradients in the disc.

Second, in the analysis of the horseshoe drag, we have taken the streamline pattern as given. Readers familiar with the restricted three-body problem will have noticed the absence of a Roche lobe and several Lagrange points. It can be shown that the configuration in the right panel of Fig. 3 is to be expected when all velocities are very subsonic (Paardekooper and Papaloizou 2009b), which essentially means that gas pressure dominates over gravity. This assumption breaks down for massive planets.

Finally, it can be shown that the scattering encounter with the planet leaves the test particles on eccentric orbits. The response of a gaseous disc is different: instead of eccentric particles we get coherent density wave emission (Ogilvie and Lubow 2002). We can therefore call the scattering torque the wave torque in gaseous discs. The left panel of Fig. 5 shows the surface density after 20 orbits of a planet with $q = 10^{-5}$ in a disc with constant $h = 0.05$. The disc response is dominated by an inner and an outer spiral density wave. Since the planet is moving in the positive φ-direction, it is clear from the picture that the gravitational field of the outer wave is trying to slow down the planet, while the inner wave is trying to speed it up. Therefore, the outer wave promotes inward migration, and the inner wave outward migration.

When doing the full calculation for isothermal discs, one usually finds negative torques and therefore inward migration (Tanaka et al. 2002), unless the surface density increases outward (Paardekooper and Papaloizou 2009a). When temperature fluctuations are allowed for, a much richer variety of migration behaviour can be seen, with possible outward migration (Masset and Casoli 2009; Paardekooper et al. 2010). This rich behaviour is almost completely due to the action of the horseshoe drag: while the wave torque is usually negative, promoting inward migration, the horseshoe drag can be of either sign and often dominate over the wave torque, depending on local disc parameters, leading to complex migration maps for low-mass planets (e.g. Bitsch et al. 2013).

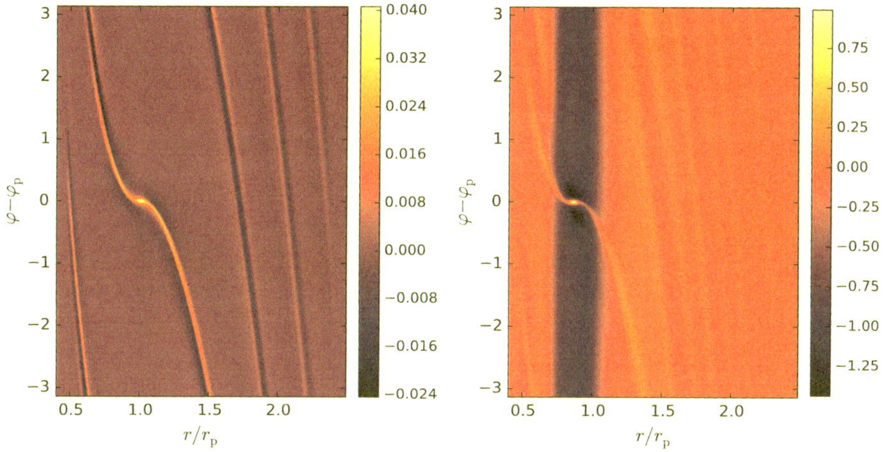

Fig. 5 Surface density ($\log_{10}(\Sigma/\Sigma_0)$) for a disc with constant $h = 0.05$, viscosity $\nu = 10^{-5}$ and an embedded planet of $q = 10^{-5}$ after 20 orbits of the planet (*left panel*) and $q = 10^{-3}$ after 100 orbits (*right panel*)

6.6 Migration Time Scale

The torque given by (73) enters the angular momentum equation of the planet,

$$\frac{d}{dt}\left(M_p r_p^2 \Omega_p\right) = \Gamma. \tag{74}$$

Assuming the planet stays on a circular orbit with $\Omega_p = \Omega_K(r_p)$, we find that the radial migration speed of the planet is given by

$$\frac{dr_p}{dt} = \frac{2\Gamma}{M_p r_p \Omega_p} = 2\gamma \frac{q}{h^2} \frac{\Sigma_p r_p^2}{M_*} r_p \Omega_p. \tag{75}$$

Ignoring factors of order unity, this leads to a migration time scale

$$\tau_m = \frac{h^2}{q} \frac{M_*}{r_p^2 \Sigma_p} \Omega_p^{-1} \tag{76}$$

For a typical protoplanetary disc this suggests migration time scales at 1 AU of $\sim 10^6$ years for Earth and $\sim 10^5$ years for Neptune. Since these time scales are shorter than the life times of protoplanetary discs, this means that Type I migration should be an important ingredient in any planet formation theory. It is clearly not easy to prevent planets from migrating very close to the central star, unless planetary growth rates are as high as in the pebble accretion scenario (see Sect. 5) so that they grow out of the Type I migration regime before migrating all the way in. This only applies to planets that end up as gas giants: the population of Super-Earths and mini-Neptunes are usually thought to be subject to Type I migration.

6.7 Recent Results

Early models of Type I migration were constructed for isothermal discs, notably the semi-analytical work of Tanaka et al. (2002). It was realised quickly that the resulting inward migration time scales as quoted above were too short to be compatible with planets ending up at several astronomical units from their star, which led to investigations of more complex discs in search of stopping and slowing down mechanisms. The discovery of rapid planet formation by pebble accretion has alleviated some of the concerns about rapid migration (Sect. 5) for gas giant planets, but the problem of Type I migration persists for Neptunes and Super-Earths. Here we briefly mention some of the progress that has been made, as well as some recent results.

One of the first extensions of the isothermal model was the inclusion of magnetic fields. This leads to additional torques on the planet (Terquem 2003; Guilet et al. 2013), but also allows the disc to be naturally turbulent because of the magnetorotational instability (MRI Balbus and Hawley 1991). Type I migration in such a turbulent disc has a stochastic character, but in the end drives planets inward on too short time scales still (Nelson and Papaloizou 2004).

Most recent progress has come from releasing the isothermal assumption, and take care of the energy budget in a more realistic way. Paardekooper and Mellema (2006) found *outward* migration in three-dimensional, radiation-hydrodynamical simulations; an effect that was later found to be due to a radial entropy gradient in the disc (Baruteau and Masset 2008; Paardekooper and Papaloizou 2008).

Recently, several groups have reported interesting effects on Type I migration originating very close to the planet itself. Fung et al. (2015) found that in three-dimensional, isothermal simulations, for low values of the viscosity there exists what they call a transient horseshoe flow, much like the green region in the right panel of Fig. 4, but three-dimensional in nature and for a static (i.e. non-migrating) planet. Material on these orbits goes very close to the planet (well within its Bondi sphere) and appears to slow down migration by a factor of 3. For three-dimensional non-isothermal simulations, Lega et al. (2014) found that very close to the planet, again well within the Bondi sphere, an asymmetry in cooling leads to an additional torque. Finally, Benítez-Llambay et al. (2015) reported that heating due to planetesimal accretion can change the sign of the torque on rapidly accreting low-mass planets.

These recent results indicate that the region very close to the planet and what happens there thermally is extremely important. Future numerical models can hopefully zoom in on the planet sufficiently with all necessary physics included (radiative cooling, self-gravity, magnetic fields) so that an ever more realistic picture of Type I migration can emerge.

7 High Mass Planet Migration (Type II)

7.1 Gap Formation

As mentioned in Sect. 6.5, gas pressure does not fundamentally alter the physics of Type I migration, and therefore we could get reasonable estimates of the migration rates through a ballistic approach. Moreover, there is the assumption that the interaction with the planet does not significantly change the properties of the disc. This works well if velocity perturbations are small compared to the sound speed. Otherwise, we expect shock waves that will make order unity changes to the region around the planet.

Setting $\Delta v_x < c_s$ in (52), together with $\epsilon = H$ as argued in Sect. 6.4, in order for Type I migration to be valid leads to a condition on the mass

$$q < \frac{3}{2}h^3, \tag{77}$$

sometimes called the *thermal criterion* for gap-opening (e.g. Bryden et al. 1999). For the canonical value of $h = 0.05$ the boundary lies at 60 Earth masses. This means that roughly any planet up to the mass of Neptune is safely in the Type I regime. Note that the thermal criterion is roughly equivalent to demanding that the Hill radius of the planet is smaller than the scale height of the disc.

In the shearing sheet, the quantity playing the role of angular momentum is

$$p_y = v_y + 2\Omega x. \tag{78}$$

Note that in the unperturbed disc, this quantity changes sign at $x = 0$. Since Δv_y changes sign as well (see (55)), this means that particles with initially $p_y > 0$ will experience an increase in p_y, while particles with initially $p_y < 0$ will experience a decrease in p_y. If these particles, by some damping process, return to equilibrium orbits with their new p_y, we have that particles with initial $x > 0$ will move to larger values of x, while particles with initial $x < 0$ will move to a position with x even more negative. In other words, the gravitational pull of the planet is repulsive: it is pushing material away from its orbit.

In the linear regime, where the waves seen in the left panel of Fig. 5 are of small amplitude, the flux of angular momentum is constant, and therefore no surface density evolution is expected unless the waves shock far away from the planet (Goodman and Rafikov 2001). If the waves start out as shock waves, as is the case for $q > h^3$, angular momentum is deposited locally leading to dramatic evolution of the surface density. Since the planet is pushing material away from its orbit, the result is an annular gap, as can be seen in the right panel of Fig. 5, which shows a Jupiter mass planet embedded in a disc with $h = 0.05$ after 100 orbits.

An estimate for the gap formation time scale can be obtained by integrating $|f_y|$ from (56) times r_p to get the total absolute torque exerted on the disc,

$$|\Gamma_\mathrm{tot}| = r_\mathrm{p} \int_{-\infty}^{\infty} |f_y| dx = \frac{2\pi}{9} \frac{q^2}{h^3} \Sigma_\mathrm{p} r_\mathrm{p}^4 \Omega_\mathrm{p}^2, \tag{79}$$

where we have substituted $\epsilon = H$ as before. Taking the half-width of the gap to be w, the total angular momentum of the gap region is initially

$$L_\mathrm{gap} \approx 2\pi r_\mathrm{p} w r_\mathrm{p}^2 \Omega_\mathrm{p} \Sigma_\mathrm{p}. \tag{80}$$

The time scale for removing this amount of angular momentum through the torque $|\Gamma_\mathrm{tot}|$ is

$$\tau_\mathrm{gap} = \frac{L_\mathrm{gap}}{|\Gamma_\mathrm{tot}|} = 18 \frac{h^3}{q^2} \frac{w}{r_\mathrm{p}} \frac{1}{\Omega_\mathrm{p}}, \tag{81}$$

which, for a gap width of $2H$, gives a gap opening time scale of ~ 100 orbits for Jupiter in a disc with $H/r = 0.05$, consistent with numerical simulations.

While $q > h^3$ is a necessary condition for local non-linear waves and therefore gap formation, if the disc is viscous it is also necessary to make sure viscosity is not able to close the gap. Suppose we impose a gap of width $2H$ in the disc. The total gravitational torque on one side of the disc is given by

$$\begin{aligned} \Gamma_\mathrm{g} &= r_\mathrm{p} \int_H^{\infty} f_y(x) dx \\ &= \frac{-r_\mathrm{p}(GM_\mathrm{p})^2 \Sigma}{2S_\mathrm{p}^2} \int_H^{\infty} \frac{dx}{(x^2+\epsilon^2)^2} \\ &\approx -0.032 \frac{r_\mathrm{p}^4 \Omega_\mathrm{p}^2 q^2 \Sigma}{h^3}. \end{aligned} \tag{82}$$

Demanding this to be larger in magnitude than the viscous torque (e.g Lin and Papaloizou 1993)

$$\Gamma_\nu = 3\pi \Sigma \nu r_\mathrm{p}^2 \Omega_\mathrm{p} \tag{83}$$

leads to the following condition on the viscosity in order for the planet to open a gap,

$$\nu < C \frac{r_\mathrm{p}^2 \Omega_\mathrm{p} q^2}{h^3}, \tag{84}$$

where $C = 0.0034$. For a Jupiter mass planet in a disc with $h = 0.05$, we must have $\nu < 2.7 \times 10^{-5}$. Therefore, for canonical disc parameters $h = 0.05$ and $\nu = 10^{-5}$, both the thermal and viscous criteria yield similar minimum planetary masses for gap opening. Combining both requirements, Crida et al. (2006) found a general criterion

$$\frac{3}{4} \frac{H}{r_\mathrm{Hill}} + \frac{50}{q\mathrm{R}} < 1 \tag{85}$$

where $r_\mathrm{Hill} = r_\mathrm{p}(q/3)^{1/3}$ is the radius of the Hill sphere, and $\mathrm{R} = r_\mathrm{p}^2 \Omega_\mathrm{p}/\nu$ denotes the Reynolds number.

7.2 Migration Inside Gaps

When a planet is able to open up a gap, its interaction with the disc changes fundamentally. Consider the very much idealised case of a planet orbiting in a completely empty gap, and that the gap is wide enough so that the planet does not interact with the disc when it is in the centre of the gap (note that the planet does need to be able to interact with the gap region itself, otherwise there would be no gap in the first place). Now displace the planet inward by a small amount dr. It can now interact with an annulus of disc material just inside the inner edge of the gap. This interaction, through the wave (or scattering) torque, will push the planet outward, back to the centre of the gap. The opposite situation arises if we displace the planet outward. In other words, the edges of the gap repel the planet.

This means that the planet is locked inside the gap: it can not move with respect to the gap. If we now bring in global disc evolution, where the disc as a whole is slowly draining onto the central star, this means that the gap will try to accrete with the rest of the disc, taking the planet with it. In effect, the planet acts as a normal disc particle, accreting onto the central star. This is called Type II migration (Lin and Papaloizou 1986; Ward 1997), and it necessarily occurs on the viscous time scale.

Several modifications have to be made to this simple picture. First of all, consider again a planet in a gap but displaced inward by a small amount dr. The annulus it interacts with at radial location r_in has total angular momentum

$$dL = \Sigma r_\mathrm{in}^2 \Omega_\mathrm{in} 2\pi r_\mathrm{in} dr \tag{86}$$

To take the planet back to the centre of the gap requires a change in angular momentum of the planet of

$$dL = \frac{1}{2} M_\mathrm{p} r_\mathrm{p} \Omega_\mathrm{p} dr. \tag{87}$$

Therefore, the annulus only has enough angular momentum to push the planet back if

$$M_\mathrm{p} < 4\pi \Sigma r_\mathrm{p}^2. \tag{88}$$

In other words, only if the local disc mass is larger than the planet mass does the planet behave like a disc particle, otherwise it *can* move with respect to the gap. This is why very massive planets migrate on time scales larger than the viscous time scale (Syer and Clarke 1995; Ivanov et al. 1999).

A second important modification concerns the implicit assumption in the simple picture of Type II migration that no material can cross the gap. Suppose for example that the annulus interacting with the inward-displaced planet as described above is not 'scattered' away by the planet, but instead is transferred outward in a single horseshoe turn, much like in the case of dynamic corotation torques. The back reaction on the planet now takes it even further inward towards the inner edge of the gap. In this case, therefore, the gap edges attract the planet rather than repel it. In reality, a combination of these two extremes probably occurs, which means that gap-opening planets do in general not migrate on the viscous time scale (Edgar 2007; Duffell et al. 2014; Dürmann and Kley 2015).

7.3 Accretion Inside Gaps

Since mass flows continuously through the gaps opened by massive planets, gap formation does not spell the end of gas accretion onto the planet. In fact, the mass flow to the planet is hardly diminished by the presence of a gap (Bryden et al. 1999; Kley 1999; Lubow et al. 1999; Lubow and D'Angelo 2006). This raises the question of what determines the final mass of a giant planet. In fact, even if a gap really meant a stop to accretion, the existence of Saturn-mass planets would still be a problem. Clearly, if the planet would be able to accept a large amount of the mass available through the flow across the gap, it would always end up far more massive than Jupiter, unless the disc disappears at exactly the right time. This seems inconsistent with the mass distribution of exoplanets, with ubiquitous Jupiter- and Saturn-mass planets.

Therefore, a mechanism is needed to limit the accretion onto a gap-opening planet. It may be that zooming in onto the planet (e.g. D'Angelo et al. 2003) with all the necessary physics included (radiation, self-gravity, realistic equation of state, magnetic fields) will reveal plausible mechanisms for stopping accretion at various points in the life of the planet. Another possibility is that accretion is regulated by the circumplanetary disc that forms around high-mass planets (Ayliffe and Bate 2009a,b; Ward and Canup 2010). These subdiscs are discussed in detail elsewhere in this volume; here we just mention that recent simulations show that the circumplanetary disc may indeed be a bottleneck (Rivier et al. 2012; Szulágyi et al. 2014).

7.4 Migration of Planets Formed by Disc Instability

Now we finally return to the disc instability scenario outlined in Sect. 3.1, where giant planets were formed directly from a gravitationally unstable disc. Since these planets have gap-opening masses, one might expect that the appropriate migration mechanism is Type II. However, an implicit assumption in the derivation of Type II migration is that the gap formation time scale is smaller than other important time scales such as the migration time scale and gas accretion time scale. With the gap formation time scale ~ 100 orbits, this is always the case in the core accretion scenario, and as a result a growing planet always has the gap consistent with its current mass and migrates at a rate that does not depend on its history (an exception is Type III migration).

In the disc instability model, planets form on a dynamical time scale. Therefore, the formation time scale is much shorter than the gap opening time scale and we have a situation of a giant planet fully embedded in a massive disc. Next, we have to compare the migration time scale across the with of the gap $\tau_{\rm m,w}$ to the gap formation time scale $\tau_{\rm gap}$. If $\tau_{\rm gap} < \tau_{\rm m,w}$, a gap will form and the planet will enter the Type II migration regime. For this to be the case we need

$$\tau_{\rm gap} = 18\frac{h^3}{q^2}\frac{w}{r_{\rm p}}\frac{1}{\Omega_{\rm p}} < \frac{h^2}{q}\frac{w}{r_{\rm p}}\frac{M_*}{r_{\rm p}^2\Sigma_{\rm p}}\frac{1}{\Omega_{\rm p}} = \tau_{\rm m,w} \tag{89}$$

Rearranging we get a condition on q,

$$q > 18h\frac{r_{\rm p}^2\Sigma_{\rm p}}{M_*} = \frac{18}{h}\frac{\Sigma_{\rm p}H^2}{M_*}. \tag{90}$$

Therefore, we need the mass of the fragment to be at least $18/h$ times bigger than expected from (30). Therefore, rather surprisingly, planets formed by disc instability migrate in the fast Type I regime (Baruteau et al. 2011). Migration time scales can indeed be very short. Using the Q parameter instead of the surface density, we get that

$$\tau_{\rm m} = \frac{h}{q}\pi Q\frac{1}{\Omega_{\rm p}} \tag{91}$$

Substituting $q = h^3$ from (31) and $h = 0.1$ as the aspect ratio in the outer parts of the disc, we expect migration time scales to be in the tens of orbits. Such fast migration is indeed observed in simulations in the context of accretion outburst (Vorobyov and Basu 2006, 2010b; Machida et al. 2011) and giant planet formation (Vorobyov and Basu 2010a; Baruteau et al. 2011). This means that the disc instability model suffers from similar migration perils as the core accretion model; in particular, it is difficult to see how to keep fragments at large radii where planets are observed (Marois et al. 2008).

8 Gas Accretion and Growth Maps

Growing protoplanets acquire gaseous envelopes that are in both hydrostastic and thermal balance. At the early growth stages of the core the gaseous envelope grows together with the core, but does not experience independent accretion. Mizuno (1980) showed that the thermal equilibrium breaks down approximately when the envelope grows more massive than the core, which occurs for nominal conditions in the protoplanetary disc at around 10 Earth masses. This breakdown is followed by a run-away contraction of the gaseous envelope as the energy released by accretion can not replace the energy lost by cooling to the surroundings.

Mizuno (1980) considered core growth rates relevant for planetesimal accretion. However, for pebble accretion the core can grow several orders of magnitude faster. This implies that the critical core mass of run-away contraction is several hundred Earth masses (Lambrechts et al. 2014). However, Lambrechts et al. (2014) also demonstrated that the flow of pebbles on to the core is stopped at a mass of around 10 Earth masses, as such a massive protoplanet carves a partial gap in the protoplanetary disc and pebbles are trapped at the outer edge of the gas where the gas orbits at the Keplerian speed. The termination of heating by the infalling pebbles allows the envelope to contract. Initially, the contraction is slow as the gravity is dominated by the mass of the solids core. Piso and Youdin (2014) considered the contraction of an envelope that is only supported by the heat of the adiabatic contraction. This yields an envelope growth rate of (Bitsch et al. 2015)

$$\frac{dM_g}{dt} = 0.00175 M_E\ \mathrm{Myr}^{-1}\ f^{-2}\left(\frac{\kappa}{\mathrm{cm}^2\,\mathrm{g}^{-1}}\right)^{-1}\left(\frac{\rho_\bullet}{5.5\ \mathrm{g\,cm}^{-3}}\right)^{-1/6}\left(\frac{M_c}{M_E}\right)^{11/3} \times \left(\frac{M_{\mathrm{env}}}{0.1 M_E}\right)^{-1}\left(\frac{T}{81\ \mathrm{K}}\right)^{-0.5}. \tag{92}$$

Here κ is the opacity of the envelope material. Contraction continues until the envelope mass M_e is equal to the core mass M_c. After that the self-gravity of the envelope leads to accretion of gas from the disc. Machida et al. (2010) calculated the accretion rate of planets using hydrodynamical simulations with nested grids. They find that the accretion rate follows the minimum of the two functions

$$\frac{dM_g}{dt} = 0.83 \Omega \Sigma_g H^2 \left(\frac{R_H}{H}\right)^{9/2}, \tag{93}$$

$$\frac{dM_g}{dt} = 0.14 \Omega \Sigma_g H^2. \tag{94}$$

Gas accretion is an area of active research. Recent findings indicate that low-mass planets do not possess bound envelopes at all, but rather experience a recycling of gas through the envelope that is more rapid than the contraction (Ormel et al. 2015). This may help explain how some super-Earths do not undergo rapid run-away accretion of gas. Nevertheless, the existing prescriptions for pebble accretion, gas accretion and planetary migration allow us to construct growth maps of planetary populations. Two examples of such maps are shown in Fig. 6, a plot taken from Bitsch and Johansen (2016). The maps show the final planetary mass and semi-major axis as a function of the starting location and starting time of the protoplanet. Figure 6 demonstrates that all the major classes of planets can form by a combination of pebble accretion and planetary migration, including hot, warm and cold gas giants (red and yellow contours), super-Earths and ice giants (green contours) and low-mass ice planets (blue contours). Migration of planets is not a nuisance in this picture, but rather a necessary (and elegant) ingredient in defining the final orbits of the planets. The price for the high migration rates is nevertheless that planets must form much further out than their final position.

9 Interaction with Water

There are several interesting links between the story of giant planet formation and migration as told above and the delivery of water to planets. First of all for the giant planets themselves: in the simplest picture, their water content is the sum of the water content of their building blocks. Therefore, we need to know at what stage in their evolution the planet gets most of its mass (and in what form, solid or gaseous), and exactly where it is in the disc at the time. Measuring carbon-to-oxygen and carbon-to-hydrogen ratios may provide important insights into the formation history of a planet (Öberg et al. 2011). One simple example could be trying to constrain planet migration by looking at its water content. When confronted by a planet rich in water, but located too close to the central star for water to have accreted locally, it seems likely that inward migration has taken place. In principle, all migration could have

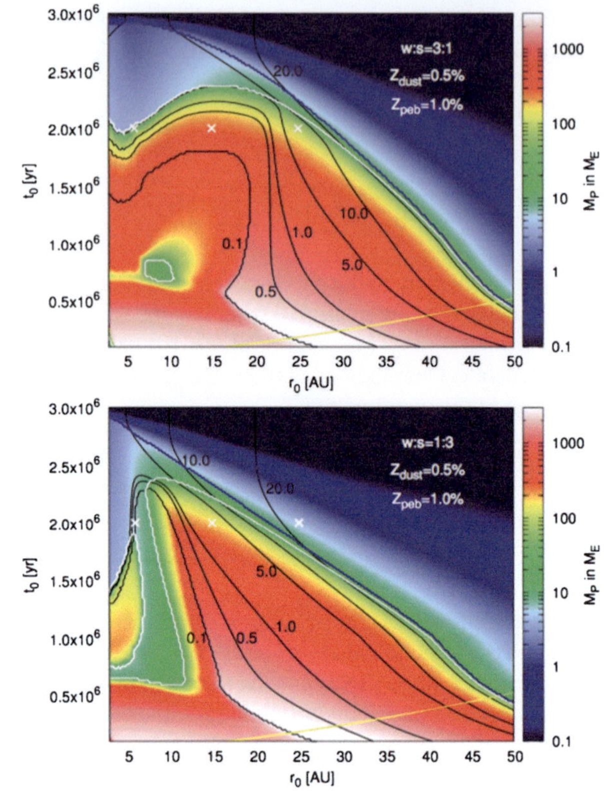

Fig. 6 Growth maps of planets that form by pebble accretion and undergo Type I and Type II, from Bitsch and Johansen (2016). The plot shows the planetary mass (*coloured contours*) and final semi-major axis (*black contour lines*) as a function of the starting position r_0 and the starting time t_0. The *top plot* has only gas giants inside of 0.5 AU—this is due to the outwards migration zone that prevents low mass cores from migrating interior of the ice line. The *bottom plot* shows the result of lowering the water contents of the disc (w:s denotes the water-to-silicates ratio). The lowered water abundance leads to a less pronounced outwards migration zone and hence transport of icy super-Earths to warm and hot orbits close to the star

happened while the building blocks were still small: inward drift of dust particles may lead to enough transport of solids to build the planet locally. However, it is unlikely that these small particles can keep their water for long enough in a water-hostile region in order to build a water-rich planet. It seems more likely that migration has taken place while the object was already of planetary-size.

However, it must be noted that this reverse engineering of the process of planet formation can be hopelessly degenerate. Madhusudhan et al. (2014) suggests a possible way to distinguish planet migration through the protoplanetary disc from migration processes after the disc has gone (e.g. Rasio and Ford 1996) might be to look at the metallicity of the atmospheres of extrasolar giant planets, since migration through the disc would lead to accretion of more solids. However, it is easy to come up with a scenario where disc migration would lead to no accretion of solids at all: since solids react strongly to any pressure gradient (see Sect. 4), when a planet starts to open up a gap, it very effectively pushes solids away from its orbit, much more efficiently than the gas (e.g. Paardekooper and Mellema 2004). Nevertheless, it can be hoped that these kind of observations can at least point towards more likely scenarios so that the story of giant planet formation can be updated and refined.

There is also a strong connection that works the other way: local abundance of water can have strong effects on the migration processes, in particular Type I migration. Because of its effect on the local temperature gradient, the water ice line is usually a region of outward Type I migration (Kretke and Lin 2007; Bitsch et al. 2014). Moreover, the overall water abundance in the disc influences the regions of outward Type I migration and therefore the final orbital distance of for example super Earths (Bitsch and Johansen 2016), as discussed in the previous section.

Finally, it is important to realise that since giant planets make up most of the mass and usually angular momentum in planetary systems, they can profoundly influence the evolution of and water delivery to other planets in the system. Any formation scenario of terrestrial or Super-Earth planets including their water content needs to include the position of any giants in the system, and therefore the migration history of these giant planets (e.g. Walsh et al. 2011; O'Brien et al. 2014).

10 Concluding Remarks

The aim of this review was to give an overview of the basic formation and migration mechanisms for giant planets. During all stages of giant planet formation, some form of migration seems to be inevitable: small solid particles undergo radial drift, low-mass planets and cores of giant planets undergo Type I migration, and giant planets undergo Type II migration. These migration types affect all stages of planet formation. Hence an increased understanding of migration will be crucial for gaining insight into the formation of the planets in the Solar System and in the multitudes of observed exoplanetary systems.

Acknowledgements S.-J. Paardekooper is supported by a Royal Society University Research Fellowship. A. Johansen is supported by the Knut and Alice Wallenberg Foundation, the Swedish Research Council (grant 2014-5775) and the European Research Council (ERC Starting Grant 278675-PEBBLE2PLANET).

References

P. Artymowicz, Migration type III, in *KITP Conference: Planet Formation: Terrestrial and Extra Solar* (2004)

B.A. Ayliffe, M.R. Bate, Circumplanetary disc properties obtained from radiation hydrodynamical simulations of gas accretion by protoplanets. Mon. Not. R. Astron. Soc. **397**, 657–665 (2009a). https://doi.org/10.1111/j.1365-2966.2009.15002.x

B.A. Ayliffe, M.R. Bate, Gas accretion on to planetary cores: three-dimensional self-gravitating radiation hydrodynamical calculations. Mon. Not. R. Astron. Soc. **393**, 49–64 (2009b). https://doi.org/10.1111/j.1365-2966.2008.14184.x

X.-N. Bai, J.M. Stone, Dynamics of solids in the midplane of protoplanetary disks: implications for planetesimal formation. Astrophys. J. **722**, 1437–1459 (2010). https://doi.org/10.1088/0004-637X/722/2/1437

X.-N. Bai, J.M. Stone, Magnetic flux concentration and zonal flows in magnetorotational instability turbulence. Astrophys. J. **796**, 31 (2014). https://doi.org/10.1088/0004-637X/796/1/31

S.A. Balbus, J.F. Hawley, A powerful local shear instability in weakly magnetized disks. I—Linear analysis. II—Nonlinear evolution. Astrophys. J. **376**, 214–233 (1991). https://doi.org/10.1086/170270

C. Baruteau, F. Masset, On the corotation torque in a radiatively inefficient disk. Astrophys. J. **672**, 1054–1067 (2008). https://doi.org/10.1086/523667

C. Baruteau, F. Meru, S.-J. Paardekooper, Rapid inward migration of planets formed by gravitational instability. Mon. Not. R. Astron. Soc. **416**, 1971–1982 (2011). https://doi.org/10.1111/j.1365-2966.2011.19172.x

C. Baruteau, A. Crida, S.-J. Paardekooper, F. Masset, J. Guilet, B. Bitsch, R. Nelson, W. Kley, J. Papaloizou, Planet-disk interactions and early evolution of planetary systems, in *Protostars and Planets VI*, (2014), pp. 667–689. https://doi.org/10.2458/azu_uapress_9780816531240-ch029

P. Benítez-Llambay, F. Masset, G. Koenigsberger, J. Szulágyi, Planet heating prevents inward migration of planetary cores. Nature **520**, 63–65 (2015). https://doi.org/10.1038/nature14277

T. Birnstiel, C.P. Dullemond, F. Brauer, Gas- and dust evolution in protoplanetary disks. Astron. Astrophys. **513**, 79 (2010). https://doi.org/10.1051/0004-6361/200913731

T. Birnstiel, H. Klahr, B. Ercolano, A simple model for the evolution of the dust population in protoplanetary disks. Astron. Astrophys. **539**, 148 (2012). https://doi.org/10.1051/0004-6361/201118136

T. Birnstiel, M. Fang, A. Johansen, Dust evolution and the formation of planetesimals. Space Sci. Rev. (2016). https://doi.org/10.1007/s11214-016-0256-1

B. Bitsch, A. Johansen, Influence of the water content in protoplanetary discs on planet migration and formation. Astron. Astrophys. **590**, 101 (2016). https://doi.org/10.1051/0004-6361/201527676

B. Bitsch, A. Crida, A. Morbidelli, W. Kley, I. Dobbs-Dixon, Stellar irradiated discs and implications on migration of embedded planets. I. Equilibrium discs. Astron. Astrophys. **549**, 124 (2013). https://doi.org/10.1051/0004-6361/201220159

B. Bitsch, A. Morbidelli, E. Lega, K. Kretke, A. Crida, Stellar irradiated discs and implications on migration of embedded planets. III. Viscosity transitions. Astron. Astrophys. **570**, 75 (2014). https://doi.org/10.1051/0004-6361/201424015

B. Bitsch, A. Johansen, M. Lambrechts, A. Morbidelli, The structure of protoplanetary discs around evolving young stars. Astron. Astrophys. **575**, 28 (2015). https://doi.org/10.1051/0004-6361/201424964

B. Bitsch, M. Lambrechts, A. Johansen, The growth of planets by pebble accretion in evolving protoplanetary discs. Astron. Astrophys. **582**, 112 (2015). https://doi.org/10.1051/0004-6361/201526463

A.C. Boley, T. Hayfield, L. Mayer, R.H. Durisen, Clumps in the outer disk by disk instability: why they are initially gas giants and the legacy of disruption. Icarus **207**, 509–516 (2010). https://doi.org/10.1016/j.icarus.2010.01.015

W.F. Bottke, D.D. Durda, D. Nesvorný, R. Jedicke, A. Morbidelli, D. Vokrouhlický, H. Levison, The fossilized size distribution of the main asteroid belt. Icarus **175**, 111–140 (2005). https://doi.org/10.1016/j.icarus.2004.10.026

F. Brauer, C.P. Dullemond, A. Johansen, T. Henning, H. Klahr, A. Natta, Survival of the mm-cm size grain population observed in protoplanetary disks. Astron. Astrophys. **469**, 1169–1182 (2007). https://doi.org/10.1051/0004-6361:20066865

G. Bryden, X. Chen, D.N.C. Lin, R.P. Nelson, J.C.B. Papaloizou, Tidally induced gap formation in protostellar disks: gap clearing and suppression of protoplanetary growth. Astrophys. J. **514**, 344–367 (1999). https://doi.org/10.1086/306917

D. Carrera, A. Johansen, M.B. Davies, How to form planetesimals from mm-sized chondrules and chondrule aggregates. Astron. Astrophys. **579**, 43 (2015). https://doi.org/10.1051/0004-6361/201425120

J. Casoli, F.S. Masset, On the horseshoe drag of a low-mass planet. I. Migration in isothermal disks. Astrophys. J. **703**, 845–856 (2009). https://doi.org/10.1088/0004-637X/703/1/845

A. Crida, A. Morbidelli, F. Masset, On the width and shape of gaps in protoplanetary disks. Icarus **181**, 587–604 (2006). https://doi.org/10.1016/j.icarus.2005.10.007

I. Crnkovic-Rubsamen, Z. Zhu, J.M. Stone, Survival and structure of dusty vortices in protoplanetary discs. Mon. Not. R. Astron. Soc. **450**, 4285–4291 (2015). https://doi.org/10.1093/mnras/stv828

G. D'Angelo, W. Kley, T. Henning, Orbital migration and mass accretion of protoplanets in three-dimensional global computations with nested grids. Astrophys. J. **586**, 540–561 (2003). https://doi.org/10.1086/367555

K. Dittrich, H. Klahr, A. Johansen, Gravoturbulent planetesimal formation: the positive effect of long-lived zonal flows. Astrophys. J. **763**, 117 (2013). https://doi.org/10.1088/0004-637X/763/2/117

P.C. Duffell, Z. Haiman, A.I. MacFadyen, D.J. D'Orazio, B.D. Farris, The migration of gap-opening planets is not locked to viscous disk evolution. Astrophys. J. Lett. **792**, 10 (2014). https://doi.org/10.1088/2041-8205/792/1/L10

C.P. Dullemond, C. Dominik, The effect of dust settling on the appearance of protoplanetary disks. Astron. Astrophys. **421**, 1075–1086 (2004). https://doi.org/10.1051/0004-6361:20040284

C.P. Dullemond, C. Dominik, Dust coagulation in protoplanetary disks: a rapid depletion of small grains. Astron. Astrophys. **434**, 971–986 (2005). https://doi.org/10.1051/0004-6361:20042080

C. Dürmann, W. Kley, Migration of massive planets in accreting disks. Astron. Astrophys. **574**, 52 (2015). https://doi.org/10.1051/0004-6361/201424837

R.G. Edgar, Giant planet migration in viscous power-law disks. Astrophys. J. **663**, 1325–1334 (2007). https://doi.org/10.1086/518591

S. Fromang, J. Papaloizou, Dust settling in local simulations of turbulent protoplanetary disks. Astron. Astrophys. **452**, 751–762 (2006). https://doi.org/10.1051/0004-6361:20054612

J. Fung, P. Artymowicz, Y. Wu, The 3D flow field around an embedded planet. Astrophys. J. **811**, 101 (2015). https://doi.org/10.1088/0004-637X/811/2/101

C.F. Gammie, Nonlinear outcome of gravitational instability in cooling, gaseous disks. Astrophys. J. **553**, 174–183 (2001). https://doi.org/10.1086/320631

P. Goldreich, S. Tremaine, Disk-satellite interactions. Astrophys. J. **241**, 425–441 (1980). https://doi.org/10.1086/158356

J. Goodman, R.R. Rafikov, Planetary torques as the viscosity of protoplanetary disks. Astrophys. J. **552**, 793–802 (2001). https://doi.org/10.1086/320572

Y. Greenzweig, J.J. Lissauer, Accretion rates of protoplanets. Icarus **87**, 40–77 (1990). https://doi.org/10.1016/0019-1035(90)90021-Z

J. Guilet, C. Baruteau, J.C.B. Papaloizou, Type I planet migration in weakly magnetized laminar discs. Mon. Not. R. Astron. Soc. **430**, 1764–1783 (2013). https://doi.org/10.1093/mnras/sts720

B. Gundlach, J. Blum, The stickiness of micrometer-sized water-ice particles. Astrophys. J. **798**, 34 (2015). https://doi.org/10.1088/0004-637X/798/1/34

P.B. Ivanov, J.C.B. Papaloizou, A.G. Polnarev, The evolution of a supermassive binary caused by an accretion disc. Mon. Not. R. Astron. Soc. **307**, 79–90 (1999). https://doi.org/10.1046/j.1365-8711.1999.02623.x

A. Johansen, H. Klahr, Dust diffusion in protoplanetary disks by magnetorotational turbulence. Astrophys. J. **634**, 1353–1371 (2005). https://doi.org/10.1086/497118

A. Johansen, P. Lacerda, Prograde rotation of protoplanets by accretion of pebbles in a gaseous environment. Mon. Not. R. Astron. Soc. **404**, 475–485 (2010). https://doi.org/10.1111/j.1365-2966.2010.16309.x

A. Johansen, A. Youdin, Protoplanetary disk turbulence driven by the streaming instability: nonlinear saturation and particle concentration. Astrophys. J. **662**, 627–641 (2007). https://doi.org/10.1086/516730

A. Johansen, A.C. Andersen, A. Brandenburg, Simulations of dust-trapping vortices in protoplanetary discs. Astron. Astrophys. **417**, 361–374 (2004). https://doi.org/10.1051/0004-6361:20034417

A. Johansen, H. Klahr, T. Henning, Gravoturbulent formation of planetesimals. Astrophys. J. **636**, 1121–1134 (2006). https://doi.org/10.1086/498078

A. Johansen, J.S. Oishi, M.-M. Mac Low, H. Klahr, T. Henning, A. Youdin, Rapid planetesimal formation in turbulent circumstellar disks. Nature **448**, 1022–1025 (2007). https://doi.org/10.1038/nature06086

A. Johansen, A. Youdin, H. Klahr, Zonal flows and long-lived axisymmetric pressure bumps in magnetorotational turbulence. Astrophys. J. **697**, 1269–1289 (2009a). https://doi.org/10.1088/0004-637X/697/2/1269

A. Johansen, A. Youdin, M.-M. Mac Low, Particle clumping and planetesimal formation depend strongly on metallicity. Astrophys. J. Lett. **704**, 75–79 (2009b). https://doi.org/10.1088/0004-637X/704/2/L75

A. Johansen, A.N. Youdin, Y. Lithwick, Adding particle collisions to the formation of asteroids and Kuiper belt objects via streaming instabilities. Astron. Astrophys. **537**, 125 (2012). https://doi.org/10.1051/0004-6361/201117701

A. Johansen, J. Blum, H. Tanaka, C. Ormel, M. Bizzarro, H. Rickman, The multifaceted planetesimal formation process, in *Protostars and Planets VI*, (2014), pp. 547–570. https://doi.org/10.2458/azu_uapress_9780816531240-ch024

A. Johansen, M.-M. Mac Low, P. Lacerda, M. Bizzarro, Growth of asteroids, planetary embryos, and Kuiper belt objects by chondrule accretion. Sci. Adv. **1**, 1500109 (2015). https://doi.org/10.1126/sciadv.1500109

H.H. Klahr, P. Bodenheimer, Turbulence in accretion disks: vorticity generation and angular momentum transport via the global baroclinic instability. Astrophys. J. **582**, 869–892 (2003). https://doi.org/10.1086/344743

W. Kley, Mass flow and accretion through gaps in accretion discs. Mon. Not. R. Astron. Soc. **303**, 696–710 (1999). https://doi.org/10.1046/j.1365-8711.1999.02198.x

K. Kratter, G. Lodato, Gravitational instabilities in circumstellar disks. Annu. Rev. Astron. Astrophys. **54**, 271–311 (2016). https://doi.org/10.1146/annurev-astro-081915-023307

K.M. Kratter, R.A. Murray-Clay, A.N. Youdin, The runts of the litter: why planets formed through gravitational instability can only be failed binary stars. Astrophys. J. **710**, 1375–1386 (2010). https://doi.org/10.1088/0004-637X/710/2/1375

K.A. Kretke, D.N.C. Lin, Grain retention and formation of planetesimals near the snow line in MRI-driven turbulent protoplanetary disks. Astrophys. J. Lett. **664**, 55–58 (2007). https://doi.org/10.1086/520718

S. Krijt, C.W. Ormel, C. Dominik, A.G.G.M. Tielens, Erosion and the limits to planetesimal growth. Astron. Astrophys. **574**, 83 (2015). https://doi.org/10.1051/0004-6361/201425222

S. Lain, D. Bröder, M. Sommerfeld, Experimental and numerical studies of the hydrodynamics in a bubble column. Chem. Eng. Sci. **54**(21), 4913–4920 (1999). https://doi.org/10.1016/S0009-2509(99)00212-2

M. Lambrechts, A. Johansen, Rapid growth of gas-giant cores by pebble accretion. Astron. Astrophys. **544**, 32 (2012). https://doi.org/10.1051/0004-6361/201219127

M. Lambrechts, A. Johansen, Forming the cores of giant planets from the radial pebble flux in protoplanetary discs. Astron. Astrophys. **572**, 107 (2014). https://doi.org/10.1051/0004-6361/201424343

M. Lambrechts, A. Johansen, A. Morbidelli, Separating gas-giant and ice-giant planets by halting pebble accretion. Astron. Astrophys. **572**, 35 (2014). https://doi.org/10.1051/0004-6361/201423814
E. Lega, A. Crida, B. Bitsch, A. Morbidelli, Migration of Earth-sized planets in 3D radiative discs. Mon. Not. R. Astron. Soc. **440**, 683–695 (2014). https://doi.org/10.1093/mnras/stu304
G. Lesur, J.C.B. Papaloizou, The subcritical baroclinic instability in local accretion disc models. Astron. Astrophys. **513**, 60 (2010). https://doi.org/10.1051/0004-6361/200913594
H.F. Levison, E. Thommes, M.J. Duncan, Modeling the formation of giant planet cores. I. Evaluating key processes. Astron. J. **139**, 1297–1314 (2010). https://doi.org/10.1088/0004-6256/139/4/1297
D.N.C. Lin, J. Papaloizou, Tidal torques on accretion discs in binary systems with extreme mass ratios. Mon. Not. R. Astron. Soc. **186**, 799–812 (1979). https://doi.org/10.1093/mnras/186.4.799
D.N.C. Lin, J. Papaloizou, On the tidal interaction between protoplanets and the protoplanetary disk. III—Orbital migration of protoplanets. Astrophys. J. **309**, 846–857 (1986). https://doi.org/10.1086/164653
D.N.C. Lin, J.C.B. Papaloizou, On the tidal interaction between protostellar disks and companions, in *Protostars and Planets III*, ed. by E.H. Levy, J.I. Lunine (1993), pp. 749–835
M.-K. Lin, A.N. Youdin, Cooling requirements for the vertical shear instability in protoplanetary disks. Astrophys. J. **811**, 17 (2015). https://doi.org/10.1088/0004-637X/811/1/17
D.N.C. Lin, P. Bodenheimer, D.C. Richardson, Orbital migration of the planetary companion of 51 Pegasi to its present location. Nature **380**, 606–607 (1996). https://doi.org/10.1038/380606a0
S.H. Lubow, G. D'Angelo, Gas flow across gaps in protoplanetary disks. Astrophys. J. **641**, 526–533 (2006). https://doi.org/10.1086/500356
S.H. Lubow, M. Seibert, P. Artymowicz, Disk accretion onto high-mass planets. Astrophys. J. **526**, 1001–1012 (1999). https://doi.org/10.1086/308045
M.N. Machida, E. Kokubo, S.-I. Inutsuka, T. Matsumoto, Gas accretion onto a protoplanet and formation of a gas giant planet. Mon. Not. R. Astron. Soc. **405**, 1227–1243 (2010). https://doi.org/10.1111/j.1365-2966.2010.16527.x
M.N. Machida, S.-i. Inutsuka, T. Matsumoto, Recurrent planet formation and intermittent protostellar outflows induced by episodic mass accretion. Astrophys. J. **729**, 42 (2011). https://doi.org/10.1088/0004-637X/729/1/42
N. Madhusudhan, M.A. Amin, G.M. Kennedy, Toward chemical constraints on hot Jupiter migration. Astrophys. J. Lett. **794**, 12 (2014). https://doi.org/10.1088/2041-8205/794/1/L12
G.R. Mamatsashvili, W.K.M. Rice, Axisymmetric modes in vertically stratified self-gravitating discs. Mon. Not. R. Astron. Soc. **406**, 2050–2064 (2010). https://doi.org/10.1111/j.1365-2966.2010.16825.x
C. Marois, B. Macintosh, T. Barman, B. Zuckerman, I. Song, J. Patience, D. Lafrenière, R. Doyon, Direct imaging of multiple planets orbiting the Star HR 8799. Science **322**, 1348 (2008). https://doi.org/10.1126/science.1166585
F.S. Masset, On the co-orbital corotation torque in a viscous disk and its impact on planetary migration. Astrophys. J. **558**, 453–462 (2001). https://doi.org/10.1086/322446
F.S. Masset, in *Planet Disk Interactions*, ed. by M.-J. Goupil, J.-P. Zahn. EAS Publications Series, vol. 29 (2008), pp. 165–244. https://doi.org/10.1051/eas:0829006
F.S. Masset, J. Casoli, On the horseshoe drag of a low-mass planet. II. Migration in adiabatic disks. Astrophys. J. **703**, 857–876 (2009). https://doi.org/10.1088/0004-637X/703/1/857
F.S. Masset, J.C.B. Papaloizou, Runaway migration and the formation of hot Jupiters. Astrophys. J. **588**, 494–508 (2003). https://doi.org/10.1086/373892
F.S. Masset, G. D'Angelo, W. Kley, On the migration of protogiant solid cores. Astrophys. J. **652**, 730–745 (2006). https://doi.org/10.1086/507515
M. Mayor, D. Queloz, A Jupiter-mass companion to a solar-type star. Nature **378**, 355–359 (1995). https://doi.org/10.1038/378355a0
F. Ménard, C. Bertout, in *The Nature of Young Solar-Type Stars*, ed. by C.J. Lada, N.D. Kylafis. NATO Advanced Science Institutes (ASI) Series C, vol. 540 (1999), p. 341
F. Meru, M.R. Bate, Non-convergence of the critical cooling time-scale for fragmentation of self-gravitating discs. Mon. Not. R. Astron. Soc. **411**, 1–5 (2011). https://doi.org/10.1111/j.1745-3933.2010.00978.x
H. Mizuno, Formation of the giant planets. Prog. Theor. Phys. **64**, 544–557 (1980). https://doi.org/10.1143/PTP.64.544
A. Morbidelli, W.F. Bottke, D. Nesvorný, H.F. Levison, Asteroids were born big. Icarus **204**, 558–573 (2009). https://doi.org/10.1016/j.icarus.2009.07.011
A. Morbidelli, M. Lambrechts, S. Jacobson, B. Bitsch, The great dichotomy of the solar system: small terrestrial embryos and massive giant planet cores. Icarus **258**, 418–429 (2015). https://doi.org/10.1016/j.icarus.2015.06.003
T.W.A. Müller, W. Kley, F. Meru, Treating gravity in thin-disk simulations. Astron. Astrophys. **541**, 123 (2012). https://doi.org/10.1051/0004-6361/201118737

S. Nayakshin, Formation of planets by tidal downsizing of giant planet embryos. Mon. Not. R. Astron. Soc. **408**, 36–40 (2010). https://doi.org/10.1111/j.1745-3933.2010.00923.x

R.P. Nelson, J.C.B. Papaloizou, The interaction of giant planets with a disc with MHD turbulence—IV. Migration rates of embedded protoplanets. Mon. Not. R. Astron. Soc. **350**, 849–864 (2004). https://doi.org/10.1111/j.1365-2966.2004.07406.x

R.P. Nelson, O. Gressel, O.M. Umurhan, Linear and non-linear evolution of the vertical shear instability in accretion discs. Mon. Not. R. Astron. Soc. **435**, 2610–2632 (2013). https://doi.org/10.1093/mnras/stt1475

K.I. Öberg, R. Murray-Clay, E.A. Bergin, The effects of snowlines on C/O in planetary atmospheres. Astrophys. J. Lett. **743**, 16 (2011). https://doi.org/10.1088/2041-8205/743/1/L16

D.P. O'Brien, K.J. Walsh, A. Morbidelli, S.N. Raymond, A.M. Mandell, Water delivery and giant impacts in the Grand Tack scenario. Icarus **239**, 74–84 (2014). https://doi.org/10.1016/j.icarus.2014.05.009

G.I. Ogilvie, S.H. Lubow, On the wake generated by a planet in a disc. Mon. Not. R. Astron. Soc. **330**, 950–954 (2002). https://doi.org/10.1046/j.1365-8711.2002.05148.x

S. Okuzumi, H. Tanaka, H. Kobayashi, K. Wada, Rapid coagulation of porous dust aggregates outside the snow line: a pathway to successful icy planetesimal formation. Astrophys. J. **752**, 106 (2012). https://doi.org/10.1088/0004-637X/752/2/106

C.W. Ormel, H.H. Klahr, The effect of gas drag on the growth of protoplanets. Analytical expressions for the accretion of small bodies in laminar disks. Astron. Astrophys. **520**, 43 (2010). https://doi.org/10.1051/0004-6361/201014903

C.W. Ormel, D. Paszun, C. Dominik, A.G.G.M. Tielens, Dust coagulation and fragmentation in molecular clouds. I. How collisions between dust aggregates alter the dust size distribution. Astron. Astrophys. **502**, 845–869 (2009). https://doi.org/10.1051/0004-6361/200811158

C.W. Ormel, J.-M. Shi, R. Kuiper, Hydrodynamics of embedded planets' first atmospheres—II. A rapid recycling of atmospheric gas. Mon. Not. R. Astron. Soc. **447**, 3512–3525 (2015). https://doi.org/10.1093/mnras/stu2704

S.-J. Paardekooper, Numerical convergence in self-gravitating shearing sheet simulations and the stochastic nature of disc fragmentation. Mon. Not. R. Astron. Soc. **421**, 3286–3299 (2012). https://doi.org/10.1111/j.1365-2966.2012.20553.x

S.-J. Paardekooper, Dynamical corotation torques on low-mass planets. Mon. Not. R. Astron. Soc. **444**, 2031–2042 (2014). https://doi.org/10.1093/mnras/stu1542

S.-J. Paardekooper, G. Mellema, Planets opening dust gaps in gas disks. Astron. Astrophys. **425**, 9–12 (2004). https://doi.org/10.1051/0004-6361:200400053

S.-J. Paardekooper, G. Mellema, Halting type I planet migration in non-isothermal disks. Astron. Astrophys. **459**, 17–20 (2006). https://doi.org/10.1051/0004-6361:20066304

S.-J. Paardekooper, J.C.B. Papaloizou, On disc protoplanet interactions in a non-barotropic disc with thermal diffusion. Astron. Astrophys. **485**, 877–895 (2008). https://doi.org/10.1051/0004-6361:20078702

S.-J. Paardekooper, J.C.B. Papaloizou, On corotation torques, horseshoe drag and the possibility of sustained stalled or outward protoplanetary migration. Mon. Not. R. Astron. Soc. **394**, 2283–2296 (2009a). https://doi.org/10.1111/j.1365-2966.2009.14511.x

S.-J. Paardekooper, J.C.B. Papaloizou, On the width and shape of the corotation region for low-mass planets. Mon. Not. R. Astron. Soc. **394**, 2297–2309 (2009b). https://doi.org/10.1111/j.1365-2966.2009.14512.x

S.-J. Paardekooper, C. Baruteau, A. Crida, W. Kley, A torque formula for non-isothermal type I planetary migration—I. Unsaturated horseshoe drag. Mon. Not. R. Astron. Soc. **401**, 1950–1964 (2010). https://doi.org/10.1111/j.1365-2966.2009.15782.x

S.-J. Paardekooper, C. Baruteau, W. Kley, A torque formula for non-isothermal type I planetary migration—II. Effects of diffusion. Mon. Not. R. Astron. Soc. **410**, 293–303 (2011a). https://doi.org/10.1111/j.1365-2966.2010.17442.x

S.-J. Paardekooper, C. Baruteau, F. Meru, Numerical convergence in self-gravitating disc simulations: initial conditions and edge effects. Mon. Not. R. Astron. Soc. **416**, 65–69 (2011b). https://doi.org/10.1111/j.1745-3933.2011.01099.x

A. Pepliński, P. Artymowicz, G. Mellema, Numerical simulations of type III planetary migration—II. Inward migration of massive planets. Mon. Not. R. Astron. Soc. **386**, 179–198 (2008). https://doi.org/10.1111/j.1365-2966.2008.13046.x

A. Pierens, Fast migration of low-mass planets in radiative discs. Mon. Not. R. Astron. Soc. **454**, 2003–2014 (2015). https://doi.org/10.1093/mnras/stv2024

A.-M.A. Piso, A.N. Youdin, On the minimum core mass for giant planet formation at wide separations. Astrophys. J. **786**, 21 (2014). https://doi.org/10.1088/0004-637X/786/1/21

N. Raettig, W. Lyra, H. Klahr, A parameter study for baroclinic vortex amplification. Astrophys. J. **765**, 115 (2013). https://doi.org/10.1088/0004-637X/765/2/115

N. Raettig, H. Klahr, W. Lyra, Particle trapping and streaming instability in vortices in protoplanetary disks. Astrophys. J. **804**, 35 (2015). https://doi.org/10.1088/0004-637X/804/1/35

R.R. Rafikov, Fast accretion of small planetesimals by protoplanetary cores. Astron. J. **128**, 1348–1363 (2004). https://doi.org/10.1086/423216

R.R. Rafikov, Can giant planets form by direct gravitational instability? Astrophys. J. Lett. **621**, 69–72 (2005). https://doi.org/10.1086/428899

F.A. Rasio, E.B. Ford, Dynamical instabilities and the formation of extrasolar planetary systems. Science **274**, 954–956 (1996). https://doi.org/10.1126/science.274.5289.954

S. Richard, R.P. Nelson, O.M. Umurhan, Vortex formation in protoplanetary discs induced by the vertical shear instability. Mon. Not. R. Astron. Soc. **456**, 3571–3584 (2016). https://doi.org/10.1093/mnras/stv2898

G. Rivier, A. Crida, A. Morbidelli, Y. Brouet, Circum-planetary discs as bottlenecks for gas accretion onto giant planets. Astron. Astrophys. **548**, 116 (2012). https://doi.org/10.1051/0004-6361/201218879

V.S. Safronov, On the gravitational instability in flattened systems with axial symmetry and non-uniform rotation. Ann. Astrophys. **23**, 979 (1960)

V.S. Safronov, *Evoliutsiia Doplanetnogo Oblaka* (1969)

S.S. Sheppard, C.A. Trujillo, The size distribution of the Neptune Trojans and the missing intermediate-sized planetesimals. Astrophys. J. Lett. **723**, 233–237 (2010). https://doi.org/10.1088/2041-8205/723/2/L233

J.B. Simon, K. Beckwith, P.J. Armitage, Emergent mesoscale phenomena in magnetized accretion disc turbulence. Mon. Not. R. Astron. Soc. **422**, 2685–2700 (2012). https://doi.org/10.1111/j.1365-2966.2012.20835.x

J.B. Simon, P.J. Armitage, R. Li, A.N. Youdin, The mass and size distribution of planetesimals formed by the streaming instability. I. The role of self-gravity. Astrophys. J. **822**, 55 (2016). https://doi.org/10.3847/0004-637X/822/1/55

M.H.R. Stoll, W. Kley, Vertical shear instability in accretion disc models with radiation transport. Astron. Astrophys. **572**, 77 (2014). https://doi.org/10.1051/0004-6361/201424114

D. Syer, C.J. Clarke, Satellites in discs: regulating the accretion luminosity. Mon. Not. R. Astron. Soc. **277**, 758–766 (1995). https://doi.org/10.1093/mnras/277.3.758

J. Szulágyi, A. Morbidelli, A. Crida, F. Masset, Accretion of Jupiter-mass planets in the limit of vanishing viscosity. Astrophys. J. **782**, 65 (2014). https://doi.org/10.1088/0004-637X/782/2/65

H. Tanaka, T. Takeuchi, W.R. Ward, Three-dimensional interaction between a planet and an isothermal gaseous disk. I. Corotation and Lindblad torques and planet migration. Astrophys. J. **565**, 1257–1274 (2002). https://doi.org/10.1086/324713

C.E.J.M.L.J. Terquem, Stopping inward planetary migration by a toroidal magnetic field. Mon. Not. R. Astron. Soc. **341**, 1157–1173 (2003). https://doi.org/10.1046/j.1365-8711.2003.06455.x

L. Testi, T. Birnstiel, L. Ricci, S. Andrews, J. Blum, J. Carpenter, C. Dominik, A. Isella, A. Natta, J.P. Williams, D.J. Wilner, Dust evolution in protoplanetary disks, in *Protostars and Planets VI*, (2014), pp. 339–361. https://doi.org/10.2458/azu_uapress_9780816531240-ch015

A. Toomre, On the gravitational stability of a disk of stars. Astrophys. J. **139**, 1217–1238 (1964). https://doi.org/10.1086/147861

E.I. Vorobyov, S. Basu, The burst mode of protostellar accretion. Astrophys. J. **650**, 956–969 (2006). https://doi.org/10.1086/507320

E.I. Vorobyov, S. Basu, Formation and survivability of giant planets on wide orbits. Astrophys. J. Lett. **714**, 133–137 (2010a). https://doi.org/10.1088/2041-8205/714/1/L133

E.I. Vorobyov, S. Basu, The burst mode of accretion and disk fragmentation in the early embedded stages of star formation. Astrophys. J. **719**, 1896–1911 (2010b). https://doi.org/10.1088/0004-637X/719/2/1896

K.J. Walsh, A. Morbidelli, S.N. Raymond, D.P. O'Brien, A.M. Mandell, A low mass for Mars from Jupiter's early gas-driven migration. Nature **475**, 206–209 (2011). https://doi.org/10.1038/nature10201

W.R. Ward, Density waves in the solar nebula—differential Lindblad torque. Icarus **67**, 164–180 (1986). https://doi.org/10.1016/0019-1035(86)90182-X

W.R. Ward, in *Horsehoe Orbit Drag*. Lunar and Planetary Science Conference, vol. 22 (1991)

W.R. Ward, Protoplanet migration by nebula tides. Icarus **126**, 261–281 (1997). https://doi.org/10.1006/icar.1996.5647

W.R. Ward, R.M. Canup, Circumplanetary disk formation. Astron. J. **140**, 1168–1193 (2010). https://doi.org/10.1088/0004-6256/140/5/1168

S.J. Weidenschilling, Aerodynamics of solid bodies in the solar nebula. Mon. Not. R. Astron. Soc. **180**, 57–70 (1977). https://doi.org/10.1093/mnras/180.1.57

F.L. Whipple, On certain aerodynamic processes for asteroids and comets, in *From Plasma to Planet*, ed. by A. Elvius (1972), p. 211

P. Woitke, I. Kamp, W.-F. Thi, Radiation thermo-chemical models of protoplanetary disks. I. Hydrostatic disk structure and inner rim. Astron. Astrophys. **501**, 383–406 (2009). https://doi.org/10.1051/0004-6361/200911821

A.N. Youdin, J. Goodman, Streaming instabilities in protoplanetary disks. Astrophys. J. **620**, 459–469 (2005). https://doi.org/10.1086/426895

A. Youdin, A. Johansen, Protoplanetary disk turbulence driven by the streaming instability: linear evolution and numerical methods. Astrophys. J. **662**, 613–626 (2007). https://doi.org/10.1086/516729

A. Zsom, C.W. Ormel, C. Güttler, J. Blum, C.P. Dullemond, The outcome of protoplanetary dust growth: pebbles, boulders, or planetesimals? II. Introducing the bouncing barrier. Astron. Astrophys. **513**, 57 (2010). https://doi.org/10.1051/0004-6361/200912976

Space Sci Rev (2018) 214:47
https://doi.org/10.1007/s11214-018-0475-8

The Delivery of Water During Terrestrial Planet Formation

David P. O'Brien[1] · **Andre Izidoro**[2,3] · **Seth A. Jacobson**[4,5,6] · **Sean N. Raymond**[3] · **David C. Rubie**[5]

Received: 12 September 2016 / Accepted: 17 January 2018 / Published online: 8 February 2018

Abstract The planetary building blocks that formed in the terrestrial planet region were likely very dry, yet water is comparatively abundant on Earth. Here we review the various mechanisms proposed for the origin of water on the terrestrial planets. Various in-situ mechanisms have been suggested, which allow for the incorporation of water into the local planetesimals in the terrestrial planet region or into the planets themselves from local sources, although all of those mechanisms have difficulties. Comets have also been proposed as a source, although there may be problems fitting isotopic constraints, and the delivery efficiency is very low, such that it may be difficult to deliver even a single Earth ocean of water this way. The most promising route for water delivery is the accretion of material from beyond the snow line, similar to carbonaceous chondrites, that is scattered into the terrestrial planet region as the planets are growing. Two main scenarios are discussed in detail. First is the classical scenario in which the giant planets begin roughly in their final locations and the disk of planetesimals and embryos in the terrestrial planet region extends all the way into the outer asteroid belt region. Second is the Grand Tack scenario, where early inward and outward migration of the giant planets implants material from beyond the snow line into the asteroid belt and terrestrial planet region, where it can be accreted by the growing planets.

The Delivery of Water to Protoplanets, Planets and Satellites
Edited by Michel Blanc, Alessandro Morbidelli, Yann Alibert, Lindy Elkins-Tanton, Paul Estrada, Keiko Hamano, Helmut Lammer, Sean Raymond and Maria Schönbächler

✉ D.P. O'Brien
obrien@psi.edu

[1] Planetary Science Institute, 1700 E. Ft. Lowell, Suite 106, Tucson, AZ 85719, USA

[2] Present address: Grupo de Dinamica Orbital Planetologia, UNESP, Universidade Estadual Paulista, Guaratingueta, CEP 12.516-410, Sao Paulo, Brazil

[3] Laboratoire d'Astrophysique de Bordeaux, CNRS, Universite de Bordeaux, B18N, Allee Geoffroy Saint-Hilaire, 33615 Pessac, France

[4] Present address: Department of Earth and Planetary Sciences, Northwestern University, Evanston, IL 60208, USA

[5] Bayerisches Geoinstitut, University of Bayreuth, 95490 Bayreuth, Germany

[6] Observatoire de la Côte d'Azur, Bd. de l'Observatoire, CS 34229, 06304, Nice Cedex 4, France

Sufficient water is delivered to the terrestrial planets in both scenarios. While the Grand Tack scenario provides a better fit to most constraints, namely the small mass of Mars, planets may form too fast in the nominal case discussed here. This discrepancy may be reduced as a wider range of initial conditions is explored. Finally, we discuss several more recent models that may have important implications for water delivery to the terrestrial planets.

Keywords Terrestrial planet formation · Water delivery

1 Introduction

The existence of water on the Earth is somewhat of a puzzle. Evidence suggests that solids that condensed around 1 AU in the solar nebula were dry and highly reduced, yet Earth clearly has water on its surface, and likely even more in its mantle. Other terrestrial planets show evidence for water as well. The large D/H ratio of Venus' atmosphere suggests an ancient ocean was lost to space (Donahue et al. 1982), while geomorphology (Baker 2007), neutron spectroscopy data (Feldman et al. 2004), and various other lines of evidence point to a wet past for Mars and significant water still present as subsurface ice. Numerous mechanisms have been suggested for delivering this water to the terrestrial planets, or allowing for incorporation from local sources. In this chapter, we review the different mechanisms and assess their plausibility, in light of available constraints and evidence.

We begin the chapter with an overview of the key geochemical and cosmochemical evidence for the origin of Earth's water in Sect. 2. In Sect. 3 we describe the different stages of terrestrial planet formation, from the growth of planetesimals and planetary embryos through the final stage of accretion, and discuss two different scenarios for late-stage accretion. We first describe the 'classical' scenario, with a disk of planetesimals and embryos extending from the terrestrial planet region into the outer asteroid belt region, and the giant planets in roughly their current locations (e.g. Chambers 2001; Raymond et al. 2004; O'Brien et al. 2006; Raymond et al. 2009; Izidoro et al. 2013, 2014). Second, we discuss the 'Grand Tack' scenario of Walsh et al. (2011), where the giant planets experience an inwards-then-outward migration that implants material from beyond the snow line into the asteroid belt and terrestrial planet region.

Beginning our discussion of water delivery, we first describe various 'in-situ' mechanisms in Sect. 4 that could allow for the incorporation of water into planetesimals local to the growing terrestrial planets or directly into the planets themselves. We then discuss the possibility of water delivery by comets in Sect. 5. Problems and difficulties with both scenarios are discussed. Finally, in Sect. 6 we discuss water delivery in the classical and Grand Tack scenarios of late-stage planet formation. Both scenarios allow for sufficient water delivery to explain the water budget of Earth. The Grand Tack scenario provides the best fit to most constraints, namely the small mass of Mars, but some issues with the timescale of planet formation remain. We summarize our discussion and point to further avenues of research on this topic in Sect. 7.

2 Geochemical and Cosmochemical Constraints

Here we discuss some of the key constraints on modeling of water delivery to the terrestrial planets, with a focus on the Earth. For further details see the other chapters in this collection

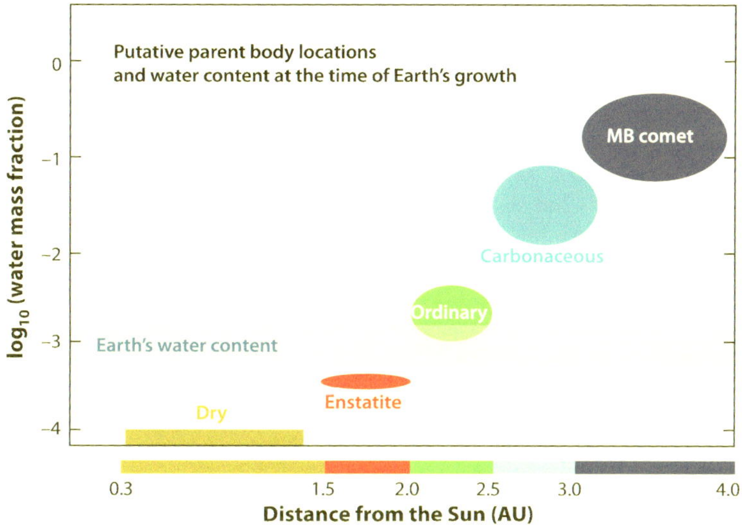

Fig. 1 Water contents of material estimated to have formed at different heliocentric distances, with the range of Earth's possible water content shown for comparison. Ordinary chondrites are spectroscopically linked to S-type asteroids, which dominate the inner asteroid belt, while carbonaceous chondrites are thought to originate from C-type asteroids, which dominate the outer belt. There is not a clear spectral link between enstatite chondrites and a known class of asteroids, although their chemistry suggests that they formed interior to the asteroid belt. Figure from Morbidelli et al. (2012b)

by Peslier et al. (2017) on water in the Earth and Alexander et al. (2018) on water in small Solar System bodies.

Lécuyer (1998) estimate the total mass of water in the Earth's crust, oceans, and atmosphere to be 2.8×10^{-4} $M_\oplus$. The water content of the mantle is more uncertain, with estimates ranging from 0.8–8 $\times 10^{-4}$ $M_\oplus$ (Lécuyer 1998) to 2×10^{-3} $M_\oplus$ (Marty 2012). The primitive mantle could have potentially contained even more, as much as 10–50 Earth oceans (Dreibus and Waenke 1989; Righter and Drake 1999; Abe et al. 2000), although there is no definitive constraint on this value. The core may also contain significant water, although that too is unclear. Nomura et al. (2014) estimate a quantity of hydrogen in the core equivalent to 80 Earth oceans of water, whereas Badro et al. (2014) find very little. From this, a reasonable lower limit for the amount of water that must be delivered to (and retained by) the Earth is 5×10^{-4} $M_\oplus$, although the actual amount of water delivered could be significantly larger.

It has traditionally been assumed that the solar nebula around 1 AU was too hot for ice to condense and be incorporated into local Earth-forming material. This is based in part on associations between different meteorite types and their asteroidal parent bodies or estimated locations of formation (Fig. 1), showing a trend towards extremely dry material near Earth's location. However, nebula models taking into account stellar irradiation, viscous heating, and time-dependent opacities (e.g. Hueso and Guillot 2005; Davis 2005; Garaud and Lin 2007; Oka et al. 2011; Bitsch et al. 2015; Baillié et al. 2015) show that the ice-condensation line (the *snowline*) can move inwards to $\sim$1 AU as the accretion of gas onto the central star slows. Morbidelli et al. (2016) show that even if the temperature around 1 AU is cold enough for ice to condense, the snowline could be fossilized at around 3 AU since the gas drifts inwards more quickly than the snowline moves, and the formation of a proto-Jupiter would block the migration of icy particles into the inner disk. Thus, there is still a problem of condensing ice and incorporating it into planetesimals in the terrestrial planet zone.

Water is clearly present in bodies in the outer asteroid belt and beyond, however. Primitive meteorites such as CM and CI chondrites can have ~5–20% water by mass, and are believed to come from asteroids such as the C-types that dominate the outer asteroid belt (Burbine et al. 2002). Ceres at ~2.7 AU contains several tens of percent water ice (e.g. McCord and Sotin 2005) and shows signs of water vapor emission (Küppers et al. 2014) and recent aqueous activity (de Sanctis et al. 2016). Water ice has been directly detected on the asteroid Themis at ~3.1 AU (Campins et al. 2010; Rivkin and Emery 2010). Some outer asteroid belt bodies called main-belt comets, which show signs of activity, likely have at least 10% water ice (Jewitt 2012). Centaurs, comets, trans-Neptunian objects, and many outer planet satellites are also ice-rich.

A key constraint on the origin of Earth's water is the deuterium/hydrogen (D/H) ratio, estimated by Lécuyer (1998) to be $\sim 150 \times 10^{-6}$, or about 6× the solar value. This value is close to the mean whole-rock value for water-rich carbonaceous chondrite meteorites (e.g. Dauphas et al. 2000; Robert 2003, 2006; Alexander et al. 2012), suggesting that material similar to carbonaceous chondrites is a potential source of Earth's water. The value for comets is often quoted as being 12× the solar value, or 2× terrestrial, although there are some exceptions to this, as described further in Sect. 5. While carbonaceous chondrites may provide a good match to the Earth's D/H ratio, their oxygen isotope ratios and volatile element compositions, when compared to the Earth, suggest that carbonaceous chondritic material may have only contributed ~2% of the Earth's mass (Drake and Righter 2002; Marty 2012).

There are geochemical constraints on the timing of water delivery as well. Studies of core formation in the Earth suggest that the early material that accreted to form Earth must have been highly reduced (Wood et al. 2008; Rubie et al. 2011), which would preclude the presence or delivery of significant water early on. As we discuss in Sect. 6, Rubie et al. (2015b) combined a core-mantle differentiation model with N-body simulations to explore this issue in further detail.

Some models (e.g. Rubie et al. 2004; Siebert et al. 2013) have proposed that core-mantle differentiation occurs under oxidizing conditions, which could be consistent with abundant water being accreted to Earth very early. In those models, accreted silicate material has a high FeO content (e.g. FeO mole fraction $X_{FeO} \approx 0.2$). During core formation, oxygen partitions into the core (as FeO) and reduces X_{FeO} of the mantle to the observed value of ~0.06. However, there is a significant problem with such models because several weight percent Si also partitions into the core, and every mole of Si that enters the core causes two moles of FeO to be transferred to the mantle. This makes it impossible for the silicate mantle to achieve the observed low final X_{FeO} value (Rubie et al. 2015b,a).

3 Stages of Terrestrial Planet Formation

The process of terrestrial planet formation begins with the condensation of the first solids ~4.567–4.568 Gyr ago (e.g. Amelin et al. 2002; Bouvier and Wadhwa 2010; Connelly et al. 2012) and can be considered complete by the end of the heavy bombardment of the inner Solar System around 4.1–3.8 Gyr ago (Tera et al. 1974; Chapman et al. 2007; Bottke et al. 2012; Marchi et al. 2012; Morbidelli et al. 2012a). The gaseous solar nebula dissipated within ~2–10 Myr of the formation of the first solids (e.g. Haisch et al. 2001; Kita et al. 2005), and during that time, the initial planetesimals and planetary embryos formed and the giant planets grew their cores and captured their gaseous envelopes. While Mars may essentially be a leftover embryo from this early phase (Dauphas and Pourmand 2011), the larger terrestrial planets experienced a much more drawn-out accretion process, lasting around 30–100 Myr (Raymond et al. 2006, 2009; Bromley and Kenyon 2006; O'Brien et al. 2006; Touboul et al. 2007; Kleine et al. 2009; Morbidelli et al. 2012b). For the Earth, the last major impact was likely that which formed the Moon (Cameron 2000; Canup and Asphaug 2001; Canup 2004; Ćuk and Stewart 2012; Canup 2012; Reufer et al. 2012). Here we give a brief overview of the different stages of terrestrial planet formation to provide a background for the rest of the chapter. There have been numerous reviews of terrestrial planet formation over the last decade or so, which the reader can consult for further details (e.g. Chambers 2004; Chambers et al. 2010; Righter and O'Brien 2011; Morbidelli et al. 2012b; Raymond et al. 2014).

The formation of terrestrial planets has often been described as consisting of three main stages, although there may be some overlap between them. The first stage, in which solid grains in the nebula accumulate into kilometer-scale or larger planetesimals is one of the most uncertain. The main problem in this stage is that the radial drift of bodies in the nebula due gas drag effects (e.g. Weidenschilling 1977, 1980) is fastest for meter-scale bodies, which are too large to be coupled to the gas, but small enough in mass that they feel a substantial drag force. Bodies of this size can drift into the Sun on a timescale $\sim$100 years, and thus any planetesimal growth process must be rapid enough to overcome this *meter size barrier*.

One proposal is that planetesimals grow by pairwise accretion of smaller grains in the nebula (e.g. Weidenschilling 1980; Weidenschilling and Cuzzi 1993; Weidenschilling et al. 1997; Wurm et al. 2001; Weidenschilling 2011). One potential issue with this scenario is that it requires a non-turbulent nebula, at least at the location and time when planetesimals were forming, which may not necessarily be the case (Armitage 2011). Otherwise, growth is slowed and bodies may not grow past the meter size barrier fast enough, and turbulence may drive collisions and erosion of bodies, rather than growth.

Another scenario is that particles settling to the midplane of the nebula may exceed a critical density such that planetesimals could grow purely by self-gravity, a process called *gravitational instability* (e.g. Goldreich and Ward 1973; Ward 2000; Youdin and Shu 2002). Like with pairwise accretion, nebular turbulence can impede this process, but even if the nebula is initially non-turbulent, shear between the midplane particle layer and the overlying gas could drive turbulence and prevent the solids from reaching a critical density (e.g. Weidenschilling and Cuzzi 1993; Weidenschilling 2003).

Some models, however, take advantage of turbulence to aid in the formation of planetesimals. In a process called *turbulent concentration*, eddies in the solar nebula may preferentially concentrate bodies of specific sizes, in particular chondrule-sized bodies, and lead to rapid growth (Cuzzi et al. 2001, 2008; Chambers 2010). Alternatively, meter-scale boulders may become concentrated in over-dense regions of the disk, and the collective drag force on the solids forces the gas to move along with them (rather than sub-Keplerian). This reduces the headwind and radial drift in a process called the *streaming instability* (Youdin and Goodman 2005; Johansen et al. 2007), and solids drifting into this region continue to accumulate until a critical density is reached.

Regardless of the specific process by which planetesimals form, they soon enter a second stage of growth in which they begin to gravitationally accrete one-another and grow into larger planetary embryos. At first, this proceeds through a process called *runaway accretion* (e.g. Greenberg et al. 1978; Wetherill and Stewart 1989; Kokubo and Ida 1996; Weidenschilling et al. 1997), in which bodies that first manage to grow larger are able to accrete

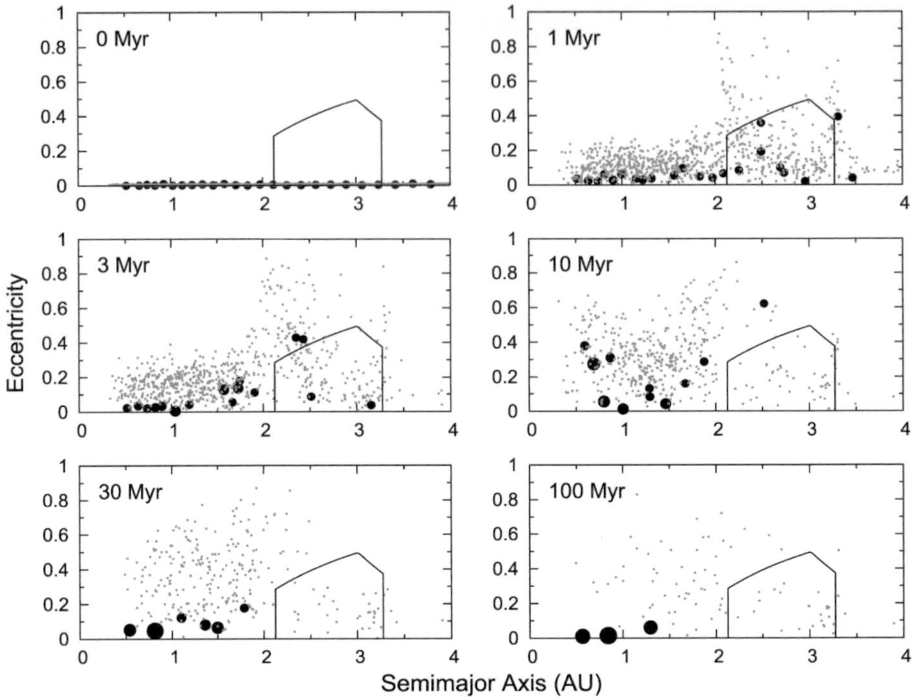

Fig. 2 Terrestrial planet formation from the EJS1 simulation of O'Brien et al. (2006). *Large black dots* are Mars-mass embryos, and *small grey dots* are planetesimal with 1/40 the mass of the embryos. The *outlined region* from ∼2–3 AU is the asteroid belt. Figure from O'Brien and Sykes (2011)

even more rapidly due to their enhanced cross sections and gravitational focusing. Once the growing embryos are large enough to significantly excite the relative velocities of surrounding planetesimals, their accretion rate slows and they enter a phase of *oligarchic growth* in which other, smaller embryos are able to catch up (Kokubo and Ida 1998, 2000). The end result of this stage is a set of Lunar- to Mars-mass embryos embedded in a swarm of remnant planetesimals.

Recently, *pebble accretion* has emerged as an extremely efficient process for the growth of planetary embryos and giant planet cores (e.g. Ormel and Klahr 2010; Lambrechts and Johansen 2012, 2014; Morbidelli and Nesvorny 2012). So-called pebbles, defined as having a stopping time due to gas drag on the order of their orbital period (generally mm- to cm-scale), are efficiently accreted onto large planetesimals due to gas dynamic effects, and those planetesimals can rapidly grow into planetary embryos. Levison et al. (2015) find that the terrestrial planets themselves can grow largely by accreting pebbles. Water delivery has yet to be quantified in such a model. Other models find that pebble accretion, at least in the terrestrial planet region, results in a swarm of embryos and remnant planetesimals much like the end result of runaway and oligarchic growth (Jacobson and Walsh 2015; Morbidelli et al. 2015; Chambers 2016).

In the final stage of terrestrial planet formation, embryos collide violently with one-another and grow into the final planets on a timescale of ∼30–100 Myr. In the classical scenario, embryos and planetesimals are spread through the terrestrial planet region, and likely into the asteroid belt zone, and the giant planets are fully-formed and located at or near their present locations (e.g. Chambers 2001; Raymond et al. 2004; O'Brien et al. 2006;

Raymond et al. 2009; Fischer and Ciesla 2014). Figure 2 shows the results of a simulation from O'Brien et al. (2006), starting with 25 Mars-mass embryos (large black dots) and 1000 planetesimals, each with 1/40 the mass of the embryos (small grey dots). The gas disk is assumed to have dissipated by this point, so gas effects are not included. While there is no gas to provide a damping effect on the growing planets, the swarm of planetesimals provides a damping effect due to *dynamical friction* (e.g. Wetherill and Stewart 1993). Terrestrial planets grow on a timescale of ~100 Myr, and the asteroid belt region, outlined in the figure, is dynamically excited and depleted by perturbations from the giant planets and the mutual scattering of embryos (Wetherill 1992; Petit et al. 2001; O'Brien et al. 2007). An important aspect of this scenario from the point of view of water delivery is that material from beyond ~2.5 AU can be scattered into the terrestrial planet region and accreted by the growing planets. This will be discussed in further detail in Sect. 6.

A major problem with this scenario is that the Mars analogues forming around 1.5 AU are typically 5–10 times more massive than Mars itself. This problem of explaining the small mass of Mars was first noted by Wetherill (1991) and persists over a wide range of parameter space, as shown by Raymond et al. (2009), Morishima et al. (2010), and Izidoro et al. (2015). Numerous models have been tested in an attempt to reconcile this issue, as well as to fit the full range of constraints on terrestrial planet formation and Solar System evolution (e.g. Nagasawa et al. 2005; Thommes et al. 2008; Kominami and Ida 2002, 2004; Ogihara et al. 2007; Morishima et al. 2008, 2010; Lykawka and Ito 2013; Izidoro et al. 2014). Often these models have included the effects of a gas disk, which can provide eccentricity and inclination damping as well drive the migration of embryos through a process called *Type I migration* (e.g. Goldreich and Tremaine 1980; Ward 1986). In general, though, such models are still not able to fit the full range of constraints, or require initial conditions such as eccentric giant planet orbits that are not consistent with Solar System formation models. For purposes of discussing water delivery in a classical scenario in Sect. 6, we will use the O'Brien et al. (2006) simulations described in the previous paragraph. For a more thorough review of the other scenarios listed above, see Morbidelli et al. (2012b) and Raymond et al. (2014).

A significant breakthrough in resolving the problem of forming a small Mars came when Hansen (2009) showed that the mass distribution of the terrestrial planets can be reproduced if the initial distribution of material is confined to a narrow annulus between ~0.7–1 AU. While they did not propose a specific mechanism for achieving this initial distribution, their work led Walsh et al. (2011) to model how the migration of the giant planets in the presence of nebular gas may allow for the truncation of the planetesimal/embryo distribution around 1 AU.[1] A large planet like Jupiter can open a gap in the nebular disk and migrate inwards via a process called *Type 2 migration*, where the planet follows the viscous evolution of the disk (Lin and Papaloizou 1986; Kley and Nelson 2012). If a second, smaller planet (e.g. Saturn) is present just exterior to the first planet, however, the profile of the gap and the resulting torque balance changes, and the inward migration can be halted and reversed (e.g. Masset and Snellgrove 2001; Morbidelli and Crida 2007; Pierens and Nelson 2008; Pierens and Raymond 2011; Pierens et al. 2014; Zhang and Zhou 2010; D'Angelo and Marzari 2012). This allows for a scenario in which Jupiter migrates into the terrestrial planet region, Saturn catches up with it, and the two planets open a mutual gap in the disk, reversing the direction of migration and moving Jupiter and Saturn back outwards to near their current locations. Walsh et al. (2011) termed this migration reversal the 'Grand Tack', in analogy with the turning of a sailboat.

[1] Several mechanisms have subsequently been proposed to explain the inner edge at ~0.7 AU, including a fossilized silicate condensation line (Morbidelli et al. 2016) and the outward migration of Jupiter's core in a model where it initially forms close to the Sun (Raymond et al. 2016).

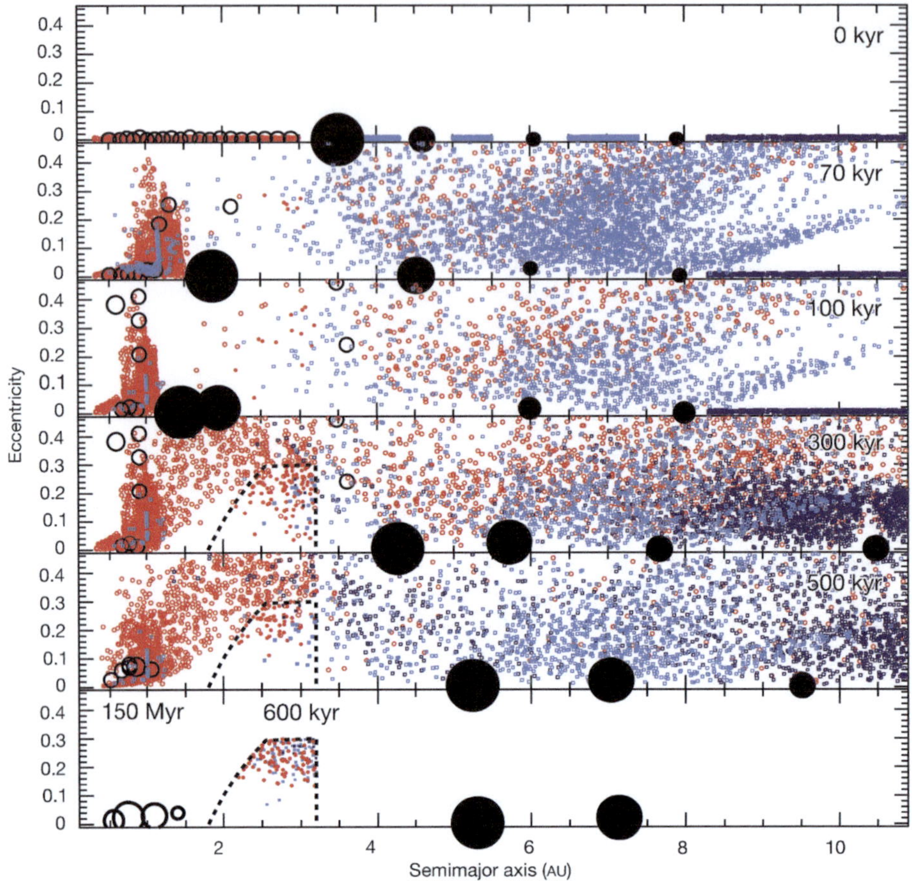

Fig. 3 Figure from Walsh et al. (2011) showing a simulation of the Grand Tack scenario. The *top three panels* show the inward migration phase of the giant planets (lasting 100 kyr), and the *last three panels* show their outward migration (lasting 500 kyr). The last panel also shows the terrestrial planets formed after 150 Myr. Material implanted into the asteroid belt is not uniformly distributed, the inner asteroid belt is dominated by material originating within a few AU of the Sun (*red dots*), and the outer belt is dominated by material scattered inwards from the giant planet region (*blue dots*), which broadly reproduces the S-Type/C-Type dichotomy seen in the asteroid belt today (e.g. Gradie and Tedesco 1982; Mothé-Diniz et al. 2003). The eccentricity distribution of asteroids after 600 kyr is higher than the observed distribution, but Deienno et al. (2016) show that it evolves to roughly match the current distribution over the subsequent 4.5 Gyr of evolution. The terrestrial planets will accrete some of the material scattered inwards from the giant planet region, which is likely water rich. This was quantified in detail by O'Brien et al. (2014)

Walsh et al. (2011) found that if Jupiter reversed direction at 1.5 AU, the inner disk would be truncated at ~1 AU, and Mars would generally form from one or a few embryos that scattered outwards and became relatively isolated from the rest of the disk, as in the simulations of Hansen (2009). Furthermore, while the passage of Jupiter through the primordial asteroid belt would be highly disruptive to the initial distribution of material there, Walsh et al. (2011) showed that the scattering effects of Jupiter as it moved inwards and outwards actually repopulated the asteroid belt with original material from the asteroid belt region as well as material scattered inwards from the giant planet region. Figure 3 shows a simulation from Walsh et al. (2011) that illustrates this process, with the top three panels showing the inward migration phase (over 100 kyr), and the last three showing outward migration (over 500 kyr). The resulting inner asteroid belt is dominated by material originating within a few AU of the Sun (red dots), while the outer belt is dominated by material scattered inwards from the giant planet region (blue dots), broadly reproducing the S-Type/C-Type dichotomy seen in the asteroid belt today (e.g. Gradie and Tedesco 1982; Mothé-Diniz et al. 2003). The eccentricity distribution of asteroids after 600 kyr is higher than the observed distribution, but will evolve to roughly match the current distribution over the subsequent 4.5 Gyr of evolution (Deienno et al. 2016). Material scattered inwards from the giant planet region would likely be water-rich and some would be accreted by the terrestrial planets, which was quantified by O'Brien et al. (2014) and will be discussed in further detail in Sect. 6. Brasser et al. (2016) performed more detailed simulations of the Grand Tack scenario and showed that a reversal of Jupiter at $\sim$2 AU might provide an even better match to some constraints than the 1.5 AU value used by Walsh et al. (2011), although the results in terms of water delivery are expected to be fairly similar.

More recently, there have been several models published besides the Grand Tack that can match many aspects of the inner Solar System. The 'low-mass asteroid belt model' proposes that only a small population of planetesimals formed beyond 1–1.5 AU (Izidoro et al. 2014, 2015; Drążkowska et al. 2016; Morbidelli and Raymond 2016; Raymond and Izidoro 2017b). This is consistent with some models of planetesimal formation by the streaming instability in evolving disks (Drążkowska et al. 2016; Surville et al. 2016). This mass deficit naturally reproduces the large Earth/Mars mass ratio much like Hansen (2009) and the Grand Tack, but requires a mechanism like chaotic excitation to match the asteroid belt (Izidoro et al. 2016). In addition, some models based on pebble accretion can match the terrestrial planets and give a depleted asteroid belt (Levison et al. 2015).

4 In-Situ/Early Water Delivery

Numerous studies have shown that water could be incorporated into olivine grains around 1 AU through adsorption directly from the gaseous nebula (Stimpfl et al. 2006; Muralidharan et al. 2008; King et al. 2010; Asaduzzaman et al. 2015). These grains would then follow the steps outlined in Sect. 3 and grow into planetesimals, potentially contributing several oceans worth of water to a growing Earth from local, early-accreted material. Several key questions remain, however. Studies show that the early material that accreted to form Earth was highly reduced (e.g. Wood et al. 2008; Rubie et al. 2011), which would not be the case if it contained significant water. Also, enstatite chondrite meteorites are drier than the Earth (see Fig. 1) and highly reduced, and may be difficult to explain if significant water-bearing material was present. An additional criticism is that this process would lead to the capture of water with a D/H ratio close to Solar, whereas the D/H ratio of the Earth's water is approximately 6× Solar. Ganguly et al. (2016) show that a fractionation process between adsorbed and nebular water may explain at least part of this enrichment.

Another possibility for the early incorporation of water-bearing material is the oxidation of an early hydrogen atmosphere by FeO in the terrestrial magma ocean (Ikoma and Genda 2006; Genda and Ikoma 2008). This requires fast growth of the planet in order to capture hydrogen from the solar nebula, which likely dissipated within 10 Myr (Haisch et al. 2001; Kita et al. 2005), in contrast to radiometric estimates for the formation timescale of the Earth that are much longer (e.g. Touboul et al. 2007; Kleine et al. 2009). The D/H ratio of the initial captured gas would be solar, not terrestrial. Hydrodynamic escape of atmosphere over billions of years could potentially lead to a terrestrial D/H ratio (Genda and Ikoma 2008).

However, this timescale is much longer than the ~100 Myr closure age of the atmosphere estimated from I-Xe isotopes (Wetherill 1975; Avice and Marty 2014).

Even if these two processes are not able to explain all of the Earth's water, or give a terrestrial D/H ratio, they may have still occurred to some degree. Deep-mantle lavas sampling relatively unmixed reservoirs show evidence for solar D/H (Hallis et al. 2015), suggesting that at least some of Earth's water could have been incorporated directly from the solar nebula.

Finally, Ciesla and Lauretta (2005) proposed that water-bearing phyllosilicate dust could drift inwards and be incorporated into planetesimals around 1 AU. This would require that water be locked in phyllosilicates while nebular gas was still around. However, many phyllosilicates found in meteorites seem to have been formed by parent-body processes after nebular gas would have dissipated.

5 Cometary Delivery

Comets have long been suggested as a source of terrestrial water, given their obvious ice-rich nature (e.g. Chyba 1987; Delsemme 1992, 1997, 1998; Owen and Bar-Nun 1995). The most common objection to this is that most comets with measured D/H ratios have about twice the terrestrial value. Initially all of those measured comets were Oort-cloud or Halley-type comets. The Jupiter-family comet 103P/Hartley 2 was recently found to have a terrestrial D/H ratio (Hartogh et al. 2011), as was 45P/Honda-Mrkos-Pajdušáková (Lis et al. 2013), hinting that JFCs could be a possible source. However, comet 67P/Churyumov-Gerasimenko, also a Jupiter-family comet, was found to have a D/H nearly 3 times terrestrial (Altwegg et al. 2015), so there is no clear trend from the current observational data.

Possibly more problematic than any geochemical constraints, however, is the significant dynamical difficulty in delivering the Earth's water primarily through comets. The probability that comets originating in the giant planet region hit the Earth is $\sim 1 \times 10^{-6}$ (Morbidelli et al. 2000). Various models of outer Solar System evolution suggest that ~30-50 $M_\oplus$ of material may have been present in a primordial cometessimal disk (e.g. Malhotra 1993, 1995; Hahn and Malhotra 1999; Gomes et al. 2004, 2005; Tsiganis et al. 2005; Morbidelli et al. 2010). Thus, if 50 $M_\oplus$ of comets were initially present and scattered by the giant planets, 5×10^{-5} $M_\oplus$ of cometary material would have hit the Earth. As discussed in Sect. 2, the Earth's water content is at least 5×10^{-4} $M_\oplus$, so even if those comets were pure ice, only ~10% of Earth's water could be provided by a cometary source.

6 Later Delivery

Morbidelli et al. (2000) and Raymond et al. (2004) showed that if the initial disk of planetesimals and embryos in the inner Solar System extended through the asteroid belt region, water-bearing material from beyond ~2.5 AU can be incorporated into the growing terrestrial planets, often relatively late in the accretion process. O'Brien et al. (2006) performed several sets of simulations with higher resolution that allowed for the analysis of this delivery mechanism in greater detail (see also Raymond et al. 2006, 2007). The O'Brien et al. (2006) simulations started with 25 Mars-mass embryos and 1000 planetesimals, each with 1/40 the mass of an embryo (so that there is equal mass in the planetesimal and embryo populations). Two configurations of Jupiter and Saturn were used: The EJS simulations, with Jupiter and Saturn on their current, eccentric orbits, and the CJS simulations, with Jupiter

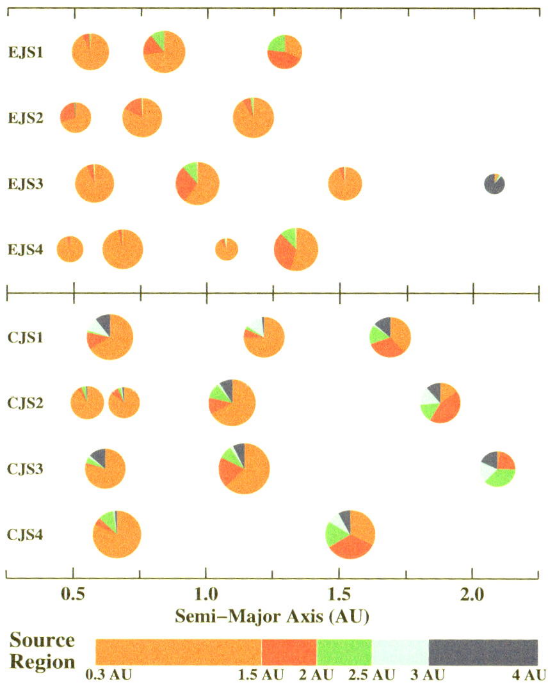

Fig. 4 Relative contributions of material from different semimajor axis zones in the simulations of O'Brien et al. (2006) (adapted from a figure in that paper). The diameter of each symbol is proportional to the diameter of the planet, assuming that they all have the same density. In the EJS simulations, Jupiter and Saturn are on their current, eccentric orbits, and in the CJS simulations, Jupiter and Saturn are on more circular and closely-spaced orbits, as predicted by the Nice Model (Tsiganis et al. 2005; Gomes et al. 2005; Morbidelli et al. 2005)

and Saturn on more circular and closely-spaced orbits, as predicted by the Nice Model (Tsiganis et al. 2005; Gomes et al. 2005; Morbidelli et al. 2005). Four simulations were run for each giant planet configuration.

Figure 4 shows the final planets in the simulations of O'Brien et al. (2006), with the contributions of material from different semimajor axis regions. 12 terrestrial planets were formed in the four CJS simulations. All of them accreted at least a few planetesimals from beyond 2.5 AU (median number of 7), and 9 of the planets contained at least one embryo that originated from beyond 2.5 AU. The fraction of each planet's total mass that originates from beyond 2.5 AU ranges from 1.6% to 38%, with a median value of 15%. On average, about 85% of the material from beyond 2.5 AU that is accreted by the planets is delivered by embryos.

In contrast, the 14 planets in the EJS simulations receive very little material from beyond 2.5 AU. None of the planets accrete an embryo from beyond 2.5 AU, although one planet (in simulation EJS3) originated as an embryo beyond 2.5 AU and never accreted any other embryos. At most, planets accrete 3 planetesimals from beyond 2.5 AU, and 5 of the planets contain no material at all from beyond 2.5 AU. The fraction of each planet's total mass that originates from beyond 2.5 AU (excluding the single-embryo planet in EJS3) has a median value of 0.3% and ranges from 0% to 1.6%. This profound disparity between the two sets of simulations is primarily due to the fact that resonances in the asteroid belt region are stronger with Jupiter and Saturn on eccentric orbits, and this leads to most material from beyond 2.5 AU being driven into the Sun or ejected from the Solar System, whereas more of that material slowly diffuses into the terrestrial planet region in the CJS simulations. Interestingly, earlier simulations incorporating only embryos (Morbidelli et al. 2000; Raymond et al. 2004) found that sufficient water could be delivered to the terrestrial planets if Jupiter and Saturn are eccentric. In the O'Brien et al. (2006) simulations, the large number of planetesimals tends to aid in the clearing of embryos from the region beyond 2.5 AU, such that little or no material from that region ends up being incorporated into the planets when Jupiter and Saturn are eccentric.

The mass fractions of material from beyond 2.5 AU in the final planets can be converted to estimates of water content, for comparison to the Earth. As noted above, the median mass fraction for planets in the CJS simulations is 15% (ranging from 1.6–38%). Assuming a mass fraction of 10% water for the material originating from outside 2.5 AU, consistent with the values for carbonaceous chondrites, and also assuming that all accreted volatiles are retained by the planet and not lost in the impact, a 15% mass fraction implies 0.015 $M_\oplus$ of water for an Earth-mass planet, or 30× the lower limit of 5×10^{-4} $M_\oplus$ from Sect. 2. The full range of values for the planets spans 3–75× the lower limit. Even assuming a more conservative mass fraction of water in the material beyond 2.5 AU of 5% and assuming that only 10% of the water is retained in the impact, the median amount of water in an Earth-mass planet is still 1.5× the lower limit. We note that if the material from beyond 2.5 AU is similar to carbonaceous chondrites, the large amount delivered in these simulations could pose a geochemical conflict. As discussed in Sect. 2, Drake and Righter (2002) and Marty (2012) argue that, based on oxygen isotope ratios and volatile element abundances, carbonaceous chondrites may have only contributed about 2% of the Earth's mass.

For the EJS simulations, the median mass fraction of material from beyond 2.5 AU that ends up in the final terrestrial planets is 0.3%, with a maximum value of 1.6%. Assuming a 10% mass fraction of water in the impactors from beyond 2.5 AU, and 100% retention of volatiles in impacts, the median water content for an Earth-mass planet would be 3×10^{-4} $M_\oplus$, less than the lower limit value of 5×10^{-4} $M_\oplus$. Hence, simulations with an eccentric Jupiter and Saturn have a much harder time forming Earth-like planets, at least in terms of water content.

O'Brien et al. (2006) also analyzed the timing of the delivery of material from beyond 2.5 AU. Figure 5 shows the timing for the planets in the CJS set of simulations. For each planet, the large black dots show the planet mass following the accretion of any embryos from beyond 2.5 AU, and the vertical bars show the planet mass when it has accreted 50% of its final number of planetesimals originating from beyond 2.5 AU. Nearly all of the embryo impacts occur well after the planet is half grown. In fact, 8 of the 9 planets that do experience at least one impact by an embryo from beyond 2.5 AU actually have that embryo as their final large impactor. Similarly, for the majority of planets, 50% of their final number of planetesimals from beyond 2.5 AU are not delivered until after the planet has grown to half its size. So whether from embryos or planetesimals, the delivery of potential volatile-carrying material tends to occur late in a planet's growth, and volatile-carrying embryos often occur as the final large impactor. Given the fact that the planets in the EJS simulations contain no embryos from beyond 2.5 AU and at most three such planetesimals, a similar plot can not be constructed for the EJS simulations.

Having water delivered relatively late in the accretion process makes it more likely to be retained, and is also consistent with the findings of Wood et al. (2008) and Rubie et al. (2011) that early material accreting to form Earth must have been highly reduced and the oxidation state increased with time. We stress, though, that while water is delivered relatively late in these simulations, it is not delivered as a *late veneer*, which is generally considered to be the mass delivered after differentiation has ceased. Differentiation and core formation would almost entirely remove highly siderophile elements from the mantle, and the fact that some do exist today suggests that $\lesssim$1% of Earth's total mass was accreted after differentiation ceased (Drake and Righter 2002). O'Brien et al. (2006) assumed that differentiation occurs as long as embryo impacts are occurring, and calculated the late veneer mass as that arriving

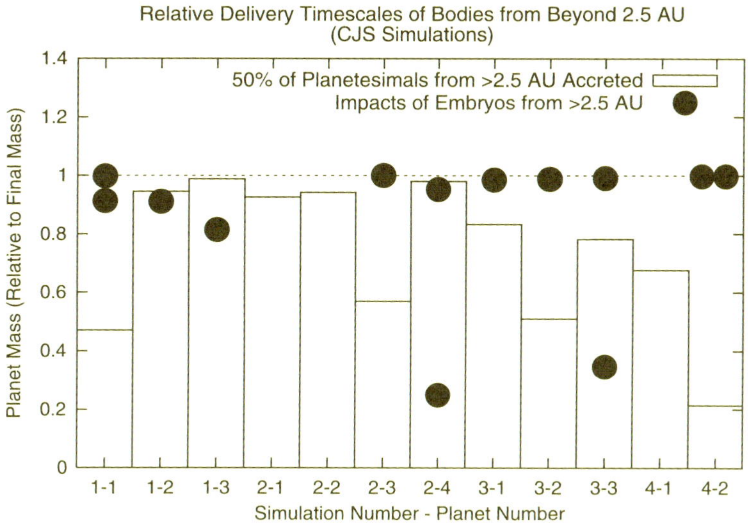

Fig. 5 Timing of the accretion of material from beyond 2.5 AU, which likely contains water and other volatiles, in the CJS simulations of O'Brien et al. (2006). Each column is for a given planet in a given simulation. The *bars* show the mass of each planet at the point when it has accreted 50% of its final number of planetesimals originating from beyond 2.5 AU. Almost all planets reach the 50% point after they are more than half grown, implying that water-bearing planetesimals are delivered preferentially late rather than early. Even more striking is the timing of accretion of embryos from beyond 2.5 AU. The *large black dots* give the planet mass following the impact of these embryos. Planet 4–2 is impacted by a body containing 2 embryos from beyond 2.5 AU, and only 3 planets do not experience the impact of an embryo from beyond 2.5 AU. In most cases, the impact of an embryo from beyond 2.5 AU is the final embryo impact on a planet. Planetesimals from beyond 2.5 AU that are 'pre-accreted' by another embryo before hitting the planet are not counted in this graph. Figure from O'Brien et al. (2006)

after the final giant impact on a given planet. For the CJS simulations, they find a median value of 1.2% for planets that are comparable in mass to the Earth, reasonably consistent with the Drake and Righter (2002) estimate. The median mass fraction of material from beyond 2.5 AU that arrives after the last giant impact, as a fraction of planet mass, is 0.7%, and ranges from 0–2.9%, whereas the total median mass fraction of material from beyond 2.5 AU is 15%. Thus, the vast majority of water-bearing material arrives prior to the late veneer in these simulations.

As can be seen in Fig. 4, the simulations of O'Brien et al. (2006) do not do a good job of reproducing Mars—even in the EJS simulations, the best Mars analogues are at least several times too massive. With larger numbers of EJS and CJS simulations (50 each), Fischer and Ciesla (2014) found that Mars-like planets could be produced in a small fraction of cases, but it is a very low-probability event, especially in the CJS case. This is an issue for nearly all simulations of the classical scenario, except in the case where Jupiter is made to be substantially more eccentric than its current value (Raymond et al. 2009). This issue was part of the motivation for the work of Hansen (2009) and the development of the Grand Tack model of Walsh et al. (2011) described in Sect. 3. O'Brien et al. (2014) extended the simulations of Walsh et al. (2011) to investigate water delivery in the Grand Tack scenario in more detail. Here we describe their key results.

The Walsh et al. (2011) simulations began with a population of embryos with 1/2 or 1/4 of a Mars mass and a swarm of planetesimals extending out to 3 AU, and populations of primitive planetesimals between the four giant planets (the *belts*) and beyond Neptune (the *disk*). The inward and outward migration of the giant planets pushes much of the inner Solar System material to inside of 1 AU, roughly reproducing the initial conditions of Hansen (2009), scatters material out of and then back into the asteroid belt, and scatters some of the material from the belt and disk populations into the terrestrial planet region as well. Walsh et al. (2011) did not explicitly track the accretion of the belt and disk material, but estimated that sufficient water would be delivered to explain the Earth's water budget. O'Brien et al. (2014) took the simulations of Walsh et al. (2011) immediately following the inward-then-outward migration of the giant planets and integrated them to 150 Myr, explicitly including the material originating from between and beyond the giant planets. 16 planetary systems were generated.

O'Brien et al. (2014) find that for planets larger than 0.75 $M_\oplus$ the median fraction of material originating from the belts population f_{belts} is 2.3%, and for the disk population f_{disk} is 0.7%. Because the masses of the belt and disk populations were normalized such that each could individually explain the mass of C-type asteroids in the main belt, the total mass fraction of material from the giant planet region that ends up in the terrestrial planets is not the sum of f_{belts} and f_{disk}, but rather a value between the two (i.e. between 0.7 and 2.3%). Note that if the primitive planetesimals are similar to carbonaceous chondrites, these values are consistent with the estimates of Drake and Righter (2002) and Marty (2012) that carbonaceous chondrites only contributed about 2% of the Earth's total mass.

For the same assumption as made for the O'Brien et al. (2006) simulations, that the primitive bodies have a water mass fraction of 10% and no water is lost in collisions, Earth-mass planets would have a median water content of 2.3×10^{-3} $M_\oplus$ if the water were delivered entirely by the belts population or 7×10^{-4} $M_\oplus$ if it were delivered entirely by the disk population (and most likely a value somewhere between the two). Both values exceed the lower limit of 5×10^{-4} $M_\oplus$ discussed in Sect. 2 (4.6 and 1.4 times larger, respectively). We note that the Grand Tack scenario has the important feature of matching both the water budget of Earth and the mass of C-type asteroids in the main belt in a self-consistent manner—a constraint that any model of water delivery by carbonaceous chondrite material must be able to match.

The 10% value assumed above for the water content of primitive planetesimals originating from the giant planet region may be conservative, given that main-belt comets may have a much larger water content than this (Jewitt 2012). The discovery of water vapor emanating from Ceres (Küppers et al. 2014) and the detection of water ice on the surface of Themis (Campins et al. 2010; Rivkin and Emery 2010) also suggest that the water content of primitive material may be larger than estimated from meteorites (in which only water bound to the silicates has survived). Thus, larger water contents are a possibility, and would not be inconsistent with the Earth's total water budget since several tens of oceans worth of water can potentially be stored in the mantle, as discussed in Sect. 2. A higher planetesimal water content could also allow for significant impact-related losses of water, while still delivering the minimum required amount.

Figure 6 shows the timing of accretion of water-bearing material. f_{belts} and f_{disk} are plotted vs. time for all planets whose final mass is larger than 0.75 $M_\oplus$. The time axis is in terms of *normalized accretion time*, which is the mass of the planet as a fraction of its final mass. If equal amounts of primitive material were accreted during each impact, all lines on the plot would be horizontal; the fact that the curves in Fig. 6 all increase with increasing time means that the primitive planetesimals preferentially arrive late in the planets' accretion. The delivery of primitive material late in the accretion process, when the planet is larger, means that water and volatiles that may be delivered by those impactors are more likely to be retained by the planet. Wood et al. (2008) and Rubie et al. (2011) show that material accreting to

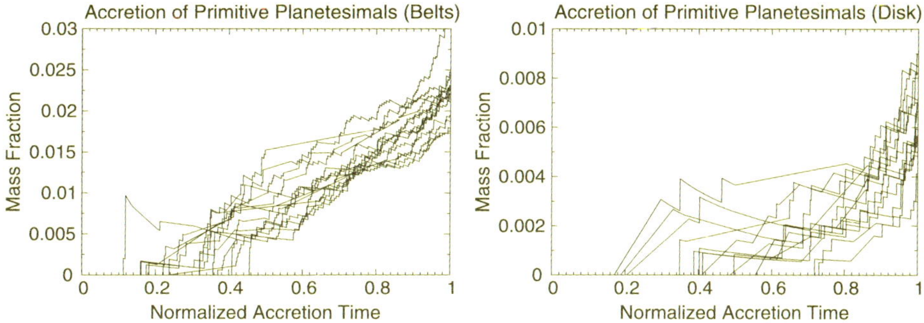

Fig. 6 Accretion curves for all planets larger than 0.75 $M_\oplus$ in the Grand Tack simulations of O'Brien et al. (2014). The mass fraction of primitive planetesimals from both the 'belts' and 'disk' populations are shown, where the belts population originally lies between the orbits of the giant planets, and the disk population initially lies beyond them. Accreted fractions are plotted vs. normalized time, which is the mass of the planet divided by its final mass. The curves all increase with time, which shows that the primitive material, and hence water and other volatiles, arrives relatively late during the accretion process. Figure from O'Brien et al. (2014)

form Earth was initially highly reduced, and the oxidation state increased with time, consistent with late addition of water to the terrestrial planets. Rubie et al. (2015b) integrated their multistage core-mantle differentiation model with actual Grand Tack simulations, and found that they can produce Earth-like planets with mantle concentrations of non-volatile elements and water that are close to the estimated values for the Earth's primitive mantle.

As with the CJS and EJS simulations described previously, the fact that primitive material is delivered late does not necessarily mean that water is all delivered as a late veneer at the tail end of accretion. Water delivery is still a relatively gradual process, with the planets continuing to grow while accreting water-rich planetesimals. The late veneer mass that is accreted in these simulations, however, tends to be significantly larger than in the CJS simulations, which were reasonably consistent with the estimate of $\lesssim 1\%$ total mass from Drake and Righter (2002). Planets larger than 0.75 $M_\oplus$ have a median late veneer mass (defined as mass accreted after the last embryo impact) of 18%, only a fraction of which is primitive water-bearing material. This high value is in part due the fact that many planets experience their last embryo impact in less than 20 Myr, and as shown by Jacobson et al. (2014), there tends to be an inverse correlation between late veneer mass and the time of last giant impact. Of the 27 planets larger than 0.5 $M_\oplus$ formed in the O'Brien et al. (2014) simulations, only 8 have last giant impacts after 20 Myr and 5 of those have late veneer masses of a few percent or less. Jacobson et al. (2014) have shown that the timing of the last giant impact and the late veneer mass can be strongly affected by the initial embryo mass and embryo/planetesimal mass ratio. This suggests a wider range of initial conditions should be explored to determine which parameters are able to best fit all available constraints.

Raymond and Izidoro (2017a) recently proposed a new scenario for water delivery by planetesimals scattered inward during the giant planets' growth and migration, which has implications for the other models discussed above. Jupiter and Saturn experienced a phase of runaway gas accretion during which the planets' masses increased dramatically, until they carved a gap in the disk and accretion slowed (e.g. Lissauer et al. 2009). This growth was likely accompanied by orbital migration, although there are a number of viable migration histories (Pierens et al. 2014). The giant planets' growth invariably destabilizes the orbits of nearby planetesimals, gravitationally scattering them in all directions onto eccentric orbits. Given the dissipative effects of aerodynamic gas drag (Adachi et al. 1976), planetesimals' eccentricities may be damped, decoupling the planetesimals from the giant planets and preventing further scattering. A fraction of planetesimals is naturally deposited interior to Jupiter's orbit, preferentially populating the outer part of the main asteroid belt (Raymond and Izidoro 2017a). Implanted planetesimals originate mainly between 4 and 10 AU (but with a tail out to 15–20 AU) and match the orbits of the C-type asteroids. In fact, Ceres is located immediately adjacent to the peak in the distribution of 1000 km-class planetesimals implanted by this mechanism (Raymond and Izidoro 2017a). Some planetesimals are scattered interior to the asteroid belt by Jupiter, crossing the orbits of the terrestrial planets. This preferentially happens for large planetesimals and later in the disk lifetime, when gas drag is weaker.

The scenario proposed in Raymond and Izidoro (2017a) would occur before any significant planetary growth in the terrestrial planet region, and hence could fit within any model of terrestrial planet formation. In the Grand Tack model, for example, there would be an additional, earlier scattering of water-rich material into the terrestrial planet region by the Raymond and Izidoro (2017a) mechanism, in addition to the later scattering of material during the inward and outward migration of Jupiter. The amount of material scattered inwards during this early phase, however, is unconstrained since it is based on the (unknown) planetesimal distribution in the primordial Jupiter-Saturn zone. Likewise, the timing of accretion of that material is unclear—even though it is implanted early, it may be accreted relatively late, especially if it is on eccentric orbits. Future modeling of terrestrial planet formation incorporating the Raymond and Izidoro (2017a) scenario, along with geochemical constraints (e.g. Rubie et al. 2015b), may help better understand the magnitude and effects of this early planetesimal scattering process.

7 Summary and Conclusions

We have reviewed the main scenarios for delivery of water to the terrestrial planets. Several early/in-situ mechanisms have been proposed, including the incorporation of water into olivine grains around 1 AU through adsorption directly from the gaseous nebula (Stimpfl et al. 2006; Muralidharan et al. 2008; King et al. 2010; Asaduzzaman et al. 2015), oxidation of an early hydrogen atmosphere by FeO in the terrestrial magma ocean (Ikoma and Genda 2006; Genda and Ikoma 2008), and incorporation of inward-drifting phyllosilicate dust into planetesimals around 1 AU (Ciesla and Lauretta 2005). As discussed in Sect. 4, each of these scenarios has its problems, for example potential inconsistencies with the measured D/H ratio of the Earth, and all run counter to work showing that the material that first accreted to form the Earth was highly reduced (Wood et al. 2008; Rubie et al. 2011). Cometary delivery has also been proposed, as discussed in Sect. 5. It too may have issues matching the measured D/H of the Earth, but a bigger issue is that the delivery efficiency of comets is very low (Morbidelli et al. 2000), such that even delivering a single Earth ocean of water is unlikely.

Later-stage delivery of water-bearing material, in the course of planetary accretion, is a more promising scenario (Sect. 6). In the classical scenario, with planetesimals and embryos extending through the outer asteroid belt region, substantial amounts of water can be delivered to the terrestrial planets if Jupiter and Saturn are on circular, low-eccentricity orbits as predicted by the Nice Model (e.g. O'Brien et al. 2006; Raymond et al. 2009). However, it is possible that so much carbonaceous-chondrite-like material could be delivered in that case that the volatile element abundances and oxygen isotope ratios of the final planets are not consistent with Earth's values (Drake and Righter 2002; Marty 2012). Such simulations are also not able to reproduce the small mass of Mars. In the Grand Tack scenario (Walsh et al. 2011; O'Brien et al. 2014), where inward and outward migration of the giant planets truncates the terrestrial planetesimal/embryo distribution and scatters material inwards from the giant planet region, the final terrestrial planet systems are a better match to the actual Solar System (with a small Mars). Sufficient water delivery occurs to explain the Earth's water budget without delivering excessive carbonaceous material that would violate the constraints of Drake and Righter (2002) and Marty (2012) regarding volatile elements and oxygen isotope ratios. Water is delivered relatively late in the accretion process, but not as a late veneer. The later addition of water is consistent with evidence that the earliest material accreted to the Earth was highly reduced, and later material was more oxidized (Wood et al. 2008; Rubie et al. 2011, 2015b). While some issues still remain with the Grand Tack scenario, specifically regarding the accretion timescales of the planets that form, this discrepancy may be reduced as a wider range of initial conditions is explored (e.g. Jacobson et al. 2014). Finally, new models of terrestrial planet formation have been proposed since the introduction of the Grand Tack scenario, such as those based on pebble accretion (e.g. Levison et al. 2015) or having an initially low mass in the asteroid belt region (Izidoro et al. 2014, 2015; Drążkowska et al. 2016; Morbidelli and Raymond 2016; Raymond and Izidoro 2017b), as well as new mechanisms for water delivery, such as early inward scattering of planetesimals during the formation of the giant planets (Raymond and Izidoro 2017a). These may be promising avenues for future research.

Acknowledgements We thank the International Space Science Institute (ISSI) for organizing and supporting the workshop "The Delivery of Water to Protoplanets, Planets and Satellites", and an anonymous reviewer for their helpful comments and suggestions. AI acknowledges financial support from FAPESP through grants number 16/12686-2 and 16/19556-7. SNR acknowledges Agence Nationale pour la Recherche grant ANR-13-BS05-0003-002 (grant MOJO). DCR and SAJ were supported by the European Research Council Advanced Grant "ACCRETE" (contract number 290568), and additional support to DCR was provided by the German Science Foundation (DFG) Priority Programme SPP1833 "Building a Habitable Earth" (Ru1323/10-1).

References

Y. Abe, E. Ohtani, T. Okuchi, K. Righter, M. Drake, Water in the early earth, in *Origin of the Earth and Moon*, ed. by R.M. Canup, K. Righter (University of Arizona Press, Tucson, 2000), pp. 413–433

I. Adachi, C. Hayashi, K. Nakazawa, The gas drag effect on the elliptical motion of a solid body in the primordial solar nebula. Prog. Theor. Phys. **56**, 1756–1771 (1976). https://doi.org/10.1143/PTP.56.1756

C.M.O.'D. Alexander, R. Bowden, M.L. Fogel, K.T. Howard, C.D.K. Herd, L.R. Nittler, The provenances of asteroids, and their contributions to the volatile inventories of the terrestrial planets. Science **337**, 721 (2012). https://doi.org/10.1126/science.1223474

C.M.O.'D. Alexander, K.D. McKeegan, K. Altwegg, Water reservoirs in small planetary bodies: meteorites, asteroids, and comets. Space Sci. Rev. (2018, in press). https://doi.org/10.1007/s11214-018-0474-9

K. Altwegg, H. Balsiger, A. Bar-Nun, J.J. Berthelier, A. Bieler, P. Bochsler, C. Briois, U. Calmonte, M. Combi, J. De Keyser, P. Eberhardt, B. Fiethe, S. Fuselier, S. Gasc, T.I. Gombosi, K.C. Hansen, M. Hässig, A. Jäckel, E. Kopp, A. Korth, L. LeRoy, U. Mall, B. Marty, O. Mousis, E. Neefs, T. Owen, H. Rème, M. Rubin, T. Sémon, C.Y. Tzou, H. Waite, P. Wurz, 67P/Churyumov-Gerasimenko, a Jupiter family comet with a high D/H ratio. Science **347**(27), 1261952 (2015). https://doi.org/10.1126/science.1261952

Y. Amelin, A.N. Krot, I.D. Hutcheon, A.A. Ulyanov, Lead isotopic ages of chondrules and calcium-aluminum-rich inclusions. Science **297**, 1678–1683 (2002). https://doi.org/10.1126/science.1073950

P.J. Armitage, Dynamics of protoplanetary disks. Annu. Rev. Astron. Astrophys. **49**, 195–236 (2011). https://doi.org/10.1146/annurev-astro-081710-102521

A. Asaduzzaman, K. Muralidharan, J. Ganguly, Incorporation of water into olivine during nebular condensation: insights from density functional theory and thermodynamics, and implications for phyllosilicate formation and terrestrial water inventory. Meteorit. Planet. Sci. **50**, 578–589 (2015). https://doi.org/10.1111/maps.12409

G. Avice, B. Marty, The iodine-plutonium-xenon age of the Moon-Earth system revisited. Philos. Trans. R. Soc. Lond. Ser. A **372**, 20130260 (2014). https://doi.org/10.1098/rsta.2013.0260

J. Badro, A.S. Côté, J.P. Brodholt, A seismologically consistent compositional model of Earth's core. Proc. Natl. Acad. Sci. **111**, 7542–7545 (2014). https://doi.org/10.1073/pnas.1316708111

K. Baillié, S. Charnoz, E. Pantin, Time evolution of snow regions and planet traps in an evolving protoplanetary disk. Astron. Astrophys. **577**, 65 (2015). https://doi.org/10.1051/0004-6361/201424987

V.R. Baker, Geomorphological evidence for water on Mars. Elements **2**(3), 139–143 (2007). https://doi.org/10.2113/gselements.2.3.139

B. Bitsch, A. Johansen, M. Lambrechts, A. Morbidelli, The structure of protoplanetary discs around evolving young stars. Astron. Astrophys. **575**, 28 (2015). https://doi.org/10.1051/0004-6361/201424964

W.F. Bottke, D. Vokrouhlický, D. Minton, D. Nesvorný, A. Morbidelli, R. Brasser, B. Simonson, H.F. Levison, An Archaean heavy bombardment from a destabilized extension of the asteroid belt. Nature **485**, 78–81 (2012). https://doi.org/10.1038/nature10967

A. Bouvier, M. Wadhwa, The age of the Solar System redefined by the oldest Pb–Pb age of a meteoritic inclusion. Nat. Geosci. **3**, 637–641 (2010). https://doi.org/10.1038/ngeo941

R. Brasser, S. Matsumura, S. Ida, S.J. Mojzsis, S.C. Werner, Analysis of terrestrial planet formation by the Grand Tack Model: system architecture and tack location. Astrophys. J. **821**, 75 (2016). https://doi.org/10.3847/0004-637X/821/2/75

B.C. Bromley, S.J. Kenyon, A hybrid N-body-coagulation code for planet formation. Astron. J. **131**, 2737–2748 (2006). https://doi.org/10.1086/503280

T.H. Burbine, T.J. McCoy, A. Meibom, B. Gladman, K. Keil, Meteoritic parent bodies: their number and identification, in *Asteroids III*, ed. by W.F. Bottke Jr., A. Cellino, P. Paolicchi, R.P. Binzel (University of Arizona Press, Tucson, 2002), pp. 653–667

A.G.W. Cameron, Higher-resolution Simulations of the Giant Impact, in *Origin of the Earth and Moon*, ed. by R.M. Canup, K. Righter, et al.(University of Arizona Press, Tucson, 2000), pp. 133–144

H. Campins, K. Hargrove, N. Pinilla-Alonso, E.S. Howell, M.S. Kelley, J. Licandro, T. Mothé-Diniz, Y. Fernández, J. Ziffer, Water ice and organics on the surface of the asteroid 24 Themis. Nature **464**, 1320–1321 (2010). https://doi.org/10.1038/nature09029

R.M. Canup, Simulations of a late lunar-forming impact. Icarus **168**, 433–456 (2004). https://doi.org/10.1016/j.icarus.2003.09.028

R.M. Canup, Forming a Moon with an Earth-like composition via a giant impact. Science **338**, 1052–1055 (2012). https://doi.org/10.1126/science.1226073

R.M. Canup, E. Asphaug, Origin of the Moon in a giant impact near the end of the Earth's formation. Nature **412**, 708–712 (2001). https://doi.org/10.1038/35089010

J.E. Chambers, Making more terrestrial planets. Icarus **152**, 205–224 (2001). https://doi.org/10.1006/icar.2001.6639

J.E. Chambers, Planetary accretion in the inner Solar System. Earth Planet. Sci. Lett. **223**, 241–252 (2004). https://doi.org/10.1016/j.epsl.2004.04.031

J.E. Chambers, Planetesimal formation by turbulent concentration. Icarus **208**, 505–517 (2010). https://doi.org/10.1016/j.icarus.2010.03.004

J.E. Chambers, Pebble accretion and the diversity of planetary systems. Astrophys. J. **825**, 63 (2016). https://doi.org/10.3847/0004-637X/825/1/63

J.E. Chambers, D.P. O'Brien, A.M. Davis, Accretion of planetesimals and the formation of rocky planets, in *Protoplanetary Dust: Astrophysical and Cosmochemical Perspectives*, ed. by D.A. Apai, D.S. Lauretta (2010), pp. 299–335

C.R. Chapman, B.A. Cohen, D.H. Grinspoon, What are the real constraints on the existence and magnitude of the late heavy bombardment? Icarus **189**, 233–245 (2007). https://doi.org/10.1016/j.icarus.2006.12.020

C.F. Chyba, The cometary contribution to the oceans of primitive earth. Nature **330**, 632–635 (1987). https://doi.org/10.1038/330632a0

F. Ciesla, D. Lauretta, Radial migration and dehydration of phyllosilicates in the solar nebula. Earth Planet. Sci. Lett. **231**, 1–8 (2005). https://doi.org/10.1016/j.epsl.2004.12.022

J.N. Connelly, M. Bizzarro, A.N. Krot, Å. Nordlund, D. Wielandt, M.A. Ivanova, The absolute chronology and thermal processing of solids in the solar protoplanetary disk. Science **338**, 651–655 (2012). https://doi.org/10.1126/science.1226919

M. Ćuk, S.T. Stewart, Making the Moon from a fast-spinning Earth: a giant impact followed by resonant despinning. Science **338**, 1047–1052 (2012). https://doi.org/10.1126/science.1225542

J.N. Cuzzi, R.C. Hogan, J.M. Paque, A.R. Dobrovolskis, Size-selective concentration of chondrules and other small particles in protoplanetary nebula turbulence. Astrophys. J. **546**, 496–508 (2001). https://doi.org/10.1086/318233

J.N. Cuzzi, R.C. Hogan, K. Shariff, Toward planetesimals: dense chondrule clumps in the protoplanetary nebula. Astrophys. J. **687**, 1432–1447 (2008). https://doi.org/10.1086/591239

G. D'Angelo, F. Marzari, Outward migration of Jupiter and Saturn in evolved gaseous disks. Astrophys. J. **757**, 50 (2012). https://doi.org/10.1088/0004-637X/757/1/50

N. Dauphas, A. Pourmand, Hf-W-Th evidence for rapid growth of Mars and its status as a planetary embryo. Nature **473**, 489–492 (2011). https://doi.org/10.1038/nature10077

N. Dauphas, F. Robert, B. Marty, The late asteroidal and cometary bombardment of earth as recorded in water deuterium to protium ratio. Icarus **148**, 508–512 (2000). https://doi.org/10.1006/icar.2000.6489

S.S. Davis, Condensation front migration in a protoplanetary nebula. Astrophys. J. **620**, 994–1001 (2005). https://doi.org/10.1086/427073

M.C. de Sanctis, A. Raponi, E. Ammannito, M. Ciarniello, M.J. Toplis, H.Y. McSween, J.C. Castillo-Rogez, B.L. Ehlmann, F.G. Carrozzo, S. Marchi, F. Tosi, F. Zambon, F. Capaccioni, M.T. Capria, S. Fonte, M. Formisano, A. Frigeri, M. Giardino, A. Longobardo, G. Magni, E. Palomba, L.A. McFadden, C.M. Pieters, R. Jaumann, P. Schenk, R. Mugnuolo, C.A. Raymond, C.T. Russell, Bright carbonate deposits as evidence of aqueous alteration on (1) Ceres. Nature **536**, 54–57 (2016). https://doi.org/10.1038/nature18290

R. Deienno, R.S. Gomes, K.J. Walsh, A. Morbidelli, D. Nesvorný, Is the Grand Tack model compatible with the orbital distribution of main belt asteroids? Icarus **272**, 114–124 (2016). https://doi.org/10.1016/j.icarus.2016.02.043

A.H. Delsemme, Cometary origin of carbon and water on the terrestrial planets. Adv. Space Res. **12**, 5–12 (1992). https://doi.org/10.1016/0273-1177(92)90147-P

A. Delsemme, The origin of the atmosphere and of the oceans, in *Comets and the Origin and Evolution of Life*, ed. by P.J. Thomas, C.F. Chyba, C.P. McKay (Springer, New York, 1997), pp. 29–67

A.H. Delsemme, The deuterium enrichment observed in recent comets is consistent with the cometary origin of seawater. Planet. Space Sci. **47**, 125–131 (1998). https://doi.org/10.1016/S0032-0633(98)00093-2

T.M. Donahue, J.H. Hoffman, R.R. Hodges, A.J. Watson, Venus was wet—a measurement of the ratio of deuterium to hydrogen. Science **216**, 630–633 (1982). https://doi.org/10.1126/science.216.4546.630

M.J. Drake, K. Righter, Determining the composition of the Earth. Nature **416**, 39–44 (2002). https://doi.org/10.1038/416039a

J. Drążkowska, Y. Alibert, B. Moore, Close-in planetesimal formation by pile-up of drifting pebbles. Astron. Astrophys. **594**, 105 (2016). https://doi.org/10.1051/0004-6361/201628983

G. Dreibus, H. Waenke, Supply and loss of volatile constituents during the accretion of terrestrial planets, in *Origin and Evolution of Planetary and Satellite Atmospheres*, ed. by S.K. Atreya, J.B. Pollack, M.S. Matthews (University of Arizona Press, Tucson, 1989), pp. 268–288

W.C. Feldman, T.H. Prettyman, S. Maurice, J.J. Plaut, D.L. Bish, D.T. Vaniman, M.T. Mellon, A.E. Metzger, S.W. Squyres, S. Karunatillake, W.V. Boynton, R.C. Elphic, H.O. Funsten, D.J. Lawrence, R.L. Tokar, Global distribution of near-surface hydrogen on Mars. J. Geophys. Res., Planets **109**, 09006 (2004). https://doi.org/10.1029/2003JE002160

R.A. Fischer, F.J. Ciesla, Dynamics of the terrestrial planets from a large number of N-body simulations. Earth Planet. Sci. Lett. **392**, 28–38 (2014). https://doi.org/10.1016/j.epsl.2014.02.011

J. Ganguly, A. Asaduzzaman, K. Muralidharan, Origin of water in Earth with high D/H ratio relative to protosolar nebula, and an explanation of its similarity with the isotopic ratios of carbonaceous chondrites and asteroid Vesta, in *79th Annual Meeting of the Meteoritical Society*. LPI Contributions, vol. 1921 (2016), p. 6055

P. Garaud, D.N.C. Lin, The effect of internal dissipation and surface irradiation on the structure of disks and the location of the snow line around Sun-like stars. Astrophys. J. **654**, 606–624 (2007). https://doi.org/10.1086/509041

H. Genda, M. Ikoma, Origin of the ocean on the Earth: early evolution of water D/H in a hydrogen-rich atmosphere. Icarus **194**, 42–52 (2008). https://doi.org/10.1016/j.icarus.2007.09.007

P. Goldreich, S. Tremaine, Disk-satellite interactions. Astrophys. J. **241**, 425–441 (1980). https://doi.org/10.1086/158356

P. Goldreich, W.R. Ward, The formation of planetesimals. Astrophys. J. **183**, 1051–1062 (1973). https://doi.org/10.1086/152291

R.S. Gomes, A. Morbidelli, H.F. Levison, Planetary migration in a planetesimal disk: why did Neptune stop at 30 AU? Icarus **170**, 492–507 (2004). https://doi.org/10.1016/j.icarus.2004.03.011

R. Gomes, H.F. Levison, K. Tsiganis, A. Morbidelli, Origin of the cataclysmic Late Heavy Bombardment period of the terrestrial planets. Nature **435**, 466–469 (2005). https://doi.org/10.1038/nature03676

J. Gradie, E. Tedesco, Compositional structure of the asteroid belt. Science **216**, 1405–1407 (1982). https://doi.org/10.1126/science.216.4553.1405

R. Greenberg, W.K. Hartmann, C.R. Chapman, J.F. Wacker, Planetesimals to planets—numerical simulation of collisional evolution. Icarus **35**, 1–26 (1978). https://doi.org/10.1016/0019-1035(78)90057-X

J.M. Hahn, R. Malhotra, Orbital evolution of planets embedded in a planetesimal disk. Astron. J. **117**, 3041–3053 (1999). https://doi.org/10.1086/300891

K.E. Haisch Jr., E.A. Lada, C.J. Lada, Disk frequencies and lifetimes in young clusters. Astrophys. J. Lett. **553**, 153–156 (2001). https://doi.org/10.1086/320685

L.J. Hallis, G.R. Huss, K. Nagashima, G.J. Taylor, S.A. Halldórsson, D.R. Hilton, M.J. Mottl, K.J. Meech, Evidence for primordial water in Earth's deep mantle. Science **350**, 795–797 (2015). https://doi.org/10.1126/science.aac4834

B.M.S. Hansen, Formation of the terrestrial planets from a narrow annulus. Astrophys. J. **703**, 1131–1140 (2009). https://doi.org/10.1088/0004-637X/703/1/1131

P. Hartogh, D.C. Lis, D. Bockelée-Morvan, M. de Val-Borro, N. Biver, M. Küppers, M. Emprechtinger, E.A. Bergin, J. Crovisier, M. Rengel, R. Moreno, S. Szutowicz, G.A. Blake, Ocean-like water in the Jupiter-family comet 103P/Hartley 2. Nature **478**, 218–220 (2011). https://doi.org/10.1038/nature10519

R. Hueso, T. Guillot, Evolution of protoplanetary disks: constraints from DM Tauri and GM Aurigae. Astron. Astrophys. **442**, 703–725 (2005). https://doi.org/10.1051/0004-6361:20041905

M. Ikoma, H. Genda, Constraints on the mass of a habitable planet with water of nebular origin. Astrophys. J. **648**, 696–706 (2006). https://doi.org/10.1086/505780

A. Izidoro, K. de Souza Torres, O.C. Winter, N. Haghighipour, A compound model for the origin of Earth's water. Astrophys. J. **767**, 54 (2013). https://doi.org/10.1088/0004-637X/767/1/54

A. Izidoro, N. Haghighipour, O.C. Winter, M. Tsuchida, Terrestrial planet formation in a protoplanetary disk with a local mass depletion: a successful scenario for the formation of Mars. Astrophys. J. **782**, 31 (2014). https://doi.org/10.1088/0004-637X/782/1/31

A. Izidoro, S.N. Raymond, A. Morbidelli, O.C. Winter, Terrestrial planet formation constrained by Mars and the structure of the asteroid belt. Mon. Not. R. Astron. Soc. **453**, 3619–3634 (2015). https://doi.org/10.1093/mnras/stv1835

A. Izidoro, S.N. Raymond, A. Pierens, A. Morbidelli, O.C. Winter, D. Nesvorny', The asteroid belt as a relic from a chaotic early Solar System. Astrophys. J. **833**, 40 (2016). https://doi.org/10.3847/1538-4357/833/1/40

S.A. Jacobson, K.J. Walsh, Earth and terrestrial planet formation. American Geophysical Union Geophysical Monograph Series **212**, 49–70 (2015). https://doi.org/10.1002/9781118860359.ch3

S.A. Jacobson, A. Morbidelli, S.N. Raymond, D.P. O'Brien, K.J. Walsh, D.C. Rubie, Highly siderophile elements in Earth's mantle as a clock for the Moon-forming impact. Nature **508**, 84–87 (2014). https://doi.org/10.1038/nature13172

D. Jewitt, The active asteroids. Astron. J. **143**, 66 (2012). https://doi.org/10.1088/0004-6256/143/3/66

A. Johansen, J.S. Oishi, M. Mac Low, H. Klahr, T. Henning, A. Youdin, Rapid planetesimal formation in turbulent circumstellar disks. Nature **448**, 1022–1025 (2007). https://doi.org/10.1038/nature06086

H.E. King, M. Stimpfl, P. Deymier, M.J. Drake, C.R.A. Catlow, A. Putnis, N.H. de Leeuw, Computer simulations of water interactions with low-coordinated forsterite surface sites: implications for the origin of water in the inner solar system. Earth Planet. Sci. Lett. **300**, 11–18 (2010). https://doi.org/10.1016/j.epsl.2010.10.019

N.T. Kita, G.R. Huss, S. Tachibana, Y. Amelin, L.E. Nyquist, I.D. Hutcheon, Constraints on the origin of chondrules and CAIs from short-lived and long-lived radionuclides, in *Chondrites and the Protoplanetary Disk*, ed. by A.N. Krot, E.R.D. Scott, B. Reipurth. Astronomical Society of the Pacific Conference Series, vol. 341 (2005), pp. 558–587

T. Kleine, M. Touboul, B. Bourdon, F. Nimmo, K. Mezger, H. Palme, S.B. Jacobsen, Q.Z. Yin, A.N. Halliday, Hf-W chronology of the accretion and early evolution of asteroids and terrestrial planets. Geochim. Cosmochim. Acta **73**, 5150–5188 (2009). https://doi.org/10.1016/j.gca.2008.11.047

W. Kley, R.P. Nelson, Planet-disk interaction and orbital evolution. Annu. Rev. Astron. Astrophys. **50**, 211–249 (2012). https://doi.org/10.1146/annurev-astro-081811-125523

E. Kokubo, S. Ida, On runaway growth of planetesimals. Icarus **123**, 180–191 (1996). https://doi.org/10.1006/icar.1996.0148

E. Kokubo, S. Ida, Oligarchic growth of protoplanets. Icarus **131**, 171–178 (1998). https://doi.org/10.1006/icar.1997.5840

E. Kokubo, S. Ida, Formation of protoplanets from planetesimals in the solar nebula. Icarus **143**, 15–27 (2000). https://doi.org/10.1006/icar.1999.6237

J. Kominami, S. Ida, The effect of tidal interaction with a gas disk on formation of terrestrial planets. Icarus **157**, 43–56 (2002). https://doi.org/10.1006/icar.2001.6811

J. Kominami, S. Ida, Formation of terrestrial planets in a dissipating gas disk with Jupiter and Saturn. Icarus **167**, 231–243 (2004). https://doi.org/10.1016/j.icarus.2003.10.005

M. Küppers, L. O'Rourke, D. Bockelée-Morvan, V. Zakharov, S. Lee, P. von Allmen, B. Carry, D. Teyssier, A. Marston, T. Müller, J. Crovisier, M.A. Barucci, R. Moreno, Localized sources of water vapour on the dwarf planet (1) Ceres. Nature **505**, 525–527 (2014). https://doi.org/10.1038/nature12918

M. Lambrechts, A. Johansen, Rapid growth of gas-giant cores by pebble accretion. Astron. Astrophys. **544**, 32 (2012). https://doi.org/10.1051/0004-6361/201219127

M. Lambrechts, A. Johansen, Forming the cores of giant planets from the radial pebble flux in protoplanetary discs. Astron. Astrophys. **572**, 107 (2014). https://doi.org/10.1051/0004-6361/201424343
C. Lécuyer, The hydrogen isotope composition of seawater and the global water cycle. Chem. Geol. **145**, 249–261 (1998). https://doi.org/10.1016/S0009-2541(97)00146-0
H.F. Levison, K.A. Kretke, K.J. Walsh, W.F. Bottke, Growing the terrestrial planets from the gradual accumulation of sub-meter sized objects. Proc. Natl. Acad. Sci. **112**, 14180–14185 (2015). https://doi.org/10.1073/pnas.1513364112
D.N.C. Lin, J. Papaloizou, On the tidal interaction between protoplanets and the protoplanetary disk. III—Orbital migration of protoplanets. Astrophys. J. **309**, 846–857 (1986). https://doi.org/10.1086/164653
D.C. Lis, N. Biver, D. Bockelée-Morvan, P. Hartogh, E.A. Bergin, G.A. Blake, J. Crovisier, M. de Val-Borro, E. Jehin, M. Küppers, J. Manfroid, R. Moreno, M. Rengel, S. Szutowicz, A Herschel study of D/H in water in the Jupiter-family comet 45P/Honda-Mrkos-Pajdušáková and Prospects for D/H Measurements with CCAT. Astrophys. J. Lett. **774**, 3 (2013). https://doi.org/10.1088/2041-8205/774/1/L3
J.J. Lissauer, O. Hubickyj, G. D'Angelo, P. Bodenheimer, Models of Jupiter's growth incorporating thermal and hydrodynamic constraints. Icarus **199**, 338–350 (2009). https://doi.org/10.1016/j.icarus.2008.10.004
P.S. Lykawka, T. Ito, Terrestrial planet formation during the migration and resonance crossings of the giant planets. Astrophys. J. **773**, 65 (2013). https://doi.org/10.1088/0004-637X/773/1/65
R. Malhotra, The origin of Pluto's peculiar orbit. Nature **365**, 819–821 (1993). https://doi.org/10.1038/365819a0
R. Malhotra, The origin of Pluto's orbit: implications for the solar system beyond Neptune. Astron. J. **110**, 420–429 (1995). https://doi.org/10.1086/117532
S. Marchi, W.F. Bottke, D.A. Kring, A. Morbidelli, The onset of the lunar cataclysm as recorded in its ancient crater populations. Earth Planet. Sci. Lett. **325**, 27–38 (2012). https://doi.org/10.1016/j.epsl.2012.01.021
B. Marty, The origins and concentrations of water, carbon, nitrogen and noble gases on Earth. Earth Planet. Sci. Lett. **313**, 56–66 (2012). https://doi.org/10.1016/j.epsl.2011.10.040
F. Masset, M. Snellgrove, Reversing type II migration: resonance trapping of a lighter giant protoplanet. Mon. Not. R. Astron. Soc. **320**, 55–59 (2001). https://doi.org/10.1046/j.1365-8711.2001.04159.x
T.B. McCord, C. Sotin, Ceres: Evolution and current state. J. Geophys. Res., Planets **110**, 5009 (2005). https://doi.org/10.1029/2004JE002244
A. Morbidelli, A. Crida, The dynamics of Jupiter and Saturn in the gaseous protoplanetary disk. Icarus **191**, 158–171 (2007). https://doi.org/10.1016/j.icarus.2007.04.001
A. Morbidelli, D. Nesvorny, Dynamics of pebbles in the vicinity of a growing planetary embryo: hydro-dynamical simulations. Astron. Astrophys. **546**, 18 (2012). https://doi.org/10.1051/0004-6361/201219824
A. Morbidelli, S.N. Raymond, Challenges in planet formation. J. Geophys. Res., Planets **121**, 1962–1980 (2016). https://doi.org/10.1002/2016JE005088
A. Morbidelli, J. Chambers, J.I. Lunine, J.M. Petit, F. Robert, G.B. Valsecchi, K.E. Cyr, Source regions and time scales for the delivery of water to Earth. Meteorit. Planet. Sci. **35**, 1309–1320 (2000). https://doi.org/10.1111/j.1945-5100.2000.tb01518.x
A. Morbidelli, H.F. Levison, K. Tsiganis, R. Gomes, Chaotic capture of Jupiter's Trojan asteroids in the early Solar System. Nature **435**, 462–465 (2005). https://doi.org/10.1038/nature03540
A. Morbidelli, R. Brasser, R. Gomes, H.F. Levison, K. Tsiganis, Evidence from the asteroid belt for a violent past evolution of Jupiter's orbit. Astron. J. **140**, 1391–1401 (2010). https://doi.org/10.1088/0004-6256/140/5/1391
A. Morbidelli, S. Marchi, W.F. Bottke, D.A. Kring, A sawtooth-like timeline for the first billion years of lunar bombardment. Earth Planet. Sci. Lett. **355**, 144–151 (2012a). https://doi.org/10.1016/j.epsl.2012.07.037
A. Morbidelli, J.I. Lunine, D.P. O'Brien, S.N. Raymond, K.J. Walsh, Building terrestrial planets. Annu. Rev. Earth Planet. Sci. **40**, 251–275 (2012b). https://doi.org/10.1146/annurev-earth-042711-105319
A. Morbidelli, M. Lambrechts, S. Jacobson, B. Bitsch, The great dichotomy of the Solar System: Small terrestrial embryos and massive giant planet cores. Icarus **258**, 418–429 (2015). https://doi.org/10.1016/j.icarus.2015.06.003
A. Morbidelli, B. Bitsch, A. Crida, M. Gounelle, T. Guillot, S. Jacobson, A. Johansen, M. Lambrechts, E. Lega, Fossilized condensation lines in the Solar System protoplanetary disk. Icarus **267**, 368–376 (2016). https://doi.org/10.1016/j.icarus.2015.11.027
R. Morishima, M.W. Schmidt, J. Stadel, B. Moore, Formation and accretion history of terrestrial planets from runaway growth through to late time: implications for orbital eccentricity. Astrophys. J. **685**, 1247–1261 (2008). https://doi.org/10.1086/590948

R. Morishima, J. Stadel, B. Moore, From planetesimals to terrestrial planets: N-body simulations including the effects of nebular gas and giant planets. Icarus **207**, 517–535 (2010). https://doi.org/10.1016/j.icarus.2009.11.038

T. Mothé-Diniz, J.M.Á. Carvano, D. Lazzaro, Distribution of taxonomic classes in the main belt of asteroids. Icarus **162**, 10–21 (2003). https://doi.org/10.1016/S0019-1035(02)00066-0

K. Muralidharan, P. Deymier, M. Stimpfl, N.H. de Leeuw, M.J. Drake, Origin of water in the inner Solar System: a kinetic Monte Carlo study of water adsorption on forsterite. Icarus **198**, 400–407 (2008). https://doi.org/10.1016/j.icarus.2008.07.017

M. Nagasawa, D.N.C. Lin, E. Thommes, Dynamical shake-up of planetary systems. I. Embryo trapping and induced collisions by the sweeping secular resonance and embryo-disk tidal interaction. Astrophys. J. **635**, 578–598 (2005). https://doi.org/10.1086/497386

R. Nomura, K. Hirose, K. Uesugi, Y. Ohishi, A. Tsuchiyama, A. Miyake, Y. Ueno, Low core-mantle boundary temperature inferred from the solidus of pyrolite. Science **343**(6170), 522–525 (2014). https://doi.org/10.1126/science.1248186

D.P. O'Brien, M.V. Sykes, The origin and evolution of the asteroid belt—implications for Vesta and Ceres. Space Sci. Rev. **163**, 41–61 (2011). https://doi.org/10.1007/s11214-011-9808-6

D.P. O'Brien, A. Morbidelli, H.F. Levison, Terrestrial planet formation with strong dynamical friction. Icarus **184**, 39–58 (2006). https://doi.org/10.1016/j.icarus.2006.04.005

D.P. O'Brien, A. Morbidelli, W.F. Bottke, The primordial excitation and clearing of the asteroid belt—revisited. Icarus **191**, 434–452 (2007). https://doi.org/10.1016/j.icarus.2007.05.005

D.P. O'Brien, K.J. Walsh, A. Morbidelli, S.N. Raymond, A.M. Mandell, Water delivery and giant impacts in the 'Grand Tack' scenario. Icarus **239**, 74–84 (2014). https://doi.org/10.1016/j.icarus.2014.05.009

M. Ogihara, S. Ida, A. Morbidelli, Accretion of terrestrial planets from oligarchs in a turbulent disk. Icarus **188**, 522–534 (2007). https://doi.org/10.1016/j.icarus.2006.12.006

A. Oka, T. Nakamoto, S. Ida, Evolution of snow line in optically thick protoplanetary disks: effects of water ice opacity and dust grain size. Astrophys. J. **738**, 141 (2011). https://doi.org/10.1088/0004-637X/738/2/141

C.W. Ormel, H.H. Klahr, The effect of gas drag on the growth of protoplanets. Analytical expressions for the accretion of small bodies in laminar disks. Astron. Astrophys. **520**, 43 (2010). https://doi.org/10.1051/0004-6361/201014903

T. Owen, A. Bar-Nun, Comets, impacts and atmospheres. Icarus **116**, 215–226 (1995). https://doi.org/10.1006/icar.1995.1122

A.H. Peslier, M. Schönbächler, H. Busemann, S.I. Karato, Water in the Earth's interior: distribution and origin. Space Sci. Rev. **212**, 743–810 (2017). https://doi.org/10.1007/s11214-017-0387-z

J. Petit, A. Morbidelli, J. Chambers, The primordial excitation and clearing of the asteroid belt. Icarus **153**, 338–347 (2001). https://doi.org/10.1006/icar.2001.6702

A. Pierens, R.P. Nelson, Constraints on resonant-trapping for two planets embedded in a protoplanetary disc. Astron. Astrophys. **482**, 333–340 (2008). https://doi.org/10.1051/0004-6361:20079062

A. Pierens, S.N. Raymond, Two phase, inward-then-outward migration of Jupiter and Saturn in the gaseous solar nebula. Astron. Astrophys. **533**, 131 (2011). https://doi.org/10.1051/0004-6361/201117451

A. Pierens, S.N. Raymond, D. Nesvorny, A. Morbidelli, Outward migration of Jupiter and Saturn in 3:2 or 2:1 resonance in radiative disks: implications for the Grand Tack and Nice models. Astrophys. J. Lett. **795**, 11 (2014). https://doi.org/10.1088/2041-8205/795/1/L11

S.N. Raymond, A. Izidoro, Origin of water in the inner Solar System: planetesimals scattered inward during Jupiter and Saturn's rapid gas accretion. Icarus **297**, 134–148 (2017a). https://doi.org/10.1016/j.icarus.2017.06.030

S.N. Raymond, A. Izidoro, The empty primordial asteroid belt. Sci. Adv. **3**, 1701138 (2017b). https://doi.org/10.1126/sciadv.1701138

S.N. Raymond, T. Quinn, J.I. Lunine, Making other earths: dynamical simulations of terrestrial planet formation and water delivery. Icarus **168**, 1–17 (2004). https://doi.org/10.1016/j.icarus.2003.11.019

S.N. Raymond, T. Quinn, J.I. Lunine, High-resolution simulations of the final assembly of Earth-like planets I. Terrestrial accretion and dynamics. Icarus **183**, 265–282 (2006). https://doi.org/10.1016/j.icarus.2006.03.011

S.N. Raymond, T. Quinn, J.I. Lunine, High-resolution simulations of the final assembly of Earth-like planets. 2. Water delivery and planetary habitability. Astrobiology **7**, 66–84 (2007). https://doi.org/10.1089/ast.2006.06-0126

S.N. Raymond, D.P. O'Brien, A. Morbidelli, N.A. Kaib, Building the terrestrial planets: constrained accretion in the inner Solar System. Icarus **203**, 644–662 (2009). https://doi.org/10.1016/j.icarus.2009.05.016

S.N. Raymond, E. Kokubo, A. Morbidelli, R. Morishima, K.J. Walsh, Terrestrial planet formation at home and abroad, in *Protostars and Planets VI*, ed. by B. Beuther, R.S. Klessen, C.P. Dullemond, T. Henning (University of Arizona Press, Tucson, 2014), pp. 595–618

S.N. Raymond, A. Izidoro, B. Bitsch, S.A. Jacobson, Did Jupiter's core form in the innermost parts of the Sun's protoplanetary disc? Mon. Not. R. Astron. Soc. **458**, 2962–2972 (2016). https://doi.org/10.1093/mnras/stw431

A. Reufer, M.M.M. Meier, W. Benz, R. Wieler, A hit-and-run giant impact scenario. Icarus **221**, 296–299 (2012). https://doi.org/10.1016/j.icarus.2012.07.021

K. Righter, M.J. Drake, Effect of water on metal-silicate partitioning of siderophile elements: a high pressure and temperature terrestrial magma ocean and core formation. Earth Planet. Sci. Lett. **171**, 383–399 (1999). https://doi.org/10.1016/S0012-821X(99)00156-9

K. Righter, D.P. O'Brien, Cosmochemistry Special feature: terrestrial planet formation. Proc. Natl. Acad. Sci. **108**, 19165–19170 (2011). https://doi.org/10.1073/pnas.1013480108

A.S. Rivkin, J.P. Emery, Detection of ice and organics on an asteroidal surface. Nature **464**, 1322–1323 (2010). https://doi.org/10.1038/nature09028

F. Robert, The D/H ratio in chondrites. Space Sci. Rev. **106**, 87–101 (2003). https://doi.org/10.1023/A:1024629402715

F. Robert, Solar system deuterium/hydrogen ratio, in *Meteorites and the Early Solar System II*, ed. by D.S. Lauretta, H.Y. McSween (University of Arizona Press, Tucson, 2006), pp. 341–351

D.C. Rubie, C.K. Gessmann, D.J. Frost, Partitioning of oxygen during core formation on the Earth and Mars. Nature **429**, 58–61 (2004). https://doi.org/10.1038/nature02473

D.C. Rubie, D.J. Frost, U. Mann, Y. Asahara, F. Nimmo, K. Tsuno, P. Kegler, A. Holzheid, H. Palme, Heterogeneous accretion, composition and core-mantle differentiation of the Earth. Earth Planet. Sci. Lett. **301**, 31–42 (2011). https://doi.org/10.1016/j.epsl.2010.11.030

D.C. Rubie, F. Nimmo, H.J. Melosh, Formation of the Earth's core, in *Treatise on Geophysics*, vol. 9, ed. by G. Schubert (Elsevier, Amsterdam, 2015a), pp. 43–79

D.C. Rubie, S.A. Jacobson, A. Morbidelli, D.P. O'Brien, E.D. Young, J. de Vries, F. Nimmo, H. Palme, D.J. Frost, Accretion and differentiation of the terrestrial planets with implications for the compositions of early-formed Solar System bodies and accretion of water. Icarus **248**, 89–108 (2015b). https://doi.org/10.1016/j.icarus.2014.10.015

J. Siebert, J. Badro, D. Antonangeli, F.J. Ryerson, Terrestrial accretion under oxidizing conditions. Science **339**(6124), 1194–1197 (2013). https://doi.org/10.1126/science.1227923

M. Stimpfl, A.M. Walker, M.J. Drake, N.H. de Leeuw, P. Deymier, An ångström-sized window on the origin of water in the inner solar system: atomistic simulation of adsorption of water on olivine. J. Cryst. Growth **294**, 83–95 (2006). https://doi.org/10.1016/j.jcrysgro.2006.05.057

C. Surville, L. Mayer, D.N.C. Lin, Dust capture and long-lived density enhancements triggered by vortices in 2D protoplanetary disks. Astrophys. J. **831**, 82 (2016). https://doi.org/10.3847/0004-637X/831/1/82

F. Tera, D.A. Papanastassiou, G.J. Wasserburg, Isotopic evidence for a terminal lunar cataclysm. Earth Planet. Sci. Lett. **22**, 1–21 (1974). https://doi.org/10.1016/0012-821X(74)90059-4

E. Thommes, M. Nagasawa, D.N.C. Lin, Dynamical shake-up of planetary systems. II. N-body simulations of solar system terrestrial planet formation induced by secular resonance sweeping. Astrophys. J. **676**, 728–739 (2008). https://doi.org/10.1086/526408

M. Touboul, T. Kleine, B. Bourdon, H. Palme, R. Wieler, Late formation and prolonged differentiation of the Moon inferred from W isotopes in lunar metals. Nature **450**, 1206–1209 (2007). https://doi.org/10.1038/nature06428

K. Tsiganis, R. Gomes, A. Morbidelli, H.F. Levison, Origin of the orbital architecture of the giant planets of the Solar System. Nature **435**, 459–461 (2005). https://doi.org/10.1038/nature03539

K.J. Walsh, A. Morbidelli, S.N. Raymond, D.P. O'Brien, A.M. Mandell, A low mass for Mars from Jupiter's early gas-driven migration. Nature **475**, 206–209 (2011). https://doi.org/10.1038/nature10201

W.R. Ward, Density waves in the solar nebula—differential Lindblad torque. Icarus **67**, 164–180 (1986). https://doi.org/10.1016/0019-1035(86)90182-X

W.R. Ward, On planetesimal formation: the role of collective particle behavior, in *Origin of the Earth and Moon*, ed. by R.M. Canup, K. Righter (University of Arizona Press, Tucson, 2000), pp. 75–84

S.J. Weidenschilling, Aerodynamics of solid bodies in the solar nebula. Mon. Not. R. Astron. Soc. **180**, 57–70 (1977). https://doi.org/10.1093/mnras/180.2.57

S.J. Weidenschilling, Dust to planetesimals—settling and coagulation in the solar nebula. Icarus **44**, 172–189 (1980). https://doi.org/10.1016/0019-1035(80)90064-0

S.J. Weidenschilling, Radial drift of particles in the solar nebula: implications for planetesimal formation. Icarus **165**, 438–442 (2003). https://doi.org/10.1016/S0019-1035(03)00169-6

S.J. Weidenschilling, Initial sizes of planetesimals and accretion of the asteroids. Icarus **214**, 671–684 (2011). https://doi.org/10.1016/j.icarus.2011.05.024

S.J. Weidenschilling, J.N. Cuzzi, Formation of planetesimals in the solar nebula, in *Protostars and Planets III*, ed. by E.H. Levy, J.I. Lunine (University of Arizona Press, Tucson, 1993), pp. 1031–1060

S.J. Weidenschilling, D. Spaute, D.R. Davis, F. Marzari, K. Ohtsuki, Accretional evolution of a planetesimal swarm. Icarus **128**, 429–455 (1997). https://doi.org/10.1006/icar.1997.5747

G.W. Wetherill, Radiometric chronology of the early solar system. Annu. Rev. Nucl. Part. Sci. **25**, 283–328 (1975). https://doi.org/10.1146/annurev.ns.25.120175.001435

G.W. Wetherill, Why isn't Mars as big as Earth? in *Lunar and Planetary Science Conference Abstracts*, vol. 22 (1991), p. 1495

G.W. Wetherill, An alternative model for the formation of the asteroids. Icarus **100**, 307–325 (1992). https://doi.org/10.1016/0019-1035(92)90103-E

G.W. Wetherill, G.R. Stewart, Accumulation of a swarm of small planetesimals. Icarus **77**, 330–357 (1989). https://doi.org/10.1016/0019-1035(89)90093-6

G.W. Wetherill, G.R. Stewart, Formation of planetary embryos—effects of fragmentation, low relative velocity, and independent variation of eccentricity and inclination. Icarus **106**, 190 (1993). https://doi.org/10.1006/icar.1993.1166

B.J. Wood, J. Wade, M.R. Kilburn, Core formation and the oxidation state of the Earth: additional constraints from Nb, V and Cr partitioning. Geochim. Cosmochim. Acta **72**, 1415–1426 (2008). https://doi.org/10.1016/j.gca.2007.11.036

G. Wurm, J. Blum, J.E. Colwell, NOTE: a new mechanism relevant to the formation of planetesimals in the solar nebula. Icarus **151**, 318–321 (2001). https://doi.org/10.1006/icar.2001.6620

A.N. Youdin, J. Goodman, Streaming instabilities in protoplanetary disks. Astrophys. J. **620**, 459–469 (2005). https://doi.org/10.1086/426895

A.N. Youdin, F.H. Shu, Planetesimal formation by gravitational instability. Astrophys. J. **580**, 494–505 (2002). https://doi.org/10.1086/343109

H. Zhang, J.L. Zhou, On the orbital evolution of a giant planet pair embedded in a gaseous disk. II. A Saturn-Jupiter configuration. Astrophys. J. **719**, 671–684 (2010). https://doi.org/10.1088/0004-637X/719/1/671

Space Sci Rev (2018) 214:76
https://doi.org/10.1007/s11214-018-0508-3

Water Partitioning in Planetary Embryos and Protoplanets with Magma Oceans

M. Ikoma[1,2] · L. Elkins-Tanton[3] · K. Hamano[4] · J. Suckale[5]

Received: 14 October 2017 / Accepted: 27 April 2018 / Published online: 8 May 2018

Abstract The water content of magma oceans is widely accepted as a key factor that determines whether a terrestrial planet is habitable. Water ocean mass is determined as a result not only of water delivery and loss, but also of water partitioning among several reservoirs. Here we review our current understanding of water partitioning among the atmosphere, magma ocean, and solid mantle of accreting planetary embryos and protoplanets just after giant collisions. Magma oceans are readily formed in planetary embryos and protoplanets in their accretion phase. Significant amounts of water are partitioned into magma oceans, provided the planetary building blocks are water-rich enough. Particularly important but still quite uncertain issues are how much water the planetary building blocks contain initially and how water goes out of the solidifying mantle and is finally degassed to the atmosphere. Constraints from both solar-system explorations and exoplanet observations and also from laboratory experiments are needed to resolve these issues.

Keywords Magma ocean · Water · Protoplanet · Habitable planet

1 Introduction

Water plays crucial roles in the climate, plate tectonics, mantle dynamics, and thermal evolution of terrestrial planets. This is partly because water vapor is a strong greenhouse gas.

The Delivery of Water to Protoplanets, Planets and Satellites
Edited by Michel Blanc, Alessandro Morbidelli, Yann Alibert, Lindy Elkins-Tanton, Paul Estrada, Keiko Hamano, Helmut Lammer, Sean Raymond and Maria Schönbächler

✉ M. Ikoma
ikoma@eps.s.u-tokyo.ac.jp

1 Department of Earth and Planetary Science, The University of Tokyo, Tokyo, Japan

2 Research Center for the Early Universe (RESCEU), The University of Tokyo, Tokyo, Japan

3 School of Earth and Space Exploration, Arizona State University, Tempe, AZ, USA

4 Earth-Life Science Institute, Tokyo Institute of Technology, Tokyo, Japan

5 Department of Geophysics, Stanford University, Stanford, CA, USA

Also, water makes a great contribution to control the CO_2 partial pressure and thus the surface temperature of the Earth through continental weathering, which occurs by reactions between water and silicate/carbonate (Walker et al. 1981). Furthermore, on the Earth, water is known to influence the lithospheric density and viscosity, which likely leads to plate tectonic (e.g., Korenaga 2013). Likewise, water affects the rheology of mantle materials and thus the thermal evolution of terrestrial planets.

The amount of water in the hydrosphere is thought to be a key factor that determines the surface environment (and habitability) of terrestrial planets. The total mass of the Earth's oceans accounts for only 0.023 wt% of the Earth's total mass. Such a small amount of water allows continents to be present on the Earth. If seawater were about three times more abundant than at present (e.g., Maruyama et al. 2013), there would be no continents. If so, continental weathering would hardly work and, consequently, the Earth's climate would be much different from the present (Abbot et al. 2012). Also, the evolution and lifetime of habitable planets is known to differ greatly, depending on the amount of seawater (Abe et al. 2011; Kodama et al. 2015).

Of great importance for understanding the formation, evolution, and habitability of terrestrial planets are, hence, not only water delivery and loss, but also water partitioning among the reservoirs such as the atmosphere/oceans, crust, mantle, and core. This is the main focus of this chapter. Water budget of the Earth is also discussed in Chap. 4. Delivery and loss processes of water are mainly discussed in Chap. 10 and in Chaps. 13 and 14, respectively. However, one can never consider water partitioning separately from the delivery (or accretion) and loss processes, because all those processes sometimes occur concurrently. Water moves into and from molten portions of a mantle, namely, magma oceans. Magma oceans are formed because of the large amount of energy brought by accretion of solids. Also, magma oceans are maintained by the strong blanketing effect of atmospheres, which accrete simultaneously with the growth of planetary embryos and are lost by intense stellar irradiation.

The key questions that we address in this chapter are, thus,

(1) how magma oceans form and solidify and how deep are they?
(2) how much water is dissolved from atmospheres into magma oceans?
(3) how is water transported in the interior of solidifying magma oceans and how the transported water goes out to atmospheres?
(4) how much water is finally retained in the hydrosphere and interior of planetary embryos and planets?

Note that partitioning of hydrogen and oxygen into the core is potentially of considerable importance (e.g., Rubie et al. 2004, 2015a). However, while hydrogen is inferred to be contained in the core (e.g., Nomura et al. 2014), it still remains quite uncertain, because partitioning of hydrogen into liquid Fe is extremely difficult to study experimentally (e.g., Okuchi 1997). Detailed discussion about water partitioning into the core is beyond the scope of this chapter.

The rest of this chapter is organized as follows. We first outline theories regarding the growth of planetary embryos and the formation and depth of magma oceans of different types in Sect. 2. In Sect. 3, we discuss the accumulation of vapor atmospheres and water partitioning between the atmosphere and magma ocean in accreting planetary embryos. In Sect. 4, we discuss the degassing of water in the solidifying mantles and the secondary formation of atmospheres on protoplanets after giant collisions. Finally we present conclusions of this chapter in Sect. 5.

2 Magma Oceans on Growing Planetary Embryos

2.1 Growth of Planetary Embryos

Planetary embryos are thought to experience a great number of collisions with objects of different sizes, ranging from 1 cm to 1000 km, in the accretion stages. First there would be no doubt that planet formation occurs in a disk-shaped gas that surrounds the central proto-star, which is in general called a protoplanetary disk or the solar nebula (or sometimes the proto-solar disk) in the case of the solar system. Observation of young stellar clusters indicates that protoplanetary disks contain dust particles of μm size. Such small particles are dynamically coupled with the ambient disk gas and, thus, are unable to accrete directly to form planets. Because of their Brownian motion, dust particles meet each other and stick together to grow to a certain size. They then start accreting by mutual collisions after being slightly decoupled from the ambient gas. The size of the initial building blocks for solid planets, however, remains a matter of debate.

Classical theories for the formation of the solar system assume that the initial building blocks, which are called *planetesimals*, are about 10 km in size (corresponding to about 10^{15} kg) around the Earth's heliocentric distance. This is based on the gravitational instability model for planetesimal formation (Safronov 1972; Hayashi 1972; Goldreich and Ward 1973), where vertical settling of dust particles forms a thin layer with high solid concentration, followed by gravitational fragmentation of the layer. The thickness of the layer determines the typical planetesimal size. Several mechanisms are, however, known to hinder the formation of such thin dust layers in protoplanetary disks. Instead of vertical settling, radial motion of dust particles may result in areas with a high dust concentration through turbulent concentration and/or streaming instability (see Johansen et al. 2014, for a recent review).

Current understanding of planetary accretion from planetesimals is as follows. First, planetesimals grow in a relatively orderly fashion through mutual collisions. When planetesimal mass is roughly 10^{20} kg, which is similar to the mass of Vesta, gravitational focusing becomes so effective that the largest planetesimal starts to grow much faster than the other smaller ones in a runaway fashion (Wetherill and Stewart 1989; Kokubo and Ida 1996). In this runaway growth phase, a bimodal size distribution of solids develops, consisting of a small number of large bodies, which are called planetary embryos, and a large number of small planetesimals. In this phase, planetary embryos experience successive collisions with small planetesimals. The runaway growth is, however, known to end relatively soon, because planetary embryos stir the surrounding planetesimals by gravitational scattering and enhance their random velocities, thus lowering the gravitational focusing effect (Ida and Makino 1993). Subsequently, several large embryos grow in an orderly fashion. This is called the oligarchic growth stage (Lissauer 1987; Kokubo and Ida 1998). Those embryos finish their growth after capturing almost all the planetesimals in their own feeding zones with a width of about 10 Hill radii. The final mass of those embryos, which is called the isolation mass, depends on the initial solid surface density. In the minimum mass solar nebula (Hayashi 1981, MMSN, hereafter), the isolation mass is almost equal to the Mars mass. Further growth requires collisions between those isolated embryos, which are often called giant collisions. This means that the Earth and Venus likely experienced several giant collisions. In the case of the Earth, the final collision is believed to be the moon-forming one (e.g., Hartmann and Davis 1975; Benz et al. 1986; Ida et al. 1997).

The timing of disk gas dispersal has a great influence on the volatile contents, magma ocean formation, and redox state of planetary embryos and planets (see Sect. 3). Also, it affects the dynamics and migration of planetary embryos. Unfortunately, we have no direct evidence for the timing of the solar nebula dispersal. Instead, astronomical observations of young stellar clusters indicate that protoplanetary disks dissipate in $\sim$5–10 million years (e.g. Haisch et al. 2001). On the other hand, according to recent Hf-W chronology, the formation ages of Mars and the Earth are about a few million (Dauphas and Pourmand 2011) and $>$ 30 million years (Halliday 2014), respectively. Thus, Earth formation is likely to have been completed well after solar nebula dispersal, whereas Mars may have formed while the solar nebula still existed. Note that some isotopic ratios, such as D/H and ^{18}Ne/^{20}Ne, of materials from the lower mantle of the Earth suggest that the Earth gained the nebular components at some point during its formation (e.g., Hallis et al. 2015). Thus, although not conclusive, a possible scenario in the framework of the above classical formation model is that the nebular gas existed until Mars-size protoplanets were formed and its dissipation took place during the giant collision stage.

Recently, there has been a movement to make a drastic re-examination of planet formation theories based on planetesimal accretion. This is motivated by the well-known theoretical difficulty in forming gas giants before disk gas dispersal. Within the classical framework based on planetesimal accretion, in a protoplanetary disk with a normal surface density like MMSN, the in-situ core growth proceeds so slowly at $>$ 5 AU that runaway gas accretion never occurs before disk gas dispersal (e.g., Pollack et al. 1996; Ikoma et al. 2000). A promising way to hasten the core growth is to consider much smaller objects of $\sim$ 1–10 cm in size, which are often termed pebbles. This is because such objects are subject to gas drag, which keeps their velocity dispersion low and enlarges their cross section for collision with proto-cores surrounded by envelopes (Inaba et al. 2003; Inaba and Ikoma 2003; Ormel and Klahr 2010). Indeed, recent planet formation theories in which pebbles are assumed to accrete to earlier-formed planetary embryos have demonstrated successfully that cores of critical mass are formed well before disk dispersal (Lambrechts and Johansen 2012, 2014). In such theories, radial drift of pebbles plays an important role in supplying materials. This could have a great influence also on terrestrial planet formation and water delivery (Morbidelli et al. 2015). The pebble accretion paradigm is, however, still developing. There is no detailed quantitative argument regarding incorporation and partitioning of water in planetary embryos and planets in the framework of the pebble accretion paradigm. Thus, our discussion below is based only on planetesimal accretion, unless otherwise stated.

2.2 Origins of Magma Oceans on Planetesimals and Planets

Three heat sources may dominate the period of planetary accretion and solidification (e.g., Elkins-Tanton 2012; Rubie et al. 2015a):

(1) *Impact*—The heat from accretionary impacts primarily affects the near-surface and surface of the planet. The energy involved in impacts between planetesimals is likely insufficient to significantly melt those small bodies; impact-induced melting becomes more important in planetary embryos, and dominant in planets (e.g., Davison et al. 2010).

(2) *Differentiation*—Metal-silicate differentiation may occur at mid-mantle depths in a magma ocean due to local differentiation between heavy metal and light silicates (Sasaki and Nakazawa 1986; Tonks and Melosh 1992) or closer to the core in larger impacts or as the metal phase gradually sinks towards the center of the impacted protoplanet (Monteux et al. 2009). In all the cases the potential energy conversion to heat likely primarily affects the deep mantle and core. Again this heat source is likely important for melting only in planetary embryos and planets.

(3) *Radiogenic decay*—Heat from radiogenic elements may be evenly dispersed unless fractional solidification of silicate liquids concentrates them. Following fractional solidification of a silicate mantle on a planet of Mars-sized or smaller, the incompatible elements, including almost all the radiogenic elements, will be concentrated in the last dregs of liquid near the surface of the planet. This material is compositionally dense and likely to sink to the bottom of the mantle, possibly leaving small pockets behind near the surface of the planet. On a larger planet, both a basal and an upper magma ocean may fractionate separately, though with the same outcome that the incompatible elements will be concentrated at the bottom in the end (Labrosse et al. 2007; Brown et al. 2014). On large planets, therefore, heating may occur simultaneously from above and below. In planetesimals, melting is almost certainly the near-exclusive domain of the short-lived radioisotope ^{26}Al, since the small mass of these bodies precludes significant melting from core formation or impacts.

Different heat sources may translate into different types of magma oceans, from (1) magma ponding at the surface of a largely solid planet, to (2) a basal magma ocean overlain by solids, to (3) a whole-mantle magma ocean. Magma oceans on planetary surfaces are mostly created by impactors. The shock pressure is nearly uniform in a spherical region around the impact, the isobaric core, and decays rapidly with distance (Croft 1982; Pierazzo et al. 1997). In the isobaric core, the kinetic energy of the impact increases the temperature by several hundred degrees depending mostly on the properties of the impactor (Monteux et al. 2007). Surface magma oceans of different sizes are likely to have existed on the early Earth at different stages of its evolution due to the varying energies of accretionary collisions (Chambers 2013) and the late heavy bombardment (Chou 1978).

A basal magma ocean could form in one of two ways: First, it could have resulted from a density inversion. The compositionally dense material remaining at the end of fractional solidification could either sink as a liquid or as a solid that melts through the heat of long-lived radiogenic isotopes (Brown et al. 2014). This possibility was explored in detail for the early Earth by Labrosse et al. (2007), who concluded that fractional solidification was important for producing deep, dense material that still might exist today. Second, a basal magma ocean may be created by the potential energy of core formation. For the early Earth, the release of gravitational energy during core formation may induce melting of the whole planet (Ricard et al. 2009), but the degree of melting depends on the mechanism of metal descent, which might range from complete core merger during a giant accretionary impact, to emulsification and slow sinking of the core of a small planetesimal, to a rain of tiny metal droplets from an undifferentiated impactor (Golabek et al. 2008; Samuel and Tackley 2008).

Probably the main mechanism for generating a magma ocean from the entire silicate fraction of a planet or protoplanet is through giant collisions (Tonks and Melosh 1993; Canup 2008; Nakajima and Stevenson 2015). While giant collisions provide enough heat for whole-mantle melting, hemispheric or local magma oceans are also possible. Tonks and Melosh (1993) showed that if the isostatic rebound time is less than the freezing time of the magma, then the return of the planet's center of mass to the center of the body would result in extrusion of melt over the planet's surface as discussed in more detail by Reese and Solomatov (2006).

3 Partitioning in Accreting Planetary Embryos

During the main accretion phase of planetary embryos, a large amount of energy is brought into their interiors through accretionary impacts of planetesimals, as described above. If all the accretion energy is converted to the thermal energy without loss, the planetary embryo has an averaged interior temperature of the order of 10^4 K, which is high enough that the planetary embryo is totally molten. However, since the cooling timescale is much shorter than the typical growth timescale of planetary embryos, the magma ocean is solidified immediately after formation. For magma oceans to be maintained, the presence of atmospheres is needed to delay the loss of energy from the planetary embryos.

Unless planetesimals are free of volatiles and grow in a vacuum, planetary embryos readily gain volatiles and have atmospheres, provided they become massive enough to bind the atmospheres gravitationally. The gravitational potential energy of an embryo of mass M_e is $GM_e/R_e = G(4\pi\bar{\rho}_e/3)^{1/3}M_e^{2/3}$, where G is the gravitational constant and R_e and $\bar{\rho}_e$ are the radius and mean density of the embryo, respectively, while the specific thermal energy of gas is $\sim kT/(\mu m_{\rm H})$, where k is the Boltzmann constant, T is the temperature, μ is the mean molecular weight, and $m_{\rm H}$ is the mass of a hydrogen atom. The condition is, thus, given with respect to embryo mass as $M_e >\sim 6 \times 10^{22}\mu^{-3/2}$ kg for a gas temperature of 300 K and a planetary mean density of 3500 kg/m^3. This indicates that an H_2O-vapor atmosphere is gravitationally bound on a planetary embryo of $\sim 8 \times 10^{20}$ kg, which is less massive than the moon. Even for an H_2 atmosphere, the lunar mass is large enough to attract it gravitationally, from the energetic point of view. The reason that the moon has no atmosphere today would be that it was subject to atmospheric erosion processes such as the solar wind-induced one.

Planetesimals, if containing water, release some fraction of their internal water upon impact, which results in forming a vapor atmosphere on an accreting planetary embryo in the following way. Impact-induced shockwaves propagate in the interior of the planetesimals. The shock heating is known experimentally to cause the internal water to evaporate off the planetesimals (e.g., Lange and Ahrens 1982). Since the impact velocity is almost the same as the escape velocity of the planetary embryo, it becomes high with increasing planetary embryo mass. When the embryo mass exceeds approximately the lunar mass, impact degassing starts to occur (Abe and Matsui 1985). This type of atmosphere is termed an impact-generated steam atmosphere.

Since water vapor absorbs infra-red radiation well, the steam atmosphere traps the accretion energy of planetesimals and exerts a strong thermal blanketing effect. According to numerical models that simulate the accumulation and energy budget of the atmosphere and mantle simultaneously (Abe and Matsui 1985, 1986), the impact-generated steam atmosphere on a Mars-mass embryo is thick enough to provide insulation that enables heat sources to cause mantle melting and magma ocean formation, as shown in Fig. 1. Provided the planetesimal accretion rate is high enough (see the solid and long-dashed lines in Fig. 1a), the surface temperature and atmospheric vapor mass are found not to increase monotonically, but to level off until just before the end of growth. As realized by Abe and Matsui (1986), this coupled system of the accumulating atmosphere and molten mantle is a self-stabilizing system. This is because water is well mixed with molten silicate (i.e., magma). An increase of atmospheric vapor enhances the greenhouse effect, which leads to extending the molten area of the surface. This enhances dissolution of water into the magma ocean and thus decreases atmospheric vapor, leading to a reduction of the surface temperature. Consequently, because of this negative feedback loop, the surface temperature remains almost constant around the melting temperature of silicate ($\sim$ 1500 K). This means that accreting embryos are always covered with magma oceans during the main accretion phase, provided the accretion rate is high enough.

Likewise, the mass of the steam atmosphere M_a remains almost constant after the embryo mass exceeds 0.01–0.1 Earth masses, depending on the planetesimal accretion rate $\dot{M}_e$,

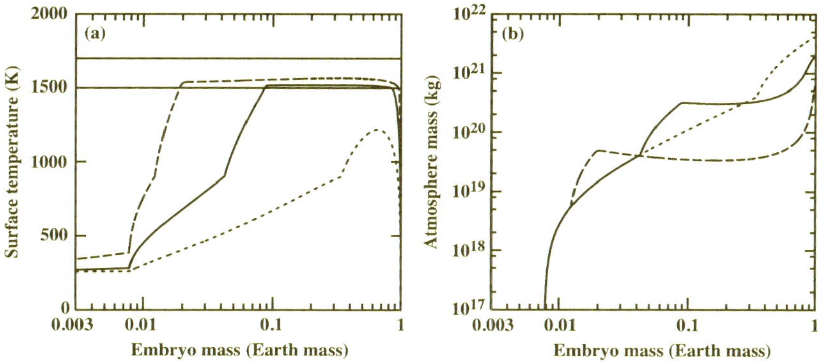

Fig. 1 Growth of an impact-generated steam atmosphere. (**a**) The surface temperature and (**b**) atmospheric vapor mass of a growing planetary embryo are plotted as a function of planetary embryo mass. The calculations have been done in a similar way to those of Abe and Matsui (1986). (Note that we have ignored the heat flux from the interior.) The solid line represents the growth with the fiducial value of planetesimal accretion rate adopted by Abe and Matsui (1986), at which the embryo grows to the Earth mass in 50 million years. The long- and short-dashed lines represent the cases with 10 and 0.1 times the fiducial value, respectively. The lower and upper horizontal solid lines in panel (**a**) indicate the assumed solidus and liquidus of silicate

as shown in Fig. 1b. This is interpreted as follows: In a radiative atmosphere, the surface temperature T_s is related to the energy flux F and the optical thickness τ as $T_s^4 \propto F\tau$. In the present case, the energy flux results from planetesimal accretion and is written as $F \simeq GM_e\dot{M}_e/4\pi R_e^3$. Also, because the opacity κ is assumed to be constant, the optical thickness is related to the pressure P such that $\tau \simeq \kappa P/g \simeq \kappa M_a/4\pi R_e^2$. Therefore, the surface temperature is related to the embryo mass, atmospheric mass, and planetesimal accretion timescale t_{acc} ($\equiv M_e/\dot{M}_e$) as

$$\sigma T_s^4 \simeq \frac{GM_e\dot{M}}{4\pi R_e^3} \cdot \frac{\kappa M_a}{4\pi R_e^2} \propto \frac{M_e^{1/3} M_a}{t_{\mathrm{acc}}}, \tag{1}$$

or

$$M_a \propto \frac{t_{\mathrm{acc}}}{M_e^{1/3}} T_s^4. \tag{2}$$

Since the surface temperature is fixed to the melting temperature of silicate, as found in Fig. 1a, the atmospheric mass is determined by the accretion timescale and the embryo mass. In the ordinary growth regime that Abe and Matsui (1986) assumed, $t_{\mathrm{acc}} \propto M_e^{1/3}$ (e.g., Safronov 1972), so that M_a is independent of M_e and constant through the growth. In the runaway growth regime, since $t_{\mathrm{acc}} \propto M_e^{-1/3}$ (Wetherill and Stewart 1989), the atmospheric mass decreases with growth in such a way that $M_a \propto M_e^{-2/3}$. In any case, from the viewpoint of water partitioning, of particular importance would be the fact that all the accumulated water other than the atmospheric vapor is incorporated in the magma ocean and the core.

Hence, the water contents of planetesimals affect the amount of water that accreting planetary embryos finally contain in their interior, since the mass of the steam atmosphere is limited to more or less the mass of the current Earth's oceans ($\simeq 0.02\%$ of the Earth mass), according to the self-stabilizing mechanism described above. A systematic investigation of the impact-generated steam atmosphere was done by Zahnle et al. (1988), who modeled the formation and growth of the atmosphere in a similar way to Abe and Matsui (1986), but treated the atmospheric blanketing effect in more detail and included interchange of water and energy between the surface and interior, and hydrogen escape by the solar extreme ultraviolet ray (EUV) and by impact erosion. They confirmed that a wide range of the model parameters yield solutions of the Abe and Matsui (1986)-type atmosphere, which are favored by moderate ($\sim 0.1\%$) water content in the planetesimals. However, if the incoming planetesimals were too dry or the EUV flux too high, very little water would accumulate at the surface, which means the magma ocean is rather dry.

If embedded in a protoplanetary disk, planetary embryos attract gravitationally hydrogen and helium from the disk. This type of atmosphere is often termed the primordial atmosphere. The primordial atmosphere is also known to affect the water and energy budgets of planetary embryos. As shown by numerical models of the structure of the primordial atmosphere of an embedded planetary embryo (Hayashi et al. 1979; Ikoma and Genda 2006), the atmospheric blanketing effect is strong enough for a magma ocean to be generated, if the embryo mass is larger than ~ 0.3 Earth masses. This is a rather robust conclusion, because the surface temperature, which is given by

$$T_s \approx \frac{G M_e \mu m_{\rm H}}{4 R_e k}, \tag{3}$$

is independent of the opacity and energy flux (or accretion rate). In contrast to atmospheres of embryos in a vacuum, changes in those quantities are compensated by that in the atmospheric mass (i.e., the total optical thickness) in the case of the atmosphere connected to the surrounding gaseous disk. For example, if the planetesimal accretion rate decreases for some reason, the atmosphere contracts gravitationally, which leads to further accretion of gas. Analytically, the atmospheric mass is given by

$$M_a \approx \frac{\pi^2 \sigma}{3 \kappa L} \left(\frac{G M_e \mu m_{\rm H}}{k} \right)^4 \ln \left(\frac{R_{\rm B}}{R_e} \right) \propto \dot{M}_e^{-1} M_e^{10/3} \mu^4, \tag{4}$$

where L is the protoplanetary luminosity, which is given approximately by $G M_e \dot{M}_e / R_e$ in the accretion phase, σ is the Stefan-Boltzman constant and $R_{\rm B}$ is the Bondi radius.

As indicated by Eqs. (3) and (4), the surface temperature and atmospheric mass depend strongly on the molecular weight μ. The primordial atmosphere is usually assumed to have solar composition; namely, the atmosphere consists mainly of hydrogen and helium and contains only $\sim 0.1\%$ H_2O. However, since oxides are available at the embryo surface, H_2O can be produced in the atmosphere through reactions with the atmospheric hydrogen. As discussed in Ikoma and Genda (2006), for the H_2O production process to work efficiently, magma oceans should be maintained. In the case of the solar-composition atmosphere, the embryo mass must be larger than $\sim 0.3 M_\oplus$. The amount of H_2O thus produced depends on the oxidation state of the mantle. For example, the iron-wüstite (IW) buffer (1.894Fe + $O_2 \leftrightarrow 2\ Fe_{0.974}O$) has the ability of producing H_2O comparable in amount with H_2 (Robie et al. 1978).

Figure 2 shows the mass of the solar abundance atmosphere (left panel) and the 50% water and 50% H/He mixed atmosphere (right panel) as functions of planetesimal accretion rate. Here the atmospheric structure has been integrated in a similar way to that of Venturini et al. (2015). Comparing the two panels, one realizes that the water fraction has a great impact on the atmospheric mass. While the water fraction in the enriched atmosphere (panel b) is larger by a factor of $\sim$500 than that in the solar-abundance atmosphere (panel a), the difference in vapor mass is larger by about four orders of magnitude. In the case of $\dot{M}_e = 1 \times 10^{-7}\ M_\oplus$/yr, for example, the atmospheric H_2O is comparable in mass with the current Earth's oceans even for a Mars-mass embryo.

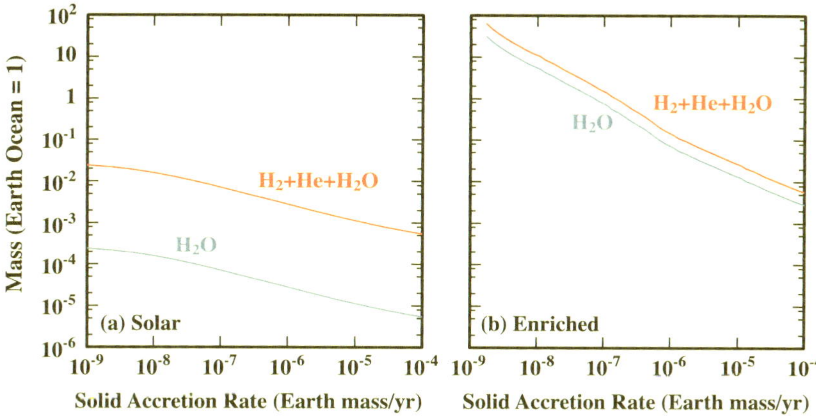

Fig. 2 Mass of the primordial atmosphere on a Mars-mass embryo embedded in a protoplanetary disk, which is plotted as a function of planetesimal accretion rate. In panel (**a**), the atmospheric composition is assumed to be solar. In panel (**b**), the atmosphere is assumed to consist of 50% H_2O and 50% H_2 + He

The thermal state of the mantle (e.g., the depth of the magma ocean) and the partitioning of the produced water between the atmosphere and magma ocean are yet to be investigated in the case of the embedded atmospheres, while those have been investigated in studies of impact-induced melting (e.g., de Vries et al. 2016) and core formation (e.g., Rubie et al. 2015a,b). As inferred from Eq. (3), the surface temperature is more sensitive to the molecular weight of the atmospheric gas than the embryo mass. Thus, it is likely that the surface temperature of a Mars-mass embryo becomes higher than the melting temperature of rock, provided the atmosphere becomes enriched in water. The initial nebula-captured atmosphere of a Mars-mass embryo is not massive enough. However, if water is added to the atmosphere for some reason (for example, impact and evaporation of relatively wet materials), the atmosphere readily becomes massive, according to Eq. (4). Then, the surface temperature may exceed the melting temperature of rock, generating a magma ocean, as found from Eq. (3). This means the H_2O production mechanism starts functioning. Consequently, the captured atmosphere becomes quite massive even on a Mars-mass embryo. Since this atmosphere contains a large amount of water, part of which is dissolved into the magma ocean, the mantle of planetary embryos may be much wetter than previously thought.

4 Partitioning in Protoplanets

The solidification and degassing processes of magma oceans set the initial interior and atmospheric volatile budgets for terrestrial planets. Since magma oceans do not degas perfectly, they provide a major mechanism for adding a small but potentially critical volatile fraction to planetary mantles; this volatile fraction will lower the melting temperature and viscosity of the mantle, encouraging the onset of thermal convection and volcanism. The interior and exterior volatile budgets are set by the degassing processes of the solidifying magma ocean.

Current models of magma ocean solidification mostly assume that volatiles degas whenever they are in excess of saturation, and demonstrate that even with nominal initial volatile concentrations, thick steam atmospheres are likely built above solidifying magma oceans, as discussed in the previous section (Abe and Matsui 1985, 1986; Zahnle et al. 1988; Elkins-Tanton 2008; Lebrun et al. 2013). Alternatively, degassing might be delayed if bubble nucleation requires a finite supersaturation pressure, as observed in the terrestrial context (Bottinga and Javoy 1990; Lensky et al. 2006; Hardiagon et al. 2013). Delayed and sudden degassing would imply supersaturated magma-ocean liquids that produce more volatile-rich mantles and comparatively late atmospheric formation. A later-forming atmosphere means faster early solidification through higher heat flux, unoccluded by a dense atmosphere. This process may also allow planets relatively closer to their stars to retain more volatiles in their interiors. Since stellar EUV radiation rapidly declines according to a power law in time, the later-forming atmosphere might avoid atmospheric escape.

4.1 Magma Ocean Solidification

A cooling protoplanet has four potential water reservoirs: an atmosphere, a magma ocean, a solid mantle, and a metallic core (Elkins-Tanton 2012). As magma ocean solidification proceeds, the water that terrestrial planets acquired during formation is partitioned among these reservoirs (e.g., Zahnle et al. 2007; Elkins-Tanton 2008). Out of these, the solid mantle is the least dynamic. Some of the water dissolved in the magma ocean can be incorporated into the solid mantle at the solidification front, but unless remelting occurs the water stored in the solid mantle has a minor influence on the evolution of the magma ocean and the atmosphere, which, on the other hand, are tightly coupled.

The liquid magma ocean is expected to contain some fraction of water (or hydroxyl) along with carbon and carbon compounds. Water, and almost all trace elements and volatiles, have extremely low solid/liquid partition coefficients and hence behave as incompatible elements on crystallization. Upon crystallization, water would thus be exsolved from solidifying cumulates and accumulate in the residual magma ocean liquids. As solidification proceeds, the magma ocean liquids become increasingly enriched in volatiles. When circulating at shallow depth, the volatiles degas into the growing atmosphere.

At pressures and temperatures of magma ocean crystallization, no hydrous or carbonate minerals will crystallize (Ohtani et al. 2004; Wyllie and Ryabchikov 2000), although reducing conditions may stabilize graphite (Hirschmann and Withers 2008). Nonetheless, water and carbon dioxide will enter solidifying minerals in very small quantities. Even the small fractions of water or hydroxyl that are incorporated into the defects and voids of nominally anhydrous minerals can produce significant reductions in melting temperature and viscosity, and thus they can greatly influence the volcanic and dynamic evolution of planets. Olivine and orthopyroxene, for example, can contain in excess of 1000–1500 ppm of hydroxyl (e.g., Hauri et al. 2006). In general water or hydroxyl is partitioned into nominally anhydrous minerals at about 1:100 in relation to the water content of the magma, but exceptions such as ringwoodite exist (see Brown et al. 2014; Elkins-Tanton 2008, supplementary material for compilations of mineral-melt partition coefficients).

Several groups have evaluated the storage rate of water, by considering the solid/liquid partitioning according to partition coefficients of water and its saturation limits for nominally anhydrous minerals, and/or by assuming the interstitial trapping of water-enriched melts (Elkins-Tanton 2008, 2011; Lebrun et al. 2013; Hamano et al. 2013). In some models, a constant mass fraction of 1% melt is assumed to be trapped in the solidifying cumulates over the solidification period. In these cases, the interstitial melts and the nominally anhydrous minerals contribute, respectively, about half the total interior reservoir (Elkins-Tanton 2008). This is because a bulk mineral-melt partition coefficient of water for lherzolite is about 1%, though the lack of experimental partitioning data makes quantitative estimation difficult under lower mantle conditions. The value of 1% for the mass fraction of trapped melts is comparable to the upper limit in the degree of melting estimated for abyssal peridotites at which melt separation starts (Johnson and Dick 1992). However, the efficiency of solid-liquid separation is highly uncertain at the bottom of the vigorously convective magma ocean, and possibly depends on the solidification rate itself.

The distribution of any elements that are incompatible in common mantle minerals is largely controlled by fractional crystallization, which in turn depends sensitively on whether growing mineral grains remain suspended or settle out (Solomatov and Stevenson 1993a,b; Solomatov et al. 1993; Solomatov 2000). For fractional solidification to dominate, mineral grains must settle out and be effectively removed from equilibrium with the magma ocean liquid. As solids are removed, remaining liquids are progressively enriched in incompatible elements including radiogenic potassium and uranium, and the volatiles hydroxyl, nitrogen, and carbon. In batch solidification, liquids and crystals remain in equilibrium throughout solidification. These two crystallization sequences hence lead to completely different predictions for the dynamics of magma oceans, the relative importance of layering or mixing, as well as the mineral assemblage, bulk composition and trace-element composition of the solidified magma oceans. The relative importance of batch vs. fractional crystallization might explain the seemingly diverse signatures of the magma ocean on the Moon as compared to the relative scarcity of magma ocean signatures on the Earth (Tonks and Melosh 1990; Solomatov and Stevenson 1993b).

Quantifying when a crystal settles or remains entrained in a solidifying magma ocean remains challenging. One line of reasoning argues that mineral grains are unable to settle if their settling velocity is small as compared to the convective flow speed in the magma ocean (Tonks and Melosh 1990). This rationale suggests that the characteristic crystal size is the most important factor for evaluating whether crystal settling occurs (Solomatov and Stevenson 1993a,b; Solomatov et al. 1993; Solomatov 2000). Analogue laboratory experiments, however, show that settling is possible along the boundaries of a rapidly convecting body even if the settling velocity of individual particles is small (Martin and Nokes 1988), but these experiments are limited to very low crystal fractions of $< 1\%$.

Recent direct numerical simulations of crystal settling in the lunar magma ocean (Suckale et al. 2012a,b) suggest that the crystal fraction, defined as the local volume fraction of crystals, rather than the crystal size might be the determining factor for assessing crystal settling. The study shows that crystals may form a network at low crystal fractions of approximately 20%, which hinders the individual motion of crystals. Below this threshold, crystals decouple from the magmatic liquid, often in the form of clusters, and settle or rise, suggesting that fractional crystallization might be a valid approximation in this regime. Above that threshold, however, crystals are trapped in a network that moves with the ambient flow, which could be indicative of batch crystallization. An interesting ramification of this study (Suckale et al. 2012a,b) is that the initial crystal fraction in the magma ocean might control its solidification history. If a magma ocean starts at a crystal fraction of 20% because it melted incompletely, batch crystallization would dominate unless there is a significant remelting event. An entirely liquid magma ocean, however, could evolve through fractional crystallization initially and transition to a batch crystallization path later once solidification has progressed further.

Evaluating the crystal fraction of a magma ocean depends both on whether initial melting was complete and on the size of the planetary body. Figure 3 shows the solidi, liquidi, and adiabats for the Moon, Mars, and Earth. In all three cases, the adiabat first intersects the solidus at the bottom of the mantle, which suggests that solidification starts from below (Walker et al. 1975). Only on the Moon, however, does the adiabat fall between liquidus and solidus for almost the entirely solidification history. The steep slope of the adiabat for the depth range characteristic of the lunar mantle hence implies that crystals are stable at all depths. For large planets such as the Earth, only a relatively small portion of the magma

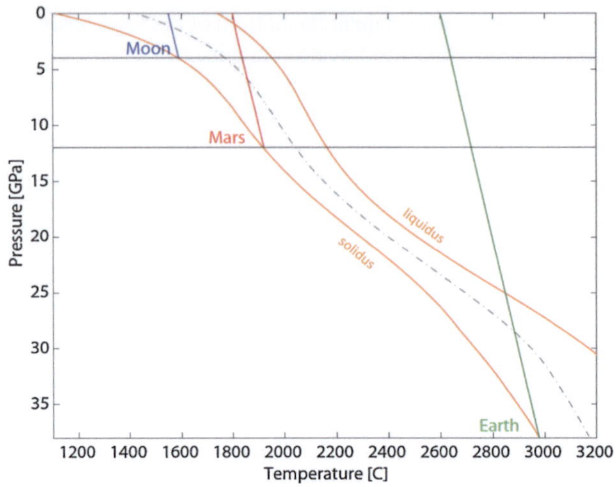

Fig. 3 Idealized adiabats for the Moon (blue), Mars (red) and Earth (green) in comparison to the solidus and liquidus (both orange) for the mantle composition KLB-1, a mantle xenolith representative of fertile peridotite. Correct adiabats deflect strongly toward the slope of the melting curves once they cross the solidus, but this figure still serves to convey the point that larger planets will have smaller depth ranges of solidification. The plotted curves are fits to experimental data from Abe (1997), Takahashi (1986), Zerr et al. (1998), and Trønnes and Frost (2002). The gray horizontal lines indicate the pressure at the bottom of a magma ocean of approximately half the depth of the mantle maximum pressures in the mantle for Moon, Mars, and Earth based on planetary size. The potential temperatures for the Moon, Mars, and Earth were chosen such that their putative whole-mantle magma oceans would be just beginning to solidify, that is, the adiabat touches the solidus at their base

ocean is likely to contain crystals for most of the solidification history as crystals may remelt when being convected upwards. Convection may thus not disperse crystals as easily as in a lunar magma ocean implying that the crystal fraction in the partially molten depth range is likely high enough to limit or impede crystal settling. The role of crystals in determining the solidification history of the magma ocean could help understand the diverse signatures of magma oceans on the Moon as compared to Earth (Tonks and Melosh 1990; Solomatov and Stevenson 1993b). For example, if a terrestrial magma ocean extends to high pressures equivalent to the lowermost mantle on the Earth, energetics indicate that solidification may begin above the bottom of the magma ocean, creating a basal magma ocean solidifying downward (Labrosse et al. 2007).

Even complete fractional solidification will seldom drive water contents of the evolving magma ocean liquids high enough to create water-saturated minerals. Under the assumption in these models, to reach saturation in an Earth-sized planet, the magma ocean must begin with at least 10 wt% of water (Elkins-Tanton and Seager 2008). Whether saturation is reached or not, later cumulates may form under higher oxygen fugacities than did early cumulates in a low-water magma ocean.

Late-solidifying minerals will have slightly higher trace and volatile element concentrations simply from partitioning out of an enriched fluid, but the incompatible element contents of the late-solidifying minerals will be tiny in comparison to the enriched interstitial fluids from the evolved magma ocean. Tonks and Melosh (1990) found that crystallization may be rapid in comparison to liquid percolation, so high fractions of retained liquid may be left in the cumulate mantle. The earliest solid mantle of the planet, then, would have a heterogeneous incompatible element composition. Interstitial melts and surface films will control the planet's internal budget of trace elements and volatiles, including hydroxyl, rare Earth elements, and noble gases.

Assuming that the solidifying mantle cumulates retain no interstitial liquids at all, Tikoo and Elkins-Tanton (2017) found that a terrestrial magma ocean that began with 0.5 wt% of water produced a cumulate mantle with between a few and 45 ppm of water, enough to lower viscosity by a factor of 5 or so. The retention of interstitial liquids, however, can double the water content at the beginning of solidification (if, for example, 1% of interstitial liquids containing 500 ppm water are retained), and can quadruple water content or more at the end of solidification (if 2% of interstitial liquids containing 7000 ppm are retained). Without going outside the range of water contents found in achondritic meteorites, these models can produce a hypothetical Earth with one or potentially multiple oceans' worth of water (Elkins-Tanton 2011).

4.2 Magma-Ocean Degassing

Previous models of magma-ocean solidification view degassing as an equilibrium process in which the volatile content in the magma ocean is assumed to follow solubility laws and the atmosphere grows continuously (Elkins-Tanton 2008; Lebrun et al. 2013). As volatile-rich magmatic liquid approaches the surface, water would be partitioned between the atmosphere and the magma ocean according to its solubility. Water could be released from the enriched magma to the atmosphere through two processes: nucleation and growth of bubbles and molecular diffusion of water across a thermal boundary layer. As long as the crystal fraction in the surface magma is sufficiently low for it to behave like a liquid (a soft magma ocean, defined by Abe 1997), the bubble formation, growth, rise and burst of bubbles is likely the dominant process. It is difficult to quantify the exact value of the crystal fraction required, because it depends on a variety of parameters such as the crystal size distribution and crystal shape. For example, plagioclase-rich magmas can form load-bearing networks at crystal fractions of 10%–20% (Philpotts et al. 1999) while other, more spherical minerals form networks at a typical rheologically critical crystal fraction of approximately 50% (Renner et al. 2000). Coalescence of bubbles during ascent could be fast enough to enable their efficient separation from the low viscous magma flow, by analogy to terrestrial volcanism (Massol et al. 2016).

If the high degree of supersaturation required for bubble nucleation, growth, and separation from magma flow is not present, then molecular diffusion is the dominant process and provides a minimum estimate for the degassing rate. The diffusion rate is controlled by gradient of the water content. As recently reviewed by Elkins-Tanton (2012), the flow field in a molten magma ocean is sufficiently rapid to reach Rayleigh numbers in the range of 10^{20}–10^{30}. For comparison, a Rayleigh number on the order of 10^7 characterizes current solid-mantle convection on the Earth. Based on scaling arguments (Solomatov 2007), the thermal boundary layer at the surface is thin, on the order of cm, and the typical circulation time in the magma ocean ranges from a few days to less than one year. Thus, molecular diffusion may keep the atmosphere and the surface magma nearly at equilibrium (Hamano et al. 2013).

Here and throughout the remainder of this chapter, we refer to continuous (equilibrium) degassing as a process where all volatiles in excess of their depth-dependent magma solubility degas instantaneously. Continuous degassing has been assumed in almost all existing magma ocean solidification models. Non-equilibrium processes are those that lead to a delay in degassing and imply a possible excess in volatiles at specific depths. Lacking observational constraints, it is currently not possible to conclusively determine the relative importance of the described continuous processes as opposed to non-equilibrium degassing in magma oceans. Here, we describe the latter mostly to complement the existing literature and to highlight potential observational signatures that could be indicative of non-equilibrium degassing. A variety of different factors could lead to non-equilibrium degassing. These include, but are not necessarily limited to, incomplete mixing of the magma ocean which could lead to an irregular distribution of volatiles horizontally and vertically, formation of a flotation crust or other solid-like surface features, incomplete melting of the magma ocean, likely most relevant if the magma ocean is created through impact cratering, a lack of mineral phases that allow for efficient nucleation of bubbles on crystal surfaces such as Fe-Ti oxides (heterogeneous nucleation), or high and/or variable crystal fraction in the magma ocean, particularly at shallow depth.

For whole-mantle magma oceans resulting from a significant melting event caused by a giant collision, we hypothesize that the two primary factors that could lead to non-equilibrium degassing in magma oceans are (1) a lack of mineral phases that allow for efficient heterogeneous nucleation, which implies that a finite supersaturation pressure at depth is required for bubble nucleation and (2) incomplete escape of formed gas bubbles through a solidifying crystal mush. We briefly review existing constraints on both of these factors.

A melt that is supersaturated in volatiles exsolves a gas phase to reestablish thermodynamic and chemical equilibrium. Homogeneous nucleation theory (Landau and Lifshitz 1980) attempts to quantify the energy required to nucleate bubbles spontaneously, but predicts unrealistically high supersaturation pressures in basaltic melts (Sparks 1978; Sparks et al. 1994). Bubble nucleation may be facilitated significantly by the existence of preferential nucleation sites such as surfactants, crystallites, or phenocrysts in the melt phase (Blander and Katz 1975; Navon and Lyakhovsky 1998). The supersaturation required for heterogeneous nucleation is determined by the surface energy of the crystal-gas as compared to the melt-gas interface (e.g., Blander and Katz 1975) and is typically much lower than for homogeneous nucleation (Hurwitz and Navon 1994; Gardner and Denis 2004). An empirical criterion for comparing the two surface energies is the wetting angle θ. Unfortunately, the experimental determination of wetting angles is very challenging (Gardner and Denis 2004), but some constraints exist (Eichelberger and Hayes 1982; Hurwitz and Navon 1994; Mangan and Sisson 2000).

Experimental studies of the supersaturation required for bubble nucleation in basaltic liquids are rare (Lensky et al. 2006; Hardiagon et al. 2013) and thus constraints for basaltic systems are often deduced from laboratory experiments performed in more silicic magma. Estimates for the supersaturation required for nucleation in rhyolitic magmas typically range from 120–200 MPa in melts exhibiting homogeneous or inefficient heterogeneous nucleation (Mourtada-Bonnefoi and Laporte 1999; Mangan and Sisson 2000; Mourtada-Bonnefoi and Laporte 2002) to 5–20 MPa in systems where heterogeneous nucleation is very efficient (e.g., Hurwitz and Navon 1994; Gardner and Denis 2004; Gardner 2007). A similar span of supersaturation pressures may be expected for basaltic magma (Bottinga and Javoy 1990) and is consistent with current estimates ranging from as little as 1–3 MPa (Harris and Anderson 1983; Cashman et al. 1994), to tens of MPa (Edmonds et al. 2013), 100–300 MPa (Lensky et al. 2006) and even 1.5 GPa (Hardiagon et al. 2013). Summarizing, existing observational, experimental and theoretical evidence suggests the possibility of non-zero supersaturation pressures of tens of MPa and possibly higher in the absence of crystals with high wetting angles.

Once bubbles have been formed, they have to rise through the convecting magma ocean to burst and degas at the surface. The ability of bubbles to degas hence depends on the flow conditions in the vicinity of the surface, the presence of an atmosphere or conductive lid,

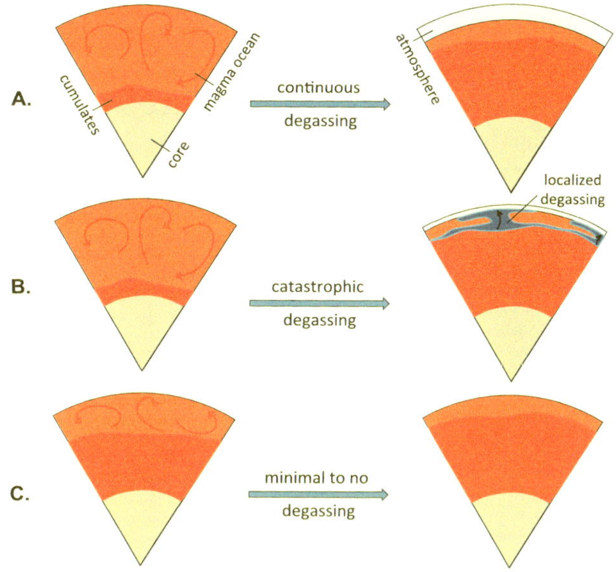

Fig. 4 Schematic illustration of the three end-member cases of magma-ocean degassing

and the crystallinity of the magma. It is interesting to note in this context that many natural examples of very efficient degassing on Earth refer to nearly aphyric mafic magmas mostly from Hawaii (Harris and Anderson 1983; Cashman et al. 1994; Edmonds et al. 2013).

Degassing at a late stage of magma ocean solidification could be complicated due to a variety of factors. Due to the high solid-like viscosity, the circulation time of convection reaches up to several millions years (Lebrun et al. 2013). Also, a viscous thermal boundary layer starts to grow and its thickness reaches the order of 1–10 km, which could prevent the volatile exchange between the atmosphere and the surface magma. In this stage, melt percolation is a dominant process for melt extraction through the partially molten boundary layer (McKenzie 1984; Solomatov 2007). The melt percolation may carry up volatiles to the surface, but only if it occurs faster than the propagation of the solidification front (Lebrun et al. 2013).

On the other hand, if fractional crystallization creates an unstable stratification due to compositional difference, a gravitational overturn can transport back and release volatiles into the deep interior (Tikoo and Elkins-Tanton 2017). If that is the case, volatiles might be transported back down into the mantle and not degas at all, leading to elevated volatile contents in portions of the solidified mantle after overturn. Another hypothetical consequence of degassing potentially occurring as late as just prior to or during overturn is that a finite bubble fraction could stabilize segments of the uppermost portion of the solidified magma ocean against overturn by reducing its density. This scenario, however, would require inefficient escape of the bubbles from the crystal mush as the time scales of solid-mantle overturn are much longer than those of gas percolation through a crystalline suspension. Since heterogeneous nucleation occurs preferentially on crystal interfaces, however, it is possible that a significant fraction of the bubbles could be trapped. The chemical and thermal structure, and dynamics of a magma ocean in the late stages of solidification is particularly important to quantify the final amount of water retained in the interior.

Synthesizing, we suggest that degassing from solidifying magma oceans could be described by three end-member cases: (A) continuous degassing, (B) catastrophic degassing, and (C) negligible degassing (Fig. 4).

4.2.1 (A) Continuous Degassing

Continuous degassing implies that degassing of volatiles in excess of saturation occurs constantly during the solidification of a magma ocean. This scenario is viable if heterogeneous nucleation requires negligible supersaturation. That might be the case, for example, if Fe-Ti oxides are abundant as small crystals in the liquid convecting magma, because oxides provide energy-efficient nucleation sites for bubbles (Hurwitz and Navon 1994). Alternatively, continuous degassing might arise if the volatile content is high enough to initiate degassing immediately after the formation of the magma ocean. It is also possible that the magma ocean may get continually replenished in volatiles by volatile-rich impactors, although impactors can play different roles depending on their size and the degree of melting upon impact and may lead to both volatile loss and replenishment.

The existence of a steam-rich atmosphere that precedes magma ocean degassing would enhance degassing by keeping the surface of the magma ocean above its solidus, and eliminating the possibility of a solid crust (Abe and Matsui 1986). Without a solid layer, liquids can ascend to the thermal boundary layer at the planetary surface and experience the lowest possible pressures. They will thus degas more thoroughly because at the lowest pressures any given liquid has its supersaturation maximized. The overall duration of the magma ocean phase would affect the total budget of water. As long as the magma ocean is sustained, all surface water is present in a vapor phase due to the hot surface temperature and therefore vulnerable to escape processes.

We note that the mere existence of an atmosphere at the time of magma ocean formation, however, may not be sufficient to ensure continuous degassing. If partial pressures are low, the main effect of a primordial atmosphere could be to slow down the solidification process but degassing could still occur discontinuously. Magma oceans that degas continuously are characterized by substantial atmospheres and comparatively low solidification speed (Zahnle et al. 1988; Elkins-Tanton 2008).

4.2.2 (B) Catastrophic Degassing

Delayed degassing could result if bubble nucleation required a finite supersaturation pressure of bubble trapping in the vicinity of the surface of the magma ocean. Even a relatively small supersaturation on the order of 10 MPa could result in delayed degassing, because it implies degassing occurring later in the solidification history when a larger portion of the magma ocean is already solidified or has a high crystal content. If finite supersaturation pressures are required for nucleation, it is possible that bubbles would form preferably in upwellings. In upwellings, compositional and thermal buoyancy reinforce each other and degassing should be possible essentially irrespective of bubble size.

In the context of volcanic degassing on Earth, it has been noticed that high surface pressures could inhibit degassing rates notably and change the degassing composition (Gaillard et al. 2011, 2015). Applied to degassing magma oceans, this finding would suggest that the presence of a substantial atmosphere with high surface pressures could reduce the efficiency of magma-ocean degassing, although thermal processes could have the opposite effect as discussed in the previous paragraph and it is difficult to know a priori how these competing processes would balance out.

If delayed degassing is related to bubble trapping, fluid and gas percolation is driven by the compaction of the cumulate pile. Degassing might hence occur preferentially in fluid-filled gas pathways. Both the preferential nucleation of bubbles in upwellings and the preferential nucleation of bubbles in fluid-filled pathways suggest that degassing occurs locally in flow regions that are conducive to nucleation events and may not be ubiquitous across the surface of the magma ocean. Delayed degassing could also entail a sequence of approximately simultaneous nucleation events throughout the magma ocean.

Given that bubble nucleation and growth happen on time scales that are small compared to the time scale of magma ocean solidification, degassing may constitute a rapid and potentially even catastrophic event rather than a continuous process. The localized and sudden nature of late-stage degassing is an important difference compared to the more commonly invoked continuous degassing in excess of saturation (Abe and Matsui 1985; Elkins-Tanton 2008; Lebrun et al. 2013). Compared to continuous degassing, magma oceans that exhibit catastrophic degassing are characterized by a less substantial early atmosphere, higher volatile content in the solidified mantle, particularly the upper mantle, and a less homogeneous distribution of volatiles throughout the mantle. The composition of the atmosphere depends on the partitioning behavior of the volatiles involved (e.g., Papale et al. 2006) and on the time scales of decompression relative to volatile diffusion. The faster the decompression, the less time is available for diffusion which could lead to substantial residual volatiles in the melt (Mangan and Sisson 2000).

4.2.3 (C) Negligible Degassing

If the total volatile content is insufficient, a solidifying magma ocean may never reach the enrichment level required for degassing. Although volatile enrichment increases rapidly during the very late stages of solidification, very shallow magma oceans might not contain sufficient volatiles to reach the supersaturation pressure required for degassing or solid-mantle overturn may initiate earlier than degassing in which case a significant portion of the volatiles may be recycled into the mantle. Two magma oceans with comparable initial volatile contents but different depths might therefore experience drastically different degassing histories. Magma oceans with limited spatial extent on the planetary surface could also form a stable conductive lid, not unlike lava lakes, which hinders efficient degassing. The consequences of incomplete degassing for the planetary interior are substantial: Magma oceans that do not degas catastrophically will be characterized by elevated volatile contents during most of its solidification, which translates into higher volatile contents in the nominally anhydrous phases as well as in trapped interstitial liquid, minimal atmospheres and rapid time scales of solidification. A single deep magma ocean on a given planet might hence degas very differently than a sequence of shallow magma oceans (Chambers 2013; Tucker and Mukhopadhyay 2014).

The style of magma-ocean degassing could also depend on the types of volatiles. For example, water and carbon dioxide may play different roles in the degassing process, because of their differing solubility behavior at depth and diffusivities.

4.3 Coupled Evolution of Magma Ocean and Atmosphere

The formation of an outgassed atmosphere affects the thermal evolution of a magma ocean. Without an atmosphere, the molten surface would emit significant thermal radiation, the flux of which is on the order of 10^6–10^7 W/m^2 due to high surface temperatures. As a result, the surface of an Earth-mass terrestrial planet would rapidly solidify in about 1000 years or less, even if a shallow, highly viscous magma ocean could continue to transport its heat effectively (Solomatov 2007). The greenhouse or blanketing effect of the outgassed atmosphere prevents heat from escaping from the surface. In general, as the atmosphere becomes

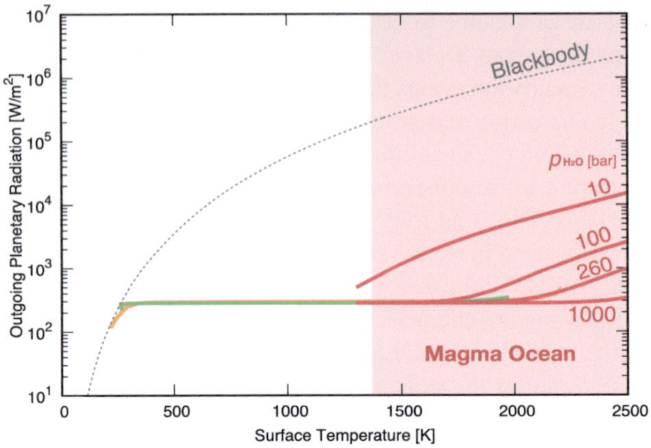

Fig. 5 Thermal radiation as a function of surface temperature for various surface pressures of the steam atmosphere. Red lines are results obtained by using a radiative transfer model of Hamano et al. (2015), compared with results by Goldblatt et al. (2013) (green) and Kopparapu et al. (2013) (orange) for a 260-bar steam atmosphere. The outgoing radiation increases at the higher surface temperature for two reasons: (1) leakage of visible and near-infrared radiation from the surface and the hotter, lower part of the atmosphere, and (2) temperature increase at altitudes, where the optical depth measured from the top of the atmosphere is around unity

massive, its greenhouse/blanketing effect reduces the outgoing radiation to space even further. Water vapor is one of the most potent greenhouse gases: Even 10 bar of a steam atmosphere, which is equivalent to about $1/3$ of the total mass of the current Earth's oceans, reduces the outgoing radiation by about 2 orders of magnitude, compared with the corresponding blackbody radiation (see Fig. 5). CO_2 is also important as an opacity source in the infrared, though its mass absorption coefficient, therefore its blanketing effect, is generally much smaller than that of water vapor (e.g. Marcq et al. 2017).

The growth rate of the atmosphere would depend on how degassing proceeds (i.e., equilibrium/non-equilibrium processes and volatile content in magmas), as described above. In the cases where super-saturation required for degassing becomes higher, degassing would occur in a more catastrophic way, leading to later formation of massive atmospheres. The outgassed atmosphere is also vulnerable to escape processes. Spectroscopic observations and atmospheric modeling of young stars have revealed that young, active stars emit higher X-ray and extreme ultraviolet radiation (e.g., Ribas et al. 2005; Linsky et al. 2014; France et al. 2016), which fuels hydrodynamic blow-off of early planetary atmospheres (e.g., Watson et al. 1981; Kasting and Pollack 1983). The hydrodynamic escape process could cause a significant loss of water from solidifying planets. This is because water vapor dissociates in the upper atmosphere by the intense EUV radiation, producing hydrogen atoms, which easily escape into space. This could affect the thermal evolution of a magma ocean, because the atmosphere grows or is depleted as a net result of the degassing and escape processes. Furthermore, the total amount of water lost during a magma ocean phase depends on its overall duration. As the magma ocean period becomes longer, the total inventory of water just after solidification becomes smaller.

Another possible process that affects the total water budget is impacts of planetary bodies. Terrestrial planets just after formation would experience a number of impacts by leftover fragments during accretion or planetesimals scattered from the outer orbital regions. These numerous impacts could erode or supply the atmosphere, depending on various parameters such as the impact velocity, the volatile content of impactors and their size distribution (e.g.,

Chyba 1990; Vickery and Melosh 1990; de Niem et al. 2012). As the net effect depends on the ambient atmospheric pressure as well, it could be worthwhile to examine the volatile budget by impacts along with the atmospheric growth during the magma ocean period in future work.

Water partitioning in protoplanets has thus been examined with the thermal evolution of magma oceans, including feedback processes such as the greenhouse/blanketing effect of outgassed atmospheres, degassing of volatile species from the interior, and loss of water associated with hydrodynamic escape of hydrogen (Elkins-Tanton 2008; Lebrun et al. 2013; Hamano et al. 2013; Salvador et al. 2017). In the case that steam-rich atmospheres are formed by degassing, terrestrial planets could have different thermal and water budget histories during solidification, depending on their orbits (Hamano et al. 2013; Lebrun et al. 2013). As a result, the solidified planets would have two distinct characteristics: Planets that maintained certain amounts of water after their short solidification period, and others that dried out during a slow solidification process.

The evolutionary dichotomy originates from the presence of a lower limit in outgoing planetary radiation. As shown in Fig. 5, the thermal radiation from the top of the atmosphere decreases along with decrease in the surface temperature or increase in the atmospheric mass, but finally approaches $\sim 280\ \mathrm{W/m^2}$ (Abe and Matsui 1988; Kasting 1988; Goldblatt et al. 2013; Kopparapu et al. 2013). This means that the outgoing radiation from the top of steam atmospheres has a lower limit under high surface temperature conditions, such as a magma ocean phase. Hereafter, we referred to it as a radiation limit, which is also called a threshold for runaway greenhouse effects. The value of the radiation limit is determined by a saturation curve of water vapor, its opacity, and the planetary gravity (Nakajima et al. 1992). The value is less sensitive to including other gaseous species such as CO_2 and N_2, as long as water vapor is the most major component in the atmosphere (Kopparapu et al. 2014; Marcq et al. 2017).

Terrestrial planets receive the net stellar radiation, which could be smaller or larger than the radiation limit, depending on their orbits. According to Hamano et al. (2013, 2015), a critical orbital distance a_{cr} is defined as the distance from the host star where the net incoming stellar radiation is equal to the radiation limit,

$$a_{cr} \approx 0.83 \left(\frac{F_{lim}}{280\ \mathrm{W\,m^{-2}}} \right)^{-1/2} \left(\frac{S_*(\tau_{GI})}{0.7 S_\odot} \right)^{1/2} \left(\frac{1-\alpha_{pl}}{1-0.2} \right)^{1/2} \mathrm{AU}, \tag{5}$$

where F_{lim} is the radiation limit, α_{pl} is the planetary albedo, and $S_\odot$ and $S_*(\tau_{GI})$ are, respectively, the bolometric fluxes at 1 AU of the current Sun and the host star at the time of giant impact. The critical distance separates the orbital regions of the two types of planet.

The presence of the radiation limit gives rise to a different water partitioning between the atmosphere and the interior, depending on the planetary orbit. For a planet formed beyond the critical distance, as in the case of the Earth in Fig. 6(a), the primary reservoir of water after solidification is the steam atmosphere. More than half of the total inventory of water has already outgassed during the soft magma ocean stage. After the crystal fraction at the surface reaches about 60%, the magma rheology abruptly changes due to a sudden jump of magma viscosity, and solid-state convection takes over from liquid-state convection (a hard magma ocean, defined by Abe 1997). The final amount of water left in the interior would be established through this late stage of solidification, as discussed in the previous section. On the assumption of continuous (equilibrium) degassing in this later stage, the water stored in the early mantle accounts for less than 10% of the total inventory, because most water has outgassed into the atmosphere.

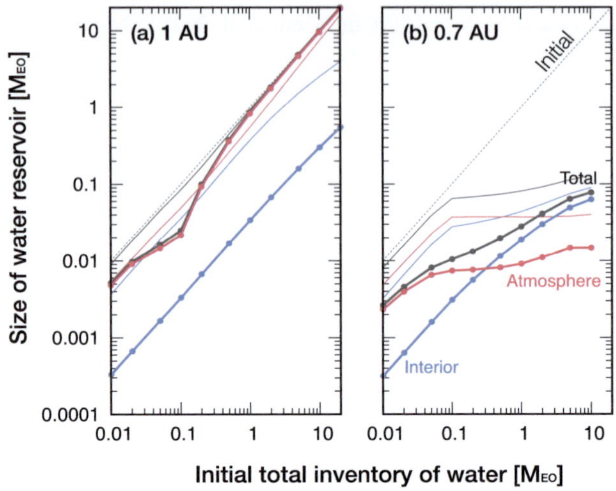

Fig. 6 The bold lines indicate the relationship between the total inventory of water and the size of its reservoirs at the end of solidification of an Earth-sized planet at 1 AU (**a**) and 0.7 AU (**b**), on the assumption of solution equilibrium at the surface and with a mass fraction of trapped melt of 1%. The thin lines indicate those at the time that the potential temperature of the interior reaches 1700 K, at which the crystal fraction in the surface magma reaches 60%, a threshold for rheological transition. This corresponds to the transition from a soft magma ocean to a hard magma ocean. The reservoir size at potential temperature of 1700 K would constitute a maximum estimate of the interior reservoir (the sum of solid mantle and magma ocean). All the results are obtained from Hamano et al. (2013)

Loss of water has little effect on the total water budget in this type of planet especially with a water content comparable to the total inventory of the present Earth. By definition, planets formed beyond the critical distance emit the outgoing radiation exceeding the incoming radiation, irrespective of the amounts of steam atmospheres. Consequently, Earth-sized planets solidify typically in less than several million years (Lebrun et al. 2013; Hamano et al. 2013), which is short enough for most of the primordial water to survive against the hydrodynamic escape. As long as the major component of the atmosphere is water vapor, the radiation limit also characterizes the cooling timescale of the atmosphere after solidification. In the case of a 300-bar steam atmosphere, water vapor would start to rain out on a timescale as short as about 1000 years after solidification, meaning that the surface water, which already outgassed by the time of complete solidification, would remain almost intact and contribute to the earliest water ocean. The following decrease in the mixing ratio of water vapor in the atmosphere would prevent further rapid loss of water (e.g. Kasting and Pollack 1983).

In contrast, if formed closer to the host star than the critical distance, a planet would become desiccated during crystallization by hydrodynamic escape. At orbits inside the critical distance, stellar radiation can keep the planetary surface in a molten state, if the planet has a steam atmosphere of about 30 bar or less, depending on the orbital distance (Fig. 5). In other words, the magma ocean would be sustained until the planet loses the steam atmosphere. The interior reservoir can account for a large fraction of the total inventory, especially in the case with the large initial total inventory of water. This is because the silicate melt has the higher water content in an early stage with the larger total inventory, whereas the amount of the steam atmosphere is regulated so that the net outgoing heat flux becomes positive.

There have been arguments that dissociation of water molecules followed by preferential escape of hydrogen may cause a build-up of oxygen in the atmosphere, especially in the context of water loss from early Venus (e.g. Kasting 1995, and references therein). An extreme build-up of oxygen could possibly shut down the escape of hydrogen itself. This may not be the case during the solidifying phase of a magma ocean, because the dissociated oxygen atoms can be consumed by oxidation of ferrous iron in the abundant surface magmas (Gillmann et al. 2009; Hamano et al. 2015). As long as the magma ocean acts as an oxygen sink, this process would have a role in suppressing a significant build-up of abiotic oxygen on molten terrestrial planets (Schaefer et al. 2016). Recently, Marcq et al. (2017) have shown that addition of CO_2 with a fixed amount of water vapor content reduces the blanketing effects. In the case of an atmosphere consisting of 270 bar of H_2O and 90 bar of CO_2, its blanketing effects would be still enough for keeping a long-lived magma ocean on early Venus, as long as the CO_2-H_2O atmosphere have the albedo as low as that of a thick H_2O atmosphere (~ 0.2 for the Sun). However, if subsequent cloud formation enhances the planetary albedo, early Venus could have started to quickly solidify and water ocean could have formed, depending on the amount of water remaining in the atmosphere (Salvador et al. 2017). If a rotation rate of early Venus was as low as the present one, formation of high thick clouds on the day side might have helped to reflect a large fraction of the incident sunlight (Yang et al. 2014; Way et al. 2016). On the other hand, clouds also have a role to enhance the blanketing effects. It would be necessary to investigate a balance of the opposing effects by clouds on planetary albedo and a blanketing effect to address early history of Venus.

The amount of water left in the interior increases with the assumed mass fraction of trapped melts. With a trapped melt fraction of 1%, the interior reservoir is the primary water reservoir in the case of the initial water inventory comparable to the total water inventory on the present Earth, 1.25–5 times the current ocean mass (e.g., Hirschmann 2006), as shown in Fig. 6. Note that solution equilibrium between the atmosphere and the magma ocean is assumed in these calculations. If degassing requires some degree of super-saturation of water in silicate melts, it leads to the later formation of the outgassed atmosphere. As a result, the solidification time of the magma ocean would become shorter. In addition, the silicate melts keep the higher volatile content during solidification, which leads to the formation of a wetter early mantle. According to calculations using a 50-fold water solubility, which are intended to mimic the effects of super-saturation of water, the amount of water remained in the interior increases by a factor of 4 at most, compared to the results with a nominal water solubility (Hamano et al. 2013).

These conclusions may be qualitatively extended to terrestrial rocky planets, ranging from Mars-sized to super-Earths, as long as heat flux from a magma ocean is limited by radiative balance at the surface. The value of the radiation limit slightly increases with planetary mass, and is about 20% larger for 5-Earth-mass planets, compared to that for Earth-mass planets (Kopparapu et al. 2014). In this case, the critical distance from the star becomes about 10% smaller. Experimental and ab initio studies of melting curves are still absent at high pressures above 150 GPa. In the case that super-Earths start to solidify with higher potential temperatures compared to the Earth, magma oceans of super-Earths could be capped by silicate atmospheres at early stages of solidification. Since the planet would emit outgoing radiation as high as 10^6 W/m^2 from the top of hot silicate clouds, this phase lasts for at most several thousand years for 5-Earth-mass planets. The atmospheric escape thus would not affect the total budget of water, but some water might be incorporated into cumulates during this short phase, depending on the efficiency of melt/solid separation. On the assumption that water in excess of the saturation limit is degassed from the surface, larger planets can retain larger fractions of water in the interior for a given bulk content of water (Elkins-Tanton 2011). In the case of a 5-Earth-mass rocky planet formed beyond the critical distance, the fraction of water retained in the final 1% of silicate melt increases by a factor of less than 2, compared to that of an Earth-mass planet.

5 Concluding Remarks

Not only the presence of oceans, but also the current mass of ocean water is essential for making the Earth habitable, as described in the Introduction. In this chapter, however, one has found it really challenging to answer the question of what determined the present total mass of the oceans on the Earth. Quite uncertain are how water/ice was distributed in the solar nebula and which processes dominated water delivery to the Earth. In addition, while magma oceans are readily formed, water partitioning between the atmosphere, magma ocean, solid mantle and core in accreting planetary embryos and protoplanets is complicated and poorly understood. The partitioning of water is also affected significantly by its accumulation and also mutually coupled with its escape. Processes relevant to water partitioning that are yet to be understood well include the solubility of water in peridotite liquid, the distribution of volatiles during crystallization, crystal settling and entrainment in solidifying magma oceans, non-equilibrium degassing such as bubble formation, and so on.

The initial amounts of water in planetary building blocks are obviously a key that determines the final amount of ocean water on terrestrial planets. As mentioned in Sect. 2, we are undergoing a paradigm shift regarding planet formation. Detection of many exoplanets with short orbital periods has compelled us to accept a dynamic picture including planetary migration and radial drift of solids, instead of the conventional, static picture. Also, rather than km-sized planetesimals, cm-sized pebbles may contribute dominantly to planetary accretion. If that is the case, since pebbles come from beyond the snowline in general and thus contain large amounts of water, water delivery and accumulation processes differ greatly from those previously considered. Such a great degree of freedom in planetary formation theories demands much more observational constraints. Fortunately, new observational constraints are expected to come from both exoplanet observations and solar system explorations in the coming years.

The first-generation exoplanet survey projects, *Kepler* and CoRoT, already confirmed that planetary systems are common. At present, the second-generation survey projects such as K2, TESS, and PLATO are under way or scheduled for detecting exoplanets orbiting nearby stars, which are suitable for further characterization. Those planets will include terrestrial planets with different amounts of water. The atmospheres of those planets will be characterized mainly via transit spectroscopy in the optical to infra-red by the upcoming space telescope JWST and the planned space telescope ARIEL. Key species such as water, methane, and oxygen will be hopefully detected in those atmospheres. Also, although being too close to host stars to be habitable, super-Earths covered with magma oceans might be detected (Ito et al. 2015). Further characterization of Earth-like planets will be challenged by the 30m-class ground-based telescopes E-ELT and TMT. With those telescopes, Earth-like planets covered with magma oceans might be detected in the future (Lupu et al. 2014; Hamano et al. 2015). For the details, see Chap. 7 in this issue (Noack et al. 2017).

The initial atmospheres of planets are determined largely by the material that accretes to create the planet. Small bodies in our solar system (in addition to meteorites) are the observable targets from which we learn more. In the wake of the highly successful Rosetta and Dawn missions, which visited a comet and the asteroids Vesta and Ceres, future missions will examine additional suites of differentiated and primitive rocky bodies. The OSIRIS-REx and Hayabusa2 missions are on their ways to the carbonaceous chondrite near-Earth asteroids Bennu and Ryugu, respectively, and will return surface samples to the Earth. In 2021 the Lucy mission will launch to the Jupiter Trojans, and will send back compositional data about these little-known but presumably ice-rich remnants of planet-building. In 2022 the Psyche mission to the metal asteroid Psyche will be launched, and will examine there what is hypothesized to be the core of a planetesimal, stripped of its silicate mantle in the first millions of years of the solar system. Through data from these missions we will be able to further constrain the starting compositions of growing planets.

Acknowledgements First of all, we would like to express our deepest appreciation to Prof. Yutaka Abe, who passed away recently. We learned a lot from discussion with him, without which we could not complete this chapter. We thank anonymous referees for their careful reading and constructive comments, which helped us improve this manuscript greatly. Also we thank the SOC members of the ISSI workshop "The Delivery of Water to Protoplanets, Planets and Satellites" for giving us the opportunity to write this chapter. M. Ikoma is supported by the Astrobiology Center Program of National Institutes of Natural Sciences (NINS) (No. AB291004) and JSPS Core-to-Core Program "International Network of Planetary Sciences". K. Hamano acknowledges JSPS KAKENHI Grant Numbers JP26800242 and JP16J06133, and MEXT KAKENHI Grant Number JP17H06457.

References

D.S. Abbot, N.B. Cowan, F.J. Ciesla, Indication of insensitivity of planetary weathering behavior and habitable zone to surface land fraction. Astrophys. J. **756**, 178 (2012). https://doi.org/10.1088/0004-637X/756/2/178

Y. Abe, Thermal and chemical evolution of the terrestrial magma ocean. Phys. Earth Planet. Inter. **100**, 27–39 (1997). https://doi.org/10.1016/S0031-9201(96)03229-3

Y. Abe, T. Matsui, The formation of an impact-generated H_2O atmosphere and its implications for the early thermal history of the Earth, in *Lunar and Planetary Science Conference Proceedings*, ed. by G. Ryder, G. Schubert. Lunar and Planetary Science Conference Proceedings, vol. 15 (1985), pp. 545–559

Y. Abe, T. Matsui, Early evolution of the Earth: accretion, atmosphere formation, and thermal history. J. Geophys. Res. **91**, 291–302 (1986). https://doi.org/10.1029/JB091iB13p0E291

Y. Abe, T. Matsui, Evolution of an impact-generated H_2O-CO_2 atmosphere and formation of a hot protoocean on Earth. J. Atmos. Sci. **45**, 3081–3101 (1988). https://doi.org/10.1175/1520-0469(1988)045<3081:EOAIGH>2.0.CO;2

Y. Abe, A. Abe-Ouchi, N.H. Sleep, K.J. Zahnle, Habitable zone limits for dry planets. Astrobiology **11**, 443–460 (2011). https://doi.org/10.1089/ast.2010.0545

W. Benz, W.L. Slattery, A.G.W. Cameron, The origin of the moon and the single-impact hypothesis. I. Icarus **66**, 515–535 (1986). https://doi.org/10.1016/0019-1035(86)90088-6

M. Blander, J.L. Katz, Bubble nucleation in liquids. AIChE J. **21**, 833–848 (1975)

Y. Bottinga, M. Javoy, Mid-ocean ridge basalt degassing: bubble nucleation. J. Geophys. Res. **95**, 5125–5131 (1990). https://doi.org/10.1029/JB095iB04p05125

S.M. Brown, L.T. Elkins-Tanton, R.J. Walker, Effects of magma ocean crystallization and overturn on the development of ^{142}Nd and ^{182}W isotopic heterogeneities in the primordial mantle. Earth Planet. Sci. Lett. **408**, 319–330 (2014). https://doi.org/10.1016/j.epsl.2014.10.025

R.M. Canup, Lunar-forming collisions with pre-impact rotation. Icarus **196**, 518–538 (2008). https://doi.org/10.1016/j.icarus.2008.03.011

K.V. Cashman, M.T. Mangan, S. Newman, Surface degassing and modifications to vesicle size distributions in active basalt flows. J. Volcanol. Geotherm. Res. **61**, 45–68 (1994). https://doi.org/10.1016/0377-0273(94)00015-8

J.E. Chambers, Late-stage planetary accretion including hit-and-run collisions and fragmentation. Icarus **224**, 43–56 (2013). https://doi.org/10.1016/j.icarus.2013.02.015

C.-L. Chou, Fractionation of siderophile elements in the Earth's upper mantle, in *Lunar and Planetary Science Conference Proceedings*. Lunar and Planetary Science Conference Proceedings, vol. 9 (1978)

C.F. Chyba, Impact delivery and erosion of planetary oceans in the early inner solar system. Nature **343**, 129–133 (1990). https://doi.org/10.1038/343129a0

S.K. Croft, A first-order estimate of shock heating and vaporization in oceanic impacts, in *Geological Society of America Special Papers*, vol. 190 (1982), pp. 143–152

N. Dauphas, A. Pourmand, Hf-W-Th evidence for rapid growth of Mars and its status as a planetary embryo. Nature **473**, 489–492 (2011). https://doi.org/10.1038/nature10077

T.M. Davison, G.S. Collins, F.J. Ciesla, Numerical modelling of heating in porous planetesimal collisions. Icarus **208**, 468–481 (2010). https://doi.org/10.1016/j.icarus.2010.01.034

D. de Niem, E. Kührt, A. Morbidelli, U. Motschmann, Atmospheric erosion and replenishment induced by impacts upon the Earth and Mars during a heavy bombardment. Icarus **221**, 495–507 (2012). https://doi.org/10.1016/j.icarus.2012.07.032

J. de Vries, F. Nimmo, H.J. Melosh, S.A. Jacobson, A. Morbidelli, D.C. Rubie, Impact-induced melting during accretion of the Earth. Prog. Earth Planet. Sci. **3**(1), 7 (2016)
M. Edmonds, I.R. Sides, D.A. Swanson, C. Werner, R.S. Martin, T.A. Mather, R.A. Herd, R.L. Jones, M.I. Mead, G. Sawyer, T.J. Roberts, A.J. Sutton, T. Elias, Magma storage, transport and degassing during the 2008-10 summit eruption at Kilauea Volcano, Hawaii. Geochim. Cosmochim. Acta **123**, 284–301 (2013). https://doi.org/10.1016/j.gca.2013.05.038
J.C. Eichelberger, D.B. Hayes, Magmatic model for the Mount St. Helens blast of May 18, 1980. J. Geophys. Res. **87**, 7727–7738 (1982). https://doi.org/10.1029/JB087iB09p07727
L.T. Elkins-Tanton, Linked magma ocean solidification and atmospheric growth for Earth and Mars. Earth Planet. Sci. Lett. **271**, 181–191 (2008). https://doi.org/10.1016/j.epsl.2008.03.062
L.T. Elkins-Tanton, Formation of early water oceans on rocky planets. Astrophys. Space Sci. **332**, 359–364 (2011). https://doi.org/10.1007/s10509-010-0535-3
L.T. Elkins-Tanton, Magma oceans in the inner solar system. Annu. Rev. Earth Planet. Sci. **40**, 113–139 (2012). https://doi.org/10.1146/annurev-earth-042711-105503
L.T. Elkins-Tanton, S. Seager, Ranges of atmospheric mass and composition of super-Earth exoplanets. Astrophys. J. **685**, 1237–1246 (2008). https://doi.org/10.1086/591433
K. France, R.O. Parke Loyd, A. Youngblood, A. Brown, P.C. Schneider, S.L. Hawley, C.S. Froning, J.L. Linsky, A. Roberge, A.P. Buccino, J.R.A. Davenport, J.M. Fontenla, L. Kaltenegger, A.F. Kowalski, P.J.D. Mauas, Y. Miguel, S. Redfield, S. Rugheimer, F. Tian, M.C. Vieytes, L.M. Walkowicz, K.L. Weisenburger, The MUSCLES treasury survey. I. Motivation and overview. Astrophys. J. **820**, 89 (2016). https://doi.org/10.3847/0004-637X/820/2/89
F. Gaillard, B. Scaillet, N.T. Arndt, Atmospheric oxygenation caused by a change in volcanic degassing pressure. Nature **478**(7368), 229 (2011)
F. Gaillard, B. Scaillet, M. Pichavant, G. Iacono-Marziano, The redox geodynamics linking basalts and their mantle sources through space and time. Chem. Geol. **418**, 217–233 (2015)
J.E. Gardner, Heterogeneous bubble nucleation in highly viscous silicate melts during instantaneous decompression from high pressure. Chem. Geol. **236**(1), 1–12 (2007)
J.E. Gardner, M.-H. Denis, Heterogeneous bubble nucleation on Fe-Ti oxide crystals in high-silica rhyolitic melts. Geochim. Cosmochim. Acta **68**, 3587–3597 (2004). https://doi.org/10.1016/j.gca.2004.02.021
C. Gillmann, E. Chassefière, P. Lognonné, A consistent picture of early hydrodynamic escape of Venus atmosphere explaining present Ne and Ar isotopic ratios and low oxygen atmospheric content. Earth Planet. Sci. Lett. **286**, 503–513 (2009). https://doi.org/10.1016/j.epsl.2009.07.016
G.J. Golabek, H. Schmeling, P.J. Tackley, Earth's core formation aided by flow channelling instabilities induced by iron diapirs. Earth Planet. Sci. Lett. **271**, 24–33 (2008). https://doi.org/10.1016/j.epsl.2008.02.033
C. Goldblatt, T.D. Robinson, K.J. Zahnle, D. Crisp, Low simulated radiation limit for runaway greenhouse climates. Nat. Geosci. **6**, 661–667 (2013). https://doi.org/10.1038/ngeo1892
P. Goldreich, W.R. Ward, The formation of planetesimals. Astrophys. J. **183**, 1051–1062 (1973). https://doi.org/10.1086/152291
K.E. Haisch Jr., E.A. Lada, C.J. Lada, Disk frequencies and lifetimes in young clusters. Astrophys. J. Lett. **553**, 153–156 (2001). https://doi.org/10.1086/320685
A.C. Halliday, The origin and earliest history of the Earth, in *Treatise on Geochemistry*, ed. by H. Holland, T. Turekian, 2nd edn. (2014), pp. 149–211
L.J. Hallis, G.R. Huss, K. Nagashima, G.J. Taylor, S.A. Halldórsson, D.R. Hilton, M.J. Mottl, K.J. Meech, Evidence for primordial water in Earth's deep mantle. Science **350**, 795–797 (2015). https://doi.org/10.1126/science.aac4834
K. Hamano, Y. Abe, H. Genda, Emergence of two types of terrestrial planet on solidification of magma ocean. Nature **497**, 607–610 (2013). https://doi.org/10.1038/nature12163
K. Hamano, H. Kawahara, Y. Abe, M. Onishi, G.L. Hashimoto, Lifetime and spectral evolution of a magma ocean with a steam atmosphere: its detectability by future direct imaging. Astrophys. J. **806**, 216 (2015). https://doi.org/10.1088/0004-637X/806/2/216
M. Hardiagon, D. Laporte, Y. Marizet, A. Provost, Co 2 degassing in a haplo-basaltic magma: an experimental approach. Mineral. Mag. **77**, 1258 (2013)
D.M. Harris, A.T. Anderson Jr., Concentrations, sources, and losses of H_2O, CO_2, and S in Kilauean basalt. Geochim. Cosmochim. Acta **47**, 1139–1150 (1983). https://doi.org/10.1016/0016-7037(83)90244-2
W.K. Hartmann, D.R. Davis, Satellite-sized planetesimals and lunar origin. Icarus **24**, 504–514 (1975). https://doi.org/10.1016/0019-1035(75)90070-6
E.H. Hauri, G.A. Gaetani, T.H. Green, Partitioning of water during melting of the Earth's upper mantle at H_2O-undersaturated conditions. Earth Planet. Sci. Lett. **248**, 715–734 (2006). https://doi.org/10.1016/j.epsl.2006.06.014

C. Hayashi, Origin of the solar system, in *Proceedings of the 5th ISAS Lunar and Planetary Symposium* (Insti. of Space and Astron. Sci., Kanagawa, 1972), pp. 13–18

C. Hayashi, Structure of the solar nebula, growth and decay of magnetic fields and effects of magnetic and turbulent viscosities on the nebula. Prog. Theor. Phys. Suppl. **70**, 35–53 (1981)

C. Hayashi, K. Nakazawa, H. Mizuno, Earth's melting due to the blanketing effect of the primordial dense atmosphere. Earth Planet. Sci. Lett. **43**, 22–28 (1979). https://doi.org/10.1016/0012-821X(79)90152-3

M.M. Hirschmann, Water, melting, and the deep Earth H_2O cycle. Annu. Rev. Earth Planet. Sci. **34**, 629–653 (2006). https://doi.org/10.1146/annurev.earth.34.031405.125211

M.M. Hirschmann, A.C. Withers, Ventilation of CO_2 from a reduced mantle and consequences for the early Martian greenhouse. Earth Planet. Sci. Lett. **270**, 147–155 (2008). https://doi.org/10.1016/j.epsl.2008.03.034

S. Hurwitz, O. Navon, Bubble nucleation in rhyolitic melts: experiments at high pressure, temperature, and water content. Earth Planet. Sci. Lett. **122**, 267–280 (1994). https://doi.org/10.1016/0012-821X(94)90001-9

S. Ida, J. Makino, Scattering of planetesimals by a protoplanet—slowing down of runaway growth. Icarus **106**, 210 (1993). https://doi.org/10.1006/icar.1993.1167

S. Ida, R.M. Canup, G.R. Stewart, Lunar accretion from an impact-generated disk. Nature **389**, 353–357 (1997). https://doi.org/10.1038/38669

M. Ikoma, H. Genda, Constraints on the mass of a habitable planet with water of nebular origin. Astrophys. J. **648**, 696–706 (2006). https://doi.org/10.1086/505780

M. Ikoma, K. Nakazawa, H. Emori, Formation of giant planets: dependences on core accretion rate and grain opacity. Astrophys. J. **537**, 1013–1025 (2000). https://doi.org/10.1086/309050

S. Inaba, M. Ikoma, Enhanced collisional growth of a protoplanet that has an atmosphere. Astron. Astrophys. **410**, 711–723 (2003). https://doi.org/10.1051/0004-6361:20031248

S. Inaba, G.W. Wetherill, M. Ikoma, Formation of gas giant planets: core accretion models with fragmentation and planetary envelope. Icarus **166**, 46–62 (2003). https://doi.org/10.1016/j.icarus.2003.08.001

Y. Ito, M. Ikoma, H. Kawahara, H. Nagahara, Y. Kawashima, T. Nakamoto, Theoretical emission spectra of atmospheres of hot rocky super-Earths. Astrophys. J. **801**, 144 (2015). https://doi.org/10.1088/0004-637X/801/2/144

A. Johansen, J. Blum, H. Tanaka, C. Ormel, M. Bizzarro, H. Rickman, The multifaceted planetesimal formation process, in *Protostars and Planets VI*, (2014), pp. 547–570. https://doi.org/10.2458/azu-uapress-9780816531240-ch024

K.T.M. Johnson, H.J.B. Dick, Open system melting and temporal and spatial variation of peridotite and basalt at the Atlantis II fracture zone. J. Geophys. Res. **97**, 9219–9241 (1992). https://doi.org/10.1029/92JB00701

J.F. Kasting, Runaway and moist greenhouse atmospheres and the evolution of Earth and Venus. Icarus **74**, 472–494 (1988). https://doi.org/10.1016/0019-1035(88)90116-9

J.F. Kasting, O_2 concentrations in dense primitive atmospheres: commentary. Planet. Space Sci. **43**, 11–13 (1995). https://doi.org/10.1016/0032-0633(94)00203-4

J.F. Kasting, J.B. Pollack, Loss of water from Venus. I—Hydrodynamic escape of hydrogen. Icarus **53**, 479–508 (1983). https://doi.org/10.1016/0019-1035(83)90212-9

T. Kodama, H. Genda, Y. Abe, K.J. Zahnle, Rapid water loss can extend the lifetime of planetary habitability. Astrophys. J. **812**, 165 (2015). https://doi.org/10.1088/0004-637X/812/2/165

E. Kokubo, S. Ida, On runaway growth of planetesimals. Icarus **123**, 180–191 (1996). https://doi.org/10.1006/icar.1996.0148

E. Kokubo, S. Ida, Oligarchic growth of protoplanets. Icarus **131**, 171–178 (1998). https://doi.org/10.1006/icar.1997.5840

R.K. Kopparapu, R. Ramirez, J.F. Kasting, V. Eymet, T.D. Robinson, S. Mahadevan, R.C. Terrien, S. Domagal-Goldman, V. Meadows, R. Deshpande, Habitable zones around main-sequence stars: new estimates. Astrophys. J. **765**, 131 (2013). https://doi.org/10.1088/0004-637X/765/2/131

R.K. Kopparapu, R.M. Ramirez, J. SchottelKotte, J.F. Kasting, S. Domagal-Goldman, V. Eymet, Habitable zones around main-sequence stars: dependence on planetary mass. Astrophys. J. Lett. **787**, 29 (2014). https://doi.org/10.1088/2041-8205/787/2/L29

J. Korenaga, Initiation and evolution of plate tectonics on Earth: theories and observations. Annu. Rev. Earth Planet. Sci. **41**, 117–151 (2013). https://doi.org/10.1146/annurev-earth-050212-124208

S. Labrosse, J.W. Hernlund, N. Coltice, A crystallizing dense magma ocean at the base of the Earth's mantle. Nature **450**, 866–869 (2007). https://doi.org/10.1038/nature06355

M. Lambrechts, A. Johansen, Rapid growth of gas-giant cores by pebble accretion. Astron. Astrophys. **544**, 32 (2012). https://doi.org/10.1051/0004-6361/201219127

M. Lambrechts, A. Johansen, Forming the cores of giant planets from the radial pebble flux in protoplanetary discs. Astron. Astrophys. **572**, 107 (2014). https://doi.org/10.1051/0004-6361/201424343

L.D. Landau, E.M. Lifshitz, *Statistical Physics* (Pergamon, New York, 1980)

M.A. Lange, T.J. Ahrens, The evolution of an impact-generated atmosphere. Icarus **51**, 96–120 (1982). https://doi.org/10.1016/0019-1035(82)90031-8

T. Lebrun, H. Massol, E. Chassefière, A. Davaille, E. Marcq, P. Sarda, F. Leblanc, G. Brandeis, Thermal evolution of an early magma ocean in interaction with the atmosphere. J. Geophys. Res., Planets **118**, 1155–1176 (2013). https://doi.org/10.1002/jgre.20068

N.G. Lensky, R.W. Niebo, J.R. Holloway, V. Lyakhovsky, O. Navon, Bubble nucleation as a trigger for xenolith entrapment in mantle melts. Earth Planet. Sci. Lett. **245**, 278–288 (2006). https://doi.org/10.1016/j.epsl.2005.11.064

J.L. Linsky, J. Fontenla, K. France, The intrinsic extreme ultraviolet fluxes of F5 V TO M5 V stars. Astrophys. J. **780**, 61 (2014). https://doi.org/10.1088/0004-637X/780/1/61

J.J. Lissauer, Timescales for planetary accretion and the structure of the protoplanetary disk. Icarus **69**, 249–265 (1987). https://doi.org/10.1016/0019-1035(87)90104-7

R.E. Lupu, K. Zahnle, M.S. Marley, L. Schaefer, B. Fegley, C. Morley, K. Cahoy, R. Freedman, J.J. Fortney, The atmospheres of earthlike planets after giant impact events. Astrophys. J. **784**, 27 (2014). https://doi.org/10.1088/0004-637X/784/1/27

M. Mangan, T. Sisson, Delayed, disequilibrium degassing in rhyolite magma: decompression experiments and implications for explosive volcanism. Earth Planet. Sci. Lett. **183**, 441–455 (2000). https://doi.org/10.1016/S0012-821X(00)00299-5

E. Marcq, A. Salvador, H. Massol, A. Davaille, Thermal radiation of magma ocean planets using a 1-D radiative-convective model of H_2O-CO_2 atmospheres. J. Geophys. Res., Planets **122**, 1539–1553 (2017). https://doi.org/10.1002/2016JE005224

D. Martin, R. Nokes, Crystal settling in a vigorously converting magma chamber. Nature **332**, 534–536 (1988). https://doi.org/10.1038/332534a0

S. Maruyama, M. Ikoma, H. Genda, K. Hirose, T. Yokoyama, M. Santosh, The naked planet Earth: most essential pre-requisite for the origin and evolution of life. Geosci. Front. **4**(2), 141–165 (2013)

H. Massol, K. Hamano, F. Tian, M. Ikoma, Y. Abe, E. Chassefière, A. Davaille, H. Genda, M. Güdel, Y. Hori, F. Leblanc, E. Marcq, P. Sarda, V.I. Shematovich, A. Stökl, H. Lammer, Formation and evolution of protoatmospheres. Space Sci. Rev. (2016). https://doi.org/10.1007/s11214-016-0280-1

D. McKenzie, The generation and compaction of partially molten rock. J. Petrol. **5**, 713–765 (1984)

J. Monteux, N. Coltice, F. Dubuffet, Y. Ricard, Thermo-mechanical adjustment after impacts during planetary growth. Geophys. Res. Lett. **34**, 24201 (2007). https://doi.org/10.1029/2007GL031635

J. Monteux, Y. Ricard, N. Coltice, F. Dubuffet, M. Ulvrova, A model of metal-silicate separation on growing planets. Earth Planet. Sci. Lett. **287**, 353–362 (2009). https://doi.org/10.1016/j.epsl.2009.08.020

A. Morbidelli, M. Lambrechts, S. Jacobson, B. Bitsch, The great dichotomy of the solar system: small terrestrial embryos and massive giant planet cores. Icarus **258**, 418–429 (2015). https://doi.org/10.1016/j.icarus.2015.06.003

C.C. Mourtada-Bonnefoi, D. Laporte, Experimental study of homogeneous bubble nucleation in rhyolitic magmas. Geophys. Res. Lett. **26**, 3505–3508 (1999). https://doi.org/10.1029/1999GL008368

C.C. Mourtada-Bonnefoi, D. Laporte, Homogeneous bubble nucleation in rhyolitic magmas: an experimental study of the effect of H_2O and CO_2. J. Geophys. Res., Solid Earth **107**, 2066 (2002). https://doi.org/10.1029/2001JB000290

M. Nakajima, D.J. Stevenson, Melting and mixing states of the Earth's mantle after the Moon-forming impact. Earth Planet. Sci. Lett. **427**, 286–295 (2015). https://doi.org/10.1016/j.epsl.2015.06.023

S. Nakajima, Y.-Y. Hayashi, Y. Abe, A study on the 'runaway greenhouse effect' with a one-dimensional radiative-convective equilibrium model. J. Atmos. Sci. **49**, 2256–2266 (1992). https://doi.org/10.1175/1520-0469(1992)049<2256:ASOTGE>2.0.CO;2

O. Navon, V. Lyakhovsky, Vesiculation processes in silicic magmas. Geol. Soc. (Lond.) Spec. Publ. **145**, 27–50 (1998). https://doi.org/10.1144/GSL.SP.1996.145.01.03

Noack et al., Space Sci. Rev. (2017, this issue). https://doi.org/10.1007/s11214-017-0413-1

R. Nomura, K. Hirose, K. Uesugi, Y. Ohishi, A. Tsuchiyama, A. Miyake, Y. Ueno, Low core-mantle boundary temperature inferred from the solidus of pyrolite. Science, 1248186 (2014). https://doi.org/10.1126/science.1248186

E. Ohtani, K. Litasov, T. Hosoya, T. Kubo, T. Kondo, Water transport into the deep mantle and formation of a hydrous transition zone. Phys. Earth Planet. Inter. **143**, 255–269 (2004). https://doi.org/10.1016/j.pepi.2003.09.015

T. Okuchi, Hydrogen partitioning into molten iron at high pressure: implications for Earth's core. Science **278**, 1781 (1997). https://doi.org/10.1126/science.278.5344.1781

C.W. Ormel, H.H. Klahr, The effect of gas drag on the growth of protoplanets. Analytical expressions for the accretion of small bodies in laminar disks. Astron. Astrophys. **520**, 43 (2010). https://doi.org/10.1051/0004-6361/201014903

P. Papale, R. Moretti, D. Barbato, The compositional dependence of the saturation surface of $H_2O + CO_2$ fluids in silicate melts. Chem. Geol. **229**(1), 78–95 (2006)

A.R. Philpotts, C.M. Brustman, J. Shi, W.D. Carlson, C. Denison, Plagioclase-chain networks in slowly cooled basaltic magma. Am. Mineral. **84**(11–12), 1819–1829 (1999)

E. Pierazzo, A.M. Vickery, H.J. Melosh, A reevaluation of impact melt production. Icarus **127**, 408–423 (1997). https://doi.org/10.1006/icar.1997.5713

J.B. Pollack, O. Hubickyj, P. Bodenheimer, J.J. Lissauer, M. Podolak, Y. Greenzweig, Formation of the giant planets by concurrent accretion of solids and gas. Icarus **124**, 62–85 (1996). https://doi.org/10.1006/icar.1996.0190

C.C. Reese, V.S. Solomatov, Fluid dynamics of local martian magma oceans. Icarus **184**, 102–120 (2006). https://doi.org/10.1016/j.icarus.2006.04.008

J. Renner, B. Evans, G. Hirth, On the rheologically critical melt fraction. Earth Planet. Sci. Lett. **181**(4), 585–594 (2000)

I. Ribas, E.F. Guinan, M. Güdel, M. Audard, Evolution of the solar activity over time and effects on planetary atmospheres. I. High-energy irradiances (1–1700 Å). Astrophys. J. **622**, 680–694 (2005). https://doi.org/10.1086/427977

Y. Ricard, O. Šrámek, F. Dubuffet, A multi-phase model of runaway core-mantle segregation in planetary embryos. Earth Planet. Sci. Lett. **284**, 144–150 (2009). https://doi.org/10.1016/j.epsl.2009.04.021

R.A. Robie, B.S. Hemmingway, J.R. Fisher, Thermodynamic properties of minerals and related substances at 298.15 K and 1 bar pressure and at higher temperature, in *Geol. Surv. Bull.*, (1978), p. 1452

D.C. Rubie, C.K. Gessmann, D.J. Frost, Partitioning of oxygen during core formation on the Earth and Mars. Nature **429**, 58–61 (2004). https://doi.org/10.1038/nature02473

D.C. Rubie, S.A. Jacobson, A. Morbidelli, D.P. O'Brien, E.D. Young, J. de Vries, F. Nimmo, H. Palme, D.J. Frost, Accretion and differentiation of the terrestrial planets with implications for the compositions of early-formed solar system bodies and accretion of water. Icarus **248**, 89–108 (2015a). https://doi.org/10.1016/j.icarus.2014.10.015

D.C. Rubie, F. Nimmo, H.J. Melosh, Formation of the Earth's core, in *Treatise on Geochemistry*, ed. by G. Schubert, 2nd edn. (2015b), pp. 43–79

V.S. Safronov, *Evolution of the Protoplanetary Cloud and Formation of the Earth and Planets* (1972)

A. Salvador, H. Massol, A. Davaille, E. Marcq, P. Sarda, E. Chassefière, The relative influence of H_2O and CO_2 on the primitive surface conditions and evolution of rocky planets. J. Geophys. Res., Planets **122**, 1458–1486 (2017). https://doi.org/10.1002/2017JE005286

H. Samuel, P.J. Tackley, Dynamics of core formation and equilibration by negative diapirism. Geochem. Geophys. Geosyst. **9**, 06011 (2008). https://doi.org/10.1029/2007GC001896

S. Sasaki, K. Nakazawa, Metal-silicate fractionation in the growing Earth—energy source for the terrestrial magma ocean. J. Geophys. Res. **91**, 9231–9238 (1986). https://doi.org/10.1029/JB091iB09p09231

L. Schaefer, R.D. Wordsworth, Z. Berta-Thompson, D. Sasselov, Predictions of the atmospheric composition of GJ 1132b. Astrophys. J. **829**, 63 (2016). https://doi.org/10.3847/0004-637X/829/2/63

V. Solomatov, Fluid dynamics of a terrestrial magma ocean, in *Origin of the Earth and Moon*, vol. 1 (2000), pp. 323–338

V. Solomatov, Evolution of the Earth, in *Magma Oceans and Primordial Mantle Differentiation*, ed. by D. Stevenson. Treatise on Geophysics 9 (Elsevier, Amsterdam, 2007), pp. 91–119

V.S. Solomatov, D.J. Stevenson, Nonfractional crystallization of a terrestrial magma ocean. J. Geophys. Res. **98**, 5391–5406 (1993a). https://doi.org/10.1029/92JE02579

V.S. Solomatov, D.J. Stevenson, Suspension in convective layers and style of differentiation of a terrestrial magma ocean. J. Geophys. Res. **98**, 5375–5390 (1993b). https://doi.org/10.1029/92JE02948

V.S. Solomatov, P. Olson, D.J. Stevenson, Entrainment from a bed of particles by thermal convection. Earth Planet. Sci. Lett. **120**, 387–393 (1993). https://doi.org/10.1016/0012-821X(93)90252-5

R.S.J. Sparks, The dynamics of bubble formation and growth in magmas: a review and analysis. J. Volcanol. Geotherm. Res. **3**, 1–37 (1978). https://doi.org/10.1016/0377-0273(78)90002-1

R. Sparks, J. Barclay, C. Jaupart, H. Mader, J. Phillips, Physical aspects of magma degassing; 1, experimental and theoretical constraints on vesiculation. Rev. Mineral. Geochem. **30**(1), 413–445 (1994)

J. Suckale, L.T. Elkins-Tanton, J.A. Sethian, Crystals stirred up: 2. Numerical insights into the formation of the earliest crust on the Moon. J. Geophys. Res., Planets **117**, 08005 (2012a). https://doi.org/10.1029/2012JE004067

J. Suckale, J.A. Sethian, J.-d. Yu, L.T. Elkins-Tanton, Crystals stirred up: 1. Direct numerical simulations of crystal settling in nondilute magmatic suspensions. J. Geophys. Res., Planets **117**, 08004 (2012b). https://doi.org/10.1029/2012JE004066

E. Takahashi, Melting of a dry peridotite KLB-1 up to 14 GPa—implications on the origin of peridotitic upper mantle. J. Geophys. Res. **91**, 9367–9382 (1986). https://doi.org/10.1029/JB091iB09p09367

S.M. Tikoo, L.T. Elkins-Tanton, The fate of water within Earth and super-Earths and implications for plate tectonics. Philos. Trans. R. Soc. Lond. Ser. A **375**, 20150394 (2017). https://doi.org/10.1098/rsta.2015.0394

W. Tonks, H. Melosh, The physics of crystal settling and suspension in a turbulent magma ocean, in *Origin of the Earth*, vol. 1 (1990), pp. 151–174

W.B. Tonks, H.J. Melosh, Core formation by giant impacts. Icarus **100**, 326–346 (1992). https://doi.org/10.1016/0019-1035(92)90104-F

W.B. Tonks, H.J. Melosh, Magma ocean formation due to giant impacts. J. Geophys. Res. **98**, 5319–5333 (1993). https://doi.org/10.1029/92JE02726

R.G. Trønnes, D.J. Frost, Peridotite melting and mineral-melt partitioning of major and minor elements at 22–24.5 GPa. Earth Planet. Sci. Lett. **197**, 117–131 (2002). https://doi.org/10.1016/S0012-821X(02)00466-1

J.M. Tucker, S. Mukhopadhyay, Evidence for multiple magma ocean outgassing and atmospheric loss episodes from mantle noble gases. Earth Planet. Sci. Lett. **393**, 254–265 (2014). https://doi.org/10.1016/j.epsl.2014.02.050

J. Venturini, Y. Alibert, W. Benz, M. Ikoma, Critical core mass for enriched envelopes: the role of H_2O condensation. Astron. Astrophys. **576**, 114 (2015). https://doi.org/10.1051/0004-6361/201424008

A.M. Vickery, H.J. Melosh, Atmospheric erosion and impactor retention in large impacts. With application to mass extinctions, in *Geol. Soc. Am. Spec. Pap.*, vol. 247 (1990), pp. 289–300

D. Walker, J. Longhi, J.F. Hays, Differentiation of a very thick magma body and implications for the source regions of mare basalts, in *Lunar and Planetary Science Conference Proceedings*. Lunar and Planetary Science Conference Proceedings, vol. 6 (1975), pp. 1103–1120

J.C.G. Walker, P.B. Hays, J.F. Kasting, A negative feedback mechanism for the long-term stabilization of the Earth's surface temperature. J. Geophys. Res. **86**, 9776–9782 (1981). https://doi.org/10.1029/JC086iC10p09776

A.J. Watson, T.M. Donahue, J.C.G. Walker, The dynamics of a rapidly escaping atmosphere—applications to the evolution of Earth and Venus. Icarus **48**, 150–166 (1981). https://doi.org/10.1016/0019-1035(81)90101-9

M.J. Way, A.D. Del Genio, N.Y. Kiang, L.E. Sohl, D.H. Grinspoon, I. Aleinov, M. Kelley, T. Clune, Was Venus the first habitable world of our solar system? Geophys. Res. Lett. **43**, 8376–8383 (2016). https://doi.org/10.1002/2016GL069790

G.W. Wetherill, G.R. Stewart, Accumulation of a swarm of small planetesimals. Icarus **77**, 330–357 (1989). https://doi.org/10.1016/0019-1035(89)90093-6

P.J. Wyllie, I.D. Ryabchikov, Volatile components, magmas, and critical fluids in upwelling mantle. J. Petrol. **41**(7), 1195–1206 (2000)

J. Yang, G. Boué, D.C. Fabrycky, D.S. Abbot, Strong dependence of the inner edge of the habitable zone on planetary rotation rate. Astrophys. J. Lett. **787**, 2 (2014). https://doi.org/10.1088/2041-8205/787/1/L2

K.J. Zahnle, J.F. Kasting, J.B. Pollack, Evolution of a steam atmosphere during Earth's accretion. Icarus **74**, 62–97 (1988). https://doi.org/10.1016/0019-1035(88)90031-0

K. Zahnle, N. Arndt, C. Cockell, A. Halliday, E. Nisbet, F. Selsis, N.H. Sleep, Emergence of a habitable planet. Space Sci. Rev. **129**, 35–78 (2007). https://doi.org/10.1007/s11214-007-9225-z

A. Zerr, A. Diegeler, R. Boehler, Solidus of Earth's deep mantle. Science **281**, 243 (1998). https://doi.org/10.1126/science.281.5374.243

Space Sci Rev (2018) 214:39
https://doi.org/10.1007/s11214-018-0473-x

Water and the Interior Structure of Terrestrial Planets and Icy Bodies

J. Monteux[1] · **G.J. Golabek**[2] · **D.C. Rubie**[2] · **G. Tobie**[3] · **E.D. Young**[4]

Received: 30 July 2016 / Accepted: 11 January 2018 / Published online: 24 January 2018

Abstract Water content and the internal evolution of terrestrial planets and icy bodies are closely linked. The distribution of water in planetary systems is controlled by the temperature structure in the protoplanetary disk and dynamics and migration of planetesimals and planetary embryos. This results in the formation of planetesimals and planetary embryos with a great variety of compositions, water contents and degrees of oxidation. The internal evolution and especially the formation time of planetesimals relative to the timescale of radiogenic heating by short-lived ^{26}Al decay may govern the amount of hydrous silicates and leftover rock–ice mixtures available in the late stages of their evolution. In turn, water content may affect the early internal evolution of the planetesimals and in particular metal-silicate separation processes. Moreover, water content may contribute to an increase of oxygen fugacity and thus affect the concentrations of siderophile elements within the silicate reservoirs of Solar System objects. Finally, the water content strongly influences the differentiation rate of the icy moons, controls their internal evolution and governs the alteration processes occurring in their deep interiors.

Keywords Water · Differentiation · Internal evolution · Terrestrial planets · Icy moons

The Delivery of Water to Protoplanets, Planets and Satellites
Edited by Michel Blanc, Alessandro Morbidelli, Yann Alibert, Lindy Elkins-Tanton, Paul Estrada, Keiko Hamano, Helmut Lammer, Sean Raymond and Maria Schönbächler

✉ J. Monteux
julien.monteux@uca.fr

[1] CNRS, IRD, OPGC, Laboratoire Magmas et Volcans, Université Clermont Auvergne, 63000 Clermont-Ferrand, France

[2] Bayerisches Geoinstitut, Universität Bayreuth, Universitätsstrasse 30, 95440 Bayreuth, Germany

[3] Laboratoire de Planétologie et Géodynamique, UMR-CNRS 6112, Université de Nantes, 44322 Nantes Cedex, France

[4] Department of Earth, Planetary, and Space Sciences, University of California Los Angeles, Los Angeles, CA 90095, USA

1 Introduction

The Solar System displays a strong dichotomy between the inner region that is characterized by relatively dry planetary objects having a very small water fraction (<0.1%), and the outer solar system where water ice constitutes a large fraction of solid material (>20%), which is inherited from accretion processes. When compared with other planetary systems, the Solar System seems however rather unusual. Exoplanet surveys have revealed that planets intermediate in mass between Earth and Neptune are surprisingly common, but are notably absent in the Solar System (Mayor et al. 2014; Howard et al. 2012; Marcy 2014). Model mass-radius relationships indicate a great diversity of interior compositions and atmospheric extents for the Super-Earth/Mini-Neptune-planet class (e.g. Howard et al. 2012), suggesting a wide range of volatile contents (including water) and compositions.

The distribution of water in the planetary system is controlled by the temperature structure in the protoplanetary disk and dynamics and migration of planetesimals and planetary embryos (Raymond et al. 2004; Cowan and Abbot 2014; O'Brien et al. 2014; Rubie et al. 2015a). This results in the formation of planetesimals and planetary embryos with a great variety of compositions and water contents. Subsequent accretionary processes can lead to the formation of planets that range from completely dry planets to "water worlds" with more than 100 Earth-oceans of water (1 ocean = 1.5×10^{21} kg H_2O) (Raymond et al. 2004; Cowan and Abbot 2014; O'Brien et al. 2014; Rubie et al. 2015a). As the evolution of water content inside a growing proto-planet is a strong function of pressure-temperature conditions, the early stages of planetary formation that govern their accretion rates and their early heat budget may have played a major role in the water distribution in the solar system. Indeed, depending on the accretion rate and on the formation time of the building blocks of the terrestrial planets relative to the timescale of radiogenic heating by the decay of short-lived ^{26}Al, water can either be incorporated as an ice–rock mixture or in the form of hydrated silicates or be absent. Aqueous alteration appears to be common in carbonaceous chondrites (Kerridge and Bunch 1979; Zolensky et al. 1989; Krot et al. 1998). The proportion of planetesimal water incorporated into rock depends critically on whether water was mobile or stagnant (e.g., Young 2001; Young et al. 1999, 2003; Bland et al. 2009; Fu et al. 2017; Castillo-Rogez and Young 2017). In turn, the water content of Solar System objects likely played a key role in the internal evolution of the solar system bodies. For example, water dissolved in silicate minerals is known to significantly reduce mantle viscosity and melting temperature (e.g. Hirth and Kohlstedt 1996; Katz et al. 2003; Hirschmann 2006). It has therefore a potential influence on the early differentiation processes and a strong feedback on the thermal state of the mantle and the vigor of convection (e.g. Korenaga 2010).

Water is also the major constituent of the moons orbiting Jupiter (except for Io), Saturn, Uranus and Neptune, and all planetary objects beyond Neptune (e.g. Hussmann et al. 2015). In these icy objects, water is likely present either as thick ice shells or deep water oceans (e.g. Hussmann et al. 2006) and in some cases is being ejected from venting plumes composed mainly of water vapor and ice particles (e.g. Porco et al. 2006 for Enceladus, Roth et al. 2014 for Europa). The combined effect of accretional heating, radiogenic decay by short-lived isotopes, tidal heating associated with despinning, and viscous heating associated with sinking negative rock diapirs may increase the internal temperature above the melting point of ice (e.g. Tobie et al. 2013). This may result in a partial to full ice–rock separation. During this differentiation as well as at the present-day, interactions between solid rock and liquid water might have promoted chemical reactions facilitated by the presence of ammonia (Schubert et al. 2010; Sohl et al. 2010). It is also possible that aqueous alteration might have led to serpentinization of olivine-rich rocks and to the formation of a highly hydrated rocky core that may result from a rapid differentiation process that implies large-scale melting events, as proposed for Saturn's moon Titan (e.g. Castillo-Rogez and Lunine 2010).

Finally, comets and comet-like objects are another class of objects in which ice has been detected as a major component (80%) (Encrenaz 2008). These bodies, with radii usually smaller than 10 km, were formed in the outer Solar System. Comets are undifferentiated and, hence, represent weakly altered remnants of the early stages of solar-system formation. Because of radiogenic heating from ^{26}Al and ^{60}Fe, the interior of comets may reach temperatures above the melting point of water and ultravolatile species may be lost (Prialnik et al. 2008; Mousis et al. 2017). In the following study, we do not consider the comets and we will only focus on objects with radii larger than 25 km (see review of Prialnik et al. 2008 for more details).

The structure of this article is as follows: We first detail the distribution of water in the protoplanetary disc and its effect on oxidation. Second, we present a discussion of the effect of internal evolution of planetesimals on water content based on numerical models. In particular we investigate the influence of planetesimal radius and formation time. We then discuss the role of water on the internal evolution of proto-planets and terrestrial bodies focusing particularly on metal/silicate separation and oxygen fugacity evolution. Finally we briefly describe the internal evolution of icy satellites from their accretion to their present-day activity and emphasize the role of water in differentiation and alteration processes.

2 Distribution of Water in the Protoplanetary Disc and Its Effect on Oxidation

Protoplanetary disks of gas and dust are by necessity characterized by radial increases in temperature and pressure towards the nascent star at their center. The inner portions of the disk are heated by two mechanisms. So-called "passive" disks are heated at their surfaces by illumination from the central star. Viscous heating in "active" accretion disks can warm up the inner portions of the disk to temperatures well above those imposed by passive illumination alone. The degree of viscous heating depends on the mechanism for angular transport (accretion onto the star) (e.g., Bai and Stone 2013). The greater the effective viscosity of the disk, the greater the heating related to accretion. In all cases, there will be specific radii corresponding to the condensation fronts for various volatile components (referred to as "snowlines"). From the perspective of volatiles and planet formation, the most important of these may be that for water. The water snowline separating the inner regions of the disk where water vapor is stable from the more distal regions of the disk where water ice is stable is thought to exert a first-order control on the chemistry of planetesimals. The snowline in an accretion disk is closest to the star at the midplane where the intervening mass of gas and dust shields the disk from light from the central star. Higher in the disk, the snowline curves to greater radial distances due to heating of the surfaces of the disk by the central star. We are most concerned with the position of the water snowline at the midplane when contemplating planetesimal formation. However it must be kept in mind that the time-dependent position of the snowline in the midplane and its effects on planetesimal formation over time are still debated (e.g., Morbidelli et al. 2016).

The bulk compositions of planetesimals and planetary embryos in the protoplanetary disc, including water and some volatile element abundances, have been constrained by core formation modelling (Rubie et al. 2015a, 2016). Mantle abundances of siderophile elements

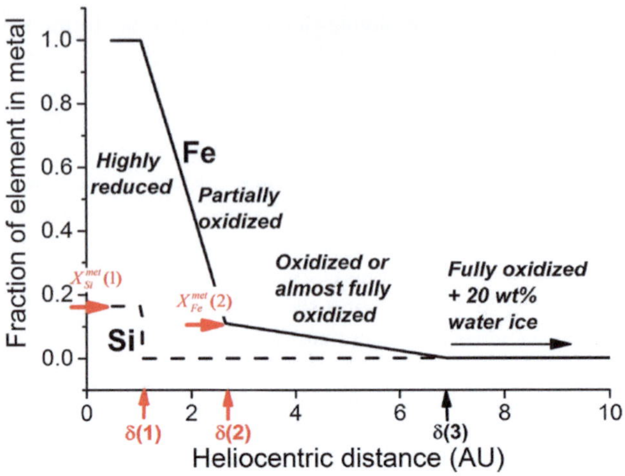

Fig. 1 The composition of initial bodies (embryos and planetesimals) as a function of heliocentric distance determined in the accretion/core formation model of Rubie et al. (2015a). At heliocentric distances less than ~1 au, 99% of all Fe is present as metal and ~18% of total Si is dissolved in the metal—which makes the bulk composition very reduced. Beyond ~1 au, compositions become increasingly oxidized with increasing heliocentric distance because the fraction of Fe present as metal decreases and the fraction of Fe present as FeO in the silicate increases. Beyond 6–7 au, compositions are fully oxidized (no metal) and contain 20 wt% water—these are bodies that originally accreted beyond the snowline. The *red arrows* indicate values of parameters that were refined by least squares

(Ni, Co, W etc.) can be reproduced in multistage core formation models when the initial 60–70% of Earth accretes from highly reduced material and the final 30–40% from relatively-oxidized material, in combination with the effects of Si partitioning into the core (Rubie et al. 2011). The late accretion of oxidized material is also consistent with the relatively late accretion of volatile elements to the Earth (e.g. Schönbächler et al. 2010; Dauphas 2017). This scenario was developed further by Rubie et al. (2015a, 2016) by formulating a combined accretion/core formation model. This was based on "Grand Tack" N-body accretion models (Walsh et al. 2011; O'Brien et al. 2014) that started with up to ~200 embryos (distributed from 0.7 to 3.0 au) embedded in a protoplanetary disc consisting of up to 4400 planetesimals (distributed from 0.7 to 9.5 au). Each accretional collision between embryos and other embryos and planetesimals was considered to result in an episode of core formation, provided the impactor was metal-bearing. The bulk composition of each of the starting bodies was defined in terms of CI ratios of non-volatile elements, with oxygen and water contents varying as a function of heliocentric distance. Based on these bulk compositions, the compositions of equilibrated metal and silicate in each core-formation event were determined by a combined mass balance/element partitioning approach, thus enabling the evolving compositions of the mantles and cores of all growing planets to be modeled. The composition of the model Earth in each simulation was fit to Earth's mantle composition by a least squares refinement of the metal-silicate equilibration pressure and several parameters that describe the compositions of initial bodies (Fig. 1). Best fits to Earth's mantle composition were obtained when compositions of bodies close to the Sun are highly reduced and those further out are relatively oxidized (Fig. 1).

The gradient in Fig. 1 requires explanation in the context of the young solar protoplanetary disk. Dynamical models have shown that radial mixing within the protoplanetary disk was likely to occur especially during the late stages of planetary formation (e.g. Chambers

Fig. 2 Schematic illustrating the enhancement of the fayalite component in olivine due to an increase in water. This scheme could have operated at moderately high temperatures in the protoplanetary disk where dust was processed

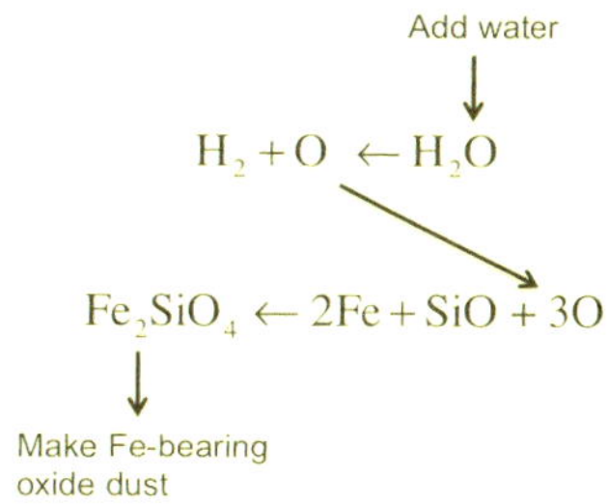

2013; Raymond et al. 2004). Variation in oxygen fugacity across the protoplanetary gaseous disk is a longstanding question. At temperatures greater than about 500 to 600 K or so, Fe_2SiO_4, representing oxidized iron, is unstable at solar oxygen fugacities. This temperature is lower than the ca. 1400 K required for diffusion and annealing of silicate grains (Figs. 2 and 3). The solution is to pump up the oxidation state of the vapor phase by infusing it with water inside of the snowline. As the protoplanetary disk evolved, small icy bodies will have drifted inwards of the snowline, resulting in addition of H_2O vapor by subsequent sublimation of water ice. The longevity of this vapor depended critically on the competing rates of inward migration across the snowline and the outward diffusive flux of water driven by freezing of water at the snowline (e.g., Cuzzi and Zahnle 2004). The result of the competing fluxes may have been a pulse of elevated H_2O/H_2 just inside the snowline (Fig. 4). Such a localized pulse of water vapor has been observed in the TW Hydra transition disk at a distance of about 4 au from the central star (Zhang et al. 2013). The significant amount of oxidized iron in grains comprising chondrite matrix, even in rocks with limited evidence for water–rock reactions, is testament to the fact that such a process must have occurred where planetesimals were forming (Grossman et al. 2012). The amount of water required can be estimated from the equilibrium relationship

$$H_2 + 1/2O_2 \Leftrightarrow H_2O$$

for which we obtain the equation for f_{O_2} (e.g., Krot et al. 2000)

$$\log(f_{O_2}) \simeq 2\log\left(\frac{H_2O}{H_2}\right) + 5.67 - \frac{256664}{T(K)}. \tag{1}$$

Inspection of this equation shows that in order to change oxygen fugacity by ~5 log units as required to shift from solar to "planetary" oxidation states, H_2O/H_2 must have increased by a factor of ~375 at inner disk temperatures of ~1400 K (Fig. 5). Geometric considerations suggest that focusing solar abundances of water into the inner 5 au of the disk by inward transport can only raise the H_2O/H_2 ratio by about a factor of 10 (where, for example, surface density varies as $1/R$ where R is the radial distance from the Sun).

The picture that emerges is one in which ice accreted within planetesimals outside of the snowline may have reacted with the host rock as the planetesimals evolved in response to heating by ^{26}Al (see also Sect. 3). Liquid water reactions with rock will have oxidized the rock. Inside of the snowline there was likely a region where inward migration of ice enhanced the H_2O/H_2 ratio in the vapor phase just inside of the snowline, but further into the disk in the region of terrestrial planet formation, the ambient conditions favored reduced iron over oxidized iron as a result of higher temperatures and a decrease in gaseous H_2O toward the star. The snowline must have migrated inward with time in response to a decrease

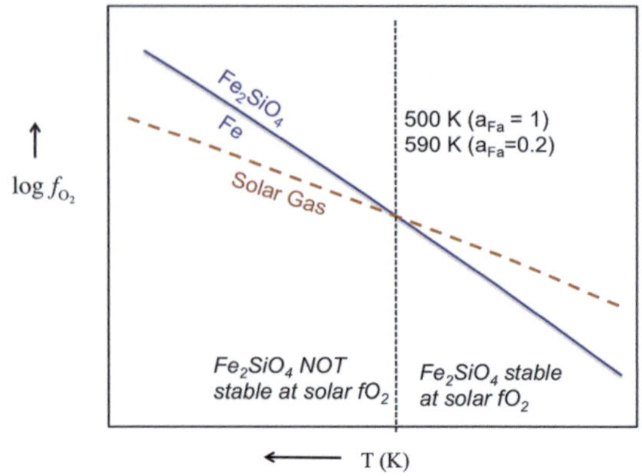

Fig. 3 Schematic diagram showing the relationship between oxygen fugacity and temperature for a solar gas and for the oxidation state of iron. At higher temperatures, reduced iron is favored over the fayalite component for a solar-like gas. At lower temperatures, Fe_2SiO_4 can be stabilized, but whether this reaction can actually occur at lower temperatures is in question. Typical activities for fayalite in olivine at two different temperatures are shown for illustration

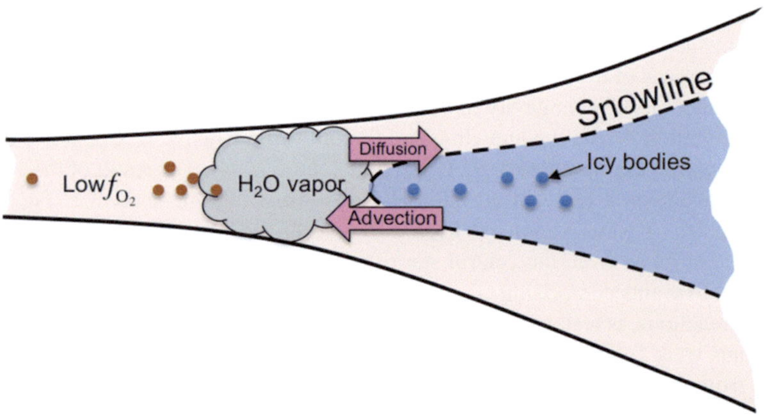

Fig. 4 Schematic diagram illustrating the competition between advection of icy material through the snowline of the protoplanetary disk and backwards net transfer of water via diffusion driven by a buildup of water vapor inside of the snowline. The process is described by Cuzzi and Zahnle (2004) and is consistent with Spitzer and Herschel observations of TW Hydra (Zhang et al. 2013)

in the stellar accretion rate, and the best fit shown in Fig. 1 suggests that its time-integrated "fossilized" location (see below) was somewhere between 6 and 2 au.

The position of the midplane water snowline as a function of time can be obtained from the temperature and pressure structure of a generic protoplanetary accretion disk. The T and P structure of the disk can be modeled using basic principles of dissipative radiative losses resulting from orbital torques. The torques are sustained by the viscosity of the disk that results in the inward spiral of gas towards the growing star. For dimensionless viscosity parameter α (e.g., Shakura and Sunyaev 1973), sound speed c_s, and scale height $H(R)$, the effective viscosity of the disk is $\upsilon = \alpha c_s H(R)$. The temperature at the midplane of the disk can be written in terms of the accretion rate of the system, $\dot{m}$, as follows

$$T_{\text{midplane}} = \left[\frac{\frac{3}{2}\Sigma(R)\kappa_R}{4} + 1\right]^{1/4} \left[\frac{3}{8\pi\sigma}\frac{GM_*\dot{m}}{R^3}\right]^{1/4} \tag{2}$$

where κ_R is the Rosseland mean opacity, $\Sigma(R)$ is the surface density of the disk that is a function of R from the central star, G is the gravitational constant, M_* is the mass of the

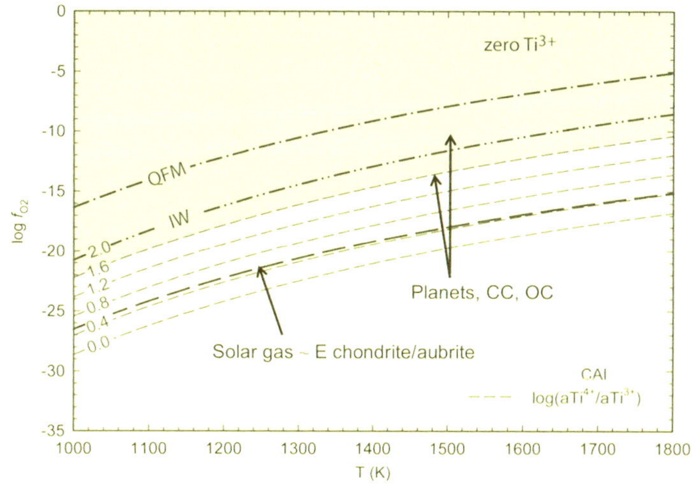

Fig. 5 The logarithm of oxygen fugacity versus temperature for various oxygen buffers relevant for the early solar system. The heavy *dashed line* shows the relationship for a solar gas. For comparison, planetary materials (Earth's mantle, basalts from various bodies) exhibit oxygen fugacities closer to the iron-wüstite (IW) and quartz–fayalite–magnetite (QFM) oxygen buffers. Note that the difference between a solar gas and those planetary materials implies an increase of more than 5 orders of magnitude in oxygen fugacity. The region shaded in *grey* represents oxidation states that favor Ti^{3+} over Ti^{4+} in calcium–aluminum-rich inclusions (CAIs) in carbonaceous chondrites. *Contours* for the activity ratio of the two forms of Ti in CAIs are also shown

central star, $\dot{m}$ is the accretion rate and σ is the Stefan–Boltzmann constant. The first term on the right-hand-side of (2) accounts for the optical depth perpendicular to the midplane and is equivalent to $T_{\text{midplane}}/T_{\text{surface}}$. The surface temperature is given by the second term. Equation (2) shows that the temperature of the disk at any radial distance R from the star depends on the rate of mass accretion through the disk. The latter has been measured in young stellar objects today and varies from about $10^{-6}\ M_\odot\ \text{yr}^{-1}$ early in the history of stellar accretion to $10^{-8}\ M_\odot\ \text{yr}^{-1}$ (Hartmann 2000) later on.

Evaluation of (2) requires evaluating the surface density with distance from the star, $\Sigma(R)$. The surface density can be written in terms of the viscosity as (Armitage 2010)

$$\Sigma(R) = \frac{\dot{m}}{3\pi\alpha c_s H(R)}, \tag{3}$$

where scale height H of the disk as a function of radial distance from the star is

$$H(R) = \frac{c_s}{\Omega} = \sqrt{\frac{k_b T R^3}{G M_* \mu}}, \tag{4}$$

and where Ω is orbital angular velocity, k_b is Boltzmann's constant, T is the temperature and μ is the average molecular mass for the gas comprising the disk. Equation (2) can be evaluated analytically with reasonable precision by assuming a constant value of 1 m^2/kg for κ_R and replacing (3) with a parameterization for $\Sigma(R)$ suggested by the mass distribution of planets in the solar system: $\Sigma(R) \sim 1200\ \text{kg/m}^2\ (R/5\ \text{au})^{-3/4}$ (compare with Chiang and Youdin 2010). The result for $10^{-7}\ M_\odot\ \text{yr}^{-1}$ is shown in Fig. 6.

The competing rates of temperature and pressure-dependent adsorption and desorption of water onto grain surfaces in the disk define the position of the water snow line. Pressure

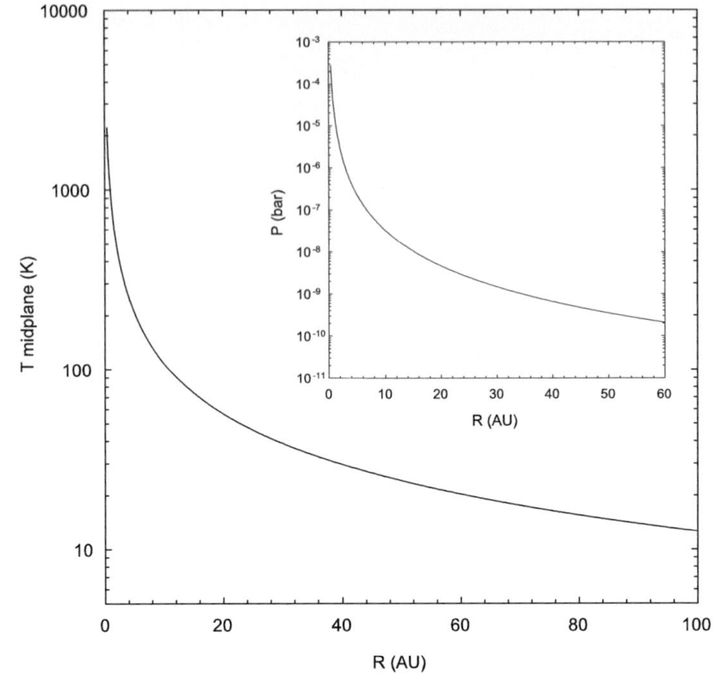

Fig. 6 Calculated temperature and pressure vs. distance from the central star for a solar-like protoplanetary accretion disk with a mass accretion rate of 10^{-7} solar masses per year

defines the total number density n_{total} of molecules since $P = n_{\mathrm{total}} k_{\mathrm{b}} T$. Midplane pressures as a function of radial distance from the star are

$$P(R) = \frac{\Sigma}{\sqrt{\pi} H(R)} \frac{k_{\mathrm{b}} T}{\mu}. \tag{5}$$

The kinetic theory of gases yields the rate constant for adsorption of molecules onto grain surfaces:

$$k_{\mathrm{ads}} = \langle \pi r_{\mathrm{grains}}^2 \rangle \hat{V}_{\mathrm{gas}} n_{\mathrm{grains}} \tag{6}$$

where k_{ads} is the rate constant (s^{-1}), r_{grains} is the average radius of the dust grains, n_{grains} is the number density of the dust grains (cm^{-3}), and $\hat{V}_{\mathrm{gas}} = \sqrt{8 k_{\mathrm{b}} T / (\pi \mu)}$ is the average gas velocity. Hasegawa et al. (1992) suggest a parameterization for the number density of grains based on total hydrogen such that $n_{\mathrm{grains}} \sim 1.33 \times 10^{-12} n_{\mathrm{H}}$ and $n_{\mathrm{H}} = 2 \times n_{\mathrm{H2}} \sim n_{\mathrm{total}}$. Substitution of this expression into (6), together with the mass of water molecules, yields in SI units

$$k_{\mathrm{ads}} = \langle \pi r_{\mathrm{grains}}^2 \rangle \sqrt{1.18 \times 10^4 T(\mathrm{K})}\, 1.33 \times 10^{-6} (P_{\mathrm{H_2}} / (k_{\mathrm{b}} T)). \tag{7}$$

Equation (7) can be evaluated using midplane temperatures and pressures ((2), (4), and (5)) and the fact that typical radii for dust grains relevant to the protoplanetary disk environment are 0.1 μm, or 1×10^{-7} m. The competing rate constant for desorption of water from grains in SI units is (Willacy et al. 1998):

$$k_{\mathrm{des}} = 1 \times 10^{12} \exp(-4815 / T(\mathrm{K})). \tag{8}$$

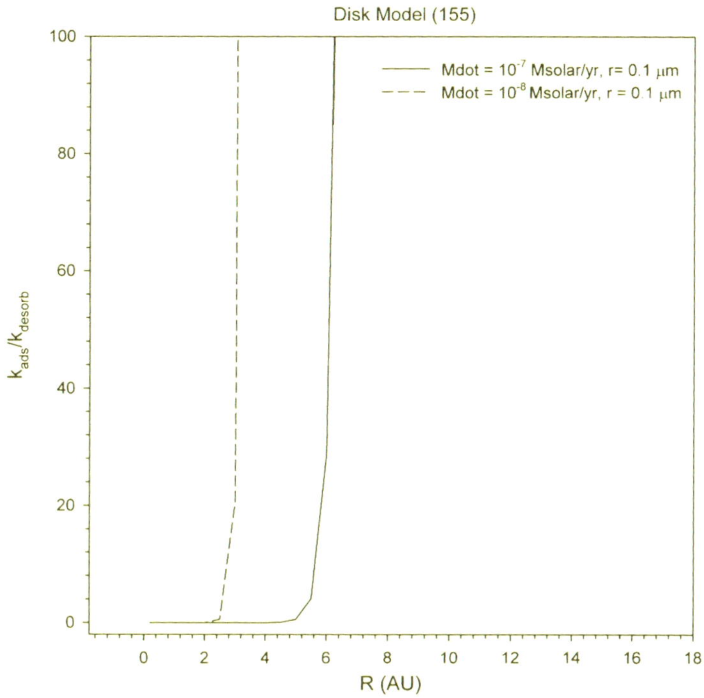

Fig. 7 Position of the water snow line as a function of mass accretion rate as defined by the ratio of the adsorption and desorption rate constants

At equilibrium rates of adsorption and desorption of water onto dust grains are equal so we have $k_{ads}n_{H_2O,gas} = k_{des}n_{H_2O,grains}$. Therefore, $k_{ads}/k_{des} \gg 1$ represents water frozen to grains and $k_{ads}/k_{des} \ll 1$ represents where most water molecules exist in the gas phase. A unit value for k_{ads}/k_{des} defines conditions corresponding to the effective position of the snow line in the midplane of the disk. Substitution of (2) and (5) into (7) and (8) yields the plot in Fig. 7 showing the calculated position of the snowline in the disk as a function of accretion rate. The effective freeze out temperature for $\dot{m} = 10^{-8}$ $M_\odot$/yr is ~190 K. The effective temperature for freeze out of water is 170 K for $\dot{m} = 10^{-7}$ $M_\odot$/yr. Most importantly, we note that basic physics of accretion disks indicates that the calculated position of the snow line varies from ~5 au to 2 au when accretion rate varies from 10^{-7} $M_\odot$/yr to 10^{-8} $M_\odot$/yr, respectively. This is the basis for the assertion that in a solar-like protoplanetary disk, we have a good idea of where the snowline should have been as a function of time.

Of course, radial drift complicates the simple picture described above. Morbidelli et al. (2016) point out that decoupling of radial drift velocities between dust and gas can result in a decoupling of "wet" and "dry" gas from the position of the snowline. The rates of icy particle drift inwards towards the star compared with the rates of planetesimal formation then become critical for understanding the compliments of water in protoplanetary materials, as described above in connection with Fig. 1. Morbidelli et al. (2016) suggested that the apparent position of the water snowline recorded by planetesimal formation may be a "fossilized" snowline imprinted on the disk by the gap opened up by a growing Jupiter. Therefore, the rough correspondence between the position of the shift in oxygen fugacity in Fig. 1 and the natural position of the water snowline for disk mass accretion rates of between 10^{-7} $M_{\odot}$/yr and 10^{-8} $M_{\odot}$/yr could be a manifestation of a locking in of the position of the snowline at these accretion rates by opening of a gap in the vicinity of the snowline by proto-Jupiter.

3 Effect of Internal Evolution on Water

During the early evolution of the terrestrial planets and icy bodies, four main sources of energy may have contributed to their internal thermal budget (e.g. Tobie et al. 2013): (i) Tidal heating associated with despinning, (ii) radiogenic heating by the decay of short-lived radioactive isotopes, (iii) impact heating and (iv) conversion of potential energy into heat via viscous dissipation during core-mantle separation. These four sources of energy likely concern different growth stages or reservoirs and hence will be separated in time and space. For planetesimals, only short-lived radiogenic (mainly ^{26}Al) heating is relevant (Yoshino et al. 2003; Rubie et al. 2015b). This source of energy was for several million times larger than the radiogenic heating rate due to the decay of long-lived isotopes currently expected on planets and moons. However, this huge heat production decayed rapidly. On the other hand on planetary embryos and fully accreted terrestrial planets, the release of potential energy and impact heating have to be taken into account in addition to radiogenic heating. The impact heating strongly depends on the impact velocity, on the masses of the target and the impactor as well as the way kinetic energy is converted into heat (Monteux et al. 2007). Hence the size of the growing body that governs the minimum impact velocity (i.e. the escape velocity) is a key factor. For large growing bodies, large impact events may generate significant melting within the target and trigger metal–rock or ice–rock separation potentially leading to runaway differentiation processes (Golabek et al. 2009; Ricard et al. 2009). Finally on icy moons long-lived radiogenic heating and tidal dissipation are the most important energy sources, while only the largest impact heating events can play a role (Monteux et al. 2014).

The composition of the building blocks from which planetary objects form as well as the early thermal budget should control the amount of water that can be stored in planetary embryos. To get a first order estimation of the amount of hydrous silicates and leftover rock–ice mixture available at the end of the thermo-mechanical evolution of planetesimals, we perform numerical models in 2D infinite cylinder geometry using the finite-difference marker-in-cell code I2ELVIS (Gerya and Yuen 2007; Golabek et al. 2014; Lichtenberg et al. 2016). The physical parameters and the setup employed are identical to those used in our previous studies (see Golabek et al. 2014; Lichtenberg et al. 2016).

Based on current planetesimal formation models (e.g. Johansen et al. 2015; Simon et al. 2016), we model planetesimals with radii R_P ranging between 25 and 230 km and consider instantaneous formation times t_{form} ranging from 0 to 3.5 Myr after formation of Ca–Al-rich inclusions (CAIs). We study here the evolution of planetesimals that formed across the snowline and were scattered later into the inner solar system (Raymond and Izidoro 2017). Thus we assume for each planetesimal a starting temperature equal to the temperature of surrounding space $T_{space} = 150$ K. Since for small objects the release of potential energy related to accretion is small (Schubert et al. 1986), the resulting temperature increase is negligible and is not considered here. Thus only short-lived radiogenic heating is relevant. For simplicity we assume that the planetesimals do not contain metal because both formation of hydrated silicates and their dehydration occur at lower temperatures than the melting temperature of the eutectic Fe–FeS mixture (1243 K) at low pressures (Brett and Bell 1969), so core formation due to percolation is expected to start only after the breakdown of hydrated silicates. Therefore we assume that at the start of the model each planetesimal is composed only of a rock–ice mixture. Here we assume that at $T = 273$ K the ice melts and hydrated silicates can form. Under the low-pressure conditions inside a planetesimal the most temperature-resistant hydrated silicates (amphibolites) break down at $T \approx 1223$ K and dehydrate (Fu and Elkins-Tanton 2014). However it should be taken into account that by considering the breakdown of amphibolites the current models give only an upper limit for the amount of hydrated silicates present, since for example more abundant hydrated silicates like serpentine-phyllosilicates break down at significantly lower temperatures (573–673 K) (Nakamura 2006; Nakato et al. 2008).

The following additional assumptions and simplifications are used in the numerical model: We ignore here reaction kinetics and assume that both the formation of hydrated silicates and their breakdown occur instantaneously at the given temperatures. Also we ignore possible water delivery or loss processes related to impact events. Additionally we do not consider pore water convection (Young et al. 2003) suggested for smaller planetesimals ($R_P < 40$–60 km) and assume that after silicate dehydration the water previously incorporated into hydrated silicates is completely lost to space. Also it should be kept in mind that even dry silicates still contain small amounts of water (see e.g. Fu and Elkins-Tanton 2014), thus even planetesimals loosing all hydrated silicates still contain some water, which is not considered in our numerical model.

The time-dependent amount of hydrated silicate material in each modeled planetesimal is tracked on the markers of the numerical model. The final 2D area covered by (a) rock–ice mixture and (b) hydrated silicates is converted for representation purposes analytically into a 3D volume (see Fig. 8a, b).

The general results show that the amount of both remaining rock–ice mixture and hydrated silicates is strongly dependent on the formation time t_{start} since this parameter controls the abundance of short-lived ^{26}Al ($t_{1/2} = 0.716$ Myr), while the radius has only a modest effect (see Fig. 8a, b). As expected late formed objects contain orders of magnitude more rock–ice mixture and hydrated silicates than early-formed planetesimals. Since formation of hydrated silicates requires some heating, more of the interior contains hydrated silicates in late-formed, but large planetesimals, while a primordial rock–ice mixture is best preserved in the smallest objects experiencing fast cooling. However the results also show that less hydrated silicates form in planetesimals that accreted very late because the reduced amount of ^{26}Al radiogenic heating is insufficient to allow for the formation of hydrated silicates throughout the bulk of the planetesimal interior. This leads to a "sweet spot" for the formation of hydrated silicates, which is narrower for small planetesimals with $R_P < 80$ km since these objects cool more efficiently than large planetesimals (see Fig. 8a).

All models start with a rock–ice mixture. Because the temperature at the center of the planetesimal reaches the maximum value, formation of hydrated silicates starts there and over time the formation front propagates outwards. For large objects ($R_P \geq 100$ km) that formed early enough ($t_{form} < 0.75$ Myr) the center is also the location where dehydration starts during later evolution and a dehydration front propagates from there towards the surface. For planetesimals with sufficient ^{26}Al heating both the formation and later the dehydration front are able to propagate throughout the entire planetesimal interior and within $\sim 10^6$ years the entire planetesimal will loose all hydrated silicates (see Fig. 9). On smaller objects full dehydration occurs only on objects that formed earlier than $t_{form} < 0.75$ Myr, while in the smallest planetesimals considered ($R_P \leq 50$ km) complete loss of hydrated silicates is not achieved even inside planetesimals formed at the same time as CAIs (see Fig. 8a).

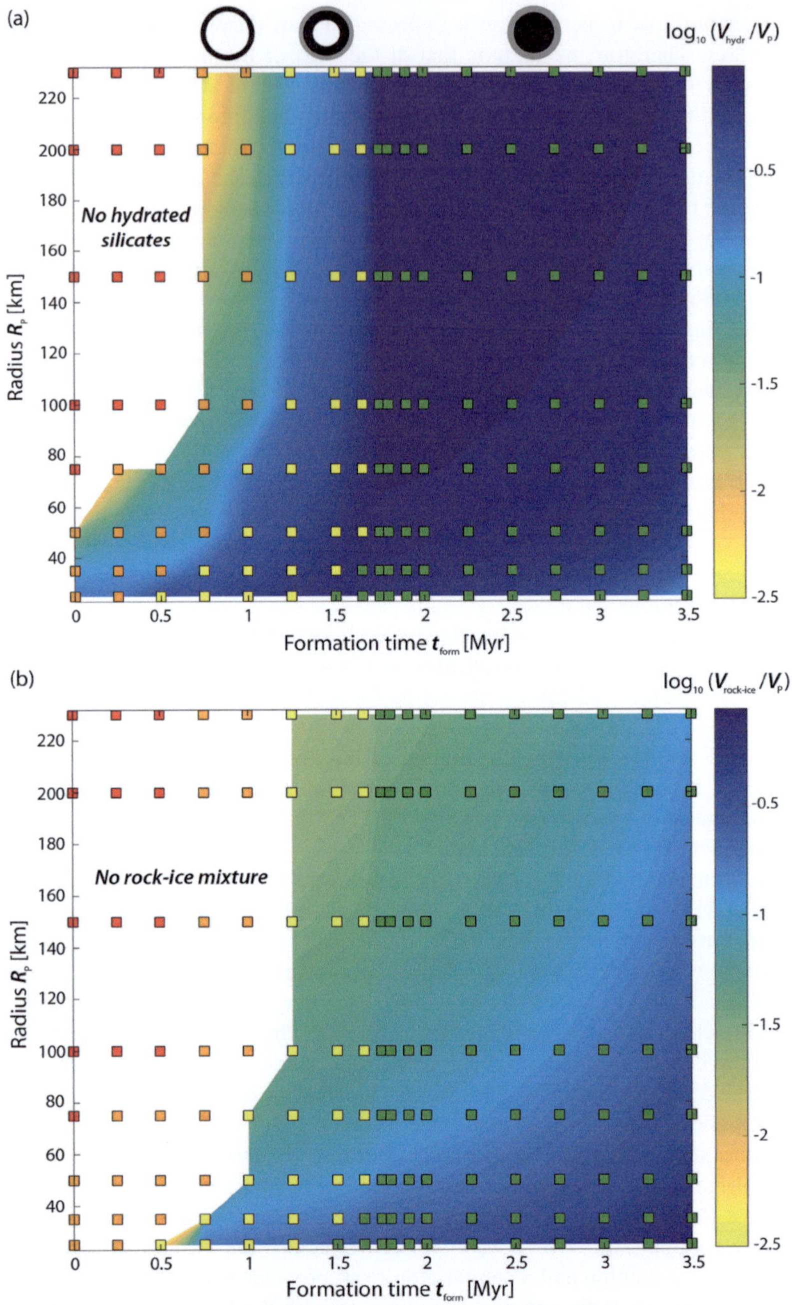

Fig. 8 (**a**) Final volume of hydrated silicates V_{hydr} and (**b**) final volume of rock–ice mixture $V_{\mathrm{rock\text{-}ice}}$. Both volumes are scaled by the total volume of the specific planetesimal V_{P} for models with different planetesimal radii R_{P} and formation times t_{form} assuming an initial temperature $T_{\mathrm{space}} = 150$ K. *Squares* represent the numerical model results. *Colours* of the squares stand for various final internal states: No hydrous silicates left (*red*), outer hydrated shell (*orange*), inner hydrated shell (*yellow*) and hydrated interior (*green*). For clarity three of the possible final outcomes are sketched on top of subfigure (a) with dehydrated silicates (*white*), hydrated silicates (*black*) and rock–ice mixture (*grey*)

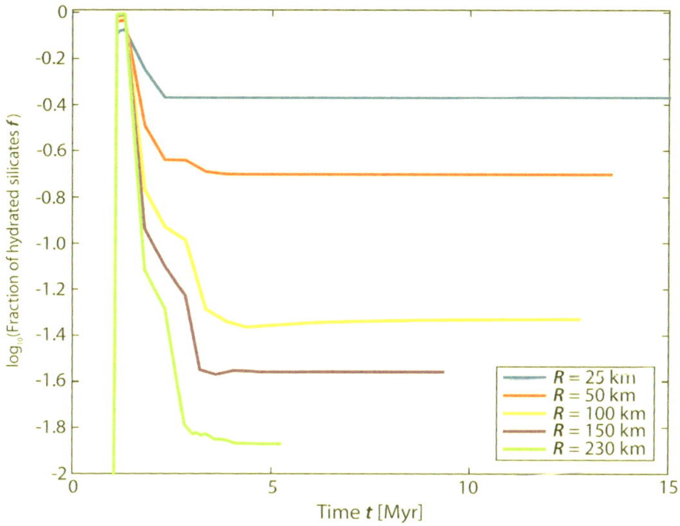

Fig. 9 Time-dependent fraction of hydrated silicates inside planetesimals with $t_{\mathrm{form}} = 1$ Myr and $R_{\mathrm{P}} = 25$–230 km.

Large, but later-formed planetesimals (0.75 Ma $\leq t_{\mathrm{form}} < 1.25$ Myr) or early-formed, but small objects exhibit a different final outcome, namely an outer hydrated shell, where the formation front reached the object's surface, while the propagation of the dehydration front ceased at depth due to insufficient ^{26}Al radiogenic heating. Inside large planetesimals that formed even later (1.25 Ma $\leq t_{\mathrm{form}} < 1.75$ Myr) neither the formation nor the dehydration front reach the surface. Under these circumstances the surface is still composed of a rock–ice mixture while the deep interior is dehydrated and an annulus at intermediate depth contains hydrated silicates. For late-formed objects (large objects with $t_{\mathrm{form}} \geq 1.75$ Myr or for smaller planetesimals with $t_{\mathrm{form}} \sim 1.5$ Myr after CAI formation) temperatures are never sufficiently high as to start dehydration and these objects display a hydrated interior and a rock–ice mixture closer to the surface (see Fig. 8a, b). This is in agreement with observations indicating that some asteroids contain even at present-day significant amounts of volatiles (Hsieh and Jewitt 2006; Jewitt et al. 2009; Küppers et al. 2014).

Based on these simple models, we can speculate that large ($R_{\mathrm{P}} > 80$ km), late-formed planetesimals ($t_{\mathrm{form}} > 1.7$ Myr), initially formed across the snowline and implanted into the main asteroid belt or scattered into the inner solar system during the accretion of the gas giants (Raymond and Izidoro 2017), can contribute significant amounts of water in the form of hydrated silicates to growing planetary embryos. This is in general agreement with both the classical model (e.g. Chambers 2013; Raymond et al. 2009) and the Grand Tack scenario (Walsh et al. 2011; Rubie et al. 2015a) suggesting that during the later stages of accretion planetary embryos will accrete material that formed at larger distances from the Sun.

In conclusion, we have shown that the internal evolution and especially the formation time of planetesimals relative to the timescale of radiogenic heating by short-lived ^{26}Al decay may govern the amount of hydrous silicates and leftover rock–ice mixture available during the late stages of their evolution. In turn, water content may affect the early internal evolution of the planetesimals and in particular the metal-silicate separation processes to be discussed in the next section.

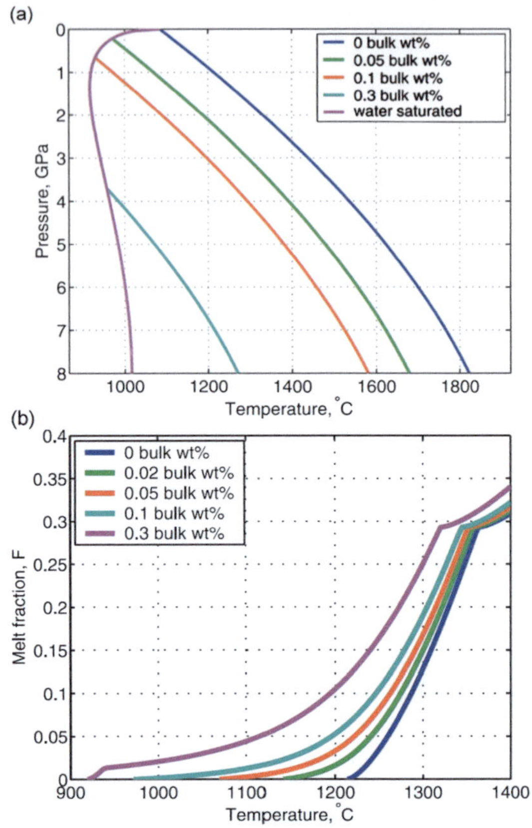

Fig. 10 Effects of water on the melting of peridotite. (**a**) Depression of the peridotite solidus temperature as a function of pressure for bulk water contents of 0, 0.05, 0.1 and 0.3 wt%. (**b**) Effect of water on melt fraction F during partial melting of peridotite shown as isobaric melting curves at 1 GPa for bulk water contents ranging from 0 to 0.3 wt%. Note that the kink at a melt fraction of ∼0.3 is caused by the disappearance of clinopyroxene from the mineral assemblage (from Katz et al. 2003). Note that equivalent temperatures for a bulk chondritic composition are lower by ∼200 °C (Fig. 7 in Asahara et al. 2004). Also the addition of any incompatible elements (e.g. the C and S that are present in chondritic meteorites) will reduce melting temperatures

4 Effects of Water on Internal Evolution

The most important process occurring during the early internal evolution of a rocky planetary body is differentiation that results in a silicate mantle and a metallic iron-rich core. The terrestrial planets, Mercury, Venus, Earth and Mars, as well some asteroids, such as Vesta, all underwent core-mantle differentiation. In addition, based on the existence of iron meteorites, many early-formed planetesimals also experienced core-mantle differentiation. Undifferentiated material, as represented by chondritic meteorites, consists of intimate mixtures of metal, sulfide, silicate and oxide grains. The formation of metallic cores in differentiated bodies therefore requires metal and silicate to separate over significant length scales (up to 3000 km in the case of Earth but on the order of 100–500 km in the case of planetesimals). Such segregation was only physically possible at high temperatures at which at least the metal and probably also the silicates were in a molten state (Stevenson 1990; Rubie et al. 2015b). High temperatures were the result of the decay of the short-lived ^{26}Al isotope in planetesimals during the first ∼3 Myr of Solar System history and were later the result of high-energy collisions that took place during planetary accretion (e.g. Rubie et al. 2015b; de Vries et al. 2016). Thus planetesimals could only differentiate early, while ^{26}Al was extant, and were largely unaffected by impact-induced heating because of their low masses (see Rubie et al. 2015b, Fig. 10). In general, core-mantle differentiation that occurred as a consequence of impact-induced melting could only happen later after bodies had grown sufficiently in mass. Physical and chemical conditions differed significantly during these respec-

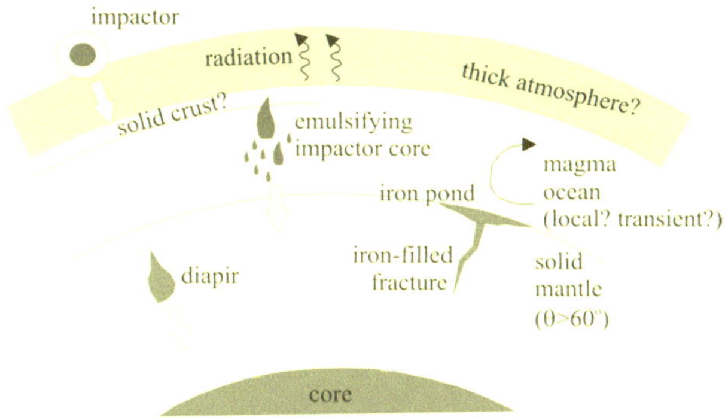

Fig. 11 Schematic section through a terrestrial planet during core formation. A global magma ocean is shown overlying solid crystalline mantle below which lies the metallic proto-core. Metal is delivered by impactors, many of which are likely to have been differentiated into core and mantle. The molten impactor cores partially or completely emulsify in the magma ocean to form small droplets that sink to the base of the magma ocean and collect there to form liquid "iron ponds". These sink through the crystalline mantle as diapirs or possibly are transported through fractures (see Rubie et al. 2015b for a review). Although a possible solid crust is shown, this may not form when an atmosphere is present because its insulating effect keeps the temperature at the surface of a magma ocean above the solidus (e.g., Matsui and Abe 1986)

tive episodes. For example, gravitational forces were much smaller during early planetesimal differentiation and such bodies likely differentiated in the presence of nebular gases, which dispersed before the formation of the terrestrial planets.

The presence of water in planetary bodies could have had a significant effect on both the physical processes of core formation and their chemical consequences. For example, the presence of even small amounts of water strongly reduces the solidus temperature of silicates (Fig. 10a). It also increases the melt fraction at a given temperature, although this effect is largest close to the solidus and becomes much smaller as the degree of melting increases (Fig. 10b).

4.1 Effects of Water on Mechanisms of Core Formation

There are several different mechanisms by which metal and silicate could have segregated (Fig. 11) and the presence of water would have affected each of these differently. Potentially the mechanisms shown in Fig. 11 could all operate during the differentiation of an Earth-mass planet. However, as discussed towards the end of this section, core-mantle differentiation on small bodies (planetesimals) likely required significant melting (with a melt fraction of at least 30–50%) as the result of heat produced by the decay of ^{26}Al (see also Scheinberg et al. 2015).

We first review the possible metal-silicate segregation mechanisms and how these would be affected by the presence of water and then discuss chemical consequences of water during core-mantle differentiation.

Percolation of Liquid Metal Through a Polycrystalline or Partially-molten Silicate Matrix Liquid iron-rich metal or sulfide, (Fe, Ni)S, can percolate through a solid polycrystalline silicate matrix by porous flow provided the liquid phase forms an interconnected network along grain boundaries or grain edges rather than isolated melt pockets. Interconnection depends upon the dihedral angle, θ, which is the angle between two solid–liquid boundaries that are intersected by a solid–solid boundary at a triple junction (Fig. 12; von

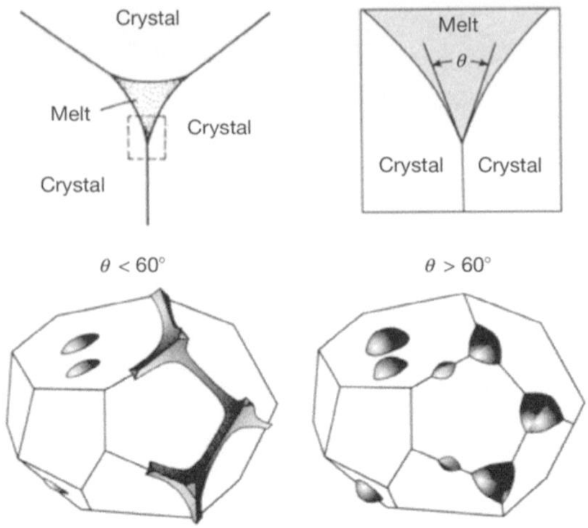

Fig. 12 The dihedral angle θ is the angle between two melt–crystal interfaces where they are intersected by a crystal–crystal grain boundary at a triple junction. When $\theta < 60°$, the melt forms an interconnected network along grain edges (*bottom left*), whereas when $\theta > 60°$ there is no interconnectivity (unless the melt fraction is high) (from Stevenson 1990)

Bargen and Waff 1986; Stevenson 1990; Rubie et al. 2015b). When θ is less than 60°, the liquid metal phase is interconnected at any melt fraction and can percolate efficiently through the crystalline matrix. When θ exceeds 60°, interconnection only occurs when a critical melt fraction (e.g. a few volume % or more) is exceeded, which means that metal-silicate segregation cannot occur efficiently. Experimental studies of both natural and synthetic systems relevant to core formation have shown that, in general, θ lies in the range 80–120°, at least up to pressures of ∼25 GPa (Rubie et al. 2015b and references therein). Therefore percolation is unlikely to be a major mechanism of metal-silicate segregation in the terrestrial planets. However, at pressures <2–3 GPa, the dihedral angle is less than 60° if the liquid Fe contains significant concentrations of oxygen and sulfur (Terasaki et al. 2005, 2008). It is therefore possible that under oxidizing conditions percolation was an important core formation mechanism in planetesimals during heating by ^{26}Al decay, especially when the metal contains sulfur (which reduces the melting temperature of the metal relative to that of the silicate).

As discussed in the previous section, dehydration of silicates is likely to leave only a few hundred ppm of water in planetesimals that have been heated by the decay of ^{26}Al. However, such H_2O concentrations would be sufficient to lower the silicate solidus temperature by 100–150 K (Fig. 10a), which means that the percolation of liquid metal or sulfide through partially molten silicates, rather than crystalline silicates, needs to be considered. Intuitively, it might be expected that a few percent silicate liquid in the system would enhance the ability of liquid metal to percolate. However, the opposite is the case and the presence of silicate melt actually reduces the interconnectivity of metallic liquids (Holzheid et al. 2000; Rushmer and Petford 2011; Holzheid 2013; Cerantola et al. 2015; Todd et al. 2016). This is the result of the high liquid metal-liquid silicate interfacial energy that causes metallic blobs to adopt a spherical geometry so that they become isolated in silicate melt. Consequently the metallic blobs become trapped in interstices in the partial molten aggregate. Thus, once silicate liquid forms by partial melting, as temperature increases, any interconnectivity of liquid metal or sulfide is destroyed. As the temperature rises further, the silicate melt fraction will eventually become large enough to enable the metallic droplets to become mobile and to segregate gravitationally. The value of the silicate melt fraction at which this happens is poorly known but is likely to be in the range 30–50% (Stevenson 1990; Minarik et al. 1996; Holzheid et al. 2000; Costa et al. 2009; Solomatov 2015). At this degree of partial melting a magma ocean forms and the segregation mechanism changes to that described below. Thus the presence of water is likely to significantly delay metal-silicate segregation in planetesimals by requiring a higher temperature to start than under anhydrous conditions.

Segregation of Liquid Metal in a Silicate Magma Ocean As described above, when the fraction of silicate melt exceeds a certain value (probably 30–50%), a magma ocean state is achieved. When the metallic cores of impacting bodies sink through a magma ocean by gravitational settling, they tend to break up into smaller masses, eventually forming small droplets through emulsification. The extent to which emulsification occurs is important because it determines the fraction of accreted metal that equilibrates chemically with silicate liquid of the magma ocean (e.g., Nimmo et al. 2010; Rudge et al. 2010; Rubie et al. 2015a). When emulsification and equilibration fail to occur, the impactor's core merges directly with the core of the target body by "core merging". The efficiency of emulsification is uncertain, especially for the cores of giant impactors (Rubie et al. 2003; Dahl and Stevenson 2010; Samuel 2012; Deguen et al. 2014; Wacheul et al. 2014; Kendall and Melosh 2016; Landeau et al. 2016). The process is extremely difficult to study numerically because the length scales involved range from 100's km (or more) to a few mm. In the case of the Earth, it has been estimated that emulsified droplets of liquid Fe have a stable diameter of $\sim$1 cm and sink with a velocity of $\sim$0.5 m/s (Stevenson 1990; Rubie et al. 2003). Such parameters are a result of molten peridotite having a very low viscosity (Liebske et al. 2005)—which also results in magma oceans convecting vigorously in a turbulent regime. Because peridotite liquid is already highly depolymerized, the addition of water is unlikely to have a significant effect on its viscosity. In addition, as shown in Fig. 10b, the presence of water will have only a small effect on the temperature of magma ocean formation when the latter is defined to have a silicate melt fraction $\geq$0.3.

Descent of Diapirs of Liquid Metal Through Crystalline Mantle It has been proposed that segregated liquid metal accumulates as a pool at the bottom of a magma ocean during the differentiation of terrestrial planets. If the base of the magma ocean is separated from the proto-core by a crystalline mantle, it is likely that the iron melt pool sinks through the latter as diapirs, perhaps 1–10 km in diameter or larger (Fig. 11; e.g. Stevenson 1990; Karato and Murthy 1997; Samuel et al. 2010). Water dissolved in crystalline upper mantle rocks has a strong weakening effect on rheology and reduces the viscosity of solid-state flow by up to 2–3 orders of magnitude. For example, Hirth and Kohlstedt (1996) show that 810 ± 490 H/10^6 Si dissolved in olivine reduces the viscosity by a factor of 500 ± 300 compared with that of dry olivine. The descent velocity of metal diapirs can be described by Stokes' Law. Since the Stokes' sinking velocity is inversely proportional to mantle viscosity, the effect of water is large and could enhance the descent velocity by up to 2–3 orders of magnitude, at least in upper mantle peridotite.

Numerical models of diapirs sinking through the solid mantle of Earth-sized bodies indicate that a wet rheology favors diapir sinking, while a dry rheology leads to preferential formation of iron-filled dikes (Golabek et al. 2009). It was shown both analytically and numerically that this changes the heat partitioning between iron and silicates (Ke and Solomatov 2009; Golabek et al. 2009). Whereas the sinking of diapirs results in heat preferentially partitioning into the silicate mantle, the propagation of iron dikes results in the preferential heating of the iron. Thus the presence of water in the crystalline mantle could affect the thermal state of a terrestrial planet after core formation.

In summary, the effect of water on the temperature and timing of magma ocean formation in accreting planetary bodies, as required for metal-silicate segregation, is very small.

In relatively oxidized planetesimals, the presence of water may eliminate the possibility of metal-silicate segregation by percolation, as proposed by Terasaki et al. (2008), so that segregation only starts at a magma ocean stage after mantle material is at least 30–50% molten. If the descent of diapirs of liquid metal through crystalline mantle occurs, water dissolved in the mantle material would greatly enhance the rate of this process. On the other hand, an absence of water could change the descent mode from sinking diapirs to diking, thus affecting the heat partitioning between iron and silicates and modifying the thermal state of the early terrestrial planet.

4.2 Effects of Water on the Chemistry of Core Formation

The main geochemical effect of metal-silicate segregation to form the core and mantle of a planetary body is that siderophile (metal-loving) elements, such as Ni, Co, V, Cr, W, Mo, etc., partition into liquid metal and are thus transported from the mantle into the core. Consequently, all siderophile elements are variably depleted in Earth's mantle, relative to solar relative abundances, by factors of up to almost 1000. The degree of depletion depends on the extent to which an element partitions into metal. This is described by the metal-silicate partition coefficient, $D_{\mathrm{M}}^{\text{met-sil}}$, which for element M is defined simply as:

$$D_{\mathrm{M}}^{\text{met-sil}} = \frac{C_{\mathrm{M}}^{\text{met}}}{C_{\mathrm{M}}^{\text{sil}}}$$

where $C_{\mathrm{M}}^{\text{met}}$ and $C_{\mathrm{M}}^{\text{sil}}$ are the molar (or wt%) concentrations of M in the metal and silicate, respectively. Most metal-silicate partition coefficients for siderophile elements depend on pressure (P) and/or temperature (T). In addition, and of great importance for the present discussion, they depend strongly on oxygen fugacity f_{O_2}, as shown by the redox reaction:

$$\underset{\text{metal}}{\mathrm{M}} + (n/4)\mathrm{O}_2 = \underset{\text{silicate liquid}}{\mathrm{MO}_{n/2}}$$

where n is the valence of element M when incorporated in silicate liquid as an oxide component. When f_{O_2} is low, this reaction shifts to the left thus increasing $D_{\mathrm{M}}^{\text{met-sil}}$, whereas when f_{O_2} is high the reaction shifts to the right thus decreasing $D_{\mathrm{M}}^{\text{met-sil}}$.

When modeling core formation, the extent to which Earth's mantle has been depleted in siderophile elements enables estimates to be made of the conditions (P, T, f_{O_2}) of metal-silicate equilibration during core formation (e.g. Wade and Wood 2005; Walter and Cottrell 2013; Rubie et al. 2011, 2015a, 2016). Traditionally, it has been assumed that core formation was a single-stage event, thus enabling unique values for P, T and f_{O_2} to be determined (Walter and Cottrell 2013; see Table 3 in Rubie et al. 2015b). However, "single-stage" core formation is clearly unrealistic because metal is delivered to a growing planet during the entire accretion process, which in the case of Earth likely lasted for up to 100 Myr (e.g. Jacobson et al. 2014). Thus, in recent years more realistic core formation models have been formulated. Firstly, in models of "continuous" core formation (Wade and Wood 2005; Wood et al. 2006), material is accreted to Earth in small (1%) increments with the metal component being equilibrated with the silicate mantle at the base of a magma ocean. In a second class of models, core formation is multi-stage and the accretion of mostly-differentiated embryos and planetesimals to growing planets is linked to episodes of metal-silicate equilibration and core formation (Rubie et al. 2011, 2015a, 2016). In both types of models, magma ocean depth increases as planets grow so that the pressure and temperature of metal-silicate equilibration both increase as accretion proceeds. In addition, the oxygen fugacity at which metal and silicate equilibrate has to increase by 2–3 orders of magnitude during accretion in order to reproduce the concentrations of siderophile elements in Earth's mantle (Wade and Wood 2005; Wood et al. 2006; Rubie et al. 2011). There are at least three mechanisms that can result in, or contribute to, this increase of f_{O_2}, two of which are intimately related to water. Note that oxygen fugacity is determined from the compositions of co-existing metal and silicate relative to that of the iron-wüstite (IW) buffer, as follows:

$$\Delta IW = 2\log\left[\frac{a_{FeO}}{a_{Fe}}\right].$$

Here ΔIW is the deviation of f_{O_2} from the IW buffer in log units, and a_{FeO} and a_{Fe} are the activities of FeO and Fe in the silicate and metal, respectively. Under low oxygen fugacities (e.g. $\Delta IW = -4$) the FeO content of the silicate is low (e.g. <1 wt%), whereas at higher oxygen fugacities (e.g. $\Delta IW = -2$) the FeO content becomes similar to that of Earth's mantle (~ 8 wt%).

Partitioning of Silicon into the Core Under reducing conditions and especially at high temperature, Si becomes siderophile (e.g. Mann et al. 2009) and this element partitions into liquid core-forming metal by the reaction

$$\underset{\text{mantle}}{SiO_2} + \underset{\text{core}}{2Fe} \rightarrow \underset{\text{core}}{Si} + \underset{\text{mantle}}{2FeO}.$$

According to this reaction, for every mole of Si that partitions into the core, two moles of FeO are added to the mantle. This reaction is therefore very efficient at increasing the oxygen fugacity, as required by the core formation models of Wade and Wood (2005) and Rubie et al. (2011, 2015b).

Oxidation of Metal by Accreted Water At the end of core formation, the oxygen fugacity of Earth's mantle was low ($\sim$IW $-$ 2) whereas it is currently several orders of magnitude higher (close to the FMQ buffer). In the upper mantle, the currently high oxygen fugacity is the result of a high Fe^{3+} concentration in some of the constituent minerals (O'Neill et al. 1993). The increase in f_{O_2} after core formation may have been caused by the disproportionation reaction

$$Fe^{2+} \rightarrow Fe^{3+} + \text{Fe metal}$$

which occurs as a result of the crystallization of the lower-mantle mineral bridgmanite (silicate perovskite) (Frost et al. 2004). Water can also potentially cause oxidization in planetary interiors, for example by the reaction:

$$H_2O + \underset{\text{metal}}{Fe} \rightarrow H_2 + \underset{\text{silicate}}{FeO},$$

for which the loss of hydrogen is required, or, in the absence of metal, by the dissociation of H_2O into oxygen and hydrogen, with the latter being lost. Sharp et al. (2013) proposed that such reactions were the cause of the increase in f_{O_2} of Earth's mantle following core formation.

All oxidation reactions that are caused by water require that hydrogen is lost from the system. In the absence of metal, this may be problematic because hydrogen is a strong reducing agent. Therefore, if oxidation is taking place deep in the Earth's interior, the released hydrogen has to migrate to the Earth's surface without causing any reduction of the material that it comes into contact with, which seems very unlikely. If metal is present, the hydrogen may be removed by dissolving in the metal, so that, during core formation, it would be transported to the core of the planet. It is uncertain how efficient this process would be because the partitioning of hydrogen between metal and silicate under core formation conditions is poorly known (Okuchi 1997). The oxidation of metal by water during core formation has been included in the accretion/core formation model of Rubie et al. (2015b) assuming that all released hydrogen dissolves in the metal. It was thus shown that the role of accreted water in oxidizing the mantle during core formation is insignificant and that all oxidation (up to the end of core formation) can be explained by the dissolution of Si into the core (as discussed above) combined with the effects of the accretion of oxidized bodies due to an oxidation gradient in the solar nebula (Fig. 1).

5 Differentiation and Evolution of Water-Rich Objects: From Icy Moons to Ocean-Planets

Most of the physical processes that are discussed to occur during the early stages of planetary formation also occur in the outer part of the Solar System during the formation of icy moons. In these bodies the mantle is composed mainly of water ice while their core is essentially made of rocky material. While the variety of internal structures among the icy moons is still debated, the essential role of the water is widely accepted. These objects are discussed in this section.

5.1 Variety of Internal Structures of Icy Moons

Differences in composition and internal structure exist among the major icy satellites of Jupiter and Saturn, suggesting distinct accretion and differentiation histories (e.g., Kirk and Stevenson 1987; Mueller and McKinnon 1988; Mosqueira and Estrada 2003; Barr and Canup 2008; Monteux et al. 2014). In this section we will focus particularly on the largest icy moons having a planetary scale, namely Jupiter's moons Ganymede and Callisto, and Saturn's moon Titan. The internal structure of these bodies has been constrained from the measurements of the gravity coefficients from spacecraft radio tracking during close flybys (e.g. Galileo, Cassini) (Anderson et al. 1996, 2001; Iess et al. 2010, 2012). The degree-2 gravity coefficients can then be used to infer the moment of inertia of the satellite, thus providing constraints on the density structure. The normalized moment of inertia about the spin axis, $C/(M_T R_T^2)$, is commonly calculated from the fluid Love number using the Radau–Darwin approximation. This formulation implies that the body's equipotential corresponds to the equilibrium shape due to fluid dynamic flattening. In this approximation, the degree-two gravity coefficients, C_{20} and C_{22}, should be equal to $-10/3$ and are proportional to the fluid Love number.

In the case of Titan, the ratio between the observed C_{20}^{obs} and C_{22}^{obs} is close to $-10/3$ (Iess et al. 2010), suggesting that Titan is relatively close to hydrostatic equilibrium. We can therefore use the degree-two gravity coefficients to first order to estimate the fluid Love number and the corresponding moment of inertia factor. Depending on whether we use the C_{20} or C_{22}

coefficients as a reference to determine the fluid Love number, the normalized moment of inertia varies between 0.335 and 0.342. For Ganymede and Callisto, the two coefficients have not been determined independently, and the $-10/3$ relationship has been prescribed when inverting the degree-two gravity coefficients. Using the degree-two coefficients, and applying the Radau–Darwin relationship, the moment of inertia factors of Ganymede and Callisto are estimated to be 0.31 and 0.355 (assuming hydrostatic equilibrium), respectively (Anderson et al. 1996, 2001). Titan's value is therefore intermediate between those of Ganymede and Callisto, suggesting a modest increase of density toward the center. The high moment of inertia factor obtained by Galileo gravity measurements ($C/(MR^2) = 0.355$) (Anderson et al. 2001) suggests an incomplete ice–rock separation within Callisto. In contrast, Ganymede has a much smaller moment of inertia ($C/(MR^2) = 0.31$) (Anderson et al. 2001) and shows signs of past endogenic activity (Pappalardo et al. 2004). Hence, Ganymede is potentially fully differentiated with an icy upper mantle, a rocky lower mantle and a metallic core, which is the origin of a relatively intense intrinsic magnetic field (Kivelson et al. 1998). With similar size and mass, Titan may be an intermediate case between Callisto and Ganymede. Based on the estimation of its moment of inertia factor, Titan's interior is more differentiated than Callisto's but probably much less than Ganymede's interior. Like Callisto, Titan might still possess a layer of ice–rock mixture between a rocky core and an outer ice-rich mantle, unless the rocky core is composed mostly of hydrated minerals (Sohl et al. 2010; Castillo-Rogez and Lunine 2010). A dense atmosphere is currently present at its surface but no intrinsic magnetic field has yet been detected on Titan. However the presence of a metallic core is plausible within Titan if enough iron was delivered during its accretion (Grasset et al. 2000). The fact that the interior of Callisto and possibly Titan may still contain a layer of ice–rock mixture suggests that the satellites have avoided significant melting during accretion and subsequent evolution.

We now discuss the apparently partially-differentiated state of Callisto and Titan, as suggested by their elevated moment of inertia factors (e.g. Anderson et al. 2001; Iess et al. 2010; Gao and Stevenson 2013). In the absence of lithospheric strength and shell thickness variations, these satellites would adopt a perfectly hydrostatic shape, corresponding to a degree-two gravity field. On Titan, the existence of a non-negligible degree-three in the gravity field as well as significant topography suggest that non-hydrostatic effects may significantly affect the estimation of the moment of inertia factor (Iess et al. 2010; Gao and Stevenson 2013; Baland et al. 2014) and that the moment of inertia factor may be significantly smaller than the value estimated from the Radau–Darwin approximation. On Callisto, similar non-hydrostatic contributions originating in the lithosphere, which could not be tested with the Galileo spacecraft data, may also affect the estimation of its moment of inertia factor (McKinnon 1997; Gao and Stevenson 2013). On these two moons, the hydrostatic dynamical flattening is relatively small as they orbit relatively far from their respective planets, and therefore the non-hydrostatic contributions need to be correctly estimated in order to accurately infer the moment of inertia factor and the density profiles of their interior. Future measurements by the JUICE mission (Grasset et al. 2013) will allow us to test how far Ganymede and Callisto are from hydrostatic equilibrium.

5.2 Differentiation Process in Water-Dominated Interiors

Several scenarios have been proposed to explain the wide range of extents of differentiation among the icy moons that resulted from radioactive and accretional heating. O'Rourke and Stevenson (2014) showed that although rock–ice separation may be delayed by double-diffusive convection in the ice–rock interior, ice melting due to progressive radiogenic heating and subsequent differentiation cannot be prevented. Previous studies showed that it was possible to avoid melting if the accumulation of accretion energy was inefficient, i.e. if the energy was radiated away at a rate comparable to the accretion rate (e.g., Schubert et al. 1981; Squyres et al. 1988; Kossacki and Leliwa-Kopystynski 1993; Coradini et al. 1995; Grasset and Sotin 1996). Based on these models, the accretion timescales should be longer than 1 Myr to avoid significant melting and hence differentiation of Callisto, while an accretion timescale as short as 10^3–10^4 yr may be possible for Ganymede. However, these timescales are dependent on the way heat deposition and cooling are treated. Barr and Canup (2010) proposed that the Ganymede–Callisto dichotomy can be explained through differences in the energy received during the Late Heavy Bombardment. Impacts would have been sufficiently energetic on Ganymede to lead to a complete ice–rock separation, but not on Callisto. More recently, Monteux et al. (2014) have developed a 3D numerical model that characterizes the thermal evolution of a satellite growing by multiple impacts, simulating the satellite growth and thermal evolution for a body radius ranging from 100 to 2000 km. Their results indicate that a satellite exceeding 2000 km in radius may accrete without experiencing significant melting only in case (i) its accretion history is dominated by small impactors (radius less than a few kilometers) and (ii) the conversion of impact energy into heat is unrealistically inefficient. In this context, they underlined that global melting for large bodies like Titan or Callisto is difficult to avoid.

Once the melting of the icy phase is initiated, ice–rock separation can proceed. This separation will lead to the accumulation of dense blocks of rock overlying a undifferentiated core consisting of a mixture of rock and ice (e.g., Friedson and Stevenson 1983; Kirk and Stevenson 1987). The dense layer of accumulated rock is gravitationally unstable, which leads to large-scale ice–rock separation. Friedson and Stevenson (1983) suggested that an increase in the silicate volume fraction from 44% for Callisto to 55% for Ganymede could have induced an increase in the internal viscosity of between 1 and 2 orders of magnitude. This increase in the internal viscosity of Ganymede might have led to a failure of its solid-state convection self-regulation. This failure would then have limited heat loss, favored partial melting of the ice and enhanced a catastrophic runaway differentiation (Kirk and Stevenson 1987). Similarly, Nagel et al. (2004) suggested that if Callisto accreted from a mixture of rock and ice and the average size of the rocks was of the order of meters to tens of meters, then the viscosity would range between 10^{13} and 10^{17} Pa s. For such viscosities, Callisto would have experienced a gradual, but incomplete unmixing of ice and rocks. This mantle/core separation potentially involves a large amount of heating due to the viscous dissipation of the potential energy (Flasar and Birch 1973). Contrary to terrestrial planets (Golabek et al. 2009; Ricard et al. 2009), this runaway separation has not yet been modeled.

5.3 Metal–Rock Separation

If a sufficient amount of native iron is present within the icy moon, an iron core may eventually form during or after the ice–rock separation. In this case, the deep interior would be comprised of a probably liquid iron core, and a convecting silicate mantle. Gravity measurements performed by the Radio Science System onboard the Cassini spacecraft suggest that the deep interior of Titan is not fully differentiated (Sotin et al. 2009a; Iess et al. 2010). However, the presence of an iron core cannot be definitely ruled out, and additional measurements are required to determine whether Titan possesses an iron core like Ganymede. On Ganymede, a second differentiation event likely occurred during or after the ice–rock separation since enough heat had been stored in the metal-silicate mixture leading to the formation of an iron core. The major observable feature of this metal-silicate separation is the presence of an intrinsic magnetic field generated in its deep interior (Kivelson

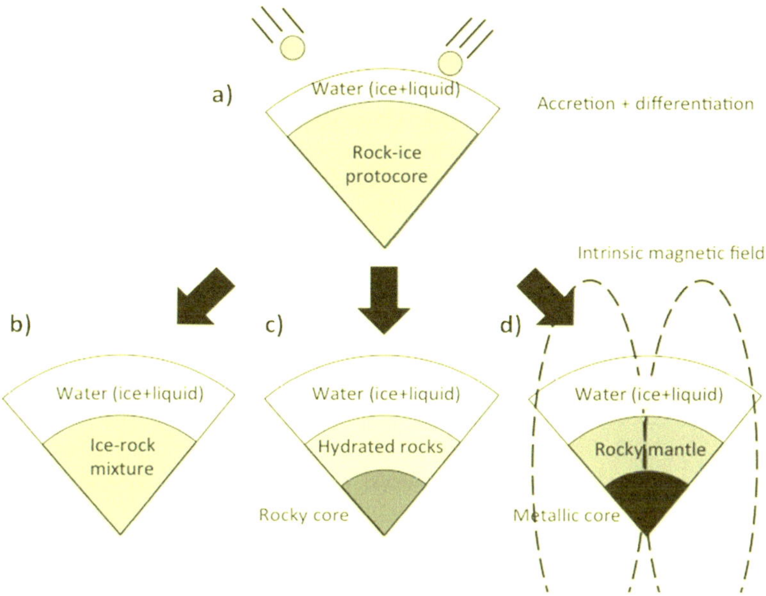

Fig. 13 Schematic representation of the structure of an icy moon (**a**) just after accretion and after differentiation leading to (**b**) a Callisto-like structure, (**c**) a Titan-like structure or (**d**) a Ganymede-like structure

et al. 1996). While the ice–rock separation has been extensively studied, the modalities of metal–rock separation on icy moons and, as a consequence, the modalities of the generation of intrinsic magnetic fields remain poorly constrained. Another major difficulty is to constrain the composition and size of the metallic core. Depending on whether the core is pure iron or a sulfur-iron alloy, it could represent 1 to 30% of the total mass of the icy moon (Anderson et al. 1996).

Although the interior of Ganymede is currently fully differentiated, the silicate rock and the iron likely coexisted and reacted with H_2O (as ice or water) before metal/silicate separation occurred. In this context, it is difficult to envision the formation of a pure iron core because oxidation of the iron would be unavoidable. This led Crary and Bagenal (1998) to suggest that Fe_3O_4-magnetite could maintain a strong remanent magnetic field within the rocky material of Ganymede. Scott et al. (2002) have shown that the presence of water could have enhanced the formation of low-density hydrated silicates and the alloying of iron with sulfur would have resulted in FeS-dominated cores. They performed high-pressure experiments that combined chondritic chemistry with water at a pressure of 1.5 GPa, temperatures between 300 °C and 800 °C and a range of oxygen fugacities. Their results indicate that sulfur may be a significant constituent of Ganymede's core but they did not observe the presence of either iron oxide species or metallic iron. Hence they concluded that magnetite is not a stable phase in the chondrite-water system in contrast with the suggestion of Crary and Bagenal (1998).

Hence, Ganymede's core probably contains not just pure iron but also lighter elements. As mentioned above a candidate for the dominant core-forming assemblage is an alloy of iron (Fe) and sulfur (S) (Hauck et al. 2006). Indeed sulfur is an element that was abundant in the proto-solar system (Hillgren et al. 2000). Within Ganymede, the core sulfur concentrations could range between 12 and 20 wt%. Under the pressure conditions of Ganymede's core ($\approx$6–10 GPa) a mixture of iron and sulfur forms a binary, eutectic system. A major consequence of adding sulfur to liquid iron is a depression of the liquidus (e.g., Usselman 1975; Fei et al. 1997). Upon freezing, nearly pure iron (Fe) (Li et al. 2001; Kamada et al. 2010) crystallizes on the Fe-rich side of the eutectic and iron sulfide (FeS) on the Fe-poor side. Various experiments have revealed that the eutectic composition of Fe–FeS becomes increasingly Fe-rich with increasing pressure (e.g., Chudinovskikh and Boehler 2007; Fei et al. 1997, 2000; Morard et al. 2008). Buono and Walker (2011) have suggested low and even negative gradients for the melting temperatures of Fe-rich Fe–FeS alloys in the pressure range of Ganymede's core. If the melting temperature has a lower gradient than the core temperature, iron will start to crystallize at the core-mantle boundary (CMB) and a growing snow zone will form during further cooling (McKinnon 1996; Hauck et al. 2006). However, the unknown oxidation state of the interior during differentiation limits our ability to constrain whether the composition of a sulfur-bearing core is on the Fe- or FeS-rich side of the eutectic composition (e.g., Scott et al. 2002; Breuer et al. 2015). Another important element could be hydrogen. The interaction of native iron with hydrated material could have led to the addition of hydrogen to the iron core resulting in the presence of iron hydride (Ohtani et al. 2005; Shibazaki et al. 2009).

5.4 Evidence for Active and Past Aqueous Processes

A range of geophysical evidence indicates that several moons of Jupiter (Europa, Ganymede, Callisto) and Saturn (Titan, Enceladus, Mimas) harbor extensive liquid water reservoirs beneath their cold icy surfaces (Khurana et al. 1998; Kivelson et al. 2000; Iess et al. 2014; Tajeddine et al. 2014; Thomas et al. 2016). Detection of ammoniated phyllosilicates on Ceres (de Sanctis et al. 2015) as well as the young age of some extensional features observed on Pluto (Hammond et al. 2016; Moore et al. 2016) suggest the existence of a subsurface ammonia-rich ocean, at least in the past, inside these two dwarf planets. Based on evolution models, many other icy bodies are also predicted to harbor internal oceans underneath their cold icy shells (Hussmann et al. 2006). These massive oceans constitute the greatest volumes of potentially habitable space in the Solar System. If chemical interactions with the rocky core and the irradiated surface provide nutriments and energy sources in sufficient amounts, the development of primitive life may be envisaged in these water-rich worlds. Several observational clues suggest that complex interactions between the subsurface oceans and their warm rocky interiors are presently occurring.

The most convincing evidence for active aqueous processes is provided by the surprising observations of eruptions of water vapor and salted icy grains at Enceladus' active south pole, indicating connections with a subsurface ocean (Postberg et al. 2009, 2011). The discovery of silicon-rich nano-particles further indicates that hydrothermal interactions are currently occurring in porous warm (100–200 °C) rock and that hydrothermal products are quickly transported to the plume source (Hsu et al. 2015; Sekine et al. 2015). Moreover, analysis of gravity, shape and libration data shows that the ice shell thickness at the south pole is thinner than 5 km, with a globally average value of not more than 21 km, requiring about 25–30 GW of internal power for such a configuration to be thermally stable (Cadek et al. 2016). This provides additional constraints in favor of a warm rocky core, powered by strong tidal dissipation. However, it remains unclear how Enceladus reached such a hyperactive state.

On Europa, although there is no direct evidence yet, the occurrence of hydrothermal processes at the seafloor, at present or in recent geological history (<100 Ma), is also likely. The observation of non-ice colored and hydrated materials in many disrupted areas on the surface of Europa points to exposure of oceanic species at the surface and to the possible existence of suboceanic hydrothermal systems (Zolotov and Kargel 2009). The likelihood of presently active hydrothermal processes on Europa mostly depends on how much heat can be generated in the interior at present. Similarly to Enceladus and Io, tidal friction may contribute to Europa's internal budget, but in absence of observational constraints, it is still unclear if tidal friction in the rocky mantle is large enough at present to maintain active hydrothermal processes (Sotin et al. 2009b).

Despite the separation of the subsurface ocean from the rocky core by a thick high pressure ice mantle, water–rock interactions at the base of the thick ice mantle and the transport of brine through the ice mantle are still possible on the largest moons like Saturn's moon Titan and Jupiter's moon Ganymede (Vance and Brown 2013; Choblet et al. 2017). It is very likely that water–rock interactions occurred on these large moons in the past, especially during the differentiation stage when a large volume of water percolated throughout rock-dominated layers toward the surface (Tobie et al. 2013) (Fig. 13). The detection of ^{40}Ar, the decay product of ^{40}K initially contained in silicate minerals, in Titan's atmosphere (Niemann et al. 2010) as well as the likely presence of salts in the internal ocean of Titan (Béghin et al. 2012; Baland et al. 2014; Mitri et al. 2014), Ganymede (Saur et al. 2015) and Callisto (Khurana et al. 1998) indicate that water–rock interaction occurred during the evolution of these large icy moons. However, it is still unclear for how long these water–rock interactions lasted and what effects they had on the thermo-chemical evolution of their interiors.

5.5 Role of Volatile Compounds in the Aqueous Alteration Within Icy Bodies

During the early evolution of Pluto, impact heating, radiogenic heat sources (short- and long-lived) and tidal heating due to the interaction with Charon have likely been significant to induce a complete differentiation of Pluto's interior (Schubert et al. 2010, Sekine et al. 2017). At present-day, because of the rapid decay of short-lived radioisotopes and since the Pluto/Charon system has reached the "dual synchronous" state in which Charon's orbital period, spin period, and Pluto's rotation period are equal, these two major heat sources are probably too weak to maintain a current water ocean within Pluto's interior (Robuchon and Nimmo 2011; Barr and Collins 2015). The surprising discovery of two quasi-circular mounds on Pluto by New Horizons, suspected to be of cryovolcanic origin (Moore et al. 2016), suggests that the interior of the dwarf planet is more active than initially anticipated and that water–volatile–rock interactions may have operated during a long period of time in its interior. The presence of significant amounts of anti-freeze compounds, such as ammonia and methanol, together with gas compounds in the form of clathrate hydrates may explain the occurrence of such activity on Pluto. This hypothesis is consistent with the detection of ammonia-rich and methane-rich areas on Charon (Grundy et al. 2016). Although ammonia has not been identified on Pluto's surface, that is dominated by N_2 ices (Grundy et al. 2016), its presence on Charon indicates that it may have played a key role in the chemical evolution of the Pluto–Charon system, as well as in those of other Kuiper belt objects (Hussmann et al. 2006).

For decades, both ammonia and methanol, have been suspected to strongly affect the melting and crystallization process in icy bodies (e.g. Lewis 1971; Kargel 1992). The role of ammonia is expected to be particularly strong during the early stage in the evolution of icy bodies, when the first aqueous melts are generated during the post-accretional differentiation process and at the end of their evolution when the internal ocean becomes strongly concentrated in solutes by progressive crystallization of ice (Hussmann et al. 2006; Malamud and Prialnik 2015). Even in the absence of significant heat sources, the presence of ammonia, as well as methanol (both are present in comparable amount in comets, Cochran et al. 2015) can sustain active aqueous processes for long periods of time. The recent detection of widespread ammoniated phyllosilicates across Ceres' surface (de Sanctis et al. 2015) may provide evidence for alteration of chondritic materials by N-bearing fluids. However, in absence of reliable experimental data on aqueous alteration by concentrated N-bearing fluids, it is difficult to investigate the conditions under which such ammoniated phyllosilicates may have formed.

6 Conclusions and Outlook

Water content and the internal evolution of terrestrial planets and icy bodies are closely linked. Indeed, the internal evolution and especially the formation time of the planetesimals and therefore the extent of radiogenic heating governs the amount of hydrous silicates and leftover rock–ice mixtures available in the late stages of their evolution. In turn, water content may affect the early internal evolution of the planetesimals and metal-silicate separation by favoring the sinking of metal diapirs over percolation processes. On the other hand, the absence of water could change the descent mode on larger terrestrial bodies from sinking diapirs to diking, thus affecting the heat partitioning between iron and silicates and modifying the thermal state of the accreting planet. Moreover, water content can contribute to an increase of f_{O_2} and thus affect the concentration of siderophile elements within the silicate reservoirs of the Solar System objects. Finally, the water content strongly influences the differentiation rate among the icy moons, controls their internal evolution and governs the alteration processes occurring in their deep interiors.

Partitioning of water between the mantle and the surface/atmosphere is also a key factor in regulating both planetary climate and internal dynamics. This partitioning is controlled by release of water by erupting melts, by hydrothermal alterations, mainly serpentinization, trapping some water in the form of hydrated minerals in the crust, and by recycling part of it deep into the mantle where it can be dissolved in nominally-anhydrous minerals. On Earth, where the water cycle is driven by plate tectonics, exactly how water partitions between surface and interior remains uncertain, but it is commonly thought to be of the order of 1. Parameterized models have recently been proposed to address the question of water cycle on other terrestrial planets, with different sizes and water contents (Crowley et al. 2011; Cowan and Abbot 2014; Schaefer and Sasselov 2015). These parameterized approaches enable some important characteristics of the water cycle to be understood, especially how surface gravity and the water volume at the surface influence the interior/surface exchange. However, they cannot describe the complexity of water transport in the mantle and its feedback on internal melting and crustal recycling. Describing correctly the coupled evolution of surface and internal reservoirs of water requires the use of thermo-chemical convection codes in 3D or at least 2D, including exchange of water between the crust and mantle, the convective transport in the mantle, its retroaction on viscosity, melting and associated outgassing.

Finally, with the discoveries of thousands of exoplanets, we are now entering a new era in comparative planetology. Even if the dataset is limited to the mass, radius and orbital characteristics of the planetary systems, it already offers a unique opportunity to test the models of planet formation and evolution. Model mass-radius relationships indicate a great diversity of interior compositions and atmospheric extents for the Super-Earth/Mini-Neptune-planet class (e.g. Howard et al. 2012), suggesting a wide range of volatile contents and compositions. Hence, planetary worlds are probably much more diverse than originally thought, with a wide range of water and other volatile contents.

Acknowledgements We thank L. Elkins-Tanton, A. Morbidelli and two anonymous reviewers for detailed and thoughtful comments that helped to improve the manuscript. D.C.R. was supported by the European Research Council Advanced Grant "ACCRETE" (contract number 290568) and additional support was provided by the German Science Foundation (DFG) Priority Programme SPP1833 "Building a Habitable Earth" (Ru 1323/10-1). J.M. was funded by the Auvergne Fellowship program. This is Laboratory of Excellence ClerVolc contribution no. 283. G.J.G. thanks Taras Gerya for providing the code I2MART. G.T. acknowledges support from the project ANR OASIS.

References

J.D. Anderson, R.A. Jacobson, T.P. McElrath, W.B. Moore, G. Schubert, P.C. Thomas, Shape, mean radius, gravity field, and interior structure of Callisto. Icarus **153**, 157–161 (2001)

J.D. Anderson, E.L. Lau, W.L. Sjogren, G. Schubert, W.B. Moore, Gravitational constraints on the internal structure of Ganymede. Nature **384**, 541–543 (1996)

P.J. Armitage, *Astrophysics of Planet Formation* (Cambridge University Press, Cambridge, 2010), 294 p.

Y. Asahara, T. Kubo, T. Kondo, Phase relations of a carbonaceous chondrite at lower mantle conditions. Phys. Earth Planet. Inter. **143–144**, 421–432 (2004)

X-N. Bai, J.M. Stone, Wind-driven accretion in protoplanetary disks. I. Supression of the magnetorotational instability and launching of the magnetocentrifugal wind. Astrophys. J. **769**, 21 (2013)

R.M. Baland et al., Titan's internal structure inferred from its gravity field, shape, and rotation state. Icarus **237**, 29–41 (2014)

A.C. Barr, R.M. Canup, Constraints on gas giant satellite formation from the interior states of partially differentiated satellites. Icarus **198**, 163–177 (2008)

A.C. Barr, R.M. Canup, Origin of the Ganymede–Callisto dichotomy by impacts during the Late Heavy Bombardment. Nat. Geosci. **3**, 164–167 (2010)

A.C. Barr, G.C. Collins, Tectonic activity on Pluto after the Charon-forming impact. Icarus **246**, 146–155 (2015)

C. Béghin, O. Randriamboarison, M. Hamelin, E. Karkoschka, C. Sotin, R.C. Whitten, J.-J. Berthelier, R. Grard, F. Simões, Analytic theory of Titan's Schumann resonance: constraints on ionospheric conductivity and buried water ocean. Icarus **218**(2), 1028–1042 (2012)

P.A. Bland et al., Why aqueous alteration in asteroids was isochemical: high porosity does not equal permeability. Earth Planet. Sci. Lett. **287**, 559–568 (2009)

R. Brett, P.M. Bell, Melting relations in the Fe-rich portion of the system Fe–FeS at 30 kb pressure. Earth Planet. Sci. Lett. **6**, 479–482 (1969)

D. Breuer et al., Iron snow, crystal floats, and inner-core growth: modes of core solidification and implications for dynamos in terrestrial planets and moons. Prog. Earth Planet. Sci. **2**, 39 (2015)

A.S. Buono, D. Walker, The Fe-rich liquidus in the Fe–FeS system from 1 bar to 10 GPa. Geochim. Cosmochim. Acta **75**, 2072–2087 (2011)

O. Cadek, G. Tobie, T. Van Hoolst, M. Massé, G. Choblet, A. Lefèvre, G. Mitri, R-M. Baland, M. Behounkova, O. Bourgeois, A. Trinh, Enceladus's internal ocean and ice shell constrained from Cassini gravity, shape, and libration data. Geophys. Res. Lett. **43**(11), 5653–5660 (2016)

J.C. Castillo-Rogez, J.I. Lunine, Evolution of Titan's rocky core constrained by Cassini observations. Geophys. Res. Lett. **37**, L20205 (2010)

J.C. Castillo-Rogez, E. Young, Origin and evolution of volatile-rich asteroids, in *Early Differentiation and Consequences for Planets*, ed. by L.T. Elkins-Tanton, B. Weiss. Cambridge Planetary Science (2017)

F. Cerantola, N. Walte, D.C. Rubie, Deformation of a crystalline olivine aggregate containing two immiscible liquids: implications for early core-mantle differentiation. Earth Planet. Sci. Lett. **417**, 67–77 (2015)

J.E. Chambers, Late-stage planetary accretion including hit-and-run collisions and fragmentation. Icarus **224**, 43–56 (2013)

E. Chiang, A.N. Youdin, Forming planetesimals in solar and extrasolar nebula. Annu. Rev. Earth Planet. Sci. **38**, 493–522 (2010)

G. Choblet, G. Tobie, C. Sotin, K. Kalousova, O. Grasset, Heat and melt transport in the high-pressure ice mantle of large icy moons. Icarus **285**, 252–262 (2017)

L. Chudinovskikh, R. Boehler, Eutectic melting in the system Fe–S to 44 GPa. Earth Planet. Sci. Lett. **257**(1), 97–103 (2007)

A.L. Cochran et al., The composition of comets. Space Sci. Rev. **197**(1–4), 9–46 (2015)

A. Coradini, C. Federico, O. Forni, G. Magni, Origin and thermal evolution of icy satellites. Surv. Geophys. **16**, 533–591 (1995)

A. Costa, L. Caricchi, N. Bagdassarov, A model for the rheology of particle-bearing suspensions and partially molten rocks. Geochem. Geophys. Geosyst. **10**, 3010 (2009)
N.B. Cowan, D.S. Abbot, Water cycling between ocean and mantle: super-Earths need not be waterworlds. Astrophys. J. **781**(1), 27 (2014).
F.J. Crary, F. Bagenal, Remanent ferromagnetism and the interior structure of Ganymede. J. Geophys. Res. **103**, 25757–25773 (1998)
J.W. Crowley, M. Gérault, R.J. O'Connell, On the relative influence of heat and water transport on planetary dynamics. Earth Planet. Sci. Lett. **310**, 380–388 (2011)
J.N. Cuzzi, K.J. Zahnle, Material enhancement in protoplanetary nebulae by particle drift through evaporation fronts. Astrophys. J. **614**, 490–496 (2004)
T.W. Dahl, D.J. Stevenson, Turbulent mixing of metal and silicate during planet accretion and interpretation of the Hf–W chronometer. Earth Planet. Sci. Lett. **295**, 177–186 (2010)
N. Dauphas, The isotopic nature of the Earth's accreting material through time. Nature **541**, 521–524 (2017)
M.C. de Sanctis et al., Ammoniated phyllosilicates with a likely outer solar system origin on Ceres. Nature **528**, 241–244 (2015)
J. de Vries, F. Nimmo, H.J. Melosh, S.A. Jacobson, A. Morbidelli, D.C. Rubie, Impact-induced melting during accretion of the Earth. Prog. Earth Planet. Sci. **3**, 7 (2016). https://doi.org/10.1186/s40645-016-0083-8
R. Deguen, M. Landeau, P. Olson, Turbulent metal–silicate mixing, fragmentation, and equilibration in magma oceans. Earth Planet. Sci. Lett. **391**, 274–287 (2014)
T. Encrenaz, Water in the solar system. Annu. Rev. Astron. Astrophys. **46**, 57–87 (2008)
Y. Fei, C.M. Bertka, L.W. Finger, High-pressure iron-sulfur compound, Fe_3S_2, and melting relations in the Fe–FeS system. Science **275**, 1621–1623 (1997)
Y. Fei, J. Li, C.M. Bertka, C.T. Prewitt, Structure type and bulk modulus of Fe_3S, a new iron–sulfur compound. Am. Mineral. **85**, 1830–1833 (2000)
F.M. Flasar, F. Birch, Energetics of core formation: a correction. J. Geophys. Res. **78**, 6101–6103 (1973)
A.J. Friedson, D.J. Stevenson, Viscosity of rock–ice mixtures and applications to the evolution of icy satellites. Icarus **56**, 1–14 (1983)
D.J. Frost, C. Liebske, F. Langenhorst, C.A. McCammon, R.G. Trønnes, D.C. Rubie, Experimental evidence for the existence of iron-rich metal in the Earth's lower mantle. Nature **428**, 409–412 (2004)
R.R. Fu, L.T. Elkins-Tanton, The fate of magmas in planetsimals and the retention of primitive chondritic crusts. Earth Planet. Sci. Lett. **390**, 128–137 (2014)
R.R. Fu, E. Young, R.C. Greenwood, L.T. Elkins-Tanton, Silicate melting and volatile loss during differentiation in planetesimals, in *Early Differentiation and Consequences for Planets*, ed. by L.T. Elkins-Tanton, B. Weiss. Cambridge Planetary Science (2017)
P. Gao, D.J. Stevenson, Nonhydrostatic effects and the determination of icy satellites' moment of inertia. Icarus **226**, 1185–1191 (2013)
T.V. Gerya, D.A. Yuen, Robust characteristics method for modelling multiphase visco-elasto-plastic thermomechanical problems. Phys. Earth Planet. Inter. **163**, 83–105 (2007)
G.J. Golabek, B. Bourdon, T.V. Gerya, Numerical models of the thermomechanical evolution of planetesimals: application to the acapulcoite-lodranite parent body. Meteorit. Planet. Sci. **49**, 1083–1099 (2014)
G.J. Golabek, T.V. Gerya, B.J.P. Kaus, R. Ziethe, P.J. Tackley, Rheological controls on the terrestrial core formation mechanism. Geochem. Geophys. Geosyst. **10**, Q11007 (2009)
O. Grasset et al., Jupiter ICy moons Explorer (JUICE): an ESA mission to orbit Ganymede and to characterise the Jupiter system. Planet. Space Sci. **78**, 1–21 (2013)
O. Grasset, C. Sotin, The cooling rate of a liquid shell in Titan's interior. Icarus **123**, 101–112 (1996)
O. Grasset, C. Sotin, F. Deschamps, On the internal structure and dynamics of Titan. Planet. Space Sci. **48**, 617–636 (2000)
L. Grossman, A.V. Fedkin, S.B. Simon, Formation of the first oxidized iron in the solar system. Meteorit. Planet. Sci. **47**, 2160–2169 (2012)
W.M. Grundy et al., Surface compositions across Pluto and Charon. Science **351**(6279), aad9189 (2016)
N.P. Hammond, A.C. Barr, E.M. Parmentier, Recent tectonic activity on Pluto driven by phase changes in the ice shell. Geophys. Res. Lett. **43**(13), 6775–6782 (2016)
L. Hartmann, in *Discs, Planetesimals, and Planets*, ed. by F. Garzon, C. Eiroa, D. de Winter, T.J. Mahoney. ASP Conf. Ser., vol. 219 (Astron. Soc. Pac, San Francisco, 2000), p. 95
T.I. Hasegawa, E. Herbst, C.M. Leung, Models of gas–grain chemistry in dense interstellar clouds with complex organic molecules. Astrophys. J. Suppl. Ser. **82**, 167 (1992)
S.A. Hauck, J.M. Aurnou, A.J. Dombard, Sulfur's impact on core evolution and magnetic field generation on Ganymede. J. Geophys. Res. **111**(E9), E09008 (2006)
V.J. Hillgren, C.K. Gessmann, J. Li, An experimental perspective on the light element in Earth's core, in *Origin of the Earth and Moon*, ed. by R.M. Canup, K. Righter (University of Arizona Press, Tucson, 2000), pp. 245–264

M.M. Hirschmann, Water, melting, and the deep Earth H_2O cycle. Annu. Rev. Earth Planet. Sci. **34**, 629–653 (2006)

G. Hirth, D.L. Kohlstedt, Water in the oceanic upper mantle: implications for rheology, melt extraction and the evolution of the lithosphere. Earth Planet. Sci. Lett. **144**, 93–108 (1996)

A. Holzheid, Sulphide melt distribution in partially molten silicate aggregates: implications to core formation scenarios in terrestrial planets. Eur. J. Mineral. **25**, 267–277 (2013)

A. Holzheid, M.D. Schmitz, T.L. Grove, Textural equilibria of iron sulphide liquids in partly molten silicate aggregates and their relevance to core formation scenarios. J. Geophys. Res. **105**, 13555–13567 (2000)

A.W. Howard et al., Planet occurrence within 0.25 AU of solar-type stars from Kepler. Astrophys. J. Suppl. Ser. **201**(2), 15 (2012)

H.H. Hsieh, D. Jewitt, A population of comets in the main asteroid belt. Science **312**, 561–563 (2006)

H.-W. Hsu, F. Postberg, Y. Sekine et al., Ongoing hydrothermal activities within Enceladus. Nature **519**, 207–210 (2015)

H. Hussmann et al., Subsurface oceans and deep interiors of medium-sized outer planet satellites and large trans-neptunian objects. Icarus **185**, 258–273 (2006)

H. Hussmann, C. Sotin, J.I. Lunine, Interiors and evolution of icy satellites, in *Physics of Terrestrial Planets and Moons Treatise on Geophysics*, vol. 10, 2nd edn. (Elsevier, Amsterdam, 2015), pp. 605–635

L. Iess et al., Gravity field, shape, and moment of inertia of Titan. Science **327**, 1367–1369 (2010)

L. Iess et al., The tides of Titan. Science **337**, 457 (2012)

L. Iess et al., The gravity field and interior structure of Enceladus. Science **344**, 78–80 (2014)

S.A. Jacobson, A. Morbidelli, S.N. Raymond, D.P. O'Brien, K.J. Walsh, D.C. Rubie, Highly siderophile elements in the Earth's mantle as a clock for the Moon-forming impact. Nature **508**, 84–87 (2014)

D. Jewitt, B. Yang, N. Haghighipour, Main-belt Comet P/2008 R1 (Garradd). Astron. J. **137**, 4313–4321 (2009)

A. Johansen, M.-M. Mac Low, P. Lacerda, M. Bizzarro, Growth of asteroids, planetary embryos, and Kuiper belt objects by chondrule accretion. Sci. Adv. **1**, 1500109 (2015)

S. Kamada, H. Terasaki, E. Ohtani, T. Sakai, T. Kikegawa, Y. Ohishi, N. Hirao, N. Sata, T. Kondo, Phase relationships of the Fe–FeS system in conditions up to the Earth's outer core. Earth Planet. Sci. Lett. **294**(1), 94–100 (2010)

S. Karato, V.R. Murthy, Core formation and chemical equilibrium in the Earth—I. Physical considerations. Phys. Earth Planet. Inter. **100**, 61–79 (1997)

J.S. Kargel, Ammonia±water volcanism on icy satellites: phase relations at 1 atmosphere. Icarus **100**, 556–574 (1992)

R.F. Katz, M. Spiegelman, C.H. Langmuir, A new parameterization of hydrous mantle melting. Geochem. Geophys. Geosyst. **4** (2003). https://doi.org/10.1029/2002GC000433

Y. Ke, V.S. Solomatov, Coupled core-mantle thermal evolution of early Mars. J. Geophys. Res. **114**, E07004 (2009)

J.D. Kendall, H.J. Melosh, Differentiated planetesimal impacts into a terrestrial magma ocean: fate of the iron core. Earth Planet. Sci. Lett. **448**, 24–33 (2016)

J.F. Kerridge, T.E. Bunch, in *Asteroids*, ed. by T. Gehrels (University of Arizona Press, Tucson, 1979), pp. 745–764

K.K. Khurana, M.G. Kivelson, D.J. Stevenson, G. Schubert, C.T. Russell, R.J. Walker, S. Joy, C. Polanskey, Induced magnetic fields as evidence for subsurface oceans in Europa and Callisto. Nature **395**, 777–780 (1998)

R.L. Kirk, D.J. Stevenson, Thermal evolution of a differentiated Ganymede and implications for surface features. Icarus **69**, 91–134 (1987)

M.G. Kivelson et al., Ganymede's magnetosphere: magnetometer overview. J. Geophys. Res. **103**, 19,963–19,972 (1998)

M.G. Kivelson, K.K. Khurana, C.T. Russell, M. Volwerk, R.J. Walker, C. Zimmer, Galileo magnetometer measurements: a stronger case for a subsurface ocean at Europa. Science **289**, 1340–1343 (2000)

M. Kivelson, K. Khurana, C. Russell, R. Walker, J. Warnecke, F. Coroniti, C. Polanskey, D. Southwood, G. Schubert, Discovery of Ganymede's magnetic field by the Galileo spacecraft. Nature **384**(6609), 537–541 (1996)

J. Korenaga, Scaling of plate-tectonic convection with pseudoplastic rheology. J. Geophys. Res. **115**, B11405 (2010)

K.J. Kossacki, J. Leliwa-Kopystynski, Medium-sized icy satellites: thermal and structural evolution during accretion. Planet. Space Sci. **41**, 729–741 (1993)

A.N. Krot et al., Progressive alteration in CV3 chondrites: more evidence for asteroidal alteration. Meteorit. Planet. Sci. **33**, 1065–1085 (1998)

A.N. Krot, B. Fegley, K. Lodders, Meteoritical and astrophysical constraints on the oxidation state of the solar nebula, in *Protostars and Planets IV*, ed. by V. Mannings, A.P. Boss, S.S. Russell (University of Arizona Press, Tucson, 2000), pp. 1019–1054

M. Küppers et al., Localized sources of water vapour on the dwarf planet (1) Ceres. Nature **505**, 525–527 (2014)

M. Landeau, P. Olson, R. Deguen, B.H. Hirsh, Core merging and stratification following giant impact. Nat. Geosci. **9**, 786–789 (2016)

J.S. Lewis, Satellites of the outer planets: their physical and chemical nature. Icarus **15**, 174–185 (1971)

J. Li, Y. Fei, H. Mao, K. Hirose, S. Shieh, Sulfur in the Earth's inner core. Earth Planet. Sci. Lett. **193(3)**, 509–514 (2001)

T. Lichtenberg, G.J. Golabek, T.V. Gerya, M.R. Meyer, The effects of short-lived radionuclides and porosity on the early thermo-mechanical evolution of planetesimals. Icarus **274**, 350–365 (2016)

C. Liebske, B. Schmickler, H. Terasai, B.T. Poe, A. Suzuki, K. Funakoshi, R. Ando, D.C. Rubie, Viscosity of peridotite liquid up to 13 GPa: implications for magma ocean viscosities. Earth Planet. Sci. Lett. **240**, 589–604 (2005)

U. Malamud, D. Prialnik, Modeling Kuiper belt objects Charon, Orcus and Salacia by means of a new equation of state for porous icy bodies. Icarus **246**, 21–36 (2015)

U. Mann, D.J. Frost, D.C. Rubie, Evidence for high-pressure core-mantle differentiation from the metal-silicate partitioning of lithophile and weakly siderophile elements. Geochim. Cosmochim. Acta **73**, 7360–7386 (2009). https://doi.org/10.1016/j.gca.2009.08.006

G.W. Marcy et al., Masses, radii, and orbits of small Kepler planets: the transition from gaseous to rocky planets. Astrophys. J. Suppl. Ser. **210**(2), 20 (2014)

T. Matsui, Y. Abe, Evolution of an impact-induced atmosphere and magma ocean on the accreting Earth. Nature **319**, 303–305 (1986)

M. Mayor, C. Lovis, N.C. Santos, Doppler spectroscopy as a path to the detection of Earth-like planets. Nature **513**, 328–335 (2014)

W.B. McKinnon, Core evolution in the icy Galilean satellites, and the prospects for dynamo-generated magnetic fields. Bull. Am. Astron. Soc. **28**, 1076 (1996)

W.B. McKinnon, Note: mystery of Callisto: is it undifferentiated? Icarus **130**, 540–543 (1997)

W.G. Minarik, F.J. Ryerson, E.B. Watson, Textural entrapment of core-forming melts. Science **272**, 530–533 (1996)

G. Mitri et al., Shape, topography, gravity anomalies and tidal deformation of Titan. Icarus **236**, 169–177 (2014)

J. Monteux, N. Coltice, F. Dubuffet, Y. Ricard, Thermo-mechanical adjustment after impacts during planetary growth. Geophys. Res. Lett. **34**, L24201 (2007)

J. Monteux, G. Tobie, G. Choblet, M. Le Feuvre, Can large icy moons accrete undifferentiated? Icarus **237**, 377–387 (2014)

J.M. Moore et al., The geology of Pluto and Charon through the eyes of New Horizons. Science **351**, 6279 (2016)

G. Morard, D. Andrault, N. Guignot, C. Sanloup, M. Mezouar, S. Petitgirard et al., In situ determination of Fe–Fe_3S phase diagram and liquid structural properties up to 65 GPa. Earth Planet. Sci. Lett. **272**, 620–662 (2008)

A. Morbidelli, B. Bitsch, A. Crida, M. Gounelle, T. Guillot, S. Jacobson, A. Johansen, M. Lambrechts, E. Lega, Fossilized condensation lines in the Solar System protoplanetary disk. Icarus **267**, 368–376 (2016)

I. Mosqueira, P.R. Estrada, Formation of the regular satellites of giant planets in an extended gaseous nebula. I: subnebula model and accretion of satellites. Icarus **163**, 198–231 (2003)

O. Mousis, A. Drouard, P. Vernazza, J.I. Lunine, M. Monnereau, R. Maggiolo, K. Altwegg, H. Balsiger, J.-J. Berthelier, G. Cessateur, J. De Keyser, S.A. Fuselier, S. Gasc, A. Korth, T. Le Deun, U. Mall, B. Marty, H. Rème, M. Rubin, C.-Y. Tzou, J.H. Waite, P. Wurz, Impact of radiogenic heating on the formation conditions of comet 67P/Churyumov–Gerasimenko. Astrophys. J. Lett. **839**, L4 (2017)

S. Mueller, W.B. McKinnon, Three-layered models of Ganymede and Callisto—compositions, structures, and aspects of evolution. Icarus **76**, 437–464 (1988)

K. Nagel, D. Breuer, T. Spohn, A model for the interior structure, evolution, and differentiation of Callisto. Icarus **169**, 402–412 (2004)

T. Nakamura, Yamato 793321 CM chondrite: dehydrated regolith material of a hydrous asteroid. Earth Planet. Sci. Lett. **242**, 26–38 (2006)

A. Nakato, T. Nakamura, F. Kitajima, T. Noguchi, Evaluation of dehydration mechanism during heating of hydrous asteroids based on mineralogical and chemical analysis of naturally and experimentally heated CM chondrites. Earth Planets Space **60**, 855–864 (2008)

H.B. Niemann et al., Composition of Titan's lower atmosphere and simple surface volatiles as measured by the Cassini–Huygens probe gas chromatograph mass spectrometer experiment. J. Geophys. Res. **115**, E12006 (2010)

F. Nimmo, D. O'Brien, T. Kleine, Tungsten isotopic evolution during late-stage accretion: constraints on Earth–Moon equilibration. Earth Planet. Sci. Lett. **292**, 363–370 (2010)

D.P. O'Brien, K.J. Walsh, A. Morbidelli, S.N. Raymond, A.M. Mandell, Water delivery and giant impacts in the 'Grand Tack' scenario. Icarus **239**, 74–84 (2014). https://doi.org/10.1016/j.icarus.2014.05.009

E. Ohtani, N. Hirao, T. Kondo, M. Ito, T. Kikegawa, Iron–water reaction at high pressure and temperature, and hydrogen transport into the core. Phys. Chem. Miner. **32**, 77–82 (2005)

T. Okuchi, Hydrogen partitioning into molten iron at high pressure: implications for Earth's core. Science **278**, 1781–1784 (1997)

H.St.C. O'Neill, D.C. Rubie, D. Canil, C.A. Geiger, C.R. Ross II., F. Seifert, A.B. Woodland, Ferric iron in the upper mantle and in transition zone assemblages: implications for relative oxygen fugacities in the mantle, in *Evolution of the Earth and Planets*, ed. by E. Takahashi, R. Jeanloz, D.C. Rubie. Geophysical Monograph, vol. 74, IUGG 14 (1993), pp. 73–88

P. O'Rourke, D.J. Stevenson, Stability of ice/rock mixtures with application to a partially differentiated Titan. Icarus **227**, 67–77 (2014)

R.T. Pappalardo, G.C. Collins, J.W. Head III., P. Helfenstein, T.B. McCord, J.M. Moore, L.M. Prockter, P.M. Schenk, J.R. Spencer, Geology of Ganymede, in *Jupiter: The Planet, Satellites and Magnetosphere*, ed. by F. Bagenal, T.E. Dowling, W.B. McKinnon. Cambridge Planetary Science (2004), pp. 363–396

C.C. Porco et al., Cassini observes the active South Pole of Enceladus. Science **311**, 1393–1401 (2006)

F. Postberg et al., Sodium salts in E-ring ice grains from an ocean below the surface of Enceladus. Nature **459**, 1098–1101 (2009)

F. Postberg, J. Schmidt, J. Hillier, S. Kempf, R. Srama, A salt-water reservoir as the source of a compositionally stratified plume on Enceladus. Nature **474**, 620–622 (2011)

D. Prialnik, G. Sarid, E.D. Rosenberg, R. Merk, Thermal and chemical evolution of comet nuclei and Kuiper belt objects. Space Sci. Rev. **138**, 147–164 (2008)

S.N. Raymond, A. Izidoro, Origin of water in the inner Solar System: planetesimals scattered inward during Jupiter and Saturn's rapid gas accretion. Icarus (2017). https://doi.org/10.1016/j.icarus.2017.06.030

S.N. Raymond, D.P. O'Brien, A. Morbidelli, N.A. Kaib, Building the terrestrial planets: constrained accretion in the inner Solar System. Icarus **203**, 644–662 (2009)

S.N. Raymond, T. Quinn, J.I. Lunine, Making other earths: dynamical simulations of terrestrial planet formation and water delivery. Icarus **168**, 1–17 (2004)

Y. Ricard, O. Šrámek, F. Dubuffet, A multi-phase model of runaway core-mantle segregation in planetary embryos. Earth Planet. Sci. Lett. **284**, 144–150 (2009)

G. Robuchon, F. Nimmo, Thermal evolution of Pluto and implications for surface tectonics and a subsurface ocean. Icarus **216**(2), 426–439 (2011)

L. Roth et al., Transient water vapor at Europa's South pole. Science **343**, 171 (2014)

D.C. Rubie, D.J. Frost, U. Mann, Y. Asahara, K. Tsuno, F. Nimmo, P. Kegler, A. Holzheid, H. Palme, Heterogeneous accretion, composition and core-mantle differentiation of the Earth. Earth Planet. Sci. Lett. **301**, 31–42 (2011). https://doi.org/10.1016/j.epsl.2010.11.030

D.C. Rubie, S.A. Jacobson, A. Morbidelli, D.P. O'Brien, E.D. Young, J. de Vries, F. Nimmo, H. Palme, D.J. Frost, Accretion and differentiation of the terrestrial planets with implications for the compositions of early-formed Solar System bodies and accretion of water. Icarus **248**, 89–108 (2015a)

D.C. Rubie, V. Laurenz, S.A. Jacobson, A. Morbidelli, H. Palme, A.K. Vogel, D.J. Frost, Highly siderophile elements were stripped from Earth's mantle by iron sulfide segregation. Science **353**, 1141–1144 (2016)

D.C. Rubie, H.J. Melosh, J.E. Reid, C. Liebske, K. Righter, Mechanisms of metal-silicate equilibration in the terrestrial magma ocean. Earth Planet. Sci. Lett. **205**, 239–255 (2003)

D.C. Rubie, F. Nimmo, H.J. Melosh, Formation of the Earth's core, in *Treatise on Geophysics*, ed. by D. Stevenson. Evolution of the Earth, vol. 9, 2nd edn. (Elsevier, Amsterdam, 2015b), pp. 43–79

J. Rudge, T. Kleine, B. Bourdon, Broad bounds on Earth's accretion and core formation constrained by geochemical models. Nat. Geosci. **3**, 439–443 (2010)

T. Rushmer, N. Petford, Microsegregation rates of liquid Fe–Ni–S metal in natural silicate-metal systems: a combined experimental and numerical study. Geochem. Geophys. Geosyst. **12** (2011). https://doi.org/10.1029/2010GC003413

H. Samuel, A re-evaluation of metal diapir breakup and equilibration in terrestrial magma oceans. Earth Planet. Sci. Lett. **313**, 105–114 (2012)

H. Samuel, P.J. Tackley, M. Evonuk, Heat partitioning in terrestrial planets during core formation by negative diapirism. Phys. Earth Planet. Inter. **290**, 13–19 (2010)

J. Saur et al., The search for a subsurface ocean in Ganymede with Hubble Space Telescope observations of its auroral ovals. J. Geophys. Res. Space Phys. **120** (2015)

L. Schaefer, D. Sasselov, The persistence of oceans on Earth-like planets: insights from the deep-water cycle. Astrophys. J. **801**(1), 40 (2015)

A. Scheinberg, R.R. Fu, L.T. Elkins-Tanton, B.P. Weiss, Asteroid differentiation: melting and large-scale structure, in *Asteroids IV*, ed. by P. Michel, F.E. DeMeo, W.F. Bottke (University of Arizona Press, Tucson, 2015), pp. 533–552

M. Schönbächler, R.W. Carlson, M.F. Horan, T.D. Mock, E.H. Hauri, Heterogeneous accretion and the moderately volatile element budget of Earth. Science **328**, 884–887 (2010). https://doi.org/10.1126/science.1186239

G. Schubert, H. Hussmann, V. Lainey, D.L. Matson, W.B. McKinnon, F. Sohl, C. Sotin, G. Tobie, D. Turrini, T. Van Hoolst, Evolution of icy satellites. Space Sci. Rev. **153**(1), 447–484 (2010)

G. Schubert, T. Spohn, R.T. Reynolds, Thermal histories, compositions and internal structures of the moons of the solar system, in *Satellites*, ed. by J.A. Burns, M.S. Matthews (University of Arizona Press, Tucson, 1986), pp. 224–292

G. Schubert, D.J. Stevenson, K. Ellsworth, Internal structures of the Galilean satellites. Icarus **47**, 46–59 (1981)

H.P. Scott, Q. Williams, F.J. Ryerson, Experimental constraints on the chemical evolution of large icy satellites. Earth Planet. Sci. Lett. **203**, 399–412 (2002)

Y. Sekine et al., High-temperature water–rock interactions and hydrothermal environments in the chondrite-like core of Enceladus. Nat. Commun. **6**, 8604 (2015)

Y. Sekine, H. Genda, S. Kamata, T. Funatsu, The Charon-forming giant impact as a source of Pluto's dark equatorial regions. Nat. Astron. **1**, 0031 (2017)

N.I. Shakura, R.A. Sunyaev, Black holes in binary systems. Observational appearance. Astron. Astrophys. **24**, 337–355 (1973)

Z.D. Sharp, F.M. McCubbin, C.K. Shearer, A hydrogen-based oxidation mechanism relevant to planetary formation. Earth Planet. Sci. Lett. **380**, 88–97 (2013)

Y. Shibazaki, E. Ohtani, H. Terasaki, A. Suzuki, K. Funakoshi, Hydrogen partitioning between iron and ringwoodite: implications for water transport into the Martian core. Earth Planet. Sci. Lett. **287**, 463–470 (2009)

J.B. Simon, P.J. Armitage, R. Li, A.N. Youdin, The mass and size distribution of planetesimals formed by the streaming instability. I. The role of self-gravity. Astrophys. J. **822**, 55 (2016)

F. Sohl et al., Subsurface water oceans on icy satellites: chemical composition and exchange processes. Space Sci. Rev. **153**, 485–510 (2010)

V.S. Solomatov, Magma oceans and primordial mantle differentiation, in *Treatise on Geophysics*, 2nd edn. (2015), pp. 81–104

C. Sotin, G. Mitri, N. Rappaport, G. Schubert, D. Stevenson, in *Titan from Cassini/Huygens*, ed. by R.H. Brown, J.P. Lebreton, J.H. Waite (2009a), pp. 61–73

C. Sotin, G. Tobie, J. Wahr, W.B. McKinnon, in *Tides and Tidal Heating on Europa*, ed. by R. Pappalardo et al., Space Science Series (University of Arizona Press, Tucson, 2009b), pp. 85–117

S.W. Squyres, R.T. Raynolds, A.L. Summers, F. Shung, Accretional heating of the satelites of Saturn and Uranus. J. Geophys. Res. **93**, 8779–8794 (1988)

D.J. Stevenson, Fluid dynamics of core formation, in *The Origin of the Earth*, ed. by H.E. Newsom, J.H. Jones (Oxford University Press, London, 1990), pp. 231–249

R. Tajeddine, N. Rambaux, V. Lainey, S. Charnoz, A. Richard, A. Rivoldini, B. Noyelles, Constraints on Mimas' interior from Cassini ISS libration measurements. Science **346**(6207), 322–324 (2014)

H. Terasaki, D.J. Frost, D.C. Rubie, F. Langenhorst, The effect of oxygen and sulphur on the dihedral angle between Fe–O–S melt and silicate minerals at high pressure: implications for Martian core formation. Earth Planet. Sci. Lett. **232**, 379–392 (2005). https://doi.org/10.1016/j.epsl.2005.01.030

H. Terasaki, D.J. Frost, D.C. Rubie, F. Langenhorst, Percolative core formation in planetesimals. Earth Planet. Sci. Lett. **273**, 132–137 (2008)

P.C. Thomas et al., Enceladus's measured physical libration requires a global subsurface ocean. Icarus **264**, 37–47 (2016)

G. Tobie, J.I. Lunine, J. Monteux, O. Mousis, F. Nimmo, The origin and evolution of Titan, in: *Titan: Surface, Atmosphere and Magnetosphere*, ed. by I. Mueller-Wodarg, C. Griffith, T. Cravens, E. Lellouch (Cambridge University Press, Cambridge, 2013)

K.A. Todd, H.C. Watson, T. Yu, Y. Wang, The effects of shear deformation on planetesimal core segregation: results from in-situ X-ray micro-tomography. Am. Mineral. **101**, 1996–2004 (2016)

T.M. Usselman, Experimental approach to the state of the core. Part I. The liquidus relations of the Fe-rich portion of the Fe–Ni–S system. Am. J. Sci. **275**, 278–290 (1975)

S. Vance, J.M. Brown, Thermodynamic properties of aqueous $MgSO_4$ to 800 MPa at temperatures from -20 to $100\,^{\circ}C$ and concentrations to 25 $mol\,kg^{-1}$ from sound speeds, with applications to icy world oceans. Geochim. Cosmochim. Acta **110**, 176–189 (2013)

N. von Bargen, H.S. Waff, Permeabilities, interfacial areas and curvatures of partially molten systems: results of numerical computations of equilibrium microstructures. J. Geophys. Res. **91**, 9261–9276 (1986)

J.-B. Wacheul, M. Le Bars, J. Monteux, J.M. Aurnou, Laboratory experiments on the breakup of liquid metal diapirs. Earth Planet. Sci. Lett. **403**, 236–245 (2014)
J. Wade, B.J. Wood, Core formation and the oxidation state of the Earth. Earth Planet. Sci. Lett. **236**, 78–95 (2005)
K.J. Walsh, A. Morbidelli, S.N. Raymond, D.P. O'Brien, A.M. Mandell, A low mass for Mars from Jupiter's early gas-driven migration. Nature **475**, 206–209 (2011)
M.J. Walter, E. Cottrell, Assessing uncertainty in geochemical models for core formation in Earth. Earth Planet. Sci. Lett. **365**, 165–176 (2013)
K. Willacy, H.H. Klahr, T.J. Millar, Th. Henning, Gas and grain chemistry in a protoplanetary disk. Astron. Astrophys. **338**, 995–1005 (1998)
B.J. Wood, M.J. Walter, J. Wade, Accretion of the Earth and segregation of its core. Nature **441**, 825–833 (2006). https://doi.org/10.1038/nature04763
T. Yoshino, M.J. Walter, T. Katsura, Core formation in planetesimals triggered by permeable flow. Nature **422**, 154–157 (2003)
E.D. Young, The hydrology of carbonaceous chondrite parent bodies and the evolution of planet progenitors. Philos. Trans. R. Soc. Lond. A **359**, 2095–2110 (2001)
E.D. Young, R.D. Ash, P. England, D. Rumble III, Fluid flow in chondrite parent bodies: deciphering the compositions of planetesimals. Science **286**, 1331–1335 (1999)
E. Young, K. Zhang, G. Schubert, Conditions for pore water convection within carbonaceous chondrite parent bodies—implications for planetesimal size and heat production. Earth Planet. Sci. Lett. **213**, 249–259 (2003)
K. Zhang, K.M. Pontoppidan, C. Salyk, G.A. Blake, Evidence for a snow line beyond the transitional radius in the TW Hya protoplanetary disk. Astrophys. J. **766**, 82 (2013)
M.E. Zolensky, W.L. Bourcier, J.L. Gooding, Aqueous alteration on the hydrous asteroids: results of EQ3/6 computer simulations. Icarus **78**, 411–425 (1989)
M.Y. Zolotov, J.S. Kargel, On the composition of Europa's icy shell, ocean and underlying rocks, in *Europa*, ed. by R. Pappalardo, W.B. McKinnon, K. Khurana (University of Arizona Press, Tucson, 2009), pp. 431–457

Space Sci Rev (2018) 214:65
https://doi.org/10.1007/s11214-018-0490-9

Water Loss from Young Planets

Feng Tian[1] · Manuel Güdel[2] · Colin P. Johnstone[2] · Helmut Lammer[3] · Rodrigo Luger[4] · Petra Odert[3]

Received: 15 October 2016 / Accepted: 21 February 2018 / Published online: 4 April 2018

Abstract Good progress has been made in the past few years to better understand the XUV evolution trend of Sun-like stars, the capture and dissipation of hydrogen dominant envelopes of planetary embryos and protoplanets, and water loss from young planets around M dwarfs. This chapter reviews these recent developments. Observations of exoplanets and theoretical works in the near future will significantly advance our understanding of one of the fundamental physical processes shaping the evolution of solar system terrestrial planets.

Keywords Exoplanet · Evolution · Water · Habitability

1 Introduction

Currently there are more than 3000 exoplanets discovered and a handful of them are potentially habitable (with mass between 0.1 and 10 Earth masses, $M_\oplus$, and within the classical liquid water habitable zone, referred to HZ in this paper). It is exciting to the science community that the nearest star to the Sun, Proxima Centauri, also has an Earth-mass planet in its HZ (Anglada-Escude et al. 2016). Existing exoplanet detection missions and facilities (TESS, PLATO 2.0, NGTS, ESPRESSO, etc.) will likely detect more potentially habitable exoplanets within the next decade. But how do we know whether these planets are indeed habitable?

The Delivery of Water to Protoplanets, Planets and Satellites
Edited by Michel Blanc, Alessandro Morbidelli, Yann Alibert, Lindy Elkins-Tanton, Paul Estrada, Keiko Hamano, Helmut Lammer, Sean Raymond and Maria Schönbächler

✉ F. Tian
tianfengco@mail.tsinghua.edu.cn

1 Department for Earth System Science, Tsinghua University, Beijing, 100084, China

2 Department of Astrophysics, University of Vienna, Türkenschanzstrasse 17, 1180 Vienna, Austria

3 Space Research Institute, Austrian Academy of Sciences, Schmiedlstrasse 6, 8042 Graz, Austria

4 Department of Astronomy, University of Washington, Box 351580, Seattle, WA 98195, USA

Observations will certainly have the final words on this. However the observation time needed to characterize Earth-mass planets (here we define them as planets with mass between 0.1 and $10M_\oplus$) in the HZ of nearby stars is long with telescopes available in the foreseeable future (Leger et al. 2015). Thus it will be important to prioritize the target list, and this is where water loss from planets becomes a central issue. Indeed a planet without water, such as our sister planet Venus, is not considered habitable in our current understanding.

Tian et al. (2013) reviewed how atmosphere escape might have influenced the climate evolution of solar system terrestrial planets. Tian (2015a, 2015b) reviewed recent developments on atmosphere escape from planets in and beyond the solar system. Massol et al. (2016) reviewed the formation and evolution of protoatmospheres, which included a section discussing the basic concepts and details of atmospheric escape. The focus of this review is on water loss from planets in their early evolution stages. We focus on hydrodynamic escape, an extreme scenario in which a planetary upper atmosphere (with total number densities lower than 10^{13} cm^{-3}, which includes both the thermosphere and ionosphere) is heated and expands to a large distance. In this situation processes occurring beyond the top of the atmosphere (the exobase) do not severely alter the total escape (Tian 2013), and thus we can focus on the physics within the upper atmosphere. In the early stage of the solar system, hydrodynamic escape from early Venus, Earth and Mars most likely played important roles in the evolution of planetary water inventories and atmospheres (e.g., Watson et al. 1981; Kasting and Pollack 1983; Chassefière 1996; Lammer et al. 2008; Lichtenegger et al. 2016). Hydrodynamic escape from planetary atmospheres is a subject that has received much attention in the past decade because of the discoveries of hot hydrogen-dominated exoplanets (e.g., Lammer et al. 2003; Yelle 2004; Tian et al. 2005; Garcia Muñoz 2007; Koskinen et al. 2013; Shaikhislamov et al. 2014; Guo and Ben-Jaffel 2016). An interesting feature of hydrodynamic escape is that heavier gases could be dragged out by lighter species, which could have implications on the composition of planetary atmospheres. This process has been discussed more than 40 years ago and recent observations of hot Jupiters provided real evidence that it actually occurs (Vidal-Madjar et al. 2003, 2004, 2013; Linsky et al. 2010; Bourrier et al. 2013).

In Sect. 2.1 we will introduce an approximation equation describing hydrodynamic escape. Readers interested in detailed discussions of planetary upper atmospheres are encouraged to check previous reviews (Tian et al. 2013; Tian 2015a, 2015b; Massol et al. 2016). In Sect. 2.2 we introduce current understanding on the XUV environment, the driving force of water loss, and in Sect. 2.3 the drag equations are summarized. In Sect. 3.1 we set the stage for water loss by discussing the capture and loss of hydrogen-dominated atmospheres and in Sect. 3.2 the loss of H_2O from planetary embryos is discussed. In Sect. 3.3 we discuss water loss from planets in the HZ of M dwarfs during the pre-main sequence phase and the possible fractionation and accumulation of oxygen (O) as a by-product of water loss.

2 Processes and Equations

2.1 Hydrodynamic Escape

Soft X-ray (wavelengths of 5 Å to $\sim$ 100 Å) and EUV ($\sim$ 100 Å to 912 Å) photons from stars are the main heating source of planetary upper atmospheres because the absorption cross sections of atmospheric species in these wavelength ranges are large (Tian 2009, 2015a, 2015b; Massol et al. 2016). FUV (912 Å to 1750 Å) photons also play a role in heating up the lowest parts of the upper atmospheres. The cross sections of photons with wavelengths < 5 Å and > 1750 Å are too small to contribute to upper atmosphere heating significantly.

A useful way to calculate the hydrodynamic escape flux (Φ, in unit of $cm^{-2}s^{-1}$) from planetary atmospheres is the energy limited formula:

$$\Phi = \frac{F_{XUV}\eta R_p}{4GM_p m K} \times \frac{R_{XUV}^2}{R_p^2} \tag{1}$$

where M_p and R_p are the mass (in gram) and radius of the planet (in cm), m the mass of escaping species in gram, G the gravitational constant in cgs unit, F_{XUV} the photon fluxes (in $erg\,cm^{-2}\,s^{-1}$) reaching the planet in the soft X-ray and EUV range, R_{XUV} the typical radius of the planet at which most of the XUV photons are absorbed (cm). η is the heating efficiency (the fraction of energy converted from absorbed XUV photons to the energy driving the hydrodynamic escape) which includes contributions from ion-neutral chemistry, radiative cooling, and dynamic processes. K is an approximate correction factor accounting for the gravity of the star (Erkaev et al. 2007):

$$\begin{cases} K = 1 - 1.5\zeta + 0.5\zeta^3 \\ \zeta = \frac{R_p}{a} \times (\frac{3M_*}{M_p})^{1/3} \end{cases} \tag{2}$$

where M_* is stellar mass and a the semi-major axis of the planet.

Detailed calculations of planetary upper atmospheres are needed in order to obtain good estimates of both η and R_{XUV} because both parameters are functions of altitude-dependent density, composition, and temperature profiles. Thus in some literature R_{XUV} is set to R_p. Under this simplification, comparisons of observed close-in exoplanets with evolution models using (1) indicate that η is $\sim 15\%$ for hydrogen dominated planetary atmospheres (Lopez et al. 2012). It is important to note that some theoretical models of the upper atmospheres of hot Jupiters (Koskinen et al. 2014) obtain heating efficiencies from lower than 10% to greater than 40% for different pressure levels in the upper atmospheres of hot Jupiters. Thus to use an overall constant heating efficiency in planetary upper atmosphere modeling is a highly simplified approach and caution is needed.

2.2 Evolution of Stellar XUV Radiation

A fundamental parameter controlling the hydrodynamic escape rate is the XUV radiation from the host star reaching the planetary body. Observations show that the ratios between the X-ray emission and the total luminosity of M dwarfs are 10–1000 times greater than that of modern Sun, which is largely an effect of the much slower activity evolution for lower-mass stars (Scalo et al. 2007; Stelzer et al. 2013; Tian 2015a, 2015b). Because the X-ray emission is emitted from the same region in stellar atmospheres as the XUV radiation, young planets around M dwarfs suffer greater dosage of XUV radiation for longer time than their counterparts around K or G dwarfs. Thus water loss from young planets around M dwarfs is an important and open scientific problem and the evolution of a planet's atmosphere and water inventory is sensitively linked to the evolution of its host star's XUV emission.

2.2.1 A Brief Research History of Stellar XUV Activity

Early theoretical and observational work in the sixties and seventies started establishing a picture of activity and rotation evolution of solar-like stars that later became known as the stellar "activity-rotation-age paradigm". Kraft (1967) found that the Ca II activity indicator is stronger for more rapidly rotating stars, which are typically younger than the average, promoting a picture in which magnetic winds decelerate stellar rotation and weaken the internal dynamo. Weber and Davis (1967) developed a relevant wind-torque model. These elements were brought together by Skumanich (1972), who reported similar evolutionary decay of the stellar rotation rate and stellar activity as seen in Ca II, both following an inverse square root law; he suggested that magnetic surface fields—responsible for activity signatures—follow similar trends.

A coherent observational picture was also developed in particular for stellar X-rays in the context of age and rotation. A brief summary of the three key elements in the rotation-activity-age relation is given below:

Activity-Rotation Relation After the initial work by Pallavicini et al. (1981) and Walter (1981), a picture emerged in which the stellar X-ray luminosity is roughly proportional to the square of the equatorial velocity, or equivalently, to the square of the stellar rotation rate (Maggio et al. 1987). The Sun occupies an area shared by the least X-ray active stars and slowest rotators. By contrast, the most active stars reach a saturation level at which $L_X \approx 10^{-3} L_{bol}$, where L_{bol} is the stellar bolometric luminosity and L_X is the X-ray luminosity. The change from an activity-rotation law to the saturation regime occurs at an L_{bol}-dependent saturation rotation period P_{sat}, which is about 1.2 days for a solar-mass star and follows $P_{sat} \approx 1.2 \times (\frac{L_\odot}{L_{bol}})^{0.5}$ (Pizzolato et al. 2003). For reference, the modern Sun has a soft X-ray (5–124 Å) luminosity between 6×10^{26} and 8×10^{27} erg s^{-1} in a typical solar cycle (Judge et al. 2003; Tian 2009). The solar luminosity in 50–1200 Å range between 1978 and 1997 is between 4×10^{27} and 1.4×10^{28} erg s^{-1} (Kazachevskaya et al. 1998; Tian 2009). Observations of *AE-E* satellite (50–1050 Å) between 1976 and 1979 also gives 2×10^{28} erg s^{-1} (Hinteregger 1981; Tian 2009). And the ratio L_X/L_{bol} for the modern Sun is $\sim 10^{-6}$.

An optimum description of this relation uses either L_X and P_{rot} (rotation period), or normalized $R_X = L_X/L_{bol}$ versus the so-called Rossby number $\mathrm{Ro} = P_{rot}/\tau$, where τ is the mass-dependent convective turnover time (Noyes et al. 1984). A modern activity-rotation diagram for a variety of stellar masses and based on stars from the field and various open clusters with different ages is shown in Wright et al. (2011).

Rotation-Age Relation As suggested by Kraft (1967) and Skumanich (1972), stars continuously spin down during their main-sequence evolution. Quantitatively, $P_{rot} \propto t^{0.6}$ on average for solar-like stars. However, the situation is complicated by the wide range of "initial conditions", i.e., the starting rotation rate at or somewhat before the zero-age main-sequence (ZAMS). Studies of co-eval samples (stars of the same age, formed at about the same time) from open clusters at ages of up to a few hundred Myr show a wide scatter of rotation rates for any given stellar mass, but the width of the rotation rate distribution narrows while shifting to larger ages and eventually converges to a near-unique rotation rate for a given age and stellar mass, typically after several hundred Myr (the convergence time increases with decreasing mass; see, for example, Soderblom et al. 1993; Gallet and Bouvier 2013; Johnstone et al. 2015). Fundamentally a star spins down because it loses its angular momentum through stellar wind. The factors determining how rapidly a star could lose its angular momentum are stellar magnetic field, stellar wind, stellar mass, stellar radius, and angular velocity of the star. Relating the spin-down rate to these parameters in a unified theory is difficult. But in principle it is reasonable that the rate at which a star spins down is primarily determined by its rotation rate, which means generally speaking a faster rotator will spin

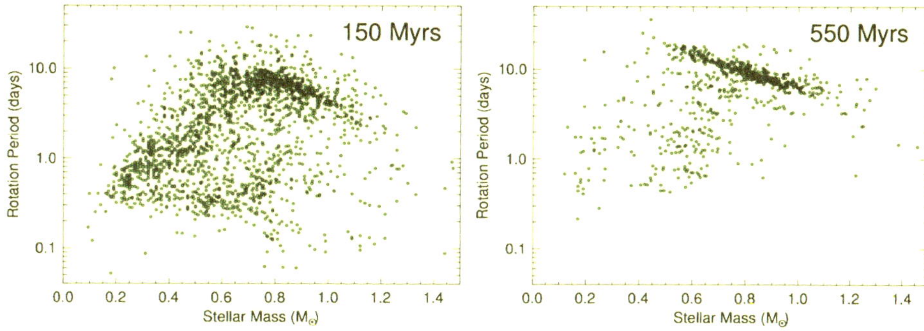

Fig. 1 Distributions of stellar rotation periods at ages of 150 Myr (left) and 550 Myr (right). Rotational data were obtained from several young clusters (Johnstone et al. 2015). All data points are green circles and the overlapping part may appear black

down more rapidly. Note that the spin-down rate is *higher* for lower-mass stars on the cool main sequence.

Activity-Age Relation The final piece of the triple correlation should then follow from the above, i.e., decreasing activity with decreasing rotation rate and with increasing age. Empirical evidence suggested, based on X-ray cluster studies, an age dependence of the form $L_X \propto t^{-1.5}$, where t is the stellar age (Maggio et al. 1987; Ayres 1997; Güdel et al. 1997). This is consistent with the other two relations. An extension of these decay laws, also including the EUV and FUV, was presented by Ribas et al. (2005). They find that all wavelength bands decay, on average, as power laws, but the power-law index increases with increasing photon energy. This implies an increasing decay factor for higher-energy radiation (factor of ≈ 1000 for X-rays, ≈ 100 for EUV, and 10–30 for the UV). Further extension of these models was presented by Claire et al. (2012).

2.2.2 A New Picture of Stellar XUV Evolution

The gross power-law relations for activity versus age summarized above are problematic because of the wide scatter of rotation periods at young ages and the activity-rotation relation. A proper description of the activity and radiation evolution must properly account for the evolution of the rotation distribution rather than using single-valued rotation periods during the first billion years.

Figure 1 shows two snapshots (at 150 Myr and 550 Myr) of the rotational distribution based on a large sample of observed stars coupled with a spin-down model compatible with measured distributions (Johnstone et al. 2015). The distribution at ~ 150 Myr includes measurement data from Pleiades, M50, M35, and NGC 2516. The distribution at ~ 550 Myr includes measurement data from M37 and Praesepe. Johnstone et al. 2015 contains more details and discussions (their Fig. 2). The convergence toward a single-valued dependence of rotation on mass and age is evident. This model can be coupled to the activity-rotation relation (Wright et al. 2011), in particular considering also the saturation level, to obtain the evolution of the activity distribution.

Figure 2 (based on Tu et al. 2015) shows the evolutionary tracks for the 10th, the 50th, and the 90th percentiles based on the initial (pre-main sequence) rotation periods derived from the rotational evolution model in Johnstone et al. (2015). It is important to understand that the XUV irradiation history of a planetary atmosphere can be largely different for hundreds

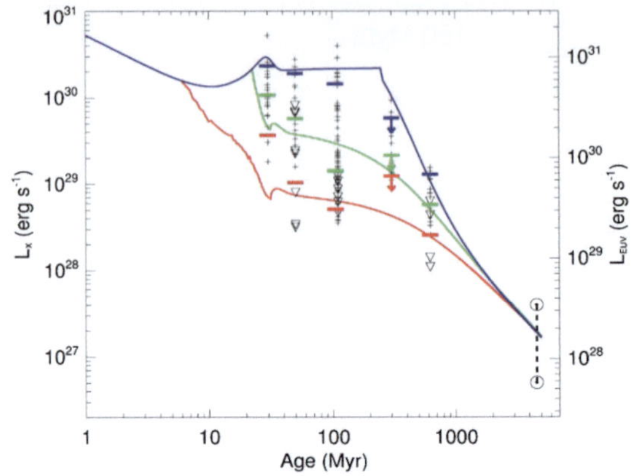

Fig. 2 X-ray and extreme-ultraviolet evolutionary tracks for solar-mass stars, based on the rotational evolution model from Tu et al. (2015). The red, green and blue tracks refer to the 10th, 50th, and 90th percentiles for the rotation rates. The black symbols show observed values in open clusters with the $\bigtriangledown$ symbols represent upper limits; the colored horizontal bars mark the percentiles of the observed distributions. The two circles at 4.5 Gyr indicate the range for the Sun, from minimum to maximum (from Tu et al. 2015)

of millions of years, with the maximum and minimum possible L_X and L_{EUV} values being as far apart as a factor of 30 and 10, respectively, during the first ~ 700 Myr for a Sun-like star.

The evolutionary tracks look qualitatively similar for K and M stars, although the evolutionary timescale is longer for K and much longer for M stars, and more of these stars reside in the saturation regime for longer time. This latter issue was investigated by West et al. (2008) who found that the magnetic activity lifetime increases from ~ 1 Gyr for early M dwarfs to ≥ 7 Gyr for late M dwarfs, relating that to the onset of full convection around the spectral type of $\sim$ M3. These extended timescales are relevant to the atmospheric escape on planets around those stars (see Sect. 3.3).

As Fig. 2 already suggests, the scales for the X-ray and the EUV part of the spectrum are different. The harder part of the spectrum generally decays more rapidly, and the high-energy stellar spectrum therefore softens with time. This correlation is important for the absorption, heating and ionization studies of planetary atmospheres given the vastly different penetration depths for photons with different energies, and thus deserves careful study.

Johnstone and Güdel (2015) found a tight correlation between the stellar X-ray surface flux and the average coronal temperature. More active stars show a hotter corona, resulting in harder spectra. The reconstruction of the entire coronal X-ray and EUV spectrum then requires calculations of atomic transitions and continuum radiation occurring in such a plasma, potentially improved for some distribution of coronal temperatures around the mean. However, extension into the EUV portion of the spectrum is challenging. While the coronal contributions, dominating in the shorter-wavelength part of this spectral range ($\sim$ 100 to 400 Å), can be estimated from line + continuum models of optically thin plasmas (e.g., Sanz-Forcada et al. 2011), there are important contributions from the stellar transition regions and chromospheres, especially at longer EUV wavelengths (400 to 912 Å), that are inaccessible on stars due to strong interstellar absorption of this spectral range. Some information is available from spectra of solar features and corresponding solar atmospheric models (Fontenla et al. 2015). Alternatively, evolutionary trends measured in the X-ray and the ultraviolet ranges can be interpolated, making use of the increasingly steeper decay laws for shorter wavelengths (Ribas et al. 2005; Claire et al. 2012). In a complementary approach, Linsky et al. (2013, 2014) developed a method to estimate the EUV luminosity based on correlations with the Lyα line flux. A database of FUV ((MUSCLE, including the Lyα line) and NUV photon fluxes of several planet-hosting M dwarfs based on HST observations exists (France et al. 2012, 2013, 2016; Loyd et al. 2016; Youngblood et al. 2016). But more observation data is still urgently needed in order to better understand the XUV environment of exoplanets.

In summary, younger stars rotate more rapidly and are therefore more active, although we need to consider rotation rate distributions at any given age, and these distributions are particularly wide up to ages of a few 100 Myrs. The activity evolution for a star of given mass is therefore not a unique function of time but depends on the initial rotation rate, which thus indirectly determines the evolution of solar system terrestrial planets (Johnstone et al. 2015; Tu et al. 2015). Future work combining evidence in multiple fronts coming from different research domains (astronomy, geosciences, and planetary sciences) will be needed to better constrain this important issue for solar system evolution. The much slower activity evolution for lower-mass stars (Scalo et al. 2007; Stelzer et al. 2013; Tian 2015a, 2015b) makes water and atmospheric loss from their planets a more severe problem in comparison to their counterparts around Sun-like stars (Luger and Barnes 2015; Tian and Ida 2015). Indeed, a similar rotation-activity relationship governing the rotation-XUV evolution of Sun-like stars can also be used for M dwarfs (Wright and Drake 2016).

2.3 Drag Equations of Heavy Species

With hydrodynamic escape of hydrogen, heavier species such as noble gases or oxygen could have been dragged along with the escaping hydrogen (e.g., Öpik and Singer 1961; Hunten 1973, 1987; Zahnle and Kasting 1986). Kasting and Pollack (1983) have shown that the slightly heavier deuterium (D) escaped less efficiently compared to atomic hydrogen (H), indicating that hydrodynamic escape can be invoked to explain the observed D/H ratio in the present Venus atmosphere. In a similar way isotopic anomalies and relative abundances of noble gases in planetary atmospheres compared to the solar composition can be generally explained (Pepin 1991).

If a planetary body experiences strong hydrodynamic atmospheric escape of a species i, which is typically atomic hydrogen, one can estimate the loss of a heavier species j that is dragged upward with the dynamic flow of the bulk atmosphere. Hunten et al. (1987) derived a simple analytic expression of the escape flux of a minor heavy species dragged along with the outflow of a major light species assuming isothermal, subsonic conditions. They defined a so-called cross-over mass m_c, which depends on the mass m_i and the upward flux Φ_i of the lighter species. For the heavier constituent j whose mass $m_j > m_i$ is lower than the cross-over mass m_c, the upward frictional force induced by Φ_i is larger than the downward directed gravitational force. In such a condition the escape flux of species j can be written as (Hunten et al. 1987; Chassefière 1996):

$$\begin{cases} \Phi_j = \frac{y_j}{y_i} \Phi_i \frac{m_c - m_j}{m_c - m_i} \\ m_c = m_i + \frac{kT\Phi_i}{b_{i,j} g y_i} \end{cases} \tag{3}$$

where y_i and y_j are the volume mixing ratios of the light and heavier species, respectively, k is the Boltzmann constant, T the atmospheric temperature, g the gravity, and $b_{i,j}$ the binary diffusion parameter of species j in i.

This simple analytic expression has been applied in several studies to investigate the loss of oxygen in hydrodynamically escaping hydrogen bulk atmospheres from early Mars and

Venus, as well as from exoplanets (e.g., Hunten et al. 1987; Chassefière 1996; Lammer et al. 2011, 2013; Erkaev et al. 2014; Luger and Barnes 2015; Tian 2015a, 2015b).

To extend the formalism to more than 2 species, Zahnle and Kasting (1986) derived a more general analytic solution. Defining $f_j = n_j/n_i$, where n_i and n_j are the number densities of species i (the primary escaping species) and j (heavier constituents such as O, C, and CO_2) respectively, the fractionation factor of a heavier species x_j, which is a measure of the ratio between escape fluxes of species i and j, can be written as

$$x_j = 1 - \frac{GM_{pl}(m_j - m_i)b_{i,j}}{\Phi_i r_0^2 kT(1 + f_j)} = 1 - \frac{g(m_j - m_i)b_{i,j}}{\Phi_i kT(1 + f_j)} \tag{4}$$

where M_{pl} is the mass of a planetary body, r_0 the planetary radius, and G the gravitational constant. The escape flux of the heavier species is $\Phi_j = \Phi_i f_j x_j$, which is equal to Eq. (2) in the limit $f_j \ll 1$. Note that the same simplifications in Hunten et al. (1987), except for the requirement that the second species is minor, are used to obtain (4).

If an additional species k is present, the fractionation factor is

$$x_k = \frac{1 - g(m_k - m_i)b_{i,k}/(\Phi_i kT) + (1 - x_j)f_j b_{i,k}/b_{i,j} + x_j f_j b_{i,k}/b_{j,k}}{1 + f_j b_{i,k}/b_{j,k}} \tag{5}$$

with a corresponding escape flux of $\Phi_k = \Phi_i f_k x_k$ (Zahnle and Kasting 1986; Zahnle et al. 1990).

These formulae can be applied to magma oceans under H_2O and CO_2 dominant steam atmospheres, which are likely to exist on planets in their early evolution stages. Considering the dissociation of H_2O by solar/stellar UV radiation, one can define species i as H, j as O, and k as CO_2. The binary diffusion parameters are $b_{H,O} = 4.8 \times 10^{17} T^{0.75}$ cm^{-1} s^{-1}, $b_{H,CO2} = 8.4 \times 10^{17} T^{0.6}$ cm^{-1} s^{-1} and $b_{O,CO2} = 7.86 \times 10^{16} T^{0.776}$ cm^{-1} s^{-1} (Zahnle and Kasting 1986). The fractionation x_k of CO_2 caused by the escape of both H and O could then be calculated based on g and T. Note that strong XUV radiation may not change the composition of CO_2-dominant early planetary atmospheres because recombination reactions, especially catalytic ones, become increasingly important in deeper and denser atmospheres (Tian et al. 2014). Thus to convert CO_2 into CO near the homopause of early planetary atmospheres requires extreme and rare stellar environments (Tian 2009).

Note that Eq. (2) is not complete. If the escape flux of species j (Φ_j) is large, the escape flux of species i (Φ_i) could be influenced. A decreased Φ_i could in turn reduce Φ_j. Such a feedback is considered in Tian (2015b). In addition, Eqs. (2)–(4) assume an isothermal upper atmosphere which probably never exists at all. To calculate a planetary upper atmosphere temperature profile is complicated and interested readers are referred to recent reviews (Tian et al. 2013; Tian 2015a; Massol et al. 2016). Nevertheless, these equations could be used to obtain a preliminary picture on loss of light and heavy species in hydrodynamic escape, while the results need to be checked by future models of planetary upper atmospheres with detailed chemistry and physics. We note that the development of multi-component models (either 1D or 3D) is still not adequately sophisticated despite many efforts to model hydrodynamic escape (Guo and Ben-Jaffel 2016; Koskinen et al. 2014; Khodachenko et al. 2017; Dong 2018). On the other hand, Loyd et al. (2017) finds an observational upper limit of $\sim 20\%$ for the transit depth of C II in the extended hydrogen envelope surrounding GJ436b, which is one order of magnitude greater than the model calculated C II transit depth. Thus in addition to more model efforts, substantial improvements in our current observation capability for the upper atmospheres of close-in exoplanets need to be fulfilled to better advance our goal to understand water loss from young planets and planet evolution.

3 Implications for Water Loss from Young Planets

3.1 Capture and Loss of the Initial Hydrogen Envelopes

In the first a few millions years after the formation of a protostar, the environment surrounding the star is dominated by a disk of mostly H and He. A planetary core that grows to ≥ 0.1 Earth mass ($M_\oplus$) during this gas phase will be able to accumulate a thick envelope of H and He (Massol et al. 2016). The presence of such atmospheres is supported by recent discoveries of low density terrestrial planets (e.g. Lissauer et al. 2011) and the trend that most high mass terrestrial planets have low densities (Marcy et al. 2014).

The lower regions of a thick hydrogen envelope are optically thick and therefore the cooling of the core is inefficient. At the surface of the solid core, the gas should be hot and convective. The temperature decreases with increasing altitude throughout the optically thick region following an adiabatic law. The envelope should be isothermal in the optically thin region (e.g., see Fig. 5 of Stökl et al. 2015; Massol et al. 2016). The temperature in the optically thin region is approximately the planetary effective temperature. Before the dissipation of protoplanetary disk, stellar XUV radiation rarely reaches planets due to absorption of dusts and gases. After the protoplanetary disk is gone, planetary upper atmospheres above the isothermal region would be heated by solar or stellar XUV radiation to temperatures of $> 10^3$ K.

Lammer et al. (2014) modelled the capture and escape of primordial atmospheres for terrestrial planets formed during the disk phase with core masses between 0.1 and $5.0M_\oplus$. In their calculations, the amount of atmosphere captured during the disk phase is strongly dependent on the core mass, with $5M_\oplus$ cores capturing 5–6 orders of magnitude more atmosphere than $0.1M_\oplus$ cores. The envelopes of the high-mass cores are too massive to be ever lost, whereas the envelopes of low-mass cores can be lost very quickly under incident XUV radiation (Lammer et al. 2014). Erkaev et al. (2014) and Stökl et al. (2015) showed that the amount of gas captured by Mars-size bodies from the protoplanetary disk is much smaller compared to more massive protoplanets, and that depending on the assumed nebula properties and the XUV evolution of host stars, planetary embryos could have lost all their hydrogen envelopes captured from the nebula in timescale of 10^4–10^6 years after the nebula gas evaporated. These results suggest that if the solid core of a planet grows to $> 1M_\oplus$ within a few Myr after the formation of the star, it will possess a thick H/He envelope for its entire lifetime. These conclusions have also been found by other theoretical studies (Luger and Barnes 2015; Owen and Mohanty 2016).

Water vapor could reach high altitudes in a hydrogen dominated atmosphere as long as the temperature profile does not present a 'cold trap' for water vapor to condense. In this scenario atomic oxygen, the dissociation product of water vapor (either through photodissociation or thermal dissociation), could be present in the upper atmosphere and be lost through the dragging effect in hydrodynamic escape. This process has been studied using analytic methods (Luger and Barnes 2015; Tian 2015b) but has not been modeled in detail yet.

3.2 Water Loss from Planetary Embryos

Most rocky planets are formed through collisions of multiple planetary embryos with radii between tens and hundreds of km, which may or may not be formed at the orbit of the planets. Objects formed beyond the snowline contain large amounts of water, 40–80% according to a number of estimations (e.g. Alibert et al. 2010). Because collisions between a few large planetary embryos are sufficient to deliver a huge amount of mass and volatiles to the growing planets during the accretion process of planets (e.g., Abe and Matsui 1985, 1986; Alibert et al. 2010), planets closer to the Sun/star may not be absolutely anhydrous (e.g., Alibert et al. 2010). Thus, the fractionation of heavy atoms by hydrodynamic escape of atomic hydrogen or other light species is important for both planetary embryos and planets.

The internal structures of large moons in the solar system are comparable to those of the terrestrial planets (e.g., Schubert et al. 2004). The silicates and metals that make up planetary bodies in general were most likely melted once or several times during their formation process (e.g., Albarède and Blichert-Toft 2007; Elkins-Tanton 2012). The melting will also be caused by 2 processes: gravitational heating which increases as the square of the radius of planetary bodies, and radiogenic heating from short-lived radioisotopes (e.g., Urey 1955; Safronov 1969; Wetherill and Stewart 1989; LaTourrette and Wasserburg 1998; Halliday et al. 2001; Albarède and Blichert-Toft 2007; Elkins-Tanton 2008, 2011, 2012). Thus larger Moon- or Mars-size planetary embryos could also form magma oceans. These magma oceans also create a compositional differentiation that affects the final volatile contents of planetary embryos.

The solidification of the magma oceans will result in the outgassing of steam atmospheres. If the steam atmosphere last longer than the nebula, it will be exposed to the strong XUV flux of the young host star (see Sect. 2.2) and frequent impacts of smaller planetesimals (Maindl et al. 2015). Dissociation of water vapor produces H and O. Because H is much lighter than O and CO_2, H will dominate the upper steam atmospheres (e.g. Hunten et al. 1987; Kasting and Pollack 1983; Chassefière 1996).

For the estimation of the loss of water from planetary embryos at different orbital distances we apply the energy-limited formula discussed in Sect. 2.1 for the total mass-loss rate $\dot{M}$, assuming a heating efficiency η of $\sim 15\%$. Note that η may be different for water-dominated steam atmospheres and also be time-dependent because the composition of the escaping atmosphere changes during its evolution. Applying Eq. (2) to an upper atmosphere made of H and O and considering the conservation of mass and energy, the mass-loss rates for H and O are:

$$\dot{M}_H = \frac{\dot{M}}{1 + \frac{m_O}{m_H}\frac{n_O}{n_H}x_O},$$

and

$$\dot{M}_O = \frac{\dot{M}}{1 + (\frac{m_O}{m_H}\frac{n_O}{n_H}x_O)^{-1}}.$$

Note that the dependence of the mass loss rates on T, gravity, and binary diffusion parameters are embedded in the calculations of $\dot{M}$ and χ_O (see previous sections). Alternative expressions for these mass loss rates can be found in Tian (2015b) with detailed derivations.

Here we apply the above formulae to a Mars-size embryo with a steam atmosphere of 50 and 260 bars pressure (corresponding to the minimum and maximum outgassing rates of H_2O in Erkaev et al. 2014), and a Moon-size embryo with a steam atmosphere of 30 bars. Assuming complete dissociation of the water vapor, Fig. 3 shows the loss timescale for the above mentioned steam atmospheres at the orbital distances of Venus (V), Earth (E), and Mars (M) for an Sun-like star with 30, 100, and 500 times higher XUV fluxes than the modern solar mean value ($4.64\ \mathrm{erg\,s^{-1}\,cm^{-2}}$; Ribas et al. 2005).

This simple analysis shows that Moon-sized planetary embryos at Venus orbit with steam atmospheres between ~ 30 to 50 bars could lose all water inventory in timescale of

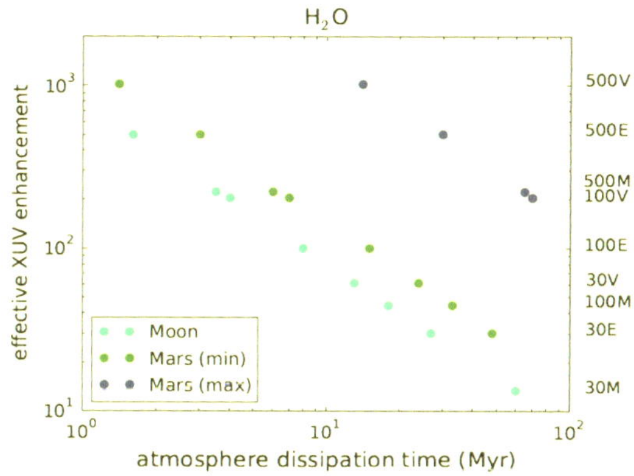

Fig. 3 Dissipation time of outgassed H_2O from Moon- and Mars-size planetary embryos with water vapor surface partial pressures of 30 bars (Moon), 52 bars (Mars-min.) and 257 bars (Mars-max.) as a function of effective XUV flux enhancement levels (left y-axis) compared to that of today's solar value at Earth's orbit. The right y-axis indicates the stellar XUV luminosity enhancement (30, 100 and 500 times stronger) and the orbit locations (Venus V, Earth E and Mars M) which the effective enhancement corresponds to

0.5–10 Myr. For Mars-size embryos with minimum initial water inventory exposed to XUV from slowly rotating young Sun-like stars, ~ 30 times modern Earth value (Sect. 2.2), water inventories could be maintained after 100 Myrs of exposure only if the embryos are on Mars orbit. But for Mars-size embryos with maximum initial water inventories, the water loss timescale is greater than 100 Myrs except for the highest XUV exposure level.

Another important factor concerning water loss is the lifetime of the steam atmospheres. Lebrun et al. (2013) modeled the thermal evolution of magma oceans and steam atmospheres on Earth-size planets by estimating the cooling timescales. These authors concluded that liquid oceans could form at the surfaces only ~ 10, ~ 1.5 and < 1 Myr after planet formation at 0.7, 1.0 and 1.5 AU, respectively. Hamano et al. (2013) used a different model and obtained longer timescales (~ 4 Myr and 100 Myrs at 1 and 0.7 AU respectively). Note that frequent impacts may extend the lifetime of the steam atmospheres.

By considering all the unknowns and uncertainties there is much evidence that wet planetary embryos moving inward from beyond the snowline can lose a large fraction of their initial volatile inventories. Planetary embryos at Venus' orbit are most likely drier in comparison to their counterparts at orbits between Earth and Mars. However, the exact degree of dehydration of planetary embryos depends on several physical processes and the sequence for them to occur, which will require further study. In addition, noble gases can also be dragged by the strong hydrogen loss during the loss of hydrogen and water. Thus the building blocks of Venus, Earth, and Mars may have had different evolution pathways in terms of isotopic fractionation, which could contribute to different $^{20}Ne/^{22}Ne$, $^{36}Ar/^{38}Ar$, on these bodies (e.g., Pepin 1991; Gillmann et al. 2009).

3.3 Water Loss from Planets in the HZ of M Dwarfs and Possible O_2 Buildup

In Sect. 2.2 we reviewed the XUV evolution of young stars. In addition to being XUV active during their pre-main sequence (PMS) phase, M dwarfs are also well known to be superluminous in this phase (Hayashi 1961; Paxton et al. 2011; Tognelli et al. 2011). As a star moves down the Hayashi track (Hayashi 1961) before arriving at the main sequence (MS), the stellar surface temperature is roughly constant but its radius decreases significantly. Since $L \sim R^2T^4$, the luminosity of PMS stars may be 1 or even 2 orders of magnitude higher than the ZAMS (Zero Age Main Sequence) value. Before reaching the ZAMS, solar-like stars reach a minimum bolometric luminosity, $L_{\rm bol} \approx 0.45L_{\odot}$ ($L_{\odot}$ = modern solar total luminosity) at an age of ≈ 13 Myr (Baraffe et al. 1998). The duration of the PMS phase scales inversely with the mass of the star; while a solar-type star arrives at the main sequence after ~ 50 Myr (Baraffe et al. 1998), M dwarfs can spend several hundreds of millions of years in the PMS phase (Reid and Hawley 2005).

The evolution trend of stellar luminosity during the PMS phase is just the opposite to that during the main sequence. After reaching the main sequence, a Sun-like star's bolometric luminosity steadily increases from $\sim 70\%$ $L_{\odot}$ to 100% $L_{\odot}$ at an age of 4.5 Gyr, mainly as a result of slightly increasing effective temperature as a consequence of the evolution of nuclear reactions in the stellar core (Sackmann and Boothroyd 2003). This luminosity increase will continue for the Sun until it reaches a luminosity of up to $\approx 3L_{\odot}$ at an age of 10 Gyrs. The significant change in $L_{\rm bol}$ implies that the classical *Habitable Zone* (HZ) around the star moves outward during this evolution. The Earth will in fact be inside the inner HZ boundary at an age of ≈ 7 Gyrs, i.e., only 2.4 Gyrs in the future (for the most pessimistic models; see Kasting et al. 1993) and it will begin to lose its water. This effect is less serious for M dwarfs because of their slower evolution.

In contrast, planets orbiting in what will be the HZ of main sequence M dwarfs are experiencing stellar radiation several to 10 times higher than that received by modern Earth during the PMS phase of their stars. If these planets are water-rich, they will be in a state known as a *runaway greenhouse* (Kasting et al. 1993), wherein liquid water is not stable on the surface, but instead forms a steam atmosphere that efficiently prevents the surface from cooling. In this situation, the *cold trap* (a local minimum in the temperature structure of the lower atmosphere that causes water to condense out of rising gas parcels, typically at 10–15 km altitude in modern Earth's atmosphere) does not operate efficiently, and water can reach the part of the atmosphere where UV photolysis can occur. Hydrodynamic escape of the photolytically-produced hydrogen can operate and drag photolytically-produced oxygen atoms, leading to a net loss of water from the planet.

This process was studied in details by 3 independent groups and a consensus of these works is that Earth-mass planets in the HZ of main sequence M dwarfs could have lost several Earth oceans (TO $\equiv 1.4 \times 10^{24}$ g) during the pre-main-sequence phase (Fig. 4; Ramirez and Kaltenegger 2014; Tian and Ida 2015; Luger and Barnes 2015).

Tian and Ida (2015) combined a planet population synthesis model and a water loss model to show that extreme water loss on rocky planets in the HZ of M dwarfs naturally leads to a bimodal distribution of the water contents. Initially planets in the HZs of M dwarfs should have a wide range of water contents (see, e.g., Mulders et al. 2015), owing to the stochastic assimilation of water-rich planetesimals migrating from beyond the ice line early on. Around solar-mass stars, this distribution is maintained once the star reaches the main sequence. However, the initial water mass distribution evolves during the pre-main sequence of M dwarfs such that a bi-modal distribution of water mass fraction is predicted (Tian and Ida 2015). Whereas the occurrence rate of Earth twins (Earth-sized planets in the HZ with water abundances similar to that of Earth) is on the order of 20% for solar-type stars, this number drops to 1% and 0.1% for $0.5M_{\odot}$ and $0.3M_{\odot}$ M dwarfs, respectively (Tian and Ida 2015).

One interesting result in Luger and Barnes (2015) is that super-Earths are likely to lose *more* water than Earth-sized planets. In the model of Luger and Barnes (2015), most of the energy in the escape flow goes toward the escape of *oxygen* atoms on Earth-sized planets, resulting in a significantly slower rate of net water loss. On super-Earths, while the

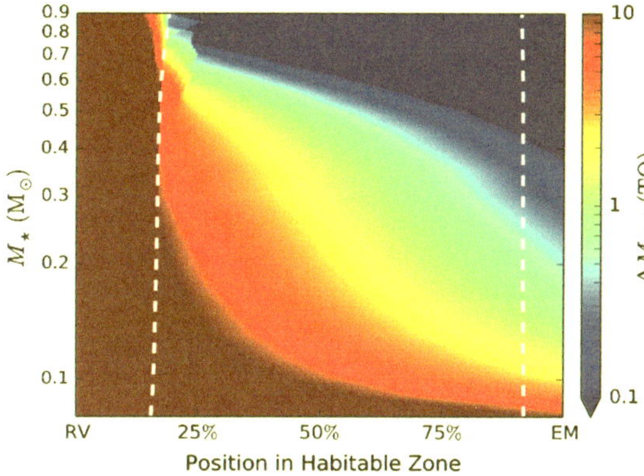

Fig. 4 Total amount of water lost (in TO) during the pre-main sequence runaway phase from an Earth-mass planet, plotted as a function of stellar mass (y-axis) and fractional position in the HZ (x-axis). RV and EM correspond to the Recent Venus and Early Mars "empirical" limits to the HZ, respectively; the dashed white lines indicate the more conservative theoretical inner and outer edges. Planets in the inner HZs of all M dwarfs lose at least 1 TO of water; around the lowest mass M dwarfs, these planets may lose in excess of 10 TO. Reproduced from Luger and Barnes (2015)

total amount of *mass* lost is lower, most of the escape is in the form of hydrogen atoms, which results in a faster water depletion. A modified consideration of fractionation during hydrodynamic escape (Tian 2015a, 2015b) results in a scenario different from that in Luger and Barnes (2015). One possible consequence of the extreme water loss process is the buildup of O_2 (up to several hundred to a few thousand bars in some cases) in these planets' atmospheres, which could be mistaken for a biosignature (Luger and Barnes 2015). Tian (2015a, 2015b) obtained similar results with more detailed calculations of the O_2 buildup rate, accounting for the changing hydrogen mixing ratio in the atmosphere as the escape proceeds, and suggested that the amount of O_2 retained in the atmospheres of rocky planets in the HZ of M dwarfs could follow a bimodal distribution. Both Luger and Barnes (2015) and Tian and Ida (2015) noted that the interactions between the atmospheres and surface/interior materials on planets experiencing rapid water loss could reduce atmospheric O_2.

Oxygen is typically unstable in planetary atmospheres in the absence of a continuous source—a planet's interior is overall reducing in comparison to its surface. Catling (2014) estimated that various surface sinks on Earth could remove oxygen at a rate of ~ 0.15 bars/Myr, although tectonic activities are required to provide a continuous source of reduced iron. Some exoplanets may either lack tectonics or have less reducing mantles, which would enable the long-term stability of oxygen in their atmospheres. This problem can become mediated during the magma ocean phase, which could last tens of Myrs for planets around pre-main-sequence M dwarfs. Schaefer et al. (2016) presented the first model considering the possible consumption of atmospheric O_2 by a magma ocean, by assuming that atmospheric O_2 can be readily embedded into the magma and eventually be used to oxidize FeO into Fe_2O_3, on early evolution history of GJ1132b and concluded that in most likely scenario the planet's atmospheric O_2 pressure should be less than a few bars of O_2.

In the interesting case of Proxima Centauri b, if the planet received approximately 200 times XUV radiation from its star during its first 100 Myrs evolution and the escape efficiency is 10%, the loss timescale of 1 Earth ocean would be ~ 25 Myrs without considering the energy consumed by the escape of O (pure hydrogen escape)—4 Earth's oceans could have been lost during the PMS phase of Proxima Centauri. The probability for Proxima Centauri b to maintain liquid water on its surface is closely linked to how much water it started with. If H and O were escaping at a 2:1 ratio, the loss timescale of 1 Earth ocean would be 150 Myrs. In this scenario the likelihood for Proxima Centauri b to maintain liquid water on its surface is much greater. However, the O_2 content in a planetary atmosphere required for H and O escape to be at a 2:1 ratio is rather close to unity for a planet similar to Proxima Centauri b (Fig. 2 in Tian 2015b). During the early PMS phase, the atmosphere should be dominated by water vapor and thus H loss would exceed that of O. Only in late PMS phase will it be possible for the O_2 content of the atmosphere to reach the H/O loss balance. And if O_2 does react with the magma ocean on Proxima Centauri b during its star's PMS phase, the amount of water allowed to survive the PMS phase would be even lower. Detailed studies are apparently needed to fully explore the possible parameter space.

Although recent studies reached a general consensus on the significance of water loss from planets in the HZ of main sequence M dwarfs during the PMS phase, the levels of XUV radiation used in these works are different. Specifically, Luger and Barnes (2015) applied a saturation time of 1 Gyrs and assumed an $L_{\rm xuv}/L_{\rm bol} = 10^{-3}$ in the saturation period for M dwarfs with mass between 0.1 and $0.3M_\odot$. Tian and Ida (2015) tested constant $L_{\rm xuv}/L_{\rm bol}$ between 10^{-5} and 10^{-3}, as well as a heating efficiency between 1% and 10%, in the PMS phase (100 Myrs) of M dwarfs with masses 0.3 and $0.5M_\odot$ respectively. Schaefer et al. (2016) applied 2 scenarios, one with an $L_{\rm xuv}/L_{\rm bol} = 10^{-3}$ and the other with a constant XUV level (10^{-3} of GJ1132's current luminosity) before 1 Gyrs of age. Apparently more detailed study based on updated understanding of the multiple evolutionary paths of $L_{\rm X}$ and $L_{\rm EUV}$ of M dwarfs (Sect. 2.2.2) will be useful future work.

4 Discussions and Summary

In this review we do not cover planet formation processes and water deliver processes which have important consequences on the final water contents on planets. These contents are covered in other chapters of this book. Recent observational discovery of the migration of the snowline around stars (Cieza et al. 2016) further complicates the picture.

One important recent progress regarding water loss from young planets is the new understanding on the complexity and diversity of the XUV evolutionary histories of Sun-like stars. Currently we still do not know whether our Sun was a rapid rotator or a slow rotator, which added a not-well-constrained parameter in our understanding of the early evolution of solar system terrestrial planets. New and interdisciplinary future works are urgently needed in order to reveal the nature of the early young Sun.

A potential issue to apply Eq. (4) and (5) in Sect. 2 to study the fate of atmospheric CO_2 is photochemistry. After H and O atoms escape efficiently from water dominated planetary atmospheres for certain time, CO_2 could become a major atmospheric component. However, under the high UV fluxes of the young Sun and stars it is likely that a significant fraction of CO_2 is also dissociated in the upper atmosphere (Tian 2009). This has two consequences: 1) IR cooling in planetary upper atmosphere will be less effective if CO_2 is dissociated efficiently because it is the most effective cooling agent for the upper atmosphere; 2) atomic C is much less massive and could participate more efficiently in hydrodynamic escape of hydrogen, similar to the escape of atomic O described earlier in this review. Finally the stability of a CO_2-dominant planetary atmosphere under strong XUV radiation is not guaranteed (Tian 2009; Tian et al. 2009). Kepler's second-light mission, $K2$, and the Transiting Exoplanet Survey Satellite, *TESS*, are likely to detect other nearby Earth-sized planets around M dwarfs, some of which could be in the HZ and more could be close-in exoplanets (Howell et al. 2014; Ricker et al. 2014). Spectroscopic follow-up of planets detected around

M dwarfs is likely to soon reveal planets that have undergone or are undergoing extreme water loss. Earth-size planets have been discovered around GJ1132 ($0.18M_\odot$, Berta-Thompson et al. 2015), TRAPPIST-1 ($0.08M_\odot$, Gillon et al. 2016, 2017), Proxima Centauri ($0.12M_\odot$, Anglada-Escude et al. 2016), and LHS 1140 ($0.15M_\odot$, Dittmann 2017).

Bolmont et al. (2016) suggests that TRAPPIST-1b and 1c could have lost 15 Earth oceans, while TRAPPIST-1d could have lost less than 1 Earth ocean. However Bourrier et al. (2017) suggests that planets within the orbit of TRAPPIST-1g might have lost more than 20 Earth oceans, while 3 Earth oceans might have been lost from TRAPPIST-1e to h. The case of TRAPPSIT1-d is especially interesting in light of the findings in recent climate models (Wolf 2017; Kopparapu et al. 2017; Bin et al. 2018) that TRAPPIST-1b, c, and d should still be in a runaway state today. Because of the large uncertainty on the initial planetary water inventories, a rapid water loss does not necessarily result in a dehydrated planet. Luckily these planets are all nearby and observations can reveal their nature. The lack of observational features in the combined transmission spectrum of TRAPPIST-1b and c has been interpreted as the lack of extended hydrogen envelopes around these planets (de Wit et al. 2016). Because a steam atmosphere would likely lead to a hydrogen envelope, the observation appears to favor the scenario of dehydrated planets and could be used to place upper limits on the initial water inventories or the efficiency of water delivery. These planets and other close-in low mass exoplanets to be discovered in the near future will be prime candidates for future follow-up observations by both space-based UV telescopes and JWST. The former observations could reveal ongoing hydrodynamic escape from rocky exoplanets, which will be crucial to observationally check our theoretical understanding of water loss and atmospheric evolution. The later observations might directly detect absorption features from major atmospheric components, including dense O_2 atmospheres as a result of extreme water loss (Luger and Barnes 2015; Tian and Ida 2015; Schwieterman et al. 2016).

In summary, good progress has been accomplished in the past a few years to better understand water loss from young planets. Opportunities on both observations and theoretical works in the near future are even better.

Acknowledgements FT is supported by Natural and Science Foundation of China (11661161014) and the Tsinghua University Initiative Science Research Program (523001028). HL and PO acknowledge the Austrian Science Fund (FWF): P27256-N27. CPJ, MG, and HL acknowledge the support of the FWF NFN project S116, FWF NFN S11604-N16, and S11607-N16. This review is a result of the ISSI workshop "Delivery of Water to Protoplanets".

References

Y. Abe, T. Matsui, The formation of an impact-generated H_2O atmosphere and its implications for the thermal history of the Earth. J. Geophys. Res. **90**, 545–559 (1985)

Y. Abe, T. Matsui, Early evolution of the Earth: accretion, atmosphere formation, and thermal history. J. Geophys. Res. **91**, E291–E302 (1986)

F. Albarède, J. Blichert-Toft, The split fate of the early Earth, Mars, Venus, and Moon. C. R. Géosci. **339**, 917–927 (2007)

Y. Alibert, C. Broeg, W. Benz, G. Tinetti, G.J. White, Origin and formation of planetary systems. Astrobiology **10**, 19–32 (2010)

G. Anglada-Escude et al., A terrestrial planet candidate in a temperate orbit around Proxima Centauri. Nature **536**, 437 (2016). https://doi.org/10.1038/nature19106

T.R. Ayres, Evolution of the solar ionizing flux. J. Geophys. Res. **102**, 1641 (1997)

I. Baraffe, G. Chabrier, F. Allard, P.H. Hauschildt, Evolutionary models for solar metallicity low-mass stars: mass-magnitude relationships and color-magnitude diagrams. Astron. Astrophys. **337**, 403 (1998)

Z.K. Berta-Thompson et al., A rocky planet transiting a nearby low-mass star. Nature **527**, 204 (2015). https://doi.org/10.1038/nature15762
J. Bin et al., New inner boundaries of the habitable zones around M dwarfs. Earth Planet. Sci. Lett. (2018 in review)
E. Bolmont et al., Water loss from terrestrial planets orbiting ultra cool dwarfs: implications for the planets of TRAPPIST-1. Mon. Not. R. Astron. Soc. (2016)
V. Bourrier et al., Temporal evolution of the high-energy irradiation and water content of TRAPPIST-1 exoplanets. Astrophys. J. (2017)
V. Bourrier, A. Lecavelier des Etangs, H. Dupuy, D. Ehrenreich, A. Vidal-Madjar, G. Hébrard, G.E. Balleste, J.-M. Désert, R. Ferlet, D.K. Sing, P.J. Wheatley, Atmospheric escape from HD189733b observed in HI Lyman-α: detailed analysis of HST/STIS September 2011 observations. Astron. Astrophys. **551**, A63 (2013)
D. Catling, The great oxidation event transition, in *Treatise on Geochemistry*, ed. by H.D. Holland, K.K. Turekian 2nd edn. (Elsevier, Oxford, 2014), pp. 177–195
E. Chassefière, Hydrodynamic escape of hydrogen from a hot water-rich atmosphere: the case of Venus. J. Geophys. Res. **101**(E11), 26039–26056 (1996)
L.A. Cieza et al., Imaging the water snow-line during a protostellar outburst. Nature **535**, 258–261 (2016)
M.W. Claire et al., The evolution of solar flux from 0.1 nm to 160 μm: quantitative estimates for planetary studies. Astrophys. J. **757**, 95–107 (2012)
J. de Wit et al., A combined transmission spectrum of the Earth-sized exoplanets TRAPPIST-1 b and c. Nature **537**, 69–72 (2016)
J. Dittmann, A temperate rocky super-Earth transiting a nearby cool star. Nature **544**, 333 (2017)
C. Dong, The influence of the extreme ultraviolet spectral energy distribution on the structure and composition of the upper atmosphere of exoplanets. Proc. Natl. Acad. Sci. USA **115**, 260–265 (2018)
L.T. Elkins-Tanton, Linked magma ocean solidification and atmospheric growth for Earth and Mars. Earth Planet. Sci. Lett. **271**, 181–191 (2008)
L.T. Elkins-Tanton, Formation of water ocean on rocky planets. Astrophys. Space Sci. **332**, 359–364 (2011)
L.T. Elkins-Tanton, Magma oceans in the inner solar system. Annu. Rev. Earth Planet. Sci. **40**, 113–139 (2012)
N.V. Erkaev, Yu.N. Kulikov, H. Lammer, F. Selsis, D. Langmayr, G.F. Jaritz, H.K. Biernat, Roche lobe effects on the atmospheric loss from "Hot Jupiters". Astron. Astrophys. **472**, 329–334 (2007)
N.V. Erkaev, H. Lammer, L.T. Elkins-Tanton, A. Stökl, P. Odert, E. Marcq, E.A. Dorfi, K.G. Kislyakova, Yu.N. Kulikov, M. Leitzinger, M. Güdel, Planet. Space Sci. **98**, 106–119 (2014)
J.M. Fontenla et al., Solar spectral irradiance, solar activity, and the near-ultra-violet. Astrophys. J. **809**, 157 (2015)
K. France, J.L. Linsky, F. Tian, C.S. Froning, A. Roberge, Astrophys. J. Lett. **750**, L32 (2012)
K. France, C.S. Froning, J.L. Linsky et al., Astrophys. J. **763**, 149 (2013)
K. France et al., The MUSCLES Treasury Survey. I. Motivation and overview. Astrophys. J. **820**, 89–113 (2016)
F. Gallet, J. Bouvier, Astron. Astrophys. **556**, A36 (2013)
A. Garcia Muñoz, Physical and chemical aeronomy of HD209458b. Planet. Space Sci. **55**, 1426–1455 (2007)
C. Gillmann, E. Chassefière, P. Lognonné, A consistent picture of early hydrodynamic escape of Venus atmosphere explaining present Ne and Ar isotopic ratios and low oxygen atmospheric content. Earth Planet. Sci. Lett. **286**, 503–513 (2009)
M. Gillon et al., Temperate Earth-sized planets transiting a nearby ultracool dwarf star. Nature **533**, 221–224 (2016). https://doi.org/10.1038/nature17448
M. Gillon, A.H.M.J. Triaud, B.O. Demory et al., Nature **542**, 456 (2017)
M. Güdel, E.F. Guinan, S.L. Skinner, Astrophys. J. **483**, 947 (1997)
J.H. Guo, L. Ben-Jaffel, The influence of the extreme ultraviolet spectral energy distribution on the structure and composition of the upper atmosphere of exoplanets. Astrophys. J. **818**, 107 (2016)
A.N. Halliday, H. Wänke, J.L. Birck, R.N. Clayton, The accretion, composition and early differentiation of Mars. Space Sci. Rev. **96**, 197–230 (2001)
K. Hamano, Y. Abe, H. Genda, Emergence of two types of terrestrial planet on solidification of magma ocean. Nature **497**, 607–610 (2013)
C. Hayashi, Stellar evolution in early phases of gravitational contraction. Publ. Astron. Soc. Jpn. **13**, 450–452 (1961)
H.E. Hinteregger, Adv. Space Res. **1**, 39 (1981)
S.B. Howell, C. Sobeck, M. Haas, M. Still, T. Barclay, F. Mullally, J. Troeltzsch, S. Aigrain, S.T. Bryson, D. Caldwell, W.J. Chaplin, W.D. Cochran, D. Huber, G.W. Marcy, A. Miglio, J.R. Najita, M. Smith, J.D. Twicken, J.J. Fortney, Publ. Astron. Soc. Pac. **126**, 398 (2014)
D.M. Hunten, The escape of light gases from planetary atmospheres. J. Atmos. Sci. **30**, 1481–1494 (1973)

D.M. Hunten, R.O. Pepin, J.C.G. Walker, Mass fractionation in hydrodynamic escape. Icarus **69**, 532–549 (1987)

C.P. Johnstone, M. Güdel, Astron. Astrophys. **578**, A129 (2015)

C.P. Johnstone, M. Güdel, I. Brott, T. Lüftinger, Stellar winds on the main-sequence II. The evolution of rotation and winds. Astron. Astrophys. **577**, A28 (2015)

P. Judge, S.C. Solomon, T.R. Ayres, Astrophys. J. **593**, 534 (2003)

J.F. Kasting, J.B. Pollack, Loss of water from Venus. I. Hydrodynamic escape of hydrogen. Icarus **53**, 479–508 (1983)

J.F. Kasting, D.P. Whitmire, R.T. Reynolds, Icarus **101**, 108 (1993)

T.V. Kazachevskaya et al., Stud. Geophys. Geod. **42**, 92 (1998)

M.L. Khodachenko et al., Lyα absorption at transits of HD 209458b: a comparative study of various mechanisms under different conditions. Astrophys. J. **847**, 126 (2017)

R.V. Kopparapu et al., Habitable moist atmospheres on terrestrial planets near the inner edge of the habitable zone around M dwarfs. Astrophys. J. **845**, 5 (2017)

T.T. Koskinen, M.J. Harris, R.V. Yelle, P. Lavvas, The escape of heavy atoms from the ionosphere of HD209458b. I. A photochemical dynamical model of the thermosphere. Icarus **226**, 1678–1694 (2013)

T.T. Koskinen, P. Lavvas, M.J. Harris, R.V. Yelle, Thermal escape from extrasolar giant planets. Philos. Trans. R. Soc. Lond. Ser. A, Math. Phys. Sci. **372**, 20130089 (2014)

R.P. Kraft, Astrophys. J. **150**, 551 (1967)

H. Lammer, F. Selsis, I. Ribas, E.F. Guinan, S.J. Bauer, W.W. Weiss, Atmospheric loss of exoplanets resulting from stellar X-ray and extreme-ultraviolet heating. Astrophys. J. **598**, L121–L124 (2003)

H. Lammer et al., Atmospheric escape and evolution of terrestrial planets and satellites. Space Sci. Rev. **139**, 399–436 (2008). https://doi.org/10.1007/s11214-008-9413-5

H. Lammer, K.G. Kislyakova, P. Odert, M. Leitzinger, R. Schwarz, E. Pilat-Lohinger, Yu.N. Kulikov, M.L. Khodachenko, M. Güdel, A. Hanslmeier, Pathways to Earth-like atmospheres: Extreme ultraviolet (EUV)-powered escape of hydrogen-rich protoatmospheres. Orig. Life Evol. Biosph. **41**, 503–522 (2011)

H. Lammer, N.V. Erkaev, P. Odert, K.G. Kislyakova, M. Leitzinger, M.L. Khodachenko, Probing the blow-off criteria of hydrogen-rich 'super-Earths'. Mon. Not. R. Astron. Soc. **430**, 1247–1256 (2013)

H. Lammer, A. Stokl, N.V. Erkaev, E.A. Dorfi, P. Odert, M. Gudel, Yu.N. Kulikov, K.G. Kislyakova, M. Leitzinger, Origin and loss of nebula-captured hydrogen envelopes from 'sub'- to 'super-Earths' in the habitable zone of Sun-like stars. Mon. Not. R. Astron. Soc. **439**, 3225–3238 (2014)

T. LaTourrette, G.J. Wasserburg, Mg diffusion in anorthite: implications for the formation of early solar system planetesimals. Earth Planet. Sci. Lett. **158**, 91–108 (1998)

T. Lebrun, H. Massol, E. Chassefiere et al., Thermal evolution of an early magma ocean in interaction with the atmosphere. J. Geophys. Res., Planets **118**, 1155–1176 (2013)

A. Leger et al., Impact of $\eta_{\rm earth}$ on the capabilities of affordable space missions to detect biosignatures on extrasolar planets. Astrophys. J. (2015)

H.I.M. Lichtenegger, K.G. Kisylakova, P. Odert, N.V. Erkaev, H. Lammer, H. Gröller, C.P. Johnstone, L. Elkins-Tanton, L. Tu, M. Güdel, M. Holmström, Solar XUV and ENA-driven water loss from early Venus' steam atmosphere. J. Geophys. Res. **121**, 4718–4732 (2016)

J.L. Linsky, H. Yang, K. France, C.S. Froning, J.C. Green et al., Observations of mass loss from the transiting exoplanet HD 209458b. Astrophys. J. **717**, 1291–1299 (2010)

J.L. Linsky, K. France, T.R. Ayres, Astrophys. J. **766**, 69 (2013)

J.L. Linsky, J. Fontenla, K. France, Astrophys. J. **780**, 61 (2014)

J.J. Lissauer et al., A closely packed system of low-mass, low-density planets transiting Kepler-11. Nature **470**, 53–59 (2011)

E.D. Lopez, J.J. Fortney, N. Miller, How thermal evolution and mass-loss sculpt populations of super-Earths and sub-Neptunes: application to the Kepler-11 system and beyond. Astrophys. J. **761**, 59 (2012)

R.O.P. Loyd, K. France, A. Youngblood et al., Astrophys. J. **824**, 10 (2016)

R.O. Loyd et al., Ultraviolet C II and Si III transit spectroscopy and modeling of the evaporating atmosphere of GJ436b. Astrophys. J. Lett. **834**, L17 (2017)

R. Luger, R. Barnes, Extreme water loss and abiotic O_2 buildup on planets throughout the habitable zones of M dwarfs. Astrobiology **15**, 119–143 (2015)

A. Maggio, S. Sciortino, G.S. Vaiana, P. Majer, J. Bookbinder et al., Astrophys. J. **315**, 687 (1987)

T.I. Maindl, R. Dvorak, H. Lammer, M. Güdel, C. Schäfer, R. Speith, P. Odert, N.V. Erkaev, K.G. Kislyakova, E. Pilat-Lohinger, Impact induced surface heating by planetesimals on early Mars. Astron. Astrophys. **574**, A22 (2015)

G.W. Marcy et al., Masses, radii, and orbits of small Kepler planets: the transition from gaseous to rocky planets. Astrophys. J. Suppl. Ser. **210**, 20–90 (2014)

H. Massol et al., Formation and evolution of protoatmospheres. Space Sci. Rev. (2016). https://doi.org/10.1007/s11214-016-0280-1
J.D. Mulders et al., The snow line in viscous disks around low-mass stars: implications for water delivery to terrestrial planets in the habitable zone. Astrophys. J. **807**, 9 (2015). https://doi.org/10.1088/0004-637X/807/1/9
R.W. Noyes, L.W. Hartmann, S.L. Baliunas, D.K. Duncan, A.N. Vaughan, Astrophys. J. **279**, 763 (1984)
E.J. Öpik, S.F. Singer, Distribution of density in a planetary exosphere. II. Phys. Fluids **4**, 221–233 (1961)
J.E. Owen, S. Mohanty, Mon. Not. R. Astron. Soc. **459**, 4088 (2016)
R. Pallavicini, L. Golub, R. Rosner, G.S. Vaiana, T. Ayres, J.L. Linsky, Astrophys. J. **248**, 279 (1981)
B. Paxton, L. Bildsten, A. Dotter, F. Herwig, P. Lesaffre, F. Timmes, Modules for experiments in stellar astrophysics (MESA). Astrophys. J. **192**, 3–37 (2011)
R.O. Pepin, On the origin and early evolution of terrestrial planet atmospheres and meteoritic volatiles. Icarus **92**, 2–79 (1991)
N. Pizzolato, A. Maggio, G. Micela, S. Sciortino, P. Ventura, The stellar activity-rotation relationship revisited: dependence of saturated and non-saturated X-ray emission regimes on stellar mass for late-type dwarfs? Astron. Astrophys. **397**, 147–157 (2003)
R.M. Ramirez, L. Kaltenegger, The habitable zones of pre-main-sequence stars. Astrophys. J. **797**, L25–L33 (2014)
I.N. Reid, S.L. Hawley, *New Light on Dark Stars* (Springer, Heidelberg, 2005)
I. Ribas, E.F. Guinan, M. Güdel, M. Audard, Evolution of the solar activity over time and effects on planetary atmospheres, I. High energy irradiances (1–1700 Å). Astrophys. J. **622**, 680–694 (2005)
G.R. Ricker, R.K. Vanderspek, D.W. Latham et al., The Transiting Exoplanet Survey satellite mission, in *American Astronomical Society Meeting Abstracts* (2014), p. 224
I.-J. Sackmann, A.I. Boothroyd, Astrophys. J. **583**, 1024 (2003)
V.S. Safronov, *Evolution of the Protoplanetary Cloud and Formation of the Earth and the Planets* (Akad. Nauk. SSSR, Moscow, 1969)
J. Sanz-Forcada, G. Micela, I. Ribas, A.M.T. Pollock, C. Eiroa, A. Velasco, E. Solano, D. Garcia-Alvarez, Astron. Astrophys. **532**, A6 (2011)
J. Scalo, L. Kaltenegger, A. Segura, M. Fridlund et al., M stars as targets for terrestrial exoplanet searches and biosignature detection. Astrobiology **7**, 85–171 (2007)
L. Schaefer et al., Predictions of the atmospheric composition of GJ 1132B. arXiv:1607.03906v1 (2016)
G. Schubert, J.D. Anderson, T. Spohn, W.B. McKinnon, Interior composition, structure and dynamics of the Galilean satellites, in *Jupiter—The Planet, Satellites and Magnetosphere*, ed. by F. Bagenal, T.E. Dowling, W.B. McKinnon (Cambridge University Press, Cambridge, 2004), pp. 281–306
E.W. Schwieterman et al., Identifying planetary biosignature impostors: spectral features of CO and O_4 resulting from abiotic O_2/O_3 production. Astrophys. J. **819**, L13–L19 (2016)
I.F. Shaikhislamov, M.L. Khodachenko, Yu.L. Sasunov, H. Lammer, K.G. Kislyakova, N.V. Erkaev, Atmospheric expansion and mass loss of close-orbit giant exoplanets heated by stellar XUV. I. Modeling of hydrodynamic escape of upper atmospheric material. Astrophys. J. **795**, 132 (2014)
A. Skumanich, Astrophys. J. **171**, 565 (1972)
D.R. Soderblom, J.R. Stauffer, K.B. MacGregor, B.F. Jones, Astrophys. J. **409**, 624 (1993)
B. Stelzer et al., The UV and X-ray activity of the M dwarfs within 10 pc of the Sun. Mon. Not. R. Astron. Soc. **431**, 2063–2079 (2013)
A. Stökl, E. Dorfi, H. Lammer, Astron. Astrophys. **576**, A87 (2015)
F. Tian, Thermal escape from super earth atmospheres in the habitable zones of M stars. Astrophys. J. **703**, 905–909 (2009)
F. Tian, Conservation of total escape from hydrodynamic planetary atmospheres. Earth Planet. Sci. Lett. **379**, 104–107 (2013)
F. Tian, Atmospheric escape from solar system terrestrial planets and exoplanets. Annu. Rev. Earth Planet. Sci. **43**, 459–476 (2015a). https://doi.org/10.1146/annurev-earth-060313-054834
F. Tian, History of water loss and atmospheric O_2 buildup on rocky exoplanets near M dwarfs. Earth Planet. Sci. Lett. **432**, 126–132 (2015b)
F. Tian, S. Ida, Water content of earth-mass planets around M dwarfs. Nat. Geosci. **8**, 177–180 (2015). https://doi.org/10.1038/NGEO2372
F. Tian, O.B. Toon, A.A. Pavlov, H. DeSterck, Tran-sonic hydrodynamic escape of hydrogen from extrasolar planetary atmospheres. Astrophys. J. **621**, 1049–1060 (2005)
F. Tian, J.F. Kasting, S.C. Solomon, Thermal escape of carbon from the early Martian atmosphere. Geophys. Res. Lett. **36**, L02205 (2009). https://doi.org/10.1029/2008GL036513
F. Tian, E. Chassefière, F. Leblanc, D.A. Brain, Atmosphere escape and climate evolution of terrestrial planets, in *Comparative Climatology of Terrestrial Planets*, ed. by S.J. Mackwell, A.A. Simon-Miller, J.W. Harder, M.A. Bullock (University of Arizona Press, Tucson, 2013), pp. 567–581

F. Tian et al., High stellar FUV/NUV ratio and oxygen contents in the atmospheres of potentially habitable planets. Earth Planet. Sci. Lett. **22**, 27 (2014)

E. Tognelli, P.G. PradaMoroni, S. Degl'Innocenti, The Pisa pre-main sequence tracks and isochrones. A database covering a wide range of *Z*, *Y*, mass, and age values. Astron. Astrophys. **533**, A109 (2011). https://doi.org/10.1051/0004-6361/200913913

L. Tu, C.P. Johnstone, M. Güdel, H. Lammer, Astron. Astrophys. **577**, L3 (2015)

H.C. Urey, The cosmic abundances of potassium, uranium, and thorium and the heat balance of the Earth, the Moon, and Mars. Proc. Natl. Acad. Sci. **41**, 127–144 (1955)

A. Vidal-Madjar, A. Lecavelier des Etangs, J.M. Désert, G.E. Ballester, R. Ferlet, G. Hébrard, M. Mayor, An extended upper atmosphere around the extrasolar giant planet HD209458b. Nature **422**, 143–146 (2003)

A. Vidal-Madjar, J.-M. Désert, A. Lecavelier des Etangs, G. Hébrard, G.E. Ballester et al., Detection of oxygen and carbon in the hydrodynamically escaping atmosphere of the extrasolar planet HD209458b. Astrophys. J. **604**, L69–L72 (2004)

A. Vidal-Madjar, C.M. Huitson, V. Bourrier, J.M. Désert, G. Ballester et al., Magnesium in the atmosphere of the planet HD 209458 b: observations of the thermosphere-exosphere transition region. Astron. Astrophys. **560**, A54 (2013)

F.M. Walter, Astrophys. J. **245**, 677 (1981)

A.J. Watson, T.M. Donahue, J.C.G. Walker, The dynamics of a rapidly escaping atmosphere—the evolution of Earth and Venus. Icarus **48**, 150–166 (1981)

E.J. Weber, L. Davis Jr., Astrophys. J. **148**, 217 (1967)

A. West et al., Astron. J. **135**, 785 (2008)

G.W. Wetherill, G.R. Stewart, Accumulation of a swarm of small planetesimals. Icarus **77**, 330–357 (1989)

E.T. Wolf, Assessing the habitability of the TRAPPIST-1 system using a 3D climate model. Astrophys. J. Lett. **839**, L1 (2017)

N.J. Wright, J.J. Drake, Solar-type dynamo behaviour in fully convective stars without a tachocline. Nature **535**, 526–530 (2016)

N.J. Wright, J.J. Drake, E.E. Mamajek, G.W. Henry, Astrophys. J. **743**, 48 (2011)

R.V. Yelle, Aeronomy of extra-solar giant planets at small orbital distances. Icarus **170**, 167–179 (2004). Corrigendum: Icarus **183**, 508 (2006)

A. Youngblood, K. France, R.O.P. Loyd et al., Astrophys. J. **824**, 101 (2016)

K.J. Zahnle, J.F. Kasting, Mass fractionation during transonic escape and implications for loss of water from Mars and Venus. Icarus **68**, 462–480 (1986)

K. Zahnle, J.F. Kasting, J.B. Pollack, Mass fractionation of noble gases in diffusion-limited hydrodynamic hydrogen escape. Icarus **84**, 502–527 (1990)

Space Sci Rev (2018) 214:34
https://doi.org/10.1007/s11214-018-0471-z

Atmosphere Impact Losses

Hilke E. Schlichting[1,2] · Sujoy Mukhopadhyay[3]

Received: 15 November 2016 / Accepted: 11 January 2018 / Published online: 23 January 2018

Abstract Determining the origin of volatiles on terrestrial planets and quantifying atmospheric loss during planet formation is crucial for understanding the history and evolution of planetary atmospheres. Using geochemical observations of noble gases and major volatiles we determine what the present day inventory of volatiles tells us about the sources, the accretion process and the early differentiation of the Earth. We further quantify the key volatile loss mechanisms and the atmospheric loss history during Earth's formation. Volatiles were accreted throughout the Earth's formation, but Earth's early accretion history was volatile poor. Although nebular Ne and possible H in the deep mantle might be a fingerprint of this early accretion, most of the mantle does not remember this signature implying that volatile loss occurred during accretion. Present day geochemistry of volatiles shows no evidence of hydrodynamic escape as the isotopic compositions of most volatiles are chondritic. This suggests that atmospheric loss generated by impacts played a major role during Earth's formation. While many of the volatiles have chondritic isotopic ratios, their relative abundances are certainly not chondritic again suggesting volatile loss tied to impacts. Geochemical evidence of atmospheric loss comes from the $^{3}He/^{22}Ne$, halogen ratios (e.g., F/Cl) and low H/N ratios. In addition, the geochemical ratios indicate that most of the water could have been delivered prior to the Moon forming impact and that the Moon forming impact did not drive off the ocean. Given the importance of impacts in determining the volatile budget of the Earth we examine the contributions to atmospheric loss from both small and large impacts.

The Delivery of Water to Protoplanets, Planets and Satellites
Edited by Michel Blanc, Alessandro Morbidelli, Yann Alibert, Lindy Elkins-Tanton, Paul Estrada, Keiko Hamano, Helmut Lammer, Sean Raymond and Maria Schönbächler

✉ H.E. Schlichting
hilke@ucla.edu

S. Mukhopadhyay
sujoy@ucdavis.edu

1 University of California, Los Angeles, 595 Charles E. Young Drive East, Los Angeles, CA 90095-1567, USA

2 Massachusetts Institute of Technology, 77 Massachusetts Avenue, Cambridge, MA 02139-4307, USA

3 University of California, Davis, One Shields Avenue, Davis, CA 95616, USA

We find that atmospheric mass loss due to impacts can be characterized into three different regimes: 1) Giant Impacts, that create a strong shock transversing the whole planet and that can lead to atmospheric loss globally. 2) Large enough impactors ($m_{cap} \gtrsim \sqrt{2}\rho_0(\pi h R)^{3/2}$, $r_{cap} \sim 25$ km for the current Earth), that are able to eject all the atmosphere above the tangent plane of the impact site, where h, R and ρ_0 are the atmospheric scale height, radius of the target, and its atmospheric density at the ground. 3) Small impactors ($m_{min} > 4\pi\rho_0 h^3$, $r_{min} \sim 1$ km for the current Earth), that are only able to eject a fraction of the atmospheric mass above the tangent plane. We demonstrate that per unit impactor mass, small impactors with $r_{min} < r < r_{cap}$ are the most efficient impactors in eroding the atmosphere. In fact for the current atmospheric mass of the Earth, they are more than five orders of magnitude more efficient (per unit impactor mass) than giant impacts, implying that atmospheric mass loss must have been common. The enormous atmospheric mass loss efficiency of small impactors is due to the fact that most of their impact energy and momentum is directly available for local mass loss, where as in the giant impact regime a lot of energy and momentum is 'wasted' by having to create a strong shock that can transverse the entirety of the planet such that global atmospheric loss can be achieved. In the absence of any volatile delivery and outgassing, we show that the population of late impactors inferred from the lunar cratering record containing 0.1% $M_\oplus$ is able to erode the entire current Earth's atmosphere implying that an interplay of erosion, outgassing and volatile delivery is likely responsible for determining the atmospheric mass and composition of the early Earth. Combining geochemical observations with impact models suggest an interesting synergy between small and big impacts, where giant impacts create large magma oceans and small and larger impacts drive the atmospheric loss.

Keywords Volatiles · Accretion · Planet formation · Water · Impacts

1 Introduction

Terrestrial planet formation can generally be characterized by three distinct stages. The first is dominated by the accretion of planetesimals, which results in the formation of planetary embryos comparable to their isolation masses. The isolation mass is simply the mass that a planetary embryo can grow to by locally accreting all the material available in its feeding zone and it is given by

$$M_{iso} = \frac{(10\pi \Sigma a^2)^{3/2}}{(3M_\odot)^{1/2}}, \tag{1}$$

where Σ is the disk surface density in solids, a the semi-major axis and $M_\odot$ the mass of the sun (e.g. Ida and Makino 1993; Weidenschilling et al. 1997; Schlichting 2014). Evaluating the isolation mass, assuming that Σ is given by the minimum-mass solar nebular (MMSN), $\Sigma_{MMSN} \simeq 10 \times (a/1\ \mathrm{AU})^{-3/2}\ \mathrm{g\,cm^{-2}}$ (Hayashi 1981), yields at 1 AU $M_{iso} \sim M_{Mars}$. This implies that Mars may indeed have formed and survived as an isolation mass, but that Earth and Venus must have undergone an additional stage of accretion. The second stage consists of a series of giant impacts of a few dozen embryos that merge to form the Earth and Venus (e.g. Agnor et al. 1999; Chambers 2001), with the Moon forming impact being the last giant impact of the Earth (e.g. Cameron and Benz 1991; Canup and Asphaug 2001; Ćuk and Stewart 2012). Giant impacts begin when the planetesimals are no longer able to efficiently damp the eccentricities that are mutually excited by the growing planetary embryos. Order of magnitude estimates that balance the stirring rates of the protoplanets with the damping rates due to dynamical friction by the planetesimal population and numerical simulations find that giant impacts set in when the total mass in protoplanets is comparable to the mass in planetesimals (Goldreich et al. 2004; Kenyon and Bromley 2006; Schlichting et al. 2012). This implies that about 50% of the total mass still resides in planetesimals when giant impacts begin and planetesimal accretion continues throughout the giant impact phase and afterwards. Furthermore, there exists geochemical evidence that a total of about 0.01 $M_{\oplus}$ of chondritic material was delivered as 'late veneer' by planetesimals to the terrestrial planets after the end of giant impacts (Warren et al. 1999; Walker et al. 2004; Walker 2009), which is the third stage of accretion. Such late veneer, if accreted in small planetesimals, has the additional advantage that it can damp, due to the absence of strong mutual excitation among the terrestrial planets by the end of the impact stage, the post giant impact eccentricities to the almost circular values that we observe today (Schlichting et al. 2012). The terrestrial planets therefore had a rich and diverse formation history with volatiles being delivered and eroded during the planets' accretion of small planetesimals and mergers between large planetary embryos. As a result, in order to understand the origin and evolution of the terrestrial planets' atmospheres one needs to examine the contribution to volatile delivery and loss by both the giant impacts and planetesimal accretion. Geochemical observations of noble gases and major volatiles determine what the present day inventory of volatiles tells us about their sources and their accretion process as well as the key volatile loss mechanisms during Earth's formation. In this chapter we combine geochemical observations and models of small planetesimal impacts and giant collisions to shed light on the volatile delivery and atmospheric loss history of the Earth. We show that the data and models point to a rich interplay between volatile accretion and atmospheric mass loss facilitated by small and large impacts. This chapter is structured as follows: Earth's geochemical constraints on the delivery of its volatiles and losses are discussed in Sect. 2. Analytical impact models ranging from small planetesimal impacts to giant impacts are presented in Sect. 3 and used to calculate the atmospheric mass loss during accretion and the delivery of the late veneer highlighting the importance and efficiency of various different impact regimes. Conclusions follow in Sect. 4.

2 Volatile Delivery to and Losses from the Earth: Constraints from Observations

The relative abundances of a suite of elements (H, C, N, halogens, noble gases) along with their isotopic compositions allow us to decipher where volatiles were derived from during Earth's accretion. Measurements in mantle-derived rock, combined with volatile abundances in the Earth's exosphere, are now providing us with insights into both the total abundance of volatiles and their initial distribution on the Earth. Recent measurements are also providing key new information on the importance of processes like hydrodynamic escape and impacts in shaping the final abundance of the volatile budget for Earth (Marty 2012; Halliday 2013; Tucker and Mukhopadhyay 2014; Sharp and Draper 2013; Hirschmann 2016).

2.1 Volatile Sources and the History of Volatile Delivery

Over the past fifteen years, the non-radiogenic neon (Ne) isotopes ($^{20}Ne/^{22}Ne$) have played an important role in providing constraints on the acquisition of volatiles. Earth could have acquired Ne from potentially three sources: i) nebular Ne with $^{20}Ne/^{22}Ne$ ratio of 13.36 ± 0.09 (Heber et al. 2012); ii) micron-sized dust grains irradiated by solar wind in the

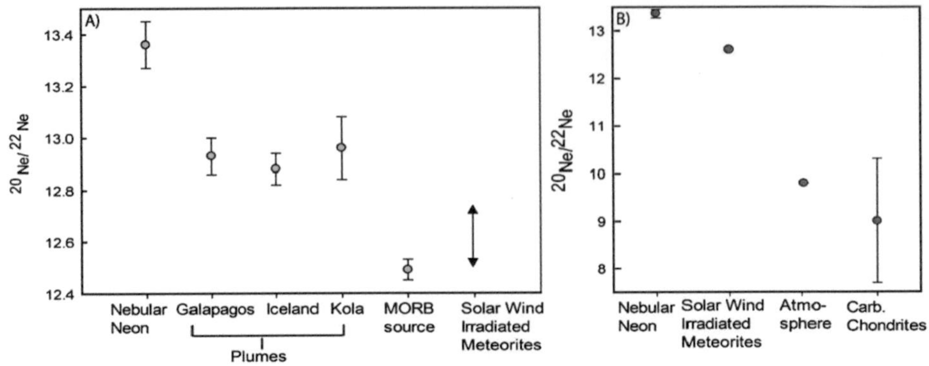

Fig. 1 The Neon isotopic composition of the mantle and atmosphere compared with that of the solar nebula, solar wind irradiated material and CI chondrites. The nebular value is from Heber et al. (2012), the range of solar wind irradiated material is from Moreira and Charnoz (2016) and the CI chondrite data is from Mazor et al. (1970). The Galapagos, Iceland and Kola plume points represent the highest measured values from these plumes and are from Péron et al. (2016), Mukhopadhyay (2012) and Yokochi and Marty (2004), respectively. The MORB source is from Holland and Ballentine (2006). So far, MORBs removed from the influence of plumes have not yielded $^{20}Ne/^{22}Ne$ ratios that are statistically different from 12.5. The large error bars for the CI chondrites reflect the inherent heterogeneity in Ne–Ar composition within this meteorite class. Error bars in both *panels* are 1 sigma and the large error bars for the CI chondrities reflect the inherent heterogeneity in Ne composition within this meteorite class

early Solar System and then incorporated into planetesimals ($^{20}Ne/^{22}Ne$ of ~12.5 to 12.7, although higher values might be possible if irradiation is not in steady state with sputtering; Moreira and Charnoz (2016)) and iii) Ne from carbonaceous chondrites ($^{20}Ne/^{22}Ne$ of ~8). The shallow mantle that is most frequently tapped during the mid ocean ridge volcanism (MORB source) appears to have a $^{20}Ne/^{22}Ne$ of 12.49 ± 0.04, similar to solar wind irradiated material that is found in gas-rich meteorites (Fig. 1; Trieloff et al. (2000), Ballentine et al. (2005), Holland and Ballentine (2006)). However, the deep mantle, sampled by plumes, appear to reach $^{20}Ne/^{22}Ne$ values close to 13.0 (Sarda et al. 2000; Yokochi and Marty 2004; Mukhopadhyay 2012; Péron et al. 2016). This value is higher than expected from solar wind irradiation of grains and significantly higher than values in carbonaceous chondrites raising the intriguing possibility that within the deepest mantle, there is a memory of Earth's acquisition of nebular volatiles.

The presence of nebular volatiles in the deep mantle plume source suggests that volatile accretion started in the presence of nebular gases while the Earth was an embryo. While nebular gases can persists for ~10 Myrs, the median lifetime of a nebula, i.e. the time at which half of all systems lost their gas disks, is ~2.5 Myrs (Hillenbrand 2008). The mean time of accretion for Earth (i.e. the time to have grown to 50% of the final mass) is ~11 Myrs (Yin et al. 2002) while the mean accretion timescale of Mars is $1.9^{+1.7}_{0.8}$ Myrs (Dauphas and Pourmand 2011; Tang and Dauphas 2014). Clearly, the entire planet could not have equilibrated with the solar nebula. However, given a median lifetime of 2.5 Myrs for the nebula, a relatively high probability event would be for one or more embryos incorporated into Earth, 1–3 times the size of Mars, to have acquired nebular volatiles.

With the exception of the deep mantle plume source, signatures of nebular gases are not found elsewhere on our planet. Ne isotopes in the MORB source are similar to solar wind irradiated meteoritic sources (e.g. Ballentine et al. 2005; Moreira 2013) and the isotopic ratios of H, N, argon (Ar), krypton (Kr) and xenon (Xe) in the convecting MORB mantle source and in the atmosphere (with the exception of Xe) are most similar to chondritic meteorites

(Alexander et al. 2012; Marty 2012; Halliday 2013; Hirschmann 2016). Furthermore, even the deep mantle values are not identical to nebular values. These observations suggest that most of the nebular volatiles acquired by the Earth embryos must have been lost and volatile elements were later replenished through delivery of chondritic volatiles after the dissipation of the nebula. We note that the isotopic compositions of the major volatiles rule out comets as the dominant source of Earth's major volatile species since these icy bodies are enriched in the heavier isotopes of H and N compared to Earth (Bockelée-Morvan et al. 1998; Meier et al. 1998; Jehin et al. 2009; Halliday 2013; Marty 2012; Marty et al. 2016)

The loss of nebular volatiles from within the Earth during its embryo stage may be a necessary consequence of the continued accretion of the Earth. After the nebula dissipates, the Earth embryo grows through impact of planetesimals and through giant impacts associated with collisions between embryos, with the Moon forming giant impact effectively terminating Earth's accretion. Since magma oceans are a consequence of giant impacts, the collisions between embryos would likely have released a large fraction of gases trapped within the solid Earth into the atmosphere. Hydrodynamic escape following the dissipation of the nebular and/or subsequent impacts must have led to loss of these nebular volatiles from the Earth embryo. Alternatively, some of the volatile elements may have been sequestered in the core (e.g. Dasgupta et al. 2013; Hirschmann 2016).

The timing of acquisition of volatiles is currently still a subject of debate (Marty 2012; Halliday 2013; Tucker and Mukhopadhyay 2014; Hirschmann 2016). The two possible hypothesis are that the majority of these volatiles were delivered prior to the Moon forming impact, or were delivered associated with the late accretion, defined as the last 0.3–1% of the Earth's mass that was accreted after the formation of the Moon (e.g. Warren et al. 1999; Walker 2009).

Radioactive isotopes, like ^{129}I and ^{244}Pu, can provide constraints on the history of volatile accretion. I is a volatile element while Pu is a refractory lithophile element. While the delivery rate of Pu should be time invariant during terrestrial accretion, I delivery would have changed with time. Thus, the I/Pu ratio of the Earth can potentially be used as a tracer of heterogeneous volatile delivery. If mantle domains have kept a memory of the different periods of terrestrial accretion, then the heterogeneous delivery of I with respect to other refractory lithophile elements could be interrogated through the $^{129}\mathrm{Xe}/^{136*}\mathrm{Xe}_{Pu}$ ratio of mantle-derived basalts. Here, ^{129}Xe is produced from ^{129}I decay only during the first 100 Myrs of solar system history and $^{136*}\mathrm{Xe}_{Pu}$ is ^{136*}Xe produced from ^{244}Pu fission during the first 500 Myrs of solar system history. The $^{129}\mathrm{Xe}/^{136*}\mathrm{Xe}_{Pu}$ ratio has been used to constrain the closure time for volatile loss of a mantle reservoir (Pepin and Porcelli 2006; Avice and Marty 2014). Since ^{129}I has a shorter half-life than ^{244}Pu, for reservoirs with the same initial I/Pu ratio, higher $^{129}\mathrm{Xe}/^{136*}\mathrm{Xe}_{Pu}$ ratios are indicative of earlier closure to volatile loss. The $^{129*}\mathrm{Xe}/^{136*}\mathrm{Xe}_{Pu}$ of the plume source mantle varies between 2.1 ± 1.6 and $2.9^{+0.4}_{-0.1}$ and the corresponding values for MORBs are between $7.8^{+3.1}_{-1.8}$ and $9.4^{+5.5}_{-2.7}$ (Fig. 2, Mukhopadhyay (2012), Pető et al. (2013), Parai and Mukhopadhyay (2015), Caracausi et al. (2016)). If the mantle had a homogeneous I/Pu ratio, the higher $^{129*}\mathrm{Xe}/^{136*}\mathrm{Xe}_{Pu}$ ratio in the MORB source would paradoxically imply that the shallow mantle became closed to volatile loss earlier than the deep mantle. A simpler explanation is that the lower $^{129*}\mathrm{Xe}/^{136*}\mathrm{Xe}_{Pu}$ reflects a lower initial I/Pu ratio for the plume source compared to the MORB source. These differences would indicate that the initial phase of Earth's accretion was volatile poor compared to the later stages of accretion. This conclusion is consistent with Pd–Ag systematics (Schönbächler et al. 2010) with the added distinction that the differences in the volatile accretion history are still preserved in the Xe isotopic composition of the mantle. As a result, the magma oceans formed during the giant impact stage of accretion did not homogenize chemical signatures acquired during the embryo stage.

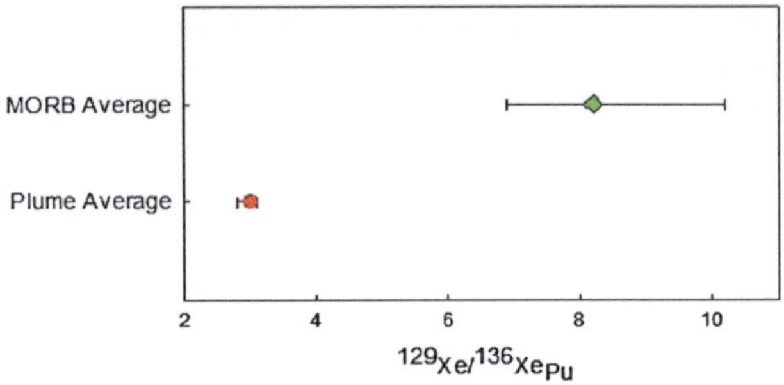

Fig. 2 Ratios of I- to Pu-derived Xe in plume source and MORB source assuming a chondritic starting composition for Xe isotopes in the mantle (Caracausi et al. 2016). The plume average was compiled from data presented in Mukhopadhyay (2012), Pető et al. (2013), and Caracausi et al. (2016). The MORB average is from Parai and Mukhopadhyay (2015)

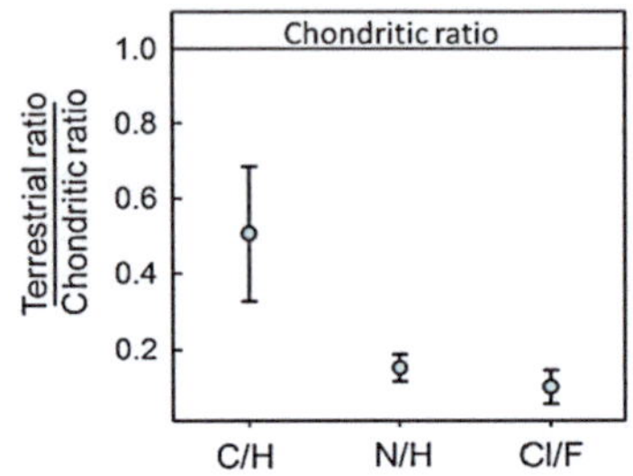

Fig. 3 Relative terrestrial abundances of C, N, H, and halogens normalized to the chondritic relative abundances. H, C and N data are from Marty (2012) and Halliday (2013). Cl data is from Sharp and Draper (2013) and F data are from McDonough (2003) and Palme and O'Neill (2014)

The relative abundances of the volatile elements also provide key insights into when volatiles must have been delivered. For example, the C/H ratio and N/H ratio on Earth is low compared to chondrites (Fig. 3), suggesting some process(es) fractionated this ratio following delivery of chondritic volatiles. For example, while the N and H isotopic ratios on Earth are chondritic, the N/H ratio does not match any known solar system object. Therefore, delivering all of the H and N associated with late accretion presents a significant challenge. Therefore Halliday (2013) concluded that water must have been delivered during the main phase of accretion. Halliday (2013) pointed out that if the entire N budget was delivered during the late accretion, the low N/H ratio of the Earth has to imply that about 70% of the Earth's hydrogen, in the form of water, must have been delivered prior to the Moon forming impact. Tucker and Mukhopadhyay (2014) argued that the low C/H and N/H ratio reflects modification of the chondritic relative abundances of these volatile elements on the Earth. Likewise, even though the Cl isotopic composition of Earth is chondritic (Sharp and Draper 2013), the extremely low Cl/F ratio of the Earth is an indication that the Cl abundance must have been modified relative to the F abundance (Tucker and Mukhopadhyay 2014). Hirschmann (2016) argued that while most of the water was delivered prior to the Moon forming giant impact, since the C/S ratio of the Earth is chondritic, and because both these elements are expected to have strongly partitioned into the Earth's core during core formation, significant C and S replenishment must have happened after cessation of core formation. That is, C and S replenishment of the Earth's mantle and exosphere must have occurred after the formation of the Moon in association with late accretion.

Hirschmann's (2016) argument about volatile delivery after the formation of the Moon is also compatible with the noble gases. The mantle and atmosphere have distinct non-radiogenic Ne, Kr and importantly, while the atmosphere is enriched in the heavier ^{22}Ne isotope relative to the mantle, it is depleted in the heavier ^{84}Kr isotope compared to the mantle (Holland et al. 2009; Marty 2012; Halliday 2013). Hence, the mantle and atmosphere cannot be related through outgassing or outgassing combined with hydrodynamic escape. The simplest explanation for the difference in mantle and atmospheric composition would be for the majority of atmospheric noble gases to be delivered after the last major mantle-atmosphere exchange—the Moon forming giant impact. Delivery of the noble gases associated with the late accretion, however presents a challenge. The atmospheric Kr inventory requires that the late accretion was CI chondrites, constituting ~4.5% of Earth's mass (Marty 2012), a number that would lead to significant overabundances in the terrestrial budget of the platinum group elements. To resolve this, delivery of a small amount of cometary ices have been invoked (Halliday 2013; Marty et al. 2016). Laboratory experiments (Bar-Nun et al. 2007; Notesco et al. 2003) and recent measurements of comet 67P/Churyumov-Gerasimenko (Altwegg et al. 2015; Marty et al. 2016) suggest that cometary ices can trap significant amounts of noble gases. If the laboratory experiments and cometary measurements are representative of ices, the budget of atmospheric noble gases could be delivered by icy planetesimals representing ~ 1% by mass of the late accretion. While the H and N isotopic compositions of comets can be significantly heavier than chondritic sources (e.g. Bockelée-Morvan et al. 1998; Meier et al. 1998; Jehin et al. 2009; Manfroid et al. 2009; Meech et al. 2009), accretion of such a small amount of cometary ice would lead to no detectable change in the H or N budget of the Earth. Therefore, the isotopic composition of these elements would not be perturbed away from chondritic values.

The compositions of the volatile elements therefore, suggest a complicated story of inheritance, loss and multiple stages of replenishment. Importantly, the budget of the different volatile elements appear to have grown at different rates, governed by the balance between delivery and loss, with loss to the core or outer space controlled by geochemical behavior of the elements. Acquisition of volatile elements started in the embryo stage, evidence of which survives in the Ne, and possibly H, isotopic compositions in the deep mantle plume source. With the dissipation of the nebula, planetary growth enters the giant impact phase of accretion—a stage marked by both addition of major volatiles with chondritic composition and volatile loss from the growing Earth. During this stage, water, and most likely all of the volatile elements, were partially retained. Following, the Moon forming giant impact, C, N, S, and halogens were partially replenished through accretion of chondritic planetesimals during the late accretion phase; the atmospheric noble gases where likely acquired from the accretion of 1×10^{22} g of icy planetesimals, which is approximately 0.1% of the late accretion mass (Marty et al. 2016). This amount of icy material is not large enough to leave its mark on the abundance of the major volatiles (H, C, S, N, halogens). The delivery of the icy planetesimals could have occurred during main phase of the late accretion or at the very tail end of late accretion (e.g. Marty et al. 2016).

2.2 What Were the Volatile Loss Processes and When Were Volatiles Lost?

2.2.1 Loss via Hydrodynamic Escape Versus Loss via Impacts

As discussed in Sect. 2.1, there is evidence in the deep mantle that nebular gases were acquired but signatures of nebular gases are not seen in the shallower mantle or in the atmosphere. This observation suggests loss of volatiles from the Earth during the main accretion phase. A simple calculation can be done based on the radiogenic Xe budget to show significant volatile loss during accretion. The abundance of ^{127}I in the bulk silicate Earth is 5–10 ppb (Lodders and Fegley 1998; Avice and Marty 2014). With an initial $^{129}I/^{127}I$ ratio of 1.0×10^{-4} (Ozima and Podosek 2002), the decay of ^{129}I to ^{129}Xe should have produced $\sim 1.5\text{–}3 \times 10^{13}$ moles of radiogenic ^{129}Xe on the Earth. The atmospheric abundance of radiogenic ^{129}Xe is $\sim 2.8 \times 10^{11}$ moles (Pepin 1991). The mantle budget is uncertain but is lower than the Xe budget of the atmosphere (Pepin and Porcelli 2002). Thus, >97% of the radiogenic ^{129}Xe has been lost from the Earth. The recent discovery that xenon isotopes in the Archean atmosphere were less fractionated than modern air suggests xenon loss during the Hadean and Archean associated with a mass fractionating process (e.g. Pujol et al. 2011). However, even after accounting for this mass fractionating loss, ~75% of the ^{129}Xe must have been lost prior to ~60 Ma, the approximate age of the Moon-forming giant impact (Avice and Marty 2014). Hence, substantial xenon loss must have occurred during the giant impact phase of accretion. The significant loss of radiogenic ^{129}Xe during the giant impact phase of accretion should not be taken to imply that all of the major volatiles were delivered after the Moon-forming giant impact. Rather, the volatile abundances could be residual to loss followed by replenishment during the late accretion phase.

Hydrodynamic escape has been invoked to explain the loss of volatiles, including Xe (Pepin 1991; Pepin and Porcelli 2006; Dauphas 2003). Hydrodynamic loss, however, requires a H-rich atmosphere, such as would be present if a nebular atmosphere was captured by the proto-Earth. In fact, models for hydrodynamic escape start off with a solar composition gas and explain the present abundances through loss of the H_2-rich atmosphere followed by resupply from the mantle and delivery of planetesimals during the late accretion (Pepin 1991; Dauphas 2003).

The present day geochemistry of volatiles, however, shows no evidence of hydrodynamic escape. Hydrodynamic escape fractionates the isotopes as the escaping H_2 preferentially drags the lighter isotope, leaving the atmosphere enriched in the heavier isotopes. Therefore, the present day chondritic isotopic composition of the H, C, N and Cl would require the H_2 flux to space to fall off at just the right time so that volatile escape is shut off exactly when chondritic isotopic composition are reached. Importantly, because the mass differences between the isotopes are different for the different elements, escape for the different volatile elements has to stop at different times and precisely at the point when the isotopic compositions have reached chondritic ratios. Such a scenario requires an extreme fine tuning of the H_2 escape flux and appears highly implausible. Alternatively, to generate the present day isotopic compositions requires adding isotopically light volatiles sequestered in the mantle to the isotopically heavy residual atmosphere that was mass fractionated through hydrodynamic escape (Pepin 1991; Pepin and Porcelli 2002). However, recent measurements of Kr isotopes indicate that the mantle is actually enriched in the heavier isotopes compared to the atmosphere (Holland et al. 2009). Therefore, hydrodynamic escape is not the primary loss signature recorded in the present day volatile budget. The lack of evidence for hydrodynamic escape in the present day volatile signatures does not imply that hydrodynamic escape never occurred. Hydrodynamic escape might have taken place on embryos shortly after the nebular gas dissipated (e.g. Erkaev et al. 2014). For our present day planet, it appears that if hydrodynamic escape occurred early in the accretion history (embryo stage), the resulting chemical signature has been completely overprinted by volatile resupply and additional loss mechanisms that did not fractionate the isotopes.

The process capable of driving loss without isotopic fractionation is impacts, both giant impacts as well as impacts of planetesimals. Atmospheric erosion associated with planetesimals has recently been shown to be quite efficient (Schlichting et al. 2015). While impact-driven loss does not fractionate the isotopes since it involves bulk ejection of the atmosphere, impact-driven losses can fractionate the elements. This is because during impacts, water oceans and crust would be retained preferentially compared to the atmosphere. Thus, elements residing in the atmosphere (nitrogen, noble gases) would be lost in preference to those in the ocean and crust (H and C). In the rest of Sect. 2.2, we will discuss constraints on magma ocean outgassing and atmospheric loss via impacts from He/Ne ratios in the mantle.

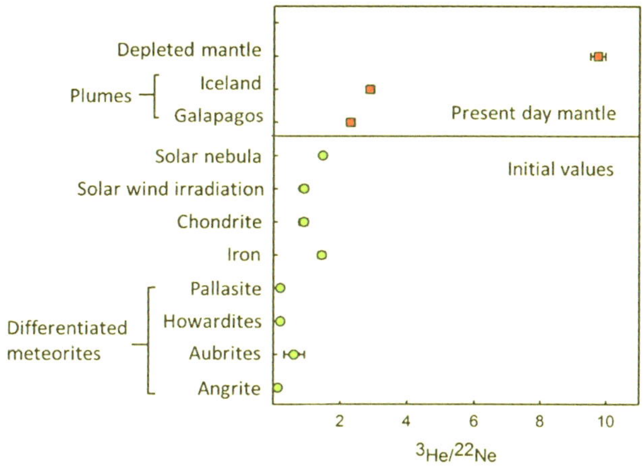

Fig. 4 $^3He/^{22}Ne$ ratios in modern terrestrial reservoirs along with the $^3He/^{22}Ne$ ratios in possible sources that may have contributed primordial He and Ne to the Earth. Figure after Tucker and Mukhopadhyay (2014), with solar nebula value update from Heber et al. (2012)

2.2.2 The He/Ne Ratio Difference Between MORB and Plume Sources

The $^3He/^{22}Ne$ ratio of the present day mantle spans a large range from ≤ 2, and possibly down to 1, to ≥ 10 (Fig. 4, Raquin and Moreira (2009), Coltice et al. (2011), Tucker and Mukhopadhyay (2014), Péron et al. (2016)). Plumes with the most primitive $^{21}Ne/^{22}Ne$ ratios from Galapagos and Iceland have $^3He/^{22}Ne$ ratios of 2 to 3 (Kurz et al. 2009; Mukhopadhyay 2012; Raquin and Moreira 2009) while the depleted MORB mantle has a $^3He/^{22}Ne$ ratio ≥ 10 with intermediate $^3He/^{22}Ne$ values due to mixing between the depleted mantle and the plume source (Tucker and Mukhopadhyay 2014). The $^3He/^{22}Ne$ ratio of the primitive reservoir sampled by plumes is higher than the nebular value by approximately a factor of ~1 to 2, with the depleted MORB source a factor of 6–10 higher than the sources of He and Ne to the Earth's mantle (Fig. 4).

Differences in the $^3He/^{22}Ne$ ratio between basalts from mid ocean ridges and basalts from mantle plumes have long been recognized (e.g. Honda and McDougall 1998; Moreira et al. 1998; Sarda et al. 2000). However, these differences were not universally accepted as reflecting a difference in the mantle sources of plumes and MORBs. Rather, magma generation and melt migration were sometimes invoked to explain the observed differences in $^3He/^{22}Ne$ ratios between MORBs and plumes (e.g. Sarda et al. 2000) with the mantle sources of basalts assumed to have similar $^3He/^{22}Ne$ ratios (e.g. Trieloff and Kunz 2005; Hopp and Trieloff 2008). More recent work however has unequivocally established that the mantle sources of plumes and MORBs have distinct $^3He/^{22}Ne$ ratios (Raquin and Moreira 2009; Kurz et al. 2009; Füri et al. 2010; Coltice et al. 2011; Marty 2012; Tucker and Mukhopadhyay 2014). Here we focus on the new interpretations of the $^3He/^{22}Ne$ ratio of mantle-derived basalts over the past few years.

2.2.3 Magma Oceans and $^3He/^{22}Ne$ *Ratios*

A striking feature of the $^3He/^{22}Ne$ ratios is that the mantle reservoirs have a higher value than even the Sun (Fig. 4). How did the Earth's interior acquire a $^3He/^{22}Ne$ ratio higher than even the solar nebula? To posit that the Earth's interior is enriched in ^3He compared to the Sun is unreasonable. More likely, the mantle must have preferentially lost ^{22}Ne. Based on the factor of two higher solubility of He compared to Ne in a basaltic liquid, Honda and McDougall (1998) argued that magma ocean outgassing must have increased the MORB source $^3He/^{22}Ne$ ratio, as magma ocean degassing would preferentially release Ne over He into the atmosphere. They also noted that the lower $^3He/^{22}Ne$ of plumes might indicate a less degassed source that was largely isolated from the MORB source for most of Earth's history.

More recently, Coltice et al. (2011) suggested that the $^3He/^{22}Ne$ ratio in plumes reflect the signature of a basal magma ocean. In the aftermath of the last giant impact, the Moon-forming impact, a global magma ocean may have existed. For a peridotite liquid, the liquidus intersects the isentrope in the mid-lower mantle (e.g. Mosenfelder et al. 2009; Thomas et al. 2012). Consequently, the magma ocean will first start crystallizing at the mid-lower mantle, which would ultimately isolate a layer of molten mantle above the core-mantle boundary—the basal magma ocean (BMO) (Labrosse et al. 2007). As crystallization in the BMO proceeds, incompatible elements, like noble gases, get enriched in the liquid. If both He–Ne partition coefficients are on order 0.01, the late crystallizing minerals could have relative high concentrations and low $^3He/^{22}Ne$ ratios. Since iron also becomes enriched in the residual liquid as the BMO crystallizes, the late crystallizing minerals would be Fe-enriched and likely to be more dense than the surrounding mantle assemblage. As a result of its high-density, the low $^3He/^{22}Ne$ reservoir would be entrained to low degrees in the convective flow, preserving the reservoir over the age of the Earth. Coltice et al. (2011) noted that the factor of 5–10 higher value in the MORB is unexplained and could reflect mantle degassing, although they did not indicate when or how this degassing occurs.

Tucker and Mukhopadhyay (2014) argued that mantle outgassing associated with the long-term plate tectonic cycling is not likely to increase the $^3He/^{22}Ne$ ratio of the mantle. Specifically, the mechanisms they investigated for raising $^3He/^{22}Ne$ ratios were (i) partial melting that generates oceanic crust and a depleted mantle residue and (ii) recycling and mixing of oceanic crust and lithosphere back into the mantle. Given current determinations of partition coefficients for He and Ne (Brooker et al. 2003; Heber et al. 2007; Jackson et al. 2013), they found that partial melting is not an efficient process for fractionating He from Ne in the mantle. Likewise, due to ubiquitous incorporation of atmospheric noble gases, both upper and lower oceanic crust have a $^3He/^{22}Ne$ of ~ 0 (Staudacher and Allègre 1988; Moreira et al. 2003). If the recycling efficiency of He and Ne back into the mantle is high, recycling of oceanic crust should decrease the $^3He/^{22}Ne$ ratio and not increase it. Therefore, Tucker and Mukhopadhyay (2014) concluded $^3He/^{22}Ne$ values of 10 for the depleted mantle could not be due to processes associated with the long-term plate tectonic cycle. They suggested that outgassing or ingassing of a magma ocean is a process that is expected to raise the mantle $^3He/^{22}Ne$ ratio. As He is more soluble in magmas than Ne, magma ocean ingassing or outgassing will raise the mantle $^3He/^{22}Ne$ ratios. The differences between the $^3He/^{22}Ne$ ratios of plume and depleted mantle sources then reflects a difference in deep versus shallow mantle magma ocean outgassing history of the Earth.

2.2.4 Multiple Magma Oceans, Giant Impacts and Atmospheric Losses

To explain the low $^3He/^{22}Ne$ ratios and solar-like $^{20}Ne/^{22}Ne$ ratios of ≥ 12.9 in plumes, Tucker and Mukhopadhyay (2014) proposed ingassing of nebular gas from a gravitationally accreted nebular atmosphere into a magma ocean, a hypothesis that was previously proposed as the mechanism responsible for acquisition of terrestrial He and Ne (Harper and Jacobsen 1996; Mizuno et al. 1980; Porcelli et al. 2001; Yokochi and Marty 2004). During nebular ingassing into a magma ocean, the $^3He/^{22}Ne$ ratio of the magma ocean would be fractionated from the nebular value of ~ 1.5 by the He/Ne solubility ratio of ~ 2 to values of ~ 2.3–3. The magma ocean ingassing occurs on an embryo, probably about the size of Mars or slightly larger.

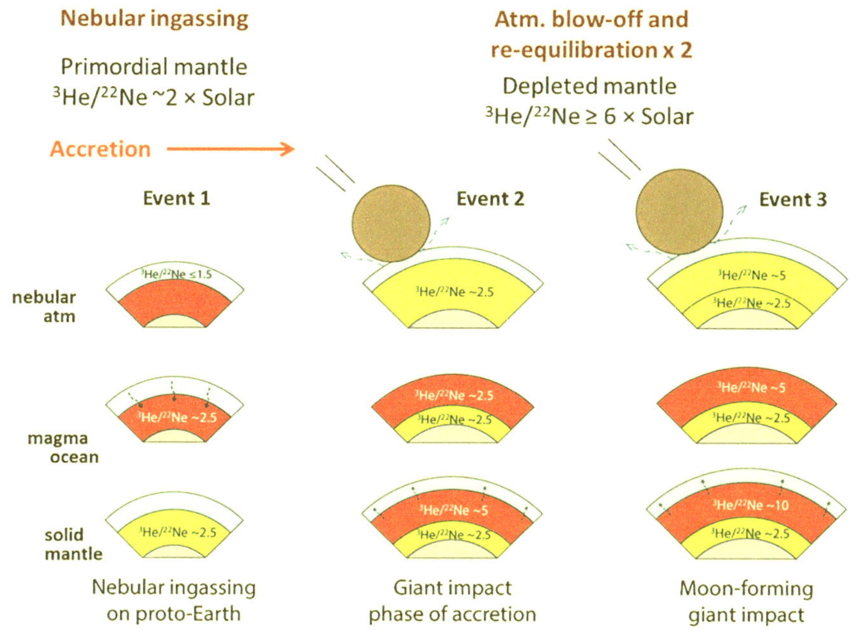

Fig. 5 A proposed chronology of events to explain the highly fractionated $^3He/^{22}Ne$ ratio of the depleted mantle. Figure after Tucker and Mukhopadhyay (2014)

The magma ocean on embryos will crystallize on a timescale on order 10^6 to 10^7 years (Zahnle et al. 2007; Elkins-Tanton 2008; Lebrun et al. 2013). The median lifetime of nebular gas is ~2.5 Myrs and the maximum observed timescale is ~10 Myr (e.g. Hillenbrand 2008). Consequently, the nebular gas will dissipate on a timescale similar to the crystallization timescale of the magma ocean. The gas in the disk has a dampening effect on planetary orbit. Consequently, dissipation of the nebular gas pumps up the eccentricity of the embryos, causing them to run into each other—the giant impact phase of terrestrial planet accretion. During the giant impact phase, magma oceans are expected as the proto-Earth grows to its present mass through violent collisions with other embryos.

Tucker and Mukhopadhyay (2014) proposed that after loss of the nebular atmosphere, and during the giant impact phase of terrestrial growth, at least two separate partial mantle magma ocean episodes raised the $^3He/^{22}Ne$ ratio of the shallower mantle from ~2 to ≥10, with the last magma ocean associated with the Moon-forming giant impact (Fig. 5). However, during this phase, the low $^3He/^{22}Ne$ ratio in the deep mantle must be preserved. Thus, Tucker and Mukhopadhyay (2014) argued that the later giant impacts including the Moon forming giant impact may not have produced a whole mantle magma ocean because the timescale for turnover of a turbulently convecting magma ocean may be a short as a few weeks (Solomatov 2000; Elkins-Tanton 2008; Pahlevan and Stevenson 2007), and hence, the magma ocean would be expected to be chemically homogeneous. Alternatively, if whole mantle magma oceans were produced, some process must have prevented mixing within the magma ocean. One possibility to prevent such mixing might be the presence of a dense basal magma ocean prior to the impact (Nakajima and Stevenson 2015).

The requirement of 2 giant impacts and magma ocean outgassing episodes comes from the He/Ne solubility ratio of ~2 in a magma ocean (Tucker and Mukhopadhyay 2014). Thus, each stage of magma ocean outgassing can only raise the residual liquid by at most a factor of 2. The degree of outgassing of a magma ocean is, however, tied to the atmospheric boundary condition above the magma ocean. If an atmosphere that had equilibrated with a previous magma ocean remained intact during a giant impact, the new magma ocean would not outgas and fractionate ^{3}He from ^{22}Ne because it would already be in equilibrium with the overlying atmosphere. Therefore, loss of the pre-existing atmosphere is a requirement for driving fractionation of He–Ne during magma ocean outgassing (Fig. 5). The loss of the nebular atmosphere in the aftermath of magma ocean ingassing on the planetary embryo could have been achieved through hydrodynamic escape (e.g. Pepin 1991), giant impacts (Genda and Abe 2003, 2005; Schlichting et al. 2015), impacts of planetesimals (Schlichting et al. 2015), or a combination of the above factors. During the giant impact phase of accretion, outgassed atmospheres are expected, which are likely to be H^2O–CO^2 rich and not H^2-rich (e.g. Heber et al. 2012). Thus, loss of an outgassed atmosphere is not likely to be driven by hydrodynamic escape. Rather, escape is more easily ascribed to giant impacts (Genda and Abe 2003) and/or impacts of planetesimals (Schlichting et al. 2015).

Complete loss of the atmosphere would lead to the largest extent of fractionation, the He/Ne solubility ratio of $\sim$2, with partial loss suppressing the extent of fractionation. Tucker and Mukhopadhyay (2014) noted that any process that leads to preferential He loss from the magma ocean or atmosphere, such as kinetically controlled degassing or hydrodynamic escape of the atmosphere, would also suppress the extent of He–Ne fractionation during degassing of the magma ocean. In addition, He/Ne fractionation associated with degassing would also be suppressed if during the giant impact phase, meteoritic He and Ne with $^{3}He/^{22}Ne$ of $\sim$0.9 were delivered to the Earth's mantle. In such cases, more than the two episodes of giant impacts may be required to drive the mantle $^{3}He/^{22}Ne$ ratio to values of $\sim$10. The proposed chronology of events in Fig. 5 should therefore, be viewed as a minimum sequence of events during the formation of the Earth.

In summary, the observation of different $^{3}He/^{22}Ne$ ratios in the plume and MORB mantle reservoirs requires the formation and preservation of two distinct mantle domains during accretion (Coltice et al. 2011; Tucker and Mukhopadhyay 2014). High $^{3}He/^{22}Ne$ ratio of $\geq$10 for the MORB source suggests multiple magma oceans and multiple atmospheric loss episodes. The lower $^{3}He/^{22}Ne$ of the plume source also appears to be associated with a magma ocean; Coltice et al. (2011) suggested that the plume values are the signature of crystallization of a basal magma ocean while Tucker and Mukhopadhyay (2014) associated it within ingassing of nebular gases into a magma ocean on the proto Earth. Tucker and Mukhopadhyay (2014) argued that given the He/Ne solubility ratio of 2, the low $^{3}He/^{22}Ne$ ratios in plumes is not likely to have its origin in a basal magma ocean produced by the last giant impact (the Moon-forming giant impact) because that magma ocean must have produced the high $^{3}He/^{22}Ne$ ratio now sampled in the depleted mantle. However, a basal magma ocean sequestering low $^{3}He/^{22}Ne$ ratios in the deep mantle could have formed from a previous whole-mantle magma ocean episode and recent work suggests that a dense basal magma ocean layer may not get mixed in with the rest of the mantle in the aftermath of a giant impact (e.g. Nakajima and Stevenson 2015).

2.2.5 Terrestrial Volatile Signatures and Atmospheric Loss

Geochemical evidence for multiple magma oceans associated with multiple giant impacts is consistent with dynamic simulations of terrestrial planet formation that suggest more than just the Moon-forming giant impact (e.g. Chambers and Wetherill 1998; Chambers 2001). The atmospheric loss episodes put forth to explain the fractionation of He/Ne during magma ocean outgassing, also provide explanations for other features of the Earth's volatile signatures, such as the depletion of carbon, and the very large depletion of nitrogen, with respect to water (Fig. 3). Carbon depletion has previously been linked to atmospheric loss or to partitioning of carbon to the core (Hirschmann and Dasgupta 2009; Hirschmann 2016). Likewise, N depletion has been linked to core formation (Marty 2012). However, since carbon appears to be more siderophile than nitrogen by factors of 25 to 100, depletion of N by core formation is unlikely because it would be accompanied by a significantly greater depletion of C (Tucker and Mukhopadhyay 2014; Hirschmann 2016). The observed moderate depletion in C and extreme depletion in N might be explained due to atmospheric loss associated with impacts. Carbon can form bicarbonate and carbonate ions and as a result, a significantly larger fraction of C can be the ocean (or crust) compared to N. Since both giant impacts and impacts of planetesimals preferentially remove the atmosphere over an ocean (Genda and Abe 2005; Schlichting et al. 2015), atmospheric loss would lead to large fractionation in the N/H ratio and a more modest fractionation in the C/H ratio (Fig. 3). Since impacts lead to bulk ejection of the atmosphere, the C and N isotopic composition of the Earth would remain chondritic.

Atmospheric loss may also explain why chlorine is depleted in the Earth relative to fluorine (Fig. 3). Compared to Cl, F is more soluble in magmas by a factor of 4 (Andrews et al. 2009). Consequently, magma ocean degassing would lead to preferential transport of Cl to the Earth's surface compared to F. If chlorine is lost due to giant impacts or impacts of planetesimals, it would explain the depletion of Cl with respect to F. On the Earth's surface, chlorine would be expected to reside primarily in the ocean and crust unless a steam atmosphere was present. Thus, loss of chlorine might also require loss of water, either in the form of partial ocean loss or through erosion of a steam atmosphere. Overall, Earth's volatile budget records the violent events during accretion. While late accretion contributed volatiles to Earth (e.g. Albarède 2009; Marty 2012; Hirschmann 2016), late accretion did not overprint the volatile characteristics acquired during the main stages of Earth's accretion (Halliday 2013; Tucker and Mukhopadhyay 2014; Dauphas and Morbidelli 2014).

3 Atmospheric Loss by Small and Large Impacts

Terrestrial planet formation consists of successive mergers of small planetesimals and larger protoplanets accumulating into final planets ranging in size from Mars to Earth. The delivery of volatiles and atmospheric erosion are intricately linked to planetary growth since planetesimal impacts and giant impacts can both deliver volatiles and lead to their loss due to atmospheric erosion.

We present here simple analytic models which we use to calculated the atmospheric mass loss due to impacts during planet formation. The aim of these models is to distill the somewhat complex and complicated nature of impacts to their essential physics, which allows us to gain an intuitive understanding of atmospheric mass loss results and to apply them over the whole range of possible impactor sizes. These models are not suitable to determine the precise outcome of a single specific impact, which for example have been investigated by Shuvalov (2009) and Liu et al. (2015), but capture the overall mass loss results in an averaged sense, e.g. once averaged over various impact geometries and angles.

For the purpose of this chapter we assume that the atmosphere is, to first order, isothermal such that its density profile is exponential and given by

$$\rho = \rho_0 \exp[-z/h], \tag{2}$$

where ρ_0 is the density on the ground, z the height in the atmosphere measured from the ground and h the atmospheric scale height. At the end of this chapter we will discuss how

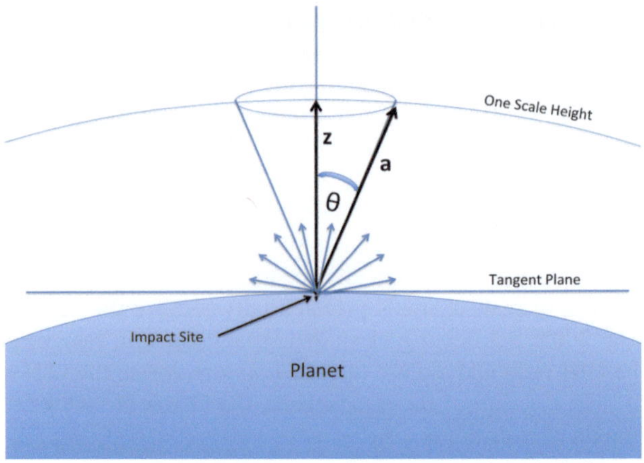

Fig. 6 Illustration of the impact geometry. Planetesimal impacts can only eject atmosphere locally. Treating their impact as a point-like explosion leading to an isotropic shock at the impact site, the maximum atmospheric mass that they can eject in a single impact is given by all the mass above the tangent plane, which is a fraction given by $h/2R$ of the total atmospheric mass. Figure after Schlichting et al. (2015)

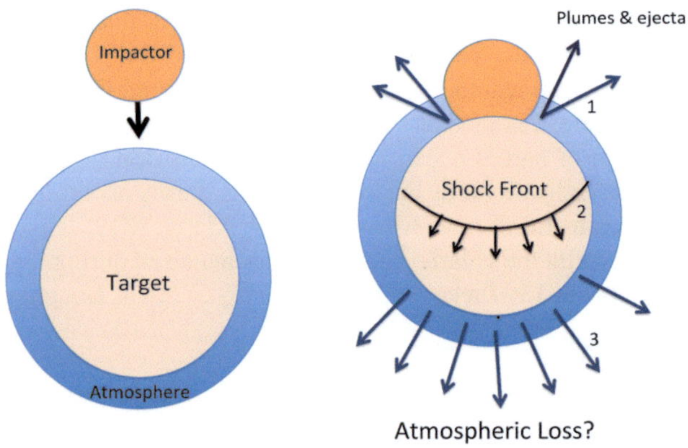

Fig. 7 Illustration of a giant impact. 1) The giant impact ejects atmosphere and ejecta close to the impact point and launches a strong shock. 2) The shock front propagates through the target causing a global ground motion. 3) This ground motion in turn launches a strong shock into the planetary atmosphere, which can lead to loss of a significant fraction of or even the entire atmosphere. Figure after Genda and Abe (2003) and Schlichting et al. (2015)

the results derived here will change for an adiabatic atmosphere in which the heat transport is facilitated by convection. We assume here that the atmosphere is planar, which is valid for the terrestrial planets since their atmospheric scale heights, h, are small compared to their radii, R. For the current Earth, $h/R \sim 0.1\%$ so the planar condition is indeed well justified for the terrestrial planets. In contrast, for close in exoplanets the planar condition breaks down and full spherical treatment should be used, see for example Inamdar and Schlichting (2015, 2016).

Impacts can lead to atmospheric losses in two distinct ways: First, the expansion of plumes generated at the impact site can erode the atmosphere locally, but not globally (see Fig. 6). Such loss can be caused by small planetesimal impacts and is discussed in detail immediately below. Second, when the target and impactor are comparable in mass, a giant impact can create a strong shock that transverses through the entire planet leading to global ground motions that in turn can launch a shock into the atmosphere (see Fig. 7). The shock accelerates the atmosphere, and fluid parcels that are accelerated to above the escape velocity of the target are lost. A simple model calculating the atmospheric mass loss due to giant impacts is described in Sect. 3.3.

3.1 Atmospheric Loss by Small Impacts

Small impacts can individually only eject planetary atmospheres locally, close to their impact sites. However, collectively they play an important role in atmospheric loss during planet formation, since they are many orders of magnitude more efficient per unit impactor mass in atmospheric erosion than giant impacts (Schlichting et al. 2015).

3.1.1 Mass Loss Regimes

The first calculations of atmospheric loss by planetesimal impacts used the Zel'dovich and Raizer (1967) solution for the expansion of a vapour plume and compared the momentum of the expanding gas and the mass of the overlying atmosphere (e.g. Melosh and Vickery 1989; Vickery and Melosh 1990; Ahrens 1993). Here we use a similar approach and compare our analytic results to atmospheric mass loss results from numerical simulations. When the Earth or a protoplanet is hit by a small impactor, the impactor's velocity is rapidly decelerated and its kinetic energy is converted into heat and pressure leading to something analogous to an explosion. We model the impact as an isotropic point-like explosion on the surface of the target. Assuming momentum conservation and impact velocities comparable to the escape velocity, v_{esc}, yields an ejecta mass comparable to the impactor mass, m_{Imp}, that propagates isotropically with a velocity of order v_{esc} into a half-sphere centered on the impact site. This implies that atmosphere is only lost where its mass per unit solid angle, m_θ, as measured from the impact site, is less than that of the ejecta, which is equal to $m_{Imp}/2\pi$.

More generally we find for an arbitrary impact velocity and momentum transfer efficiency in the impact, η, that atmosphere is only ejected within a solid angle, θ, where

$$m_\theta \leq \eta \left(\frac{v_{Imp}}{v_{esc}} \right) \frac{m_{Imp}}{2\pi}. \tag{3}$$

For the density profile given in (2) we have

$$m_\theta = 2\pi \rho_0 \int_0^\infty \exp\left[-\left(a^2 + 2aR\cos\theta\right)/2Rh\right] a^2 da, \tag{4}$$

where z is the height in the atmosphere above the ground and it is related to a, the distance from the impact site to the top of the atmosphere (see Fig. 6), by $z = (a^2 + 2aR\cos\theta)/2R$. Note, the integration in (4) is only over a and not θ since the explosion at the impact site is assumed to be isotropic (Schlichting et al. 2015). This leads to two interesting limits:

1) There is a lower limit on the impactor size that can eject any atmosphere.
2) There is a lower limit on the impactor size that can eject all the atmospheric mass above the tangent plane.

The first limit can be derived by equating the mass residing in the vertical column above the impact site to (3). Integrating (4) for $\theta = 0$ yields a minimum impactor mass of

$$m_{min} = 4\pi \rho_0 h^3 \eta^{-1} \left(\frac{v_{esc}}{v_{Imp}} \right), \tag{5}$$

which reduces to the result obtained by Schlichting et al. (2015) for $v_{imp} \sim v_{esc}$ and $\eta \sim 1$, and yields $r_{min} = (3\rho_0/\rho_{imp})^{1/3} h \sim 1$ km for the current Earth (Fig. 8a).

The second limit is obtained by integrating (4) for $\theta = \pi/2$ which yields the minimum impactor mass that can eject all the mass above the tangent plane, m_{cap}, and is given by

$$m_{cap} = \sqrt{2}\rho_0(\pi h R)^{3/2}\eta^{-1}\left(\frac{v_{esc}}{v_{Imp}}\right), \tag{6}$$

which again reduces to the result obtained by Schlichting et al. (2015) for $v_{Imp} \sim v_{esc}$ and $\eta \sim 1$, and yields $r_{cap} = (\rho_0/\rho_{imp})^{1/3}(hR)^{1/2} \sim 25$ km for the current Earth (Fig. 8c). All impactors above this size will eject all the atmospheric mass above the tangent plane but not more than that, unless they are so large that they are in the giant impact regime which is discussed in Sect. 3.3.

3.1.2 Mass Loss Efficiencies

Next we turn to calculating the mass loss efficiencies for the various impactor sizes. The mass loss efficiency is given by the amount of atmospheric mass lost divided by the impactor mass needed to achieve the loss. These two quantities are not equal because we model the explosion at the impact site as isotropic whereas the atmospheric mass follows an isothermal or adiabatic density profile (see (2)). As a result the atmospheric mass directly vertically above the impact site is easiest to eject, since it has the lowest column density and the atmospheric mass along the tangent plane is hardest to eject since it has the largest column density. The atmospheric mass residing inside a cone subtended by an angle θ measured from the normal to the impact site (see Fig. 8) is

$$\mathcal{M}_{Eject,\theta} = 2\pi\rho_0 \int_{a=0}^{a=\infty}\int_{\theta'=0}^{\theta'=\theta} \exp[-z/h]\sin\theta' a^2 d\theta' da. \tag{7}$$

The mass loss efficiency is given by dividing (7) by (4) and evaluating it for various impactor sizes. Figure 9 shows the mass loss efficiency as a function of impactor size. Small impactors with $r_* = \sqrt{3}r_{min}$ are the most efficient impactors per unit mass in ejecting the atmosphere. For the current Earth this corresponds to bodies with $r \sim 2$ km. The value of $\mathcal{M}_{Eject}/m_{Imp}$ decreases rapidly for larger planetesimals because the atmosphere along the tangent plane is harder to eject, due to its larger column density. In addition, once impactors are larger than r_{cap}, each impactor only ejects the mass residing above the tangent plane, which reduces the mass loss efficiency per unit impactor mass.

Integrating over the whole cap, i.e. from $\theta = 0$ to $\theta = \pi/2$, yields a total cap mass of

$$\mathcal{M}_{cap} = 2\pi\rho_0 h^2 R, \tag{8}$$

in the limit that $R \gg h$, which applies for the terrestrial planets. This is the maximum atmospheric mass that a single planetesimal impact can eject and is given by all the mass above the tangent plane of the impact site. The ratio of the mass in the cap compared to the total atmospheric mass is therefore $\mathcal{M}_{cap}/M_{atmos} = h/2R$.

3.1.3 Impactor Size Distribution

Figure 9 shows the ratio of atmospheric mass ejected to impactor mass. To understand the atmospheric mass loss over time we calculate the atmospheric loss due to a population of planetesimal impactors of various sizes. The impactors are assumed to follow a power-law size distribution such that the cumulative impactor flux can be parametrized as $N(>r) =$

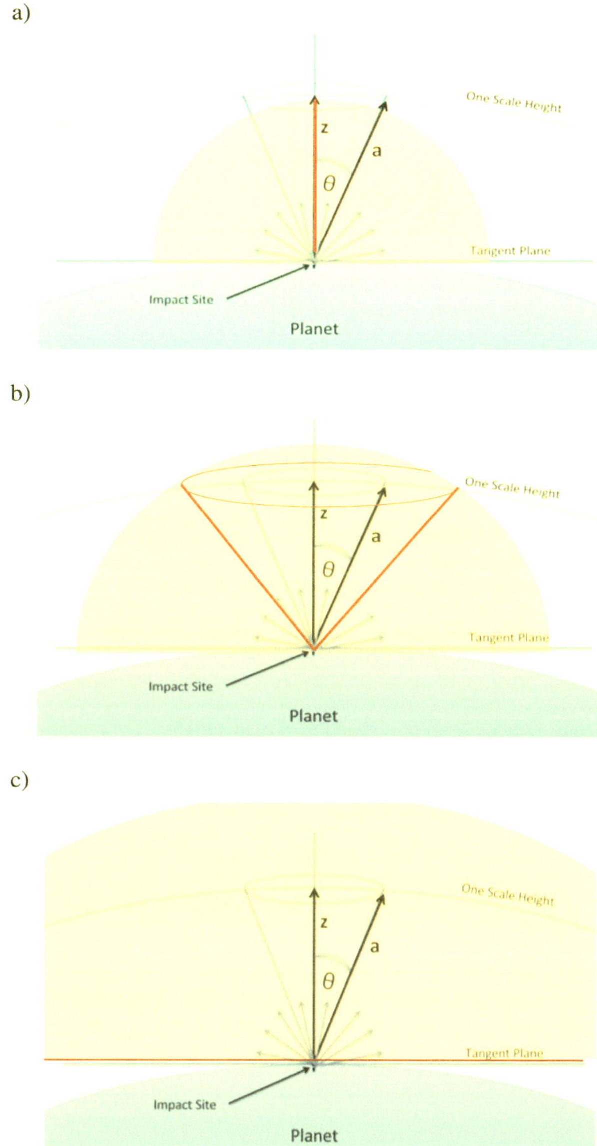

Fig. 8 Different planetesimal impact regimes. The size of the shaded hemisphere is scaled to the impactor mass (assuming an impact velocity comparable to the escape velocity) such that the mass per unit solid angle in the hemisphere presented by the shaded region is given by $m_{Imp}/2\pi$ as measured radially outward from the impact site. Panel **a**) illustrates the limit in which the impactor is able to only eject the atmosphere vertically above the impact site, which corresponds to $m_{min} = 4\pi\rho_0 h^3$. Panel **b**) corresponds to an intermediate regime, where the impactor is able to eject all the atmospheric mass residing in the red cone. Panel **c**) displays the limit in which an impactor is large enough to eject all the atmospheric mass residing above the tangent plane of the impact site, which corresponds to a minimum impactor mass of $m_{cap} = \sqrt{2}\rho_0(\pi h R)^{3/2}$

$N_0(r/r_0)^{-q+1}$, where q is the differential power law index and N_0 is normalized to impactors of size r_0 and is the impactor flux (number per unit time per unit area). The atmospheric mass loss rate is then given by

$$\frac{dM_{atmos}}{dt} = -\pi R^2 \frac{N_0(q-1)}{r_0} \int_{r_{min}}^{r_{max}} \left(\frac{r}{r_0}\right)^{-q} \mathcal{M}_{Eject}(r)dr, \tag{9}$$

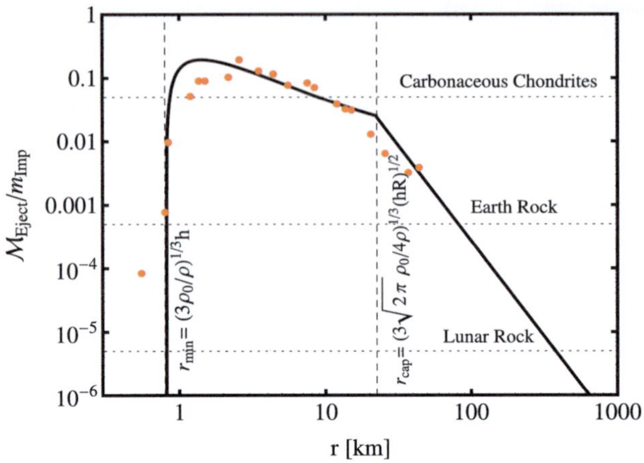

Fig. 9 Ratio of atmospheric mass ejected to impactor mass, $\mathcal{M}_{Eject}/m_{Imp}$. Numerical values are scaled to the current Earth's atmosphere and shown for $v_{imp}\eta/v_{esc} \sim 1$. Small impactors with $r_* = \sqrt{3}r_{min}$ are the most efficient impactors per unit mass in ejecting the atmosphere and the ejection efficiency decreases rapidly for larger planetesimals. Whether or not planetesimal impacts will lead to a net loss of planetary atmospheres depends on the impactor sizes distribution, their volatile budget and the amount of outgassing their impacts can initiate. The *three dotted horizontal lines* correspond to volatile contents of 5 wt.% (representative of some of the most water rich carbonaceous chondrites), 0.05 wt.% (representative of the average water content in the bulk Earth excluding the hydrosphere) and 0.0005 wt.% corresponding to an estimate of the minimum water content of the bulk moon (McCubbin et al. 2010). For comparison, data from oblique impact simulations for escape velocities of 11.2 km/s and impact velocities of 30 km/s from Shuvalov (2009) are shown by the orange points. Figure after Schlichting et al. (2015)

where $\mathcal{M}_{Eject}(r)$ is the atmospheric mass ejected due to an impactor with radius r. Solving (9) yields

$$M_{atmos}(t) = M_0\left(1 - \frac{t}{t^*}\right)^{3/(q-1)}, \tag{10}$$

where M_0 is the initial atmospheric mass at $t = 0$ and t_* is the time it takes to erode the entire atmosphere (Melosh and Vickery 1989). Equation (9) shows that the whole atmosphere is eroded in a finite time, which is due to the fact that as the atmosphere is lost, its density decreases and even smaller impactors, which are more numerous, start to contribute to the mass loss. For $1 < q < 3$, the value for t^* is

$$t^*_{q<3} = \frac{6}{\pi(q-1)CRhN_0}\left(\sqrt{\frac{\pi h}{8R}}\frac{M_0}{m_0}\right)^{(q-1)/3}, \tag{11}$$

and for $q > 3$ it is

$$t^*_{q>3} = \frac{3(q-3)}{\pi(q-1)h^2N_0}\left(\left(\frac{h}{R}\right)^2\frac{M_0}{m_0}\right)^{(q-1)/3}, \tag{12}$$

where $m_0 = 4\pi\rho_{imp}r_0^3/3$ (Schlichting et al. 2015). We get two different expressions for t^* depending on the power-law index of the planetesimal size distribution because for $1 < q < 3$ atmospheric mass loss is dominated by impactors that remove the whole atmospheric mass above the tangent plane and for $q > 3$ the atmospheric mass loss is dominated by impactors that each only eject part of the atmospheric mass residing above the tangent plane.

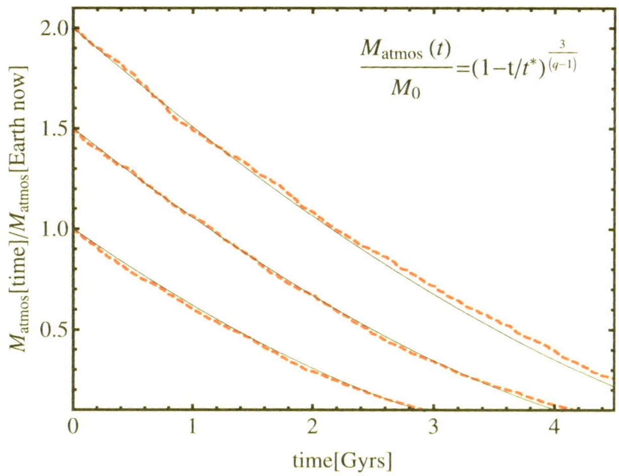

Fig. 10 Atmospheric mass loss over time for three different initial atmospheric masses normalized to the current Earth's atmosphere. The impactor population contains a total mass of 0.1% Earth's masses, consistent with the mass inferred for the late veneer, and has a power-law size distribution with a differential power-law index of $q = 2.8$ as inferred from the lunar cratering record by Neukum et al. (2001). The largest impactor is 300 km in size and the smallest 1 m. The impactor flux is assumed to be uniform over 4.5 Gyrs, but the total atmospheric mass loss would remain the same for a varying impactor flux over time as long as the total impactor mass remains unchanged. When calculating the mass loss efficiency we assumed $v_{imp}\eta/v_{esc} \sim 1$. The *thick-red dashed lines* are from numerical Monte Carlo simulations that calculate the atmospheric loss due to each individual impact and the thin *black lines* are the analytic solutions given in (10)

The constant C in (11) parameterizes any additional contribution to the atmospheric loss by impactors that are smaller than those that eject all the mass above the tangent planet. Melosh and Vickery (1989) investigated the mass loss due to impactors that remove all the atmosphere above the tangent plane (i.e. the $1 < q < 3$ regime) but they assumed that $\mathcal{M}_{cap} = m_{cap}$ (instead of $\mathcal{M}_{cap} = m_{cap}(2h/\pi R)^{1/2}$ see (6) and (8)), ignoring the fact that the atmospheric column density is larger in the direction along the tangent plane than directly vertically above the impact site, and they also neglected the numerical coefficient C discussed above.

Figure 10 shows atmospheric erosion over time due to an impactor size distribution with $q = 2.8$ and an impactor flux that bombards the Earth uniformly over 4.5 Gyrs with a total of 0.1% of the mass of the Earth. The largest body is 300 km and the smallest 1 m in radius. The thin black lines correspond to the analytic solution given in (10) and the red dotted lines to numerical simulations that calculate the atmospheric loss after each impact and determine the evolution of the atmospheric mass over time. As expected, the numerical and analytical results are in excellent agreement. They demonstrate that small impactors that contain about 0.1% of the mass of the Earth, which is about the mass inferred for the late veneer, can erode the entire Earth's atmosphere, if the Earth's initial atmospheric mass was similar to its current one (Schlichting et al. 2015). Whether such a bombardment by small impactors leads to a net loss or gain in atmospheric mass depends on the volatile content of the impactors themselves, how much of their volatiles end up in the atmosphere and the amount of volatiles released from the impact melt pools that are created by the impacts.

3.2 Atmospheric Loss by Giant Impacts

Giant impacts are believed to be the last major assembly stage of terrestrial planet formation (e.g. Chambers 2001) and may also play a major role in the formation of close-in exoplanets that have masses between that of Earth and Neptune (e.g. Raymond et al. 2008; Hansen and Murray 2012; Inamdar and Schlichting 2015). The atmospheric survival during a giant impact has been examined by several groups both in the context of terrestrial planet formation (Genda and Abe 2003, 2005; Stewart et al. 2014; Schlichting et al. 2015) and in that of close-in exoplanets (Inamdar and Schlichting 2015, 2016). These works integrate the hydrodynamic equations of motion of the planetary atmosphere and determine the amount of atmospheric mass loss for a given ground velocity that launches a strong shock into the atmosphere. Recently Liu et al. (2015) presented the first results of three dimensional hydrodynamic simulations of atmospheric mass loss that model the giant impact as well as the atmospheric mass loss together. Here we calculate the atmospheric loss due to a given ground motion. We assume that the atmosphere is planar. In addition we assume that radiative cooling can be neglected such that the flow is adiabatic. The hydrodynamic equations describing the flow are then given by

$$\frac{1}{\rho}\frac{D\rho}{Dt} + \frac{\partial u}{\partial z} = 0, \tag{13}$$

$$\frac{Du}{Dt} + \frac{1}{\rho}\frac{\partial p}{\partial z} = 0, \tag{14}$$

$$\frac{1}{p}\frac{Dp}{Dt} - \frac{\gamma}{\rho}\frac{D\rho}{Dt} = 0, \tag{15}$$

where γ is the adiabatic index and D/Dt the ordinary Stokes time derivative. The solutions to the hydrodynamic equations above can be separated into their time-dependent and spatial parts allowing for analytic self-similar solutions (Raizer 1964; Grover and Hardy 1966; Schlichting et al. 2015). Specifically we have,

$$\begin{aligned} &\rho(z,t) = \rho_0 \exp\big[-Z(t)/h\big]G(\zeta), \qquad u(z,t) = \dot{Z}U(\zeta), \\ &p(z,t) = \rho_0 \exp\big[-Z(t)/h\big]\dot{Z}^2 P(\zeta) \end{aligned} \tag{16}$$

where $Z(t)$ is the position of the shock front and $\zeta = (z - Z(t))/h$. The similarity variables for the density, velocity and pressure are given by $G(\zeta)$, $U(\zeta)$ and $P(\zeta)$, respectively. Using the expressions in (16) and substituting them into the hydrodynamic equations (13)–(15) yields for the spatial parts

$$\frac{1}{G}\frac{dG}{d\zeta}(U-1) + \frac{dU}{d\zeta} = 1, \tag{17}$$

$$(U-1)\frac{dU}{d\zeta} + \frac{1}{G}\frac{dP}{d\zeta} = -\frac{U}{\alpha}, \tag{18}$$

$$(U-1)\left(\frac{1}{P}\frac{dP}{d\zeta} - \frac{\gamma}{G}\frac{dG}{d\zeta}\right) = -\frac{2}{\alpha} - \gamma + 1, \tag{19}$$

and a time dependent part given by

$$\frac{\dot{Z}^2}{\ddot{Z}} = \alpha h. \tag{20}$$

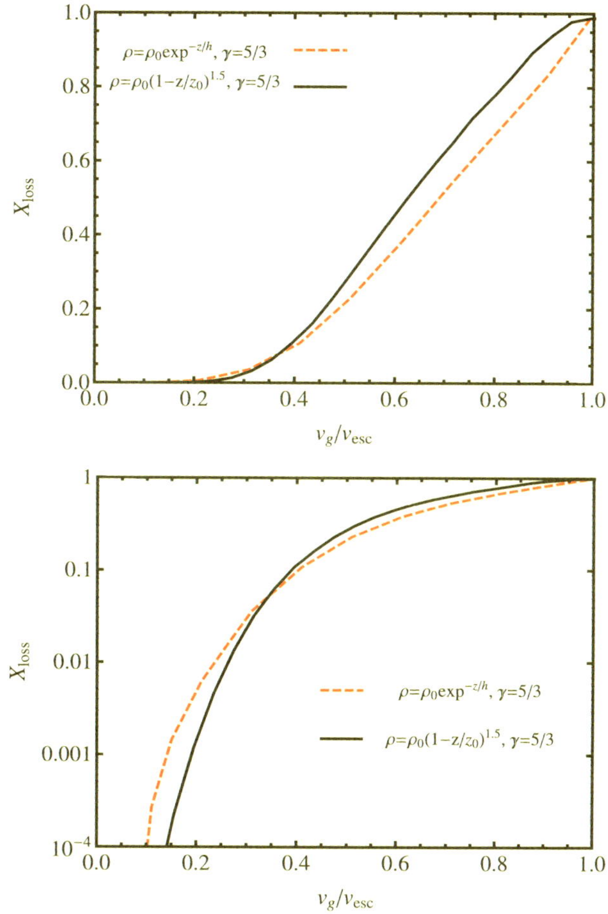

Fig. 11 Mass loss fraction, χ_{loss}, as a function of v_g/v_{esc} for an adiabatic atmosphere in black and an isothermal atmosphere shown as *red-dashed line*. Both loss *curves* correspond to and adiabatic index of $\gamma = 5/3$

Using the strong shock conditions we have

$$G(0) = \frac{\gamma+1}{\gamma-1}, \qquad U(0) = \frac{2}{\gamma+1}, \qquad P(0) = \frac{2}{\gamma+1}. \tag{21}$$

We refer the reader to Schlichting et al. (2015) for a full derivation of the self-similar solutions to Equations (17)–(20). Figure 11 shows the atmospheric mass loss fraction as a function of the ground velocity with which the shock is launched into the atmosphere, where the atmospheric mass loss fraction is defined as

$$\chi_{loss} = \exp[-z_{esc}/h] \tag{22}$$

where z_{esc} is the initial height in the atmosphere of the fluid element that has been accelerated to a velocity equal to the escape velocity by the shock, such that the atmosphere at $z \geq z_{esc}$ will be lost.

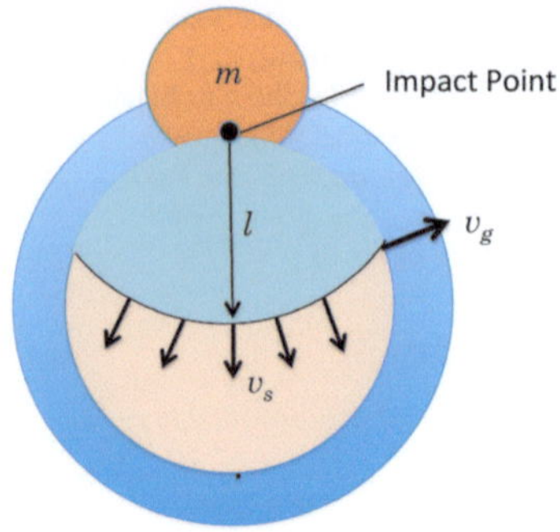

Fig. 12 Illustration of the impact geometry. An impactor of mass, m, and impact velocity, v_{imp}, impacts a target with mass, M, and radius, R. Assuming momentum conservation, we calculate the shocked fluid velocity, v_s, and the component of the ground velocity normal to the surface, v_g, as a function of the distance from the impact point. Figure after Schlichting et al. (2015)

3.2.1 Global Mass Loss

Above we calculated the atmospheric mass loss due to a given ground velocity. In order to calculate the total global atmospheric mass loss, we need to relate the ground velocity to the impactor's mass and impact velocity. We do this by using a simple model following Schlichting et al. (2015). We approximate the impact as a point like explosion on the planet's surface, which results in a self-similar solution of the second type (Zel'dovich and Raizer 1967). As the shock propagates through the planet, its velocity must fall off faster than dictated by energy conservation but slower than required by momentum conservation. This is because as the shock propagates it must lose energy, but its momentum is increased by the nonzero pressure in the target. Catastrophic impact simulations find scaling laws that are close to ones resulting from momentum conservation (Love and Ahrens 1996; Benz and Asphaug 1999; Leinhardt and Stewart 2009). Assuming momentum conservation and a constant density target, the velocity of the shocked fluid traveling through the protoplanet is

$$v_s = v_{Imp}\left(\frac{m}{M}\right)\frac{1}{(l/2R)^3(4-3(l/2R))}, \tag{23}$$

where l is the distance of the shock travelled from the impact point, (see Fig. 12). The ground velocity with which the shock is launched into the atmosphere is due to the component of the shocked fluid velocity that is perpendicular to the planet's surface, such that $v_g = v_s l/(2R)$, which yields

$$v_g = v_{Imp}\left(\frac{m}{M}\right)\frac{1}{(l/2R)^2(4-3(l/2R))}. \tag{24}$$

Using this ground velocity we can now calculate the global atmospheric mass loss by summing the loss from all the location on the planet for the various ground velocities. The resulting atmospheric loss as a function of $(v_{Imp}/v_{esc})(m/M)$ for an isothermal atmosphere is shown in Fig. 13. The total atmospheric loss consists of two components: The first is from the area of the planet's surface where the ground motion is large enough such that locally all the atmosphere is lost (dashed line in Fig. 13), the second component corresponds to the region of the planet where the local ground velocity is small enough such that only part of the atmosphere is lost (thin solid line in Fig. 13) (Schlichting et al. 2015). Independent of the exact value of the adiabatic index, we find that the global atmospheric mass loss fraction is for an isothermal atmosphere well approximated by

$$X_{loss} = 0.4\left(\frac{v_{Imp}m}{v_{esc}M}\right) + 1.4\left(\frac{v_{Imp}m}{v_{esc}M}\right)^2 - 0.8\left(\frac{v_{Imp}m}{v_{esc}M}\right)^3, \tag{25}$$

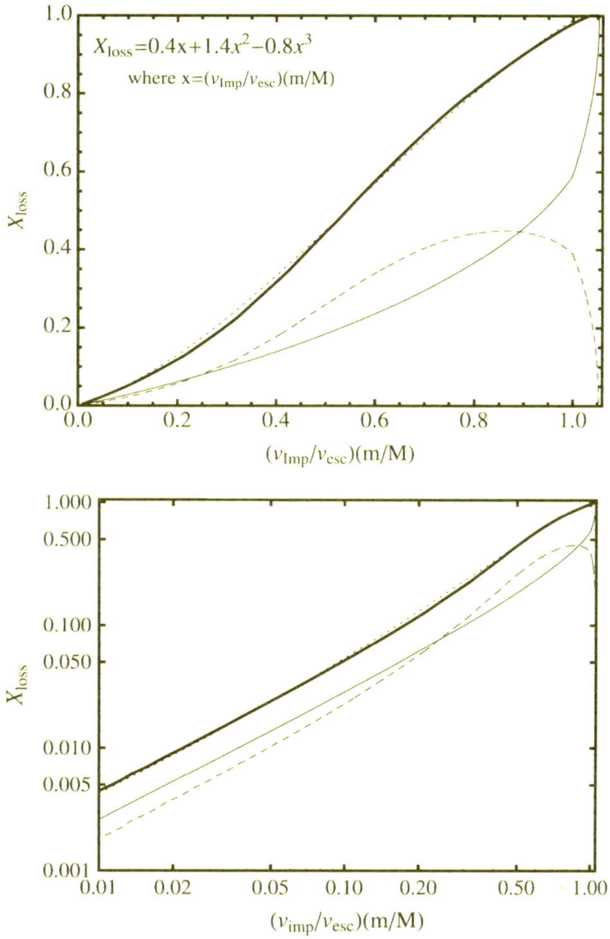

Fig. 13 Global mass loss fraction (*thick solid line*), calculated by taking into account the different ground velocities across the planet's surface. The total atmospheric loss consists of two components: the first is from the area of the planet's surface where the ground motion is large enough such that locally all the atmosphere is lost (*dashed line*) and the second component corresponds to the regions of the planet's surface where the local ground velocity is too small to lead to complete loss but large enough such that a fluid parcel residing higher in the atmosphere is accelerated to velocities exceeding the escape velocity such that the upper parts of the atmosphere are lost (*thin solid line*). The analytic fit given by the equation in the *top panel* is shown as *dotted line*. Figure after Schlichting et al. (2015)

whereas for an adiabatic atmosphere we find

$$X_{loss} = 0.4\left(\frac{v_{Imp}m}{v_{esc}M}\right) + 1.8\left(\frac{v_{Imp}m}{v_{esc}M}\right)^2 - 1.2\left(\frac{v_{Imp}m}{v_{esc}M}\right)^3. \tag{26}$$

In both cases, $X_{loss} \simeq 0.4(m/M)(v_{Imp}/v_{esc})$ in the limit that $(v_{Imp}/v_{esc})(m/M) \ll 1$. We find a mass loss fraction of 6% for an Mars-sized impactor hitting an $0.9M_{\oplus}$ protoplanet with $v_{imp} \sim v_{esc}$, which is about a factor of 2 lower than estimates by Genda and Abe (2003) who assumed a single averaged ground velocity of 4–5 km/s for all locations on the planet. The presence of an ocean can significantly enhance the atmospheric mass loss (Genda and Abe 2005). Lock et al. (2014) find that with a large ocean up to 30% of the atmosphere can be lost in the canonical Moon-forming impact.

3.3 Comparison of Atmospheric Loss Across All Impact Regimes

We now turn to comparing the three different atmospheric mass loss regimes derived above. First we turn to the regime, where the mass loss is due to small impactors that can only eject a fraction of the atmospheric mass above the tangent plane, i.e. $r_{min} < r < r_{cap}$. In this regime the total mass in impactors needed, assuming that all bodies have the same size, to remove the current Earth's atmosphere is (Schlichting et al. 2015)

$$M_T = \frac{M_{atmos} m_{Imp}}{\mathcal{M}_{Eject}} = \frac{2r}{r_{min}} \left(1 - \left(\frac{r_{min}}{r}\right)^2\right)^{-1} M_{atmos}. \tag{27}$$

This is strictly speaking an overestimate of the total impactor mass needed because as the atmospheric mass is decreasing with each impact, each new impactor is able to remove an ever larger fraction of the atmospheric mass above the tangent plane.

Next we turn to the second regime, in which impactors are large enough such they can eject all the atmosphere above the tangent plane, such that $r_{cap} < r < r_{GI}$. In this regime, the total mass in impactors needed to erode the atmosphere is

$$M_T = \frac{M_{atmos} m_{Imp}}{\mathcal{M}_{Eject}} = \frac{4\pi}{3} \rho_{imp} r^3 \frac{2R}{h}. \tag{28}$$

Since impactors with $r_{cap} < r < r_{gi}$ are always limited to eroding all the atmospheric mass above the tangent plane, atmospheric mass loss does not become easier or more efficient over time as the atmospheric mass declines.

Finally we have giant impacts, which define the third regime. Following Schlichting et al. (2015), we define the transition from planetesimal impacts to giant impacts at an impactor radius at which a giant impact ejects more atmosphere than a smaller impactor that erodes all the mass above the tangent plane. This yields giant impacts for bodies with $r > r_{gi} \simeq (2hR^2)^{1/3}$. For the values of the current Earth $r_{gi} \simeq 900$ km. In the giant impact regime, the total mass in impactors needed to eject the atmosphere is

$$M_T = \frac{M_{atmos} m_{Imp}}{\mathcal{M}_{Eject}} \simeq 4M = \text{constant}, \tag{29}$$

where we assumed that $X_{loss} \ll 1$ in a single giant impact (see Schlichting et al. 2015 for details).

Figure 14 summarizes the three distinct impact regimes discussed above. It displays the total mass in impactors needed, i.e. $M_T = M_{atmos}/\mathcal{M}_{Eject}$, to lose the whole atmosphere as a function of impactor radius. Figure 14 impressively shows that, per unit impactor mass, small impactors with $r_{min} < r < r_{cap}$ are the most efficient in eroding the atmosphere. In fact for the current atmospheric mass of the Earth, they are more than five orders of magnitude more efficient (per unit impactor mass), than giant impacts. For Venus like atmospheres, they are still more than three orders of magnitude more efficient than giant impacts. The enormous atmospheric mass loss efficiency of small impactors is due to the fact that most of their impact energy and momentum is directly available for local mass loss, where as in the giant impact regime a lot of energy and momentum is 'wasted' by having to create a strong shock that transverses the entirety of the planet such that global atmospheric loss can be achieved. This explains intuitively why for giant impacts $M_T/M_\oplus \sim 1$ whereas for small planetesimal impacts $M_T/M_\oplus \sim M_{atmos}$.

4 Summary and Conclusions

Geochemical measurements and analytic impact models demonstrate that the formation of the terrestrial planets consisted of a rich interplay between volatile accretion and atmospheric mass loss facilitated by small and large impacts throughout the planet formation

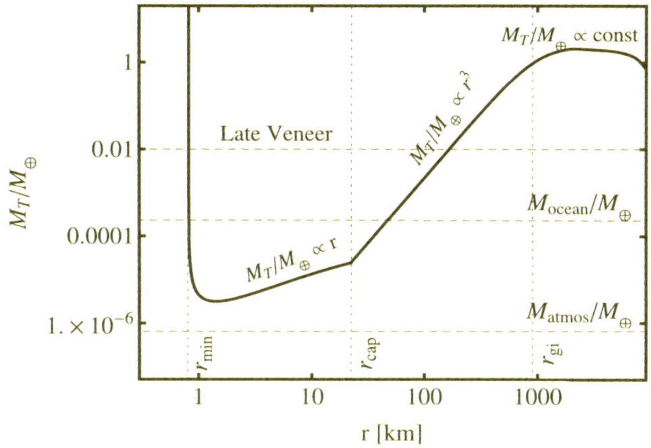

Fig. 14 Total impactor mass, M_T, needed to eject the atmosphere as a function of impactor radius, r. Three distinct ejection regimes are apparent: 1) For small $r_{min} < r < r_{cap}$ (i.e., 1 km $\lesssim r \lesssim$ 25 km), the number of bodies needed scales roughly as r^{-2}. 2) For intermediate impactor sizes (i.e. 25 km $< r <$ 1000 km), N is constant, because each impact ejects the whole atmospheric cap, and to eject the entire atmosphere one needs $N = M_{atoms}/\mathcal{M}_{cap} = (2R/h)$ number of impacts. 3) For larger impactor radii (i.e., $r >$ 1000 km) the impactors are large enough to initiate a shock wave traveling through the entire Earth and launching a shock into the atmosphere globally such that N tends to 1 as r tends to R_{Earth}. In the giant impact regime, $N \sim (R/r)^3$. Impactors with $r < r_{min} \sim$ 1 km are not able to eject any atmosphere. For comparison, the *upper*, *middle*, and *lower dashed lines* correspond to the mass ratio of the late veneer to the Earth's mass, the Earth's oceans to its total mass, and the Earth's atmosphere to its total mass, respectively. Figure after Schlichting et al. (2015)

process. A visual summary of this chapter and the accretion history of the terrestrial planets is given in Fig. 15. The following chronology of terrestrial planet formation and volatile accretion emerges: The solar nebula persisted for a few Myrs (Wang et al. 2017), during which dozens to hundred planetary embryos comparable to the size of Mars formed. Since these bodies formed in large part during the presence of the gas disk, they had the chance to accrete a primordial H/He dominated atmosphere from the gas disk. Much of this primordial atmosphere was lost. Nebular Ne in the deep mantle might be a fingerprint of this early accretion, but most of the mantle does not remember this signature implying that volatile loss occurred during accretion likely due to a combination of collisions, photo-evaporation and hydrodynamic escape. Once the growing planetary embryos contain collectively more mass than the planetesimal population, their mutual gravitational excitation can no longer be efficiently damped by dynamical friction provided by the planetesimal population. As a result, the planetary embryos experience eccentricity growth, orbit crossing and giant impacts. Geochemical data indicate that most of the water could have been delivered prior to the Moon forming impact and that the Moon forming impact did not drive off the ocean. However, present day geochemistry of volatiles shows no evidence of hydrodynamic escape as the isotopic composition of most volatiles are chondritic. The relative abundances of the volatiles are however not in chondritic proportions, suggesting that impacts played a major role in sculpting volatile budgets. Geochemical evidence of atmospheric loss comes from the $^{3}He/^{22}Ne$, halogen ratios (e.g., F/Cl) and low H/N ratios. Global magma oceans create an outgassed atmosphere during the giant impact phase. The outgassed atmospheres are eroded through a combination of large and small impacts, with smaller impacts being more efficient in eroding the atmosphere. Finally, during late accretion material is added to the Earth consisting of a mix of chondritic and icy planetesimals is added to the Earth. These impactors

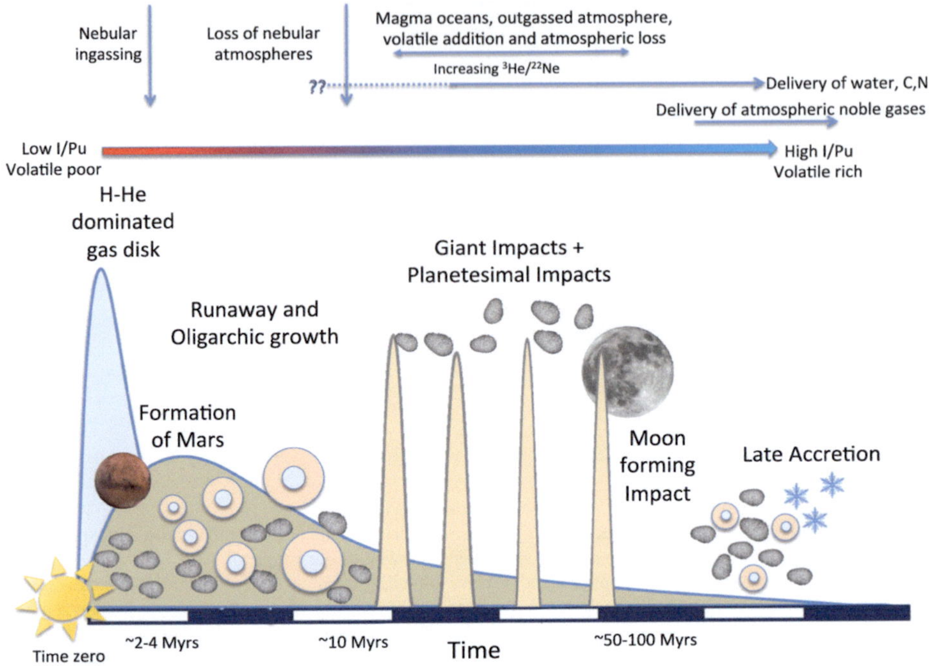

Fig. 15 Chronology of the terrestrial planets' accretion and volatile delivery. The solar system's gas disk (represented by the *blue curve*) dissipated after a few Myrs, which is comparable to the timescale for runaway growth resulting in the formation of Mars and dozens to hundred similar sized or larger planetary embryos (represented as differentiated cores). Planetary embryos continue to grow during oligarchic growth by accreting planetesimals (represented as grey boulders) and merging with comparable sized bodies. Some of these embryos form early enough to accrete a primordial H/He dominated atmosphere from the gas disk, however much of this primordial atmosphere was likely lost due to collisions, photo-evaporation and hydrodynamic escape. Although nebular Ne and possible H in the deep mantle might be a fingerprint of this early accretion, most of the mantle does not remember this signature implying that volatile loss occurred during accretion. Once the growing planetary embryos contain collectively more mass than the planetesimal population, (the planetesimal population's decease in mass is shown by the *brown curve*), the mutual gravitational excitation of the planetary embryos can no longer be efficiently damped by dynamical friction provided by the planetesimal population. As a result, the planetary embryos undergo eccentricity growth, orbit crossing and giant impacts. Global magma oceans create an outgassed atmosphere during the giant impact phase. In addition, volatiles are delivered and atmospheres are eroded by planetesimal impacts and giant impacts. Geochemical ratios indicate that most of the water could have been delivered prior to the Moon forming impact and that the Moon forming impact did not drive off the ocean. Finally, during late accretion material is added to the Earth consisting of a mix of chondritic and icy planetesimals. These late impactors deliver volatiles, lead to outgasing and atmospheric loss. The interplay of these processes determined the atmospheric conditions of the early Earth

could have supplemented the existing volatile inventory. The icy planetesimals may have been particularly important for the delivery of the atmospheric noble gases as these were likely added after the Moon-forming impact. While these late impactors may have delivered volatiles, they could have also lead to outgasing and atmospheric loss. Whether impactors lead to a net loss or gain in atmospheric mass depends on the volatile content of the impactors themselves, how much of their volatiles end up in the atmosphere and the amount of volatiles released from the impact melt pools that are created by the impacts. Combining geochemical observations with impact models suggest an interesting synergy between small and big impacts, where giant impacts create large magma oceans and small impacts dominate the atmospheric loss. In fact, since small planetesimal impacts tend to be much more efficient (per unit impactor mass) in driving atmospheric loss than giant impacts, they likely dominated the atmospheric loss during terrestrial planet formation whereas giant impacts likely dominated the outgassing by creating large magma oceans. The early Earth's atmosphere was therefore likely determined by an interplay between volatile delivery, outgassing and atmospheric erosion.

References

C.B. Agnor, R.M. Canup, H.F. Levison, On the character and consequences of large impacts in the late stage of terrestrial planet formation. Icarus **142**, 219–237 (1999). https://doi.org/10.1006/icar.1999.6201

T.J. Ahrens, Impact erosion of terrestrial planetary atmospheres. Annu. Rev. Earth Planet. Sci. **21**, 525–555 (1993). https://doi.org/10.1146/annurev.ea.21.050193.002521

F. Albarède, Volatile accretion history of the terrestrial planets and dynamic implications. Nature **461**, 1227–1233 (2009). https://doi.org/10.1038/nature08477

C.M.O.. Alexander, R. Bowden, M.L. Fogel, K.T. Howard, C.D.K. Herd, L.R. Nittler, The provenances of asteroids, and their contributions to the volatile inventories of the terrestrial planets. Science **337**, 721 (2012). https://doi.org/10.1126/science.1223474

K. Altwegg, H. Balsiger, A. Bar-Nun, J.J. Berthelier, A. Bieler, P. Bochsler, C. Briois, U. Calmonte, M. Combi, J. De Keyser, P. Eberhardt, B. Fiethe, S. Fuselier, S. Gasc, T.I. Gombosi, K.C. Hansen, M. Hässig, A. Jäckel, E. Kopp, A. Korth, L. LeRoy, U. Mall, B. Marty, O. Mousis, E. Neefs, T. Owen, H. Rème, M. Rubin, T. Sémon, C.-Y. Tzou, H. Waite, P. Wurz, 67P/Churyumov-Gerasimenko, a Jupiter family comet with a high D/H ratio. Science **347**(27), 1261952 (2015). https://doi.org/10.1126/science.1261952

S.M. Andrews, D.J. Wilner, A.M. Hughes, C. Qi, C.P. Dullemond, Protoplanetary disk structures in Ophiuchus. Astrophys. J. **700**, 1502–1523 (2009). https://doi.org/10.1088/0004-637X/700/2/1502

G. Avice, B. Marty, The iodine-plutonium-xenon age of the Moon–Earth system revisited. Philos. Trans. R. Soc. Lond. Ser. A, Math. Phys. Sci. **372**, 20130260 (2014). https://doi.org/10.1098/rsta.2013.0260

C.J. Ballentine, B. Marty, B. Sherwood Lollar, M. Cassidy, Neon isotopes constrain convection and volatile origin in the Earth's mantle. Nature **433**, 33–38 (2005). https://doi.org/10.1038/nature03182

A. Bar-Nun, G. Notesco, T. Owen, Trapping of N_2, CO and Ar in amorphous ice: application to comets. Icarus **190**, 655–659 (2007). https://doi.org/10.1016/j.icarus.2007.03.021

W. Benz, E. Asphaug, Catastrophic disruptions revisited. Icarus **142**, 5–20 (1999). https://doi.org/10.1006/icar.1999.6204

D. Bockelée-Morvan, D. Gautier, D.C. Lis, K. Young, J. Keene, T. Phillips, T. Owen, J. Crovisier, P.F. Goldsmith, E.A. Bergin, D. Despois, A. Wootten, Deuterated water in comet C/1996 B2 (Hyakutake) and its implications for the origin of comets. Icarus **133**, 147–162 (1998). https://doi.org/10.1006/icar.1998.5916

R.A. Brooker, Z. Du, J.D. Blundy, S.P. Kelley, N.L. Allan, B.J. Wood, E.M. Chamorro, J.-A. Wartho, J.A. Purton, The 'zero charge' partitioning behaviour of noble gases during mantle melting. Nature **423**, 738–741 (2003)

A.G.W. Cameron, W. Benz, The origin of the moon and the single impact hypothesis. IV. Icarus **92**, 204–216 (1991). https://doi.org/10.1016/0019-1035(91)90046-V

R.M. Canup, E. Asphaug, Origin of the Moon in a giant impact near the end of the Earth's formation. Nature **412**, 708–712 (2001)

A. Caracausi, G. Avice, P. Bernard, E. Furi, B. Marty, Chondritic xenon in the Earth's mantle. Nature **553**, 82–85 (2016)

J.E. Chambers, Making more terrestrial planets. Icarus **152**, 205–224 (2001). https://doi.org/10.1006/icar.2001.6639

J.E. Chambers, G.W. Wetherill, Making the terrestrial planets: N-body integrations of planetary embryos in three dimensions. Icarus **136**, 304–327 (1998). https://doi.org/10.1006/icar.1998.6007

N. Coltice, M. Moreira, J. Hernlund, S. Labrosse, Crystallization of a basal magma ocean recorded by helium and neon. Earth Planet. Sci. Lett. **308**, 193–199 (2011). https://doi.org/10.1016/j.epsl.2011.05.045
M. Ćuk, S.T. Stewart, Making the Moon from a fast-spinning Earth: a giant impact followed by resonant despinning. Science **338**, 1047 (2012). https://doi.org/10.1126/science.1225542
R. Dasgupta, H. Chi, N. Shimizu, A.S. Buono, D. Walker, Carbon solution and partitioning between metallic and silicate melts in a shallow magma ocean: implications for the origin and distribution of terrestrial carbon. Geochim. Cosmochim. Acta **102**, 191–212 (2013). https://doi.org/10.1016/j.gca.2012.10.011
N. Dauphas, The dual origin of the terrestrial atmosphere. Icarus **165**, 326–339 (2003). https://doi.org/10.1016/S0019-1035(03)00198-2
N. Dauphas, A. Morbidelli, Geochemical and planetary dynamical views on the origin of Earths atmosphere and oceans, in *The Atmosphere—History*, ed. by D.E. Canfield, J. Farquhar, J.F. Kasting. Treatise on Geochemistry, 2nd edn., vol. 13 (2014). https://doi.org/10.1016/B978-0-08-095975-7.01301-2. Chap. 6.1
N. Dauphas, A. Pourmand, Hf-W-Th evidence for rapid growth of Mars and its status as a planetary embryo. Nature **473**, 489–492 (2011). https://doi.org/10.1038/nature10077
L.T. Elkins-Tanton, Linked magma ocean solidification and atmospheric growth for Earth and Mars. Earth Planet. Sci. Lett. **271**, 181–191 (2008). https://doi.org/10.1016/j.epsl.2008.03.062
N.V. Erkaev, H. Lammer, L.T. Elkins-Tanton, A. Stökl, P. Odert, E. Marcq, E.A. Dorfi, K.G. Kislyakova, Y.N. Kulikov, M. Leitzinger, M. Güdel, Escape of the martian protoatmosphere and initial water inventory. Planet. Space Sci. **98**, 106–119 (2014). https://doi.org/10.1016/j.pss.2013.09.008
E. Füri, D.R. Hilton, S.A. Halldórsson, P.H. Barry, D. Hahm, T.P. Fischer, K. Grönvold, Apparent decoupling of the He and Ne isotope systematics of the Icelandic mantle: the role of He depletion, melt mixing, degassing fractionation and air interaction. Geochim. Cosmochim. Acta **74**, 3307–3332 (2010). https://doi.org/10.1016/j.gca.2010.03.023
H. Genda, Y. Abe, Survival of a proto-atmosphere through the stage of giant impacts: the mechanical aspects. Icarus **164**, 149–162 (2003). https://doi.org/10.1016/S0019-1035(03)00101-5
H. Genda, Y. Abe, Enhanced atmospheric loss on protoplanets at the giant impact phase in the presence of oceans. Nature **433**, 842–844 (2005). https://doi.org/10.1038/nature03360
P. Goldreich, Y. Lithwick, R. Sari, Final stages of planet formation. Astrophys. J. **614**, 497–507 (2004). https://doi.org/10.1086/423612
R. Grover, J.W. Hardy, The propagation of shocks in exponentially decreasing atmospheres. Astrophys. J. **143**, 48 (1966). https://doi.org/10.1086/148476
A.N. Halliday, The origins of volatiles in the terrestrial planets. Geochim. Cosmochim. Acta **105**, 146–171 (2013). https://doi.org/10.1016/j.gca.2012.11.015
L.J. Hallis, G.R. Huss, K. Nagashima, G.J. Taylor, S.A. Halldórsson, D.R. Hilton, M.J. Mottl, K.J. Meech, Evidence for primordial water in Earth's deep mantle. Science **350**, 795–797 (2015). https://doi.org/10.1126/science.aac4834
B.M.S. Hansen, N. Murray, Migration then assembly: formation of Neptune-mass planets inside 1 AU. Astrophys. J. **751**, 158 (2012). https://doi.org/10.1088/0004-637X/751/2/158
C.L. Harper Jr., S.B. Jacobsen, Noble gases and Earth's accretion. Science **273**, 1814–1818 (1996). https://doi.org/10.1126/science.273.5283.1814
C. Hayashi, Structure of the solar nebula, growth and decay of magnetic fields and effects of magnetic and turbulent viscosities on the nebula. Prog. Theor. Phys. Suppl. **70**, 35–53 (1981)
V.S. Heber, R.A. Brooker, S.P. Kelley, B.J. Wood, Crystal melt partitioning of noble gases (helium, neon, argon, krypton, and xenon) for olivine and clinopyroxene. Geochim. Cosmochim. Acta **71**, 1041–1061 (2007). https://doi.org/10.1016/j.gca.2006.11.010
V.S. Heber, H. Baur, P. Bochsler, K.D. McKeegan, M. Neugebauer, D.B. Reisenfeld, R. Wieler, R.C. Wiens, Isotopic mass fractionation of solar wind: evidence from fast and slow solar wind collected by the genesis mission. Astrophys. J. **759**, 121 (2012). https://doi.org/10.1088/0004-637X/759/2/121
L.A. Hillenbrand, Disk-dispersal and planet-formation timescales. Phys. Scr. T **130**(1), 014024 (2008). https://doi.org/10.1088/0031-8949/2008/T130/014024
M.M. Hirschmann, Constraints on the early delivery and fractionation of Earth's major volatiles from C/H, C/N and C/S ratios. Am. Mineral. **101**, 540–553 (2016)
M.M. Hirschmann, R. Dasgupta, The H/C ratios of Earth's near-surface and deep reservoirs, and consequences for deep Earth volatile cycles. Chem. Geol. **262**, 4–16 (2009)
G. Holland, C.J. Ballentine, Seawater subduction controls the heavy noble gas composition of the mantle. Nature **441**, 186–191 (2006). https://doi.org/10.1038/nature04761
G. Holland, M. Cassidy, C.J. Ballentine, Meteorite Kr in Earth's mantle suggests a late accretionary source for the atmosphere. Science **326**, 1522 (2009). https://doi.org/10.1126/science.1179518
M. Honda, I. McDougall, Primordial helium and neon in the Earth: a speculation on early degassing. Geophys. Res. Lett. **25**, 1951–1954 (1998). https://doi.org/10.1029/98GL01329

J. Hopp, M. Trieloff, Helium deficit in high-3He/4He parent magmas: predegassing fractionation, not a "helium paradox". Geochem. Geophys. Geosyst. **9**, 03009 (2008)
S. Ida, J. Makino, Scattering of planetesimals by a protoplanet—slowing down of runaway growth. Icarus **106**, 210 (1993). https://doi.org/10.1006/icar.1993.1167
N.K. Inamdar, H.E. Schlichting, The formation of super-Earths and mini-Neptunes with giant impacts. Mon. Not. R. Astron. Soc. **448**, 1751–1760 (2015). https://doi.org/10.1093/mnras/stv030
N.K. Inamdar, H.E. Schlichting, Stealing the gas: giant impacts and the large diversity in exoplanet densities. Astrophys. J. Lett. **817**, 13 (2016). https://doi.org/10.3847/2041-8205/817/2/L13
C.R.M. Jackson, S.W. Parman, S.P. Kelley, R.F. Cooper, Constraints on light noble gas partitioning at the conditions of spinel-peridotite melting. Earth Planet. Sci. Lett. **384**, 178–187 (2013). https://doi.org/10.1016/j.epsl.2013.09.046
E. Jehin, J. Manfroid, D. Hutsemékers, C. Arpigny, J.-M. Zucconi, Isotopic ratios in comets: status and perspectives. Earth Moon Planets **105**, 167–180 (2009). https://doi.org/10.1007/s11038-009-9322-y
S.J. Kenyon, B.C. Bromley, Terrestrial planet formation. I. The transition from oligarchic growth to chaotic growth. Astron. J. **131**, 1837–1850 (2006). https://doi.org/10.1086/499807
M.D. Kurz, J. Curtice, D. Fornari, D. Geist, M. Moreira, Primitive neon from the center of the Galápagos hotspot. Earth Planet. Sci. Lett. **286**, 23–34 (2009). https://doi.org/10.1016/j.epsl.2009.06.008
S. Labrosse, J.W. Hernlund, N. Coltice, A crystallizing dense magma ocean at the base of the Earth's mantle. Nature **450**, 866–869 (2007). https://doi.org/10.1038/nature06355
T. Lebrun, H. Massol, E. ChassefièRe, A. Davaille, E. Marcq, P. Sarda, F. Leblanc, G. Brandeis, Thermal evolution of an early magma ocean in interaction with the atmosphere. J. Geophys. Res., Planets **118**, 1155–1176 (2013). https://doi.org/10.1002/jgre.20068
Z.M. Leinhardt, S.T. Stewart, Full numerical simulations of catastrophic small body collisions. Icarus **199**, 542–559 (2009). https://doi.org/10.1016/j.icarus.2008.09.013
S.-F. Liu, Y. Hori, D.N.C. Lin, E. Asphaug, Giant impact: an efficient mechanism for the devolatilization of super-Earths. Astrophys. J. **812**, 164 (2015). https://doi.org/10.1088/0004-637X/812/2/164
S.J. Lock, S.T. Stewart, S. Mukhopadhyay, Was the atmosphere lost during the Moon-forming giant impact? in *Lunar and Planetary Science Conference*. Lunar and Planetary Inst. Technical Report, vol. 45 (2014), p. 2843
K. Lodders, B. Fegley, *The Planetary Scientist's Companion* (1998)
S.G. Love, T.J. Ahrens, Catastrophic impacts on gravity dominated asteroids. Icarus **124**, 141–155 (1996). https://doi.org/10.1006/icar.1996.0195
J. Manfroid, E. Jehin, D. Hutsemékers, A. Cochran, J.-M. Zucconi, C. Arpigny, R. Schulz, J.A. Stüwe, I. Ilyin, The CN isotopic ratios in comets. Astron. Astrophys. **503**, 613–624 (2009). https://doi.org/10.1051/0004-6361/200911859
B. Marty, The origins and concentrations of water, carbon, nitrogen and noble gases on Earth. Earth Planet. Sci. Lett. **313**, 56–66 (2012). https://doi.org/10.1016/j.epsl.2011.10.040
B. Marty, G. Avice, Y. Sano, K. Altwegg, H. Balsiger, M. Hässig, A. Morbidelli, O. Mousis, M. Rubin, Origins of volatile elements (H, C, N, noble gases) on Earth and Mars in light of recent results from the ROSETTA cometary mission. Earth Planet. Sci. Lett. **441**, 91–102 (2016). https://doi.org/10.1016/j.epsl.2016.02.031
E. Mazor, D. Heymann, E. Anders, Noble gases in carbonaceous chondrites. Geochim. Cosmochim. Acta **34**, 781–824 (1970). https://doi.org/10.1016/0016-7037(70)90031-1
F.M. McCubbin, A. Steele, E.H. Hauri, H. Nekvasil, S. Yamashita, R.J. Hemley, Nominally hydrous magmatism on the Moon. Proc. Natl. Acad. Sci. **107**, 11223–11228 (2010). https://doi.org/10.1073/pnas.1006677107
W.F. McDonough, Compositional model for the Earth's core, in *Treatise on Geochemistry*, vol. 2 (2003), p. 568. https://doi.org/10.1016/B0-08-043751-6/02015-6
K.J. Meech, J. Pittichová, A. Bar-Nun, G. Notesco, D. Laufer, O.R. Hainaut, S.C. Lowry, D.K. Yeomans, M. Pitts, Activity of comets at large heliocentric distances pre-perihelion. Icarus **201**, 719–739 (2009). https://doi.org/10.1016/j.icarus.2008.12.045
R. Meier, T.C. Owen, D.C. Jewitt, H.E. Matthews, M. Senay, N. Biver, D. Bockelee-Morvan, J. Crovisier, D. Gautier, Deuterium in comet C/1995 O1 (Hale-Bopp): detection of DCN. Science **279**, 1707 (1998). https://doi.org/10.1126/science.279.5357.1707
H.J. Melosh, A.M. Vickery, Impact erosion of the primordial atmosphere of Mars. Nature **338**, 487–489 (1989). https://doi.org/10.1038/338487a0
H. Mizuno, K. Nakazawa, C. Hayashi, Dissolution of the primordial rare gases into the molten Earth's material. Earth Planet. Sci. Lett. **50**, 202–210 (1980). https://doi.org/10.1016/0012-821X(80)90131-4
M. Moreira, Noble gas constraints on the origin and evolution of Earth's volatiles. Geochem. Perspect. **2**, 229–403 (2013)

M. Moreira, S. Charnoz, The origin of the neon isotopes in chondrites and on Earth. Earth Planet. Sci. Lett. **433**, 249–256 (2016). https://doi.org/10.1016/j.epsl.2015.11.002

M. Moreira, J. Kunz, C. Allegre, Rare gas systematics in popping rock: isotopic and elemental compositions in the upper mantle. Science **279**, 1178 (1998). https://doi.org/10.1126/science.279.5354.1178

M. Moreira, J. Blusztajn, J. Curtice, S. Hart, H. Dick, M.D. Kurz, He and Ne isotopes in oceanic crust: implications for noble gas recycling in the mantle. Earth Planet. Sci. Lett. **216**, 635–643 (2003). https://doi.org/10.1016/S0012-821X(03)00554-5

J.L. Mosenfelder, P.D. Asimow, D.J. Frost, D.C. Rubie, T.J. Ahrens, The MgSiO3 system at high pressure: thermodynamic properties of perovskite, postperovskite, and melt from global inversion of shock and static compression data. J. Geophys. Res. **14**, 01203 (2009)

S. Mukhopadhyay, Early differentiation and volatile accretion recorded in deep mantle neon and xenon. Nature **486**, 101–104 (2012)

M. Nakajima, D.J. Stevenson, Melting and mixing states of the Earth's mantle after the Moon-forming impact. Earth Planet. Sci. Lett. **427**, 286–295 (2015). https://doi.org/10.1016/j.epsl.2015.06.023

G. Neukum, B.A. Ivanov, W.K. Hartmann, Cratering records in the inner solar system in relation to the lunar reference system. Space Sci. Rev. **96**, 55–86 (2001)

G. Notesco, A. Bar-Nun, T. Owen, Gas trapping in water ice at very low deposition rates and implications for comets. Icarus **162**, 183–189 (2003). https://doi.org/10.1016/S0019-1035(02)00059-3

M. Ozima, F.A. Podosek, *Noble Gas Geochemistry* (2002), p. 286

K. Pahlevan, D.J. Stevenson, Equilibration in the aftermath of the lunar-forming giant impact. Earth Planet. Sci. Lett. **262**, 438–449 (2007). https://doi.org/10.1016/j.epsl.2007.07.055

H. Palme, H.S.C. O'Neill, Cosmochemical estimates of mantle composition, in *Treatise on Geochemistry, vol. 2*, ed. by R.W. Carlson, H.D. Holland, K.K. Turekian (Elsevier, New York, 2014), pp. 1–38. https://doi.org/10.1016/B978-0-08-095975-7.00201-1

R. Parai, S. Mukhopadhyay, The evolution of MORB and plume mantle volatile budgets: constraints from fission Xe isotopes in Southwest Indian Ridge basalts. Geochem. Geophys. Geosyst. **16**, 719–735 (2015)

R.O. Pepin, On the origin and early evolution of terrestrial planet atmospheres and meteoritic volatiles. Icarus **92**, 2–79 (1991). https://doi.org/10.1016/0019-1035(91)90036-S

R.O. Pepin, D. Porcelli, Origin of noble gases in the terrestrial planets. Rev. Mineral. Geochem. **47**, 191–246 (2002)

R.O. Pepin, D. Porcelli, Xenon isotope systematics, giant impacts, and mantle degassing on the early Earth. Earth Planet. Sci. Lett. **250**, 470–485 (2006). https://doi.org/10.1016/j.epsl.2006.08.014

S. Péron, M. Moreira, A. Colin, L. Arbaret, B. Putlitz, M.D. Kurz, Neon isotopic composition of the mantle constrained by single vesicle analyses. Earth Planet. Sci. Lett. **449**, 145–154 (2016). https://doi.org/10.1016/j.epsl.2016.05.052

M.K. Pető, S. Mukhopadhyay, K.A. Kelley, Heterogeneities from the first 100 million years recorded in deep mantle noble gases from the Northern Lau back-arc basin. Earth Planet. Sci. Lett. **369**, 13–23 (2013). https://doi.org/10.1016/j.epsl.2013.02.012

D. Porcelli, D. Woolum, P. Cassen, Deep Earth rare gases: initial inventories, capture from the solar nebula, and losses during Moon formation. Earth Planet. Sci. Lett. **193**, 237–251 (2001). https://doi.org/10.1016/S0012-821X(01)00493-9

M. Pujol, B. Marty, R. Burgess, Chondritic-like xenon trapped in Archean rocks: a possible signature of the ancient atmosphere. Earth Planet. Sci. Lett. **308**, 298–306 (2011). https://doi.org/10.1016/j.epsl.2011.05.053

Y.P. Raizer, Propagation of a shock wave in an inhomogeneous atmosphere in the direction of decreeing density. Zh. Prikl. Mekham. i Tekhn. Fiz. **4**, 49–56 (1964)

A. Raquin, M. Moreira, Atmospheric ^{38}Ar/ ^{36}Ar in the mantle: implications for the nature of the terrestrial parent bodies. Earth Planet. Sci. Lett. **287**, 551–558 (2009). https://doi.org/10.1016/j.epsl.2009.09.003

S.N. Raymond, R. Barnes, A.M. Mandell, Observable consequences of planet formation models in systems with close-in terrestrial planets. Mon. Not. R. Astron. Soc. **384**, 663–674 (2008). https://doi.org/10.1111/j.1365-2966.2007.12712.x

P. Sarda, M. Moreira, T. Staudacher, Rare gas systematics on the southernmost Mid-Atlantic ridge: constraints on the lower mantle and the Dupal source. J. Geophys. Res. **105**, 5973–5996 (2000)

H.E. Schlichting, Formation of close in super-Earths and mini-Neptunes: required disk masses and their implications. Astrophys. J. Lett. **795**, 15 (2014). https://doi.org/10.1088/2041-8205/795/1/L15

H.E. Schlichting, R. Sari, A. Yalinewich, Atmospheric mass loss during planet formation: the importance of planetesimal impacts. Icarus **247**, 81–94 (2015). https://doi.org/10.1016/j.icarus.2014.09.053

H.E. Schlichting, P.H. Warren, Q.-Z. Yin, The last stages of terrestrial planet formation: dynamical friction and the late Veneer. Astrophys. J. **752**, 8 (2012). https://doi.org/10.1088/0004-637X/752/1/8

M. Schönbächler, R.W. Carlson, M.F. Horan, T.D. Mock, E.H. Hauri, Heterogeneous accretion and the moderately volatile element budget of Earth. Science **328**, 884 (2010). https://doi.org/10.1126/science.1186239

Z.D. Sharp, D.S. Draper, The chlorine abundance of Earth: implications for a habitable planet. Earth Planet. Sci. Lett. **369**, 71–77 (2013). https://doi.org/10.1016/j.epsl.2013.03.005

V. Shuvalov, Atmospheric erosion induced by oblique impacts. Meteorit. Planet. Sci. **44**, 1095–1105 (2009). https://doi.org/10.1111/j.1945-5100.2009.tb01209.x

V.S. Solomatov, Fluid dynamics of a terrestrial magma ocean, in *Origin of the Earth and Moon*, ed. by R. Canup, K. Righter (University of Arizona Press, Tucson, 2000), pp. 323–338

T. Staudacher, C.J. Allègre, Recycling of oceanic crust and sediments: the noble gas subduction barrier. Earth Planet. Sci. Lett. **89**, 173–183 (1988). https://doi.org/10.1016/0012-821X(88)90170-7

S.T. Stewart, S.J. Lock, S. Mukhopadhyay, Atmospheric loss and volatile fractionation during giant impacts, in *Lunar and Planetary Science Conference*. Lunar and Planetary Science Conference, vol. 45 (2014), p. 2869

H. Tang, N. Dauphas, ^{60}Fe–^{60}Ni chronology of core formation in Mars. Earth Planet. Sci. Lett. **390**, 264–274 (2014). https://doi.org/10.1016/j.epsl.2014.01.005

C.W. Thomas, Q. Liu, C.B. Agee, P.D. Asimow, R.A. Lange, Multi-technique equation of state for Fe2SiO4 melt and the density of Fe-bearing silicate melts from 0–161 GPa. J. Geophys. Res. **117**, 10206 (2012)

M. Trieloff, J. Kunz, D.A. Clague, D. Harrison, C.J. Allègre, The nature of pristine noble gases in mantle plumes Science **288**, 1036–1039 (2000)

M. Trieloff, J. Kunz, Isotope systematics of noble gases in the Earth's mantle: possible sources of primordial isotopes and implications for mantle structure. Phys. Earth Planet. Inter. **148**, 13–38 (2005)

J.M. Tucker, S. Mukhopadhyay, Evidence for multiple magma ocean outgassing and atmospheric loss episodes from mantle noble gases. ArXiv e-prints (2014)

A.M. Vickery, H.J. Melosh, Atmospheric erosion and impactor retention in large impacts, with application to mass extinctions, in *Geological Society of America*, vol. 247 (1990), pp. 289–300

R.J. Walker, Highly siderophile elements in the Earth, Moon and Mars: update and implications for planetary accretion and differentiation. Chem. Erde **69**, 101–125 (2009). https://doi.org/10.1016/j.chemer.2008.10.001

R.J. Walker, M.F. Horan, C.K. Shearer, J.J. Papike, Low abundances of highly siderophile elements in the lunar mantle: evidence for prolonged late accretion. Earth Planet. Sci. Lett. **224**, 399–413 (2004). https://doi.org/10.1016/j.epsl.2004.05.036

H. Wang, B.P. Weiss, X.-N. Bai, B.G. Downey, J. Wang, J. Wang, C. Suavet, R. Fu, M.E. Zucolotto, Lifetime of the solar nebula constrained by meteorite paleomagnetism. Science **355**(6325), 623–627 (2017). https://doi.org/10.1126/science.aaf5043

P.H. Warren, G.W. Kallemeyn, F.T. Kyte, Origin of planetary cores: evidence from highly siderophile elements in Martian meteorites. Geochim. Cosmochim. Acta **63**, 2105–2122 (1999). https://doi.org/10.1016/S0016-7037(99)00156-8

S.J. Weidenschilling, D. Spaute, D.R. Davis, F. Marzari, K. Ohtsuki, Accretional evolution of a planetesimal swarm. Icarus **128**, 429–455 (1997). https://doi.org/10.1006/icar.1997.5747

Q. Yin, S.B. Jacobsen, K. Yamashita, J. Blichert-Toft, P. Télouk, F. Albarède, A short timescale for terrestrial planet formation from Hf-W chronometry of meteorites. Nature **418**, 949–952 (2002). https://doi.org/10.1038/nature00995

R. Yokochi, B. Marty, A determination of the neon isotopic composition of the deep mantle. Earth Planet. Sci. Lett. **225**, 77–88 (2004). https://doi.org/10.1016/j.epsl.2004.06.010

K. Zahnle, N. Arndt, C. Cockell, A. Halliday, E. Nisbet, F. Selsis, N.H. Sleep, Emergence of a habitable planet. Space Sci. Rev. **129**, 35–78 (2007). https://doi.org/10.1007/s11214-007-9225-z

Y.B. Zel'dovich, Y.P. Raizer, *Physics of Shock Waves and High-Temperature Hydrodynamic Phenomena* (1967)

Space Sci Rev (2017) 212:877–898
DOI 10.1007/s11214-017-0413-1

Water in Extrasolar Planets and Implications for Habitability

Lena Noack[1,2] · Ignas Snellen[3] · Heike Rauer[4,5]

Received: 31 July 2016 / Accepted: 22 August 2017 / Published online: 5 September 2017

Abstract Exoplanet detection missions have found thousands of planets or planet candidates outside of the Solar System—some of which are in the habitable zone, where liquid water is possible at the surface. We give an overview of the recent progress in observations of water-rich exoplanets, detection of water in the atmosphere of gas giants and less-massive targets, and modelling of the interior and evolution of water layers in exoplanets. We summarise the possible habitability of water-rich planets and discuss the potential of future missions and telescopes towards the detection of water in the atmosphere of low-mass exoplanets or on their surface.

Keywords Exoplanets · Water · Detection · Atmosphere spectroscopy · Habitability

1 Introduction

Water is abundant not only in the Solar System, but in the universe in general. The search for water on exoplanets is mainly driven by its implications for the potential habitability of these

The Delivery of Water to Protoplanets, Planets and Satellites
Edited by Michel Blanc, Alessandro Morbidelli, Yann Alibert, Lindy Elkins-Tanton, Paul Estrada, Keiko Hamano, Helmut Lammer, Sean Raymond and Maria Schönbächler

✉ L. Noack
lena.noack@fu-berlin.de

I. Snellen
snellen@strw.leidenuniv.nl

H. Rauer
heike.rauer@dlr.de

1 Royal Observatory of Belgium, Brussels, Belgium

2 Present address: Free University Berlin, Berlin, Germany

3 University of Leiden, Leiden, The Netherlands

4 Institute of Planetary Research, German Aerospace Center, Berlin, Germany

5 Center for Astronomy and Astrophysics, Technical University of Berlin, Berlin, Germany

planets, since life as we know it from Earth needs access to liquid water. This constraint led to the definition of the habitable zone (e.g. Hart 1979; Walker et al. 1981; Kasting et al. 1993; Kopparapu et al. 2013), which describes the range of distances from a star where liquid water can exist at the surface of a planet depending on its atmosphere (greenhouse gases like CO_2, H_2O or He, e.g. Seager 2013).

When searching for a possible habitable exoplanet, the focus often lies on planets with a density indicating a rocky bulk composition. Only few low-mass planets have been detected so far, for which the mass and radius could be measured, and such observations are currently only possible for close-in planets. Detected planets even include small objects in the habitable zone of solar-like stars (e.g. Kepler 425b with 1.6 Earth radii, Jenkins et al. 2015), even though their mean densities are not yet known.

The past measurements show that exoplanets are much more diverse than previously expected, including super-Earths (planets with masses up to 10 Earth masses, Valencia et al. 2007), hot Jupiters (Mayor and Queloz 1995; Marcy and Butler 1996) and mini-Neptunes (Charbonneau et al. 2009; Léger et al. 2009). The latter contain large amounts of water and may migrate inward, possibly into the habitable zone (Kuchner 2003; Léger et al. 2004). These ocean planets have added to the speculation of habitable worlds. Several studies have addressed the possible interior structures and evolution of these low-mass planets in the past years (Sotin et al. 2007; Grasset et al. 2009; Alibert 2014; Noack et al. 2016).

Planets with a small density do not necessarily indicate large amounts of water, but may also contain a thick and broad volatile envelope similar to gas giants. Observations of exoplanets (Rogers 2015) show that there is a threshold between terrestrial planets and planets with an extended atmosphere at a limit of about 1.6 Earth radii (or 6 Earth masses, Marcy et al. 2014).

Direct measurements of water vapour in exoplanet atmospheres have already been obtained for about a dozen objects, mainly hot Jupiters. Such observations are expected to be expanded to planets of significantly lower mass and temperature with upcoming space and ground-based observatories such as the James Webb Space Telescope (JWST) and the next generation of Extremely Large Telescopes.

In this paper we give a summary of recent advances in detection of exoplanets, as well as observation and modelling of water in exoplanets, and discuss possible implications on the habitability of these planets.

2 Past Detection of Exoplanets and Their Bulk Compositions

Up to today, most extrasolar planets are detected by indirect methods. The so far most successful detection methods, the radial velocity (RV) technique and the transit method, fortunately can provide us with the bulk planet parameters orbit, radius, mass, and mean density. Masses can also be derived from astrometric techniques, which will gain significant impact in particular for gas planets when results from ESA's GAIA mission become available. Direct imaging methods currently assess (young) gas planets and can provide good targets for atmosphere spectroscopy, but concerning bulk parameters (radii, masses) need model assumptions. Here, we focus on RV and transit detections because they provide the largest statistics for planet characterisation available today.

2.1 Techniques to Measure Planet Densities

Determining planet mean densities requires accurate measures of the planetary radius and mass. The only means we have to directly measure exoplanetary radii is via the transit method. Planetary masses are determined via RV data or by transit timing variations (TTVs) of transiting planets.

The RV technique measures the wobble of a star induced by an orbiting planet around their common centre of mass. The radial velocity component of the stellar motion is measured by high-resolution spectrographs from the induced spectral line shifts via the Doppler-effect. From the periodicity, amplitude, and shape of the signal, the planetary orbit parameters as well as a lower limit of the planetary mass, $m \sin i$ (i: inclination of the orbital plane), can be derived.

The transit method measures the dimming of the stellar flux when an orbiting planet passes through the line-of-sight towards the star. The dimming is proportional to the projected disk of the planet divided by the disk of the star ($k = r_p^2/r_s^2$). The periodicity allows deriving the planetary orbit parameters via Kepler's third law. Spectroscopic follow-up observations, e.g., during transit, of the secondary eclipse, and of reflected or emitted light as the planet revolves around its host star, provide access to further planet properties, like atmosphere composition, effective temperature, and albedo (see Sect. 4).

The two techniques, RV and transit method, complement each other. Planets which can be observed by both methods allow us to remove the $\sin i$ uncertainty in the mass determination and hence provide the true planetary mass, the radius and therefore mean planetary density.

The transit method provides an additional tool to derive planetary masses. Additional planets in a planetary system can induce Transit Timing Variations (TTVs), that means shifts in the occurrence time of an observed transit event. If several transiting planets in a system show TTVs, their masses can be derived with high accuracy. If only one transiting planet is observed and others are indicated only indirectly via their TTVs induced on the observed planet, their masses can still be constraint but with a larger error bar. The derivation of planetary masses for non-transiting planets via TTV is a tricky issue because their orbital parameters (e.g. eccentricity) are often only poorly constraint.

A pre-requisite to the derivation of planet parameters is the good knowledge of the host star. The RV method measures $m \sin i$ of the planet relative to the mass of the host star. The transit signal depends on the ratio of planet to host star radius. Highly accurate determinations of host star parameters are therefore curial for our understanding of the planet properties. Often it is the uncertainty of stellar parameters which limits our knowledge of the orbiting planet. Future instruments to investigate exoplanet parameters must therefore be combined with techniques to derive accurate stellar parameters (see Sect. 6).

2.2 Planet Density as Constraint to Planet Interior

Today about 2000 exoplanets are known (exoplanet.eu). NASA's Kepler mission has published in addition about 7000 planet candidates which are very useful for statistical studies. However, out of the known planets only about 200 are small/low-mass planets with masses $< 30\ M_{earth}$ (excluding planet candidates and planets with only mass upper/lower limits), and only about 60 out of these have a measured mean density (exoplanet.eu). The mean uncertainty in their density is about 40% (where the uncertainty comes from the uncertainty in mass of about 20% and radius of about 5% to the power of three). Only few small planets have measured densities with uncertainty <10%. The number of detected and characterised planets continuously grows, but today's snapshot illustrates the difficulty we have to derive accurate planetary densities with current instrumentation.

Figure 1 displays the density-mass relationship for exoplanets (red and orange dots) and compares it to the Solar System planets and large moons (blue dots). The figure also shows model results for planets with different compositions (after Wagner et al. 2012) to guide our

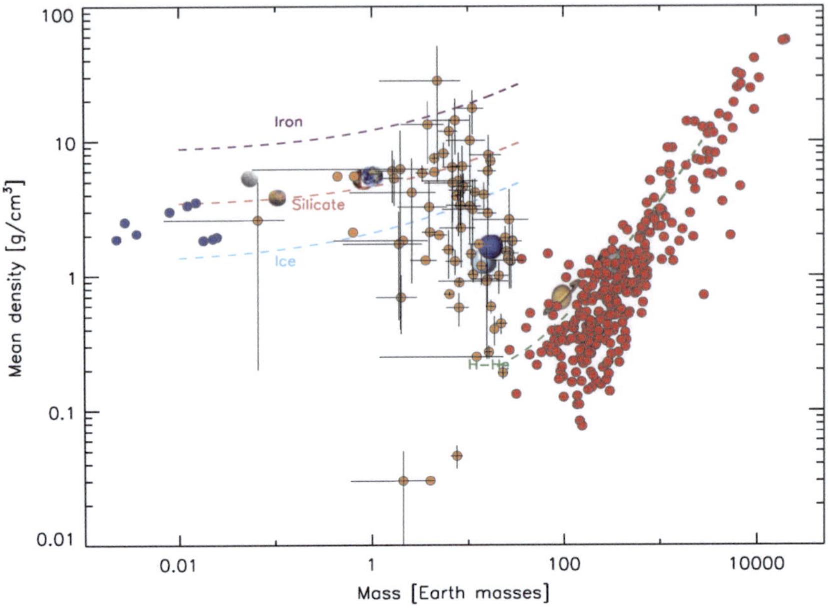

Fig. 1 Mean planet density versus mass. *Dashed lines* show density-mass relationships for planets with certain composition (*green*: Jupiter-like, *violet*: iron, *pink*: silicate, *light blue*: water ice). The Solar System planets are indicated by their images and the large Solar System moons by *blue dots*. *Red dots*: exoplanet gas giants (no error bars given for clarity), *orange dots*: low-mass exoplanets

understanding of the planet distribution seen. Planets with Jupiter-like composition fall onto the green-dashed line. This branch of the diagram contains the gas giant planets with composition dominated by hydrogen, similar to Jupiter and Saturn in our Solar System. Their mean density grows with increasing mass. Gas giants falling below the green-dashed line are larger than expected by the standard model and often called "inflated" planets. Planets above the green-dashed line require significantly more heavy elements than Jupiter and may have large rocky cores (e.g. Cabrera et al. 2010; Deleuil et al. 2012). At $\sim$0.3 $M_{jupiter}$ we find a minimum in the mean density versus mass distribution, separating the low-mass planets with significant rocky/icy composition from the gas giants (Hatzes and Rauer 2015; Laughlin and Lissauer 2015). Planet synthesis models (e.g. Mordasini et al. 2012a,b) show that we expect such planet to be numerous.

Low-mass planets in a density range which falls below the light-blue line (ice composition) have significantly extended atmospheres, resulting in large planetary radii and low mean densities. Examples in our Solar System are Neptune and Uranus. We point out that in contrast to the ice planets in the Solar System, up to today basically all small exoplanets with known mean density orbit close to their stars and belong to the class of hot planets. Some of them may have significantly extended atmospheres which make the planet appear very large, in particular in the presence of high-altitude clouds (Lammer et al. 2016). To disentangle from radius and mass observations alone the size of a rocky core from an extended atmosphere is difficult. We take the planet GJ1214b to illustrate the problem. One of the currently best studied exoplanets is GJ1214b (Charbonneau et al. 2009). It is a planet with 6.26 ± 0.91 M_{earth} and 2.8 ± 0.24 R_{earth} (Anglada-Escudé et al. 2013; Charbonneau et al. 2009), hence belongs to the planets often called "super-Earths". Its mean density is 1.56 ± 0.61. GJ1214b orbits a nearby (13 pc) cool M dwarf star (M4.5) and was subject to intense observations trying to determine the atmospheric composition of the planet. However, no atmosphere absorptions could be detected which was finally attributed to clouds in the upper atmosphere, inhibiting the detection of atmospheric features (Kreidberg et al. 2014b). GJ1214b was also subject to studies to derive its composition and internal structure. In principle three scenarios were studied. All scenarios assume a planet with an iron core and a silicate mantle layer, but they differ in the assumed atmosphere composition. Atmospheres could be dominated by hydrogen on top of the silicate mantle, or on top of a water ice layer. A third atmosphere scenario assumes a water steam atmosphere, resulting from e.g. evaporation of a thick ice layer when the planet migrated towards the star during its early evolution. It was shown that all three scenarios are able to meet the measured planetary radius and mass (Rogers and Seager 2010), but require different amounts of hydrogen and water. This was confirmed in similar studies, showing that it needs $<10\%$ hydrogen above a rocky core to explain the measured radius and mass, or alternatively large amounts of water in the atmosphere (Rogers and Seager 2010; Nettelmann et al. 2011; Valencia et al. 2013). GJ1214b illustrates that from radius and mass measurements alone it is not possible to disentangle this ambiguity. Detection of atmospheric compositions in planets could help. However, it will also require a better understanding of planet formation and evolution scenarios to build arguments to constrain probable planet compositions in this planet parameter range.

Mean densities of rocky terrestrial planets similar to Earth are dominated by their iron cores and silicate mantles. The atmosphere contributes only little. These planets fall between the violet and pink lines in Fig. 1. We can see that Mars, Venus and Earth line-up on the silicate composition line. The figure also shows that there seems to be an upper mass limit for such silicate-dominated planets. The highest-mass silicate planets observed so far have masses of 10–20 M_{earth} (Rauer et al. 2014). This is in agreement with the core-accretion model of planet formation which expects larger planets, beyond the critical core mass, to accrete hydrogen and quickly grow in size (and hence decrease in mean density, falling below the light-blue line or moving to the right part of the diagram of gas giants). Also Mercury-like planets which require large iron cores to explain their mean density are found. However, error bars in this low-mass part of the diagram are large, and a higher accuracy of measured mean densities is urgently needed to at least minimise the ambiguities from observations. However, also planets in this low-mass parameter range are subject to compositional degeneracy (see Sect. 3).

The examples above have shown that inferring the water content of planets from their bulk properties, radius, mass, and density, is a challenging task due to compositional degeneracies between core, mantle (silicate and/or ice) and atmosphere contributions. Nevertheless, statistical methods to disentangle terrestrial planet composition from radius and mass data, taking into account measurement and model uncertainties, are currently investigated (Dorn et al. 2015). These studies show that constraints may become available with an increasing data base of measured planet (radius, mass, mean density, atmosphere) and host star (composition) parameters. While the determination of a precise composition and water content of a particular, individual exoplanet will likely remain a challenge, improved data quality and especially large sets of observations will provide a new parameter input into the analysis of statistical properties of terrestrial planet composition in future.

3 Interior Structures from Bulk Compositions

When detecting small-massive exoplanets, a rough characterisation of the bulk composition can be obtained when comparing the observed mass and radius to predicted relationships

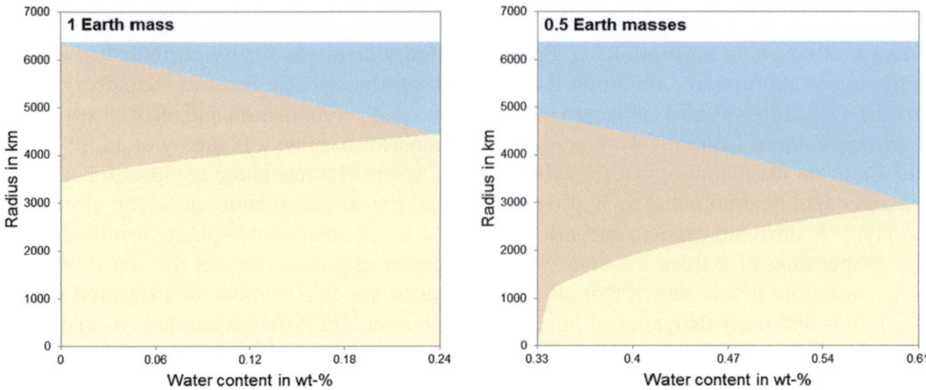

Fig. 2 Possible interior structure variations for an Earth-size planet of 1 (*left*) or 0.5 (*right*) Earth masses composed of an iron core (*yellow area*), a silicate mantle (*brown*) and water/ice (*blue*)

for different planets types—typically ranging from pure water spheres over silicate shells to iron-rich bodies (e.g. Valencia et al. 2006; Wagner et al. 2012; Dressing et al. 2015; Noack et al. 2016). An exoplanet that would lie on the mass-radius curve of an Earth-like structure would therefore be assumed to be rocky and similar to Earth. However, interior structure of planets and composition is never unambiguous.

Figure 2a shows a simplified example for different interior structures that are possible for an Earth-size planet with a mass of exactly Earth's mass (neglecting measurement inaccuracies here). The blue areas denote a pure water layer (with a surface temperature of 300 K and consisting of liquid water and possibly high-pressure ice), brown areas are the silicate mantle (containing only the minerals Mg_2SiO_4, $MgSiO_3$ and MgO for simplicity) and yellow indicates the iron core (solid in this example). Earth's surface water content is about 0.023 wt-% (Schubert et al. 2001). However, if Earth would have a larger water content (here up to 24 wt-% or appr. 1000 Earth oceans), larger amounts of iron could compensate the effect of the less-dense water on the mass of the planet and yield the same observed mass and radius as Earth. Note that a pure water-iron planet lacking any silicate minerals is unlikely, but variations in the Mg-Fe-Si ratios have already been observed for several stars—which may indicate related variations in planet compositions and can help to constrain possible interior structures (e.g. Grasset et al. 2009; Santos et al. 2015; Dorn et al. 2015; Unterborn et al. 2016).

The right plot in Fig. 2 on the other hand shows a planet of one Earth size but only half the mass of Earth. As long as an extended atmosphere, which would yield a larger observed than actual planet radius (e.g. Lissauer et al. 2011), can be excluded (for example for low-massive planets up to Earth's mass Lammer et al. 2014), such a small bulk density can only be explained for substantial amounts of water. For such planets a rough characterisation based on their mass-radius relationship is possible. For planets discovered with Earth-like densities, large amounts of water cannot be excluded, and the detected exoplanet may have either a water-rich or a rocky composition.

In Fig. 3 example profiles of a water-rich planet throughout the interior are plotted for the density, pressure, specific heat capacity, gravitational acceleration, adiabatic reference temperature and thermal expansion coefficient after Noack et al. (2016). This model planet of one Earth mass contains 35 wt-% iron and 25 wt-% water. Distinct variations of several ther-

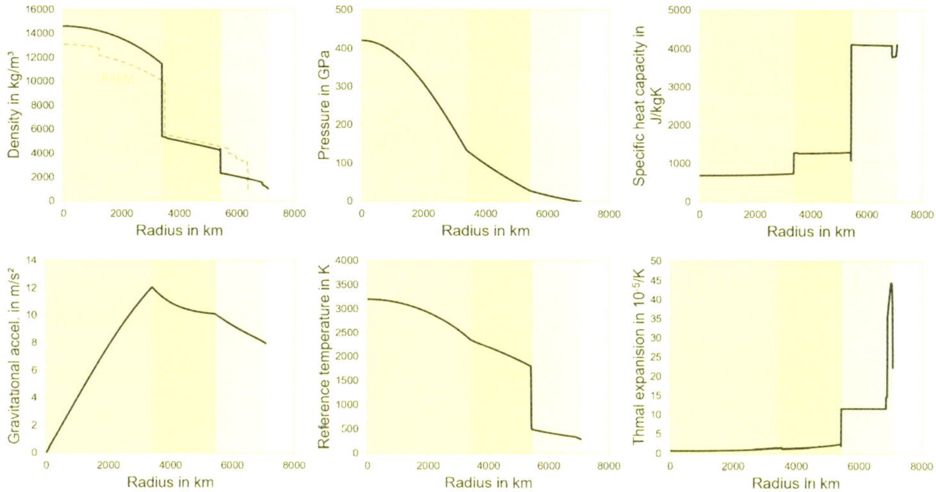

Fig. 3 Radial profiles of a water-rich planet for density, pressure, and different thermodynamic properties for a one-Earth-mass planet composed of 35 wt-% iron, 25 wt-% water and 40 wt-% silicates from Fig. 2. The *yellow colour* denotes the iron core, *brown* the silicate mantle, *grey* the high-pressure ice and *blue* the liquid water layer at the surface. Possible melting of the ice layer from below is neglected here for simplicity. The *grey dashed line* shows the density of the preliminary reference Earth model (PREM, Dziewonski and Anderson 1981)

modynamic properties for the different layers (water–mantle–core) can be observed for the density and the thermal properties. For comparison, the density profile of the Earth following the preliminary reference Earth model (PREM) by Dziewonski and Anderson (1981) is given. The main differences between the density profiles can be observed in the upper first 1000 km, where the large water content of the modelled planet leads to smaller densities in the high-pressure ice layer than in the silicate upper mantle of Earth. The density of Earth's core is reduced with respect to the model planet due to existence of lighter elements in Earth's core, which were not considered in the model planet. While mass and radius of rocky planets can be observed from space, the interior structure as well as information on thermodynamic properties of the different layers in the interior cannot be observed. Models of possible planet compositions and differentiation processes can help to better understand the thermal evolution of the interior of rocky planets, which influences the subaqueous surface, for example via volcanism and plate tectonics (Noack et al. 2016, 2017).

4 Characterisation of Exoplanet Atmospheres and Observing Water

Decades of Solar System exploration have revealed an immense complexity and diversity of planet atmospheres. Gas giants' atmospheres are molecular hydrogen dominated and exhibit strong zonal flows with thick clouds and storms. The rocky planets harbour secondary atmospheres, with a super-rotating CO_2-based atmosphere with opaque sulphuric acid clouds in the case of Venus, a N_2 based partially clear atmosphere with a significant biotic oxygen content in the case of the Earth, and a tenuous CO_2 based atmosphere with yet unexplained trace amounts of methane in the case of Mars (Webster et al. 2015). Arguably even more mysterious is the thick N_2-based atmosphere of Saturn's moon Titan, with opaque methane/ethane clouds.

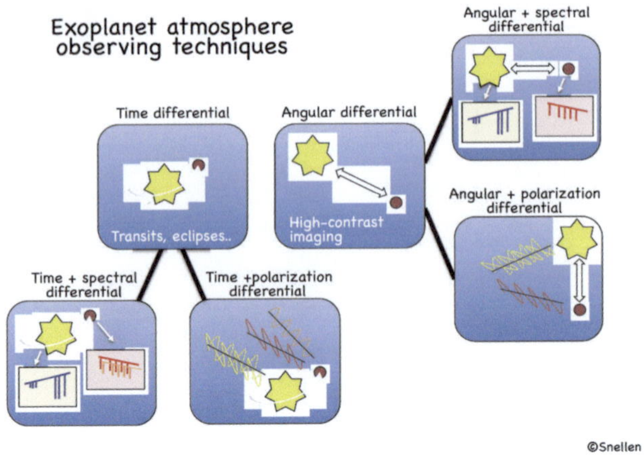

Fig. 4 Exoplanet atmosphere observations are all about separating the planet light from that of the star. This can be done using time differential measurements (using transits, eclipses, or phase curves), angular differential techniques using high-contrast imaging, and either of these combined with high-dispersion spectroscopy or polarisation observations

Scientifically, the daunting task is to understand this wide variety in atmospheres in the frame work of planet formation and evolution. Since the atmospheres of the Solar System planets are so different from each other, it indicates that it will require detailed investigations of extrasolar planets to even begin to recognise patterns, relations, and/or families between different types of planets. Planet types such as hot Jupiters, super-Earths and mini-Neptunes do not exist in our Solar System.

On the other hand, studying the atmospheres of extrasolar planets will always be significantly limited compared to what we can do for Solar System planets. The latter we can send probes to that dig and drill to understand their geology, or that perform in situ atmospheric measurements. Artificial satellites can accurately measure a planet's gravity field, and landers can collect seismic data, or in the future even collect samples that could be returned to Earth. All these important and basic observables will never be available for exoplanet studies.

Fortunately, a wide variety of information can be obtained from remote sensing alone. Moreover, the composition, thermal structure and dynamics of a planet atmosphere can be studied using a wide range of observational techniques (see Fig. 4).

4.1 Techniques to Measure Exoplanet Atmospheres

In the first two decades since the first discoveries more than 1000 exoplanets have been found. Virtually all of these planets have been discovered without identifying even one single photon from the planet itself. Unfortunately, this can no longer be the case for atmospheric studies, and such observations are about separating the planet signal from that of the star, with the latter often many orders of magnitude brighter.

Practically all aspects of astronomical observations, which measure the radiation (and polarisation) as function of time, direction, and wavelength, can be used to filter out the light from the planet:

1: *Time domain—transits and eclipses:* The way a planet is observed can be a dramatic function of time. This is particularly the case for transiting systems. Not only does the planet at given times block off part of the stellar surface, but a small fraction of star light filters through the planet atmosphere leaving a spectral imprint of absorption and scattering processes. By observing the system as function of time, the planet's atmospheric transmission spectrum can be isolated from the stellar spectrum (Charbonneau et al. 2002).

This works best when the observability of the planet changes on short time scales relative to the duration of the observation, such as in the case of transits, or eclipses—the moment the planet vanishes behind the star and the missing radiation is a measure of the planet's thermal emission spectrum and/or reflectance (Charbonneau et al. 2005; Deming et al. 2005). Along a planet orbit we also see the planet at different viewing angles, showing varying fractions of its day and night side. The time scales of such phase curve measurements are longer with the changes in planet signal less abrupt making this significantly more difficult to observe (Knutson et al. 2007).

The planet transmission spectra probe the atmosphere at the day/night terminator region. The atmospheric signal largely depends on the scale height and the abundance of spectroscopically active species and scattering or absorption by aerosols. A secondary eclipse spectrum or phase curve measurement probes a combination of star light reflected off a planet's atmosphere and its intrinsic thermal spectrum—the latter largely depending on the atmospheric temperature structure and abundances of molecular species on the planet's day-side. Note that this type of exoplanet atmospheric measurements, which depend on time differential filtering, is mainly constrained to planets in close-in orbits, since both the transit probability and transit frequency scale down with orbital distance, but also because these planets are very warm due to the close proximity of their host star.

2: *Spatial domain—high contrast imaging:* The small angular separation between star and planet can also be used to isolate the planet signal, called *direct* or *high contrast imaging (HCI)*. Since the theoretical angular resolution is proportional to the telescope diameter, best results so far are being reached with the large ground-based telescopes making use of adaptive optics systems aimed to cancel out the negative effects of Earth-atmospheric turbulence (seeing), coronography to optimally darken regions directly around the star, and utilising smart analysis algorithms that further remove telescope-induced systematics. The hot Jupiters that are being probed with transit measurements are inaccessible with direct imaging since they are too close to the host star. Mature planets that are located at such large orbital distance that they could be angularly separated from their host star are extremely faint, both in thermal as in reflected light, with planet/star contrasts of 10^{-8} or lower. This means that they are also beyond reach of current instrumentation. However, very young systems with planets at tens or more astronomical units *can* be observed (Marois et al. 2008; Lagrange et al. 2010). Although these planets receive only little energy form their host stars, they are still warm from their formation resulting in a much more favourable planet-to-star contrast. In particular young A-stars, which are 2–3 times more massive than the Sun, sometimes exhibit gas giant planets at large orbital distances of which a dozen or so are now known. HCI immediately allows the production of a planet thermal spectrum by combining measurements at different wavelengths (Currie et al. 2010).

Both the filtering in the time-domain as in the spatial domain can be used in combination with filtering on other aspects of astronomical observations. E.g. high-dispersion spectroscopy (HDS) can be used in conjunction with transit or phase-curve observations. At $\lambda/\Delta\lambda \approx 100000$, a planet molecular band (such as from CO or H_2O) resolves into tens or hundreds of individual lines. Since the orbital motion of hot Jupiters is large (up to 150 km/s) the radial component of the orbital velocity of the planet can change by tens of km/s during an observation—adding a time differential component to the observation. This allows the planet also to be separated in wavelength-space by filtering out all stationary components in the observed spectra (including the spectral features from the star and Earth atmosphere). The signals from the individual molecular lines can be subsequently combined using cross-correlation techniques (Cameron et al. 1999; Snellen et al. 2010).

High-dispersion spectroscopy can also be combined with high contrast imaging (HDS+HCI). While the planet-star contrast obtained with HCI may not be sufficient to detect the planet, the remaining starlight at the planet position may have a significantly different spectrum than that of the planet. The stellar spectrum is well sampled and can be filtered out, leaving the planet signal only at the planet position of which the signal of individual lines can be combined again by using cross-correlation techniques (Sparks and Ford 2002; Riaud and Schneider 2007; Snellen et al. 2015). In this way the spin-rotation of the directly imaged planet β Pictoris b was determined for the first time (Snellen et al. 2014).

Similar techniques can also be used in conjunction with polarisation. E.g. it is expected that reflected light from a planet is partially polarised, while that of the star is not. Hence, by isolating the polarising flux, the contrast between the planet and star can be further enhanced. The Zimpol instrument, recently installed on the Very Large Telescope, has been designed to pursue this (Milli et al. 2013).

4.2 Observations of Water in Exoplanet Atmospheres

The earliest report of the detection of water vapour in the atmosphere of an exoplanet is from Tinetti et al. (2007) using the Spitzer Space Telescope, who attributed the observed variation in transit depth between three infrared wavelengths as a signature of water absorption in the hot Jupiter HD189733. The same planet was also shown to have water using observations with the NICMOS near-infrared spectrograph at the Hubble Space Telescope (HST, Swain et al. 2008), along with a heavily disputed claim of methane absorption. Charbonneau et al. (2008) and Grillmair et al. (2008) showed that the broad-band Spitzer spectrum of the thermal emission spectrum (obtained through secondary eclipse measurements) is also consistent with absorption by water vapour. With the installation of the WFC3 camera on the HST and the utilisation of innovative observational techniques that minimise its systematic effects (drift scanning), the 1.4–1.7 micron water absorption band has been detected for about a dozen hot Jupiters in their transmission spectrum (Deming et al. 2013; Mandell et al. 2013; McCullough et al. 2014; Wakeford et al. 2013), and in the thermal phase curve of WASP-43b (Kreidberg et al. 2014a).

While the strength of the water absorption feature shows planet to planet variations by up to an order of magnitude, Sing et al. (2016) show that this is anti-correlated with the strength of the UV-optical blueward slope of these planets' transmission spectra, generally attributed to Rayleigh scattering of haze particles in the atmospheres. Therefore it seems that it is not a variation in water abundance that is causing the variation in absorption strength, but the presence or not of hazes that partly block the 1.4–1.7 micron water feature.

Pushing this technique to smaller and cooler planets has been a challenge. For several planets (e.g. Kreidberg et al. 2014b; Knutson et al. 2014) the transmission spectrum seems to be dominated by clouds, basically hiding the water absorption signature (note that in such hot planets these are not water-clouds). Fortunately, at least one Neptune size planet at <1000 K has a spectrum that is not dominated by clouds (Fraine et al. 2014). However, it is clear that clouds may be an important obstacle when searching for absorption features in transmission spectra of cooler planets.

Ground-based HDS of hot Jupiters has also resulted in several water detections. Water absorption was detected in the dayside thermal spectrum of hot Jupiter HD189733 using CRIRES at the Very Large Telescope (Birkby et al. 2013), and in that of tau Bootis using NIRSPEC at Keck (Lockwood et al. 2014)—both at 3–3.5 μm. Water absorption was also observed at shorter wavelengths (e.g. Brogi et al. 2014; Schwarz et al. 2015).

HCI has also revealed water absorption in young massive gas giants at large orbital distances. Both planets b and c that orbit the young A-giant HR8799 have been found to exhibit a clear water absorption signature, together with carbon monoxide and methane, and carbon monoxide respectively (Barman et al. 2015; Konopacky et al. 2013). As expected, water vapour is also common in the atmospheres of young, self-luminous gas giant planets.

5 Planetary Evolution and Habitability

The basic definition of habitability, based on life as we know it, is the existence of liquid water as prerequisite—even though other factors like the existence of building blocks of life, energy sources, and nutrients are necessary, as well. Looking at the planets and moons in the Solar System, and at the data of the so far discovered exoplanets, suggests three different groups of bodies that could host life:

- terrestrial planets with shallow, regional water oceans,
- icy bodies with subsurface niches (either subsurface access to water as suggested for Mars or subsurface oceans like for Europa and Enceladus), or
- water-worlds with water-steam atmospheres and liquid water at the surface.

In the following, we shortly discuss studies that investigated the habitability potential of terrestrial and icy bodies, followed by a section on water-worlds.

5.1 Water on Terrestrial Exoplanets

The holy grail in exoplanet research is to discover a potential second Earth. While based on the geophysical definition an Earth-like planet (or rocky planet) would solely require a density indicating a composition of mainly silicate minerals and metals, the "second Earth" is typically expected to have surface oceans like on Earth—and therefore needs to be in the habitable zone. Just the right mass and radius would not justify to call the detected planet a "second Earth". Some scientists would in addition request the existence of a magnetic field as well as plate tectonics, in other words, a planet that might be able to host complex, Earth-like life (either in its original or present-day form).

The existence of small amounts of water on terrestrial planets cannot be detected from the mass-radius relationship (see Sects. 2, 3). On the other hand, when searching for complex life like it evolved on Earth, the existence of surface land (volcanic islands or continents) may be crucial, and only shallow ocean layers may be considered as habitable (Maruyama et al. 2013).

The detection of water and continents at the surface of a terrestrial exoplanet could be possible at some point. Water in the atmosphere can be detected from its spectrum, see Sect. 4. Observation of surface coverage of oceans and land might be possible due to interpretation of light curves and albedo measurements (Montanes-Rodriguez et al. 2006; Fujii et al. 2010) for non-transiting planets.

Another question concerning water at the surface of terrestrial planets is its survivability. During the long-term evolution of Sun-like stars (F- or G-type stars), the solar flux strongly increases with time. The standard model for the adaption of the atmosphere to changing solar radiation, as used in the calculation of habitable zone boundaries (e.g. Hart 1979; Kasting et al. 1993; Kopparapu et al. 2013), includes the formation of carbonates to extract CO_2 from the atmosphere to compensate for the increasing solar flux. This process may already need the existence of plate tectonics to subduct carbonates produced at the continental shelves, which may be limited by several factors including for example the planet mass (still under debate, e.g. Valencia et al. 2007; van Heck and Tackley 2011; Noack and Breuer 2014), interior heat (Stein et al. 2013; Stamenković and Breuer 2014) or composition (Noack et al. 2014). Inefficient atmosphere adaptation to increasing solar flux would lead to a runaway greenhouse effect and loss of surface water. On the other hand, limitations to the maximal potential amount of greenhouse gases in the atmosphere could have severe consequences for the existence of liquid surface water at the outer boundary of the habitable zone (Noack et al. 2014).

Another interesting contribution of plate tectonics is the recycling of water into the mantle and its replenishing at volcanic regions (Parai and Mukhopadhyay 2012; Höning et al. 2014; Foley 2015), which may be able to lead to stable, shallow ocean layers—even for slightly larger amounts of water than on Earth leading to an enrichment of water in the mantle instead of deeper surface ocean layers for small-massive planets (Schaefer and Sasselov 2015).

5.2 Water in Icy Bodies

While early Mars had periods of liquid water flowing locally at the surface (McKay and Davis 1991; Halevy and Head 2014; Helbert et al. 2015), present-day Mars can rather be considered as an icy world (apart from few regions at the equator where temperatures rise above zero degrees during the day). In the subsurface of Mars, access to liquid water may be possible, and if life did exist during the earlier stages of the evolution, it may have penetrated into the ground, where it is also shielded from surface radiation (Boston et al. 1992; Gold 1992).

The more traditional view of icy bodies considers the water-rich moons in the outer Solar System covered by an ice layer. These moons formed outside of the snowline. Icy moons are typically considered to be habitable, if a subsurface ocean exists, which is in contact with the silicate mantle. Hydrothermal activity or serpentinisation processes within cracks of the rocky core (e.g. Vance et al. 2016) are needed as energy source, and access to nutrients is furthermore essential for long-time habitability of these subsurface oceans. Subsurface oceans are believed to exist within Ganymede (Kivelson et al. 2002; Saur et al. 2015; Vance et al. 2014), Titan (Iess et al. 2012; Grindrod et al. 2008) and Callisto (Zimmer et al. 2000; Khurana et al. 1998), but at depth high-pressure ice forms, which hinders the direct access of the ocean to the rocky core (even though for Ganymede an upper and lower water layer have been suggested, Vance et al. (2014)). The moons experience negligible radioactive heating from within and only limited tidal heating, such that melting of the entire high-pressure ice layer from beneath is not expected. Therefore, in the Solar System, typically only Europa and Enceladus are considered as habitable icy moons, where water-silicate contact is possible (e.g. Roth et al. 2014; Khurana et al. 1998; Thomas et al. 2016; Van Hoolst et al. 2016).

5.3 Habitability of Ocean Planets

The detection of low-dense, massive exoplanets (see also Sect. 2) has started the discussion about the habitability of large water-rich planets, often called ocean planets or mini-/sub-Neptunes (Léger et al. 2004; Grasset et al. 2009; Sotin et al. 2007; Alibert 2014; Luger et al. 2015; Lammer et al. 2009; Noack et al. 2016). These planets need to contain large amounts of water and/or extended atmospheres to explain their low density (see Sect. 3).

Of interest have been especially water worlds in the habitable zone, where the hydrogen-helium atmospheres allow for liquid water at the surface of these planets. Sub-Neptunes outside the snowline could be potentially habitable, as well, although observations will be

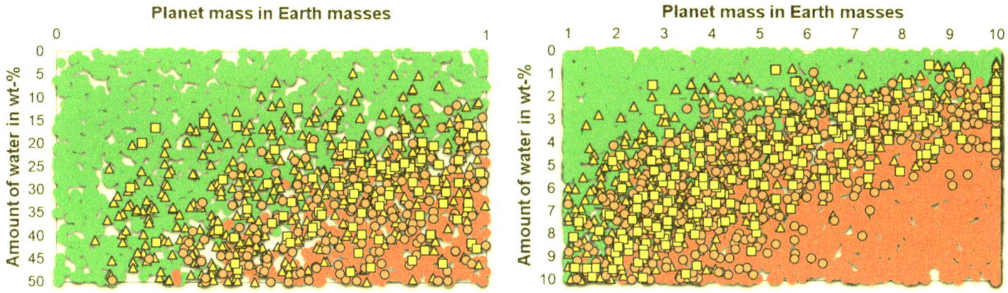

Fig. 5 Possible habitability evaluation for planets of variable masses, compositions and interior parameters. The *colours* refer to liquid water layers (*green*), lower ocean layers (*yellow*), episodic lower ocean layers (*orange*) and high-pressure ice layers that are not molten at the bottom (*red*). Based on Noack et al. (2016)

extremely challenging due to the enormous star/planet contrast ratio (if they can be observed, at all, and are not scattered out of their star system, Stevenson 1999). Furthermore, biosignatures in the atmosphere indicating life would be enhanced on planets without a surface ice crust, where life could extend towards the surface.

Alibert (2014) modelled the maximal possible habitable planet radius for water worlds of different masses considering different mantle-core composition. For large water amounts, high-pressure ice can form, which is typically considered to rule out habitability of the ocean, as the ocean is not in contact with the silicate part of the planet. Similar studies have investigated the maximal liquid ocean depth (e.g. Sotin et al. 2007), also under the aspect of different surface temperatures (Noack et al. 2016).

The recent study by Noack et al. (2016) looked closer at water-rich planets to evaluate the decrease in habitability potential with increasing water amount. The study found that, depending on several factors such as initial energy budget, amount of radioactive heat sources, or composition, high-pressure ice can be molten from beneath due to the high heat flow from silicate mantle into the ice layer. In some cases, only an episodic second, lower ocean layer appears above the silicate mantle. Such a lower ocean layer (episodic or not), would classify as a habitable niche inside water-rich planets.

To understand how the possible habitability of a deep water layer (possibly including high-pressure ice and a second, lower ocean above the silicate mantle) depends on the amount of water and mass of a planet, we model the thermal evolution of 7500 model planets over several billion years assuming arbitrary values for composition, initial parameters (as for example the amount of radioactive heat sources or initial temperature profile), crustal production parameters and different silicate lithosphere regimes (stagnant lid or plate tectonics) following a Monte Carlo approach as in Noack et al. (2016).

Figure 5 shows the habitability potential of the model planets separated into a low-mass range (left plot) and super-Earth mass range (right plot). For each planet, we determine if high-pressure ice forms during the evolution of the planet, and if the heat from the interior would be sufficient to melt the ice layer from beneath. The colours in Fig. 5 refer to the possible habitability of the water-rich planet. The green region is the region where no high-pressure ice forms, whereas in the yellow and orange cases a lower ocean forms (only episodically for planets with orange dots). For model planets with a red dot, a high-pressure ice layer forms, which extends to the surface of the silicate shell. These planets are likely uninhabitable Noack et al. (for more information on the habitability classification we refer the reader to 2016). When taking into account planets with a second, lower ocean in the classification of possibly habitable water-rich planets, the range of water amounts that still allow for habitable worlds is extended to much larger values than compared to the classical

Fig. 6 Tree of life awareness. The different-coloured branches indicate different evolution steps of the search for life, from the first thoughts on the history of life (*blue branch*) to search for life on other planets or moons in the solar system (*red branch*) and in the universe (*green branch*)

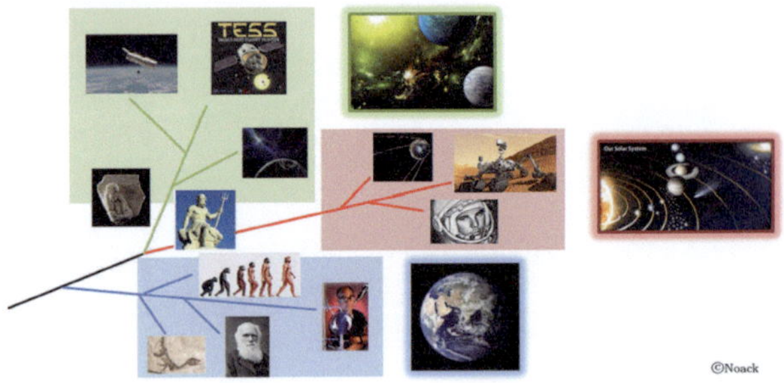

restriction of having no high-pressure ice forming in the water layer. Note, however, that the CO_2 feedback cycling between interior and atmosphere would be hindered, which may have additional consequences for the atmosphere stability and habitability (Kitzmann et al. 2015). For small massive planets (from Mars' to Earth' mass), immense water contents up to several tens of wt-% could still allow for the origin and evolution of life as we know it, whereas for massive super-Earths, even small amounts of water in the range of few wt-% would make the planets uninhabitable.

5.4 Searching for Life Outside of Earth

Figure 6 shows a different tree of life than the well-known biological tree of life (Darwin 1859; Hinchliff et al. 2015). The figure summarises the evolution of the search for life, starting with the first thoughts on the history of life, which is indicated in the lower, blue region. Archaeological findings and the fundamental works of Charles Darwin and others led to the discovery of life's origin on Earth, including the ground-breaking Miller-Urey experiment creating amino acids under early-Earth conditions in the laboratory and thus creating organic compounds, that are essential for life as we know it.

A different branch of life awareness, considering life on other planets, goes back to Greek philosophers such as Democritus around 400 BC, who claimed that the universe must contain more than one planet sustaining life. The red area in Fig. 6 refers to planets and moons in the Solar System. Mars, the moon and Venus were believed to be inhabited (see also review in Noack and Breuer 2013), until the area of space missions revolutionised the idea of life on other planets.

In the last two decades, after not having found life on another body in the Solar System, and after the first exoplanets were discovered in the early 1990s (Wolszczan and Frail 1992; Mayor and Queloz 1995), the universe is in the focus of the search for life (green region). The hunt for finding possible habitable exoplanets started a decade ago and continues with numerous upcoming detection and characterisation missions such as TESS (Ricker et al. 2015), CHEOPS (Broeg et al. 2013), JWST (Sabelhaus and Decker 2004) and PLATO (Rauer et al. 2014), see also Sect. 6. One goal of these missions is to find a second habitable world—a planet containing liquid water that life can access, and where it can flourish.

6 Outlook Missions

6.1 Bulk Planet Characterisation: Radii, Masses, Density and Ages

Figure 7 provides an overview of past and future missions relevant for exoplanet studies. The first dedicated space mission for exoplanets CoRoT (CNES, Baglin et al. 2006) detected

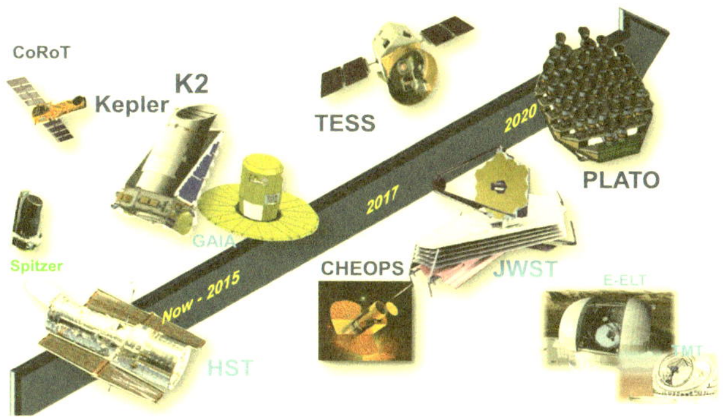

Fig. 7 Overview of past and future missions relevant for exoplanet characterisation

32 exoplanets so far, all with RV follow-up observations and hence measured radii and masses. CoRoT provided us with the first secured terrestrial planet, CoRoT-7b (Léger et al. 2009). NASA's Kepler mission (Koch et al. 2010) detected thousands of planet candidates, including ~80 planets for which RV follow-up was possible and hence mean planet densities could be derived, including the small terrestrial planet Kepler-10b (Batalha et al. 2011). Both missions stopped nominal operation in 2012. However, the Kepler satellite continuous to take data, now as K2 mission (Howell et al. 2014). K2 observes several target fields near the ecliptic plane for durations of up to about 80 days each.

CoRoT, Kepler and K2 benefited from the availability of space observatories such as HST and the IR telescope Spitzer, since planets around sufficiently bright stars could be followed up with space-based spectroscopy to study their atmospheres (see below).

NASA's exoplanet transit survey TESS (Ricker et al. 2014) will provide the first all-sky survey to search for transiting planets. Most target fields will be observed for periods around ~1 month, but a small continuous viewing zone allows for observations of up to one year. TESS is planned for launch in 2018 and will hence be able to provide targets for JWST. TESS targets will be bright and radii, masses, and mean densities will be known for most of its detections.

In Europe, ESA will launch its first small mission CHEOPS (Broeg et al. 2013) in 2018. CHEOPS is a follow-up mission and will determine accurate radii for planets detected previously by the RV method, but also refine radii of TESS targets and ground-based transit detections. In 2025, ESA is planning to launch its M3 mission PLATO (Rauer et al. 2014). PLATO's main target range is small planets, including characterised planets in the habitable zone. PLATO will provide not only accurate radii and masses (from RV follow-up), but also accurate stellar parameters from asteroseismology. This technique was first applied to space photometric data by CoRoT and then successfully applied to the brightest Kepler targets. PLATO will now perform asteroseismology of host stars on a large scale to obtain not only accurate radii and masses, but also well-known ages of a large number of planetary systems. PLATO targets will be observable in a JWST extended mission phase and come in time for the 30m-class ground-based telescopes E-ELT (ESO) (Ramsay et al. 2014) and TMT (Skidmore et al. 2015).

In the next two decades we will see a wealth of data of planets characterised for their radius, mass and mean density, allowing for statistical studies of characterised planet samples. Many of these will be suitable for spectroscopic follow-up to measure their atmospheric composition.

6.2 Atmospheric Observations: Technological and Observational Developments

Current state of affairs is that for gas giants that are hot (>1000 K), either through irradiation of their host star or because they are young and still contracting, water absorption features can be readily observed. First attempts to push this to smaller planets and those with lower temperatures were so far only partially successful, hampered by the occurrence of clouds, which in particular in the case of transmission spectroscopy can mask absorption features.

The JWST, which will be launched in 2018, will constitute an enormous enhancement over the HST. It's collecting area is a factor 7 higher, and importantly the JWST will be flown to L2, meaning that in stark contrast to the HST it will be able to observe most targets for long periods in time (compared to the low-Earth 95-minute orbit of HST during which the Earth eclipses a target for significant periods). Stability is key for transmission, eclipse and phase curve observations and the JWST will provide this. In the case of clear atmospheres, the JWST is expected to be able to reach detection of water vapour for planets significantly lower in mass and lower in temperature than the HST can provide, possibly in the super-Earth or even Earth-size regime. However, clouds may spoil the party. In that case, observations at mid-infrared wavelength with MIRI may provide a solution—probing wavelength long enough such that observations through the clouds may be made.

Ground-based high-dispersion spectroscopy may be another solution to deal with cloudy atmospheres. The high-dispersion means that observations probe the atmospheres at significantly higher altitudes, being much more sensitive to water absorption features above any (non-water) cloud layers. New or upgraded spectrographs at existing 4–10 m class telescopes will significantly push this method forward, with as ultimate instruments the upcoming extremely large telescopes such as the E-ELT, both for high-dispersion and high-contrast imaging, of which the observing speed scales with the telescope diameter to the 4th power. Planned high-dispersion spectrographs such as METIS in the mid-infrared and HIRES in the optical near-infrared will revolutionise this field and may be able to characterise the atmospheres of Earth-like planets in the habitable zones of nearby stars.

7 Discussion and Summary

The detection of water (as well as its amount) is ambiguous when using mass and radius measurements alone. On the one hand, the current mean uncertainty of measured mean densities for small planets is about 40%. On the other hand, even if the exact mass and radius of a planet would be known, its composition cannot be derived unambiguously from the data alone. Together with information on the star's composition and the planet's surface temperature, statistical methods can suggest likely compositions (Dorn et al. 2015; Santos et al. 2015). But even for the planets in the Solar System, this information gives only a rough understanding of the interior structure of the rocky planets, and seismic measurements are needed for an improved understanding of the planets' compositions. This is not possible for exoplanets and will always limit our understanding of an individual planet. However, exoplanets provide a statistical information over a wide range of mean densities which is not assessable for Solar System bodies. This statistical sample will significantly constrain future planet formation models and will lead to a better understanding of planet evolution. It is in this field where exoplanets probably will provide the largest impact.

Concerning water in a particular exoplanet, fortunately, water vapour can be measured directly in exoplanet atmospheres, as has been shown by recent detections of water in the atmospheres of close-in hot Jupiters using transmission spectroscopy, and in young gas giants at large orbital distances using high-contrast imaging. The detection of water in the atmosphere of terrestrial planets is however challenging and awaits future instrumentation.

Water is essential for life on Earth, and is often seen as necessary condition for the origin of life (Westall and Brack 2017, this issue; Cockell et al. 2016). Detecting water either indirectly (via mass and radius relationship) or directly (atmosphere spectroscopy) in exoplanets is therefore an important first step in the direction of finding life outside of the Solar System.

Water vapour in the atmosphere does not necessarily correlate with liquid water at the surface. Indeed, the past detections of water vapour in exoplanet atmospheres were only possible for hot Jupiters and one Neptune-like exoplanet, which would not be considered as good candidates for life (at least not Earth-like life). Even for smaller, terrestrial planets, water vapour does not need to refer to surface water, as can be seen in the case of Venus (e.g. Bullock and Grinspoon 2001).

On the other hand, liquid surface water also does not implicitly include habitability, as deep water layers (i.e. less-dense planets) may be uninhabitable (especially on massive planets, see Sect. 5.3). For the origin of life, the right amount of water matters.

In this paper we summarised the recent findings on the existence of water in exoplanets. The detection of low-density exoplanets gives an indirect detection of large amounts of water. Direct methods can be used to identify water in the atmosphere, and future missions will enable detection of water in the atmosphere of super-Earths (see Sect. 4), giving a clearer indication for the possible habitability of these planets. Ocean planets, hence planets without significant land masses, are shown to be less suitable for life due to negative CO_2 feed-back (Kitzmann et al. 2015). Subsurface habitability for planets with large amounts of water is not necessarily ruled out, as even underneath deep ice layers life could flourish, as long as enough heat is produced in the rocky interior of the planet to melt the ice layer at the bottom. It will, however, be difficult to detect subsurface ocean life on exoplanets in the foreseeable future. Future missions therefore focus on the detection and characterisation of Earth-like planets which can be better compared to the evolution of life on our home planet.

Acknowledgements L. Noack has been funded by the Interuniversity Attraction Poles Programme initiated by the Belgian Science Policy Office through the Planet Topers alliance.

References

Y. Alibert, On the radius of habitable planets. Astron. Astrophys. **561**, 41 (2014)

G. Anglada-Escudé, B. Rojas-Ayala, A.P. Boss, A.J. Weinberger, J.P. Lloyd, Gj 1214 reviewed-trigonometric parallax, stellar parameters, new orbital solution, and bulk properties for the super-Earth gj 1214b. Astron. Astrophys. **551**, 48 (2013)

A. Baglin, M. Auvergne, L. Boisnard, T. Lam-Trong, P. Barge, C. Catala, M. Deleuil, E. Michel, W. Weiss, CoRoT: a high precision photometer for stellar ecolution and exoplanet finding, in *36th COSPAR Scientific Assembly*, vol. 36 (2006), p. 3749

T.S. Barman, Q.M. Konopacky, B. Macintosh, C. Marois, Simultaneous detection of water, methane, and carbon monoxide in the atmosphere of exoplanet HR8799b. Astrophys. J. **804**, 61 (2015). doi:10.1088/0004-637X/804/1/61

N.M. Batalha, W.J. Borucki, S.T. Bryson, L.A. Buchhave, D.A. Caldwell, J. Christensen-Dalsgaard, D. Ciardi, E.W. Dunham, F. Fressin, T.N. Gautier III et al., Kepler's first rocky planet: Kepler-10b based in part on observations obtained at the W.M. Keck Observatory, which is operated by the University of California and the California Institute of Technology. Astrophys. J. **729**(1), 27 (2011)

J.L. Birkby, R.J. de Kok, M. Brogi, E.J.W. de Mooij, H. Schwarz, S. Albrecht, I.A.G. Snellen, Detection of water absorption in the day side atmosphere of HD 189733 b using ground-based high-resolution spectroscopy at 3.2 μm. Mon. Not. R. Astron. Soc. **436**, 35–39 (2013). doi:10.1093/mnrasl/slt107

P.J. Boston, M.V. Ivanov, C.P. McKay, On the possibility of chemosynthetic ecosystems in subsurface habitats on mars. Icarus **95**(2), 300–308 (1992)
C. Broeg, A. Fortier, D. Ehrenreich, Y. Alibert, W. Baumjohann, W. Benz, M. Deleuil, M. Gillon, A. Ivanov, R. Liseau et al., CHEOPS: a transit photometry mission for ESA's small mission programme, in *EPJ Web of Conferences*, vol. 47 (EDP Sciences, London, 2013), p. 3005
M. Brogi, R.J. de Kok, J.L. Birkby, H. Schwarz, I.A.G. Snellen, Carbon monoxide and water vapor in the atmosphere of the non-transiting exoplanet HD 179949 b. Astron. Astrophys. **565**, 124 (2014). doi:10.1051/0004-6361/201423537
M.A. Bullock, D.H. Grinspoon, The recent evolution of climate on venus. Icarus **150**, 19–37 (2001). doi:10.1006/icar.2000.6570
J. Cabrera, H. Bruntt, M. Ollivier, R. Díaz, S. Csizmadia, S. Aigrain, R. Alonso, J.-M. Almenara, M. Auvergne, A. Baglin et al., Transiting exoplanets from the CoRoT space mission—XIII. CoRoT-13b: a dense hot jupiter in transit around a star with solar metallicity and super-solar lithium content. Astron. Astrophys. **522**, 110 (2010)
A.C. Cameron, K. Horne, A. Penny, D. James, Probable detection of starlight reflected from the giant planet orbiting τ Boötis. Nature **402**, 751–755 (1999). doi:10.1038/45451
D. Charbonneau, T.M. Brown, R.W. Noyes, R.L. Gilliland, Detection of an extrasolar planet atmosphere. Astrophys. J. **568**, 377–384 (2002). doi:10.1086/338770
D. Charbonneau, L.E. Allen, S.T. Megeath, G. Torres, R. Alonso, T.M. Brown, R.L. Gilliland, D.W. Latham, G. Mandushev, F.T. O'Donovan, A. Sozzetti, Detection of thermal emission from an extrasolar planet. Astrophys. J. **626**, 523–529 (2005). doi:10.1086/429991
D. Charbonneau, H.A. Knutson, T. Barman, L.E. Allen, M. Mayor, S.T. Megeath, D. Queloz, S. Udry, The broadband infrared emission spectrum of the exoplanet HD 189733b. Astrophys. J. **686**, 1341–1348 (2008). doi:10.1086/591635
D. Charbonneau, Z.K. Berta, J. Irwin, C.J. Burke, P. Nutzman et al., A super-Earth transiting a nearby low-mass star. Nature **462**, 891–894 (2009)
C. Cockell, T. Bush, C. Bryce, S. Direito, M. Fox-Powell, J. Harrison, H. Lammer, H. Landenmark, J. Martin-Torres, N. Nicholson et al., Habitability: a review. Astrobiology **16**(1), 89–117 (2016)
T. Currie, P. Hinz, Y. Itoh, M. Fukagawa, V. Bailey, T.J. Rodigas, D. Fabrycky, R. Murray-Clay, Direct imaging of the brown dwarf/planet-mass companions to HR 8799 and GJ 758 from 1 to 5 microns: constraints on atmospheric properties, in *In the Spirit of Lyot 2010* (2010)
C. Darwin, *On the Origin of Species* (John Murray, London, 1859), p. 490
M. Deleuil, A. Bonomo, S. Ferraz-Mello, A. Erikson, F. Bouchy, M. Havel, S. Aigrain, J.-M. Almenara, R. Alonso, M. Auvergne et al., Transiting exoplanets from the CoRoT space mission—XX. CoRoT-20b: a very high density, high eccentricity transiting giant planet. Astron. Astrophys. **538**, 145 (2012)
D. Deming, S. Seager, L.J. Richardson, J. Harrington, Infrared radiation from an extrasolar planet. Nature **434**, 740–743 (2005). doi:10.1038/nature03507
D. Deming, A. Wilkins, P. McCullough, A. Burrows, J.J. Fortney, E. Agol, I. Dobbs-Dixon, N. Madhusudhan, N. Crouzet, J.-M. Desert, R.L. Gilliland, K. Haynes, H.A. Knutson, M. Line, Z. Magic, A.M. Mandell, S. Ranjan, D. Charbonneau, M. Clampin, S. Seager, A.P. Showman, Infrared transmission spectroscopy of the exoplanets HD 209458b and XO-1b using the Wide Field Camera-3 on the Hubble Space Telescope. Astrophys. J. **774**, 95 (2013). doi:10.1088/0004-637X/774/2/95
C. Dorn, A. Khan, K. Heng, J.A.D. Connolly, Y. Alibert, W. Benz, P. Tackley, Can we constrain the interior structure of rocky exoplanets from mass and radius measurements? Astron. Astrophys. **577**(183), 1–18 (2015)
C.D. Dressing, D. Charbonneau, X. Dumusque et al., The mass of Kepler-93b and the composition of terrestrial planets. Astrophys. J. **800**(2), 135 (2015)
A.M. Dziewonski, D.L. Anderson, Preliminary reference Earth model. Phys. Earth Planet. Inter. **25**(4), 297–356 (1981)
B.J. Foley, The role of plate tectonic–climate coupling and exposed land area in the development of habitable climates on rocky planets. Astrophys. J. **812**(1), 36 (2015)
J. Fraine, D. Deming, B. Benneke, H. Knutson, A. Jordán, N. Espinoza, N. Madhusudhan, A. Wilkins, K. Todorov, Water vapour absorption in the clear atmosphere of a Neptune-sized exoplanet. Nature **513**, 526–529 (2014). doi:10.1038/nature13785
Y. Fujii, H. Kawahara, Y. Suto, A. Taruya, S. Fukuda, T. Nakajima, E.L. Turner, Colors of a second Earth: estimating the fractional areas of ocean, land, and vegetation of Earth-like exoplanets. Astrophys. J. **715**(2), 866 (2010)
T. Gold, The deep, hot biosphere. Proc. Natl. Acad. Sci. USA **89**(13), 6045–6049 (1992)
O. Grasset, J. Schneider, C. Sotin, A study of the accuracy of mass–radius relationships for silicate-rich and ice-rich planets up to 100 Earth masses. Astrophys. J. **693**, 722–733 (2009)

C.J. Grillmair, A. Burrows, D. Charbonneau, L. Armus, J. Stauffer, V. Meadows, J. van Cleve, K. von Braun, D. Levine, Strong water absorption in the dayside emission spectrum of the planet HD189733b. Nature **456**, 767–769 (2008). doi:10.1038/nature07574

P.M. Grindrod, A.D. Fortes, F. Nimmo, D.L. Feltham, J.P. Brodholt, L. Vocadlo, The long-term stability of a possible aqueous ammonium sulfate ocean inside Titan. Icarus **197**, 137–151 (2008)

I. Halevy, J.W. Head III, Episodic warming of early mars by punctuated volcanism. Nat. Geosci. **7**(12), 865–868 (2014)

M.H. Hart, Habitable zones about main sequence stars. Icarus **37**, 351–357 (1979). doi:10.1016/0019-1035(79)90141-6

A.P. Hatzes, H. Rauer, A definition for giant planets based on the mass–density relationship. Astrophys. J. Lett. **810**(2), 25 (2015)

J. Helbert, E. Hauber, D. Reiss, Water on the terrestrial planets, in *Treatise on Geophysics, Volume 10: Physics of Terrestrial Planets and Moons*, 2nd edn. (Springer, Dordrecht, 2015), pp. 367–409. doi:10.1016/B978-0-444-53802-4.00174-3

C.E. Hinchliff et al., Synthesis of phylogeny and taxonomy into a comprehensive tree of life. Proc. Natl. Acad. Sci. USA **112**(41), 12764–12769 (2015). doi:10.1073/pnas.1423041112

D. Höning, H. Hansen-Goos, A. Airo, T. Spohn, Biotic vs. abiotic Earth: a model for mantle hydration and continental coverage. Planet. Space Sci. **98**, 5–13 (2014)

S.B. Howell, C. Sobeck, M. Haas, M. Still, T. Barclay, F. Mullally, J. Troeltzsch, S. Aigrain, S.T. Bryson, D. Caldwell et al., The k2 mission: characterisation and early results. Publ. Astron. Soc. Pac. **126**(938), 398 (2014)

L. Iess, R.A. Jacobson, M. Ducci, D.J. Stevenson, J.I. Lunine, J.W. Armstrong, S.W. Asmar, P. Racioppa, N.J. Rappaport, P. Tortora, The tides of Titan. Science **337**(6093), 457–459 (2012). doi:10.1126/science.1219631

J.M. Jenkins, J.D. Twicken, N.M. Batalha, D.A. Caldwell, W.D. Cochran, M. Endl, D.W. Latham, G.A. Esquerdo, S. Seader, A. Bieryla et al., Discovery and validation of Kepler-452b: a 1.6 R? Super Earth exoplanet in the habitable zone of a g2 star. Astron. J. **150**(2), 56 (2015)

J.F. Kasting, D.P. Whitmire, R.T. Reynolds, Habitable zones around main sequence stars. Icarus **101**, 108–128 (1993). doi:10.1006/icar.1993.1010

K. Khurana, M. Kivelson, D. Stevenson, G. Schubert, C. Russell, R. Walker, C. Polanskey, Induced magnetic fields as evidence for subsurface oceans in Europa and Callisto. Nature **395**(6704), 777–780 (1998)

D. Kitzmann, Y. Alibert, M. Godolt, J.L. Grenfell, K. Heng, A. Patzer, H. Rauer, B. Stracke, P. von Paris, The unstable CO_2 feedback cycle on ocean planets. Mon. Not. R. Astron. Soc. **452**(4), 3752–3758 (2015)

M. Kivelson, K. Khurana, M. Volwerk, The permanent and inductive magnetic moments of Ganymede. Icarus **157**(2), 507–522 (2002)

H.A. Knutson, D. Charbonneau, L.E. Allen, J.J. Fortney, E. Agol, N.B. Cowan, A.P. Showman, C.S. Cooper, S.T. Megeath, A map of the day-night contrast of the extrasolar planet HD 189733b. Nature **447**, 183–186 (2007). doi:10.1038/nature05782

H.A. Knutson, D. Dragomir, L. Kreidberg, E.M.-R. Kempton, P.R. McCullough, J.J. Fortney, J.L. Bean, M. Gillon, D. Homeier, A.W. Howard, Hubble Space Telescope Near-IR transmission spectroscopy of the super-Earth HD 97658b. Astrophys. J. **794**, 155 (2014). doi:10.1088/0004-637X/794/2/155

D.G. Koch, W.J. Borucki, G. Basri, N.M. Batalha, T.M. Brown, D. Caldwell, J. Christensen-Dalsgaard, W.D. Cochran, E. DeVore, E.W. Dunham et al., Kepler mission design, realized photometric performance, and early science. Astrophys. J. Lett. **713**(2), 79 (2010)

Q.M. Konopacky, T.S. Barman, B.A. Macintosh, C. Marois, Detection of carbon monoxide and water absorption lines in an exoplanet atmosphere. Science **339**, 1398–1401 (2013). doi:10.1126/science.1232003

R.K. Kopparapu, R. Ramirez, J.F. Kasting, V. Eymet, T.D. Robinson, S. Mahadevan, R.C. Terrien, S. Domagal-Goldman, V. Meadows, R. Deshpande, Habitable zones around main-sequence stars: new estimates. Astrophys. J. **765**, 131 (2013). doi:10.1088/0004-637X/765/2/131

L. Kreidberg, J.L. Bean, J.-M. Désert, M.R. Line, J.J. Fortney, N. Madhusudhan, K.B. Stevenson, A.P. Showman, D. Charbonneau, P.R. McCullough, S. Seager, A. Burrows, G.W. Henry, M. Williamson, T. Kataria, D. Homeier, A precise water abundance measurement for the Hot Jupiter WASP-43b. Astrophys. J. Lett. **793**, 27 (2014a). doi:10.1088/2041-8205/793/2/L27

L. Kreidberg, J.L. Bean, J.-M. Désert, B. Benneke, D. Deming, K.B. Stevenson, S. Seager, Z. Berta-Thompson, A. Seifahrt, D. Homeier, Clouds in the atmosphere of the super-Earth exoplanet GJ1214b. Nature **505**, 69–72 (2014b). doi:10.1038/nature12888

M.J. Kuchner, Volatile-rich Earth-mass planets in the habitable zone. Astrophys. J. **596**, 105–108 (2003)

A.-M. Lagrange, M. Bonnefoy, G. Chauvin, D. Apai, D. Ehrenreich, A. Boccaletti, D. Gratadour, D. Rouan, D. Mouillet, S. Lacour, M. Kasper, A giant planet imaged in the disk of the young star β Pictoris. Science **329**, 57 (2010). doi:10.1126/science.1187187

H. Lammer, J.H. Bredehöft, A. Coustenis et al., What makes a planet habitable? Astron. Astrophys. **17**, 181–249 (2009)

H. Lammer, A. Stökl, N.V. Erkaev, E.A. Dorfi, P. Odert et al., Origin and loss of nebula-captured hydrogen envelopes from 'sub'- to 'super-Earths' in the habitable zone of Sun-like stars. Mon. Not. R. Astron. Soc. **439**(4), 3225–3238 (2014)

H. Lammer, N. Erkaev, L. Fossati, I. Juvan, P. Odert, P. Cubillos, E. Guenther, K. Kislyakova, C. Johnstone, T. Lüftinger et al., Identifying the "true" radius of the hot sub-Neptune CoRoT-24b by mass loss modelling. Mon. Not. R. Astron. Soc. **461**(1, slw095), 62–66 (2016)

G. Laughlin, J.J. Lissauer, Exoplanetary geophysics—an emerging discipline (2015). arXiv:1501.05685

A. Léger, F. Selsis, C. Sotin et al., A new family of planets? 'Ocean-Planets'. Icarus **169**(2), 499–504 (2004)

A. Léger, D. Rouan, J. Schneider, P. Barge, M. Fridlund et al., Transiting exoplanets from the CoRoT space mission. VIII. CoRoT-7b: the first super-Earth with measured radius. Astron. Astrophys. **506**, 287–302 (2009). doi:10.1051/0004-6361/200911933

J.J. Lissauer, D.C. Fabrycky, E.B. Ford et al., A closely packed system of low-mass, low-density planets transiting Kepler-11. Nature **470**, 53–58 (2011). doi:10.1038/nature09760

A.C. Lockwood, J.A. Johnson, C.F. Bender, J.S. Carr, T. Barman, A.J.W. Richert, G.A. Blake, Near-IR direct detection of water vapor in Tau Boötis b. Astrophys. J. Lett. **783**, 29 (2014). doi:10.1088/2041-8205/783/2/L29

R. Luger, R. Barnes, E. Lopez, J. Fortney, B. Jackson, V. Meadows, Habitable evaporated cores: transforming mini-Neptunes into super-Earths in the habitable zones of M Dwarfs. Astrobiology **15**, 57–88 (2015). doi:10.1089/ast.2014.1215

A.M. Mandell, K. Haynes, E. Sinukoff, N. Madhusudhan, A. Burrows, D. Deming, Exoplanet transit spectroscopy using WFC3: WASP-12 b, WASP-17 b, and WASP-19 b. Astrophys. J. **779**, 128 (2013). doi:10.1088/0004-637X/779/2/128

G.W. Marcy, R.P. Butler, A planetary companion to 70 Virginis. Astrophys. J. Lett. **464**, 147 (1996). doi:10.1086/310096

G.W. Marcy, H. Isaacson, A.W. Howard, J.F. Rowe, J.M. Jenkins et al., Masses, radii, and orbits of small Kepler planets: the transition from gaseous to rocky planets. Astrophys. J. Suppl. Ser. **210**, 20–70 (2014)

C. Marois, B. Macintosh, T. Barman, B. Zuckerman, I. Song, J. Patience, D. Lafrenière, R. Doyon, Direct imaging of multiple planets orbiting the star HR 8799. Science **322**, 1348 (2008). doi:10.1126/science.1166585

S. Maruyama, M. Ikoma, H. Genda et al., The naked planet Earth: most essential pre-requisite for the origin and evolution of life. Geosci. Front. **4**, 141–165 (2013)

M. Mayor, D. Queloz, A Jupiter-mass companion to a solar-type star. Nature **378**, 355 (1995). doi:10.1038/378355a0

P.R. McCullough, N. Crouzet, D. Deming, N. Madhusudhan, Water vapor in the spectrum of the extrasolar planet HD 189733b. I. The transit. Astrophys. J. **791**, 55 (2014). doi:10.1088/0004-637X/791/1/55

C.P. McKay, W.L. Davis, Duration of liquid water habitats on early Mars. Icarus **90**(2), 214–221 (1991)

J. Milli, D. Mouillet, D. Mawet, H.M. Schmid, A. Bazzon, J.H. Girard, K. Dohlen, R. Roelfsema, Prospects of detecting the polarimetric signature of the Earth-mass planet α Centauri B b with SPHERE/ZIMPOL. Astron. Astrophys. **556**, 64 (2013). doi:10.1051/0004-6361/201321881

P. Montanes-Rodriguez, E. Palle, P.R. Goode, F.J. Martin-Torres, Vegetation signature in the observed globally integrated spectrum of Earth considering simultaneous cloud data: applications for extrasolar planets. Astrophys. J. **651**(1), 544 (2006)

C. Mordasini, Y. Alibert, H. Klahr, T. Henning, Characterization of exoplanets from their formation—I. Models of combined planet formation and evolution. Astron. Astrophys. **547**, 111 (2012a)

C. Mordasini, Y. Alibert, W. Benz, H. Klahr, T. Henning, Extrasolar planet population synthesis—IV. Correlations with disk metallicity, mass, and lifetime. Astron. Astrophys. **541**, 97 (2012b)

N. Nettelmann, J. Fortney, U. Kramm, R. Redmer, Thermal evolution and structure models of the transiting super-Earth gj 1214b. Astrophys. J. **733**(1), 2 (2011)

L. Noack, D. Breuer, Interior and surface dynamics of terrestrial bodies and their implications for the habitability, in *Habitability on Other Planets and Satellites: The Quest for Extraterrestrial Life*, ed. by J.-P. de Vera, F. Seckbach (Springer, Dordrecht, 2013), pp. 203–233

L. Noack, D. Breuer, Plate tectonics on rocky exoplanets: influence of initial conditions and rheology. Planet. Space Sci. **98**, 41–49 (2014). doi:10.1016/j.pss.2013.06.020

L. Noack, M. Godolt, P. von Paris, A. Plesa, B. Stracke, D. Breuer, H. Rauer, Constraints for planetary habitability from interior modeling. Planet. Space Sci. **98**, 14–29 (2014)

L. Noack, D. Höning, A. Rivoldini, C. Heistracher, N. Zimov, B. Journaux, H. Lammer, T.V. Hoolst, J.H. Bredehöft, Water-rich planets: how habitable is a water layer deeper than on Earth? Icarus **277**, 215–236 (2016). doi:10.1016/j.icarus.2016.05.009

L. Noack, D. Höning, A. Rivoldini, C. Heistracher, N. Zimov, B. Journaux, H. Lammer, T. Van Hoolst, J.H. Bredehöft, Water-rich planets: how habitable is a water layer deeper than on Earth? Icarus **277**, 215–236 (2017)

R. Parai, S. Mukhopadhyay, How large is the subducted water flux? New constraints on mantle regassing rates. Earth Planet. Sci. Lett. **317**, 396–406 (2012)

S. Ramsay, M. Casali, J. González, N. Hubin, The E-ELT instrument roadmap: a status report, in *SPIE Astronomical Telescopes+ Instrumentation* (SPIE, Bellingham, 2014), p. 91471

H. Rauer, C. Catala, C. Aerts et al., The PLATO 2.0 mission. Exp. Astron. **38**(1-2), 249–330 (2014). doi:10.1007/s10686-014-9383-4

P. Riaud, J. Schneider, Improving Earth-like planets' detection with an ELT: the differential radial velocity experiment. Astron. Astrophys. **469**, 355–361 (2007). doi:10.1051/0004-6361:20077085

G.R. Ricker, J.N. Winn, R. Vanderspek, D.W. Latham, G.Á. Bakos et al., Transiting exoplanet survey satellite (tess). Proc. SPIE **9143**(914320), 15 (2014)

G.R. Ricker, J.N. Winn, R. Vanderspek, D.W. Latham, G.Á. Bakos, J.L. Bean, Z.K. Berta-Thompson, T.M. Brown, L. Buchhave, N.R. Butler et al., Transiting exoplanet survey satellite. J. Astron. Telesc. Instrum. Syst. **1**(1), 014003 (2015)

L.A. Rogers, Most 1.6 Earth-radius planets are not rocky. Astrophys. J. **801**(1), 41 (2015)

L. Rogers, S. Seager, Three possible origins for the gas layer on gj 1214b. Astrophys. J. **716**(2), 1208 (2010)

L. Roth, J. Saur, K.D. Retherford, D.F. Strobel, P.D. Feldman, M.A. McGrath, F. Nimmo, Transient water vapor at Europa's South pole. Science **343**(6167), 171–174 (2014)

P.A. Sabelhaus, J.E. Decker, An overview of the James Webb Space Telescope (JWST) project, in *SPIE Astronomical Telescopes+ Instrumentation* (SPIE, Bellingham, 2004), pp. 550–563

N.C. Santos, V. Adibekyan, C. Mordasini, W. Benz, E. Delgado-Mena, C. Dorn, L. Buchhave, P. Figueira, A. Mortier, F. Pepe, A. Santerne, S.G. Sousa, S. Udry, Constraining planet structure from stellar chemistry: the cases of CoRoT-7, Kepler-10, and Kepler-93. Astron. Astrophys. **580**(L13), 1–5 (2015)

J. Saur, S. Duling, L. Roth, X. Jia, D.F. Strobel, P.D. Feldman, U.R. Christensen, K.D. Retherford, M.A. McGrath, F. Musacchio et al., The search for a subsurface ocean in Ganymede with Hubble Space Telescope observations of its auroral ovals. J. Geophys. Res. Space Phys. **120**(3), 1715–1737 (2015)

L. Schaefer, D. Sasselov, The persistence of oceans on Earth-like planets: insights from the deep-water cycle. Astrophys. J. **801**(1), 40 (2015)

G. Schubert, D.L. Turcotte, P. Olson, *Mantle Convection in the Earth and Planets* (Cambridge University Press, New York, 2001)

H. Schwarz, M. Brogi, R. de Kok, J. Birkby, I. Snellen, Measuring the spin of the directly imaged sub-stellar companion GQ Lupi b, in *AAS/Division for Extreme Solar Systems Abstracts*, vol. 3 (2015), pp. 104–114

S. Seager, Exoplanet habitability. Science **340**, 577–581 (2013)

D.K. Sing, J.J. Fortney, N. Nikolov, H.R. Wakeford, T. Kataria, T.M. Evans, S. Aigrain, G.E. Ballester, A.S. Burrows, D. Deming, J.-M. Désert, N.P. Gibson, G.W. Henry, C.M. Huitson, H.A. Knutson, A.L.D. Etangs, F. Pont, A.P. Showman, A. Vidal-Madjar, M.H. Williamson, P.A. Wilson, A continuum from clear to cloudy hot-Jupiter exoplanets without primordial water depletion. Nature **529**, 59–62 (2016). doi:10.1038/nature16068

W. Skidmore, TMT International Science Development Teams, TMT Science Advisory Committee, Thirty meter telescope detailed science case: 2015. Res. Astron. Astrophys. **15**(12), 1945 (2015). http://stacks.iop.org/1674-4527/15/i=12/a=1945

I.A.G. Snellen, R.J. de Kok, E.J.W. de Mooij, S. Albrecht, The orbital motion, absolute mass and high-altitude winds of exoplanet HD209458b. Nature **465**, 1049–1051 (2010). doi:10.1038/nature09111

I.A.G. Snellen, B.R. Brandl, R.J. de Kok, M. Brogi, J. Birkby, H. Schwarz, Fast spin of the young extrasolar planet β Pictoris b. Nature **509**, 63–65 (2014). doi:10.1038/nature13253

I. Snellen, R. de Kok, J.L. Birkby, B. Brandl, M. Brogi, C. Keller, M. Kenworthy, H. Schwarz, R. Stuik, Combining high-dispersion spectroscopy with high contrast imaging: probing rocky planets around our nearest neighbors. Astron. Astrophys. **576**, 59 (2015). doi:10.1051/0004-6361/201425018

C. Sotin, O. Grasset, A. Mocquet, Mass–radius curve for extrasolar Earth-like planets and ocean planets. Icarus **191**(1), 337–351 (2007)

W.B. Sparks, H.C. Ford, Imaging spectroscopy for extrasolar planet detection. Astrophys. J. **578**, 543–564 (2002). doi:10.1086/342401

V. Stamenković, D. Breuer, The tectonic mode of rocky planets: Part 1—Driving factors, models & parameters. Icarus **234**, 174–193 (2014)

C. Stein, J.P. Lowman, U. Hansen, The influence of mantle internal heating on lithospheric mobility: implications for super-Earths. Earth Planet. Sci. Lett. **361**, 448–459 (2013)

D.J. Stevenson, Life-sustaining planets in interstellar space? Nature **400**(6739), 32 (1999)

M.R. Swain, G. Vasisht, G. Tinetti, The presence of methane in the atmosphere of an extrasolar planet. Nature **452**, 329–331 (2008). doi:10.1038/nature06823
P. Thomas, R. Tajeddine, M. Tiscareno, J. Burns, J. Joseph, T. Loredo, P. Helfenstein, C. Porco, Enceladus's measured physical libration requires a global subsurface ocean. Icarus **264**, 37–47 (2016)
G. Tinetti, A. Vidal-Madjar, M.-C. Liang, J.-P. Beaulieu, Y. Yung, S. Carey, R.J. Barber, J. Tennyson, I. Ribas, N. Allard, G.E. Ballester, D.K. Sing, F. Selsis, Water vapour in the atmosphere of a transiting extrasolar planet. Nature **448**, 169–171 (2007). doi:10.1038/nature06002
C.T. Unterborn, E.E. Dismukes, W.R. Panero, Scaling the Earth: a sensitivity analysis of terrestrial exoplanetary interior models. Astrophys. J. **819**(1), 32 (2016)
D. Valencia, R.J. O'Connell, D.D. Sasselov, Internal structure of massive terrestrial planets. Icarus **181**, 545–554 (2006)
D. Valencia, R.J. O'Connell, D.D. Sasselov, Inevitability of plate tectonics on super-Earths. Astrophys. J. **670**, 45–48 (2007)
D. Valencia, T. Guillot, V. Parmentier, R.S. Freedman, Bulk composition of gj 1214b and other sub-Neptune exoplanets. Astrophys. J. **775**(1), 10 (2013)
H.J. van Heck, P.J. Tackley, Plate tectonics on super-Earths: equally or more likely than on Earth. Earth Planet. Sci. Lett. **310**, 252–261 (2011)
T. Van Hoolst, R.-M. Baland, A. Trinh, The diurnal libration and interior structure of Enceladus. Icarus **277**, 311–318 (2016)
S. Vance, M. Bouffard, M. Choukroun, C. Sotin, Ganymede's internal structure including thermodynamics of magnesium sulfate oceans in contact with ice. Planet. Space Sci. **96**, 62–70 (2014)
S. Vance, K. Hand, R. Pappalardo, Geophysical controls of chemical disequilibria in Europa. Geophys. Res. Lett. **43**(10), 4871–4879 (2016)
F.W. Wagner, N. Tosi, F. Sohl, H. Rauer, T. Spohn, Rocky super-Earth interiors structure and internal dynamics of CoRoT-7b and Kepler-10b. Astron. Astrophys. **541**, 103 (2012)
H.R. Wakeford, D.K. Sing, D. Deming, N.P. Gibson, J.J. Fortney, A.S. Burrows, G. Ballester, N. Nikolov, S. Aigrain, G. Henry, H. Knutson, A. Lecavelier des Etangs, F. Pont, A.P. Showman, A. Vidal-Madjar, K. Zahnle, HST hot Jupiter transmission spectral survey: detection of water in HAT-P-1b from WFC3 near-IR spatial scan observations. Mon. Not. R. Astron. Soc. **435**, 3481–3493 (2013). doi:10.1093/mnras/stt1536
J.C.G. Walker, P.B. Hays, J.F. Kasting, A negative feedback mechanism for the long-term stabilization of the Earth's surface temperature. J. Geophys. Res. **86**, 9776–9782 (1981)
C.R. Webster, P.R. Mahaffy, S.K. Atreya, G.J. Flesch, M.A. Mischna, P.-Y. Meslin, K.A. Farley, P.G. Conrad, L.E. Christensen, A.A. Pavlov et al., Mars methane detection and variability at Gale crater. Science **347**, 415–417 (2015). doi:10.1126/science.1261713
F. Westall, A. Brack, The importance of water for life. Space Sci. Rev. (2017, in review)
A. Wolszczan, D.A. Frail, A planetary system around the millisecond pulsar psr1257 + 12. Nature **355**, 145–147 (1992)
C. Zimmer, K.K. Khurana, M.G. Kivelson, Subsurface oceans on Europa and Callisto: constraints from Galileo magnetometer observations. Icarus **147**(2), 329–347 (2000)

MIX
Papier aus verantwortungsvollen Quellen
Paper from responsible sources
FSC® C105338

If you have any concerns about our products,
you can contact us on
ProductSafety@springernature.com

In case Publisher is established outside the EU,
the EU authorized representative is:
Springer Nature Customer Service Center GmbH
Europaplatz 3, 69115 Heidelberg, Germany

Printed by Libri Plureos GmbH
in Hamburg, Germany